ASPECTS OF PALYNOLOGY

ASPECTS OF PALYNOLOGY

ROBERT H. TSCHUDY RICHARD A. SCOTT
Editors

WILEY-INTERSCIENCE
A Division of John Wiley & Sons
New York • London • Sydney • Toronto

10 9 8 7 6 5 4 3 2 1

Library of Congress Catalogue Card Number: 73-84968

SBN 471 89220 3

Printed in the United States of America

PREFACE

Palynology, although a relatively new field, already has earned a prominent place in paleobotany. The ubiquity and abundance of palynomorphs in diverse kinds of rocks provide a source of material that has enormous potential in documenting the geologic record of plants. The pioneering efforts in the field have been rapidly and broadly supplemented as new applications of palynology have been recognized in geology and botany. In oil geology particularly the phenomenal growth and expansion of palynology has been stimulated by the successful practical application of the results of research.

Growth in palynological knowledge has not, however, been accompanied by a commensurate body of published information in the form of comprehensive accounts or syntheses. This lack is felt alike by beginning students, teachers, and those engaged in pure or applied research. The omission is particularly acute because palynology in combining aspects of geology and botany attracts interest from those whose backgrounds may be incomplete in one or the other of these disciplines.

The present volume was conceived as an attempt to summarize in one book the nature, scope, and applications of the study of fossil pollen and spores. Informal discussions about this kind of summary were conducted with a number of palynologists. The overwhelming consensus was that, in order to present a discussion of principles combined with adequate coverage of the field, a collaborative volume would be best.

The collaborative approach has led to an extended preparation time for this book. Inevitably some chapters were completed early, whereas others have been submitted only recently. Consequently in some chapters references to the most recent literature are not complete. In addition, subsequent publications altered our original plans; for example, the publication (in *Handbook of Paleontological Techniques*, edited by Kummel and Raup, W. H. Freeman and Co., San Francisco and London, 1965) of a compilation of palynological techniques by Jane Gray made superfluous a similar chapter in this book. Furthermore, little space has been allocated to the Quaternary, because the recent textbook by Knut Faegri and Johannes Iversen (*Textbook of Pollen Analysis*, Munksgaard, Copenhagen, 1964) covers Pleistocene and Recent palynology more fully than would be possible in the present volume.

This book in effect is subdivided into four sections: chapters 1 through 5 provide a general background. Following an introductory chapter, the plant kingdom and the role of pollen and spores in it are reviewed. The morphology and description of pollen and spores are then presented, along with a discussion of the microstructure of exines as revealed by the electron and scanning electron microscopes. The section concludes with discussions of the problems of palynological taxonomy and of the relationships of palynomorphs to sedimentary processes. Chapters 6 and 7 deal with the interpretation of the fossil record. This section, which contains chapters on sample reliability and applied problems, includes a short discussion of paleoecology. Chapters 8 through 17 are concerned with plant microfossils in time. The section begins with a summary of the geological history of plants as revealed by plant megafossils. Subsequent chapters treat the record of plant life as represented by palynomorphs from the first vestiges of life to the late Tertiary and Quaternary. The last section, Chapter 18, deals with acritarchs and dinoflagellates, significant microfossils often found in palynological preparations.

As the fossil record is incomplete, so also is this book incomplete. Accounts of significant new discoveries are appearing in the literature at so rapid a rate that it is impossible for a book of this kind to be up to date. This book is meant to serve as a point of departure for students of palynology. It is hoped that it will also provide useful material and provoke discussions among those with more than an elementary knowledge of the subject.

The various authors differ considerably in emphasis and outlook. Indeed some of the opinions are at variance with those of the editors. This we believe to be advantageous. Diverse viewpoints of workers in the field provide the student with broader views of palynology than would an attempt at conformity imposed by the editors; diversity rather than uniformity of opinion is actually more representative of the field of palynology. Undoubtedly the test of time will alter many of the concepts expressed in this book. It is only the open-minded acceptance of new facts that will allow a new science such as palynology to progress.

Robert H. Tschudy
Denver, Colorado Richard A. Scott

CONTENTS

ASPECTS OF PALYNOLOGY

1

INTRODUCTION

R. H. Tschudy and R. A. Scott

DEFINITION AND BASIS

Palynology, a word coined by Hyde and Williams (1944), was defined by them as "the study of pollen and other spores and their dispersal, and applications thereof." The term includes both modern and fossil pollen and spores. Although certain other microfossils also are considered, this book is concerned chiefly with the study of pollen and spores from fossil plants, perhaps more precisely termed paleopalynology. This field is a subdivision of paleobotany, the study of the plant life of the past.

The underlying assumptions of palynology are not different from those of other areas of paleontology. They include the familiar trust in the present as indicative of the past and a belief that fossils are elements of a continuum of once-living organisms whose succession was shaped by organic evolution.

Palynology depends mainly on four characteristics of pollen and spores: their greater resistance to degradation than most other plant parts, thus facilitating their survival as fossils; their small size, mostly less than 200 microns, so that they are transported and deposited as sedimentary particles; their morphological complexity, so that they can be distinguished and characterized; and their production in enormous numbers, which facilitates recovery of statistically significant assemblages.

HISTORY

The historical development of palynology has been reviewed by Wodehouse (1935), Erdtman (1943), Pokrovskaya (1958), Faegri and Iversen (1964), and others, and need not be repeated here. It is sufficient to add that, since about 1950, the science of palynology has grown remarkably and, in a sense, has matured. Research laboratories have been established throughout the world by universities, governmental agencies, commercial firms, and private individuals. At present there undoubtedly are more people occupied with the varied aspects of palynology than with the parent science—the paleobotany of megafossils.

PALYNOLOGICAL MILESTONES IN PLANT EVOLUTION

Fossil plants have been found in rocks ranging in age from the Precambrian to the Recent. Spores are among the earliest structurally preserved remains of plant life and accompany material that is probably derived from bacteria, algae, and perhaps fungi (Tyler and Barghoorn, 1954; Roblot, 1963; Schopf and others, 1965). (See also chap. 9.)

The first unequivocal plant spores bearing trilete sutures are found in rocks of Silurian age. The advent of vascular tissue, a most significant step in land-plant evolution, occurred at about the same time. Because of this coincidence, and because some early and structurally uncomplicated vascular land plants had simple trilete spores, there is a tendency to correlate these two events. This is probably a mistake. Vascular and reproductive structures may have evolved more or less concurrently, but the two developments probably were essentially independent of one another.

Table 1-1 Geologic Time Scale

Era	System(s) or Period	Series (Epoch)	Stage or Age	Beginning of Interval[1] (million years) Kulp (1961)	Beginning of Interval[1] (million years) Holmes (1965)	Duration (million years) Holmes (1965)
Cenozoic	Quaternary	Recent				
		Pleistocene		1	2 or 3	2 or 3
	Tertiary	Pliocene		13	12	9 or 10
		Miocene		25	25	13
		Oligocene		36	40	15
		Eocene	upper[2]	45		
			middle[2]	52		
			lower[2]	58	60	20
		Paleocene		63	70	10
Mesozoic	Cretaceous	Upper (Late)	Maestrichtian	72		
			Campanian			
			Santonian	84		
			Coniacian			
			Turonian	90		65
			Cenomanian	110		
		Lower (Early)	Albian	120		
			Aptian			
			Neocomian	135	135	
	Jurassic	Upper (Late)				
		Middle (Middle)	Bathonian	166		
			Bajocian			
		Lower (Early)		181	180	
	Triassic	Upper (Late)		200		
		Middle (Middle)				
		Lower (Early)		(230)	225	
Paleozoic	Permian	Upper (Late)				
				260		45
		Lower (Early)		280	270	
	Carboniferous: Pennsylvanian					
	Carboniferous: Mississippian		Visean	320		
			Tournasian	345	350	
	Devonian	Upper (Late)		(365)		
		Middle (Middle)		390		
		Lower (Early)		405	400	
	Silurian			(425)	440	40
	Ordovician	Upper (Late)				
			Trenton	445		
		Middle (Middle)				60
		Lower (Early)		500	500	
	Cambrian	Upper (Late)		530		
		Middle (Middle)				100
		Lower (Early)			600	

[1] Parentheses indicate interpolated dates.

[2] Informal subdivisions of the Eocene series.

Heterospory (the development of megaspores and microspores) is first noted in the fossil record of the Devonian Period. No megaspores of Early Devonian age have been found, but heterospory is known in at least three separate groups of plants from the Middle and Late Devonian. Heterospory is the prologue to development of the seed.

Two important structural types, monosulcate and bisaccate pollen, first appear in the Pennsylvanian Period. Both pollen types probably originated in a pteridospermous complex (Staplin, Pocock, and Jansonius, 1967). Monosulcate pollen became established as characteristic in the cycadophytes; bisaccate pollen is common in the conifers. A significant aspect of both monosulcate and bisaccate pollen is that the aperture for the emergence of the prothallial tissue is on the distal surface. In trilete and monolete spores the aperture is along the suture on the proximal side of the spore.

The last prominent evolutionary milestone was the appearance of recognizable, true angiospermous pollen. Although the time of the origin of the angiosperms is controversial, the palynological record is thus far consistent in first showing recognizable angiosperm pollen in the late Early Cretaceous.

The evolution of pollen and spores, like that of other plant organs, has proceeded in general from the simple to the complex. However, many examples are known of trends toward reduction and simplification. There also has been development of pollen and spore forms with similar morphology in widely separated taxonomic groups. The systematic affinities of dispersed fossil pollen and spores must be evaluated with these possibilities in mind.

GEOLOGIC TIME SCALE

There are three major groups of phenomena from which inferences concerning the age of rocks can be drawn: sediments and processes of sedimentation, the record of evolution of life, and rates of radioactive decay (Teichert, 1958). Utilizing the first two methods, geologists have arrived at a workable but somewhat empirical time scale. With the discovery that radioactive elements decay at constant rates yielding other elements as decay products, a method for calculating a radiometric geologic time scale was born. As data were accumulated Holmes (1947, 1959) constructed a time scale based on a framework of radioisotope ages.

We have combined in table 1-1 information from two of the more recent compilations of time scales utilizing additional and more refined radiometric determinations (Kulp, 1961; Holmes, 1965). This chart provides a base of reference for the discussions in chapters 8 through 17.

FUTURE OF PALYNOLOGY

Palynology, a relatively new member of the paleontological sciences, is a young science that cuts across traditional disciplines. Palynological studies can be approached from the viewpoint of botany, with emphasis on the plant relationships, or from the geological perspective, with emphasis on biostratigraphy. These approaches are not mutually exclusive; indeed, adequacy requires a knowledge of both fields.

Until lately comparatively few people have been active in palynology and only an insignificant portion of the sediments of the earth's crust has been examined. Despite these limitations, significant scientific contributions have demonstrated the value of palynology, and its future expansion is unquestionable. This expansion will yield further contributions in the fields of plant systematics, plant geography, paleoclimatology, and a better understanding of the history of the plant kingdom. Palynology also can be expected to furnish more refined and more extensive geologic and stratigraphic data as its coverage enlarges.

References

Erdtman, Gunnar, 1943, An introduction to pollen analysis: Waltham, Mass., Chronica Botanica Co., 239 p.

Faegri, Knut, and Iversen, Johannes, 1964, Text-book of modern pollen analysis, 2d ed., revised: Copenhagen, Munksgaard, 237 p.

Holmes, Arthur, 1947, The construction of a geological time-scale: Glasgow Geol. Soc. Trans., v. 21, pt. 1, p. 117-152.

———1959, A revised geological time-scale: Edinburgh Geol. Soc. Trans., v. 17, pt. 3, p. 183-216.

———1965, Principles of physical geology, 2d ed., revised: New York, The Ronald Press Co., 1288 p.

Hyde, H. A., and Williams, D. A., 1944, The right word (letter to Paul B. Sears, dated July 15, 1944): Pollen Analysis Circular (Oberlin, Ohio; mimeographed), no. 8, p. 6, October 28, 1944.

Kulp, J. L., 1961, Geologic time scale: Science, v. 133, no. 3459, p. 1105-1114.

Pokrovskaya, I. M., 1958, Analyse Pollenique: [France] Bur. recherches géol., géophys. et minières serv. inf. géol., v. 24, 435 p. (French translation).

Roblot, Marie-Madeleine, 1963, Découverte de sporomorphes dans des sédiments antérieurs à 550 M.A. (Briovérien): Acad. sci. [Paris] Comptes rendus, v. 256, pt. 1, p. 1557-1559.

Schopf, J. W., Barghoorn, E. S., Maser, M. D., and Gordon, R. O., 1965, Electron microscopy of fossil bacteria two billion years old: Science, v. 149, no. 3690, p. 1365-1367.

Staplin, F. L., Pocock, S. J., and Jansonius, Jan, 1967, Relationships among gymnospermous pollen: Rev. Paleobotany and Palynology, v. 3, p. 297-310.

Teichert, Curt, 1958, Some biostratigraphical concepts: Geol. Soc. America Bull., v. 69, no. 1, p. 99-119.

Tyler, S. A., and Barghoorn, E. S., 1954, Occurrence of structurally preserved plants in pre-Cambrian rocks of the Canadian Shield [Ontario]: Science, v. 119, no. 3096, p. 606-608.

Wodehouse, R. P., 1935, Pollen grains; their structure, identification, and significance in science and medicine: New York, McGraw-Hill Book Co., Inc., 574 p.

2

THE PLANT KINGDOM AND ITS PALYNOLOGICAL REPRESENTATION

R. H. Tschudy

CLASSIFICATION OF PLANTS

For a group as large and diversified as the plant kingdom to be dealt with, it must be divided into manageable units. Many systems of classification have been proposed on the basis of various criteria (Lawrence, 1951); for example, classifications based on chemical similarities among plants have been proposed. For the purpose of interpreting the fossil record the most satisfactory systems are those that are phylogenetically oriented and that utilize morphological criteria derived from the form and structure of plant organs. Schemes of classification change as previously acquired information is reinterpreted.

A primary breakdown of the plant kingdom on a structural basis yields two great divisions—the nonvascular and the vascular plants—and further subdividing is clearly necessary. The following categories for additional subdivision of the plant kingdom and their relative ranks are designated by the "International Code of Botanical Nomenclature":

Division (phylum)	Genus
Class	Section
Order	Series
Family	Species
Tribe	Variety
	Form

Publication authorized by the Director of the U.S. Geological Survey.

Further supplementary ranks such as subfamily may be intercalated or added provided that no confusion or error is introduced in the process. Understandably, different workers will have different ideas concerning the proper placement of a given group of plants; for example, some may place the bacteria in the phylum Schizomycophyta, whereas others may place them in the class Schizomycetes.

The classification of plants used by Fuller and Tippo (1949) is given in table 2-1. Henry Andrews (1961) is of the opinion that a classification system modeled after that of Harold C. Bold (1957) is more appropriate for fossil plants. This system, shown in the right-hand column is exclusive of the algae and fungi, since Andrews in his book was concerned chiefly with vascular plants. The paleobotanical literature may refer to one or the other of the classification systems or modifications of them, hence both are included.

FOSSILS OF NONVASCULAR PLANTS

It may be helpful to consider some of the microscopic types of fossil remains of nonvascular plants that have been discovered (this discussion excludes the better known megascopic calcareous algae). Some of these forms are problematic. They do not yet fit into the classification scheme because they are imperfectly known or because they are only fragmentary remains of a whole organism—or of an extinct organism. Other forms, because of their morphologic simi-

Table 2-1. **Two Systems of Plant Classification**

Fuller and Tippo (1949)	*Andrews (1961)*
Subkingdom Thallophyta	
Phylum Cyanophyta—blue-green algae	ALGAE
Phylum Euglenophyta—euglenoids	
Phylum Chlorophyta—green algae	
Phylum Chrysophyta—yellow-green algae, golden-brown algae, and diatoms	
Phylum Pyrrophyta—cryptomonads and dinoflagellates	
Phylum Phaeophyta—brown algae	
Phylum Rhodophyta—red algae	
Phylum Schizomycophyta—bacteria	FUNGI
Phylum Myxomycophyta—slime molds	
Phylum Eumycophyta—true fungi	
Subkingdom Embryophyta	
Phylum Bryophyta (or Atracheata)	
Class Musci—mosses	Division Bryophyta
Class Hepaticae—liverworts	Division Hepatophyta
Class Anthocerotae—hornworts	
Phylum Tracheophyta[1]—vascular plants	
Subphylum Psilopsida	
Class Psilophytineae	Division Psilophyta
Order Psilophytales[2]	
Order Psilotales	
Subphylum Lycopsida—clubmosses	
Class Lycopodineae	Division Lycopodophyta
Order Lycopodiales—clubmosses	
Order Selaginellales—small clubmosses	
Order Lepidodendrales[2]—giant clubmosses	
Order Pleuromeiales[2]	
Order Isoetales—quillworts	
Subphylum Sphenopsida—horsetails	
Class Equisetineae	Division Arthrophyta
Order Hyeniales[2]	
Order Sphenophyllales[2]	
Order Equisetales—horsetails	
Subphylum Pteropsida	
Class Filicineae—ferns	Division Pterophyta
Order Coenopteridales[2]	
Order Ophioglossales	
Order Marattiales	
Order Filicales	
Class Gymnospermae—conifers and their allies	
Subclass Cycadophytae	
Order Cycadofilicales[2,3]—seed ferns	Division Pteridospermophyta
Order Bennettitales[2,4]	Division Cycadophyta
Order Cycadales—cycads	
Subclass Coniferophytae	
Order Cordaitales[2]	Division Coniferophyta
Order Ginkgoales—maidenhair tree	Division Ginkgophyta
Order Coniferales—conifers	Division Coniferophyta
Order Gnetales	Division Gnetophyta
Class Angiospermae—flowering plants	Division Anthophyta
Subclass Dicotyledoneae	
Subclass Monocotyledoneae	

[1]Also known as Tracheata.
[2]Known only as fossils.
[3]Also known as Pteridospermae.
[4]Also known as Cycadeoidales

larity to an extant plant or plant part, can easily be placed in an appropriate class, family, genus, or even species.

Fresh-water algae from oil shale in the Green River Formation were reported by Bradley (1929). In a later publication (1931) he showed photographs of a variety of organic remains including pollen, fern spores, plant hairs, insects, insect hairs, insect larvae, and unicellular and multicellular algae belonging to the Cyanophyta, Euglenophyta, and Chlorophyta. He illustrated and reported bacteria and a variety of fungal spores also.

Some problematic fossil algae from western Australia—including the genera *Schizocystia*, *Lecaniella*, and *Horologinella*—were reported by Cookson and Eisenack (1962). They suggested a possible affinity of *Lecaniella* to the green alga *Phacotus* and also suggested that *Horologinella* may be an algal aplanospore. *Horologinella* is similar morphologically to the fossil *Tetraporina* of Naumova (1950). The striking resemblance of *Tetraporina* to the modern unicellular green alga *Tetraëdron* was noted by Scott, Barghoorn, and Leopold (1960).

Representatives of most, if not all, of the thallophyte phyla have been found as fossils, and most of the phyla have been recognized as microscopic remains in palynological preparations. A few of these are shown on plate 2-1.

The spore known as *Tasmanites* is of interest because it occurs abundantly in the coallike or kerogenlike "white coal" known as tasmanite, so called because this "white coal" as originally described was from Tasmania. Spores from this material were described by Newton (1875), who gave them the name *Tasmanites punctatus*. Spores that were isolated from some Tasmanian "white coal" (furnished by Harry Tourtelot of the U.S. Geological Survey) are shown on plate 2-1, figures 3 and 4.

Thin sections of an organic deposit resembling oil shale from the northern flank of the Brooks Range, Alaska (also furnished by Tourtelot) show that this material is made up almost entirely of compressed *Tasmanites* spores. (See pl. 2-1, figs. 1 and 2.) In composition this Brooks Range material is very similar to the tasmanite from Tasmania.

The affinity of *Tasmanites* to the algae, based largely on the absence of haptotypic structures, has been suggested (Schopf, Wilson, and Bentall, 1944; Sommer, 1956). The morphological similarity of the planktonic green alga *Pachysphaera pelagica* to the fossil genus *Tasmanites* was shown by Wall (1962). Specimens of *Pachysphaera pelagica* were provided by Mary Parke of the Marine Biological Laboratory, Plymouth, England. Photographs of two specimens are shown on plate 2-1, figures 5 and 6.

Coenobia of *Pediastrum* are not uncommon in palynological preparations. I have found them in Tertiary rocks from Venezuela and in Cretaceous and Tertiary rocks from the United States. Four new species of fossil *Pediastrum* were described by Wilson and Hoffmeister (1953) from the Neogene of Sumatra, and Cookson (1953) reported the fossil from the Cenozoic of Australia. *Pediastrum* may be of importance as a facies indicator. A modern *Pediastrum* is shown on plate 2-1, figure 8.

An interesting fossil was isolated from the Frontier Formation of Montana. This fossil (pl. 2-1, fig. 7) probably represents the conjugation and resultant zygospore of some member of the order Conjugales of the green algae. Filamentous algae of this type are rarely found as fossils.

Botryococcus (pl. 2-1, fig. 9) is another alga that is often found in palynological assemblages. It has been reported from rocks at least as old as Ordovician. An excellent account of the morphology and occurrence of this alga in the fossil record is given by Traverse (1955).

Boghead coal is made up largely of an alga similar to, if not identical with, *Botryococcus*. This alga is living today in fresh-water lakes and brackish-water localities such as the region of the Coorong along the southeastern coast of south Australia. The substance coorongite is the peaty equivalent of boghead coal. An attribute of living *Botryococcus* is its ability to produce large quantities of oil. The position of *Botryococcus* within the Thallophyta is not precisely known. It is placed by some authors in the Chlorophyta and by others in the Chrysophyta. The validity of the genus, however, is not in dispute.

Fossil representatives of the Cyanophyta, Euglenophyta, and Chlorophyta have been mentioned above as occurring in the Green River shales. Diatoms, belonging to the Chrysophyta,

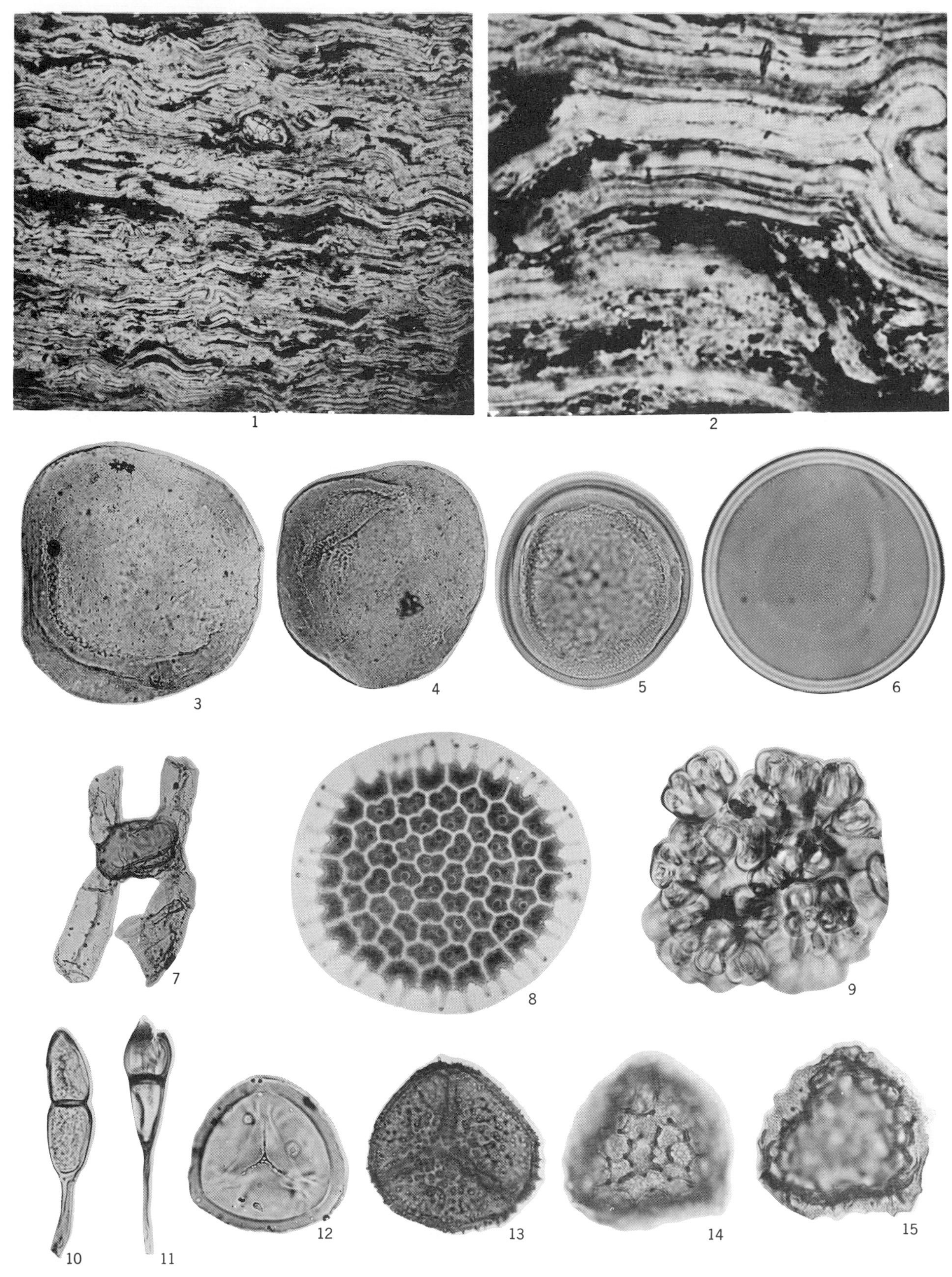

8

are common fossils in many rocks, but owing to the common use of hydrofluoric acid in preparing palynological samples, the siliceous frustules are usually not seen. The study of the siliceous skeletons of fossil diatoms is a specialized branch of paleontology and is usually given little attention by palynologists.

The phylum Pyrrophyta, which includes the dinoflagellates, is very well represented in palynological preparations. Dinoflagellates are considered in detail in chapter 18.

The two remaining algal phyla, the Phaeophyta and Rhodophyta (with the exception of calcareous forms), like the Cyanophyta possess plant bodies that are commonly difficult to preserve. Representatives of these phyla possibly were preserved only under special conditions and may be encountered as fossils only in certain types of rocks such as the Green River oil shales.

The remaining phyla of the Thallophyta do not possess chlorophyll and may be considered under the broad heading of bacteria and fungi. Bacteria have been recorded as fossils. Electron micrographs of bacteria 2 billion years old from the Precambrian Gunflint chert are shown in a recent report by Schopf and others (1965). More recently bacteria 3.1 billion years old from the Precambrian Fig Tree Series of South Africa have been found by Barghoorn and Schopf (1966). To my knowledge no fossils attributable to the Myxomycophyta or slime molds have been reported. Mycelia of the Eumycophyta, or true fungi, are common accompaniments of palynologic assemblages. Spores similar to the teliospores of rusts are also fairly common. (See pl. 2-1, figs. 10 and 11.) A few other fungal remains such as *Phragmothyrites* have been reported (Potonié and Venitz, 1934). In spite of the common occurrence of fungi in palynological preparations, comparatively little attention has been paid to them. Until more work is reported the fungi appear to have little value stratigraphically.

In the subkingdom Embryophyta only one phylum, the Bryophyta, does not possess vascular tissue. This phylum includes the mosses, liverworts, and hornworts. It is the only nonvascular phylum that produces thick-walled spores in tetrads. These spores, when separated from the tetrad, commonly display a trilete suture. Consequently, as dispersed fossil spores, they are difficult or impossible to distinguish from spores of the advanced phylum Tracheophyta. Spores representing members of the three classes of the phylum Bryophyta are shown on plate 2-1, figures 12–15. Figure 12 is a moss spore of the modern genus *Sphagnum*. Figure 13 is a spore of the modern hornwort *Anthoceros*, and figures 14 and 15 represent a spore of the modern liverwort *Riccia*. Undoubtedly many fossil spores of the bryophytes have been assigned to artificial, or form, genera. Some may have been assigned erroneously to tracheophyte genera; for example, in morphology the spore of *Anthoceros* is not too unlike some spores of the fern genus *Osmunda*. Until we are sure of the affinities of fossil spores the least confusing procedure is to assign them to a form genus. Lundblad (1954, 1955) found spores associated with the fossil liverwort *Ricciopsis*. Although she strongly suspected that these were spores belonging to the fossil plant, they were not obtained from reproductive organs of the fossil. She placed the spores in the new genus *Ricciisporites*. Lundlblad found these fossil liverworts in rocks of Mesozoic age and suggested that the major groups of the hepatics may have been differentiated in the Paleozoic.

The occurrence of the Bryophyta in the fossil record has been summarized by Jovet-Ast (1967). She suggests that the bryophytes probably originated in the Silurian, were present in the Devonian, and were well represented in the fossil record at the beginning of the Carboniferous.

Plate 2-1. Examples of spores and other structures from nonvascular plants: **1.**—thin section of Alaskan tasmanitelike organic rock (× 200); **2.**—thin section of Alaskan tasmanitelike organic rock (× 1,000); **3.**—*Tasmanites punctatus* from Tasmania (× 200); **4.**—*T. punctatus* from Tasmania (× 200); **5.**—*Pachysphaera pelagica*, modern (× 500); **6.**—*P. pelagica*, modern (× 500); **7.**—fossil conjugatae from Frontier Formation, Montana (× 1,000); **8.**—*Pediastrum* sp., modern (× 1,000); **9.**—*Botryococcus* sp. McNairy Formation, Kentucky (× 1,000); **10.**—*Puccinia malvacearum* teliospore, modern (× 1,000); **11.**—*P. graminis* teliospore, modern (× 1,000); **12.**—*Sphagnum tenellum*, modern (× 1,000); **13.**—*Anthoceros* sp., modern (× 1,000); **14.**—*Riccia natans*, modern (× 1,000); **15.**—*R. natans*, modern (× 1,000).

SPORES AND POLLEN FROM VASCULAR PLANTS

The subphylum Psilopsida embraces two orders, the Psilophytales and the Psilotales. The Psilophytales are known only as fossils. Beginning in 1917 Kidston and Lang (1917-1921) described four species of a primitive group of vascular plants from the Rhynie chert of Devonian age and placed them in the genera *Rhynia, Hornea* (now *Horneophyton*), and *Asteroxylon*. An abundance of trilete spores was found in the sporangia of *Rhynia major*. The plants probably produced only isospores. The earliest vascular plants known to possess trilete spores are *Baragwanathia* (originally considered as belonging to the Psilopsida but now placed in the Lycopsida by many workers) from rocks of Silurian age in Australia. The Psilotales are represented in the modern flora by the genera *Psilotum*, with two species, and *Tmesipteris*, with only a single species. One of the monolete spores of *Psilotum nudum* is shown on plate 2-2, figure 1.

Two of the five orders belonging in the subphylum Lycopsida are known only from fossils. They are the Lepidodendrales, or giant clubmosses, and the Pleuromeiales, which might be called giant quillworts. The Lepidodendrales were all trees. They appeared first in the Devonian and persisted to the end of the Carboniferous. Most of these arborescent lycopods produced bisporangiate cones that yielded megaspores and microspores. These plants were among the dominant elements in Carboniferous forests. The Pleuromeiales, on the other hand, attained a height of only 2 meters and never were a dominant part of any flora. The single genus *Pleuromeia* is known only from the Triassic. It has been suggested (Andrews, 1961) that modern *Isoetes* is the end member of an evolutionary series in which a stem originally like that of *Pleuromeia* became progressively reduced. *Pleuromeia*, like *Isoetes*, is heterosporous.

The Lycopodiales, or modern clubmosses, are generally herbaceous and of worldwide distribution. Some species are known from the Arctic, whereas others are found in temperate and tropical regions. Both genera of the Lycopodiales, *Lycopodium* and *Phylloglossum*, are homosporous; that is, they produce but one type of spore. A wide diversity in the sculpturing is exhibited by the spores of the many species of *Lycopodium*. Some are smooth or nearly so, others have pitted, rugulate, or reticulate exine patterns. A spore of *Lycopodium phyllanthum* is shown on plate 2-2, figure 6.

The order Selaginellales is represented by one living genus, *Selaginella*. These small, herbaceous plants are widely distributed in temperate and tropical localities. The cones bear numerous microspores in the axils of the microsporophylls in the terminal portions of the stems. The basal sporophylls bear axillary megasporangia, each containing four large megaspores. Megaspores of *Selaginella densa* are shown on plate 2-2, figures 2-3 & 5. Microspores of *S. conduplicata* and *S. weatherbiana* (pl. 2-2, figs. 4 and 11) illustrate two different types of microspores within the genus. *Selaginella conduplicata* has trilete microspores with prominent spines, and microspores of *S. weatherbiana* possess what appears to be a well-developed equatorial cingulum but is actually a perisporial membrane. The microspores of *S. densa* also possess perisporia.

Plate 2-2. Examples of spores from the phylum Tracheophyta: **1.**—*Psilotum nudum* modern (× 1,000); **2.**—*Selaginella densa* megaspore, modern (× 200); **3.**—*S. densa* megaspore, modern (× 200); **4.**—*S. conduplicata* microspore, modern (× 1,000); **5.**—*S. densa* megaspore, modern (× 200); **6.**—*Lycopodium phyllanthum*, modern (× 1,000); **7.**—*Isoetes bolanderi* microspore, modern (× 1,000); **8.**—*I. bolanderi* microspore, modern (× 1,000); **9.**—*I. bolanderi* megaspore, modern (× 200); **10.**—*Equisetum hyemale*, with elaters, modern (× 100); **11.**—*Selaginella weatherbiana* microspore, modern (× 1,000); **12.**—*Ophioglossum fulcatum*, modern (× 1,000); **13.**—*Botrychium lanceolatum*, modern (× 1,000); **14.**—*Elaterites triferens*, Pennsylvanian (× ca. 600) (photo courtesy of R. W. Baxter and S. A. Leisman); **15.**—*Danaea jenmenii*, modern (× 1,000); **16.**—*Todea barbara*, modern (× 1,000); **17.**—*Schizaea pusilla*, modern (× 500); **18.**—*S. pusilla*, modern (× 500); **19.**—*Lygodium articulatum*, modern (× 250); **20.**—*Anemia phyllitidis*, modern (× 1,000); **21.**—*Anemia villosa*, modern (× 1,000); **22.**—*Arcellites disciformis*, Late Cretaceous (× 200); **23.**—*Hemitelia karsteniana*, modern (× 1,000); **24.**—*Ginkgo biloba*, modern (× 1,000); **25.**—*G. biloba*, modern (× 1,000); **26.**—*Cycas revoluta*, modern (× 1,000); **27.**—*C. revoluta*, modern (× 1,000); **28.**—*Gnetum gnemon*, modern (× 1,000); **29.**—*Welwitschia mirabilis*, modern (× 1,000); **30.**—*Ephedra viridis*, modern (× 1,000); **31.**—*E. trifurca*, modern (× 1,000).

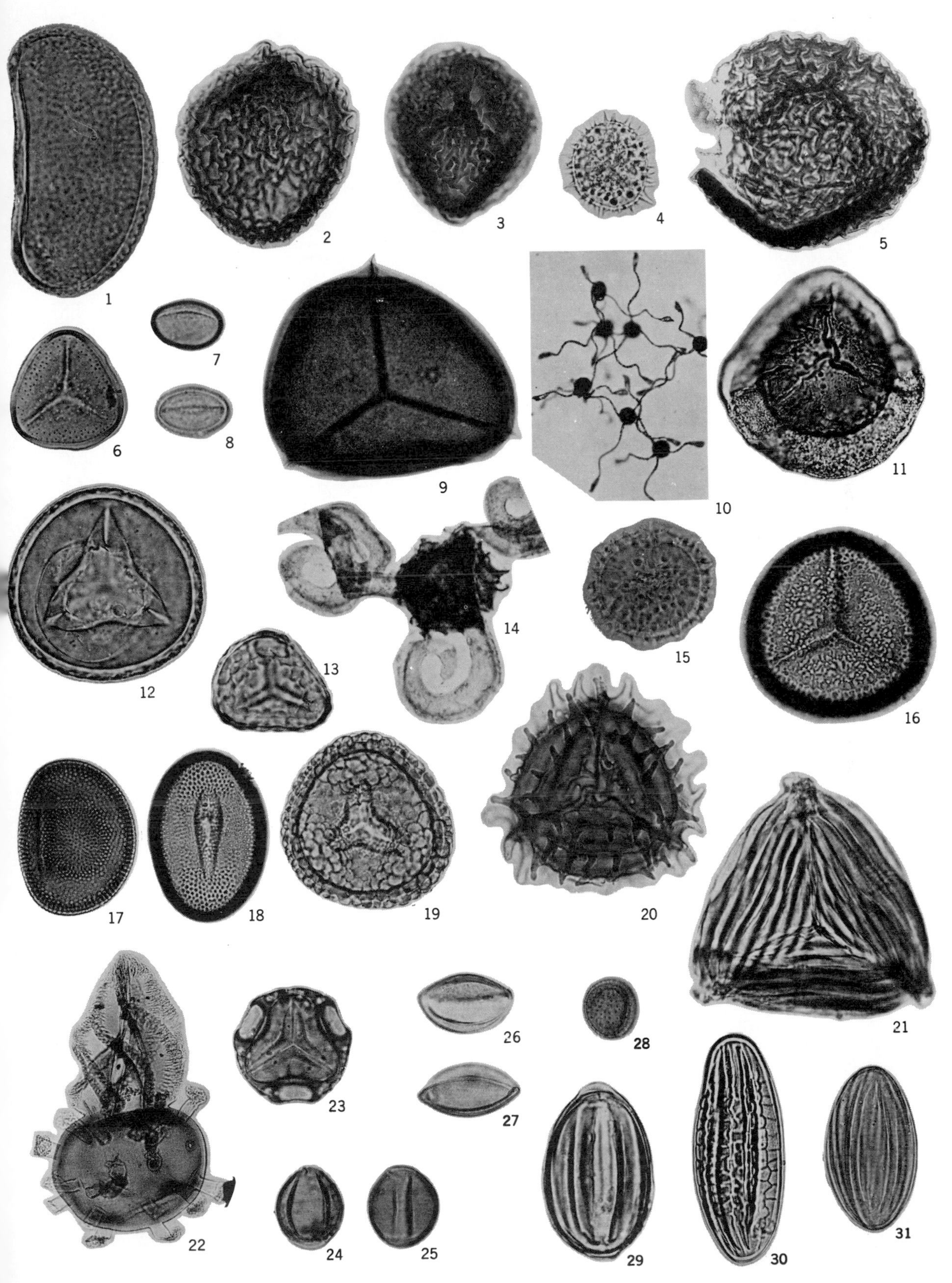
1
2
3
4
5
6
7
8
9
10
11
12
13
14
15
16
17
18
19
20
21
22
23
24
25
26
27
28
29
30
31

11

The order Isoetales is represented by two living genera—*Stylites*, recently described by Amstutz (1957), and *Isoetes*. This latter genus is worldwide in distribution and is found commonly in shallow lakes or ponds. It is a small plant, 50 or more centimeters high, and consists of a tuft of quill-like leaves arising from a cormlike tuberous axis. The outer leaves are megasporophylls, bearing single megasporangia that contain 50 to 300 megaspores. The inner leaves bear microsporangia containing many thousands of microspores. A megaspore of *Isoetes boulanderi* is shown on plate 2-2, figure 9. Characteristically megaspores of *Isoetes* have an outer siliceous coating; however, the megaspore shown on plate 2-2 lacks the outer coating because it was removed during treatment of the specimen with hydrofluoric acid. The megaspores are trilete, but the much smaller microspores are monolete. (See pl. 2-2, figs. 7 and 8.)

The Lycopsida may have evolved from the Psilopsida. They attained their greatest diversity during the Carboniferous, and, with the exception of the few living genera, most of them disappeared by the end of the Paleozoic.

The subphylum Sphenopsida contains one class, the Equisetineae, consisting of three orders—the Hyeniales, the Sphenophyllales, and the Equisetales. The first two orders are represented only as fossils. The Hyeniales were probably homosporous, whereas both homospory and heterospory are known in the Sphenophyllales and Equisetales.

The order Equisetales consists of two families—one contains the fossil genus *Calamites*, the other contains only the living genus *Equisetum*. The modern *Equisetum* looks much like a diminutive Paleozoic calamite. Both homosporous and heterosporous species are known within *Calamites*. Schopf, Wilson, and Bentall (1944) stated: "One feature stands out in the sporogenous sequence of the calamite alliance. More evidence of incipient heterospory is observable there than in any other fossil group. Apparently normal spores of two sizes but otherwise similar in appearance are found in adjacent sporangia. Most noteworthy is the fact that large spores exceeding half a millimeter in diameter show characters very similar to those of less than 100 microns in diameter."

Equisetum possesses spores that are almost unique. Each mature spore is invested with two hygroscopic elaters that coil and uncoil with changes in humidity. (See pl. 2-2, fig. 10.) (Each elater is attached at its center point to the spore; the four free ends give the appearance of four elaters.) Scott (1960) demonstrated convincingly that the supposed elaters on *Equisetosporites chinleana* Daugherty were the thickened ektexinous ridges similar to those found on many *Ephedra* species. The Pennsylvanian spore *Elaterites* Wilson 1943 is the only other spore known to me that possesses elaterlike structures. *Elaterites* possesses three long, coiled elaters, which arise from the splitting of the perispore and are joined together on the distal side of the spore by a triangular truss. They apparently functioned in much the same manner as do *Equisetum* elaters. Mature and immature *Elaterites* spores in sporangia of a Pennsylvanian calamitean cone were found by Baxter and Leisman (1967). A portion of the coal ball that contained a cone was dissolved in hydrofluoric acid. After washing with water the liberated spores were transferred to 50 percent alcohol. In this medium the elaters were observed to coil and uncoil rapidly. A photograph of *Elaterites triferens* kindly supplied by Baxter and Leisman is seen on plate 2-2, figure 14.

The subphylum Pteropsida contains all the remaining plants in the plant kingdom. The class Filicineae is subdivided into four orders. The first of these is the Coenopteridales, known exclusively as fossils. They apparently originated in the Devonian and persisted at least through the Permian. Most of the genera recognized as belonging to the Coenopteridales are known from stem and petiole anatomy. The genus *Botryopteris* is known also from its fructifications. All of the known species of the Coenopteridales from which spores have been obtained except *Stauropteris burntislandica* are homosporous and have spores of the trilete type. The heterosporous *Stauropteris burntislandica* possesses megaspores that are unique. Chaloner (1958) described the megaspore tetrads from this species. They consist of two large united megaspores and two minute (presumably abortive) spores. Chaloner gave this fossil-spore tetrad the name of *Didymosporites scotti*.

The Ophioglossales are the adder's tongue and grape ferns. In these ferns the sporangia usually are borne on fertile spikes that arise from a common stipe with the leafy segments. The Marattiales possess some characters indicating a more advanced phylogenetic position. However, together with Ophioglossales and Coenopteridales, they are usually considered to be more primitive than the Filicales. Mamay (1950) suggested the possible derivation of the Marattiaceae from the Coenopteridales. The Marattiales are homosporous and produce both the trilete and monolete types of spore. The Ophioglossales are also homosporous and, as far as I can determine, produce only the trilete type of spore.

The more advanced order of true ferns, the Filicales, is a large group containing about 132 genera. Both the monolete and trilete spore types are found in this order. In "Genera Filicum" Copeland (1947) arranged the families of modern ferns in a suggested phylogenetic order from the more primitive to the advanced. The Osmundaceae and Schizaeaceae are two of the more primitive families. Plate 2-2, figure 16, shows a spore of *Todea barbara* in the Osmundaceae, and figures 17-18, 19, and 20-21 represent spores of *Schizaea*, *Lygodium*, and *Anemia*, respectively, in the Schizaeaceae. The variety of spore types found in the Filicales is too large to be discussed here. Harris (1955), Knox (1938), Selling (1946), Erdtman (1943, 1957), and Tschudy and Tschudy (1965) have published pictures and drawings of a great number of modern fern spores. These treatments are by no means exhaustive, but they provide a starting point for the examination of modern fern-spore types. A unique spore type is produced by the modern tree-fern genus *Hemitelia*. These trilete spores possess pseudopores in the interradial areas. (See pl. 2-2, fig. 23.) This morphological arrangement is also found in the fossil genus *Kuylisporites*, and there is ample reason to believe that the *Kuylisporites* spore type represents the modern genus *Hemitelia*.

All the families of the Filicales are homosporous except the Marsiliaceae and the Salviniaceae. Both of these families are so-called water ferns. The genera are *Pilularia*, *Regnellidium*, and *Marsilia* in the Marsiliaceae, and *Salvinia* and *Azolla* in the Salviniaceae. Members of these families had already developed at least by Cretaceous time and have persisted to the present. A fossil representative of the water-fern group is *Arcellites*. A megaspore of this genus is shown on plate 2-2, figure 22.

The members of the class Gymnospermae, or conifers and their allies, are all heterosporous, although there may be little difference in size between the megaspores and the microspores, or pollen, containing the male gametophytes. Palynologists are primarily concerned with the pollen because it is commonly the structure that is dispersed and preserved.

The subclass Cycadophytae contains the orders Cycadofilicales and Bennettitales, known only as fossils; and the Cycadales, represented by modern cycads. The Cycadofilicales, or seed ferns, may have originated in the Late Devonian, attained their acme of development in the Carboniferous, and some may have persisted into the Cretaceous. Bisaccate pollen assigned to the genus *Vitreisporites* is almost identical with the pollen obtained from cones of the fossil cycadofilicinean genus *Caytonia*. The pollen of several species of *Crossotheca* is indistinguishable from the trilete fern-spore type. *Stephanospermum* produced microspores with an outer reticulate membrane similar to that seen in the Paleozoic spore genus *Florinites*. Microspores of *Whittleseya* and *Dolerotheca* are bilateral and monolete. The distal surface of these large microspores, or pollen grains, commonly possesses two prominent grooves extending the length of the spore on the margins of an elongate depression. This distal depression may be the forerunner of the sulcus seen in the more advanced Bennettitales and Cycadales. These spores belong to the dispersed-spore genus *Monoletes* (Ibrahim, 1933) Schopf, Wilson, and Bentall 1944 or *Schopfipollenites* Potonié and Kremp 1954.

The Bennettitales (Cycadeoidales) possibly were derived from the Cycadofilicales in the Carboniferous. They became extremely abundant in the Jurassic and probably became extinct in the Cretaceous. "The grains of the Bennettitales are, with one or two exceptions, scarcely different from those of the cycads. They are boat shaped and provided with a single longitudinal furrow . . ." — Wodehouse (1935). The grains are smaller than those of the Cycadofilicales and tend to be larger than those of the cycads and ginkgos. In some respects their morphology is intermediate between that of *Monoletes* and *Cycas*.

The modern representatives of the Cycadales are mostly limited to the tropics and subtropics. The pollen is consistently of the monosulcate type.

The Cordaitales, an extinct order, is perhaps the oldest of several orders of the subclass Coniferophytae. The pollen grains are ellipsoidal, large, about 100 microns in length, and are provided with a broad, deep furrow.

The order Ginkgoales, once widespread and made up of many genera, is now represented by only one genus and species, *Ginkgo biloba*. The pollen—shown on plate 2-2, figures 24 and 25—is of the monosulcate type. The male gametophyte is reduced to two vegetative cells, one generative cell, and one tube cell. So far as is known the plants belonging to the Ginkgoales and Cycadales are the first to have developed the pollen tube. They still retain, however, the multiciliate swimming sperm. In the succeeding orders the male generative cells are reduced to nonciliated nuclei.

The order Coniferales is represented by such well-known plants as pine, fir, juniper, and spruce. The plants in this group first appeared in the Permian and were dominant in Jurassic and Triassic times. With the advent of angiosperm proliferation in the Late Cretaceous they declined to their present status. The pollen of members of the Coniferales is not limited to one type, yet the bisaccate "bladdered" conifer pollen is by far the most common type. In general the Pinaceae, Abietineae, and Podocarpineae possess pollen grains with two or three sacs, whereas the Taxodineae, Cupressineae, and Taxaceae are without sacs. They possess small grains with a thin area or a papilla representing the place where the pollen tube will emerge. Exceptions, however, occur—*Larix* and *Pseudotsuga* in the Abietineae produce large pollen grains of the taxodiaceous or cupressaceous type devoid of sacs.

The Gnetales, the most advanced order of the subclass Coniferophytae, is represented at the present time by the three genera *Welwitschia*, *Gnetum*, and *Ephedra*. The phylogenetic position of the Gnetales is somewhat uncertain and is considered by some to be one of convenience. The pollen of *Gnetum* does not resemble that of any of the conifers and according to Wodehouse (1935) is "obviously" reduced. A spheroidal, echinate grain of *Gnetum gnemon* is shown on plate 2-2, figure 28. Relationship between the pollen of *Welwitschia* and *Ephedra* can be postulated on morphologic grounds. Both possess pollen unlike that from any other known group of plants. *Welwitschia* possesses a single furrow, or sulcus, and multiple heavy ribs. (See pl. 2-2, fig. 29.) The furrow is absent in *Ephedra*, and we can theorize that the polyplicate condition in the latter genus permits the exit of the pollen tube from between any of the ribs, rather than from the single furrow as in *Welwitschia*. Pollen grains of the genera *Vittatina* and *Costapollenites* from the Permian are somewhat similar to those of *Welwitschia*, and *Ephedra*-like pollen has been found in rocks ranging in age from the Triassic to the present.

The class Angiospermae, or flowering plants, divided into the subclasses Dicotyledoneae and Monocotyledoneae, will not be discussed here other than to mention that, so far as is presently known, no unequivocal angiosperm pollen from rocks older than the Aptian has been reported.

The pollen and spore types known from the orders discussed are summarized in table 2-2.

PLANT LIFE CYCLES

The life cycles of plants typically consist of two stages, a gametophyte generation with a single complement of chromosomes (n) and a sporophyte generation with a double complement ($2n$). An examination of the life cycles of a few plants demonstrate the evolutionary trends from simple life forms to the most complex—the angiosperms. (See fig. 2-1.) The gametophyte generation in most lower plants is physically the larger plant of the two generations. In some algae, for example, *Spirogyra*, the sporophyte generation is represented by a single cell—the zygote. When growth of the zygote begins reduction division takes place immediately, and the gametophyte generation reappears. The converse is

Table 2-2. Spore and Pollen Types in the Bryophyta and Tracheophyta

Plant Group	*Spore or Pollen Type*	*Homosporous or Heterosporous*
Bryophyta	Trilete, inaperturate	Homosporous
Psilophytales	Trilete, inaperturate	Homosporous
Psilotales	Monolete, trilete	Homosporous
Lycopodiales	Trilete	Homosporous
Selaginellales	Trilete	Heterosporous
Lepidodendrales	Trilete	Heterosporous
Pleuromeiales	Trilete	Heterosporous
Isoetales	Monolete, trilete	Heterosporous
Hyeniales	Trilete?	Homosporous
Sphenophyllales	Trilete?	Homosporous, heterosporous
Equisetales	Trilete, inaperturate	Homosporous, heterosporous
Coenopteridales	Trilete	Homosporous, heterosporous
Ophioglossales	Trilete	Homosporous
Marattiales	Monolete, trilete	Homosporous
Filicales	Monolete, trilete	Homosporous, heterosporous
Cycadofilicales	Trilete, monolete, monosulcate	Heterosporous
Bennettitales (Cycadeoidales)	Monosulcate	Heterosporous
Cycadales	Monosulcate	Heterosporous
Cordaitales	Monosulcate	Heterosporous
Ginkgoales	Monosulcate	Heterosporous
Coniferales	Monosulcate, inaperturate	Heterosporous
Gnetales	Monosulcate, inaperturate	Heterosporous
Angiospermae	Various	Heterosporous

true in the angiosperms in which the gametophyte generation is confined to the pollen-grain tube and to the few cells of the female gametophyte, hidden in the ovule enclosed in an ovary. The larger plant is the sporophyte.

That pollen grains have evolved from spores is evident. The spore, such as the disseminule of ferns, has a nucleus that has undergone reduction division in the formation of the spore. Consequently the spore represents the beginning of the gametophyte generation and on germination and growth produces the gametophyte generation. In some heterosporous genera, such as *Isoetes* and *Selaginella*, the female gametophyte develops within the spore coat. In both the gymnosperms and the angiosperms the megagametophyte is entirely enclosed within the tissues of the sporophyte. We are concerned chiefly with the analogy between the spore and the pollen grain. Pollen grains differ from spores in being multinucleate young male gametophytes, whereas spores are uninucleate and develop into gametophytes outside the spore coat. Pollen, although commonly different in external form, is essentially a spore in which development of a male gametophyte has proceeded before liberation from the sporangium (anther).

The megagametophyte produces eggs; one of these is fertilized by a sperm nucleus brought into the female gametophyte in the pollen tube. The fertilized egg develops into a young sporophyte within the tissues of the ovule. A seed, such as one developed within the pod (ovary) of a pea plant, consists of sporophytic tissue, the integuments, the nucellus, the endosperm, the megagametophyte and the young sporophyte or embryo ($2n$). The endosperm develops from the fusion of one sperm nucleus and the fused polar nuclei (a tissue containing a triple complex of chromosomes ($3n$). Commonly the young embryo, by the time the seed is mature, has taken up all the food material originally present within the nucellus and endosperm. Consequently such a seed then consists of the young embryo and only remnants of the megagametophyte, nucellus, and endosperm, all enclosed in the ovule integuments.

The male gametophyte within the pollen grain of angiosperms is reduced to three nuclei—the tube nucleus and the two sperm nuclei.

KINDS OF SPORES

The cells of a tetrad may be arranged in a linear series or in almost any other geometric configuration; however, the more common forms are the tetrahedral tetrad and the tetragonal tetrad. The plants whose spores form in tetrahedral tetrads commonly produce spores of the trilete type. The laesurae are the common lines of contact of the cells in the tetrad. The tetragonal tetrad gives rise to the monolete spore type. Spores that show neither type of laesura probably separated early and became spherical before any ektexinous material was deposited on the spore wall. Some

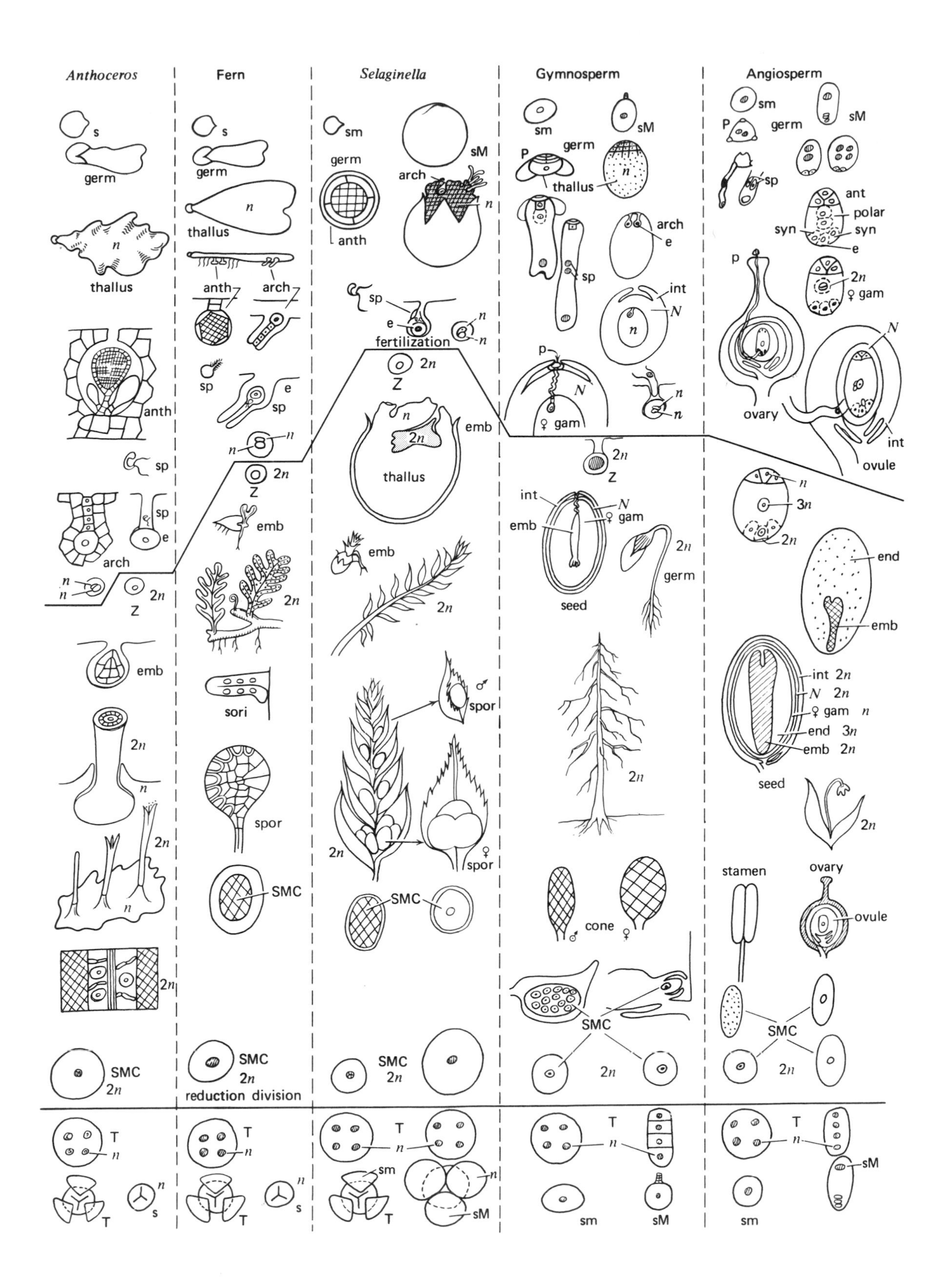
Anthoceros
Fern
Selaginella
Gymnosperm
Angiosperm
s
germ
thallus
anth
sp
arch
e
Z
emb
SMC
2n
n
T
sm
sM
fertilization
sori
spor
reduction division
P
int
N
♀ gam
seed
cone
p
ant
polar
syn
ovary
ovule
end
stamen
3n

of the thin-walled spores, such as those of *Equisetum*, may be of this type.

FUNCTION OF SPORES

The primary function of spores is to distribute and reproduce the plant. A secondary function is the protection of the spore contents during transport and before germination. The development of thick walls on some spores probably has arisen as a response to the survival advantage in prevention of excessive drying. Spores of many members of the Hymenophyllaceae, a fern family living in wet tropical rain forests, are very thin walled. Spores of *Pityrogramma* (a fern) possess exine ornamentation in the form of heavy ridges surrounding the spore at the equator and rugulate or reticulate proximal and distal ornamentation. This arrangement of the thick portions of the spore wall may allow for a certain amount of desiccation without rupture of the wall. *Pityrogramma* grows in comparatively dry localities. A structure that accommodates a semirigid exine to changes in volume has been termed "harmomegathous" by Wodehouse (1935); volume-change accommodation is expressed by the term "harmomegathy."

FUNCTION OF POLLEN GRAINS

The function of pollen grains is to accomplish the transport of the male gametophyte to the female flower so that fertilization can take place. During Cretaceous and post-Cretaceous times a host of modifications that assist in this function developed in angiosperms. Reduction in size and abundant production of pollen are modifications that aid in insuring fertilization of anemophilous plants. The development of viscin threads as in the Oenotheraceae or the specialized pollinia of some orchids are examples of special modifications for effecting the pollination of entomophilous plants. In the Oenotheraceae the viscin threads tend to entangle groups of pollen grains that stick together and are transported to another plant by insects. In the second example pollinia consist of the entire contents of the anther, or many pollen grains that are not separate, and the entire pollinium is transferred from one flower to another by insects.

Function of Exines of Pollen Grains

The function of the exine of pollen grains is threefold. It serves to protect the protoplasm from excessive desiccation and mechanical injury, it provides for the emergence of the pollen tube at the time of fertilization, and it provides for size accommodation as the grain loses or absorbs moisture with changing humidity. The means for the emergence of germ tubes is provided by the germ pore or, in its absence, by the furrows, or colpi. Wodehouse (1935) considered that the function of elongate furrows common to the pollen of many plants is almost entirely harmomegathic. Wodehouse also suggested that the many pores of polyporate grains, in addition to providing loci for the emergence of the pollen tube, also serve a harmomegathic function. Under conditions of high humidity and the accompanying absorption of moisture the intine bulges outward through the pores; during drying conditions the pores permit the intine to be drawn inward. A special modification, possibly evolved to prevent excessive desiccation, is the development of distally inclined sacs in bisaccate pollen, such as that of *Pinus*. During desiccation the two wings are brought together on the distal side of the grain, effectively covering the thin distal sulcus and thus hampering further evaporation.

Figure 2-1. Diagrammatic life cycles of *Anthoceros*, a generalized fern, *Selaginella*, a gymnosperm, and an angiosperm. Legend: s—spore; sm—microspore; sM—megaspore; germ—germination; anth—antheridium; arch—archegonium; e—egg; Z—zygote; n—gametophyte, gametophyte generation; $2n$—sporophyte, sporophyte generation; $3n$—endosperm; emb—embryo; SMC—spore mother cell; T—tetrad; sp—sperm or sperm nuclei; spor—sporangium; ♂—male; ♀—female; ♀ gam—female gametophyte; P—pollen; int—integument(s); ant—antipodal cells; syn—synergids; end—endosperm; N—nucellus.

MORPHOLOGICAL DESCRIPTION OF SPORES AND POLLEN

The rapid growth of palynology has resulted in a proliferation of terms used to describe the structure and sculpture of spores and pollen. This situation is perhaps a natural result of many people working independently. Such a situation is more confusing than helpful to the beginning student of palynology. It would be impossible here to discuss all the more common terms in use or partial use at the present time. The reader is referred to the illustrated "Morphologic Encyclopedia of Palynology" by Kremp (1965) for a more or less complete discussion of most of the terms that have appeared in the palynological literature. Only a few of the terms are discussed here, but it is hoped that this partial treatment will provide sufficient background so that the student can select the most suitable terms. A few of the terms used in the description of spores and pollen grains and their orientation are graphically explained in figure 2-2. Some of these illustrations were redrawn from Couper and Grebe (1961), and the others are original. These should be supplemented by reference to more complete descriptions (Faegri and Iversen, 1950; Couper and Grebe, 1961; Erdtman, 1952; Kremp, 1965).

FORM AND STRUCTURE OF SPORES

The simplest spore type is the alete, or inaperturate, type. Such spores are found in fossil preparations, but it is difficult or impossible to determine whether these spores have affinities with algae, the Bryophyta, the fernlike Tracheophyta, or the Angiospermae.

The monolete type (*Laevigatosporites*) of spore, which occurs in rocks from the Paleozoic to the present, is a common type. These spores occur as simple monolete spores; as monolete spores with a cingulum, such as *Pericutosporites;* and as a monolete spores with a perisporium, such as the modern *Asplenium* spores. In most modern monolete spores the perisporium is destroyed when subjected to chemical treatment. Selling (1946) wrote that in the fossil state the perispore is absent, presumably having been destroyed during the fossilization process.

The trilete type of spore has developed many more morphological variations in structure as well as in sculpture than has the monolete group. Most of these morphological variations are shown in figures 2-3, 2-4, and 2-5.

The primary shapes of spores are shown in figure 2-3, but it must be remembered that all intergradations in shape occur. A single fossil-spore genus may include spores that in outline are circular, convex, or even concave. Spores may develop modifications such as the equatorial crassitude, or zone (fig. 2-3); radial and interradial crassitudes, or interrupted zones (fig. 2-4); proximal or distal crassitudes; or combinations of these.

A complete or partial envelope around the spore is characteristic of many Paleozoic genera. *Endosporites* is a genus possessing a trilete suture on the spore body and a surrounding sac with internal reticulations. The genus *Florinites* is inaperturate, and the spore genus *Potonieisporites* is monolete, with a sac enclosing the distal hemisphere.

Common modern gymnospermous types include bisaccate and trisaccate pollen. These should not be confused with some Paleozoic multisaccate spore types; for example, spores of *Alatisporites* are multisaccate and possess a trilete aperture on the proximal surface. Modern gymnosperm pollen commonly has monosulcate apertures that are located on the distal surface. Some of the species of the bisaccate Paleozoic genus *Illinites* possess well-developed proximal trilete marks, other species are characterized by having one of the rays of the trilete mark shorter than the others.

In discussing Paleozoic spores, and to a lesser degree some Mesozoic and Cenozoic spores, we need additional terms to describe features of the trilete suture. Some of these terms are defined in figure 2-5. As the tectum over the trilete suture develops it may form large leaflike fused extensions, as in *Capulisporites*. In the larger Paleozoic megaspores these fused extensions of the lips of the trilete suture are referred to as massa, or gula. Somewhat similarly in the Mesozoic genus *Ariadnaesporites* the margos of the commissure develop leaflike extensions that are free rather than fused. These leaflike extensions of the margos have been termed acrolamellae by Tschudy (1966). (See fig. 2-5.) Paleozoic spores are discussed further in chapters 10 to 13.

ANGIOSPERMOUS POLLEN GRAINS

The protoplasmic contents of all pollen grains are covered with a more or less elastic membrane known as the intine. Upon the death of the cell this layer is, in common with the cell contents, easily destroyed by bacterial action. Very few species of modern plants produce pollen grains with only this intine layer; for example, *Zostera marina* and *Naias flexilis* do not possess

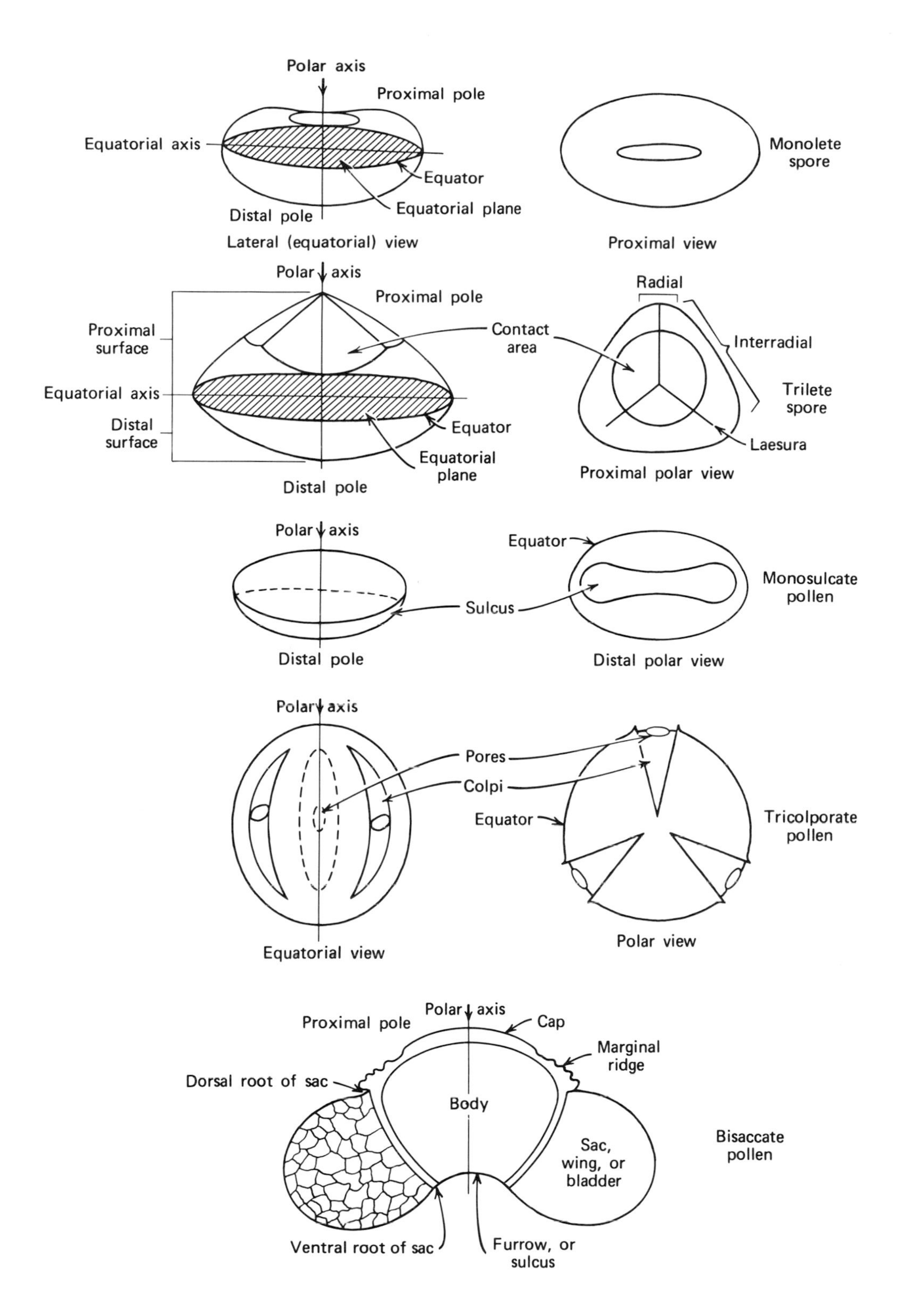

Figure 2-2. General terms used in describing pollen and spores.

Figure 2-3. General terms for shape and structure of spores, with examples.

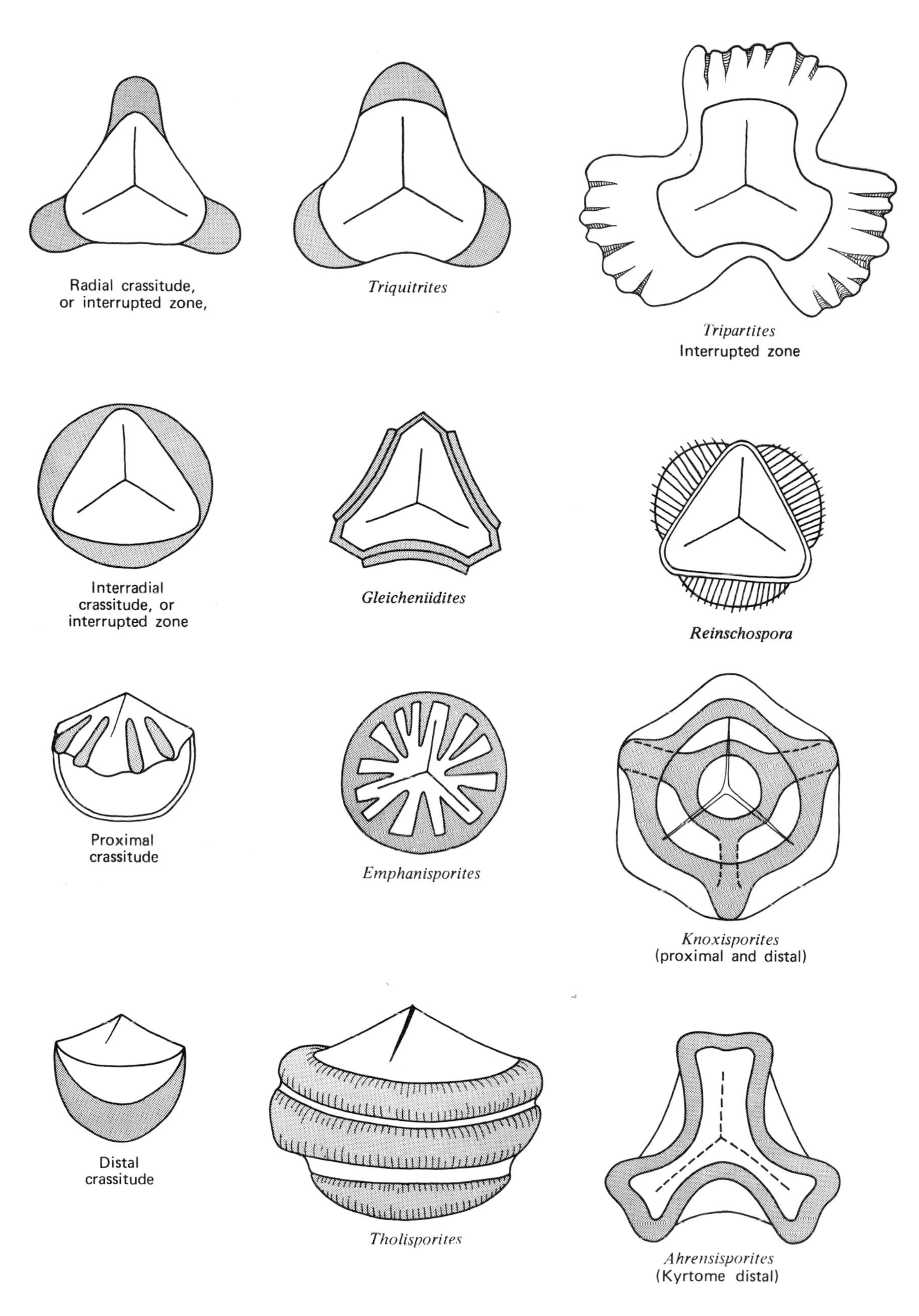

Figure 2-4. Different types of proximal, distal, and equatorial structures.

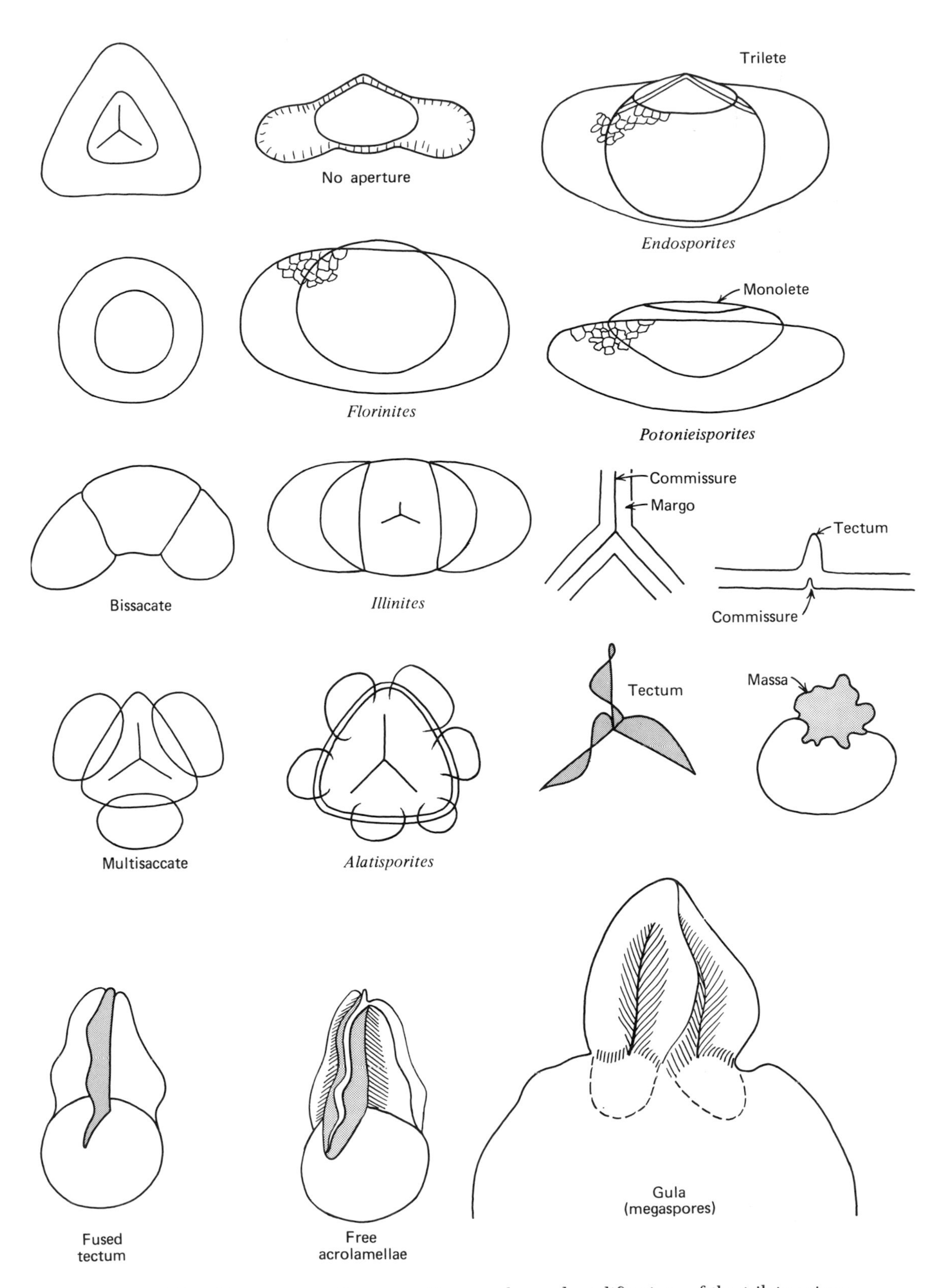

Figure 2-5. Different types of saccus, with examples, and modifications of the trilete suture.

a recognizable outer layer, or exine. The pollen of these two species is developed and shed under water. Such pollen has not been recognized in fossil material, possibly because it would not have been preserved.

By far the majority of angiospermous plants produce pollen with an additional outer coat, the exine. The exine may be thick, as in species of *Geranium*, or thin, as in some species of *Populus*. It may consist of one layer or several layers. The exine is commonly made up of an outer layer, the ektexine, and an inner layer, the endexine. In *Juniperus* the ektexine is reduced to isolated granules on the surface of the endexine. In many members of the Compositae the ektexine itself is made up of several structured layers. (See fig. 2-6.) It is this outer layer, or exine, that is preserved in the fossil state. As a consequence the exine is the part of the pollen grain that is of greatest concern to palynologists.

FORM OF POLLEN GRAINS

The form of pollen grains is influenced by two factors—heredity and position in the tetrad. The hereditary features were termed emphytic by Wodehouse (1935), and the features that are due to contact with neighboring cells during development are termed haptotypic.

The apertures of pollen grains may be furrows (colpi) or pores, or both. In general, the presence of a single furrow (sulcus) characterizes pollen of the monocotyledons, the gymnosperms, or the primitive dicotyledonous type. The presence of a greater number of colpi indicates more advanced dicotyledons. In the fossil record, the first pollen that can be assigned unequivocally to the angiosperms is the tricolpate form. In triporate and colporate pollen the germ pores are situated within a furrow or colpus. In those instances in which a colpus is not recognizable, such as in chenopodiaceous pollen, it is postulated that the colpus has been reduced to coincide with the germ pore.

The Structure and Sculpture of Pollen Exines

The terminology applied to the various subdivisions of the exine is somewhat varied, as shown in the table from Larson, Skvarla, and Lewis (1962).

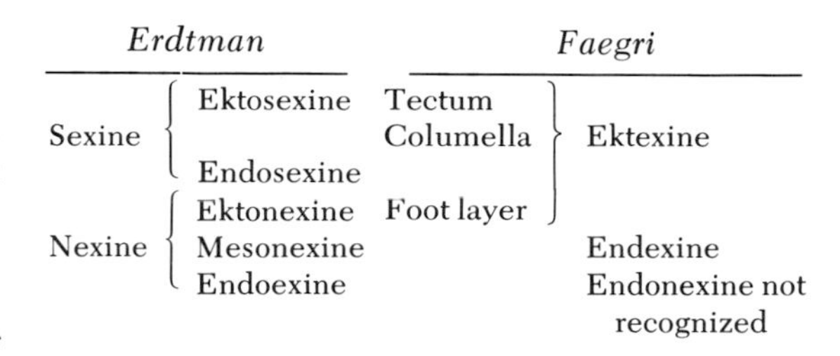

Erdtman		*Faegri*	
Sexine	Ektosexine	Tectum	Ektexine
		Columella	
	Endosexine		
Nexine	Ektonexine	Foot layer	
	Mesonexine		Endexine
	Endoexine		Endonexine not recognized

Although not all these layers are universally present, the ektexine and endexine are commonly recognizable. Most endexine layers are uniform in structure and are undifferentiated. The ektexine usually is not uniform structurally; differentiation often develops in the form of irregularities on the exterior, which are known as sculpture patterns. Figure 2-6 shows diagrams of sections of the pollen wall illustrating the common types of sculpture. Combinations and intergradations of these forms are common.

The sculpture elements and patterns found on Paleozoic (and Recent) spores are essentially the same as those found on pollen grains. In an attempt to bring about uniformity in the description of Paleozoic spores an international committee was organized. In the committee's report (Couper and Grebe, 1961) recommendations regarding terminology were made. These terms do not entirely coincide with the sculptural terms applied to pollen grains. The table below is an attempt to equate the two sets of terms.

Pollen Grains	*Paleozoic Spores*
Verrucate	Verrucate
Gemmate	Pilate
Baculate	Baculate
Echinate	With spinae or coni
	With capilli (forked or fimbriate bacula)
	Rugulate
Rugulate	
	Cristate (tops of elements serrate)
Foveolate	Foveolate
Fossulate	
	Vermiculate
Canaliculate	

The structure of pollen grains is outlined in figure 2-6. In simple pollen the wall is composed of only two layers—an inner uniformly structured endexine and a single-layered ektexine

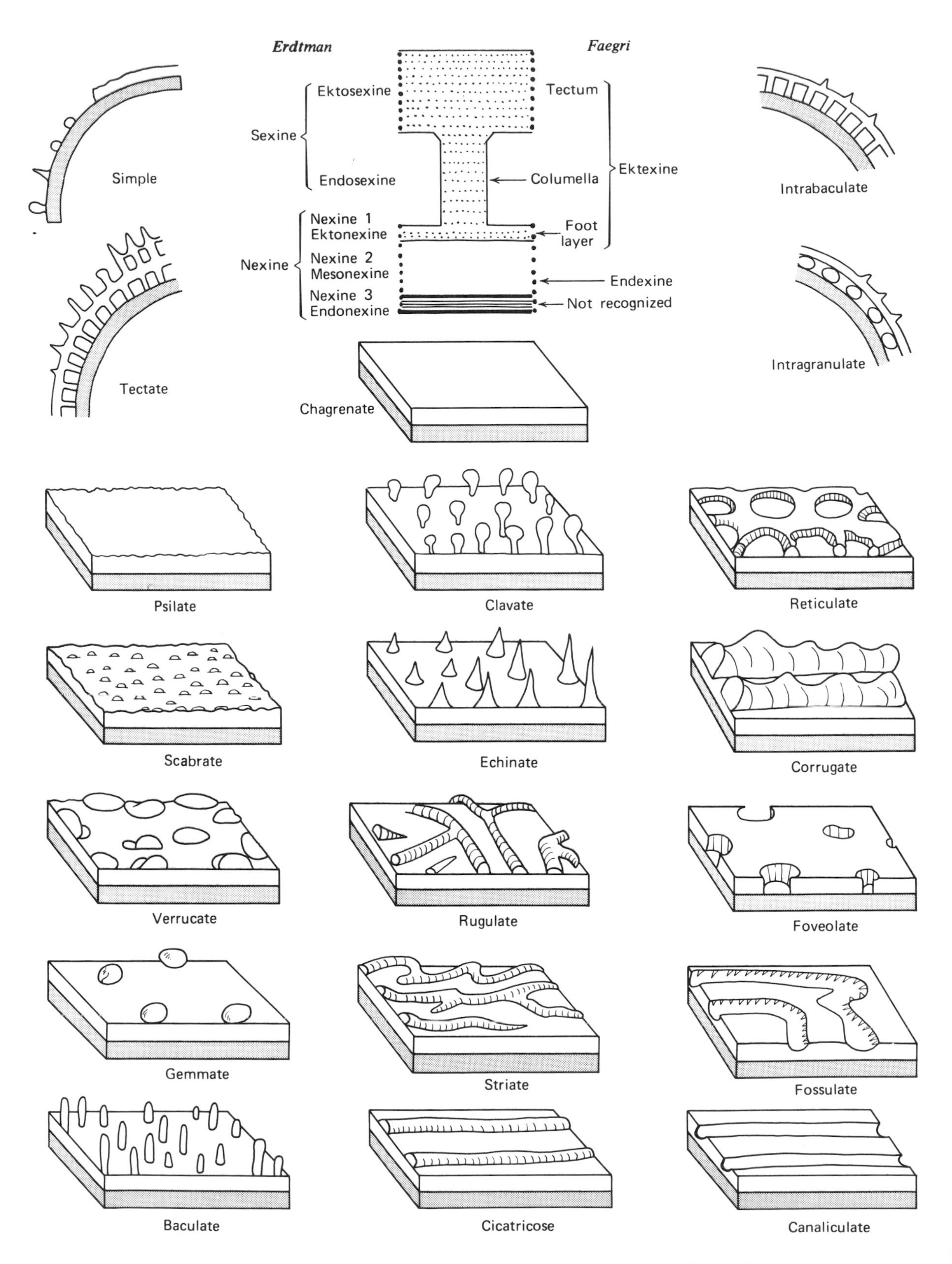

Figure 2-6. Exine layers and examples of different types of surface sculpture.

that may have external sculpture or may be smooth.

The tectate (roofed) wall is made up of three or more layers—the endexine, columellae or granules fused at their apices to form a "roof," and sculptural elements deposited on this layer.

The sculpture types that are applicable to both pollen and spores are described below:

Chagrenate From Turkish *Gaghri*—untanned leather. Shiny smooth, translucent.

Psilate From Greek *psilos*—smooth. A matlike surface that may even possess small depressions less than 1 micron in diameter, as in *Betula*.

Scabrate From Latin *scaber*—flecked. Very small projections less than 1 micron in diameter, as in *Artemisia* or *Quercus*.

Verrucate From Latin *verruca*—a wart. A rounded projection not constricted at the base.

Gemmate From Latin *gemma*—a bud. A spherical projection with constricted base, as in *Juniperus*.

Baculate From Latin *baculum*—a rod. Rodlike projections whose greatest diameter is less than the height, as in *Nymphaea*.

Clavate From Latin *clava*—a club. A clublike projection whose apex is greater in diameter than the base, as in *Ilex*.

Echinate From Latin *echinatus*—prickly. Spiny projections and tapering from base to a sharp point, as in *Taraxacum*.

Rugulate From Latin *ruga*—a wrinkle. Irregularly distributed elongate ridges or wrinkles, as in *Nymphoides peltatum* or *Selaginella densa* megaspore pl. 2-2.

Striate From Latin *stria*—striped. More or less parallel ridges that are wider than the spaces separating them, as in *Menyanthes*.

Cicatricose From Latin *cicatrix*—a scar. More or less parallel ridges that are narrower than the spaces separating them, as in *Anemia phyllitidis*.

Reticulate From Latin *rete*—a net. Ridges forming a net. The walls (muri) narrower than the diameter of the spaces (lumina), as in *Illicium*.

Corrugate From Latin *corrugatus*—wrinkled. Projections are ridges with regular or irregular radial humps or bulges, as in *Riccia natans* pl. 2-1 figs. 14-15.

Foveolate From Latin *fovea*—a pit. Circular pits whose diameter is less than the diameter of the ridges separating them, as in *Lycopodium phyllanthum* pl. 2-2 fig. 6.

Fossulate From Latin *fossa*—a ditch. Possessing a negative reticulum. Surface with irregular grooves, as in *Pteris tripartia* or *Ledum palustre*.

Canaliculate From Latin *canalis*—a channel, canal. More or less parallel cavities or channels narrower than the bastions between them, as in *Anemia simii*.

The common forms of pollen grains are shown on figure 2-7. These diagrams show in a generalized form the position, type, and number of apertures. Inaperturate, sulcate, syncolpate, colpate, porate, colporate, brevicolporate, and syncolporate grains are shown. Also shown are diagrammetic representations of dyad, tetrad, and polyad pollen types. Pollen form is further discussed by Faegri and Iversen (1950), Kremp (1965), and Wodehouse (1935).

Most of the pollen grains encountered in Eocene or younger rocks can be fitted easily into the categories outlined in figure 2-7. Some other pollen grains, however, particularly some of those from the Cretaceous and Paleocene, cannot be placed in these categories. In this group of pollen, which has been termed the Normapolles group by Pflug (1953), are the pollen grains that have so-called bizarre structural elements. Pflug has attempted to place these forms into form genera and to describe their structure. Krutzsch (1959, 1962) has added several new Normapolles genera. Undoubtedly, as more Cretaceous and lower Tertiary rocks are studied, more genera of the Normapolles type will be found.

THE NORMAPOLLES GROUP

The complex morphological features of the Normapolles group have caused difficulties in interpretation. In an attempt to provide some assistance to those who have to work with this group the remainder of this chapter is devoted to a key, definitions, and illustrations.

Pflug (1953) provided a key to the Normapolles group as follows (terms are defined at the end of the key):

I. Germinal apparatus in fives, three equatorial germinals and one at each pole—*Pentapollis*.

II. Germinals all equatorial, in fours or fives, interloculum present—*Tetrapollis*.

III. Germinals in threes:
   - A. With a double Y-mark—*Sporopollis*.
   - B. Germinals subequatorial with endannuli—*Interporopollenites*.
   - C. Pores at equator:
     1. With oculi that form equatorial relief—*Oculopollis*.

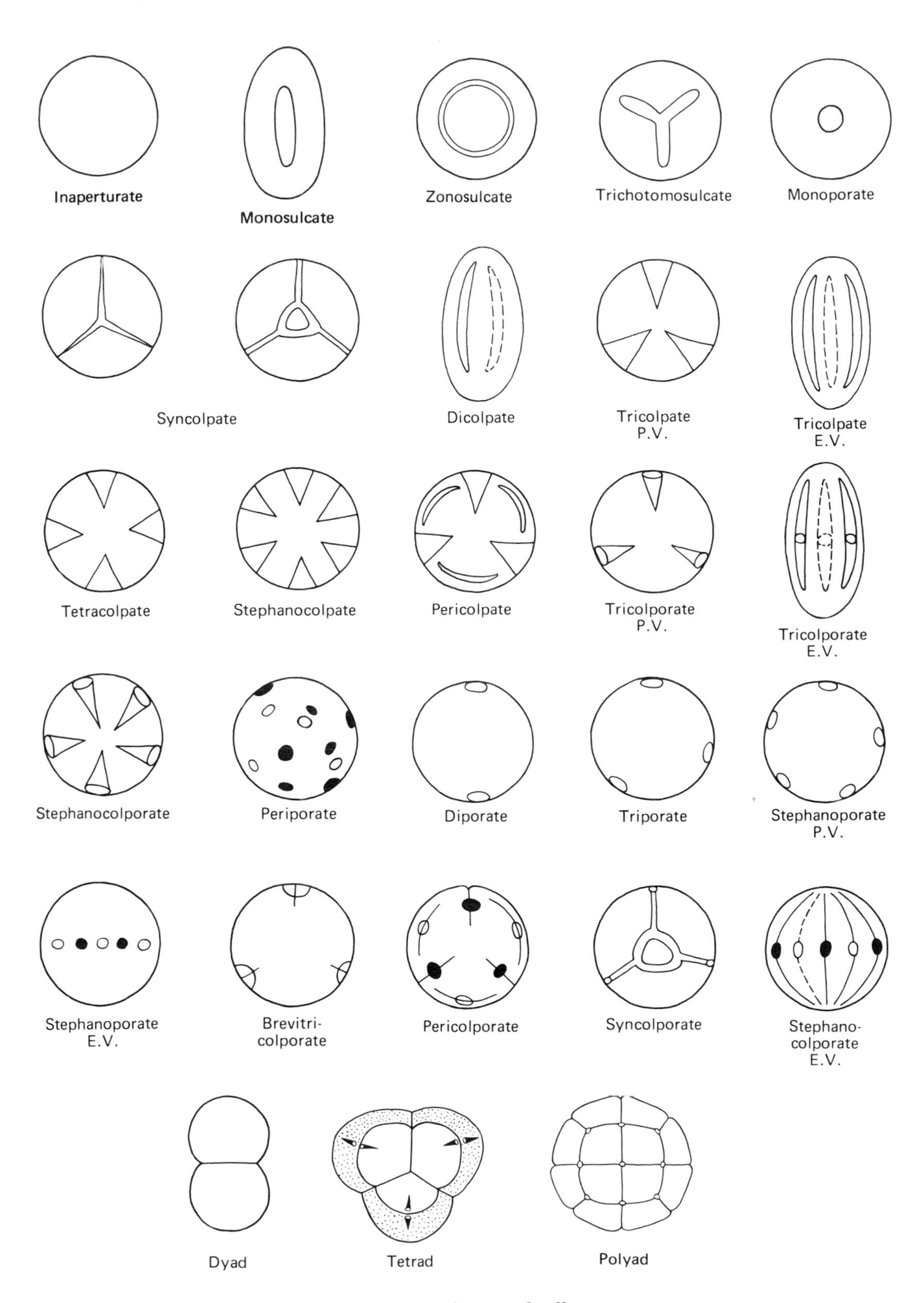

Figure 2-7. Principal types of pollen grains.

2. With pappilus polaris; thick lens shape (axis index between 2/3 and 3/4)—*Papillopollis*.
3. Pore-canal index above 0.25:
   a. With conclave and atrium; no annulus, no interloculum, no vestibulum—*Vacuopollis*.
   b. With conclave, annulus, and atrium: no interloculum, no vestibulum—*Conclavipollis*.
   c. Polyannulate, with vestibulum and praevestibulum; no atrium, no conclave, no interloculum—*Basopollis*.
   d. Polyannulate, no vestibulum, no interloculum, no conclave; endexine very scanty—*Nudopollis*.
   e. Polyannulate, with interloculum; pore-canal index above 0.3—*Extratriporopollenites*.
4. Pore-canal index under 0.25:
   a. With an interloculum—*Trudopollis*.
   b. With strong endoplicae, with vestibulum, without an interloculum—*Plicapollis*.

Definition of Terms used by Pflug

Interloculum An apparent space between the ektexine and endexine (fig. 2-8*h*).

Endannulus A thickening around the endopore, an annulus on the endexine (fig. 2-8*e*).

Double Y-mark A Y-mark or trilete suture on both the proximal and distal surfaces of the grain (fig. 2-8*i*).

Oculus A swelling of the region immediately surrounding a pore. In polar view a dark circular spot or thickening lying on the pore region (fig. 2-8*f*).

Papillus polaris A swelling or protuberance in the polar region ("polpapillus") (fig. 2-8*l*).

Pore-canal index Ratio of length of pore canal to diameter of pollen measured in the axis of the pore canal.

Conclave A roomy atrium that in cross section appears to be filled with rodlike elements at an acute angle to the wall of the atrium (fig. 2-8*c*).

Atrium Endexine absent in the pore region. Diameter of endopore at least 3 times greater than that of exopore (fig. 2-8*a*).

Vestibulum A chamber or space formed by the separation of the ektexine and endexine in the vicinity of the pore. Diameter of endopore less than 3 times as large as that of the exopore (fig. 2-8*b*).

Polyannulate The centrifugal extension of an enlarged annulus forming a long pore canal (fig. 2-8*g*). (Pflug postulates the union of many lamellae, each with an annulus, to form such a structure.)

Praevestibulum A chamber formed in the ektexine and not contiguous to the endexine (fig. 2-8*m*). (Pflug's definition: an interlamellar separation of the ektexine that communicates with the pore.)

Endoplicae Folds in the endexine. Detachment of the ektexine and endexine accompanied by swelling of the endexine, forming tubelike structures that always extend to atria (fig. 2-8*k*).

Even with the key and definition of terms it is difficult to interpret some of Pflug's genera. This is partly due to the extremely brief generic diagnoses and partly due to Pflug's interpretation of intergrading characters in several genera. Also the rather poor pictures of various species make interpretation of structure difficult; for example, the complete generic diagnosis of *Oculopollis* is, *Germinal apparatus in threes, pores lying equatorially. Large oculus complex that forms a relief in the contour.*

The single distinguishing feature of *Oculopollis* is the presence of an oculus. However, Pflug says that a weak oculus is present in the genus *Extratriporopollenites* and in some species of *Trudopollis*. Another example that lends confusion to the genera described by Pflug is presented by the key that indicates that an endannulus is present in the genus *Interporopollenites*, a genus erected by Weyland and Krieger (1953). The genotype description states, "mit Endanulus"—yet the description of the type species, *Interporopollenites proporus*, includes the statement, "kein Endanulus". Pflug propagated this error, in spite of his key definition, by placing a new species without an endannulus in *Interporopollenites*.

It becomes abundantly clear that the Normapolles group of Pflug and others is badly in need of emendation. This book, however, is not the place for such an undertaking.

I have attempted to show, by diagrams, the principal characteristics that I believe Pflug was referring to in his original work. I hope that these diagrams will assist in the interpretation and recognition of some of the already described Normapolles genera. The drawings shown in figure 2-9 are my own interpretations. The characteristics of each genus are given below.

Normapolles Genera and Characteristics Described by Pflug

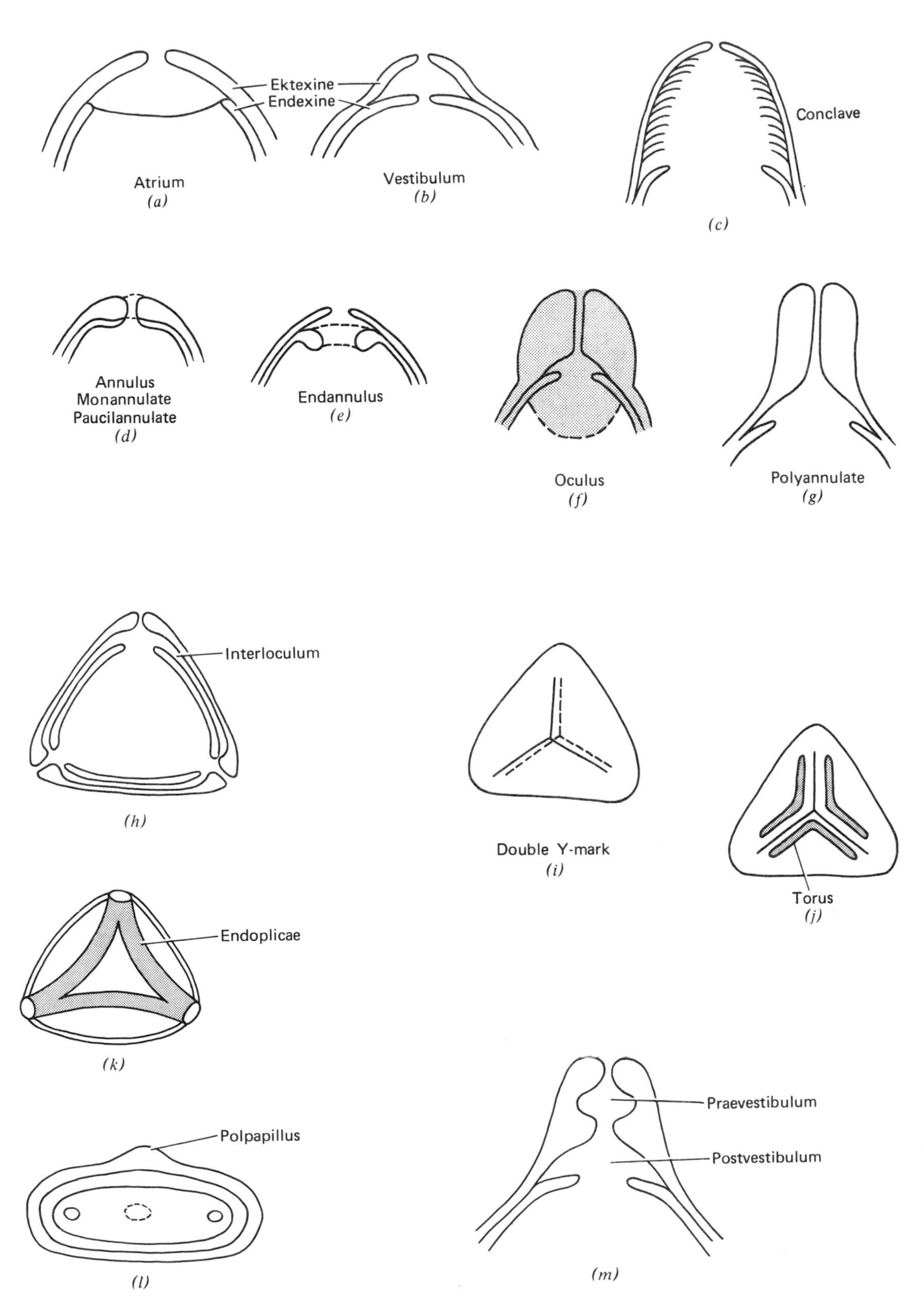

Figure 2-8. Explanation of terms used in describing pollen grains.

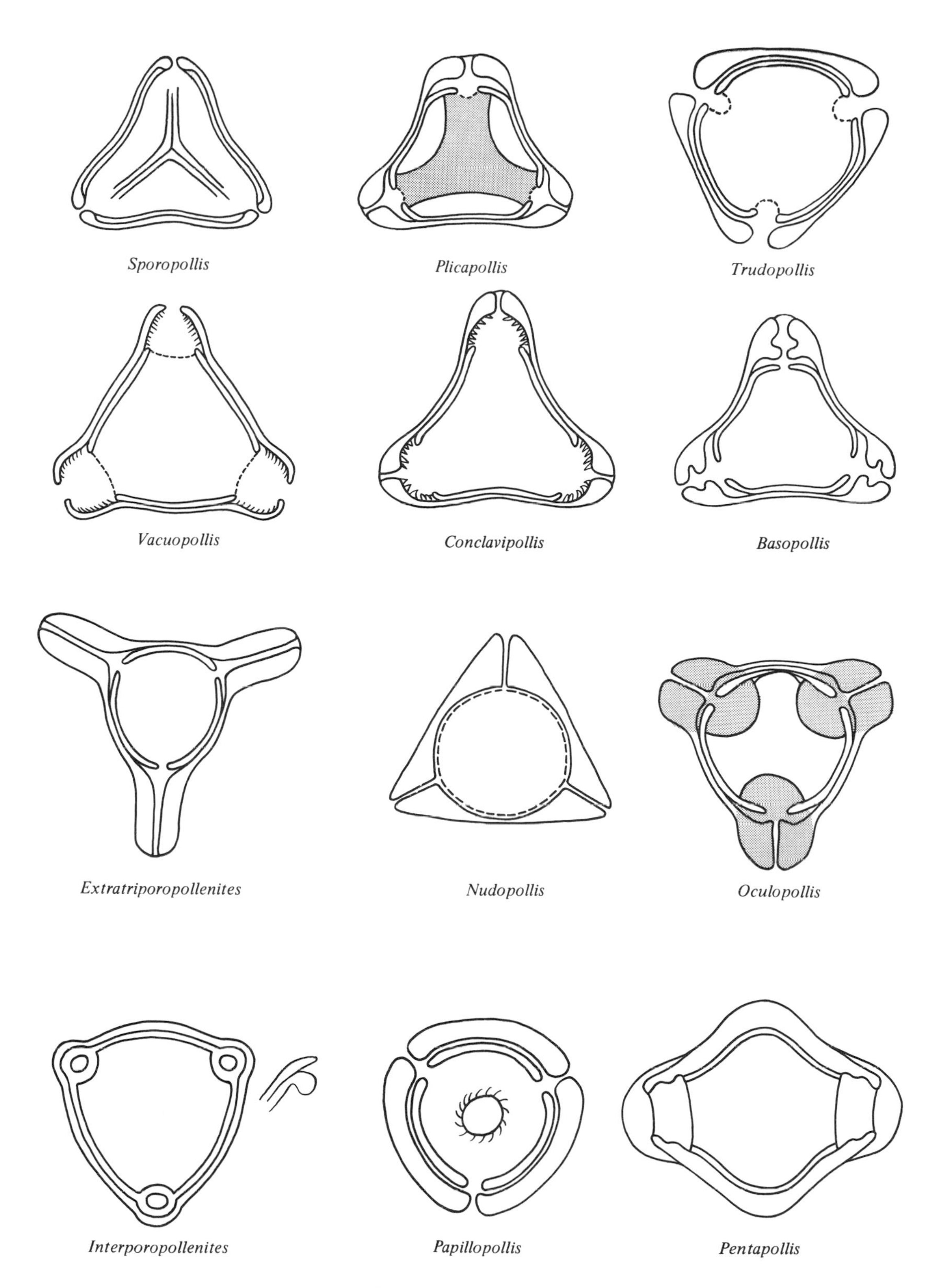

Figure 2-9. Diagrams of principal Normapolles genera (Pflug, 1953).

Sporopollis Proximal and distal Y-marks and torus. (From Pflug's photographs it is not possible to discern whether a Y-mark is present in addition to the tori parallel to its putative position.)

Plicapollis Endoplicae and vestibulum.

Trudopollis Always with an interloculum. Interloculum may be present in other genera such as *Oculopollis* and *Extratriporopollenites*. However, other genera with this characteristic have additional distinguishing features, such as an oculus or the "polyannulate" condition.

Vacuopollis A conclave—no annulus—and an atrium with a wide endoporus.

Conclavipollis An annulus, a conclave, and atrium with a wide endoporus; distinguished from *Vacuopollis* by the presence of an annulus.

Basopollis Praevestibulum and postvestibulum.

Extratriporopollenites Polyannulate; pore-canal index over 0.3; with an interloculum.

Nudopollis Polyannulate; pore-canal index over 0.25; without an interloculum. This genus is distinguished from *Extratriporopollenites* by the very thin endexine, often reduced to the point where it is difficult to recognize.

Oculopollis The presence of bulging oculi.

Interporopollenites Subequatorial pores with endannuli.

Papillopollis The presence of an apical papilla or protuberance.

Pentapollis The presence of bipolar prominences or protuberances. (Pflug says that there are five germinals but that the two polar germinals are not fully developed. Krutzsch (1962) in a later study demonstrates that *Pentapollis* possesses three equatorial germinals and two polar prominences.)

Tetrapollis Same as for *Trudopollis* except that the genus is four pored.

Krutzsch (1959) described eight new Normapolles genera, the characteristics of which are briefly described below. These characteristics are not meant to stand as substitutes for the complete generic diagnoses but are shown merely to assist in elucidating the diagrams shown in figure 2-10, which were redrawn from Krutzsch.

Characteristics of Normapolles Genera Described by Krutzsch

Latipollis Stable position equatorial rather than polar, mostly with atria; exopores with vertical slit.

Monstruosipollis Very long pore canals with a constriction and local bulge; apparent inner and outer atrium; center atrium narrow—scarcely wider than a pore canal.

Turonipollis Long pore canals, germinals two or more, layered to form a prominent praevestibulum; small atrium (postatrium).

Complexiopollis Praevestibulum with conclave; atrium and postatrium, 10 exine layers postulated (and drawn) by Krutzsch. (It is doubtful that 10 layers can be differentiated without resort to sectioning and the use of an electron microscope.)

Pflugipollis Both wall layers thick; proximal pole folded, forming tori; inner-wall boundary in the form of a Y-mark.

Emscheripollis Tori on proximal and distal poles; annulus in inner part of outer-wall layer; interloculum always present; vestibulum; endopore circular and larger than exopore.

Minorpollis Small—8 to 15 microns; three-cornered pollen with annulus, labrum or tumescence; atrium or vestibulum. This genus needs clarification to separate it from *Triatriopollenites*.

Quedlinburgipollis Inner and outer germinals nonsymmetrical; exopore on the sides of the swollen equatorial angles; annulus; endopore with tumescence or endannulus; atrium or vestibulum present.

THE POSTNORMAPOLLES GROUP

The Postnormapolles group does not possess the so-called bizarre germinal development, double Y-mark, torus, or conclave. The genera *Triatriopollenites, Trivestibulopollenites, Multiporopollenites, Stephanoporopollenites, Polyatriopollenites, Intratriporopollenites, Subtriporopollenites,* and *Triporopollenites* were apparently envisioned by Thomson and Pflug (1953) and Pflug (1953) as genera that would encompass all non-Normapolles pollen. Earlier validly published genera were largely ignored. Consequently, before these genera are used, the literature must be searched to determine if any of the genera are later homonyms. An example of the difficulties that may arise may be seen by an examination of the genus *Tricolporopollenites* erected in 1953 by Thomson and Pflug, who ignored an earlier validly published genus, *Tricolporites* Cookson 1947. Furthermore, Pflug and Thomson selected the species *Tricolporites dolium* (alias *Pollenites dolium* Pot. 1931) as the type species. Potonié (1960) stated that *Tricolporopollenites* is not valid because the type species *T. dolium* is morphologically so close to

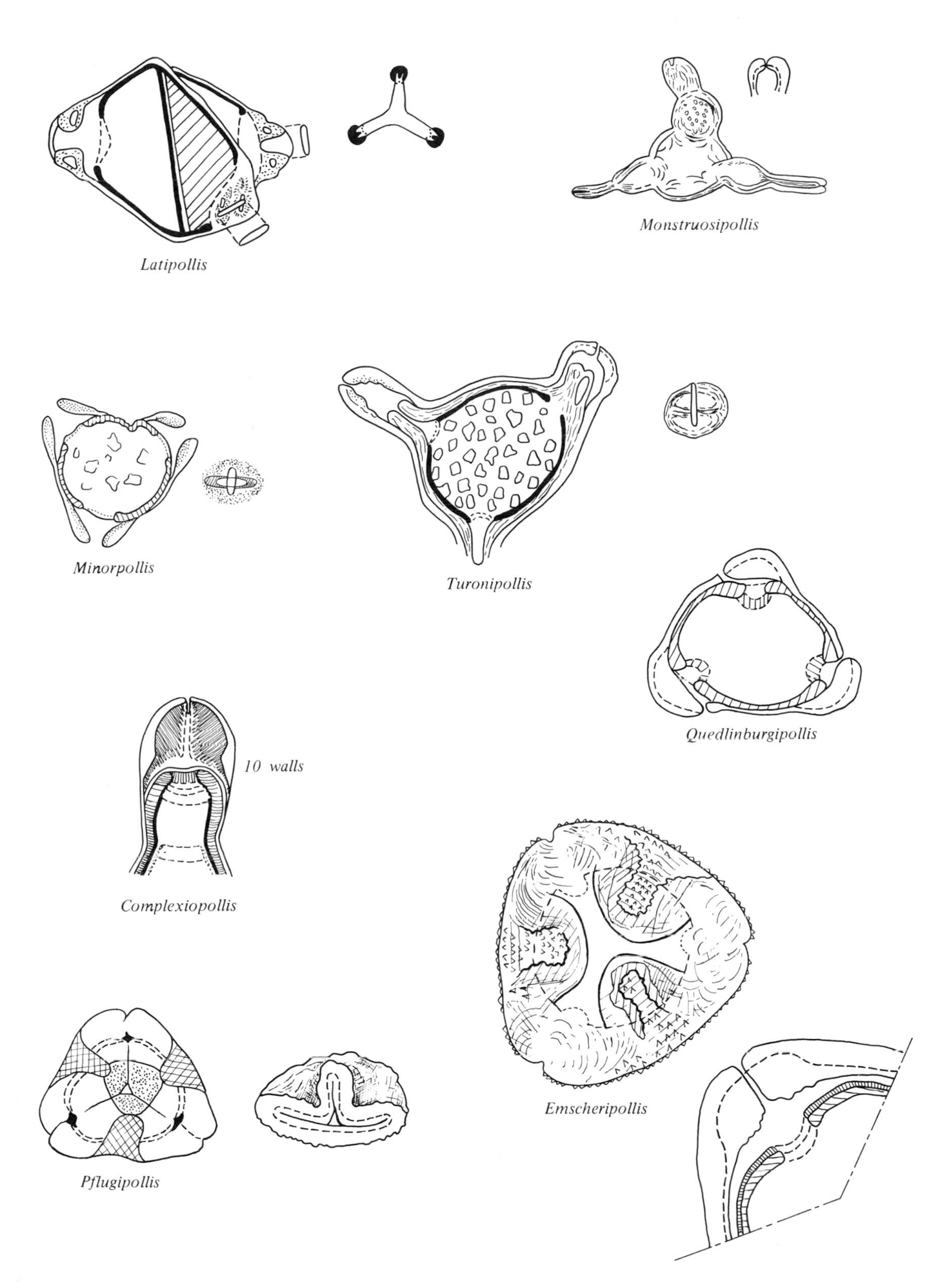

Figure 2-10. Principal Normapolles genera (Krutzsch, 1959).

the type species of *Rhoipites* Wodehouse 1933. Potonié placed the species *T. dolium* in the genus *Rhiopites*, leaving *Tricolporopollenites* a nomen nudum.

Moreover, Pflug chose the type species of *Ulmipollenites* Wolff 1934, *U. undulosus*, as the type species of his new genus *Polyporopollenites*. Because *Ulmipollenites* was validly published by Wolff in 1934. Pflug's treatment was entirely arbitrary and against the rules of botanical nomenclature. Consequently the genus *Polyporopollenites* cannot stand.

References

Amstutz, Erika, 1957, *Stylites*, a new genus of Isoetaceae: Missouri Bot. Garden Annals, v. 44, p. 121-123.

Andrews, H. N., Jr., 1961, Studies in paleobotany; with a chapter on palynology, by C. J. Felix: New York, John Wiley & Sons, Inc., 487 p.

Barghoorn, E. S., and Schopf, J. W., 1966, Microorganisms three billion years old from the Precambrian of South Africa: Science, v. 152, no. 3723, p. 758-763.

Baxter, R. W., and Leisman, G. A., 1967, A Pennsylvanian Calamitean cone with *Elaterites triferens* spores: Am. Jour. Botany, v. 54, no. 7, p. 748-754.

Bold, H. C., 1957, Morphology of plants: New York, Harper & Brothers, 669 p.

Bradley, W. H., 1929, Fresh-water algae from the Green River formation of Colorado: Torrey Bot. Club Bull., v. 56, no. 8, p. 421-428.

———1931, Origin and microfossils of the oil shale of the Green River formation of Colorado and Utah: U.S. Geol. Survey Prof. Paper 168, 58 p.

Chaloner, W. G., 1958, Isolated megaspore tetrads of *Stauropteris burntislandica*: Annals of Botany, new ser., v. 22, no. 86, p. 197-204.

Cookson, I. C., 1953, Records of the occurrence of *Botryococcus braunii*, *Pediastrum*, and the Hystrichosphaerideae in Cainozoic deposits of Australia: Natl. Mus. Victoria Mem. 18, p. 107-122.

Cookson, I. C., and Eisenack, A., 1962, Some Cretaceous and Tertiary microfossils from western Australia: Royal Soc. Victoria Proc., new ser., v. 75, pt. 2, p. 269-273.

Copeland, E. B., 1947, Genera filicum, the genera of ferns: Waltham, Mass., Chronica Botanica Co., 247 p.

Couper, R. A., and Grebe, Hilde, 1961, A recommended terminology and descriptive method for spores, *in* Comptes rendus Reunion de la Commission Internationale de Microflore du Paleozoïque, 3d, Krefeld, May 11-13, 1961, Group 16 report: Creil, Cerchar de France.

Erdtmann, Gunnar, 1943, An introduction to pollen analysis: Waltham, Mass., Chronica Botanica Co., 239 p.

———1952, Pollen morphology and plant taxonomy; Angiosperms (an introduction to palynology, v. I): Stockholm, Almqvist and Wiksell, 539 p.

———1957, Pollen and spore morphology—plant taxonomy; Gymnospermae, Pteridophyta, Bryophyta (an introduction to palynology, v. II): Stockholm, Almqvist and Wiksell, 151 p.

Faegri, Knut, and Iversen, Johannes, 1950, Text-book of modern pollen analysis: Copenhagen, Munksgaard, 168 p.

Fuller, H. J., and Tippo, Oswald, 1949, College botany: Henry Holt & Co., Inc., 993 p.

Harris, W. F., 1955, A manual of the spores of New Zealand Pteridophyta: Wellington, New Zealand Dept. Sci. and Indust. Research Bull. 116, 186 p.

Jovet-Ast, Suzanne, 1967, Bryophyta, p. 17-186 *in* Traité de paleobotanique, v. II, Boreau, E., ed.: Paris, Masson et Cie.

Kidston, Robert, and Lang, W. H., 1917-1921, On Old Red Sandstone plants showing structure, from the Rhynie chert bed, Aberdeenshire; pts. 1-5: Royal Soc. Edinburgh Trans., v. 51-52.

Knox, E. M., 1938, The spores of the Pteridophyta, with observations on microspores in coals of Carboniferous age: Bot. Soc. Edinburgh Trans. and Proc., v. 32, pt. 3, p. 438-466.

Kremp, G. O. W., 1965, Morphologic encyclopedia of palynology: Arizona Univ. Press, 185 p., 38 pl.

Krutzsch, Wilfried, 1959, Einige neue Formgattungen und -arten von Sporen und Pollen aus der mitteleuropäischen Oberkreide und dem Tertiär: Palaeontographica, v. 105, Abt. B., no. 5-6, p. 125-155.

——— 1962, Mikropaläontologische (sporenpaläontologische) Untersuchungen in der Braunkohle des Geiseltales. II – Die Formspezies der Pollengattung *Pentapollenites* Krutzsch 1958: Paläont. Abh., v. 1, no. 2, p. 71-103.

Larson, D. A., Skvarla, J. J., and Lewis, C. W., Jr., 1962, An electron-microscope study of exine stratification and fine structure: Pollen et Spores, v. 4, no. 2, p. 233-246.

Lawrence, G. H. M., 1951, Taxonomy of vascular plants: New York, The Macmillan Co., 823 p.

Lundblad, Britta, 1954, Contributions to the geological history of the Hepaticae; fossil Marchantiales from the Rhaetic-Liassic coal mines of Skromberga (province of Scania), Sweden: Svensk Botanisk Tidskr., v. 48, pt. 2, p. 381-417.

——— 1955, Contributions to the geological history of the Hepaticae. II. On a fossil member of the Marchantiineae from the Mesozoic plant-bearing deposits near Lago San Martin, Patagonia (Lower Cretaceous): Botaniska Notiser, v. 108, no. 1, p. 22-39.

Mamay, S. H., 1950, Some American Carboniferous fern fructifications: Missouri Bot. Garden Annals, v. 37, no. 3, p. 409-477.

Naumova, S. N., 1950, Pollen of angiosperm type in deposits of the lower Carboniferous [in the Moscow basin] [in Russian]: Akad. Nauk SSSR Izv. Ser. Geol., no. 3, p. 103-113.

Newton, E. T., 1875, On "Tasmanite" and Australian "White Coal": Geol. Mag., ser. 2, v. 2, no. 8, p. 337-342.

Pflug, H. D., 1953, Zur Entstehung und Entwicklung des angiospermiden Pollens in der Erdgeschichte: Palaeontographica, v. 95, Abt. B, no. 4-6, p. 60-171.

Potonié, Robert, 1960, Synopsis der Gattungen der Sporae dispersae, III Teil, Nachträge Sporites, Fortsetzung Pollenites, mit Generalregister zu Teil I-III: Geol. Jahrb. Beihefte 39, 189 p.

Potonié, Robert, and Kremp, Gerhard, 1954, Die Gattungen der paläozoischen Sporae dispersae und ihre Stratigraphie: Geol. Jahrb., v. 69, p. 111-193 [1955].

Potonié, Robert, and Venitz, Herbert, 1934, Zur Mikrobotanik der Kohlen und ihrer Verwandten I. Zur Mikrobotanik des miocänen Humodils der niederrheinischen Bucht: Arb. Inst. Paläob. Petrog. Brennsteine, Preuss. geol. Landesanst. Abh. v. 5, p. 5-54.

Schopf, J. W., Barghoorn, E. S., Maser, M. D., and Gordon, R. O., 1965, Electron microscopy of fossil bacteria two billion years old: Science, v. 149, no. 3690, p. 1365-1367.

Schopf, J. M., Wilson, L. R., and Bentall, Ray, 1944, An annotated synopsis of Paleozoic fossil spores and the definition of generic groups: Illinois Geol. Survey Rept. Inv. 91, 73 p.

Scott, R. A., 1960, Pollen of *Ephedra* from the Chinle Formation (Upper Triassic) [Arizona] and the genus *Equisetosporites*: Micropaleontology, v. 6, no. 3, p. 271-276.

Scott, R. A., Barghoorn, E. S., and Leopold, E. B., 1960, How old are the angiosperms?: Am. Jour. Sci., v. 258-A (Bradley volume), p. 284-299.

Selling, O. H., 1946, The spores of the Hawaiian pteridophytes, pt. 1 *of* Studies in Hawaiian pollen statistics: Bernice P. Bishop Mus. Spec. Pub. 37, 87 p.

Sommer, F. W., 1956, South American Paleozoic sporomorphae without haptotypic structures: Micropaleontology (Am. Mus. Nat. History), v. 2, no. 2, p. 175-181, pls. 1 and 2.

Thomson, P. W., and Pflug, H. D., 1953, Pollen und Sporen des mitteleuropäischen Tertiärs; Gestamtübersicht über die stratigraphisch und paläontologisch wichtigen Formen: Palaeontographica, v. 94, Abt. B, no. 1-4, 138 p.

Traverse, A. F., Jr., 1955, Occurrence of the oil-forming alga *Botryococcus* in lignites and other Tertiary sediments [Vermont]: Micropaleontology, v. 1, no. 4, p. 343-349, pl. 1, text figs. 1 and 2.

Tschudy, R. H., 1966, Associated megaspores and microspores of the Cretaceous genus *Ariadnaesporites* Potonié, 1956, emend., p. D76-D82 *in* U.S. Geol. Survey research 1966: U.S. Geol. Survey Prof. Paper 550-D.

Tschudy, R. H., and Tschudy, B. D., 1965, Modern fern spores of Rancho Grande, Venezuela [fern taxonomy by Volkmar Vareschi]: Acta Botanica Venezuelica, v. 1, no. 1, p. 9-71.

Wall, D., 1962, Evidence from Recent plankton regarding the biological affinities of *Tasmanites* Newton 1875 and *Leiosphaeridia* Eisenack 1958: Geol. Mag., v. 99, no. 4, p. 353-362.

Weyland, Hermann, and Krieger, Wilhelm, 1953, Die Sporen und Pollen Aachener Kreide und ihre Bedeutung für die Charakterisierung des mittleren Senons: Palaeontographica, v. 95, Abt. B., no. 1-3, p. 6-27.

Wilson, L. R., 1943, Elater-bearing spores from the Pennsylvanian strata of Iowa: Am. Midland Naturalist, v. 30, no. 2, p. 518-523.

Wilson, L. R., and Hoffmeister, W. S., 1953, Four new species of fossil *Pediastrum*: Am. Jour. Sci., v. 251, no. 10, p. 753-760.

Wodehouse, R. P., 1935, Pollen Grains, their structure, identification, and significance in science and medicine: New York, McGraw-Hill Book Co., Inc., 574 p.

Wolff, Herbert, 1934, Mikrofossilien des pliocänen Humodils der Grube Freigericht bei Dettingen a.M. und Vergleich mit älteren Schichten des Tertiärs sowie posttertiären Ablagerungen: Arb. Inst. Paläob. Petrol. Brennsteine. Preuss. Geol. Landesanst., Abh., v. 5, p. 55-86.

3

WALL STRUCTURE AND COMPOSITION OF POLLEN AND SPORES

A. Orville Dahl

The pollen grain is commonly released from the plant at maturity as a separate cellular unit composed of a bi- or tri-nucleate protoplast invested by a remarkably complex cell wall whose fine structure, chemistry, and development are still subjects of increasingly refined researches. It should be noted that each of the nuclei involved is surrounded by its own discrete membrane-limited mass of cytoplasm with the consequence that each complex made up of nucleus and associated mantle of cytoplasm is basically a cell. Thus the mature pollen grain illustrates the unusual circumstance of a biological unit composed of either a cell (generative) or cells (male gametes), as the particular case may be, within a larger, vegetative or tube cell, all enclosed by the relatively massive pollen wall.

Ordinarily the pollen cell wall is subdivisible into two general zones: an outer *exine* (Fritzsche, 1837), which was demonstrated long ago to be comparatively high in electron-scattering capacity (Fernandez-Moran and Dahl, 1952) and an inner zone, the *intine* (Fritzsche, 1837), differentiated by its marked transparency to electrons. During the past century many approaches have been employed to improve resolution of the fine structure of the component layers of both zones as they are variously represented in pollen grains and spores of different taxa. It is the exine of pollen and spores that is notable for its resistance to chemical and morphological degradation over thousands of years (Ehrlich and Hall, 1959; Faegri and Iversen, 1964). Consequently, until comparatively recently most studies have focused attention primarily on the wall or membranes of pollen and spores rather than considering the entire protoplast in its intricate association with the investing wall.

Orthodox visual light microscopy involving optimally corrected objectives of high numerical aperture, a superior light source, and appropriate optical environments (mounting media) for the specimens (Andersen, 1960; Chamot and Mason, 1938; Christensen, 1954; Dahl and Rowley, 1965; Michel, 1940), remains important in critical investigations of pollen structure. Many specialized techniques have been tried, sometimes with operational difficulty; they have yielded varying amounts of new or significant information. Examples include analyses based on polarized-light microscopy (Freytag, 1954; Sitte, 1960), interference microscopy (Gullvåg, 1964), soft-x-ray absorption (Dahl and others, 1957), ultra-violet-light microscopy (Erdtman, 1959), and freeze-etching procedure (Hess and Stocks, 1967).

Electron microscopy was first applied to thinly sectioned pollen grains in 1950 (Fernandez-Moran and Dahl, 1952) and has since been employed in increasingly refined ways by numerous workers (Gullvåg, 1966b). For the basic details of fixation, embedding, and the production of ultra-thin sections of pollen and spores for electron microscopy summaries by Pease (1964), Mollenhauer (1964), and Skvarla (1966) are most helpful. Electron microscopy of carbon replicas of pollen and spore surfaces is of considerable value in providing high-resolution structural details; these have the important potential of adding a third dimension to the electron micrographs derived from ultrathin sections (Bradley, 1958; Chanda and Rowley, 1967; Mühlethaler, 1955; Rowley, 1960; Rowley and Flynn, 1966). (See pl. 3-1, fig. 1.)

Although the principles of the scanning electron microscope have been known for decades, it is only relatively recently that newly developed scanning electron microscopes have been applied to metallurgy and certain biological materials where surface features are of particular interest. The resolution of the scanning electron microscope currently is not as great as that of the conventional, transmission-type electron microscope, but it is noteworthy for its greater depth of focus and production of striking three-dimensional images of whole, unsectioned specimens. These images are produced by secondary, or deflected, or backscattered electrons derived from the specimen being scanned by a finely focused electron probe. Impressive electron micrographs of pollen of various taxa have already been published (Erdtman and Dunbar, 1966; Thornhill, Matta, and Wood, 1965; Echlin, 1968). (See pl. 3-1, fig. 2; pls. 3-3 and, 3-4.)

The mature pollen grain (fig. 1 of pl. 3-2) in being the immature male gametophyte obviously relates to the process of fertilization (Linskens, 1964). However, a functional interpretation of the fine structural details, particularly with regard to the elements of the complex wall or sporoderm, is not so obviously attained (Gullvåg, 1966b), as will be clear from the descriptive summary that follows.

INTACT POLLEN GRAIN

Most pollen grains may be classified as being either without differentiated pores or furrows (inaperturate) or with preformed openings or thin areas (aperturate) (Erdtman, 1952; Faegri and Iversen, 1964). In the aperturate class of pollen grains there is tremendous variation in the number, size, distribution, and structure of the apertures. Critical analysis at microscopical (Erdtman, 1952; Faegri and Iversen, 1964; Van Campo, 1961) and submicroscopical (Dahl, 1964; Roland, 1966; Rowley, 1964; Rowley and Dahl, 1962; Rowley and Southworth, 1967; Skvarla and Turner, 1966) levels have yielded data of systematic, diagnostic, and cytological significance. In many diverse taxa investigated the apertural portion of the sporoderm appears usually as a distinct area that is either devoid of exine or with the component much reduced or modified. (See pl. 3-2.) In the latter instance exinous structures may be systematically or randomly distributed over the surface of the apertural membrane and persist after fossilization or acetolysis because of anchoring strands or systems of exine associated with the contiguous nonapertural exine (Chanda and Rowley, 1967; Erdtman, 1960; Roland, 1966; Rowley, 1964; Rowley and Dahl, 1962). In many taxa the apertural membrane, exclusive of any exinous components that may be present, may be markedly and abruptly thickened. Such areas have been designated Zwischenkörper (Ehrlich, 1958; Fritzsche, 1837) or onci (Ehrlich, 1958; Hyde, 1955) and, although usually not fossilized, are of complex structure (Ehrlich, 1958). In *Liriodendron* the prominent apertural membrane consists of an inner, relatively smooth-textured, cellulosic layer that is continuous with the intine and an outer, prominently expanded, pectic-callosic layer of channeled, spongelike structure. (See pl. 3-2, figs. 4, 5, 6, and 7.) Studies of the early development of this structure indicated that such channels, or canals, were points of direct association through cytoplasmic strands emanating from the protoplast of the developing pollen grain (Dahl and Rowley, 1965; Stone, Reich, and Whitfield, 1964). In some taxa certain of these strands appear to be replaced by exinous materials since they are resistant to acetolysis. An apertural membrane of such structure would appear to have great potential for imbibition and swelling. This could be the physical basis for the elegantly distinctive, but impermanent, ruthenium-red reaction described by Mangin (1893)

Plate 3-1. Electron micrographs of ultra thin sections.
(Plate 3-1) 1. *Liriodendron tulipifera* L., Botanical Garden, University of Minnesota. Electron micrograph of carbon replica of mature-pollen-grain surface, shadowed with chromium (× ca. 17,500)
(Plate 3-1) 2. *Alnus japonica* (Thub.) Steud. Scanning electron micrograph (× 6000). (Photograph provided by courtesy of Jeolco (U. S. A.), Inc. (Japan Electron Optics Laboratory Co., Ltd.), 477 Riverside Ave., Medford, Mass. 02155.)

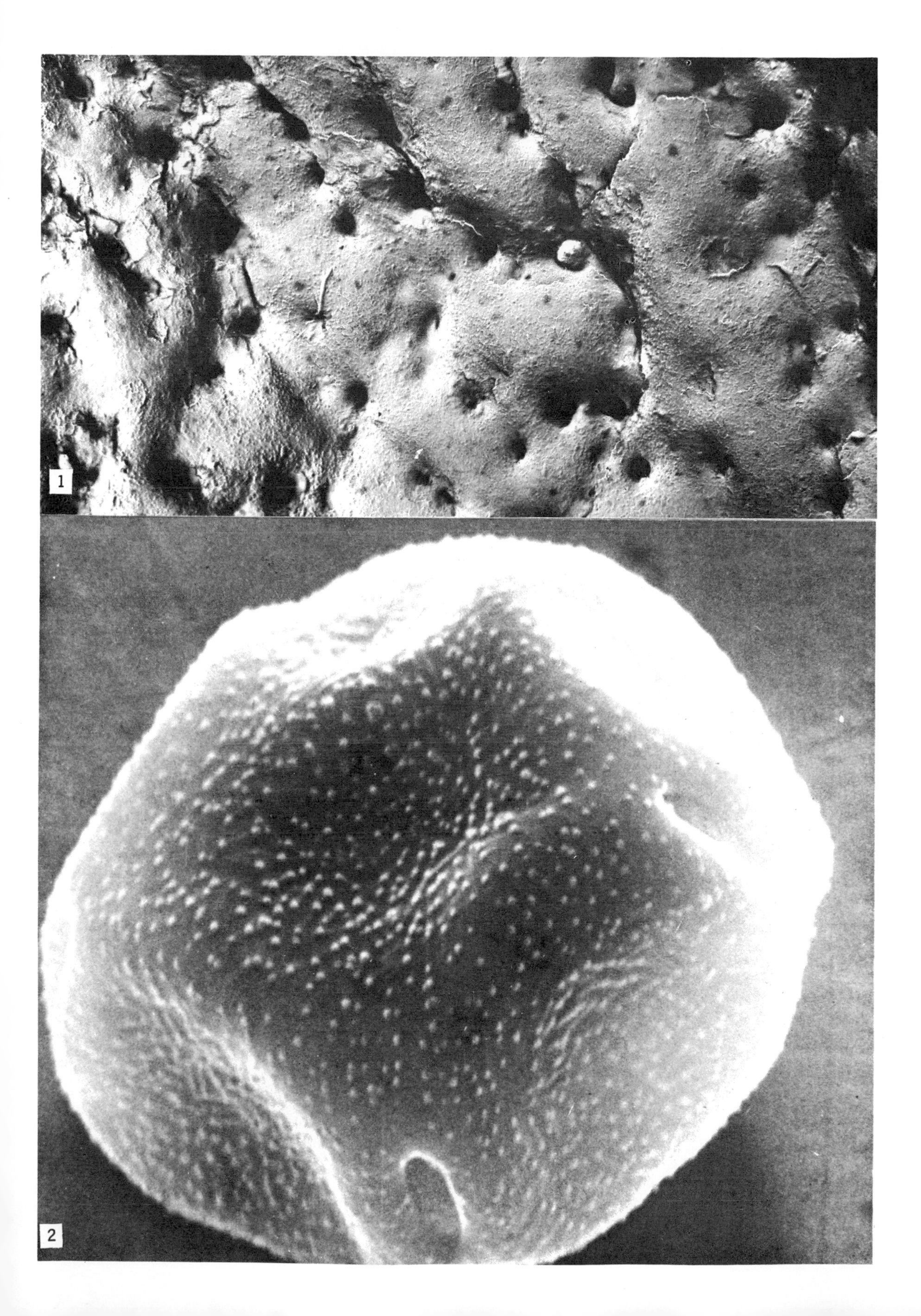
1
2

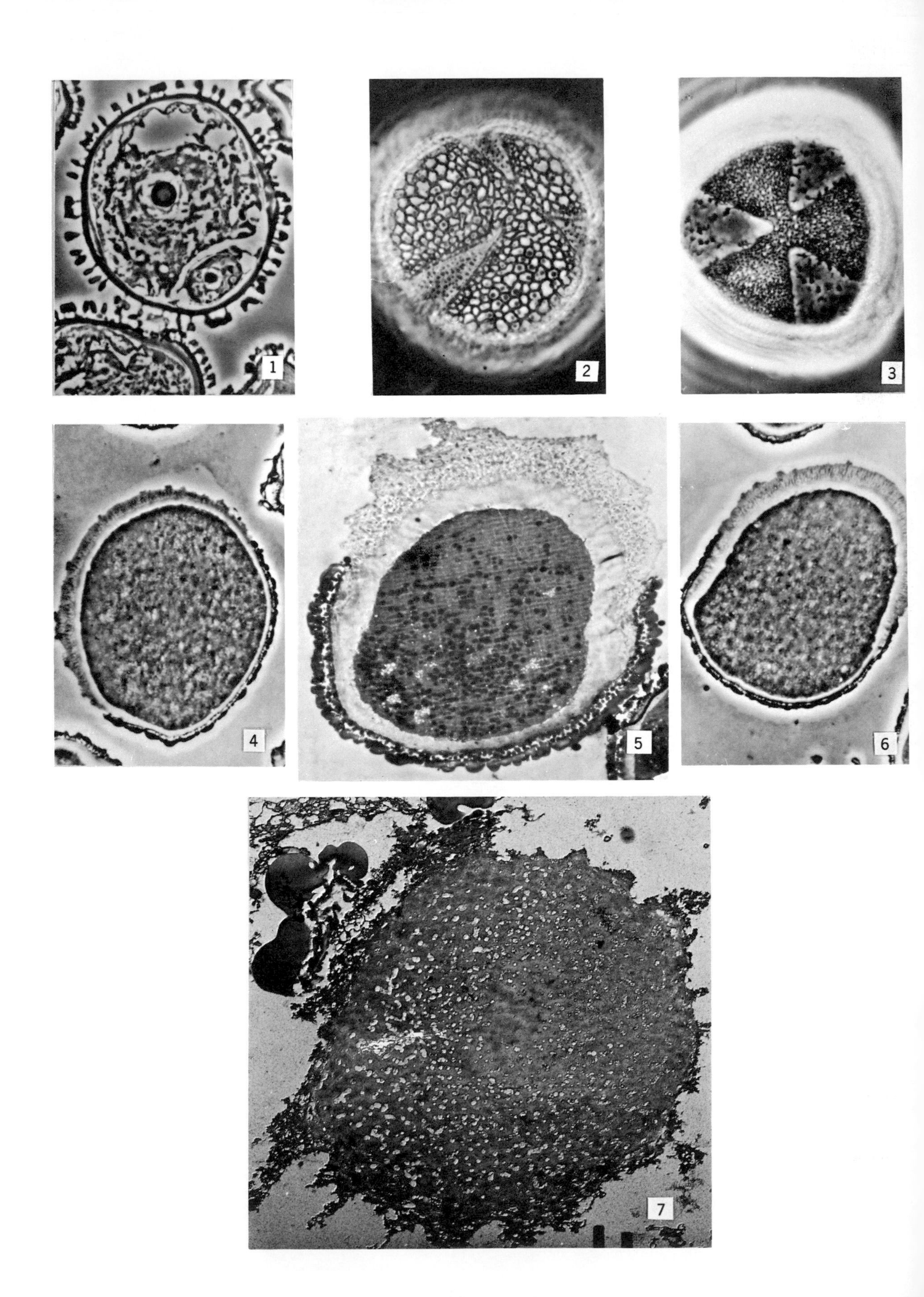
1
2
3
4
5
6
7

and Bailey (1960) in the observation of apertural membranes.

In general apertures have been related to the basic functions of (a) provision of a place of emergence for the developing pollen tube and (b) accommodation to the significant volume changes (harmomegathi – Wodehouse, 1935) that occur in the pollen grain as a result of rapidly changing humidities.

The nonapertural exine is the outer, remarkably resistant layer of the sporoderm. Its surface configuration can be extraordinarily intricate and taxonomically distinctive (Erdtman, 1952; Faegri and Iversen, 1964; Wodehouse, 1935). The presence of ornate structuring of the exine surface has been correlated in many instances with insect pollination (Faegri and Van der Pijl, 1966; Wodehouse, 1935). Visual-light microscopy, electron microscopy, and other techniques have revealed a subdivided structure within most exines. Complexity and some confusion have appeared in the widely scattered literature with respect to nomenclature and interpretation of the various layers of exine. A summary will be presented in a later paragraph. Functionally, nonapertural exine has been associated with protection against excessive water loss, irradiation, and mechanical injury (Wodehouse, 1935).

The chemistry of the highly resistant exine is still undergoing investigation. Long ago Zetzsche and Liechti (1937) related the resistance of exine to a fraction designated sporonin or pollenin, depending on source, for which he suggested the empirical formula of approximately $C_{3x}H_{4.3x}O_x$. Shaw and Yeadon (1964, 1966) and Freytag (1967) have recently confirmed some of the earlier observations but added that sporonin involves a ligninlike fraction and lipid. They explain the absence of reactivity of the ligninlike fraction as due to the nature of its physical association with cellulose and the lipid fraction. The lipid fraction, although a major constituent in both *Lycopodium* and *Pinus*, needs to be more fully investigated.

In the maturation of the sporoderm the intine is the last formed zone, or layer, immediately adjacent to the protoplast. Peripheral portions of the intine may exhibit complex intergrading relationships with the innermost zones of the exine. Since it is ordinarily devoid of sporopollenin (Faegri and Iversen, 1964), it is usually absent in fossilized or acetolyzed specimens. Frequently its inner zone is relatively high in cellulose, with pectic substances, polyuronides, or callose being more prominent in the outer, or first formed, layer of the intine (Bailey, 1960). Structurally the intine layers associate directly with apertural structures.

THE DEVELOPMENT OF THE SPORODERM

During the past decade an increasing number of investigations with specialized techniques have been directed toward improved understanding of the details of the sporoderm development,

---

Plate 3-2. Mature pollen grains.

(Plate 3-2) 1. *Helleborus orientalis* Lamb. (*H. kochii* Schiffn.), Botanical Garden, Copenhagen, Denmark. Mature pollen fixed in formalin-osmium tetroxide. Photomicrograph, Leitz phase contrast, of thin section intact in methacrylate embedding medium mounted in *n*-butanol (refractive index = 1.40). Note nuclei of generative and tube cells. (× ca. 1170.)

(Plate 3-2) 2. *H. abchasicus* A. Br., Botanical Garden, Copenhagen. Intact, mature pollen grain fixed in 1% lactic acid in 30% ethanol and mounted in 4:1 lactic acid-*n*-butanol. Intermediate focal plane of lower hemisphere of pollen grain. Note comparatively massive reticulation of exine (seen in section in fig. 1) and small exinous islets on apertural membranes. Polar view, phase contrast. (× ca. 1170.)

(Plate 3-2) 3. *H. niger* L., Botanical Garden, Copenhagen. Mature pollen grain, fixed in 1:3 glacial acetic acid-*n*-butanol. Lower focus of upper hemisphere of pollen grain. Note relatively fine foveation of exine and comparatively large islets of exine in apertural areas. Polar view, phase contrast. (× ca. 1170.)

(Plate 3-2) 4,5,6. *Liriodendron tulipifera* L. Mature pollen grain; osmium tetroxide fixation; methacrylate embedding medium. Thin sections showing ektexine, endexine, intine, and complex nature of apertural region.

(Plate 3-2) 4. Phase contrast (Leitz) photomicrograph, with methacrylate embedding medium intact, mounted in tricaproin (trihexanoin), r.i. = 1.44. Sagittal section through apertural region. (× ca. 1250.)

(Plate 3-2) 5. Electron micrograph, transverse section through aperture. (× ca. 1250.)

(Plate 3-2) 6. Same as Fig. 4, except transverse section through aperture. (× ca. 1250.)

(Plate 3-2) 7. *L. tulipifera*, mature pollen grain. Electron micrograph. Thin tangential section through outer portion of apertural membrane showing numerous channels (more or less circular, low-density structures) in transverse section. (Magnification ca. 4000 x.)

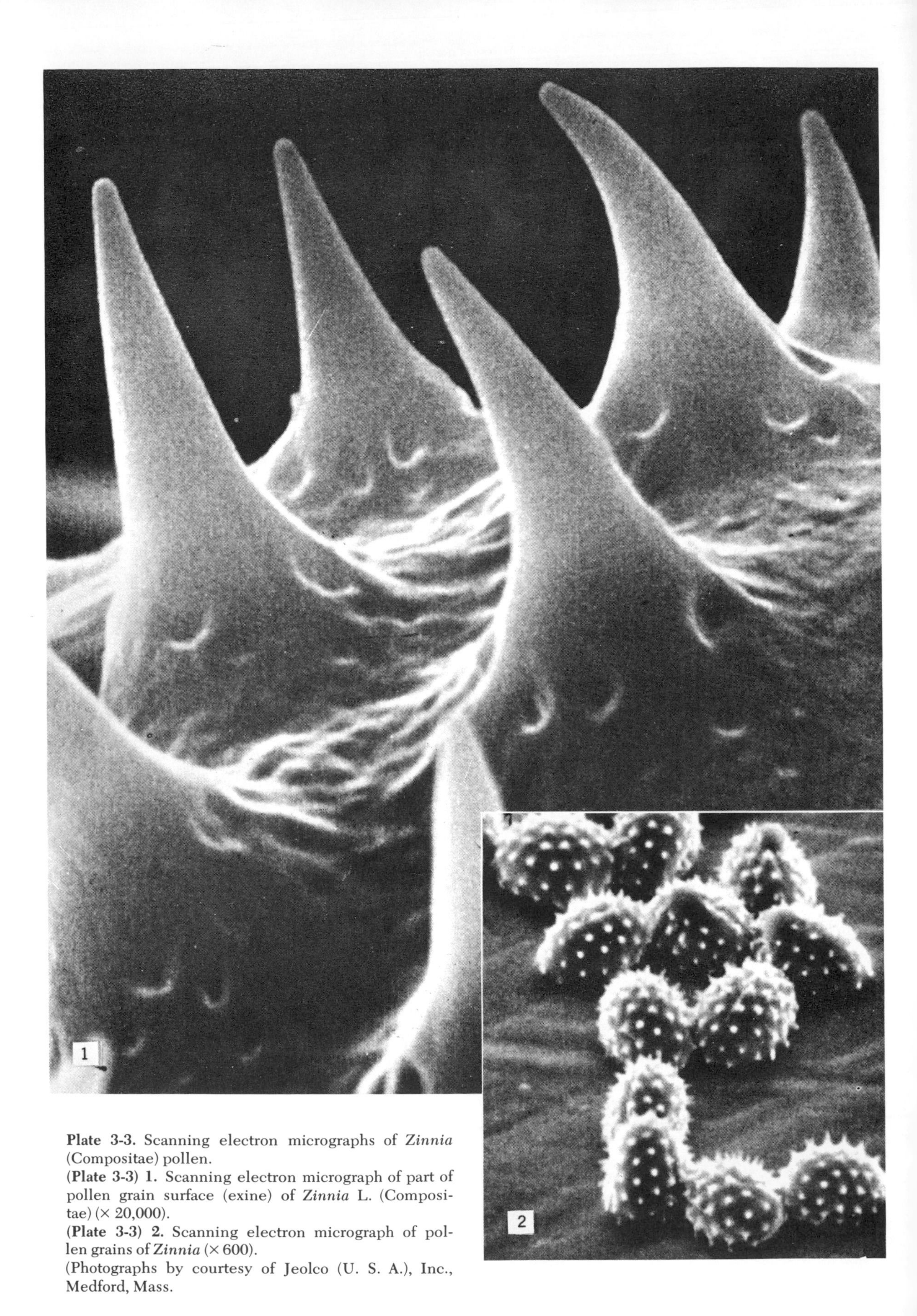

Plate 3-3. Scanning electron micrographs of *Zinnia* (Compositae) pollen.
(Plate 3-3) 1. Scanning electron micrograph of part of pollen grain surface (exine) of *Zinnia* L. (Compositae) (× 20,000).
(Plate 3-3) 2. Scanning electron micrograph of pollen grains of *Zinnia* (× 600).
(Photographs by courtesy of Jeolco (U. S. A.), Inc., Medford, Mass.

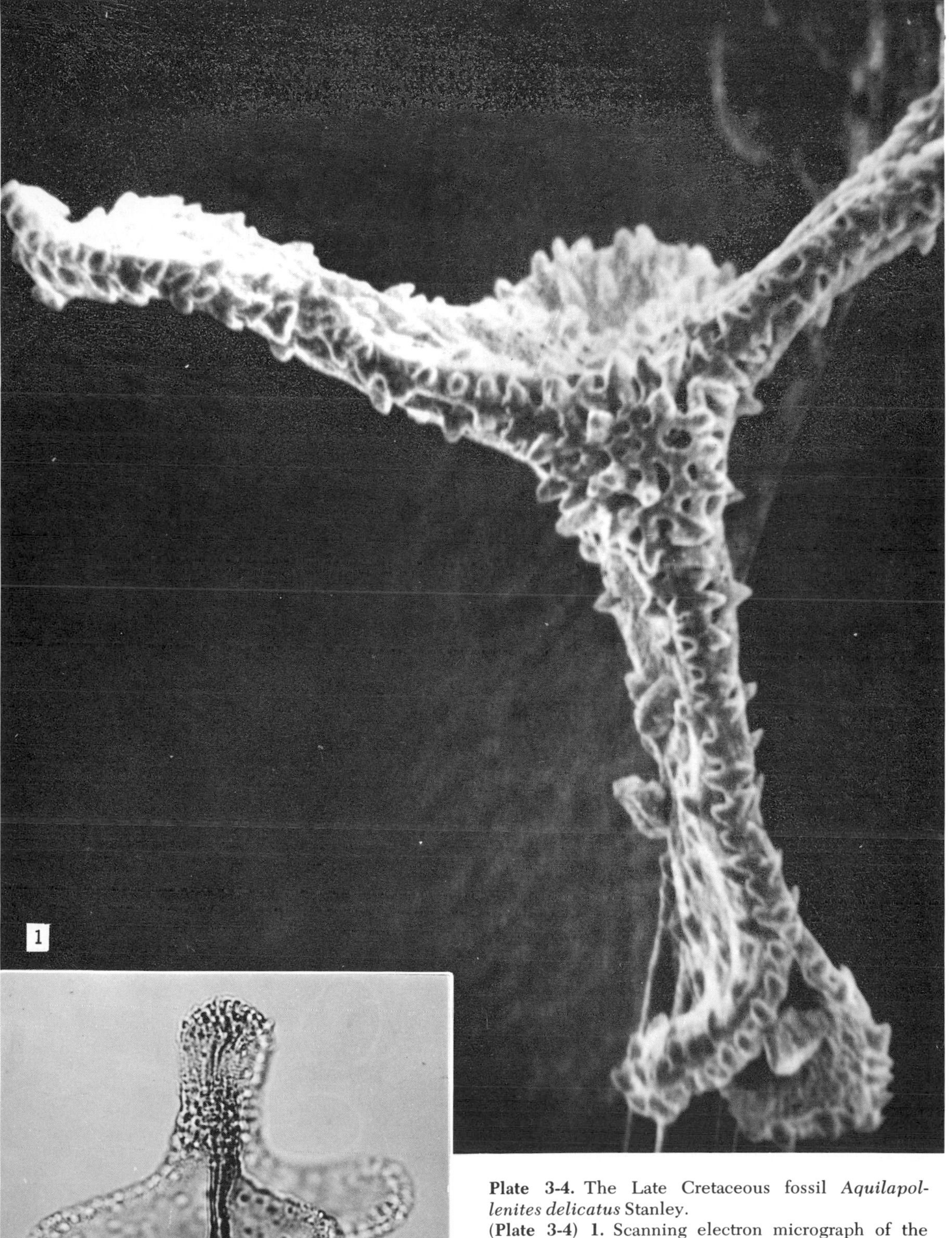

Plate 3-4. The Late Cretaceous fossil *Aquilapollenites delicatus* Stanley.

(**Plate 3-4**) **1.** Scanning electron micrograph of the Late Cretaceous fossil *Aquilapollenites delicatus* Stanley (× 4300).

(**Plate 3-4**) **2.** Same species as in fig. 1 taken with conventional light microscope (× 1000).

(Figure 1 was provided through the courtesy of Charles M. Drew, Chemical Kinetics Branch, Department of the Navy, Naval Weapons Center, China Lake, Calif. Figure 2 was supplied by Bernadine D. Tschudy, U. S. Geological Survey, Denver, Colo.)

with particular reference to the origin of the exine. Despite considerable progress (cf. Gullvåg, 1966b) much challenge remains for future research activity.

Usually at the close of telophase II of meiosis a tetrad of four microspores is produced. The microspores are enclosed by a special callose wall, formed within the original pollen mother-cell wall. Differentiation of these microspores into pollen grains is initiated while they are *inside* the intact and usually massive callose wall. Some authors give emphasis to the special callose wall as functioning to isolate physically each of the tetrads of microspores within the locule of the anther (Heslop-Harrison, 1964, 1966a, 1966b; Jones, 1964; Skvarla and Larson, 1966; Waterkeyn, 1961). Over the years there has been much discussion of the possible schemes of development of the usually intricate sporoderm of the mature pollen grain or spore. Commonly two patterns of development have been contrasted; in one the major control of sporoderm development is associated with the microspore protoplast, whereas the other pattern associates sporoderm growth largely with external phenomena involving material of tapetal origin *after* the microspores are released as individual cells from the investing special callose wall. There would appear to remain the prospect of both patterns of development being successively involved during the total range of phases from premeiosis to mature pollen grain.

Although information concerning the genetic basis for the determination of exine patterns is seriously deficient, it is noteworthy that exine structure is usually uniform (i.e., does not display segregation) within a tetrad. It is not known how such uniformity of the exine may relate to the existence of plasmodesmata and subsequently more massive cytoplasmic connections *between* pollen mother cells during early meiotic stages (Ehrlich, 1958; Gates, 1911; Heslop-Harrison, 1966b, 1968). During these stages cytoplasmic interconnections would appear to provide for ready exchange of materials, including maternal gene products, throughout the entire population of pollen nother cells. Because intensive research projects are still under way, particularly with reference to tapetum-microspore interrelationships, currently it appears premature to eliminate one or the other of the chief hypotheses (Skvarla and Larson, 1966). However, special attention is directed to several meritorious studies that have been published during the past 10 years (Chambers and Godwin, 1961; Ehrlich, 1958; Heslop-Harrison, 1966a; Rowley and Erdtman, 1967; Skvarla and Larson, 1966; Rowley, 1967). From these thought-provoking analyses there emerges the following sequence of events that in general relate to sporoderm development in the taxa investigated. At the start, the surface (plasma membrane) of each of the microspores of the tetrad enclosed within the special callose wall takes on a distinctive configuration that is somewhat suggestive of pinocytosis (Chapman-Andresen, 1962; Heslop-Harrison, 1964), which serves as a template or primexine, for subsequent exine (sporopollenin) development (Ehrlich, 1958; Rowley, 1963; Rowley and Southworth, 1967; Skvarla and Larson, 1966). With enzymatic breakdown of the special callose wall the microspores now with templated surfaces are released into the locule of the anther where they are freely exposed to a variety of substances, including factors from the tapetum. Striking particulates of sporopollenin (Ubisch bodies) in close association with tapetal cells and developing pollen grains have been described by a number of authors. The primexine (template) of the microspore gives way to the mature exine pattern through relatively rapid deposition of the resistant sporopollenin. Finally, after completion of the various layers, or zones, of exine, the intine appears between the innermost layer of the exine (e.g., endexine) and the plasma membrane. (See pls. 3-5 and 3-6.)

Plate 3-5. Pollen grain of *Populus tremula*.

(Plate 3-5) 1. Pollen grain, just prior to anther dehiscence, of *Populus tremula* L. in cross section. Some lamellae of the sporoderm are accentuated by presence of lipid droplets. Note the nearly mature, now thickened and structured intine. The plasma membrane is visible at the inner margin of the intine. (× ca. 65,000.)

(Plate 3-5) 2. Mature pollen of *P. tremula*. Note the massive intine showing low-density (light) areas comparable to those visible in fig. 1. These low-density areas are similar in translucence to polysaccharides. Granular material on ektexine is probably locular material accentuated by oxidative phases of fixation, etc. (× ca. 50,000.)

(Plate 3-5) 3. Same as fig. 2. Lines separate cytoplasm, intine and exine.

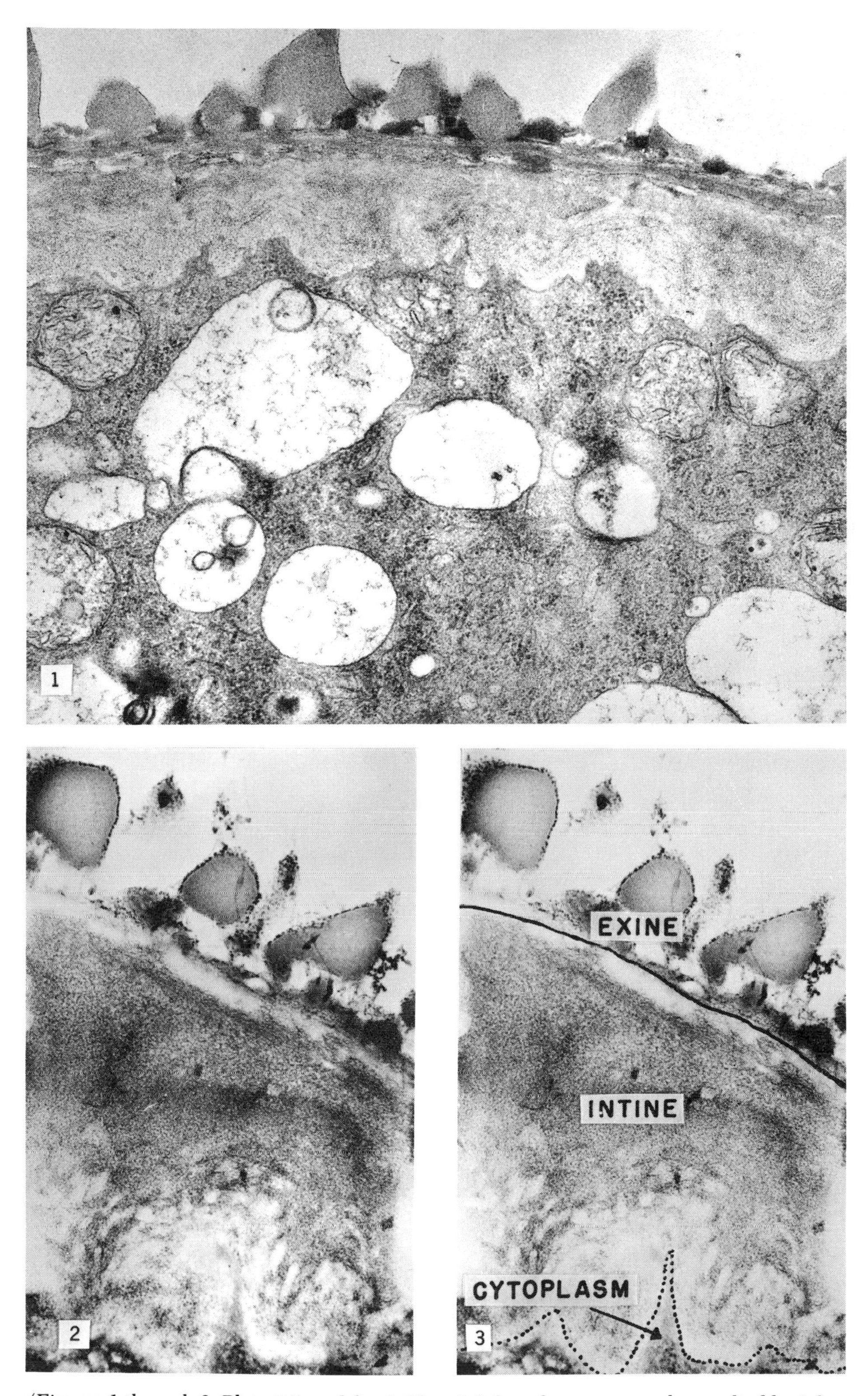

(Figures 1 through 3, Plate 3-5, and fig. 1, Plate 3-6, have been generously supplied by John R. Rowley and are based on specimens fixed in glutaraldehyde with postfixation in osmium tetroxide.)

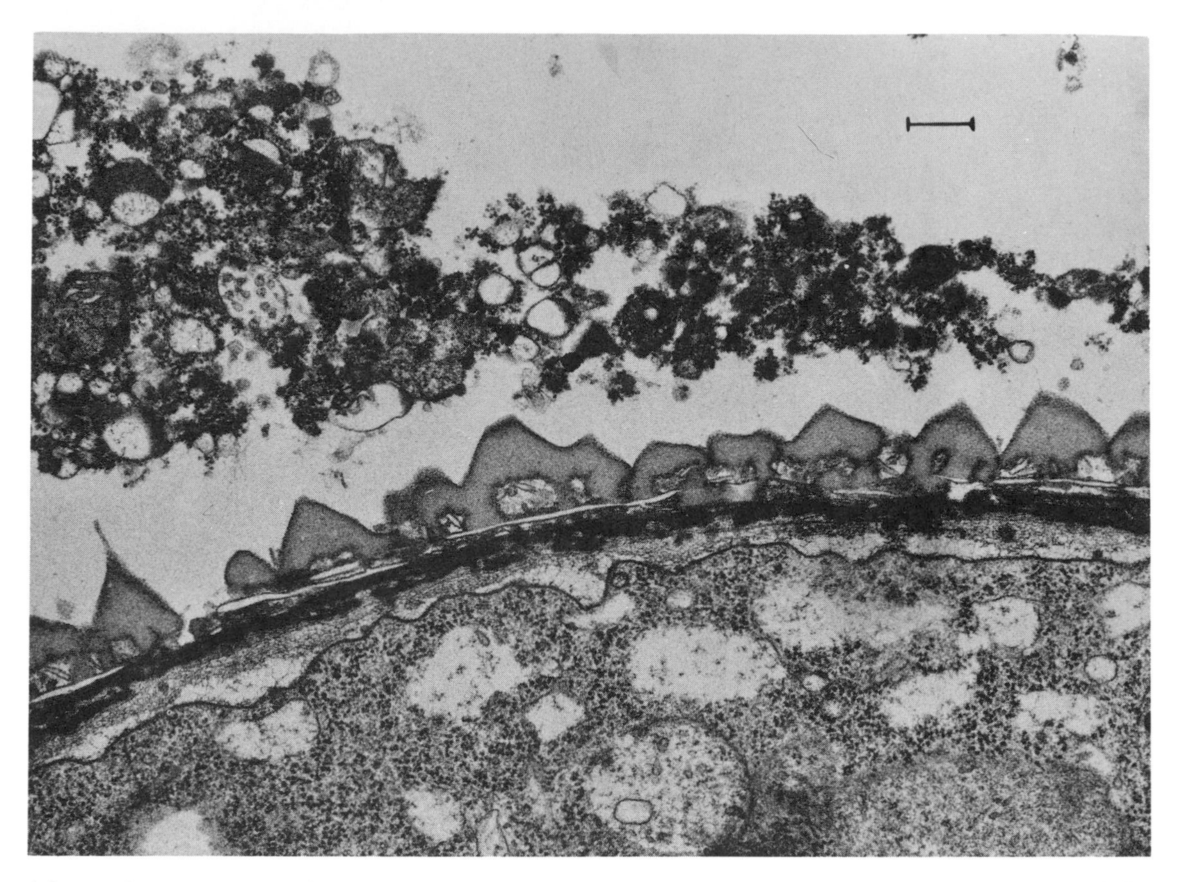

(Plate 3-6) 1. Microspore of *Populus tremula* L. Thin section through developing exine and intine and the peripheral portion of the protoplast. This electron micrograph illustrates the stage during which the intine is forming between the plasma membrane and the innermost endexine (nexine 2). Note that the nexine 2 appears to be formed on membranes (see Rowley, 1967) that are first visible as lines of low density. Locular (tapetal) substance is evident outside of the ektexine. Note the lamellar bodies resembling those of the inner endexine (nexine) in the columellate interstices of ektexine. (The black line signifies ca. 0.1 μ.)

STRATIFICATION OF THE MATURE SPORODERM

Standard comprehensive texts (Erdtman, 1952; Faegri and Iversen, 1964; Wodehouse, 1935) give impressive summarization of taxon-distinct variations in the detailed organization and character of the walls of pollen grains and spores. Critical interpretation of these structures has contributed significantly to a wide range of researches in palynology, ecology, taxonomy, archeology, and allergology. Since there is much variation in the detailed zonation of the sporoderm with reference to the taxon involved, a complex and sometimes confusing technical nomenclature has appeared in the extensive literature of the past two decades (Erdtman, 1952, 1966; Faegri, 1956; Faegri and Iversen, 1964, 1966; Larson, Skvarla, and Lewis, 1962; Van Campo and others, 1966–67). It follows that no single tabular summary will necessarily apply to a particular species under investigation. However, a generalized guide to sporoderm stratification in mature pollen grains and spores is presented below.

In most taxa, during the development of the sporoderm, collumellae appear first in ontogenetic time, followed by the tectum and the foot layer, with the endexine of varied texture, if present, usually developing just prior to the appearance of the intine.

In the application of the basic terms we must bear in mind that prior to maturity and after germination both the morphology and the chemistry of components of the sporoderm can be vastly altered (Rowley and Dunbar, 1967). In cases where one or more of the exinous layers listed above (e.g., the foot layer) are not visibly present other guides summarizing sporoderm zonation may be found applicable—for example, that of Erdtman (1952). In these classifications the outer, tectate, and columellate zones of the ektexine are together designated the sexine, and the remaining, inner components of the exine are referred to as the nexine. The nexine itself may be made up of two or more subdivisions, as revealed by electron microscopy.

Stratification of the Mature Sporoderm

I. Outer surface of mature pollen grain or spore.
- **A.** Perine if present (substance or structure deposited on outermost exine).
- **B.** Exine.
  - **1.** Ektexine.
    - **a.** Tectum.
    - **b.** Columella.
    - **c.** Foot layer.
  - **2.** Endexine.
- **C.** Intine.

II. Inner boundary of sporoderm, in contact with plasma membrane of protoplast.

The above outline of sporoderm stratification would be misleading if attention were not drawn to the many variations in lamellation that occur in various taxa. Gullvåg (1966a) provides an interesting demonstration of two-layered exines in pollen of gymnosperms, the outer layer being orbicular or granular and the inner layer laminated.

Finally I would point out that the mature sporoderm with its fascinating, often taxon-distinct array of systems of zones, layers, or specialized components represents a visual record of changing syntheses and related phenomena that took place during the course of development.

(I am much indebted to Professor John R. Rowley, University of Massachusetts, for his critical reading of the manuscript.)

References

Andersen, S. T., 1960, Silicone oil as a mounting medium for pollen grains: Danmarks geol. unders., v. 4, rk. 4, p. 1.

Bailey, I. W., 1960, Some useful techniques in the study and interpretation of pollen morphology: Jour. Arnold Arb., v. 41, p. 141-148.

Bradley, D. E., 1958, The study of pollen grain surfaces in the electron microscope: New Phytol., v. 57, p. 226-229.

Chambers, T. C., and Godwin, H., 1961, The fine structure of the pollen wall of *Tilia platyphyllos*: New Phytol., v. 60, p. 393-399.

Chamot, E. M., and Mason, C. W., 1938, Handbook of chemical microscopy, v. I, 2nd ed.: New York, John Wiley & Sons, Inc.

Chanda, S., and Rowley, J., 1967, Apertural types in pollen of the Restionaceae and Flagellariaceae: Grana Palynol., v. 7, p. 16-36.

Chapman-Andresen, C., 1962, Studies on pinocytosis in amoebae: Compt. rendus trav. Lab. Carlsberg, v. 33, p. 73-264.

Christensen, B. B., 1954, New mounting media for pollen grains: Danmarks geol. unders., 2. rk. 80, p. 7.

Dahl, A. O., 1964, The fine structure of pollen, with special reference to apertural and nonapertural surfaces: Internat. Bot. Cong., 10th, Edinburgh 1964, Abstracts of papers, p. 221.

Dahl, A. O., and Rowley, J. R., 1965, Pollen of *Degeneria vitiensis:* Jour. Arnold Arb., v. 46, p. 308-323.

Dahl, A. O., Rowley, J. R., Stein, O. L., and Wegstedt, L., 1957, The intracellular distribution of mass during ontogeny of pollen in *Tradescantia L.*: Exptl. Cell Res., v. 13, p. 31-46.

Echlin, P., 1968, Pollen: Sci. Amer. v. 218, p. 80-90.

Ehrlich, H. G., 1958, Electron microscope studies of *Saintpaulia ionantha* Wendl. pollen walls: Exptl. Cell Res., v. 15, p. 463-474.

Ehrlich, H. G., and Hall, J. W., 1959, The ultrastructure of Eocene pollen: Grana Palynol., v. 2, p. 32-35.

Erdtman, Gunnar, 1952, Pollen morphology and plant taxonomy: Stockholm, Almqvist & Wiksell.

———1959, Ultraviolet micrographs and photomicrographs from the Palynological Laboratory, Stockholm-Solna: Grana Palynol., v. 2, p. 36-39.

———1960, The acetolysis method: Sv. Bot. Tidskr., v. 54, p. 561-564.

———1966, Sporoderm morphology and morphogenesis. A collocation of data and suppositions: Grana. Palynol., v. 6, p. 317-323.

Erdtman, Gunnar, and Dunbar, A., 1966, Notes on electron micrographs illustrating the pollen morphology in *Armeria maritima* and *Armeria sibirica*: Grana Palynol., v. 6, p. 338-354.

Faegri, Knut, 1956, Recent trends in palynology: Bot. Rev., v. 22, p. 639-664.

Faegri, K., and Iversen, Johannes, 1964, Textbook of pollen analysis: New York, Hafner Publishing Co.

———1966, Terminology in palynology: Pollen et Spores, v. 8, p. 407-408.

Faegri, Knut, and Van der Pijl, L., 1966, The principles of pollination ecology: New York, Pergamon Press.

Fernandez-Moran, H., and Dahl, A. O., 1952, Electron microscopy of ultrathin frozen sections of pollen grains: Science, v. 116, p. 465-467.

Freytag, K., 1964, Polarisationsmikroskopische Beobachtungen an Nexinen von Malvaceen-Pollen: Grana Palynol., v. 5, p. 277-288.

———1967, Ueber doppelbrechende Lipoide in der Exine und im Pollenöl: Grana Palynol., v. 7, p. 3-9.

Fritzsche, C. J., 1837, Ueber den Pollen: Mem. Sav. Etrang. Acad. Sci. Petersb., 3.

Gates, R. R., 1911, Pollen formation in *Oenothera gigas:* Ann. Bot., v. 25, p. 909-940.

Gullvåg, B. M., 1964, Morphological and quantitative investigations of pollen grains and spores by means of the Francon-Johansson interference microscope: Grana Palynol., v. 5, p. 3-23.

———1966a, The fine structure of some gymnosperm pollen walls: Grana Palynol., v. 6, p. 435-475.

———1966b, The fine structure of pollen grains and spores: a selective review from the last twenty years of research: Phytomorph., v. 16, p. 211-227.

Heslop-Harrison, J., 1964, Cell walls, cell membranes and protoplasmic connections during meiosis and pollen development, *in* Linskens, H. F. (ed.), Pollen physiology and fertilization: Amsterdam, North-Holland Publishing Co., p. 39-47.

———1966a, Cytoplasmic connexions between angiosperm meiocytes: Ann. Bot., v. 30, p. 221-230.

———1966b, Cytoplasmic continuities during spore formation in flowering plants: Endeavor, v. 25, p. 65-72.

———1968, Pollen wall development: Science, v. 161, p. 230-237.

Hess, W. M., and Stocks, D. L., 1967, Surface structures and organelle characterization of frozen-etched *Aspergillus* conidiospores: Am. Jour. Bot., v. 54, p. 637-638

Hyde, H. A., 1955, Oncus, a new term in pollen morphology: New Phytol., v. 54, p. 255-256.

Jones, K., 1964, Pollen structure and development in *Drosera*: Jour. Linn. Soc. (Bot.), v. 59, p. 81-87.

Larson, D. A., Skvarla, J. J., and Lewis, C. W., Jr., 1962, An electron microscope study of exine stratification and fine structure: Pollen et Spores, v. 4, p. 233-246.

Linskens, H. F. (ed.), 1964, Pollen physiology and fertilization: Amsterdam, North-Holland Publishing Co.

Mangin, L., 1893, Sur l'emploir du rouge de ruthenium en anatomie vegetale: Acad. sci. [Paris] Comptes rendus, v. 116, p. 653-656.

Michel, K., 1950, Die Grundlagen der Theorie des Mikroskops: Stuttgart, Wiss. Verl. Ges. M. B. H.

Mollenhauer, H. H., 1964, Plastic embedding mixtures for use in electron microscopy: Stain Technology, v. 39, p. 110-114.

Mühlethaler, K., 1955, Die Struktur einiger Pollenmembranen: Planta, v. 46, p. 1-13.

Pease, D. C., 1964, Histological techniques for electron microscopy, 2nd ed.: New York, Academic Press.

Roland, F., 1966, Etude de l'ultrastructure des apertures; pollens à pores: Pollen et Spores, v. 8, p. 409-419.

Rowley, J. R., 1960, The exine structure of "cereal" and "wild" type grass pollen: Grana Palynol., v. 2, p. 9-15.

———1963, Ubisch body development in *Poa annua*: Grana Palynol., v. 4, p. 25-36.

———1964, Formation of the pore in pollen of *Poa annua*, *in* Linskens, H. F. (ed.), Pollen physiology and fertilization: Amsterdam, North-Holland Publishing Co., p. 59-69.

———1967, Fibrils, microtubules and lamellae in pollen grains: Rev. Palaeobotan. Palynol., v. 3, p. 213-226.

Rowley, J. R., and Dahl, A. O., 1962, The aperture of the pollen grain in *Commelinantia*: Pollen et Spores, v. 4, p. 221-232.

Rowley, J. R., and Dunbar, A., 1967, Sources of membranes for exine formation: Sv. Bot. Tidskr., v. 61, p. 49-64.

Rowley, J. R., and Erdtman, Gunnar, 1967, Pollen wall development in *Populus:* Am. Jour. Bot., v. 54, p. 629.

Rowley, J. R., and Flynn, J. J., 1966, Single-stage carbon replicas of microspores: Stain Technology, v. 41, p. 287-290.

Rowley, J. R., and Southworth, D., 1967, Deposition of sporopollenin on lamellae of unit membrane dimensions: Nature, v. 213, p. 703-704.

Shaw, G., and Yeadon, A., 1964, Chemical studies on the constitution of some pollen and spore membranes: Grana Palynol., v. 5, p. 247-252.

———1966, Chemical studies on the constitution of some pollen and spore membranes: Jour. Chem. Soc. (London) (C, organic), 16-22.

Sitte, P., 1960, Die optische Anisotropie von Sporodermen: Grana Palynol., v. 2, p. 16-38.

Skvarla, J. J., 1966, Techniques of pollen and spore electron microscopy. Part I. Staining, dehydration, and embedding: Okla. Geol. Notes, v. 26, p. 179-186.

Skvarla, J. J., and Larson, D. A., 1966, Fine structural studies of *Zea mays* pollen I: Cell membranes and exine ontogeny: Am. Jour. Bot., v. 53, p. 1112-1125.

Skvarla, J. J., and Turner, B. L., 1966, Systematic implications from electron microscopic studies of Compositae pollen-a review: Ann. Missouri Bot. Gard., v. 53, p. 220-256.

Stone, D. E., Reich, J., and Whitfield, S., 1964, Fine structure of the walls of *Juglans* and *Carya* pollen: Pollen et Spores, v. 6, p. 379-392.

Thornhill, J. W., Matta, R. K., and Wood, W. H., 1965, Examining three-dimensional microstructures with the scanning electron microscope: Grana Palynol., v. 6, p. 3-6.

Tsukada, M., and Rowley, J. R., 1964, Identification of modern and fossil maize pollen: Grana Palynol., v. 5, p. 406-412.

Van Campo, M., 1961, Mecanique aperturale: Grana Palynol., v. 2, p. 93-97.

Van Campo, M., Bronckers, F., Guinet, P., 1966-67, Electron microscopy's contribution to the knowledge of the structure of acetolysed pollen grains: Palynological Bull., Lucknow (India), v. 2 and 3, suppl., p. 1-21.

Waterkeyn, L., 1961, Les parois microsporocytaires de nature callosique chez *Helleborus* et *Tradescantia*: La Cellule, v. 62, p. 225-255.

Wodehouse, R. P., 1935, Pollen grains: New York, McGraw-Hill Book Co., Inc.

Zetzsche, F., and Liechti, J., 1937, XII. Biochemisch veränderte Sporopollenine: Brennstoff-Chem. v. 18, p. 280.

4

SYSTEMATICS AND NOMENCLATURE IN PALYNOLOGY

James M. Schopf

We are now witnessing an unparalleled, almost explosive increase in the number of new names proposed for use in paleobotany, chiefly names referring to plant microfossils. The names are applied in diverse ways, and about the only element they seem to have in common is a binary form that generally corresponds to that traditionally applied in plant systematics. However, modern plant systematics has an underlying philosophy of phylogeny and relationship; the names applied to plant microfossils rarely seem to involve such a consideration. Neverthless, selection may be made from works of different authors to illustrate all shades of phyletic implication.

Many authors responsible for numerous names are little concerned with phylogeny. Some systems seem almost to have been devised to confuse and confound any application of nomenclature for elucidation of phylogeny in a misguided understanding of utilitarian reasons. Thus there exists a problem concerning the appropriate function of scientific names and their relation to systems of classification. Unless some means of discrimination is employed, the very abundance of names is likely to confuse and delay a clarification of issues that are more substantial than those concerned merely with method and procedure.

Some analogy may be drawn from the established laws of economics if we consider the circulation of scientific names in any way comparable with the circulation of money. In a way scientific names do represent a type of currency in thought. Sir Thomas Gresham (1519-1579), founder of the Royal Exchange, is reputed to have first explained how "bad money drives out good"[1]; the worst form of currency in circulation regulates the value of the whole currency and drives all other forms of currency out of circulation. The question, bluntly phrased, is, will the "bad" names drive out of circulation the "good"?

In the first place let us clearly differentiate between "good" and "bad" names. The usage of names, as such, is solely a matter of nomenclature, not taxonomy. We are not contrasting "good" genera, for example, with "bad" genera, although the distinction may be a subtle one. For purposes of this discussion a "good" name is one that is established fully in accordance with the spirit and philosophy of phylogenetic taxonomy reflected by the "International Code of Botanical Nomenclature;" a "bad" name is one that ostensibly fulfills the code regulations but does not reflect the philosophy. In this sense the nomenclature that is applied to many sporomorphs is "bad." Names established on this basis are doubly confusing because they are most difficult to apply consistently. The present generation of plant-microfossil specialists has been drawn from highly diversified lines of training, and it is unlikely that for some time to come all would

Publication authorized by the Director of the U.S. Geological Survey.

[1]MacLeod, H. D., "Gresham's Law": Encyclopedia Britannica, 1965 ed., v. 10, p. 917.

agree that consistency is necessary. Plant-microfossil studies are conducted principally to serve geologic purposes. If the purpose of correlation is served, a geologist usually does not care whether the names applied to the fossils are "good" or "bad" or even illegitimate; his sole concern is with the accuracy of the stratigraphic conclusions. The eventual solution of nomenclatural problems caused by "bad" names is likely to result from improvements instituted by palynologists themselves. Gradually "good" names will prevail.

Most paleobotanists believe that accurate stratigraphic conclusions depend primarily on accurate identification. Identification consists of *establishing the identity* between newly discovered material and material known previously. Identification is always made *with* a previously recognized group. If the previously known material can be placed in relation to its stratigraphic occurrence, an inference of relative contemporaneity may be justified. Accordingly much depends on the accuracy of identification. The stratigraphic significance of any taxonomic group depends on accuracy rather than on the ease of determination.

One of the factors causing proliferation of names is the difficulty in determining identity. Since the context of phyletic relationship is difficult to establish, the identification of a *new* taxon is more readily justified than identification with a previously defined species. However, the stratigraphic significance of any fossil depends on recognition of alliance with fossils that are well enough known to possess an implication of age. If similar fossils are designated by different names in each report, even though the similarities are apparent, it is difficult to bring the information together for any purpose. The potential significance of the material is obscured by difficulties of identification. Beyond this, stratigraphic ranges are apt to be spotty and of questionable reliability because of lack of uniformity in identification if consideration is restricted to occurrence reports based on lists of names. Whether or not application of identical names is justified in study of new material, a superior insight as to stratigraphic significance can be derived from inferences that are based on phyletic interpretation.

When microfossils have been identified without reference to the phylogenetic philosophy of taxonomy, revision embodying this thought becomes an excessive task. Obviously, if the original identification was not taxonomic in the sense of natural relationships, the record has no phyletic value. Hence published determinations cannot be used indiscriminately. In order to utilize certain records it is necessary to attempt to supply, or interpret, some common denominator of systematic thought. For this reason the field is inherently a very difficult one for a novice.

Although plant microfossils are likely (a) to provide an extensive supplement to information gained from plant megafossils about distribution of some groups of ancient plants and (b) to improve the possibilities of stratigraphic correlation of beds commonly regarded as unfossiliferous, these potentialities have not been fully realized. The field is in a state of great activity, flux, and confusion. At present no one can know whether he has applied the most appropriate treatment in reporting new material. The present literature must be reviewed and classification extensively revised before plant microfossils become easy to use. The confused state of the literature is a further deterrent to publication. Authors publishing in this field today need courage and confidence in the merit of their work. Reports are needed that are more amply illustrated and that provide an adequate factual background so that significant revisions can later be accomplished. Under these circumstances it seems most important now to concentrate on the fundamentals of systematics wherein specific ranges of characteristics are studied.

It is questionable whether nonphylogenetic taxonomy can be sustained; indeed, if the analogy with Gresham's law holds true, this type of taxonomy must also have a corollary to the effect that a currency does not remain permanently debased. Eventually revaluation occurs, and an adequate support is derived for the currency. There is no way of knowing whether the excess of recently proposed microfossil names will "debase the currency" for a long or a short time. There is no question that the recent tendency has been toward application of names on a less meaningful and more superficial basis than is commonly practiced in taxonomy. For the present a practical approach to the problem of microfossil classification seems to be essential. Probably a more rapid approach to a "sound" taxonomy will be facilitated if scientific data can be accumulated without regard to nomenclature but according to practices designed not to conflict or compete with the "International Code of Botanical Nomenclature" regulating scientific names. Certainly it seems we need more facts and can do with fewer names.

DEFINITION OF TERMS

Following is a list of terms, many of which are so commonly used in systematic study that their meaning is taken for granted. A review of their implications for paleobotany may be nonetheless desirable.

1. **Terminology** is the designation of scientific and technical concepts, such as anatomical parts, methods of procedure, or objectives of study. Although priority is a consideration, convenience for communication is the paramount objective of scientific terminology. No formal standards apply to terminology, and usage differs according to language (Hitchcock, 1929, p. 1434).

2. **Nomenclature** is the formal treatment of names according to a code, or set of rules, adopted by agreement and acknowledged as a general standard in scientific work. Priority of valid publication is generally regarded as a cardinal consideration for purposes of nomenclature.

3. **Taxonomy** is the scientific and systematic classification of plants and animals according to the principles of evolution and descent (phylogenetic relationship). Taxonomic arrangements differ according to emphasis and interpretation of phylogeny by taxonomists. Because of the absence of compelling evidence and for purposes of convenience, taxonomic arrangements are, in part, arbitrary and artificial. A taxonomist is one who makes decisions about the organization of information in the plant or animal kingdoms.

4. **Systematics** refers to the manner or practice of classifying taxa according to a nomenclatural system. Commonly the term "systematics" is used as a synonym for taxonomy, but some authors give preference to systematics as having a broader connotation. The term is used in the broader sense in this chapter and should be taken to include not only taxonomy but other ancillary areas needed to support classificatory objectives.

5. **Taxon** (plural, *taxa*) is a group concept reinforced with certain evidences from physical reality. It may refer to any group that is differentiated by taxonomy, such as a given species or genus or family, or to the plant kingdom. The taxon always is to be regarded as an *abstract group concept*, a population; taxa are not finite objects. The definition of a taxon depends on interpretative inference, the estimation or expression of a degree of affinity (phyletic relationship) between subordinate groups, individuals, or specimens included. Inferences are based on evidence derived from physical specimens; the specimens *represent* or *illustrate*, but do not literally or logically *constitute*, the taxon. Specimens identified with a taxon constitute the hypodigm.

6. **Valid**, or **validity**, has a special meaning in systematic studies in which it is applied to *publication* of names according to precepts of a code of nomenclature. (See Lanjouw and others, 1966, art. 6.) The term "validity" can only refer to names of taxa and not to groups of organisms. A prescribed manner of publication is valid, but taxa literally are not "published" nor are they "valid"; the term has reference only to names.

7. **Biocharacter** is a feature that is attributable to hereditary influence or control and is therefore presumed to be independent of environment or mode of preservation.

8. **Circumscription** is the limit of any population *ascribed to* a taxon. Intention of an author, in part, must be inferred since any physical evidences cited can never extend fully to the periphery of an intended circumscription.

9. **Hypodigm** is the part of a population *that has been identified* with a taxon. (See Simpson, 1940.)

10. **Nomen nudum** means literally "bare name" and refers to a name published without information necessary to establish validity. Commonly these omissions are technical and not necessarily very informational, but for systematic purposes technical requirements must be observed.

11. Determination is a taxonomic decision as to identity.

12. Identification is the notation or other indication of agreement between an unassigned specimen (identity *with*) and elements included within a circumscription concept.

13. Assignment is the act of identifying the position of an unassigned specimen within some larger taxonomic group (assignment *to*).

14. Individual plant, for nomenclatural purposes, includes all stages of a life cycle. An individual plant is not a specimen because any specimen represents the physical existence of an organism at just one instant within the sequence of the life cycle. There is no justification for applying different names to different stages of a life cycle, *unless mutual relationship (botanical correlation) of stages is in doubt.* Special provisions are included in the botanical code of nomenclature for practical convenience in classifying vegetative stages of fungi and dissociated elements preserved as fossils *that might be, but cannot be shown to be, mutually related* (intercorrelated) with different stages of the life cycle included in a species.

15. A species, for nomenclatural purposes, is a taxon of basic rank. It is directly referenced to, and is represented by, one or more physical specimens that *you believe*[2] should be specifically designated by a scientific name.

16. A genus is a taxon that, in the light of nomenclatural requirements, includes one species (monotypic) or two or more related species that *you believe*[2] should be designated by the same substantive name. A genus is only *indirectly* referenced to an actual physical specimen (the holotype of an included species).

17. A family is a taxon almost completely abstract in nomenclatural reference and synthetic in its attributes. A family must include one or more related genera for which separate familial recognition seems appropriate *to you.*[2] The basis for belief in familial attribution depends on critical consideration by the original or some subsequent author.

[2]Many motives may be responsible for an author's belief in, or for a reader's acceptance of, names of taxa; but whether one proposes or merely accepts the name someone else has proposed, the use of the name is still a matter of personal responsibility. Scientific names should not be taken lightly.

18. Natural classification is a contradiction in terms—all classification systems are artifices designed for purposes of mankind. The term formerly implied an arrangement divinely ordained and came later (post "Origin of Species") to be interpreted in the phylogenetic sense of descent. Recently some taxonomists have used "natural classification" in the sense of a nonphyletic system that is based on the total complement of characters attributed to any organism, regardless of the manner in which those characters originated. The term probably should now be avoided as confusing. If classification is phyletically oriented, it should be so designated.

GENERAL CONSIDERATIONS

Application of the scientific method to systematic studies requires ability to deal with abstract concepts. Taxonomy involves systematic treatment of group concepts that are generalizations and hence abstractions. No one can *see* or feel a species; species do not grow in swamps or have long hairs, regardless of colloquial scientific expression. Abstractions are involved when we deal with species or any other taxon, regardless of any system of classification that is employed.

According to Benson (1943), the real goal of systematic botany is organization—the arrangement of the myriads of living plants into a readily understandable system made up according to their phyletic (genetic, or "blood") relationships. This goal should be extended to cover the plant kingdom and to include all identifiable plants, both fossil and modern.

In this account the term "phyletic (or phylogenetic) system" is used in the sense in which some authors use the term "natural system." Actually there is nothing very "natural" about manmade systematic arrangements of the plant or animal kingdoms. The term "natural system" was also used in a typologic sense for a long time before phylogeny was even vaguely recognized. The term "phyletic system" seems definitely less subject to misunderstanding, even though proof or evidence of many of the important phylogenetic details may remain obscure. Taxonomy in general depends on the uniformitarian principle of heredity and evolutionary descent, regardless of whether the organisms are known to us as modern or as fossil. Accordingly the best system is one that is considered to most closely approximate the actual line of descent. If we adopt this point of view, we can share a common orientation for purposes of subsequent discussion.

All plants, fossil or modern, *have had ancestors*; that is to say, all plant fossils, large or small, have been derived phylogenetically back to the point of initial organismal differentiation. If this is accepted as the real biogenetic law, it follows logically that a systematic search for probable phylogeny is a most fundamental objective. Taxonomy should be based on knowledge of the evolutionary history and derivation of fossils to the fullest extent possible. For this reason phyletic considerations take precedence in taxonomy of plants. In such a context taxonomy acquires a fully oriented scientific purpose, conducive to the establishment of accurate stratigraphic control. There is nothing new in this point of view. Orientation is implicit in all adequate systematic treatments, but it is commonly unstated and may on occasion be overlooked.

Species consist of populations projected in time. In a historical or phyletic sense species should not be regarded as fixed, because most species include a broad range of variants and distinctions between closely related species are rarely clear-cut. Commonly the populations are as variable as human beings, or as dogs, or any other example that might be chosen, because at any point in time there may be several lines of incipient evolution within them. Intergradation of major populations, such as is well illustrated among microfossils, is so nearly complete that borderlines between taxa are difficult to distinguish. Whether a fossil is long or broad, or smooth or ornamented, is clearly incidental to the purpose of understanding the relationships of plants. The homologous organs of higher modern plants exhibit many morphologic variations that are of minor or insignificant value in determining phyletic relationship. Useful taxonomic distinctions must be based on criteria that appear in historic perspective to have value in classification. Names applied to plant and animal populations should represent recognizable taxa, but it is not realistic to ignore forms that intergrade between related species.

All phyletic systems are tentative because the information on which they are based is incomplete. Benson (1943) indicated that no organism —not even *Drosophila*, which has been exhaustively studied genetically—has ever been thoroughly known. No organism, fossil or modern, is ever likely to be. Every taxonomist who draws on field, herbarium, museum, or laboratory material for data determines so far as he himself is able that which is significant, but no one has a complete story. Every taxonomist, paleontologist and neontologist alike, must draw tentative conclusions from incomplete data or else draw none at all.

NOMENCLATURE

None of the processes of organization—that is, classification, taxonomy, or systematics—depends in any way on nomenclature. However, it is necessary to apply scientific names to the taxa after a satisfactory scheme of classification has been formulated if we want to discuss the classification or communicate in any way regarding the plant groups being studied. There are many more scientific names for groups that have been studied for a long time than there are kinds of plants. Benson (1943) indicated that some 4,000 species of grasses have acquired something like 10,000 to 20,000 names during the last two centuries. Sometimes names have been applied because of ignorance of work done by others or as a reflection of disagreement on classification or rules of nomenclature. One species of foxtail millet, *Setaria geniculata*, has been named 22 times, and various epithets for it have been used in 74 different combinations, either as species or as varieties, within seven different genera. Probably none of the fossils has yet been so treated, chiefly because very few types of fossils have been critically studied by as many people. Regrettable as synonyms may be, they must be dealt with in systematic study.

As a further example of synonyms in modern plants, DeWolf (1964) estimated that for higher plants in general the average ratio of valid or generally accepted names to synonyms is about 1 to 5. For each "accepted" binomial there are usually 3 to 10 synonyms. He suggested that some of the larger plant families, and plants of some geographic areas, have been overnamed or overdescribed, and he aptly remarked that "names are not species." Without entering into the problem of "good" species, he nevertheless suggested that the world flora comprises no more than 280,000 species of [modern] angiosperms.

If the fossil record is taken into account, one wonders how many more species of angiosperms have existed since this important class of plants first became differentiated. Eventually some corollary information about angiosperm origin might be obtained through an attempt to answer this question. Paleobotanists are concerned with an actively differentiating assemblage of angiospermous plants that has existed for more than 130 million years. They also are concerned with differentiation of the other groups of plants, some of which have become extinct and do not contribute to the modern flora. These considerations are paramount and names serve only to facilitate discussion. Even though paleobotanists lament their problems of synonymy, it will be fairly evident that from the standpoint of the actual numbers of names in use they are well short of the neobotanists.

One obvious means of eliminating factors of confusion that lead to multiplication of names is to agree on the mechanics of treating names for scientific purposes. Rules that apply solely to the mechanics of handling names of taxa are given in "The International Code of Botanical Nomenclature" (Lanjouw, 1966). International agreement on methods for treating names so that at least the proper emphasis might be attached to substantial divergence of opinion, as distinguished from divergence in procedure, was initiated by the Rules approved by the Paris Congress in 1867, compiled largely under the direction and leadership of Alphonse DeCandolle. This code has been published in eight different revisions, the latest of which derived from the Edinburgh Congress and was published under the editorship of Lanjouw in 1966.

Although the provisions of the code are relatively simple and straightforward, decision as to the correct name for any taxon depends not only on adherence to approved nomenclatural procedure but also on the circumscription, rank, and position to which the group (taxon) is assigned. Priority is a most important principle for determining the name of any taxon of a particular position, rank, and circumscription. The name of any taxon, either fossil or recent, thus becomes a technical matter that depends on the chronology of study as well as on the quality of information and evaluation of physical evidence. In any given instance the chronologic history may become complicated and differences in interpretation are possible. Regardless of any personal bias, divergence in systematic opinion, whether nomenclatural or taxonomic, demands respect. For this reason the writer does not plan to devote attention to determination of the correct name of any group. The discussion in this book must be interpreted in the general frame of reference already established in general treatments. Names provided by different authors as a means of reference will be used without concern as to whether the names should be regarded as correct names or as synonyms.

It is of primary importance to distinguish clearly between what are matters of systematic and classificatory *judgment* and what is objectively demonstrable in fact of publication. The first question is (a) whether a *name* (which cannot be ignored and *must* be taken into consideration) has actually been legitimately established in the published record and is to be dealt with according to priority or other regulations agreed on in nomenclature or (b) whether the name is not to be so treated. Each legitimate name that is attached to its nomenclatural type and "found" within the circumscription of the taxon with which one is concerned must be regarded either as the correct name or as a synonym. Circumscription depends on taxonomic decision, but the nomenclatural decision is automatic and depends on the date at which verifiable requirements have been met to entitle a name to legitimate treatment. The criteria for the legitimacy of species names thus become the *minimum* of essential requirements for validating the name of a species. It has been agreed that "Every individual plant is treated as belonging to a number of taxa among which the rank of species is basic." (See Lanjouw, 1966, art. 2.) This means that every identifiable specimen *may be taken to represent* an individual plant. It does not mean, however, that *every* specimen must be identified or that every specimen *can be* identified. Perhaps of all the opportunities of exercising taxonomic judgment, the decision to identify or not to identify is most important.

It is most important to differentiate (a) the

names of the concepts from the concepts themselves or objects actually discussed (b) the biometric limits or theoretical criteria used to characterize or distinguish species and (c) the elements that are *treated* as species in nomenclature. In nomenclatural consideration we need not enter into the taxonomic problem of whether a taxon "deserves" assignment to species rank. What we are concerned about is whether the requirements have been met for recognition of a species *name*. We must presume that according to one or another theoretical basis the taxon can be justified because the bases for *taxonomic* recognition are subjective; they are for this reason beyond the range of nomenclatural regulation; they are taxonomic.

Nomenclature involves the philosophy of precision in scientific communication. Those who have had slight training in systematic botany commonly have a misapprehension about nomenclature, which may take either of two forms.

Students may note differences in names that refer to similar concepts and feel at a loss to evaluate or explain the discrepancy. So far as they are concerned the fine points of communication, for the aid of which scientific nomenclature presumably is designed, have broken down. These students are likely to be at first disgusted and next to attempt to devise their own simple code of rules; such a code probably would successfully serve their own limited purpose, aimed at a concise application, *if* everyone else would use it. Unfortunately not all persons are able to approach a nomenclatural problem from the same point of view. Legitimate purposes and applications do differ. General problems in nomenclature are not to be solved by a quick slash that cuts the Gordian knot.

A second reaction to nomenclature is not as naive as the first in some respects, but it becomes equally misdirected because it commences with a basic misconception as to definition. To some persons the term "nomenclature" means the same as "taxonomy" and "classification" (see p. 51); they consider that all that pertains to giving or using names is nomenclature. These persons are likely to attribute all the advantages of classification to the mystic or authoritative ability to use "right" words. They do not differentiate well between different systems, and their preferences are principally based on what they regard as authoritative recommendation. If the forms they prefer are actually correct, the preference is almost always rationalized for the wrong reasons. If only amateurs shared this point of view, it would be less disturbing. In paleobotany, however, there are professionals who do useful work and make reliable observations but lack understanding of the meaning of nomenclature. They have been so busy recording "facts" that they have avoided learning how the facts may best be reported.

There is probably a third class of "nomenclators" who worry about nomenclature to excess; in a way this zeal for nomenclature may serve subconsciously as an excuse for meager reporting. These nomenclators are the persons who have uncompleted manuscripts and abundant criticism for more productive and less meticulous colleagues. Rather obviously the attitude of overemphasis on nomenclature is wanting in certain respects. The problem is simply to utilize nomenclature in its appropriate context; but, since for one reason or another nomenclature itself may not be applied consistently by the people who use it, some degree of charity must be cultivated. This need not mean that one compromises principle, but it does mean that one adopts the practice of trying to understand the source of error or misstatement and of striving to rationalize a consistent interpretation for apparent divergences that are difficult to rationalize. If one discovers mistakes or unfortunate practices, one should always give preference to the *minimum* change that will serve to rectify the error or disadvantage.

It is possible to determine which material relates to nomenclature and which to taxonomy or classification. Systematical science, which includes both taxonomy and nomenclature as construed here, has had its skilled practitioners for a long time. The Code of Nomenclature is a product of their accumulated experience. Its numerous editions reflect the evolution of principles and practices that are beneficial to the science. Continuing efforts have made the practices recommended by the Code an improved vehicle for concise, accurate, scientific communication. The arrangement of taxa, on the other hand, is in the field of taxonomy and is quite apart from the mechanism whereby names are applied. In matters of arrangement a divergence of interpretation, whether or not we agree with it, is always permissible.

The appropriate use of nomenclature is important. In this day of digital computers, "numerical taxonomy," and information retrieval, one may easily overlook the fact that nomenclature still provides the most important system for recording and recalling scientific information. The correct names of groups of organisms, arranged according to a well-reviewed and revised taxonomic system, undoubtedly are the most effective condensation symbols for this type of knowledge. No other system is so widely comprehended. It is true that symbols of all grades of informational content are available. The most important system, however, is based on taxonomy that differentiates by its inherent construction between a high level of symbolic information transfer and a symbolism of lower order. Regardless of the level of transfer, nomenclatural legitimacy is essential. A great deal of the higher level of symbolism must be allied with knowledge of phyletic relevance. Symbols that involve lower levels of taxonomic implication will remain available for any identification that must necessarily be reported in those terms, but, as their level of information becomes superfluous, the symbols will be relegated to disuse. All of this of course is quite apart from the superficial trappings of etymology, interpretation of "inherent" meaning, or clever mnemonic construction of words. The most effective use of systematics transcends the superficial appearance of the terms.

THE SPECIES CONCEPT IN FOSSIL PLANTS

Some differences of opinion are expressed as to whether fossils properly should be classified and named as groups of organisms or whether, like minerals and chemical compounds, they should more simply be classified as "interesting" objects. There really should be no difference of opinion on this subject. Fossils are not now living, but their claim to taxonomic classification is based on the point of view that they *represent*, and may be used as a basis for interpretation of, *organisms* once living that are comparable to those of the present day. Plant microfossils, including fossil spores and pollen and any other determinable microscopic objects, first should be regarded as the representatives of plants. As such, the microfossils are (though more difficult to identify) just as amenable to systematic treatment as plant megafossils of various types or, for that matter, as herbarium specimens that, though fragmentary, may be assigned to definite plant taxa. The taxonomic purpose is concerned with the relationships of groups of organisms.

There are two means of designating any kind of fossil specimen. One designation indicates its taxonomic position and the other designates its morphology. Nomenclatural considerations apply only to the taxonomic designation, and the use of a taxonomic designation should signify that the fossils are considered as representative of an abstract taxonomic group, such as a species or genus. The morphologic designation indicates primarily the section or stage of the plant life cycle that is represented by a specimen. A single specimen, or any specimen (*all specimens* suitable for preservation, fossil and modern, are detached; even microscopic algae are detached from their environment), obviously can represent only a momentary part of a contemporary stage. A species concept, however, involves theoretical recognition of the sequence of all life-cycle stages, many of which must be generalized for want of specific information. A species represents a taxon of plants, even though the species may be primarily defined only on the restricted basis of biocharacters morphologically represented in customary specimens.

The proposal to recognize a new species depends on, among other matters, legitimate establishment of a new species name. The new name, as it appears in the literature, may thus be taken to signify a serious and legitimate taxonomic intention. It suggests (a) that no other group of plants could be identified with the fossil specimens on which the species is based and (b) that the fossils are sufficiently distinctive, in the author's opinion, to signify a different group of plants, and for this reason they deserve a name. These two elements are basic to the problem of what a species *is*. The initial task in systematic work is that of arranging the legitimate symbols (names) of the elements that are *treated*. Their subsequent taxonomic disposal depends on additional criteria variously available and variously described.

Many arguments may be advanced to justify the recognition of the new taxon and to distinguish it from others of a similar morphologic nature or approximate relationship. Some species may be proposed ostensibly for classificatory convenience. Even in such a case the argument concerning recognition of a new group of plants still holds because, if a full identification could have been established with some group already named, recognition of a new taxon and a new name obviously would be superfluous and devoid of legitimate purpose. In one form or another the two enumerated elements of systematic purpose always are involved. Recognition of a new taxon always involves lack of identity or at least some element of uncertainty of relationship with taxa previously established. If this should not be the case, the taxon would not be "new."

Some names, however, have been proposed and published as avowed synonyms. Usually the ostensible justification for such a mistake involves confusion between morphologic terminology and taxonomic nomenclature. Authors have been known to propose new taxonomic names to apply to specimens whose taxonomic assignment seemed definite and chiefly showed the need of additional morphologically descriptive terms. A reasonable policy would seem to deny an author the privilege of assigning the same specimen to more than a single species within the pages of the same publication. An author's opinion about specific relationship may change in even a short time, but the readers of scientific literature deserve at least this initial element of "decent" consistency in treatment from an author. Unless publication consistency is adhered to and seriously regarded by others, even the publication date of priority for the new name becomes uncertain. If a species is regarded seriously as new and if other students are to regard the new taxon as legitimate, the specimen on which the new taxon is based cannot *at the same time* be stated to be conspecific with another. An author must clearly adopt a policy by which the new element can be consistently treated. (See Lanjouw, 1966, art. 34.)

"LUMPERS" VERSUS "SPLITTERS"

Taxa that deserve to be named obviously should be as consistent in their botanical significance from one group to another as information permits. Benson (1943) indicated that the groups should not be so finely subdivided that a similar system applied to human beings would call for assignment of a brown-eyed man to a species different from his blue-eyed mother or father. He suggested that persons who adopt such narrow and unnatural divisions are "splitters." Taxa should not be so broadly defined, however, that they include elements that are not actually a part of the same genetic population. Benson cited lions and tigers as examples of organisms that are obviously closely related since they have too many characteristics in common for the coincidence to be merely a matter of chance. To consider them conspecific, however, would be absurd, because in no sense are they a part of the same modern genetic population. Benson regarded the persons who adopt such broad and unnatural divisions as "lumpers." Arguments in taxonomy, which have a basis apart from matters of technicality, all are essentially resolvable as questions of policy involving one or another aspect of the "lumper" or "splitter" philosophy. If students recall this precept, they may avoid distractions caused by semantic variation or the incidental use of unfamiliar terms, and they may better understand classificatory problems. Benson's identification of both "splitters" and "lumpers" as advocates of "unnatural" taxa is a most satisfying definition. Unfortunately it is a distinction that is hard to apply in practice because standards of splitting and lumping vary in different groups of plants.

Most systematists consider that the only natural classification is a phylogenetic one. It will be observed, however, that all taxonomists are advocates of one "natural" system or another, but because of divergent interpretations they actually may hold some quite different concepts.

Moreover, both broad and narrow interpretations of taxa may be equally defensible within the context of phylogenetic classification. In part perhaps the decision should depend on practical considerations. Beyond being natural in a phylogenetic sense, the fossil taxon should be defined with reference to systematic consistency—taxa of about the same "value" should be represented in each group. Due regard for potential stratigraphic utility as well as other considerations ought to be taken into account. Persons who favor broad interpretations, may be regarded as "lumpers," and those who favor narrow interpretation, as "splitters," but both their interpretations can be perfectly sound phylogenetically. In either interpretation the taxa merely represent different policies of subdivision (organization and evaluation) within a larger biogenetic group. The stigma of "unnaturalness" is extremely hard to apply if a paleontologist is concerned with phylogenetic implications in a broad range of taxonomic groups; however, he should apply a reasonable consistency in treatment. A clear distinction between "unnatural" splitters and lumpers may in fact be possible only in some ranges of neobotany where a superficial view can be adopted about the differentiation of groups of plants in the course of geologic time.

PHYLETIC DIVERSITY AND PHYLETIC UNITY

Many indirect types of evidence may provide evidence of phyletic diversity; for example, if spores of virtually identical morphology are genetically related, they are not likely to show a consistently disjunct stratigraphic occurrence as, for example, in beds of Late Cretaceous and Mississippian age. Conversely, if spores of somewhat disjunct stratigraphic occurrence are placed within a common species or genus, according to a phyletic system, we may reasonably infer that the responsible paleontologist considered that the true stratigraphic range was probably continuous. An example may be cited concerning the taxa known as *Cirratriradites*, a genus allied with the Selaginellales, and a group of plants identified from Cretaceous rocks as *Aequitriradites*, a fossil which shows a similar morphology. (See Upshaw, 1964.) The taxonomists who believe that a *Selaginella* relationship is represented by these fossil spores in the Cretaceous will find no reason to identify the material with anything but *Cirratriradites*—the morphology is very similar. Those who believe that the similarity has been arrived at homoplastically and that a group must be represented that is distinct from the selaginellas, will class the fossils with *Aequitriradites*. Study of fertile parts of associated megafossils might readily disclose which of the contrasting relationships has a greater probability.

All groups of true phyletic relationship have a continuing stratigraphic range from the time of their inception to the time of their extinction or diversification. A truly disjunct range in time is an impossibility for any group that has phyletic unity. However, in any region a local stratigraphic disjunction may occur due to migratory retreat or, conversely, invasion. Modifications in geographic range may be due to many causes. An understanding of distribution in time and space must be regarded as important evidence for the differentiation of any taxon that is constructed according to a phylogenetic system. A coherent understanding of distribution will of course be impossible if phyletic relationship does not exist.

The examples suggest that a phyletic approach to taxonomy is more meaningful because it usually involves several generalizations and conclusions of potential utility. If selaginellaceous plants were abundantly represented in many Cretaceous deposits, we would be closer to understanding the environments of deposition, and the records could be used to check on consistency of ecologic relations for the plants. These deductive avenues are largely blocked as long as phyletic alliance is disregarded. Of course the knowledge that permits such scientific generalization usually cannot be accumulated rapidly. Usually it cannot be assembled solely from information available at any one time and place but depends on a consistent effort from many reports and several sources. Probably this fund of knowledge never reaches perfection or completion. Working knowledge at any time depends on a considered evaluation of the reports available, and all reports represent stages of progress toward a more accurate and a more complete scientific understanding.

CONVERGENCE IN CHARACTERISTICS

The same functions can be served and virtually similar morphology can be achieved by different methods of growth. The fewer the characteristics involved and the simpler the nature of the function to be served, the more likely it is that apparent duplication by different modes of origin has occurred. Dispersal of disseminules is a very simple and very fundamental function, requisite to the success of virtually all plants. Wind and water are comparable media for dispersal. Under these circumstances it is not so surprising that some different groups of plants seem to converge in producing a common, efficient type of dispersal product. The more surprising fact is that, in spite of functional analogies, the disseminules of different groups differ as much as they do. We might be led to conclude that many ornamental features have a rather incidental value for species survival. Nevertheless, some of the differences among different genera are very subtle, and for determination of plants on the basis of isolated spores we must face the fact that convergent similarities occur. Few heritable features are disclosed by study of only the resistant spore or pollen coat available in many fossils. The fewer the characters concerned, the greater are the chances that convergence will cause taxonomic differences to be concealed or to be easily confused.

There are great differences in the extent of phylogenetic convergence. Theoretically the degree of convergence depends first of all on the maximum divergence in morphology that occurred between the two ancestral lines. In other words, it is necessary to consider the maximum amount of morphologic difference, the presumed maximum degree of phenotypic isolation, between the lineages that have again converged with regard to a few confusable characteristics. Even though the phyletic separation may be of ancient origin, the spores of the respective plants may have never been greatly divergent in character, even though other features may have evolved very differently. Persistent similarities of function in spores of both groups of plants would seem to favor conservatism in characteristics of the spores. Consequently the degree or duration of phyletic isolation and the degree of anatomical divergence during the period of phyletic isolation must be considered in studying convergent evolution. Simulation of characteristics may be found in groups that are as far separated as algae and anthophytes. We are more likely to presume that convergence among features of allied genera should be termed character conservatism rather than convergence; but there is no reason to question that convergence can occur after almost any degree of previous divergence.

The farther separated the two convergent lines have become in ecologic character, the more important it is that convergent features be recognized. Recognition of convergence may not be as critical for pollen grains from two grass species that are more or less adapted to a similar habitat. The "conservative" pollen features would still be interpretable as a general habitat indicator, which would not cause difficulty. If an algal spore is confused with grass pollen, a great difference in interpretation may result. Some grass pollen and algal spores are sufficiently similar in morphology for them to be confused. They could be assigned under the same form genus if only pure morphology is taken into account.

Most palynologists probably would avoid confusion of such extreme examples. Spores of soil fungi, however, have been confused with remains of marine microplankton, and, although examples are hard to detect in the literature, they may not be as rare as we would like to believe. Certainly the likelihood of convergence remaining undetected is greater when a palynologist emphasizes pure morphology in classification than when he is forced constantly and critically to question whether apparent similarity in structure means, or is reasonably interpreted as, phyletic relationship.

CONSERVATISM

Some types of spores show long stratigraphic ranges that probably indicate the continuing existence of a particular group of plants. Disseminules or sporelike microfossils referred to the genus *Tasmanites* apparently range in marine environments from Ordovician to Recent. Wall (1962) concluded that the modern *Pachysphaera* is very similar and may be the same as *Tasmanites* Newton 1875 or *Leiosphaeridia* Eisenack 1958. Felix (1965) reported other occurrences in Neogene deposits. Of course these microfossils may represent cysts and are only sporelike in morphology; they do, however, probably signify an unbroken chain of descent in a consistent environment, little affected by time.

Another example of conservatism relates to ornamentation of schizaeaceous spores (Hughes and Moody-Stuart, 1966) in which ridges usually are ornamented by a characteristic tuberculation. Many other variations have been noted, and several genera have been distinguished for this reason. Of course not all "knobby" types of spores belong to this family, but there is evidently a conservatism of spore ornamentation that started during the late Paleozoic (see Radforth, 1938; Schopf, Wilson, and Bentall, 1944, p. 55) still maintained within this family of plants.

Most commonly conservatism is apparent among plants that seem to have existed in a persistently uniform environment. However, a different weight in evidence must be recognized, depending on the particular group of plants that is concerned. Prokaryotic (asexual) organisms, such as bacteria and the blue-green algae, may have an exceptional range of environmental tolerance. These plants include the fossil iron bacteria that seem to have maintained specialized form and function from at least the Middle Pennsylvanian to the present. One of these bacterial types is quite comparable to a bacterium from Precambrian deposits about 2 billion years old, and isolated types of bacilliform organisms have existed for at least 3 billion years. *Tasmanites* probably represents an extreme example of conservatism among eukaryotic organisms that possess a form of genetic interchange in reproduction. Its environment seems to have been about the same all through the Phanerozoic.

Hystrichosphaerid types have not been as closely identified because many authors have been skeptical that forms that look alike are actually related. The similarity of form between tetragonal types assigned to *Veryhachium* and that of the modern green alga *Tetraëdron* may be mentioned. One of the reasons these forms are distinguished is that *Veryhachium* disseminules occur in, and are characteristic of, marine deposits, whereas *Tetraëdron* represents, just as characteristically, a form found in fresh water.

ORGAN GENERA

Organ genera consist of groups of plants allied within the same plant family that are defined by functionally related and commonly connected "sets" of biocharacters. Organ genera are not genera of organs—such an expression involves a contradiction in terms and is nonsense. Genera are taxa, or groups of allied species of plants; organs do not have any independent organic existence. Everyone knows, for example, that a leaf includes a host of organs. Even the epidermis includes separate organs with biocharacters having demonstrable taxonomic significance. No one pretends that a leaf can be discretely defined morphologically in many primitive vascular plants, although foliar functions are represented. Even among advanced plants of broadly divergent ancestry the foliar homologies are none too certain; for example, a monocotyledonous leaf has been homologized with a dicotyledonous petiole, but this interpretation is generally regarded as a theory that is hard to substantiate in detail. A similar situation exists for all the other morphologic categories, but this need not greatly concern us if we are able to establish *some* valid basis for more immediate comparison. The problems of the palynologist and paleobotanist are intensified by variations in the preservability of plant remains and by the vicissitudes of sedimentary transport, burial, and the diagenetic agencies that subsequently affect the fossil fragments. Consequently the morphologic categories that serve as convenient means for organizing morphological information are not a very satisfactory basis for practical taxonomy. Organs or not, certain standard types of fossils recur, and it is these common varieties of fossils, rather than morphologic organs, that have come to be placed in "organ genera."

The terms "organ genus" and "form genus" are so closely related in usage that they must be explained together. As far as nomenclature is concerned, names of organ genera, form genera, and names of any other taxa of generic rank all conform to the same regulations and all have the same protection of priority. The distinction between the two is in the *taxonomic manner* in which they are regarded.

The term "form genera" is applied to genera for which we have no proof of precise botanical affinity. The degree of precision that is involved is more explicitly stated in the Code (Lanjouw, 1966, art. 3): "A form genus is a genus unassignable to a family, but it may be referable to a taxon of higher rank . . . Form genera are artificial in varying degree." Conversely, "an organ genus is a genus assignable to a family." Thus the nomenclature of organ genera is normal according to regulations for taxa of generic rank.

Obviously some palynomorph fossils are assignable to organ genera, but in many instances, especially in the more ancient types of fossils, our taxonomic information is not that good and only a more generalized (form genus) assignment is possible. The interpretation of material may be subsequently improved for some assignments, and, with emendation if appropriate, a taxon originally instituted as a form genus may later be properly assigned to a family as an organ genus. Taxonomic unanimity, however, may not be immediate. Some taxonomists may accept the validity of a familial assignment of a taxon, and other taxonomists may feel that the evidence remains inconclusive. Thus, according to different interpretations, a taxon may be regarded as a form genus by one person and an organ genus by another. The category designation serves to indicate legitimate divergences in taxonomic opinion. We cannot expect all taxonomists to reach the same conclusions even though all have access to the same sources of evidence. Moreover, the evaluation of evidence may change as we continue to view fossil plants in better perspective. The designations of organ genus or form genus suggest the manner in which taxa were regarded by *some person*, at a *given time*, and they continue to provide assistance for interpreting literature in which such information is recorded.

SYNTHETIC ATTRIBUTES OF FAMILIES

The families of plants are based, like other taxa, on classification proposals of competent systematists. Appropriate familial classification depends on an acute sense of proportion and judgment, tempered by a reasonable concession to taxonomic tradition based on previous studies of the group. Authors who propose familial taxa must determine (subjectively on the basis of experience) which features are critical. Only subsequent studies can show whether emphasis has been placed properly. In a geochronologic sense all the modern families, and all superior taxa of plants, have been in existence for a long time and must be thought of as projected into the geologic time scale where they are represented by fossils. Many ancient fossils, however, may be best referred to families now extinct. Even among family classifications that are based principally on modern material great divergences of interpretation occur. Many families that are now proposed are monotypic, but such a disposition rarely solves a taxonomic problem—usually the problem, a phyletic one, is simply transposed to a taxon of higher rank. The manner in which family names should be constructed and the nature of their types are prescribed by the Code. A genus serves to typify a family, but this relationship, although permanently fixed, is entirely conceptual. Phyletic alliance, a matter of personal appraisal, must be the governing factor. It is most important to recognize that the families of plants are derived by conceptual syntheses of authors.

In paleobotany general alliance is indicated by discoveries that are still sometimes spectacular. The alliance of *Archaeopteris* and *Callixylon*, recently proved by the studies of Beck (1960), must at least be signified by assignment of both genera to the same family. Difficulties in the precise correlation of sufficient characters of the axis that are diagnostic of species of *Callixylon* with characters of the foliage that are diagnostic of species of *Archaeopteris* suggest that at some point in the taxonomic arrangement of the group, two alliances will have to be separated for reasons of practical convenience. Since most of the fossils consist of either foliage or of wood, they provide specific information of only one type. The separation might be made at the section level, or subgenus, if that seems appropriate, but it probably will be most convenient to maintain the several different species within the genera *Callixylon* and *Archaeopteris*, much the same as these groups have been treated in the past. In any event their reference to the same family of plants must surely be beyond dispute, and the necessary rearrangement principally affects assignment at a suprageneric level.

A question arises whether isolated spores may also be assigned to this family. Arnold (1939) has illustrated sporangial groups of both megaspores and microspores from fertile specimens of *Archaeopteris* that provide a basis for judgment. The spores by themselves are not very distinctive and probably would have to be assigned to a more generalized taxon (a form genus) if similar spores were found individually dispersed. However, features shown by sporangial aggregates of spores of both types probably would be more diagnostic. Dissociated sporangial aggregations of spores might not be directly referable to any species of *Archaeopteris*, but their more general alliance with the same family of plants might reasonably be inferred. If the occurrences of such fossils seem important, it might be desirable to assign the sporangial aggregates of spores within the same family to a genus not yet designated. The use of an additional name would be involved, however, and perhaps some alternative method of reporting would serve the purpose. Alternative possibilities should be considered before a new taxon is proposed.

Most systematists recognize that families and other taxa of superior rank serve an important function for convenience in verbal reference that is somewhat apart from the more substantial arguments of phylogeny and emphasis that distinguish the families of plants. Names of well-known groups, such as the "Amentiferae," are still referred to in discussion—even though the group has been essentially dismantled in phyletic consideration—simply because of a general comprehension of the concept referred to. It is unfortunate that family alliances have not yet been carefully systematized for fossil plants. Many modern families may be utilized for fossils of Cretaceous and Tertiary age, but older fossils are not as easily assigned, even when their relationship is well established, because familial classification has not received enough attention. Studies should be devoted to familial alliance of plant microfossils with the groups based on megafossils with which they are related. Lack of attention to the familial category in paleobotany still hampers reference convenience and systematic retrieval of information about general aspects of progression and history of the plant kingdom.

INFORMAL AND MORPHOLOGIC SYSTEMS

Artificial systems of classification do not constitute as accurate or as soundly conceived bases for generalization as do phyletic systems, but they may be more readily applied by students with lesser training or experience. Requirements of artificial systems are more easily met and more rapidly applied; their most successful range of application is local—within a particular geologic basin or province of sedimentation. If taken within so limited a frame of geologic reference for the primary purposes of stratigraphic correlation, almost any consistent system of microfossil classification will provide relatively rapid and useful results. It would therefore seem desirable to adopt for any program of microfossil studies a system of classification according to (a) the state of preexisting knowledge, and (b) the nature of the main objectives of the work. It should be realized that at least some of the results made available by use of an artificial system may contribute knowledge with phyletic implications that can be utilized after it has again been reviewed, revised, and rearranged. Repeated revision and correction is an inherent part of the systematic method.

The writer (Schopf, 1964, p. 50) has already commented on the possibility of applying informal systems of classification to plant microfossils for practical purposes. Although a phylogenetic system is of the greatest fundamental importance, informal systems based on various kinds of plant microfossils have been applied successfully for stratigraphic correlation. Such practices may be most advantageous for studies of a provincial nature in which emphasis must necessarily be placed on age relations rather than on the geological history of plants.

Precedents have long been well established for the morphologic (special) classification in modern plants of leaves (simple, compound, etc.), fruits (pome, drupe, berry, etc.), seeds (atropous, anatropous, campylotropous, etc.), and pollen (inaperturate, monoporate, syncolpate, etc.). Clarity requires that terms denoting anatomical systems of classification be differentiated from formal names denoting the taxa of plants. Fundamentally, morphologic systems of classification are not taxonomic. For morphologic purposes convenience governs rather than priority; morphologic systems employ terminology rather than nomenclature. (See definitions, p. 18.) Tschudy (1957) has suggested a symbolic coding system based on morphology that seems to be convenient and advantageous for practical application. If coded "taxa" in morphologic systems were consistently identified, it would be possible to assign many of the elements to formally named species within the established system of taxonomy at any later time. The writer is convinced that such a procedure would often serve the purely practical, mostly local, purposes of stratigraphy more satisfactorily than a taxonomic treatment that deserves more critical attention.

Much of the need for time-consuming search of taxonomic literature may be deferred if an informal system is used for practical objectives. An informal system has the further advantage of making available practical results of palynologic study during early phases of microfossil exploration when announcement of formal taxonomic results probably would be premature. Taxonomic classification, which variously reflects phylogenetic relationship to whatever extent evidence supports it, is not necessarily limited to reports of microfossils; it requires painstaking evaluation of all the potential sources of information and additional bibliographic search. It is a fact that fundamental scholarly results are difficult to produce under pressure.

The weak point in the strictly morphologic (morphographic) approach to plant-microfossil classification lies in its distinct disregard of phylogeny and phyletic relationship. Two reasons may be advanced in explanation for morphologic resemblances of fossils; namely, phylogeny and convergence. (See p. 58-59.)

Two fossils may be similar because the same factors of heredity controlled their growth; thus resemblance is due to common ancestry. Such an explanation would apply to all spores formed within a sporangium or to those formed within different sporangia of the same plant. The amount of variation noted under such conditions would be regarded as normal incidental variation, which is of no consequence for taxonomic determination of individual fossil spores, although parameters of normal variation are important for definition of species. Similarities in morphology depend on the nature and extent of hereditary control of biocharacters, and when their control is recognized the morphologic features are significant for identification of plants.

A clear differentiation between anatomical-morphological and taxonomic concepts is particularly essential in the systematic study of fossil plants, although this aspect has rarely been emphasized in standard treatments. Spore and pollen morphology seems so deceptively concise and simple that a great temptation exists in the study of microfossils to substitute a special classification, based simply on morphology. Taxonomy that is based on microfossils is actually more difficult than that based on megafossils because of difficulty in distinguishing features that arose by convergence from those that indicate close hereditary alliance. Systematic science really depends on the integration of historical and inferred hereditary information about the groups of plants. Special schemes of morphologic classification, however, tend to deemphasized hereditary principles and exert an uncoordinated and divisive influence in taxonomic work.

In an artificial (special) system of classification all definable biocharacters are treated very much alike. Organisms are classified according to resemblance in form. Unrelated plants may resemble one another in a few features of architecture or anatomy. Should these features be so situated that they occur apart from biocharacters that serve as more reliable indications of affinity and become individual dispersed fossils, then all would tend to be classified together. According to a system in which form is arbitrarily accorded precedence no consistent basis could then be justified for separating the fossils in classification, whether or not their heterogeneous affinity were recognized. Obviously this is wrong. Phyletic distinctions should be recorded whenever the distinctions can be noted.

According to a phyletic system, if there is *any* reasonable basis for recognizing the heterogeneous elements, these elements can be separated in taxonomy. The degree of taxonomic separation should be approximately proportional to the degree of phyletic separation and can be easily expressed by conventional classificatory devices. Representatives of species are always considered to be phyletically related. If any species should have a heterogeneous derivation, its stratigraphic or other use would be seriously compromised. However, any information that suggests that heterogeneous elements are included in a species also is likely to provide a basis for segregating the heterogeneous elements. At least this information provides a warning to qualify specific stratigraphic implications.

The species included within genera are subject to similar consideration, but evidence of phyletic relationship between genera is subject to more latitude of interpretation.

The contrast between an artificial system of classification and one reflecting phylogeny is best illustrated by differences in treating homoplasy, the results of so-called convergent evolution. Palynologists who work with fossils that show a restricted range of definable biocharacters must be greatly concerned with this problem since this type of morphologic resemblance is particularly common among spores and pollen. The chances for duplication of form in functionally similar organs among plants of different families are great. Demonstrable similarities commonly occur in plant groups that are unrelated in any narrow sense, such as the photosynthetic shoots of *Casuarina* and *Equisetum*, or the habit similarity of *Aloe*, *Agave*, and *Yucca*. The similarity of pollen from some species among the Casuarinaceae, Betulaceae, Juglandaceae, Myricaceae, and Moraceae is noteworthy, and even familial assignment of certain fossil forms is difficult. The determination of fossils of questionable affinity should be always qualified so that the presence of their names in a list of fossils does not obscure the significance of associated elements that can be more accurately identified.

UNORTHODOX SYSTEMS

Difficulty arises because authors who use a morphologic system and rarely mention accepted groups of the plant kingdom usually are not explicit in stating their fundamental basis of procedure. Van der Hammen (1956), however, has been exceptionally frank in using modern pollen of known derivation as the type basis for his system of nomenclature. Clearly, any materials conspecific or congeneric with the nomenclatural types so defined are systematically related to the modern taxa from which the types were derived. If the different morphologic names of "species" of spores and pollen are to be taxonomically regarded, all the new names must be later homonyms (superfluous names) for modern plant taxa. Pierce (1961) has adopted a somewhat similar procedure, although some of the names used have been different. He calls his classification system "eclectic," and properly so, because he has commonly chosen to disregard the normal systematic procedures. However, not all taxa proposed by Pierce are illegitimate. His inconsistency does not simplify the problem.

Even though the practices of Van der Hammen (1956) are in flagrant disregard of accepted nomenclatural rules and principles, some advantages in his treatment are apparent. If objections to a nomenclature that is completely illegitimate can be cast aside, it is possible to devise a system that is logical and easy to apply. The tie with systematic botany, although "illegitimate," is nonetheless real. There is no question about the status of the names Van der Hammen used, and hence there can be little difference of opinion or confusion regarding them. It may be that this system, which is not as subject to misinterpretation as many others, has advantages that ought to be more widely considered for applications of a temporary nature.

The illegitimate nomenclature used by Van der Hammen and by Pierce (1961) is written, however, as if the accepted rules of nomenclature had always been applied. An incidental observer, therefore, might labor under misapprehension. It would seem most appropriate, if one really believes that convenience in classification should supersede adherence to the Code, to express the units of such a classification by a different kind of symbol. A system should be either distinctly indicated as a convenient, morphologic scheme or else, if the format and regulations of taxonomic nomenclature are accepted, as regularly taxonomic and nomenclatural. For this reason the format of nomenclature used by Van der Hammen, and commonly by Pierce, is not desirable. It would seem that the code system advocated by Tschudy (1957) and discussed earlier in this chapter provides at least one easy and appropriate means of avoiding this difficulty.

The eclectic quality of morphologic classification systems constitutes both their strength and their weakness. Although by selection we may construct what appears to be a logical and consistent scheme of classification of pollen and spore configurations, these classification systems hold no promise of stability. The most basic principle of taxonomic nomenclature, that of priority, cannot be adhered to if morphologic convenience is adopted. The basic distinction between taxonomic nomenclature and morphologic terminology is that the latter is based on expediency and convenience. If we desire an appropriate name and temporary convenience, we may choose between the multitude of morphologic systems that have been proposed or devise a system of our own to suit our own standards. Another recent system of this sort is the one proposed by Corsin and his colleagues (1962). Whatever advantage is gained in morphologic consistency and in the descriptive quality of terms applied through a morphologic classification is lost through divergency of attack and discrepancy of interpretation by different authors. All systems that attribute inherent meaning to names are, contrary as it may seem, a source of real difficulty in systematic communication.

A basic uncertainty at the level of species and genus concerns whether the authors agree that species and genera represent groups of plants—collective abstract concepts in systematic botany—or whether the species and genus names they use symbolize only spores or grains of pollen. This is by no means the only source of uncertainty because it seems evident in a great many instances that only the most superficial approach to systematic botany has been involved. It is, however, a fundamental one.

Undoubtedly one of the major difficulties in achieving reasonable application of nomenclature to fossil plants is a tendency to adopt too superficial an attitude toward systematics. Now, more than ever before, we should give greater attention to the substance and philosophy of the Code of Nomenclature as well as to the courteous formalities of reporting. Adherence to the Code simplifies interpretation for the reader. The present Code has been derived as a means of avoiding ambiguity and confusion in systematic expression in the light of previous difficulties and mistakes. *Ideas* about classification are much more important than any unfamiliar innovation in the means adopted for expressing them. Ideas leading to improvement in classification may have merit that is more easily overlooked if an author adopts an unorthodox mode of expression. Constructive ideas receive most serious consideration if a reader is not distracted by the unfamiliar way in which they are stated.

SUGGESTIONS TO STUDENTS

In the process of becoming familiar with the methods of palynology somewhat more general knowledge about systematic biology should be acquired than that required for specific application. Palynologists commonly (but not invariably!) have a more diversified and a more fragmental type of material to deal with than do neobotanists. The point of greatest fundamental contrast is that the fossils cannot be subjected to functional experimentation and that knowledge of physiologic relations must be derived entirely by inference. However, this approach is not so different from the practices in some kinds of neobotanical science as it might initially seem. Although an experimental approach is possible when modern plants are studied, anatomists commonly are able to infer function on the basis of structural relationships rather than by experiments. Experimentation has its greatest value when one is concerned with subspecific taxa. Valid inferences of function can be derived for structures observed in a great many fossil plants. Of course pitfalls of teleology and wishful thinking are real enough and should be avoided. It is at least equally important to appreciate that effective functional arrangements were in operation during life for the plants now indicated as fossil. Pollen really *did* function during the Cretaceous (and at other times!). The fossils *do* represent organisms that lived. The grains *should not* be interpreted as "sports of Creation"!

Most groups of microorganisms present difficulty (for different reasons) in the correlation of information that is characteristic of some problems with fossil plants. The classificatory devices required by practical mycologists and bacteriologists in dealing with incomplete information about their groups of microorganisms provide a parallel for the students of palynology.

Students of modern microorganisms have defined their systematic concepts and framed their classificatory concepts more realistically with respect to difficult problems of classification than has always been necessary for students dealing with higher plants. Students of higher plants have accepted methods for comparison and reporting in spermatophyte taxonomy that are virtually traditional. In this context a thorough analysis of the classificatory process may not be essential because many fundamental concepts seem very obvious and too self-evident for statement. However, some of the self-evident considerations in systematics are not nearly so evident when the problem is that of classifying microfossils or microbionts (which must also be guarded from contamination). Benefit is to be derived by reading some of the broadly based works in systematic microbiology, such as the recent (seventh) edition of Bergey's "Manual of Determinative Microbiology" or Ward and

Whipple's "Fresh-Water Biology" (1959). The annotations included in the "Bacteriological Code" (Buchanan and others, eds., 1958) are most helpful since the significance of provisions is commonly stated and comparison is made with the manner in which botanical and zoological codes have treated common classification problems.

Practical solution of the problem of classifying many fossils of questionable relationship involves, first of all, recognition of inherent taxonomic difficulties. This might be taken to suggest that for some time to come valid uncertainties in classification will remain. Some species or groups of species have been described and classified according to principles that have no bearing on phyletic relationships. These groups must be remodeled or reassigned or else pass into disuse. There is, however, no established classificatory tradition to fall back on among palynomorphs. One scheme of classification may be as valid as another *if it accords equally well* with principles of organic evolution and stratigraphic geology. Even within this range of restriction many equally plausible taxonomic arrangements are possible.

A great deal of difficulty in classifying palynomorphs is psychological and arises from the desire to propose a nominal solution for problems for which no substantial solution is available. This tendency has caused a proliferation of schemes and names that are confusing *if* we attempt to apply them just as the various authors seem to have intended, because the underlying concepts are inconsistent and to some extent conflicting. A great deal of unnecessary complication is introduced when incidental variation is accorded an unwarranted taxonomic status. Some authors have proposed new species without considering variations induced by preparation, differences in preservation, or inherent and incidental variability. Genera seem to be proposed on the basis of insignificant or unproved differences without regard to inheritance of characters or possible functional implications. Supra-generic groups commonly seem to be based on any type of superficial similarity. A revisionary approach is commonly required in order to make effective use of earlier literature.

Effective plant taxonomy depends on the differentiation of incidental variation from differences that are genetically controlled. Genetic control can be directly proved only by means of experimental procedures. Consequently in paleontology characters must be treated according to a reasonable *presumption* of heritability or *interpretation* of an incidental causation. This distinction is related to the significance of morphologic features. Only significant variations should be accorded a rating as evidence in taxonomy.

The sources of incidental variation are particularly numerous for fossil spores. It may be presumed through analogy with modern plants that the spores and pollen grains of fossil plants show morphologic abnormalities caused by abortion, shriveling, and malformation along with the variations that are most characteristic and functional. Since spores are an indispensable link in the chain of life cycles, the more numerous potentially functional spores or grains will be perpetuated with the species. Malfunctional types will be sporadic in occurrence and relatively infrequent. In particular the malfunctional types should not be readily perpetuated in time since the functions of pollen or spores have direct survival value. On the other hand, a malfunction may be repeated, as in the case of hybridization between species with a genetic constitution that induces complete or partial spore or pollen abortion. Such an occurrence therefore might be expected wherever these species are in contact. Particular kinds of pollen of aborted configuration might be expected to have a highly peculiar and erratic stratigraphic distribution, indicative of a very strangely defined taxon of plants. Some taxa of this character may eventually be recognized.

Fortunately most of the abundant fossil spores and pollen may be safely presumed to be functional. Functionalism still implies a range of incidental variation for all features. Genetic control of characters is always to some degree subject to environmental influence and related to nutrition and position. No two pollen grains are ever identical in ultramicroscopic detail, but the differences are commonly so slight as to be insignificant. At least the differences are insignificant in relation to an ordinary classificatory purpose. The range of incidental variation in itself, since it signifies a *degree* of genetic control, is a most significant feature in definition based on spores, of any group of plants.

Spores and similar palynomorphs are also subject to different environments of preservation. Pollen analysts have noted that pollen of genera such as *Larix* is not readily preserved. Some species produce pollen or spores in which the more delicate features are usually lost during fossilization. The chaffy perisporium of some ferns serves as an example. A few years ago L. R. Wilson (1943) described a singular spore type with peculiar appendages from a coal-ball petrifaction as representative of a new equisetoid genus, *Elaterites*. Recently additional similar spores have been found in strobilar association by Baxter and Leisman (1967) to confirm and extend our knowledge of these plants. The genus has not, however, been identified from maceration residues, although the spore morphology is distinctive. This may be either a result of technique or environment of preservation. Taxa that can be recognized only from material preserved in these exceptional environments are less helpful in routine palynologic work, though they have great potential value in explaining plant relationships.

The morphology of spores commonly influences the plane of compression and affects the normal aspect of the fossils. Spores of *Densosporites* have a substantial cingulum that almost invariably leads to proximo-distal compression, and this morphology has influenced description and recognition of taxa. Assiduous search has been rewarded by the finding of laterally compressed spores of *Densosporites*, and examples of the tetrads (Kosanke, 1950), but most students have distinguished species of *Densosporites* only by features visible in proximo-distal aspect. No doubt this practice serves as a practical expedient, but it also signifies a certain limitation in morphologic understanding. More recently additional information, complementary to that afforded by the proximo-distal appearance, has been provided by use of microtome sections (Hughes, Dettmann, and Playford, 1962). Types of spores and pollen that show no discernible preferential plane of preservational compression are easier to describe, but the more stereotyped features they present are commonly less distinctive.

As in environmental control of original preservation, the changes induced by geologic vicissitudes may also be detected. Some spores are apparently more readily affected by heat and pressure within the earth's crust than others, and certain types of distinctive features also may be differentially altered. Spores in sediments adjacent to igneous intrusions are devolatilized, rendered nearly opaque in spite of a general reduction in volume, and may appear perforate although perforations were initially lacking. Devolatilized spores are more fragile, and some of the damage caused by igneous heating may actually result in greater difficulty in preparation. Some kinds of spores are more easily damaged than others. (See Chapter 5).

The question of whether ornamental features were originally absent, destroyed in deposition or in geologic alteration, or just rendered more difficult for preparation, is a common source of difficulty because one must rely on interpretation. Different maceration techniques sometimes yield sufficiently contrasting results to suggest that some biocharacters are effectively erased by improper procedure as material is prepared for study. No amount of subsequent study can compensate for the poor preservation or poor preparation of material.

All this leads to one conclusion. In order to be reasonably certain of identifying a biologic unit that is called a species, one must depend on "safety in numbers." The numbers of specimens provide a basis for evaluating incidental variation of the biocharacters (heritable characters) and for distinguishing empirically what is genetically significant. An extension of this thought relates to the numbers of horizons that should be sampled and studied in order to outline stratigraphic distribution. Stratigraphic distribution may in fact be regarded as the truest criterion for species recognition, because if the fossils occur in a predictable pattern in time and space consistent with their taxonomic evaluation, one comes closest to identifying the essence of any species. There are few species, however, that have been identified in this manner.

REPORTING

Taxonomic assignment should be a means of indicating an author's evaluation of phyletic affinity. As has been pointed out (p. 58), morphologic similarity may not signify close affinity. For a long time many plant microfossils may not be assignable to family; many may not be, with confidence, assignable within even higher taxa, and all we may be able to say is that they might represent one or another of the divisions of the plant kingdom. Even assignment at this level will reflect a stage of progress in systematic reporting. Assignment to some major group, at some level of the taxonomic hierarchy, should be made.

Microfossils still deserve description, regardless of the level at which they are assigned, because even those that have to be assigned on a most empirical basis are of use for purposes of local correlation. Needless to say, however, by themselves they have no significance in habitat determination, and their generalized assignment shows that they have little discernible bearing on the phyletic history of the plant kingdom. Forms about which we have little knowledge should not be indiscriminately arranged in a report among those whose significance is clear. Different grades of information are represented and should be separated for purposes of emphasis. The forms that are most significant botanically and ecologically should be reported in a significant manner. The system used in classification should be that which enhances all positive elements of information and knowledge.

All authors will develop their own preferences in presenting descriptive material. Descriptions should be adequate and cover all of the significant characters. They should be carefully and consistently organized. A telescoped style of writing that omits verbs and articles is commonly favored because it is concise and clear. All characters intergrade and interrelate in a historical sense, however, and not all can be included. Critical faculties must be utilized to achieve a useful and appropriate selection. Benson (1943) quoted the thought-provoking definition of intelligence as "the ability to recognize the significant elements in a situation." The test of an adequate description depends directly or indirectly on the recognition of significant elements and the ability to present them for benefit of a reader.

THE SYSTEMATIST'S RESPONSIBILITIES

A cardinal principle of taxonomic organization is the arrangement of taxa according to what *you* believe is phyletically most probable. If you *believe* that quite different groups of plants are represented by microfossils of similar morphology, you should not place the groups together. To do so would suggest a real confusion of purpose in taxonomy. A statement of evidence should serve to explain the scientific reason for believing that the groups of similar fossils are different and should justify taxonomic separation of the groups. However, if physical evidence actually is lacking, arrangements for convenience unquestionably are necessary. We may always hope that the phyletic arrangements can be improved and supplemented by access to new information and that, concurrently, at least some of the forms classed within the taxa of convenience can be transferred to a position of greater botanical significance.

Proposals of taxa leave much unstated, so that the reader must work with minimum information. The author virtually leaves it to posterity to decide whether these taxa were well formed intentionally or simply fell in line accidentally with a probable or plausible phylogeny. Care should be taken to provide all the pertinent information that is available. We may assume that any taxon formally treated has had the benefit of phyletic reasoning behind it because this is the kind of arrangement for which formal taxonomic groups are proposed. If the author supposed a taxon was unusual in its composition, we expect he would have called attention to it, but it would be reassuring if we knew this was true. Certainly, whether we agree with the reasoning or not, the presumption is that a proposed genus or species is normal and coordinate with other taxa of the same rank unless there is indication to the contrary. Anyone who differs from this presumption must carry the burden of the argument and indicate his source of evidence.

Formal nomenclature is not designed to reflect morphologic resemblance – in fact quite the contrary: it is meant to serve science as a conceptual "handle" for the treatment of plant groups that are regarded as phyletically allied or phyletically separated – regardless of their morphology. If convergent evolution can be recognized, or if it seems more reasonable by reason of extreme stratigraphic or geographic separation or for any other evidence to consider that different groups have converged in their morphology, they should be taxonomically separated. It is not that they *must* be separated. The only real obligation that a scientist *must* honor is *his own conviction* about alliance or relationship.

If a scientist is convinced that a proper genetic (phyletic) alliance exists, he *must* attempt to be consistent in expressing this conviction; if he believes that the evidence is more reasonably interpreted against the phyletic relationship, no correspondence in simple morphologic features, which may be evolved to a point of apparent duplication, should tempt him to set down a determination that is inconsistent with his belief. Should the evidence seem equivocal, a consistent policy should be determined for that time and the taxon should be treated on this basis. In a subsequent report, if necessary, the author may claim the privilege of changing his mind. If properly employed, the usage of names indicates the nature of treatment. We often observe textual statements that are diametrically opposite to the policy that has been implied by usage of names. Such a conflict conveys a printed record of confusion and uncertainty that negates the purpose of publication. The use of nomenclature should reflect the scientist's serious convictions. Anything less than this is misleading and is a misuse of the systematic method. Nothing more is involved than a serious attempt at expression of his evaluation of evidence.

Because of the inherent variability of biological material, taxonomy is not an exact science, and for this reason no solutions are unique. No one can be assured that his is the only possible, or even necessarily the best, taxonomic solution to any systematic problem. No legitimate taxonomy, in any case, is rejected by the Code; all taxonomic proposals are considered, and, on the basis of evidence initially presented and evidence brought out subsequently, proposals are accepted, modified, or rejected by persons concerned with the groups of plants that are dealt with. If a taxonomic solution seems cumbersome, impractical, or erroneous and is rejected, another solution that seems to be an improvement is required to take its place. Many taxonomic proposals are workable and are accepted, not because they are necessarily the best solution but for want of an available alternative that is an improvement. Most botanists who are not systematists and who make use of taxonomy whenever they have occasion to refer to a particular group of plants, do not pause for a review of evidence pertinent to classification and nomenclature. They adopt the names that are readily available and that they presume can be understood with enough certainty to serve their purpose. Even a completely arbitrary system of classification could serve temporarily for this means of reference, for cataloging information, and for aiding information retrieval. A purely arbitrary system, however, lacks all but arbitrary means of correction or supplementation. Arbitrary systems are not generally acceptable to systematists if any scientific alternatives are available. The phyletic system is the only one that is sufficiently fundamental to be generally acceptable and widely understood.

Emphasis should be placed on the personal responsibilities of scientists who do taxonomic work. One such responsibility is to work according to the spirit of the Code. This includes conformity in matters of context as well as in those matters that are stated more explicitly. A wide range of taxonomic categories of subordinate, intermediate, or superfamilial rank may be utilized for classification. A list of those regularly authorized is given in Table 4-1. Additional unspecified categories also may be used if needed, provided that their relative position is made clear. Within its range of context the Code does not rule out any proposed taxonomic revision or modification—it does not prescribe motives that will justify a taxonomic proposal. Doubtless motives may differ, but acceptable ones are those that are consistent with and, in fact, inherent in the context of the Code. Perhaps conformity with the context is even more important for orderly systematic progress than absolute agreement with Code provisions that are explicitly recommended.

AIDS IN TAXONOMIC STUDIES

In taxonomic study, according to Knight (1941), "The curious fact is that from a practical viewpoint one begins best the study of the group with a thoroughgoing survey of the generic names already in the literature. One first tries to discover all the names and, still working only with words, he tries to discover what species is actually the valid genotype of each name." This is library work. During the progress of this work

Table 4-1. **Categories of Taxa, Proscribed Epithets and Terminations, and Prescribed Suffixes Indicating Rank**[1]

Category	
1. INDIVIDUUM	(represented by specimens)
2. Forma specialis (microparasites)	
3. Subforma	*Inadmissible Epithets* "typicus" "originalis" "originarius" "genuinus" "verus" "veridicus" (applies to 3–7)
4. FORMA	
5. Subvarietas	
6. VARIETAS	
7. Subspecies	
8. SPECIES	
9. Subseries	
10. SERIES	
11. Subsectio	*Terminations Forbidden* "-oides"; "-opsis"; *and prefix* "Eu-" (applies to 11–12)
12. SECTIO	
13. Subgenus	
14. GENUS	

Category	*Suffix*
15. Subtribus	-inae
16. TRIBUS	-eae
17. Subfamilia	-oideae
18. FAMILIA	-aceae
19. Subordo	-ineae *when based on stem of name of a family*
20. ORDO	-ales
21. Subclassis	-phycidae = algae; -mycetidae = fungi; -idae = cormophytes
22. CLASSIS	-phyceae = algae; -mycetes = fungi; opsida = cormophytes
23. Subdivisio	-mycotina = fungi; -phytina = algae and cormophytes
24. DIVISIO (PHYLUM)	-mycota = fungi; -phyta = algae and cormophytes
25. Subregnum	
26. REGNUM VEGETABILE	

[1]If this series of ranks is insufficient, supplementary ranks may be intercalated, provided that the relative order of ranks specified is not altered and that the sequence does not cause confusion. (See Lanjouw, 1966, arts. 4 and 5.)

one becomes familiar with the origin and history of pertinent concepts and thus acquires some of the competence needed to deal with real material. Without such a background no one can effectively describe to other scientists what he has found. Not only is such background information necessary for understanding material, it is also essential for reporting.

Much time can be saved by consulting compilations of names that are generally known as indexes. Some indexes, however, are incomplete or inconsistent in treatment, and therefore the listings must be regarded with caution and each statement verified. In addition some tend to be organized for the purpose of promoting only a particular scheme of classification. An index of names is also invaluable as a bibliographic aid and a very important secondary source of information. No index, however, can ever serve as a substitute for original reference material.

Good indexes are not uniformly available for all types of material studied by palynologists or paleobotanists. The "Index Kewensis" (1893–) lists names of modern seed plants but does not include fossils. This and other modern plant indexes should be checked before proposing a new name in order to avoid the possibility of inadvertent duplication. Andrews' (1955) "Index of Generic Names of Fossil Plants" is especially valuable, although species are not included. A second edition is now in press. Unfortunately names of many genera based on plant microfossils have not been included in the new edition, although many published prior to 1950 have been given. Undoubtedly additional useful compendia will continue to appear. The following aids that deal particularly with plant microfossils are mentioned to advise students of their availability.

Schopf. J. M., Wilson, L. R., and Bentall, Ray, 1944, An annotated synopsis of Paleozoic fossil spores and the definition of generic groups: Illinois Geol. Survey, Rept. Inv. no. 91, 72 p., 3 pls., 5 figs.

This was the first attempt to provide a general index of fossil spores. Limited essentially to taxa occurring in the Paleozoic (some competing names of taxa occurring in the Triassic also are included), species were allocated among 23 genera and listed, with their nomenclatural types, in the order of their nomenclatural priority. Five new genera were proposed. Genera of some phyletic significance are distinguished from genera of convenience. In general, broader generic circumscriptions were adopted than has been characteristic of later work, and desirable revisions are now evident. However, the broader concepts have had some advantages in practical application. Guiding principles given on pages 7-10 still seem to be valid and pertinent.

Potonié, Robert, 1956, *Sporites*, pt. 1 *of* Synopsis der Gattungen der Sporae dispersae: [Hannover], Beihefte Geol. Jahrb., v. 23, 103 p., 11 pls.

——1958, *Sporites* (Nachträge), *Saccites*, *Aletes*, *Praecolpates*, *Polyplicates*, *Monocolpates*, pt. 2 *of* Synopsis der Gattungen der Sporae dispersae: [Hannover], Beihefte Geol. Jahrb., v. 31, 114 p., 11 pls.

——1960, Nachträge *Sporites*, Fortsetzung *Pollenites* mit Generalregister (to Pts. 1-3), pt. 3 *of* Synopsis der Gattungen der Sporae dispersae; [Hannover], Beihefte Geol. Jahrb., v. 39, 189 p., 9 pls.

——1966, Nachträge zu allen Gruppen (Turmae), pt. 4 *of* Synopsis der Gattungen der Sporae dispersae: [Hannover], Beihefte Geol. Jahrb., v. 72, 244 p., 15 pls.

The Robert Potonié *Synopsis* lists genera based on isolated (dispersed) spores and gives date, references, type species, source locality, and the deposit listed by each author. Usually the illustration of the holotype of the type species has been copied and is reproduced as a line drawing. An effort has been made to determine validity (legitimacy) of names. Helpful notes and comments are included. A clear statement regarding nomenclatural status usually is omitted. In part 4 some names regarded as invalid (according to an interpretation of article 42 in the prior edition of the Code) must evidently be reinstated now that the wording of article 42 has been clarified. Consequently, one must draw his own conclusions as to establishment of validity. Dates as given by Potonié apply to bibliographic references, and they do not necessarily indicate a date for priority. Entries are arranged according to a system of artificial categories (the "turma" system)—anteturma, turma, subturma, infraturma—the designations of which are treated as if they were names applied to suprageneric taxa. Generic names and comments follow infraturmae. This system seems relatively complicated since the differential features have been dispersed throughout the text instead of having been presented in the usual condensed form of an artificial key.

Many of the suprageneric designations would seem to duplicate and conflict with corresponding and well-established botanical nomenclature. (See Potonié, 1962.) Thus the fancy names for suprageneric groups probably could be dropped—in large part they must be synonyms, and all are of undefined rank. They could be treated as informal terms and arranged for more efficient use in simplified form as an artificial key.

Thanks to inclusion of detailed contents, tables, and a good index, however, the volumes may be used effectively without memorizing the system. Introductory discussion to the first volume (Potonié, 1956) is most concerned with establishing the turma system; that in volumes 2 (Potonié, 1958) and 3 (Potonié, 1960), with validity of generic and specific names, particularly with reference to nomenclatural types as required by the "International Code of Botanical Nomenclature." Potonié has attempted to treat genera and species formally so that they deserve status in priority, but the turmae serve only as do the "parataxa" in zoology, and parataxa have never been sanctioned by either the botanical or the zoological codes.

Potonié's dispersed-spore synopsis is an exceedingly valuable guide to original spore literature, particularly the literature on vascular plants. It does not cover many of the so-called phytoplankton or acritarchs.

Potonié, Robert, 1962, Synopsis der Sporae in situ. Die Sporen der fossilen Fruktifikationen (Thallophyta bis Gymnospermophyta) im natürlichen System und im Vergleich mit den Sporae dispersae: [Hannover], Beihefte Geol. Jahrb., v. 52, 204 p., 19 pls.

This volume attempts to summarize information about spores in sporangia before they have been shed, to serve as a complementary volume to the four volumes concerned with dispersed spores previously discussed. The original material is so varied and the original reports are so variable that unfortunate limitations are imposed. Most authors of the original reports had little conception of the detail with which spores could be regarded. Restudies and additional preparations need to be made for many of these spores. Potonié's volume is a useful summary of what the literature can provide.

Traverse, Alfred [and various coeditors], 1957-continuing, Catalog of fossil spores and pollen: University Park, Pa., Pennsylvania State Univ., v. 1-31.

"The Catalog of Fossil Spores and Pollen" was begun at Pennsylvania State University in August 1957 by G. O. W. Kremp, W. Spackman, Jr., Tate Ames, and members of their supporting staff with the intent of bringing together all original specific and generic descriptions based on specimens of fossil spores and pollen, with references, supplemental information, and illustrations. Alfred Traverse became associated as editor of the catalog in 1967. Twenty-eight volumes of about 150 sheets each were originally projected in 1957, with one species entry per sheet, to total about 4,200 descriptions. Once this catalog becomes current with published literature, it will prove to be invaluable for undertaking any possible taxonomic arrangement of material. The volumes are all looseleaf to readily permit any desired rearrangement.

Thirty-one volumes have now been issued, together with an index of sources of the first 20 volumes and partial coverage of stratigraphic, geographic, and taxonomic terms. Two translation volumes have also been prepared covering two of the important source references in Russian literature. During the 12 years of the catalog's existence, however, the literature has continued to grow at an increasing rate, and considerably more than 31 volumes will be needed to achieve the primary objective. The catalog has been supported by grants from oil companies and from the National Science Foundation and by sustained subscription of many libraries, organizations, and private individuals.

Deflandre, Georges, and Deflandre, Marthe, 1962, Dinoflagellés III—Peridinida à tabulation conservée, *in* Fichier Micropaléont., sér. 11: Centre Natl. Recherche Sci., no. 383, pts. 1-5, p. 1751-1947.

Essentially this volume and the following three volumes by the same authors consist of reproductions of 5- by 8-inch file cards (each of which is one fiche), reproduced two to a page.

Species entries are arranged in alphabetical order under 12 genera of which *Gonyaulax* is much the largest. Name, reference, age, and illustration of type material (often with an interpretative diagram) is given. Practical problems of classification and distinction between hystrichospheres and dinoflagellates are briefly discussed in the introduction.

Deflandre, Georges, and Deflandre, Marthe, 1964, Acritarches I—Polygonomorphitae-Netromorphitae *pro parte*—Appendice: Genres *Deflandrastrum* Combaz et *Wilsonastrum* Jansonius, *in* Fichier Micropaléont., sér. 12: Centre Natl. Recherche Sci., no. 392, pts. 1-10, p. 1948-2172.

Names of taxa with authority citations and age are given, usually with copies of illustrations of type material, with some illustrations that are original. Species are arranged alphabetically within genera on each half-page (fiche). The genera and species are listed in the introductory pages, together with notes and comments on a parataxon classification.

Deflandre, Georges, and Deflandre, Marthe, 1965a, Acritarches II. Acanthomorphitae 1, Genre *Micrhystridium* Deflandre *sens. lat.*, *in* Fichier Micropaléont., sér. 13: Centre Natl. Recherche Sci., no. 402, pts. 1-5, p. 2176-2521.

This volume is intended to cover all "microhystrichospheres," according to the size-inclusive concept for differentiating forms called *Micrhystridium.* Each fiche (half-page) gives name, citation, age, and illustrations of type material. The entries are arranged alphabetically in two sections and two appendices: section I (the largest) includes forms with simple appendages, most of which had been originally assigned to *Micrhystridium*; section II includes forms with forked or branched appendages that usually had been originally assigned to another "genus," mostly *Baltisphaeridium*, and are transferred under the new name combination ("hic = here"). Appendix A includes a dozen species that are transferred from *Micrhystridium* to other genera, and appendix B includes nine forms assigned to *Echinum*, some of which had been previously included in *Micrhystridium.*

Deflandre, Georges, and Deflandre, Marthe, 1965b, Chitinozoaires, *in* Fichier Micropaléont., sér. 7, 2me ed.: Centre Natl. Recherche Sci., no. 238, pts. 1-4, p. 1020-1095.

In the double-card format with name, figure, and citation, the volume also includes a considerable number of new illustrations of material Eisenack had distributed to Deflandre before 1939. Neotypes are designated for types lost (during World War II).

Deflandre, Georges, and Deflandre, Marthe, 1965c, Ciliés (Infusores) I. Tintinnoidea (incl. *Calpionella* auct.) et Ciliatae incertae, *in* Fichier Micropaléont., sér. 9: Centre Natl. Recherche Sci., no. 302, pts. 1-4, p. 1186-1293.

Eleven genera (one new) are briefly described. Descriptions of fossils are based chiefly on thin sections of the lorica as preserved in limestone.

Eisenack, Alfred, and Klement, K. W., 1964, Dinoflagellaten, v. 1 *of* Katalog der fossilen Dinoflagellaten, Hystrichosphären und verwandten Mikrofossilien: Stuttgart, E. Schweizerbart'sche Verlagsbuchhandlung (Nägele u. Obermiller), 895 p., 9 pls., 420 figs.

The introductory part of the volume includes a review of 17 families into which the genera are arranged according to systematics of recent dinoflagellates. It also provides a discussion of dinoflagellates and their organization in general, discussion of methods of preparation of fossil material, and prefatory remarks. Genera and species are listed in alphabetical order and illustrated by nine full-page plates of excellent halftone illustrations.

Genera and species are treated in alphabetic order, types are cited for all fossil taxa, and diagnoses are quoted; the holotype of each species is illustrated. Additional notes are given to aid taxonomic interpretation. A few mistakes have been noted, and the review of Tappan and Loeblich (1967) should be consulted regarding minor inconsistencies.

The pages are looseleaf in a binder of special design. Although convenient for consulting and comparing individual pages, the binder is somewhat fragile, and the volume must be used with care. A second volume is planned to deal with hystrichosphaerids and allied forms. This reference is an invaluable aid in taxonomy of these groups.

The pages completing volume 1 have recently been issued (Eisenack,1967) and reviewed by Loeblich (1968) in the Journal of Paleontology.

Norris, Geoffrey, and Sarjeant, W. A. S., 1965, A descriptive index of genera of fossil Dinophyceae and Acritarcha: New Zealand Geol. Survey Paleont. Bull. 40, 72 p.

The names of 241 genera regarded as valid are printed in boldface type; other names are listed in italics. Type species are cited. Diagnoses are given, and, when necessary, translated into English. Most entries are annotated, and literature is quoted as an aid to understanding the author's viewpoint. Tappan and Loeblich (1966b) indicate some inconsistencies that occur in this otherwise very useful treatment.

Downie, Charles, and Sarjeant, W. A. S., 1964, Bibliography and index of fossil Dinoflagellates and Acritarchs: Geol. Soc. America Mem. 94, 180 p., 2 figs.

About 665 references are given, and about 258 "valid" generic names are indexed. A list is presented of some 550 names of species and 33 generic names that are rejected, chiefly as a result of taxonomic interpretation. Thus the volume serves as a personal check list for those who wish to accept the present status of Downie and Sarjeant's interpretation. Not enough material is provided to enable a reader to draw his own conclusions. Type species, when fossil rather than modern, are indicated by an asterisk. The work has been reviewed by Tappan and Loeblich (1966a). A very useful index of stratigraphic units from which these microfossils have been recorded is included.

Loeblich, Alfred R., Jr., and Loeblich, Alfred R., III, 1966, Index to the genera, subgenera, and sections of the Pyrrhophyta: Miami Univ. Inst. Marine Sci., Studies in tropical oceanography, no. 3, 94 p., 1 pl.

Of interest to paleontologists, paleoecologists, palynologists, micropaleontologists, oceanographers, fisheries researchers, hydrobiologists, algologists, physiologists, cytologists, biochemists, and marine biologists, "This array of organisms is united by motile reproductive cells of a distinct dinoflagellate form." A remarkably informative introduction summarizes the morphology and biology of the forms included. Original references to nearly a thousand names are listed, with indication of their original nomenclatural status, and their type species, of which about 40 are designated for the first time here. Attention is directed to numerous homonyms, nomina nuda, and other errors discovered in the literature.

A supplement to the original index has recently been issued (Loeblich and Loeblich, 1968). The authors plan to issue additional supplements on an approximately annual basis.

Loeblich, A. R., Jr., and Tappan, Helen, 1966, Annotated index and bibliography of the calcareous nannoplankton: Phycologia, v. 5, nos. 2-3, p. 81-216.

———1968, Annotated index and bibliography of the calcareous nannoplankton II: Jour. Paleontology, v. 42, no. 2, p. 584-598.

This index is thorough; it is, however, limited to calcareous forms. It includes an excellent bibliography. Introductory notes treat the nomenclatural problems that are encountered for taxa assigned under the slightly differing codes of nomenclature as plants or as animals. Illegitimate or unavailable names are distinguished by italic type, and sources of nomenclatural error generally are indicated. Very briefly the practices under the two codes are compared, and all the minimum requirements for legitimacy are indicated. Two highly confused examples are summarized in the hope " . . . that the pointing out of past errors will lead to their avoidance in future studies . . . . " This publication is well done and is recommended as an example for indexing and nomenclature.

Most of the microplankton indexes have been carefully reviewed by Tappan and Loeblich (1966a, b; 1967). In addition to the errors that these authors have detected, their further comments deserve consideration. Recently Sarjeant (1968) has replied to a number of points raised by the Loeblichs that appear to him to be debatable. A student who is interested in the relation of nomenclature to taxonomy of any of these groups will find these review discussions of considerable interest.

References

Andrews, Henry N., Jr., 1955, Index of generic names of fossil plants, 1820-1950: U.S. Geol. Survey Bull. 1013, 262 p.

Arnold, C. A., 1939, Plant remains from the Catskill delta deposits of northern Pennsylvania and southern New York, pt. 4 *of* Observations on fossil plants from the Devonian of eastern North America: Ann Arbor, Michigan Univ. Press, Mus. Paleontology Contr., v. 5, no. 11, p. 271-314.

Baxter, R. W., and Leisman, G. A., 1967, A Pennsylvanian calamitean cone with *Elaterites triferens* spores: Am. Jour. Botany, v. 54, p. 748-754.

Beck, C. B., 1960, The identity of *Archaeopteris* and *Callixylon*: Brittonia, v. 12, no. 4, p. 351-368.

Benson, Lyman, 1943, Goal and methods of systematic botany: Cactus and Succulent Plant Jour., v. 15, p. 99-111.

Bergey, D. H., 1957, Manual of determinative bacteriology, 7th ed.: Baltimore, The Williams & Wilkins Co., 1094 p.

Buchanan, R. E., and others, eds., 1958, International code of nomenclature of bacteria and viruses (bacteriological code): Ames, Iowa, Iowa State Univ. Press, 186 p.

Corsin, Paul, Carette, Josaine, Danzé, Jacques, and Laveine, Jean-Pierre, 1962, Classification des spores et des pollens du Carbonifère au Lias: Acad. sci. [Paris] Conptes rendus, v. 254, pt. 2, p. 3062-3065.

Deflandre, Georges, and Deflandre, Marthe, 1962, Dinoflagelles III—Peridinida à tabulation conservée, *in* Fichier Micropaléont., sér. 11: Centre Natl. Recherche Sci., no. 383, pts. 1-5, p. 1751-1947.

———1964, Acritarches I—Polygonomorphitae-Netromorphitae *pro parte*—Appendice: Genres *Deflandrastrum* Combaz et *Wilsonastrum* Jansonius, *in* Fichier Micropaléont., sér. 12: Centre Natl. Recherche Sci., no. 392, pts. 1-10, p. 1948-2172.

———1965a, Acritarches II. Acanthomorphitae 1, Genre *Micrhystridium* Deflandre *sens. lat.*, *in* Fichier Micropaléont., sér. 13: Centre Natl. Recherche Sci., no. 402, pts. 1-5, p. 2176-2521.

———1965b, Chitinozoaires, *in* Fichier Micropaléont., sér. 7, 2d ed.: Centre Natl, Recherche Sci., no. 238, pts. 1-4, p. 1020-1095.

———1965c, Ciliés (Infusores) I. Tintinnoidea (incl. *Calpionella* auct.) et Ciliatae incertae, *in* Fichier Micropaléont., sér. 9: Centre Natl. Recherche Sci., no. 302, pts. 1-4, p. 1186-1293.

DeWolf, G. P., Jr., 1964, On the sizes of floras: Taxon, v. 13, no. 5, p. 149-153.

Downie, Charles, and Sarjeant, W. A. S., 1964, Bibliography and index of fossil Dinoflagellates and Acritarchs: Geol. Soc. America Mem. 94, 180 p., 2 figs.

Eisenack, Alfred, 1958, *Tasmanites* Newton 1875 und *Leiosphaeridia* N. G. als Gattungen der Hystrichosphaeridea: Palaeontographica, Abt. A., v. 110, nos. 1-3, p.1-19, pls. 1-2.

Eisenack, Alfred, 1967, Dinoflagellaten, v. 1, supp. no. 1, *of* Katalog der fossilen Dinoflagellaten, Hystrichosphären und verwandten Mikrofossilien: Stuttgart, E. Schweizerbart'sche Verlagsbuchhandlung (Nägele u. Obermiller), 158 p., 92 figs.

Eisenack, Alfred, and Klement, K. W., 1964, Dinoflagellaten, v. 1 *of* Katalog der fossilen Dinoflagellaten, Hystrichosphären und verwandten Mikrofossilien: Stuttgart, E. Schweizerbart'sche Verlagsbuchhandlung (Nägele u. Obermiller), 895 p., 9 pls., 420 figs.

Felix, C. J., 1965, Neogene *Tasmanites* and Leiospheres from southern Louisiana, U.S.A.: Palaeontology, v. 8, pt. 1, p. 16-26, pls. 5-8.

Hitchcock, A. S., 1929, The relation of nomenclature to taxonomy, *in* Duggar, B. M., ed., Internat. Cong. Plant Sci. Proc., Ithaca, N.Y., 1926: Menasha, Wis., George Banta Pub. Co., v. 2, p. 1434-1439.

Hughes, N. F., Dettmann, M. E., and Playford, Geoffrey, 1962, Sections of some Carboniferous dispersed spores: Palaeontology, v. 5, p. 247-252, pls. 37 and 38.

Hughes, N. F., and Moody-Stuart, Judith, 1966, Descriptions of schizaeaceous spores taken from early Cretaceous microfossils: Palaeontology, v. 9, p. 274-289, pls. 43-47.

Index Kewensis (1893-1953), Index Kewensis plantarum phanerogamarum: Oxford, Clarendon Press, 2 v., 11 supp.

Knight, J. B., 1941, *Review of* Lang, W. D., Smith, Stanley, and Thomas, H. D., 1940, Index of Paleozoic coral genera (London, British Mus. Nat. History): Jour. Paleontology, v. 15, no. 2, p. 178-180.

Kosanke, R. M., 1950, Pennsylvanian spores of Illinois and their use in correlation: Illinois Geol. Survey Bull. 74, 128 p., 16 pls.

Lanjouw, J., and others, eds., 1966, International code of botanical nomenclature: Utrecht, Netherlands, Internat. Bur. Plant Taxonomy and Nomenclature, 402 p.

Loeblich, A. R., Jr., 1968, *Review of* Eisenack, Alfred, 1967, Dinoflagellaten, v. 1, supp. no. 1, *of* Katalog der fossilen Dinoflagellaten, Hystrichosphären und verwandten Mikrofossilien: Jour. Paleontology, v. 42, no. 2, p. 607-608.

Loeblich, A. R., Jr., and Loeblich, A. R., III, 1966, Index to the genera, subgenera, and sections of the Pyrrhophyta: Miami Univ. Inst. Marine Sci., Studies in tropical oceanography, no. 3, 94 p., 1 pl.

______1968, Index to the genera, subgenera, and sections of the Pyrrhophyta II: Jour. Paleontology, v. 42, p. 210-213.

Loeblich, A. R., Jr., and Tappan, Helen, 1966, Annotated index and bibliography of the calcareous nannoplankton: Phycologia, v. 5, nos. 2-3, p. 81-216.

______1968, Annotated index and bibliography of the calcareous nannoplankton II: Jour. Paleontology, v. 42, no. 2, p. 584-598.

Newton, E. T., 1875, On "Tasmanite" and Australian "White Coal": Geol. Mag., ser. 2, v. 2, no. 8, p. 337-342.

Norris, Geoffrey, and Sarjeant, W. A. S., 1965, A descriptive index of genera of fossil Dinophyceae and Acritarcha: New Zealand Geol. Survey Paleont. Bull. 40, 72 p.

Pierce, R. L., 1961, Lower Upper Cretaceous plant microfossils from Minnesota: Minnesota Geol. Survey Bull. 42, 86 p., 3 pls., 1 map, 3 tables.

Potonié, Robert, 1956, *Sporites*, pt. 1 *of* Synopsis der Gattungen der Sporae dispersae: [Hannover], Beihefte Geol. Jahrb., v. 23, 103 p., 11 pls.

______1958, *Sporites* (Nachträge), *Saccites, Aletes, Praecolpates, Polyplicates, Monocolpates*, pt. 2 *of* Synopsis der Gattungen der Sporae dispersae: [Hannover], Beihefte Geol. Jahrb., v. 31, 114 p., 11 pls.

______1960, Nachträge *Sporites*, Fortsetzung *Pollenites* mit Generalregister to pts. 1-3, pt. 3 *of* Synopsis der Gattungen der Sporae dispersae: [Hannover], Beihefte Geol. Jahrb., v. 39, 189 p., 9 pls.

______1962, Synopsis der Sporae in situ. Die Sporen der fossilen Fruktifikationen (Thallophyta bis Gymnospermophyta) im natürlichen System und im Vergleich mit den Sporae dispersae: [Hannover], Beihefte Geol. Jahrb., v. 52, 204 p., 19 pls.

______1966, Nachträge zu allen Gruppen (Turmae), pt. 4 *of* Synopsis der Gattungen der Sporae dispersae: [Hannover], Beihefte Geol. Jahrb., v. 72, 244 p., 15 pls.

Radforth, N. W., 1938, An analysis and comparison of the structural features of *Dactylotheca plumosa* Artis sp. and *Senftenbergia ophiodermatica* Göppert sp.: Royal Soc. Edinburgh Trans., v. 59, pt. 2, p. 385-396.

Sarjeant, W. A. S., 1968, The Tappan-Loeblich reviews on dinoflagellate and acritarch publications—a reply: Jour. Paleontology, v. 42, no. 2, p. 599-611.

Schopf, J. M., 1964, Practical problems and principles in study of plant microfossils, *in* Cross, A. T., ed., Palynology in oil exploration—symposium, San Francisco, 1962: Soc. Econ. Paleontologists and Mineralogists Spec. Pub. no. 11, p. 29-57.

Schopf, J. M., Wilson, L. R., and Bentall, Ray, 1944, An annotated synopsis of Paleozoic fossil spores and the definition of generic groups: Illinois Geol. Survey, Rept. Inv. no. 91, 72 p., 3 pls., 5 figs.

Simpson, G. G., 1940, Types in modern taxonomy: Am. Jour. Science, v. 238, p. 413-431.

Tappan, Helen, and Loeblich, A. R., Jr., 1966a, *Review of* Downie, Charles, and Sarjeant, W. A. S., 1964, Bibliography and index of fossil Dinoflagellates and Acritarchs; (Geol. Soc. America Mem. 94, 180 p., 2 figs.): Jour. Paleontology, v. 40, no. 4, p. 977-978.

______1966b, *Review of* Norris, Geoffrey, and Sarjeant, W. A. S., 1965, A descriptive index of genera of fossil Dinophyceae and Acritarcha; (New Zealand Geol. Survey Paleont. Bull. 40, 72 p.): Jour. Paleontology, v. 40, no. 4, p. 978-980.

______1967, *Review of* Eisenack, Alfred, and Klement, K. W., 1964, Dinoflagellaten, v. 1 *of* Katalog der fossilen Dinoflagellaten, Hystrichosphären und verwandten Mikrofossilien; (Stuttgart, E. Schweizerbart'sche Verlagsbuchhandlung (Nägele u. Obermiller), 895 p., 9 pls., 420 figs.): Jour. Paleontology, v. 41, p. 525-526.

Traverse, Alfred [and various coeditors], 1957-continuing, Catalog of fossil spores and pollen: University Park, Pa., Pennsylvania State Univ., v. 1-31.

Tschudy, R. H., 1957, Pollen and spore formulate—a suggestion: Micropaleontology, v. 3, no. 3, p. 277-280.

Upshaw, C. F., 1964, Palynological zonation of the Upper Cretaceous Frontier Formation near Dubois, Wyo., *in* Cross, A. T., ed., Palynology in oil exploration—symposium, San Francisco, 1962: Soc. Econ. Paleontologists and Mineralogists Spec. Pub. 11, p. 153-168.

Van der Hammen, Thomas, 1956, A palynological systematic nomenclature: Bol. Geologico [Bogotá], v. 4, no. 2-3, p. 63-101.

Wall, David, 1962, Evidence from Recent plankton regarding the biological affinities of *Tasmanites* Newton 1875 and *Leiosphaeridia* Eisenack 1958: Geol. Mag., v. 99, no. 4, p. 353-362, pl. 17.

Ward & Whipple [Edmondson, W. T., ed.], 1959, Fresh-water biology, 2d ed.: New York, John Wiley & Sons, Inc., 1248 p., illus.

Wilson, L. R., 1943, Elater-bearing spores from the Pennsylvanian strata of Iowa: Am. Midland Naturalist, v. 30, no. 2, p. 518-523.

5

RELATIONSHIP OF PALYNOMORPHS TO SEDIMENTATION

Robert H. Tschudy

Fossil pollen and spores obtained from sedimentary matrices should be considered in relation to a number of factors that have exerted an influence on them during the time between their dispersion and their recovery from the rock. It is axiomatic that the more one knows about the history of a fossil pollen grain, a fossil pollen-and-spore assemblage, or a rock matrix that contains plant microfossils, the more reliable the inferences derived from this knowledge will be.

Different sedimentary sites, different ecological conditions during the time a pollen was being incorporated into the sediment, primary and secondary changes taking place in the sediment as it becomes rock, and differences in the chemical composition of the original pollen may all exert an influence on a particular fossil or fossil assemblage. Thus a whole series of factors may have acted on pollen grains before and during their conversion to fossils. Alone or in combination these influences affect the state of preservation of the fossils and their ultimate recovery.

Pollen and spores are almost never accumulated in the pure state as a fossil deposit. (As an exception, some specimens of tasmanite are made up almost entirely of spores.) Deposits of pollen and spores are almost universally accompanied by various amounts of other organic material, and by clay, sand, inorganic salts, or other rock-forming clastic materials. Adequate evaluation of the fossils cannot be made without considering the influences of the sedimentation process, the history of the host rock, and the chemical and physical attributes of the fossil itself.

Publication authorized by the Director of the U. S. Geological Survey.

SOURCES OF PALYNOMORPHS

The principal pollen- and spore-producing plants are those of land origin. These plants include angiospermous and gymnospermous trees, shrubs, herbs, ferns and fern allies, and, to a lesser degree, mosses and fungi. In addition to these, flagellates, algae, and fragments of animals and plants living in the water of the depositional basin may and often do contribute palynomorphs to the organic fraction of the sedimentary complex. These latter accessory organisms or materials, which provide recognizable fossils, are extremely useful additions to the spore-pollen complex. Data derived from their presence in the rocks are an aid not only in making age determinations but also in the interpretation of ecology at the deposition site.

Most of the fossil pollen and spores were derived from land plants that grew originally outside the site of deposition rather than from plants that grew within the area of deposition. They may have been derived from plants growing in highlands either at some distance from the deposition area or adjacent to it; or they may have been derived from lowland plants and then eventually have become fossilized. The disseminules of plants from the various original growth sites were carried to the place of deposition almost entirely by wind or by water.

SOURCES OF SEDIMENTS

Sediments are derived from the breakdown of rocks into particles, or ultimately into chemical compounds, and from the residues of animal and plant life. Sedimentology includes the study of the various processes responsible for the formation of sedimentary rocks. The principal processes normally involved in the formation of sedimentary rocks are erosion, transportation, deposition, diagenesis, and consolidation of particles or aggregates (Trask, 1950, 1955, p. 428-453; Krumbein and Sloss, 1951). Detrital sediments are those solid particles or aggregates that are in suspension or that have finally come to rest; any included organic particles are thus a part of the sediment.

Parent rocks, the ultimate source of many sediments, may be broken down into particles of transportable size. This breakdown is accomplished either by mechanical or by chemical agencies or by both. Mechanical agencies are most important in arid regions; they include gravity cracking and slumping of rocks, breakdown and spalling caused by frost action, glacial scouring, wind pitting, and abrasion during transportation. Chemical breakdown is dominant in moist regions; it includes solution as well as disaggregation caused by differential chemical action. Organisms, particularly soil organisms, and the byproducts of the life processes of these organisms, also play an important chemical role in the ultimate breakdown of rocks.

The product of this mechanical or chemical breakdown may consist of both altered and unaltered rock particles, plus an incorporated organic fraction, which is made up of organic matter in various stages of decay plus soil organisms, bacteria, fungi, and microscopic animal life. To this organic fraction is commonly added the fallout of pollen and spores from the atmosphere.

Within a depositional area other types of sediment may originate. As a result of the action of organisms or evaporation, soluble salts in the water may become too concentrated to remain in solution and will therefore form precipitates. The most important of these deposits are carbonates, sulfates, and chlorides. Significant volumes of such sediments have accumulated during various stages of geologic time.

TRANSPORTATION

Sediments, including the organic residues, commonly are moved from source areas by wind and water. Other agencies of transport—such as ice, volcanic explosions, or birds—are relatively minor and need not be considered here.

The transport of the organic fraction, as well as the mineral particles, may involve appreciable distances. Alder pollen from the Andes of Venezuela and Colombia has been traced by Muller (1959) down the Orinoco River to a deposition site in the sea along the northeastern coast of Venezuela, a distance of more than 500 miles. This transport was against the prevailing wind direction, so it must have been accomplished by the flow of the Orinoco River from source areas to the mouth of the river and thence by ocean currents. Long-distance wind transport of pollen across the North Atlantic has been documented by Erdtman (1943). Although such long-distance transport has been well established, experiments on modern pollen rains suggest that probably most pollen that occurs as fossils has been transported relatively short distances before becoming incorporated into sedimentary deposits.

Most sediments are finally deposited under water. Exceptions include deposits such as loess and dune sand. Once the sediments arrive in the sea or other aqueous deposition sites, many of the particles may be transported long distances before finally coming to rest. Wave action, bottom currents, turbulence, or mass movement of water may act to keep some of the fine particles, which include most pollen and spores, in suspension for a long time. During this phase of transport particularly, winnowing, or separation, of the fine and light particles occurs. The average size of particles that are transported decreases with increasing distance. The organic fraction, which for the most part is relatively small and light, is commonly separated from the coarser detritus and deposited with the fine-grained clastics. For this reason well-washed beach sand is nearly devoid of pollen and spores.

Plants growing within a basin or an area of deposition may drop their pollen or spores in situ. The pollen will then commonly be moved about only by such currents as may be active within the area. Plants growing in, and forming the vegetation of, a peat bog contribute relatively enormous amounts of pollen and spores to the sediments that may eventually be transformed into coal. Wind or water may also carry in pollen and spores from source areas outside the bog. As a rule, however, the contribution of the bog plants to the total pollen-and-spore concentration is so overwhelming that the presence of any nonbog pollen and spores may be difficult to detect. These foreign materials commonly will be represented by only a fraction of a percent.

Distribution of Palynomorphs in Bottom Deposits of Lake Maracaibo, Venezuela[1]

An examination of bottom muds from Lake Maracaibo, Venezuela, provides information on the relative effects of winds, rivers, water circulation in the lake, and salt-water influx on the distribution of palynomorphs in modern sediments.

Lake Maracaibo is not a true lake because it has a narrow opening to the Caribbean Sea to the north. It is slightly less than 200 kilometers long and is about 100 kilometers wide. It is comparatively shallow, only a small area attaining a depth of more than 30 meters (fig. 5-1a). It is mainly a body of fresh water, but it is brackish in the northern part, becoming progressively more brackish toward its mouth and contact with the sea. During the dry season, particularly, sea water flows into the lake over a shallow bar and, because it has a greater density than the lake water, sinks to the bottom and accumulates in the deeper parts at the north end of the lake. During the rainy season an influx of fresh water flushes out much of the brackish water at the north end.

The Rio Negro and Rio Catatumbo, tributaries from the southwest, contribute the greatest amount of fresh water and suspended matter to the lake. Smaller rivers flowing from the Andes Mountains, which are to the south and southeast of the lake, also contribute fresh water, particularly during the rainy season.

The trade winds, blowing from the east, are probably the source of most of the wind-distributed pollen carried into the lake.

In the Lake Maracaibo study bottom samples from more than 50 stations were examined for palynomorphs (fig. 5-1b). Data obtained from three groups of palynomorphs are presented here. The first group consisted of pollen grains 24 microns or less in diameter. These small grains probably were carried into the lake by wind. Figure 5-1c shows percentage contours obtained from these data. The contours suggest a fairly uniform deposition, although the percentages are somewhat greater in midlake than nearshore.

The second group consisted of the appreciably larger trilete fern spores, many of them from the swamp fern *Achrostichum*. The percentage contours for this group are shown in figure 5-1d. The greatest percentages were found directly off the mouth of the Rio Catatumbo. The contours definitely indicate a comparatively high concentration in the southern and the eastern lake areas and an almost total absence of spores from the northwestern part of the lake. These spores were undoubtedly carried into the lake by the Rio Negro and Rio Catatumbo, which flow for many kilometers through a swampy delta region, the source of the fern spores. The circulation of the lake water is in a counterclockwise direction; this accounts for the distribution in the southern and eastern parts of the lake.

The third group, the microforaminifera, was found at the stations shown in figure 5-1*a*. The numbers refer to the individuals found in a count of 100 spores and pollen grains. The greatest concentration appears in the northern part of the lake where sea water has accumulated on the bottom. Only one of the samples from the southernmost stations yielded microforaminifera, and then only a single specimen was recorded.

This résumé of the Lake Maracaibo study provides evidence that the relatively small, wind-borne pollen is somewhat evenly distributed throughout this large lake. The higher percentage of these grains in the midlake area may reflect a lessening effect there of the circulatory current, permitting somewhat more rapid settling of pollen grains. The distribution of trilete fern spores clearly suggests the effect of fresh-water influx as well as the influence of the circu-

[1]Approval to use these data from the files of the Creole Petroleum Corp., Caracas, Venezuela, is gratefully acknowledged.

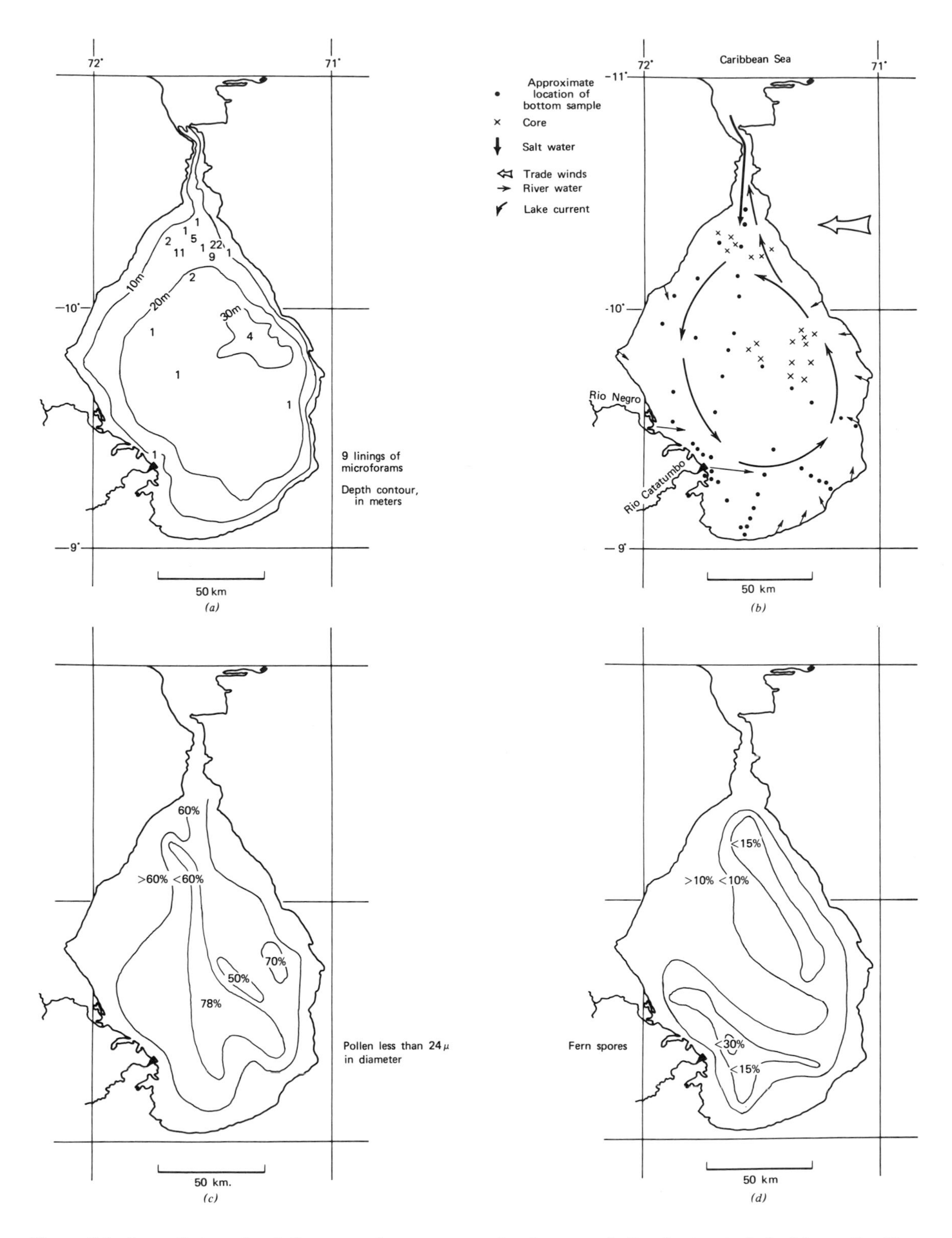

Figure 5-1. Some factors that influence sedimentation and palynomorph distribution in Lake Maracaibo, Venezuela: (*a*) depth contours and occurrence of microforaminifera; (*b*) sample localities and physical features; (*c*) isopollen contours, pollen less than 24 microns in diameter; (*d*) isofern spore contours.

latory motion of the water of the basin as a whole. Concentration of microforaminifera in the northern part of Lake Maracaibo reflects the brackish nature of water in that area.

Effect of Transport on Pollen and Spores

During the time that palynomorphs are being transported they may be subjected to various agencies whose effects may change their distribution, concentration, and state of preservation. The original distribution and concentration of palynomorphs in a sediment may be altered by sorting, flotation, stirring, mixing, and resettling —and by factors that affect palynomorph preservation. This alteration may result in qualitative or quantitative change in the recovered pollen-and-spore assemblage. The state of preservation may be adversely affected by abrasion, chemical action (decomposition), or the activity of animals, particularly those that feed on organic matter. The state of the specimens may range from only slightly altered to corroded and abraded. Differential destruction of less resistant forms could radically alter the composition of the assemblage. Relatively little is known of the effect of time and ecological, chemical, and physical factors on pollen and spores during the transport stage. Research on such problems undoubtedly would be rewarding.

Mechanical (Abrasive) Effects. Abrasion of pollen grains is usually minimal. The most common observable effect of stream or wave abrasion is the detachment of the sacs from the bodies of conifer pollen. Abrasion of fine-grained inorganic detritus is also slight; for example, only minor changes in the fine-grained sands occur during 1,100 miles of transport by the Mississippi River (Russell, 1955, p. 32-47), and experiments by Ziegler (1911) show that sand grains less than 0.75 millimeter in diameter cannot be rounded by abrasion in water. Most pollen grains and spores are much smaller than 0.75 millimeters (750 μ). This resistance to abrasion may account for the fact that whole assemblages of broken and obviously mechanically comminuted pollen and spore fragments are seldom encountered. An exception may exist in palynomorphs obtained from some wind-transported loess deposits.

Chemical Effects. Chemical action (including biochemical degradation) is the most destructive agent in the degradation of pollen and spores during transport. Oxidation, or biological attack, commonly corrodes surfaces or entirely destroys structure.

Repeated stirring and resettling of sediments in a river provide the mechanics by which effective sorting of the fine-grained particles takes place. Mud may be temporarily deposited on a bar or in an eddy, and in successive years it may be moved nearer to the mouth of the stream and the area of ultimate deposition. During this time chemical agencies have ample time to work on the organic fraction. The acid-insoluble fraction of the organic content of phytoplankton and zooplankton is about 5 percent, according to Trask (1955, p. 428-453), whereas the acid-insoluble organic content of marine sediments is more than 30 percent. This increase in concentration is the result of the earlier decomposition and destruction of all but the most resistant (acid-insoluble) fraction, including pollen and spores. Even the acid-insoluble fraction, however, may be totally destroyed if subjected for a long enough time to the effects of aerobic bacteria and fungi or to atmospheric oxygen.

DEPOSITION

The factors that influence the deposition of inorganic particles also influence the deposition of palynomorphs. These factors include particle size, shape, density, coagulability, and the physical conditions at the site of deposition. Some of the physical conditions of significance are density of water, turbulence, salinity, and bottom topography.

Hjulström's Diagram

A set of curves prepared by Hjulström (1955, p. 5-31) (see fig. 5-2) shows the interrelationships of particle size and water velocity to erosion, transportation, and deposition. The upper series of curves, A (a series of bands rather than a single line because of uncertainty of data) show that fine sand (0.3 to 0.6 mm in diameter) is relatively easy to erode and that both the smaller and the larger sized particles, silt or clay and coarse sand or gravel, require higher fluid velocities to accomplish the same amount of erosion as for fine sand. Curve B represents the lowest transportation velocity. If the velocity for a particular grain

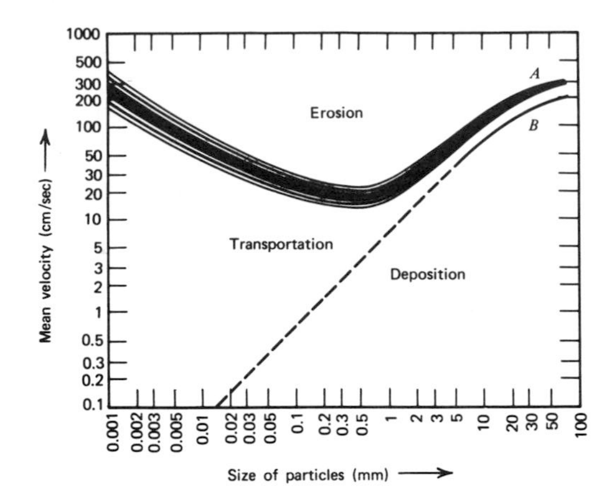

Figure 5-2. Approximate curves for erosion and deposition of uniform material (logarithmic scale). From Hjulström (1955).

size falls below values expressed by this line, deposition will take place.

Material carried into an area of deposition may have been transported only a relatively short distance and therefore not have been winnowed appreciably. This lack of winnowing may cause retention of the heterogeneous character of a deposit in which little sorting has taken place. Pollen and spore concentrations are usually low in such deposits because the fine detrital material including pollen and spores has not been separated from the coarser fraction. Deposits that have been winnowed may have no pollen in the sand fraction, but a significantly increased pollen concentration may occur in the fraction that has been removed and deposited in quiet water.

Sites of Deposition

At the deposition site a complex of physical, chemical, and biological factors influences the characteristics of the sediment. Terrestrial nonaqueous deposits are mostly either from arid regions or from glaciers; they also include those transported by gravity as talus and those developed in situ.

Aquatic depositional sites may be classified as continental, transitional, and marine. Continental deposits include fluvial, lacustrine, and paludal. The latter two are of greatest interest palynologically. These environments commonly provide reducing conditions under which pollen and spores may be very well preserved.

Transitional sites include deltaic, lagoonal, and littoral. The deltaic and lagoonal sites are more likely to yield palynomorphs than are the littoral sites. Shoreline deposits commonly have the fine-grained material, which includes palynomorphs, winnowed from them and deposited elsewhere.

Normal marine environments include the neritic, the bathyal, and the abyssal. Neritic environments are those of relatively shallow water on the margins of the sea, in general overlying the continental shelf. The bathyal environment is generally from depths of about 600 to 6,000 feet on the continental slopes. The abyssal zone is confined to the oceanic deeps. The neritic zone, particularly off deltas, may yield excellent palynomorph assemblages (Woods, 1955). Bathyal and abyssal sediments are likely to be impoverished in organic content from land-based plants but to be enriched in oceanic organic material.

Concentration in Sediments

Pollen grains and spores are rarely so abundant that they make up most of the volume of an organic deposit. Tasmanite and some cannel coals, however, represent such extremes. Some lignites have yielded more than 2 million pollen grains from a 1-centimeter cube of sediment. If we assume the pollen to be isodiametric with all dimensions 30 microns and if we assume no compression, there is room for 37 million pollen grains in a 1-centimeter cube. Even in such a rich sample as the 2-million-grain sample at an average diameter of 30 microns per grain, at least fourteen-fifteenths of the cube is composed of materials other than pollen. If we do assume some compression, the relative proportion of the pollen to the residue is appreciably less. Some rocks, however, may yield an exceedingly sparse pollen and spore flora. Most samples examined palynologically lie between the two extremes of virtual absence and extreme abundance.

That the concentration of pollen decreases rapidly as distance from shore increases has been demonstrated by Woods (1955). In some areas, at least, the absolute fossil pollen and spore concentration per unit of sediment may provide an estimate as to whether the sediment was accumulated off shore or near shore. This method was proposed by Hoffmeister (1954) as a means of determining the position of ancient shorelines. It should be used with extreme caution, however, as factors other than distance from shoreline may affect the total pollen and spore concentration.[2]

Orientation of Palynomorphs in Sediments

Ample evidence is available to indicate that the shape of a palynomorph influences the orientation of it in the sedimentary matrix. In general palynomorphs come to rest with their greatest diameter parallel to the bedding plane. Nearly all fossil grains of the genera *Carya* and *Cirratriradites* are flattened at right angles to a line running through their proximal and distal poles. In *Carya* the pollen grains when shed are oblate; that is, the polar axis is significantly shorter than the other axes. We assume the same to have been true for *Cirratriradites*. Difficulty in interpreting the shape and thickness of the cingulum of such genera as *Cirratriradites* and *Densosporites* has been experienced because the spores are rarely preserved in such a manner as to give a lateral view of the cingulum. The discovery of tetrads of *Densosporites* (Kosanke, 1950) has aided in the interpretation of the cingulum of this genus. Only recently (Hughes, Dettman, and Playford, 1962) have attempts been made to section spores so that the features that are rarely seen in fossil preparations could be reliably reported and interpreted. Optical sectioning methods for the elucidation of features of several *Densosporites*-like genera from the Paleozoic have been utilized by Staplin and Jansonius (1964).

Nearly all prolate pollen grains present an equatorial view, and, conversely, most oblate grains present a polar view. If, however, particles are spherical or nearly so, as is true for chenopodiaceous pollen or for *Calamospora* spores, they have no preferred orientation and therefore may be seen in any conceivable view as fossils.

[2]For an accurate determination absolutely synchronous samples must be used because shorelines fluctuate with time. Lithotypes must be the same. Pollen and spore content of a siltstone will be different, other things being equal, from that obtained from a shale or mudstone. If reliable data are being sought, rocks being compared should have had equivalent histories —poorly preserved pollen from one sample should not be compared with well-preserved pollen from another.

Effect of Pressure from Overburden

The first noticeable effect of the weight of overburden on palynomorphs is compression, or flattening. In ancient sediments, particularly, a nonreversible flattening is evident. As seen in thin sections of coal, for example, megaspores, small spores, and pollen are flattened parallel to the bedding planes. This flattening is observable also in dispersed pollen and in spores isolated from shales. If these grains are mounted in a viscous liquid, such as glycerine, and the cover glass is moved, some of the grains can be made to turn over. In the process edge-on views of compressed grains can be seen. Pollen and spores are compressed with the same attitude as that in which they came to rest, and their shapes are the determining factors in their preferred orientation.

Pollen and spores embedded in clays are flattened to a maximum degree. Increase in pressure, once flattening has occurred, does not induce additional lateral distortion. Such distortion is prevented by the confining pressures of the matrix itself. Compaction reduces water content and also the volume of sediment. This process continues with increasing load. A maximum compaction measured by maximum shale density is reached after about 6,000 feet of burial (Athy, 1930). That overburden is of more significance than time in its effects on sediments is attested to by the fact that some Upper Cretaceous deposits from the Mississippi Embayment region are less consolidated than rocks of younger age from some parts of the Rocky Mountain region. Uncompressed Late Cretaceous plant microfossils have been obtained from the Mississippi Embayment region, and a compressed, completely flattened assemblage has been obtained from the Eocene Green River Formation of Colorado.

Effect of the Matrix on Palynomorphs

If the matrix has not been compacted, the palynomorphs, where preserved, will retain their original shape. This preservation of shape is independent of whether deposition is in a clay, silt, sand, or an organic matrix (bog). Preservation with no evident distortion has been observed in chert, and hystrichospheres were first observed as three-dimensional fossils in thin chips of transparent chert. By dissolving chert in hydrofluoric acid, three-dimensional pollen with apparently all features intact has been recovered from the Clarno formation (Eocene) of Oregon (R. A. Scott, personal communication, 1965). In this chert silicification of palynomorphs evidently took place without compaction.

Some evidence of distortion of fossils by pressure from sand grains during compaction has been observed. Megaspores in a sandy matrix may have the imprint of sand grains impressed on their surfaces. Palynomorphs obtained from a sandy lithofacies are more likely to be distorted and to have surface features changed or eliminated than are those from finer grained sediments.

Perhaps one reason for the good preservation of palynomorphs in some shales and coals is that these matrices provide a protective cushion during compaction, thus permitting only a minimum of distortion aside from flattening.

DIAGENESIS

The rapid alterations or physico-chemico-biological changes that operate during deposition and within the first few feet of burial are defined as early diagenesis. Subsequent alterations of longer duration and lesser intensity that occur during or after lithification but prior to metamorphism are defined as late diagenesis (Ginsburg, 1957).

Before or during early diagenesis the least resistant parts of palynomorphs are destroyed, probably by bacterial action. The cell protoplasm disappears. The inner layer of the pollen grain coat, the intine, normally does not persist. Usually the perisporium, particularly the thin-walled perisporia of such monolete genera as *Asplenium*, is readily destroyed and therefore does not appear in the fossilized state (Selling, 1946).

Coals are characterized by a high content of lignin and humic acids. These substances are the relatively resistant residues left after partial selective bacterial and fungal decomposition of the original vegetable matter (Kuznetsov, Ivanov, and Lyalikova, 1963; Barghoorn, 1952). The resistant, original spore and pollen coats are preserved, typically as compression fossils. Petrified pollen and spores are known, however; the silicified, calcified, or pyritized petrifactions appear to have been produced by infiltration in the same manner as the petrified vegetative or sporangial tissue that may occur with them. Perhaps the best known is the holotype of *Pityosporites antarctica* Seward. Furthermore, large, silicified *Monoletes* microspores are present in thin sections of *Dolerotheca* material in the Renault Collection, Museum of Natural History, Paris (J. M. Schopf, personal communication, 1967). Two examples of the occurrence of isolated, petrified megaspores are known to me. Silicified megaspores have been found as discrete entities from the Knee Hills Tuff, Edmonton Formation, Canada (Late Cretaceous age). Pyritized, three-dimensional megaspores have been recovered from Pennsylvanian coal-ball material by Mamay (personal communication, 1968).

During late diagenesis changes in the matrix and in the organic fraction continue at a retarded rate as additional sediment accumulates. Interstitial water is gradually squeezed out. After expulsion of most of the interstitial water by compaction, diagenesis probably ceases; for example, according to Tourtelot (1962), in the Pierre Shale, after expulsion of interstitial water, diagenesis was greatly retarded if not completed, and further chemical changes took place very slowly if at all.

Eh and pH

The oxidation-reduction potential (Eh) of sediments is intimately related to and perhaps more important than hydrogen-ion concentration (pH) for the preservation of palynomorphs in sediments. The chart from Garrels (1960) (see fig. 5-3) shows the approximate position of some natural environments in relation to Eh and pH.

This chart indicates clearly that normal marine waters are oxidizing and that only in an euxinic marine environment is the Eh low enough to provide a reducing environment. Furthermore it is obvious that reducing conditions exist only in environments isolated from the atmosphere.

Confined waters, particularly in the presence of organic matter, rapidly lose their oxygen content. Hydrolysis of silicates causes this environment to become alkaline as well as reducing. Biochemical reactions initiated by microorganisms rapidly remove oxygen and at the same time produce carbon dioxide and hydrogen sul-

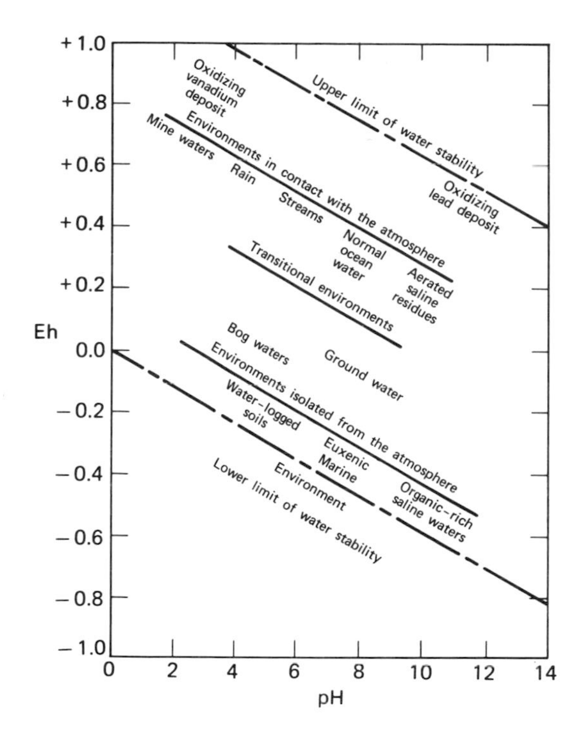

Figure 5-3. Approximate position of some natural environments as characterized by Eh and pH. From Garrels (1960).

fide, resulting in a lowering of pH. Some anaerobic bacteria actually release hydrogen, which causes strongly negative Eh potentials to be developed, and as a result strongly reducing conditions are created.

The deposition of chemical sediments—such as carbonates, sulfates, chlorides, iron and magnesium compounds, phosphates, and to some extent organic deposits—has been shown by Krumbein and Garrels (1952) to be largely controlled by Eh and pH. In their chart (fig. 5-4) they have established limits, or "fences," to separate environments characterized by the deposition of different kinds of chemical sediments.

"The boundaries of the 10 fields in the diagram serve as approximate limits to the associations shown" (Krumbein and Garrels, 1952). The most abundant members are indicated by large capital letters, those present in appreciable amounts by small capitals, and the accessory members by lower case letters. When an aqueous environment becomes concentrated to a salinity exceeding about 200 parts per thousand, those compounds shown in brackets (those to the extreme right on the chart) begin to precipitate as evaporites.

An example will serve to demonstrate the utility of this chart as an aid in the interpretation of depositional environments. The presence of primary pyrite, unaccompanied by calcite, in coal or a highly organic shale indicates that the deposit was laid down under strongly reducing conditions and at a pH less than 7.8.

"The next field below the upper right represents euxinic environments with partial stagnation. The field is limited by the limestone fence, the organic matter fence, and the Fe-Mn oxide-carbonate fence. The limestone formed in this environment, while still dominant, is partly bituminous but does not contain primary pyrite. As the succeeding fields below this are encountered, the Eh becomes more negative, the amount of organic matter may increase, and primary pyrite becomes increasingly important" (Krumbein and Garrels, 1952).

More complete discussions of the implications of this chart are in the original publication of Krumbein and Garrels (1952).

Preservation

It is to be expected that some species of pollen will be more readily attacked by oxygen and by bacterial action than will others. If this occurs to the extent that some species are totally destroyed, then the original representation of the various pollen species will be altered by differential preservation. Data presented by Havinga (1964) show that this hypothesis is correct and furthermore that there is an inverse relationship between the percentage of sporopollenine in the pollen or spore and the susceptibility to oxidation. (See table 5-1.)

Thin-walled pollen species belonging to such families as Cannaceae, Musaceae, and Zingiberaceae possibly will not survive fossilization or subsequent chemical treatment and thus do not appear in the fossil record. Another thin-walled pollen, that of *Populus*, is readily destroyed and also does not appear in the fossil record (Erdtman, 1943). A contrasting viewpoint is held by Iversen (1946: cited in Faegri and Iversen, 1959, p. 84), who says: "As a matter of fact they [*Populus tremula* grains] can be recognized and counted like those of other forest trees, even though they are delicate and easily overlooked,

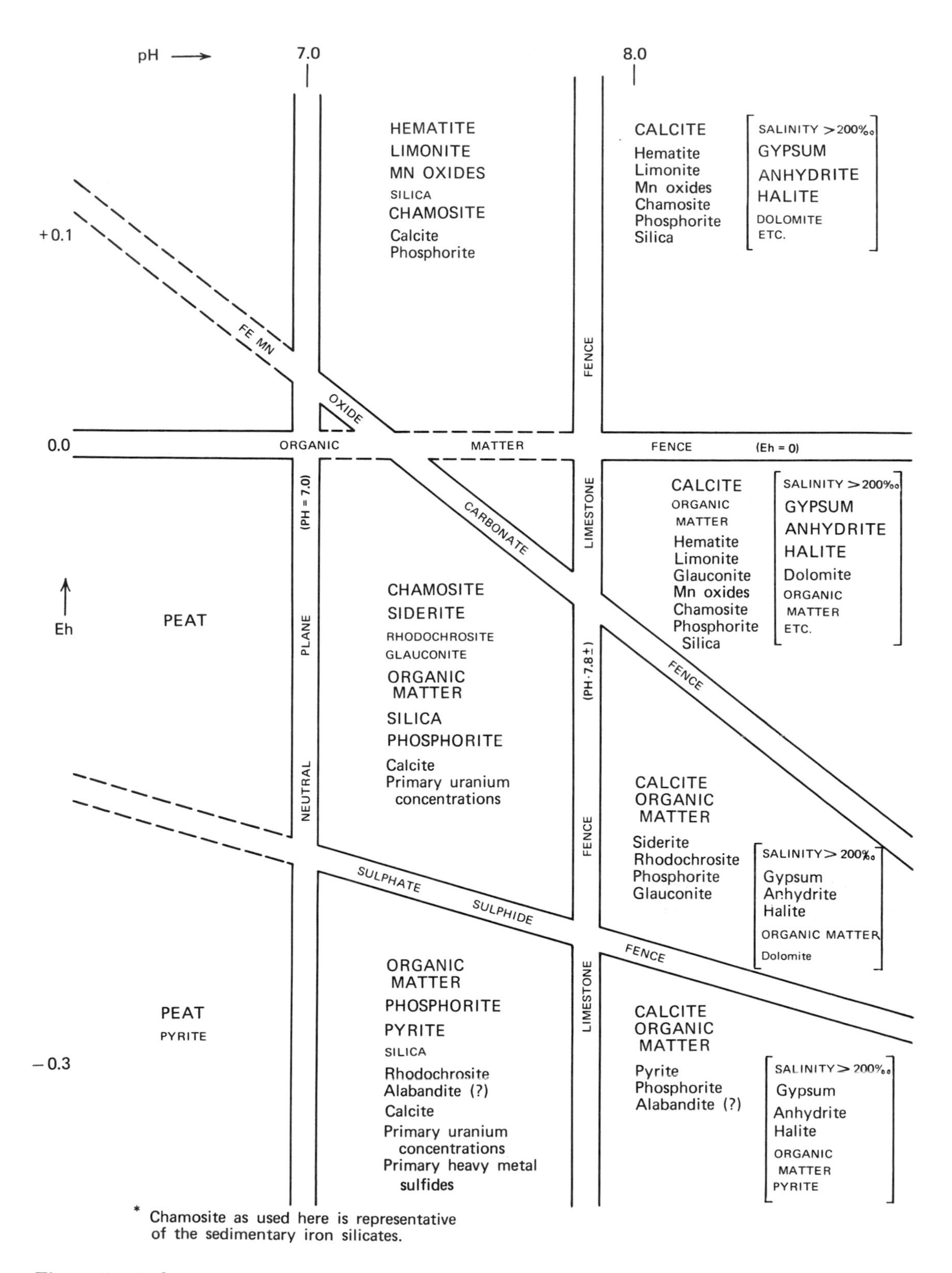

Figure 5-4. Sedimentary chemical end-member association in their relations to environmental limitations imposed by selected Eh and pH values. Associations in brackets refer to hypersaline solutions. From Krumbein and Garrels (1952).

especially if the preparation is not stained." This observation, however, does not negate the fact that some structures are less resistant than others and therefore that differential preservation is sometimes to be expected.

Acidic Environments. Well-preserved palynomorphs are more likely to be found in rocks deposited under conditions of low pH and negative Eh; that is, in a reducing acidic environment. Such conditions are commonly developed in bogs, the bottoms of lakes, and the depths of closed basins.

In an acidic bog environment the oxygen available for bacterial degradation of the organic residue decreases progressively with depth. The antiseptic properties of humic substances may be an even more limiting factor than the absence of oxygen. Phenols, quinones, phenyl carboxylic acids, and phenol glucosides are present in peat deposits (Swain, 1963, p. 87-147).

In ocean-bottom sediments quantity of bacteria and bacterial action decrease very rapidly from the water-sediment interface downward (ZoBell, 1946, 1955, p. 416-427). A few centimeters below this interface the sediments contain no free oxygen. Living aerobic bacteria decrease at a logarithmic rate as depth increases below 5 to 10 centimeters. Even anaerobic bacteria show a similar death curve below 40 to 60 centimeters Ancient sediments contain almost as much organic matter as the topmost 30 to 40 centimeters of recent sediments, suggesting that bacterial activity, which would reduce the organic content, is slight below this depth. When cores of ancient sediments are stored wet, the number of bacteria increase enormously, oxygen is consumed, and carbon dioxide is produced. Even the resistant humic or organic residue in ancient sediments is slowly decomposed in the laboratory when oxygen suddenly becomes abundant (Waksman, 1933).

Alkaline Environments. In Florida Bay the pH of sea water ranges from a high of 9.8 in bright sunlight to a low of about 8.0 after dark (Williams and Barghoorn, 1963). Photosynthesis during the day reduces carbon dioxide and increases oxygen content. Respiration, dominant at night, consumes oxygen and liberates carbon dioxide, thus decreasing the pH. Photosynthetic activity of algae is also responsible for the precipitation of carbonates from sea water, according to Williams and Barghoorn. In the presence of organic acids such alkaline precipitates redissolve. Consequently the acid content of these sediments is thereby decreased as the acid is neutralized. Most lime muds are precipitated in shallow water commonly under conditions characterized by effective oxygenation. Such sediments commonly have a low content of pollen and spores, and the pollen in sediments examined from Florida Bay commonly exhibited some evidence of corrosion. A palynological examination of the interlaminated marls and peats from Florida might indicate the effect of alkaline deposits on the preservation of pollen (Davis, 1946).

Table 5-1. **Relationship Between Sporopollenine Content and Susceptibility to Oxidation of Selected Species**[1]

	Plant Species	*Sporopollenine Content (percent)*
Increasing Susceptibility to Oxidation ↓	*Lycopodium clavatum*	23.4
	Pinus sylvestris	19.6
	Tilia sp.	14.9
	Corylus avellana	8.2
	Carpinus betulus	8.2
	Ulmus sp.	7.5

[1]Modified from Havinga (1964).

The lack of abundant well-preserved palynomorphs in most calcareous rocks probably is not a function of the alkalinity but rather is a function of the relatively long period that they have been in contact with an oxygenated environment or of the original lack of availability of palynomorphs to be preserved. Many calcareous deposits are laid down at some distance from shore, and commonly little or no sediment of terrigenous origin is present; the only palynomorphs that are present are those indigenous to the deposition site or a comparative few carried there by wind. Many calcareous deposits yield such palynomorphs as dinoflagellates, hystrichospheres, chitinozoans, and *Tasmanites*, among which preservation is excellent. That a calcareous environment is not per se responsible for corrosion or destruction is attested by the excellent preservation of plant material in coal balls (Mamay and Yochelson, 1962).

Red Beds. The common scarcity of pollen and spores in red deposits, or the absence of them, has been attributed to the oxidization and destruction of organic material. The red color in the sediments is caused by the oxidized state of the iron contained in the matrix; this indicates that the matrix was either deposited under already oxidized conditions or was later subjected to oxidation by ground water containing oxygen or by exposure to the atmosphere. Terrestrial deposits in deserts or alluvial plains may periodically or continually be exposed to atmospheric oxygen. This prolonged oxidation may tend to eliminate the slight amount of organic matter that would otherwise be present.

Two hypotheses have been advanced to explain the color of red beds. One hypothesis is that the red color is due to hematite that was formed in lateritic soils of moist tropical source areas, that this residual soil was then transported and deposited in sites in arid regions without changing color. This hypothesis, which claims that the sediments were red at the time of deposition, may be referred to as the *detrital* hypothesis (Baker, 1916). It is supported by Van Houten (1948) as follows: " . . . it is postulated that the sediments of both red and drab layers in varicolored formations were derived from red upland soils and deposited on flood plains in a lowland region. Red mud accumulated when little organic matter was incorporated at the place of deposition, whereas drab deposits resulted from reduction of ferric oxide in the presence of plant debris."

A contrasting hypothesis offers convincing evidence that red color is due to in situ formation of hematite after deposition (Walker, 1967a, b). The iron is derived from the interstitial alteration of iron-bearing detrital grains, such as hornblende and biotite, as a result of prolonged weathering on well-drained upland surfaces. This hypothesis may be referred to as the *diagenetic* hypothesis.

Either hypothesis would account for the general paucity of organic material in red beds. Fine-grained Permian red samples from Oklahoma examined by the author showed the absence not only of pollen and spores but of any organic matter whatsoever. An abundant Permian assemblage, however, occurs in the Flowerpot Formation of Oklahoma (Wilson, 1962), which consists of interbedded dolomite; gray-green, brown, and red shale; and siltstone. The assemblage examined by Wilson was derived from an olive-gray fissile shale, rather than from beds showing the red color associated with either detrital or diagenetic hematite.

The absence of pollen from some oxidized environments was demonstrated clearly by Kuyl, Muller, and Waterbolk (1955). They found in Venezuela that the oxidized pollen- and spore-free zone extends to a depth of 40 feet in the arid porous deposits near Lake Maracaibo. In the Andes, where rapid erosion is occurring, the oxidized zone may not extend more than 3 feet in depth.

SECONDARY CHANGES – METAMORPHISM

Metamorphism consists of the postdiagenetic alterations in consolidated rock that are brought about principally by pressure, heat, and introduction of new chemical substances. Organic sediments, such as peat, undergo successive changes with depth of burial and with time. These changes may be referred to as "coalification," a term proposed by Barghoorn (1952, p. 182), who realized the need for a name "to designate all physical and chemical alteration starting with the initial unaltered plant substance and ending, perhaps, with graphites." If the changes are minimal, the coal is termed lignitic. Increasing alteration produces, successively, subbituminous, bituminous, and anthracitic coal. These changes involve an increase in fixed carbon content and a decrease in volatile matter and moisture content. Coalification can be subdivided into a diagenetic stage, which ends at soft lignite, and into a metamorphic series, which leads to the formation of bright lignite and successively harder coals (Kuyl and Patijn, 1961).

The alteration of organic material into coal is accompanied by compaction and a decrease in volume. Early estimates of the amount of compaction ranged from 10 : 1 to 20 : 1. The extent of compression is still not accurately known.

Classifications of the various ranks of coal are given in table 5-2.

A fivefold thickness decrease in volume of coal during compaction is suggested by Trueman (1954). A compaction ratio of less than 3 : 1 is estimated by Wanless (1952), based on compac-

Table 5-2. Classification of Coals by Rank[1,2]

Class	Group	Fixed Carbon Limits (percent) (Dry, Mineral-Matter-Free Basis): Equal to or Greater Than	Fixed Carbon Limits: Less Than	Volatile Matter Limits (percent) (Dry, Mineral-Matter-Free Basis): Greater Than	Volatile Matter Limits: Equal to or Less Than	Calorific Value Limits (Btu/lb) (Moist,[3] Mineral-Matter-Free Basis): Equal to or Greater Than	Calorific Value Limits: Less Than	Agglomerating Character
I. Anthracitic	1. Meta-anthracite	98	...	...	2	...	...	Nonagglomerating
	2. Anthracite	92	98	2	8	...	...	
	3. Semianthracite[4]	86	92	8	14	...	...	
II. Bituminous	1. Low volatile bituminous coal	78	86	14	22	...	...	
	2. Medium volatile bituminous coal	69	78	22	31	...	...	
	3. High volatile *A* bituminous coal	...	69	31	...	14,000[5]		Commonly agglomerating[6]
	4. High volatile *B* bituminous coal	...	...	...	...	13,000[5]	14,000	
	5. High volatile *C* bituminous coal	...	...	...	...	11,500	13,000	
						10,500	11,500	Agglomerating
III. Subbituminous	1. Subbituminous *A* coal	...	...	...	...	10,500	11,500	
	2. Subbituminous *B* coal	...	...	...	...	9,500	10,500	
	3. Subbituminous *C* coal	...	...	...	...	8,300	9,500	Nonagglomerating
IV. Lignitic	1. Lignite *A*	...	...	...	...	6,300	8,300	
	2. Lignite *B*	...	...	...	...	...	6,300	

[1] From American Society of Testing and Materials (1966, p. 74).
[2] This classification does not include a few coals, principally nonbanded varieties, which have unusual physical and chemical properties and which come within the limits of fixed carbon or calorific value of the high-volatile bituminous and subbituminous ranks. All of these coals either contain less than 48 percent dry, mineral-matter-free fixed carbon or have more than 15,500 moist, mineral-matter-free British thermal units per pound.
[3] Moist refers to coal containing its natural inherent moisture but not including visible water on the surface of the coal.
[4] If agglomerating, classify in low-volatile group of the bituminous class.
[5] Coals having 69 percent or more fixed carbon on the dry, mineral-matter-free basis shall be classified according to fixed carbon, regardless of calorific value.
[6] It is recognized that there may be nonagglomerating varieties in these groups of the bituminous class, and there are notable exceptions in high volatile *C* bituminous group.

tion of coal layers adjacent to sandstone dikes (which undergo very little compaction). Coals or parts of coal seams derived from attrital material have been much more compressed than those derived from woody or sclerotic tissues. Much coal is formed from highly degraded organic material. The original volume of organic material required to eventually form a unit of coal would undoubtedly represent a ratio significantly greater than 3 : 1. As concluded by Wanless, the time required for completion of the volume shrinkage involved in the change from peat to coal is roughly equivalent to that necessary to form 30 feet of overlying sediments in terms of present sedimentary thickness.

An increase in depth of burial, time, and, to a degree, diastrophic forces forms coals of different ranks. The processes involved in coal formation are discussed in detail by Schopf (1948). According to his concept fusinized (fossil charcoal?) materials are formed very early in the diagenetic stage and are imperceptibly affected by metamorphism. The waxy-resinous elements remain essentially unchanged until the coal has been altered beyond the high-volatile-bituminous rank. The principal cellulosic or lignocellulosic constituents of plants are altered chiefly by processes induced by metamorphism. These are included under the term *vitrinization*.

Coalification (metamorphism) may change the coal to a rank so high that chemical agents will not solubilize or disperse the coal matrix, thus inhibiting separation of the spores or pollen grains from the matrix. Although outlines of the spores may still be seen in thin sections, they cannot be successfully isolated for study. Some pollen and spores, however, may be obtained from the shales adjacent to such coal. In Antarctica spores have been obtained (Schopf, 1962) more readily from carbonaceous shales than from coals. It is not known whether this is due to the fact that the shale matrix can be dissolved away whereas that of the high-rank coals cannot, or whether a protective clay film prevents devolatilization at the same rate as that occurring in adjacent coals.

Depth of burial or diastrophism is known to alter the organic content of shales, perhaps in much the same manner that the organic content is altered in coals of increasing rank. In discussing this phenomenon in shales Kuyl, Muller, and Waterbolk (1955) stated that "no well preserved flora is ever found again anywhere below a 'black zone'." As depth increases plant-tissue fragments and, gradually also, pollen and spores first lose their light color, becoming increasingly dark and finally black, with no evidence of structure other than their outline. Below this level only amorphous carbon flecks can be obtained.

The inference that the carbonizing effect is caused by overburden is made because samples from all deep bore holes (15,000 ft or deeper) exhibit some degree of this type of metamorphism. In several wells from Venezuela the author has seen the change from well-preserved pollen to amorphous carbon take place within a 2,000-foot interval. In samples of surface sediments or of young sediments with no appreciable overburden in which amorphous carbon occurs we may assume a previous much greater depth of burial with subsequent erosion that removed much of the original overburden. We can and should also examine the area for evidence of intrusives or diastrophism. The importance of elevated temperatures, associated with depth of burial or nearness to intrusives, and of time to account for the metamorphism of coal is stressed by Teichmüller and Teichmüller (1954, 1947) and Teichmüller (1952). Tectonics increase pressures and temperatures in rocks. Apparently these forces influence preservation in much the same manner as does depth of burial. Curves by Karweil (1955-1956) show the relationship of time and temperature to degree of coalification.

Indurated shales or slates that have been metamorphosed will commonly not yield palynomorphs. Obvious evidence of metamorphism is sufficient reason for rejecting a sample. However, many hard, indurated, compact shales may not be metamorphosed and may yield an abundance of well-preserved spores or pollen grains.

During the process of metamorphism, particularly in its earlier stages, spores or pollen grains may exhibit evidence of carbonization and crystallization before other evidences appear in the rock. Microcrystals may form, and during their growth they may push the somewhat plastic exines aside. After chemical isolation palynomorphs show the imprint of crystals. Commonly blackening and carbonization of the organic fraction will be encountered a thousand or so feet below such alteration; but this is not always true. The manner in which growing pyrite crystals impress their shape on palynomorphs has been shown by Neves and Sullivan (1964). The growth of other crystals may likewise perforate and alter the appearance of palynomorphs—sometimes to the extent that the organic fragment is unrecognizable. Figure 5-5 shows the imprint of crystals on fossil palynomorphs.

Leaching

Leaching of rocks by ground water may, particularly in porous rocks, carry sufficient oxygen through the rocks to oxidize or destroy any organic matter that is present. Leaching is commonly in evidence in sandstone that immediately underlies a coaly layer. Acidic water from the original coal swamp, in passing through the sandstone, removes the iron—in places redepositing it as ironstone concretions lower in the sandstone body (Young, 1955). These sandstones stand out in outcrop as white ledges. This kind of white leached interval may sometimes be used to locate coaly horizons that are hidden by talus or otherwise covered.

Redeposition

Redeposition is a constant source of concern to palynologists as it involves the mixing of palynomorphs of different ages in the same sample. Although usually not serious, it occurs with sufficient frequency to warrant constant vigilance. Redeposition of Devonian spores, for example,

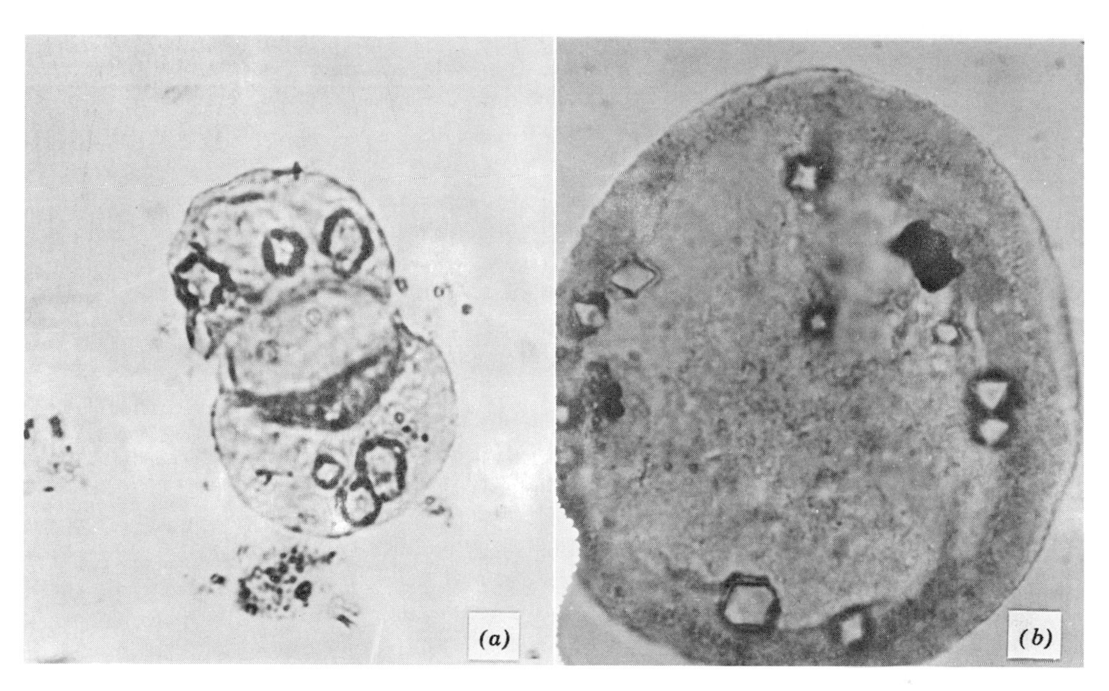

Figure 5-5. Crystal imprints on palynomorphs (× 1000): (*a*) from the Permian of Texas; (*b*) from the Mississippian of Montana.

in Cretaceous rocks is not at all serious. If the palynologist is familiar with the fossil sequence in the stratigraphic column, the Devonian fossils usually are easily identifiable as foreign to the deposit. Furthermore such redeposited spores commonly show much greater evidence of corrosion and destruction than spores from the younger deposit.

This preservational difference may be attributable to the fact that on prolonged exposure to air "sporonine" from certain pollen grains at least takes up oxygen with an increase in oxygen content of from 25 to 70 percent (Erdtman, 1943); this increase in oxygen content apparently renders the sporonine more readily soluble in weak alkali.

Spores and pollen grains preserved in a rock, on the disintegration and erosion of that rock, are again subjected to the effects not only of atmospheric oxygen but of aerobic bacteria. The chemical properties of these fossils were certainly changed somewhat during the process of fossilization. These fossils may be more susceptible to attack by organisms and by oxygen than are freshly shed pollen and spores. Redeposition of well-preserved pollen and spores is much less common than might at first be anticipated.

Further work not only on the chemistry of pollen and spore walls but also on the changes that take place during diagenesis and compaction are needed. Many other aspects of preservation could profitably be studied; for example, what is the effect on pollen grains of passing through the alimentary tracts of animals such as earthworms, copepods, or corals? Such lines of research are virtually untouched.

References

American Society for Testing and Materials, 1967, Gaseous fuels; coal and coke, pt. 19 *of* 1967 Book of ASTM standards: ASTM, 486 p.

Athy, L. F., 1930, Density, porosity and compaction of sedimentary rocks: Am. Assoc. Petroleum Geologists Bull., v. 14, no. 1, p. 1-24.

Baker, C. L., 1916, Origin of Texas red beds: Texas Univ. Bull. 29, 8 p.

Barghoorn, E. S., 1952, Degradation of plant materials and its relation to the origin of coal, Nova Scotia Dept. Mines, 2d Conf., Origin and constitution of coal, 1952, p. 181-207.

Davis, J. H., Jr., 1946, The peat deposits of Florida; their occurrence, development, and uses: Florida Geol. Survey Bull. 30, 247 p.

Erdtman, Gunnar, 1943, An Introduction to pollen analysis: Waltham, Mass., Chronica Botanica Co., 239 p.

Faegri, Knut, and Iversen, Johannes, 1950, Text-book of modern pollen analysis: Copenhagen, Munksgaard, 168 p.

Garrels, R. M., 1960, Mineral equilibria—At low temperature and pressure: New York, Harper & Brothers, 254 p.

Ginsburg, R. N., 1957, Early diagenesis and lithification of shallow-water carbonate sediments in South Florida, p. 80-99 *in* R. J. LeBlanc and J. G. Breeding, eds., Regional aspects of carbonate deposition — a symposium: Soc. Econ. Paleontologists and Mineralogists Spec. Pub. 5.

Havinga, A. J., 1964, Investigation into differential corrosion susceptibility of pollen and spores: Pollen et Spores, v. 6, no. 2, p. 621-635.

Hjulström, Filip, 1955, Transportation of detritus by moving water, p. 5-31 *in* Trask, P. D., ed., Recent marine sediments—a symposium: Soc. Econ. Paleontologists and Mineralogists Spec. Pub. 4.

Hoffmeister, W. S., 1954, Microfossil prospecting for petroleum: U.S. Patent 2,686,108.

Hughes, N. F., Dettmann, M. E., Playford, G., 1962, Sections of some Carboniferous dispersed spores: Palaeontology, v. 5, pt. 2, p. 247-252, pls. 37 and 38.

Iversen, Johannes, 1946, Geologisk datering af en senglacial Boplads ved Bromme: Aarb. f. Nordisk Oldk. og Hist., 198 p. [in Danish].

Karweil, J., 1955-1956, Die Metamorphose der Kohlen vom Standpunkt der physikalischen Chemie: Deutsch Geol. Gesell. Zeitschr., v. 107-108, p. 132-139.

Kosanke, R. M., 1950, Pennsylvanian spores of Illinois and their use in correlation: Illinois Geol. Survey Bull. 74, 128 p.

Krumbein, W. C., and Garrels, R. M., 1952, Origin and classification of chemical sediments in terms of pH and oxidation-reduction potentials: Jour. Geology, v. 60, no. 1, p. 1-33.

Krumbein, W. C., and Sloss, L. L., 1951, Stratigraphy and sedimentation: San Francisco, W. H. Freeman & Co., 497 p.

Kuyl, O. S., Muller, J., and Waterbolk, H., 1955, The application of palynology to oil geology with reference to western Venezuela: Geologie en Mijnbouw, new ser. no. 3, v. 17, p. 49-75.

Kuyl, O. S., and Patijn, R. J. H., 1961, Coalification in relation to depth of burial and geothermic gradient: Cong. Avanc. Études Stratigraphie et Géologie Carbonifère, 5th, v. 2, p. 357-365.

Kuznetsov, S. I., Ivanov, M. V., Lyalikova, N. N., 1963, Introduction to geological microbiology [translation by P. T. Broneer]: New York, McGraw-Hill Book Co., Inc., 252 p.

Mamay, S. H., and Yochelson, E. L., 1962, Occurrence and significance of marine animal remains in American coal balls: U.S. Geol. Survey Prof. Paper 354-I, p. 193-224.

Muller, Jan, 1959, Palynology of Recent Orinoco delta and shelf sediments: Micropaleontology, v. 5, no. 1, p. 1-32.

Neves, R., and Sullivan, H. J., 1964, Modification of fossil spore exines associated with the presence of pyrite crystals: Micropaleontology, v. 10, no. 4, p. 443-452.

Russell, R. D., 1955, Effects of transportation on sedimentary particles, p. 32-47 *in* Trask, P. D., ed., Recent marine sediments—a symposium: Soc. Econ. Paleontologists and Mineralogists Spec. Pub. 4.

Schopf, J. M., 1948, Variable coalification; the processes involved in coal formation: Econ. Geology, v. 43, no. 3, p. 207-225.

______1962, A preliminary report on plant remains and coal of the sedimentary section in the central range of the Horlick Mountains, Antarctica: Ohio State Univ. Inst. Polar Studies Rept. 2, 61 p.

Selling, O. H., 1946, Studies in Hawaiian pollen statistics; pt. I, The spores of the Hawaiian pteridophytes: Bernice P. Bishop Mus. Spec. Pub. 37, 87 p.

Staplin, F. L., and Jansonius, Jan, 1964, Elucidation of some Paleozoic densospores: Palaeontographica, v. 114, sec. B, no. 4-6, p. 95-117.

Swain, F. M., 1963, Geochemistry of humus, p. 87-147 *in* Breger, I. A., ed., Organic geochemistry: New York, Pergamon Press, 658 p.

Teichmüller, Marlies, and Teichmüller, Rolf, 1954, Die stoffliche und strukturelle Metamorphose der Kohle: Geol. Rundschau, v. 47, no. 2, p. 265-296.

Teichmüller, Rolf, 1952, Zur Metamorphose der Kohle: Cong. Avanc. Études Stratigraphie et Géologie Carbonifère, 3d, v. 2, p. 615-623.

Teichmüller, Rolf, and Teichmüller, Marlies, 1947, Inkohlungsfragen im Ruhrkarbon: Deutsch geol. Gesell. Zeitschr., v. 99, p. 40-75.

Tourtelot, H. A., 1962, Preliminary investigation of the geologic setting and chemical composition of the Pierre Shale, Great Plains region: U.S.Geol. Survey Prof. Paper 390, 74 p.

Trask, P. D., ed., 1950, Applied sedimentation: New York, John Wiley & Sons, Inc., 707 p.

______1955, Organic content of recent marine sediments, p. 428-453 *in* Trask, P. D., ed., Recent marine sediments—a symposium: Soc. Econ. Paleontologists and Mineralogists Spec. Pub. 4.

Trueman, A. E., ed., 1954, Coalfields of Great Britain: London, Edward Arnold, Ltd., 396 p.

Van Houten, F. B., 1948, Origin of red-banded early Cenozoic deposits in Rocky Mountain region: Am. Assoc. Petroleum Geologists Bull., v. 32, no. 11, p. 2083-2126.

Waksman, S. A., 1933, The distribution of organic matters in the sea bottom and the chemical nature and origin of marine humus: Soil Sci., v. 36, p. 125-147.

Walker, T. R., 1967a, Color of recent sediments in tropical Mexico—A contribution to the origin of red beds: Geol. Soc. America Bull., v. 78, no. 7, p. 917-920.

______1967b, Formation of red beds in modern and ancient deserts: Geol. Soc. America Bull., v. 78, no. 3, p. 353-368.

Wanless, H. R., 1952, Studies of field relations of coal beds, p. 148-180 *in* Nova Scotia Dept. Mines Conf., 2d, Origin and constitution of coal, 1952, p. 148-180.

Williams, Milton, and Barghoorn, E. S., 1963, Biogeochemical aspects of the formation of marine carbonates, p. 596-604 *in* Breger, I. A., ed., Organic geochemistry: New York, Pergamon Press, 658 p.

Wilson, L. R., 1962, Permian plant microfossils from the Flowerpot Formation, Greer County, Oklahoma: Oklahoma Geol. Survey Circ. 49, 50 p.

Woods, R. D., 1955, Spores and pollen – a new stratigraphic tool for the oil industry: Micropaleontology, v. 1, no. 4, p. 368-375.

Young, R. G., 1955, Sedimentary facies and intertonguing in the Upper Cretaceous of the Book Cliffs, Utah-Colorado: Geol. Soc. America Bull., v. 66, no. 2, p. 177-201.

Ziegler, Victor, 1911, Factors influencing the rounding of sand grains: Jour. Geology, v. 19, p. 645-654.

ZoBell, C. E., 1946, Marine microbiology, a monograph on hydrobacteriology: Waltham, Mass., Chronica Botanica Co., 240 p.

——— 1955, Occurrence and activity of bacteria in marine sediments, p. 416-427 *in* Trask, P. D., ed., Recent marine sediments – a symposium: Soc. Econ. Paleontologists and Mineralogists Spec. Pub. 4.

obtained from wells drilled with air or gas than with mud. The material circulates to the top of the hole in a fraction of the time taken by mud. Also caving is less of a problem because there is no swelling and sloughing of soft shales as a result of wetting.

Much the same comments apply to samples from other rotary-drill holes—for example, those used for seismic shots. The principal difference is that holes drilled for these are not so deep, and consequently there are not the complications imposed by depth. If a shot hole is drilled to a depth of only 50 feet there is no danger of caving from sediments 1000 feet up the hole.

Cable-tool drilling is increasingly rare, but it is still used for shallower oil wells in some areas and for water wells. In cable-tool drilling a weighted bit is repeatedly lifted and dropped to crush the rock being penetrated. The crushed rock is removed from the hole by a hollow cylinder known as a bailer. Cores can be taken by dropping a special cylinder which cuts out a circular section of the rock at the bottom of the hole (these are known as "biscuit" cores). Cable-tool cuttings are in general more reliable than those taken in rotary drilling. First, one knows the exact level from which the cuttings come. Second, cable-tool wells are generally cased throughout the full depth as they are drilled; this eliminates caving.

SOURCES OF CONTAMINATION

Sample contamination results from a variety of causes operating at all stages from the sample in situ to the final preparation of a permanent microscope slide. In-laboratory contamination can occur even in the best equipped and operated laboratories.

Contamination from Air

To know that pollen can be carried by air we have only to think of the yellow "dust" that causes allergic reactions. Air contaminants are most commonly modern material. Occasionally, however, samples may be contaminated during processing by fossils from other samples in the laboratory. It has already been mentioned that dust in mines, particularly coal mines, can contaminate samples with fossil material.

The ease with which modern contaminants can be recognized depends, of course, on the material with which one is working. If the material is Recent or near-Recent the problem of differentiating modern material is more difficult than when working with older specimens. The worker should be familiar with what is likely to turn up from the local flora. Some workers put out greased plates which they examine occasionally to see what is in the air. An abundance in samples of types found on the plates would obviously signal air contamination.

When working with older sediments the problem of identifying modern contaminants is not so difficult. Also, modern material generally has a different sheen from that of fossils and often takes stains differently. Most older sediments have been compacted, and in the process the acid-insoluble microfossils have been flattened. There are exceptions to this flattening; for example, spores and pollen from Cretaceous clays in Virginia and Maryland show little or no flattening. Also most microfossils from cherts are unflattened.

Contamination by fossils from other samples is a somewhat more vexing problem, but fortunately happens very seldom. We should keep abreast of what is passing through the processing laboratory and what colleagues are working on. If something that is obviously not modern, but seemingly out of place, turns up, we may be able to track it down.

Underlying the above considerations is the greatest factor of all: good judgment. If we find *Ulmus* in the Paleozoic, we should not rush into print with an early record for angiosperms. This admonition may sound ridiculous, but just such announcements have been published.

Contamination in Shipping

Shipping contamination can be almost impossible to cope with when dealing with disaggregated samples. Regrettably many people do not know how to pack samples to be used in palynological studies. Bags may be of loosely woven cloth that permits dust to seep through. Samples may reach the laboratory in loosely tied bags placed neck to neck; spilling into each other may have occurred at every jostle of the packing crate. Another problem is that cloth or paper bags may rot in transit. This is especially true in humid tropical areas where inefficient transportation operates in league with climate.

We can only hope that collectors and shippers may become better educated and equipped as time goes on. Hard, well-consolidated samples require the least care in packing, but finely divided or friable samples require great care. Samples should be thoroughly dried before packing and placed in leakproof containers. If the shipping container is likely to become wet after packing, it should be lined with oilcloth, oiled paper, or plastic.

Contamination from Water Supply

The water supply, particularly if stored in open reservoirs, can contain modern pollen, spores, diatoms, dinoflagellates, desmids, etc., and in rare cases fossil ones. A classic example (in Dawson, 1886) records *Sporangites* (now *Tasmanites*) in the city water supply of Chicago. To guard against this kind of contamination many laboratories use filtered or distilled water for palynological processing. If a diatomite filter is used, the filter itself can contribute diatom fragments.

Contamination from Laboratory Equipment

The apparatus used for crushing samples before they are processed can be a source of contamination unless it is carefully cleaned between samples. Fossils can be released to circulate in the air, but more likely contamination occurs between consecutively crushed samples. The use of standard grinding machines is to be discouraged because of the difficulty in cleaning them. Mortars and pestles should be carefully scrubbed after each sample and should be discarded at the first sign of pitting. The same can be said for the metal plates and hammers commonly in use. A partial answer to contamination of metal plates is to beat the sample on several layers of heavy aluminum foil, discarding the foil after each sample is crushed.

Unless properly washed, glassware can be a source of contamination through the adherence of fossils. This is particularly true of pipets whose small diameter makes the cleaning of the inside difficult; chromic-sulfuric acid cleaning solution is recommended for them. An ultrasonic cleaner helps prevent contaminated glassware.

Contamination from Caving

Caving helps to make well cuttings one of the least reliable types of sample. Caving results when pieces of a higher rock stratum break off and intermingle with particles of the rock being circulated up from the bottom of the hole. It can also occur (uncommonly, however) at the top of a conventional core as a result of the breaking off of a chunk of a higher stratum by the lowering core barrel.

If caving is suspected, helpful information can be obtained from the drilling report. Trips, stuck holes, reaming, and fishing are common causes of increased caving. On the contrary, knowing when and where casing was set or cementing done allows us to eliminate certain sediments as sources of caving. It should be kept in mind that the usual cutting fragment is ¼ inch or less in size; larger fragments are usually caved. If several lithologies are present, they can be sorted and processed separately; differences in fossil content suggest mixed cuttings. A method that is sometimes used to obtain relatively uncontaminated samples is to shut down drilling for a time and to circulate the mud. Samples are then taken as soon as drilling is commenced again. A knowledge of the section and how the various strata react to drilling is invaluable. In general the softer, less silicified shales tend to swell more, and consequently cave more, than the harder ones. Also the palynologist has to exercise judgment and common sense. The sudden appearance of Tertiary fossils in a Mesozoic sequence signals caving.

Contamination from Drilling Mud

Drilling mud can carry contamination in the form of fine particles from formations other than the one being drilled. Contamination can also come from additives containing ground lignite—for example, Carbonox, CC-16, or XP-20. In my experience these have always been Paleocene lignites from the Rocky Mountain region. The problem has been discussed by Traverse, Clisby, and Foreman (1961). Samples should be washed as free as possible from drilling mud before processing. It has already been pointed out that the mud cake can contaminate sidewall cores; also that fossils suspended in drilling mud can be forced into porous or fractured sediments.

Contamination from Recirculation Material

Circulation is lost when the drill penetrates a bed that is sufficiently permeable to drain off the circulation fluids. This most often occurs in highly porous sandstones, fractured limestones, and the like. When circulation is lost, anything that by a stretch of the imagination might plug the troublesome bed may be put down the hole. In addition to special preparations, the material can be walnut hulls, hay, or cotton-seed husks. Some of these materials can contain pollen and spores of their own; they can also be very exasperating to separate from cutting samples.

DEPOSITION OF NONCONTEMPORANEOUS FOSSILS

Reworking

Reworking is the redeposition of older fossils into younger beds; thus it is an inherent part of sedimentation. It can present extremely difficult problems and can appear in any type of sample, regardless of how collected. Its one redeeming feature is that it shows the source of a sediment.

The smaller microfossils can be transported and redeposited either in particles of their original sedimentary matrix or, less commonly, free of matrix. When free of matrix, they are usually corroded to some extent and less well preserved than fossils contemporaneous to sedimentation. On the other hand, if they are enclosed in particles of the original matrix, they may be redeposited in coarser clastic sediments than they otherwise would be. In such sediments contemporary material may have been completely removed by the winnowing action of currents or, if left, be badly corroded. Thus assemblages can be composed entirely of reworked fossils or may contain a mixture of reworked and contemporary fossils. The problem has been discussed by Wilson (1964).

In some cases reworked fossils cannot be detected on palynological evidence alone. The palynologist should gather all available geologic information and discuss the problem with geologists who are familiar with the area. Coarse-grained samples should always be suspected. Thin sections can be extremely helpful because in them it is often possible to see the reworked fossiliferous rock fragments. In samples that the palynologist knows to be in place—for example, cores or outcrops—older fossils that are stratigraphically above younger ones probably got where they are through reworking. Alternative explanations are that the younger fossils got into the older beds by stratigraphic leakage or faulting. Reworking also results in assemblages that are mixtures of younger and older fossils.

In ditch samples reworked fossils combined with cavings can present an insoluble problem, especially in an unfamiliar area.

An example of reworking that is easy to detect has been noted in the Fort Union Formation (Paleocene) from certain parts of Wyoming. It is not uncommon to find reworked Late Cretaceous fossils near the top of the Fort Union, but they are underlain by a long sequence of Paleocene assemblages. A considerably more difficult problem is encountered in Venezuela, where in one locale the Eocene directly overlies the uppermost Cretaceous and contains nothing but reworked latest Cretaceous microfossils. Here it is impossible to establish an Eocene-Cretaceous boundary by palynology; it must be determined on other bases. An extreme case, also from Venezuela, has Late Cretaceous fossils reworked into other, only slightly younger, Upper Cretaceous sediments.

Stratigraphic Leakage

Stratigraphic leakage is the opposite of reworking in that it is the deposition of younger fossils into older beds. It takes place when younger sediments are deposited in cracks, fissures, or solution channels of older ones. It occurs most commonly in limestones, which are particularly susceptible to solution channels. An example has been reported by Upshaw and Creath (1965), who found an assemblage of 69 taxa of Pennsylvanian age in a cavern fill deposit in the Middle Devonian Callaway Limestone of Missouri.

Stratigraphic leakage is much less common than reworking and seldom presents a problem to palynologists. One should be wary, however, of either collecting or processing samples that have filled cracks of any sort. Guennel (1963) reported on well-preserved Devonian spores in crack fillings in a core from the Middle Silurian Tilden Reef in Illinois.

References

Dawson, William, 1886, On Rhizocarps in the Erian (Devonian) period in America: Chicago Acad. Sci. Bull., p. 105-118.

Guennel, G. K., 1963, Devonian spores in a Middle Silurian reef: Grana Palynologica, v. 4, no. 2, p. 245-261.

Traverse, Alfred, Clisby, K. H., and Foreman, Frederick, 1961, Pollen in drilling-mud "thinners," a source of palynological contamination: Micropaleontology, v. 7, no. 3, p. 375-377.

Upshaw, C. F., and Creath, W. B., 1965, Pennsylvanian miospores from a cave deposit in Devonian limestone, Callaway County, Missouri: Micropaleontology, v. 11, no. 4, p. 431-448, pls. 1-4.

Wilson, L. R., 1964, Recycling, stratigraphic leakage, and faulty techniques in palynology: Grana Palynologica, v. 5, no. 3, p. 425-436.

7

APPLIED PALYNOLOGY

Robert H. Tschudy

The application of palynology to geologic or stratigraphic problems involves the definition and delineation of specific strata, or segments of the stratigraphic column, in terms of the palynomorphs derived from these rocks. The concept that some stratigraphic segments can be identified and distinguished from other segments is based on the fact that plants have undergone evolutionary change during geologic time. Evolutionary change is reflected in the parts preserved—pollen, spores, and some other structures—as well as in the plants as a whole. The preserved remains of plants will reliably identify and distinguish segments of the geologic column. Paleontologists since the time of William Smith have found that the position of rocks in the geologic column and the fossil content of the rocks are interrelated, and that the fossil biota presents a reliable and practical method for dating rocks.

Ecological factors, including climatic and edaphic, may also be reflected by changes in the floras of successive rock layers. Floral changes caused by these factors may be difficult to distinguish from changes resulting from evolution. Nevertheless constant effort should be directed toward recognizing and distinguishing such effects.

The principal applications of palynology are to the correlation of strata and to the determination of the relative ages of strata. Age determination must, at least for the present, be based on the correlation of palynomorph assemblages of a particular stratigraphic section with the palynomorph assemblages from a similar section that has previously been reliably dated by some other means.

Initially dating is done by comparison of palynomorph assemblages with vertebrate or invertebrate fossils of the same rocks. Well-known vertebrate and invertebrate assemblages commonly provide excellent comparative data for dating palynomorphs contained in the same rocks. Sometimes the principle of interpolation is employed; for example, a continental bed, which yields a plant-microfossil assemblage, can be given an approximate date if the overlying and underlying strata have been dated by some means other than pollen and spores.

PALYNOLOGIC CORRELATION

Correlation is the process of determining that geologic events in two or more areas are contemporaneous. In palynology it implies the establishment of qualitative and quantitative similarity between the plant-microfossil assemblages derived from two or more segments of rock strata. This type of correlation may involve the recognition of similar assemblages from two rock samples of only a gram or so each, or it may involve similarity of fossil content from stratigraphic sequences of two or more sections involving hundreds of feet.

Historically, palynologic correlation was first used by Von Post in 1916 in studies of fossil pollen from Swedish bogs (Erdtman, 1943). Application of spores to correlation of coal seams was

Publication authorized by the Director of the U.S. Geological Survey.

initiated by Thiessen and Voorhees (1922) and Thiessen and Stand (1923). Microscopic examination of thin sections of coal disclosed that individual coal beds contained characteristic spores and other plant structures. These spore assemblages differed vertically from bed to bed, but no significant variation in spore type was observed laterally. An individual coal bed such as the "Thick Freeport" coal, or the Pittsburgh, Redstone, and Sewickley coals could be identified over a wide area by means of its characteristic spores. Detection of this variety led to the conclusion that it was possible to identify each coal bed by means of the spores and other structures observed in thin sections.

With the introduction of a solution of nitric acid and potassium chlorate (Schulze's solution) as a means of breaking down coal and freeing spores from it, instead of depending on finding spores in thin sections, correlation of coals began to be common both in the United States and in Europe. At present the spore contents of coals are accepted as reliable criteria for identifying individual coal beds. This application of palynology, however, has largely been limited to regional, rather than to interregional or intercontinental problems (Kosanke, 1950; Guennel, 1958).

The step from the recognition of the abundance and usefulness of plant microfossils in coals to a search for them in shales was, in retrospect, a short one. Possibly because of the comparative abundance of spores and pollen in coals, the relative ease of separating such structures from coals, and the fact that there was, and still is, an enormous reservoir of coal available for study, this step was greatly delayed.

As early as 1925 Lang obtained and studied spores from the Old Red Sandstone of Devonian age by dissolving the rock with hydrofluoric acid. The widespread application of palynology to lithotypes other than peats and coals, however, did not begin until about 1950. The impetus to this new application was provided to a large extent by the petroleum industry.

The need for additional tools for correlation in long, otherwise nonfossiliferous, rock sequences led directly to the rediscovery that palynomorphs were present in some shales and other noncoaly lithotypes, and that such rock therefore offered a material of great potential usefulness (Wilson, 1946; Grayson, 1960; Gutjahr, 1960; Hoffmeister, 1960). This potential is now being realized. Many rocks previously considered unfossiliferous are now yielding an abundance of plant microfossils. These microfossil assemblages may provide all the information required for correlation or environmental interpretation, or they may serve to supplement and refine conclusions based on other fossil types.

As a result of the discovery of well-preserved palynomorphs in noncoaly lithotypes and the assistance thus made available to the correlation of otherwise unfossiliferous shale sequences, several oil companies immediately established palynological laboratories. In the decade from 1950 to 1960 practically all major oil companies followed suit, and numerous other institutions recognized the need for such laboratories. Many foreign countries have incorporated palynological laboratories into their national and provincial geological institutions, and numerous universities are establishing palynological curricula in their botanical or geological departments.

Collection of data

Prior to data collection a clear concept of the problem to be resolved should be at hand. If the dating of an individual sample is desired and adequate control[1] material is available, the solution to the problem is straightforward. We need only to prepare the sample, list the species, and compare the assemblage quantitatively and qualitatively with assemblages available for the control sections. Frequently, however, adequate comparative material is not available or stratigraphic coverage is incomplete. We must then either amplify the control material or make a species-by-species comparison with published material from other localities. The most suitable solution is the amplification of the control material. A comparison with published material from other regions may yield information of value but should not be expected to provide information that is as reliable as that derived from nearby control samples.

Two steps usually are involved in the collection of data: (a) preliminary examination of the prepared material followed by a qualitative list-

[1]Palynologic standards of comparison, usually obtained from type localities or from reference collections.

ing of the pollen and spore flora and (b) a quantitative determination of the dominant palynomorphs present.

Qualitative Determinations. The first, or qualitative, examination of the prepared slides should provide information on the reliability of the sample. Are the palynomorphs sufficiently numerous to provide reliable comparative data? Are they so corroded or distorted that consistently reliable species determinations cannot be made? Is the sample contaminated with either modern species or redeposited species? If the sample is unreliable for any of these reasons, time should not be spent in attempting determinations.

If a sample is reliable, the most useful information normally gained by a qualitative examination is a record of the total composition of the flora. Virtually complete qualitative information permits meaningful data to be accumulated, and this will contribute to knowledge of the range in time of specific taxa. The vertical range of species is especially significant and useful. Accessory information may be obtained relative to such subjects as floral origin, evolution, and the nature of the facies and the climate at the time of deposition.

During the examination of any sample morphologic variations within species – either previously described species or new – should be noted. Such procedure provides data permitting the individual worker to decide whether more than one species is represented or whether all variations noted should be included within a single species. Further, this procedure provides a firm basis for the recognition of this species in other samples.

When the qualitative examination of a particular sample has been completed the data normally are compiled as a list of genera or a list of species.

Quantitative Determinations. Quantitative counts in the form of absolute figures, percentages, or both, have proved to be extremely useful, particularly in local correlation problems. Such counts provide estimates of the dominant species or forms present. The dominant forms may differ with climatic or edaphic changes, even though no evolutionary changes can be recognized.

In coals, which represent a bog or swamp environment, ecological or edaphic changes gave rise to changes in the relative composition of the bog flora. Successive coal layers, many of them only inches thick, commonly present significantly different percentage compositions even on the generic level. Sometimes little or no change in taxa is represented, even though percentages differ greatly. These percentage differences are commonly uniform throughout the lateral extension of coal beds, and in many cases, but not all, represent time units within a given basin. They obviously cannot be relied on for interbasinal or intercontinental correlations.

The utility of the quantitative approach has been so thoroughly established that virtually all stratigraphic presentations today involve quantitative data. It is only in purely taxonomic reports that one may find no or only brief references to the percentage abundance of the taxa found in the sample or series of samples.

Many palynologists for practical reasons have settled on a count of about 200 entities to provide data for the analysis of an average sample. This figure is a compromise as it is not practical to count all of the units in a sample that yields thousands to millions of plant microfossils. Related factors in the composition of a fossil assemblage and the number of recurring individuals of the same species have been analyzed by Wilson (1959) (see fig. 7-1) in an attempt to determine, other than empirically, the number of palynomorphs that should be counted in a given preparation in order to arrive at a reliable percentage figure.

A sample yielding only a few species will require a smaller count than one that yields many species. Some samples contain a few exceedingly abundant species and many relatively scarce species. Such samples can be evaluated by obtaining first an overall count that will give the percentages of the dominant species, followed by a second count, ignoring the dominant species already counted. Thus a sample that accidently or otherwise has a very high representation of one species, such as would occur if an anther were incorporated into that sample, would be counted twice. The double-count method provides an estimate of the overall palynomorph representation.

The method proposed by Wilson (1959) is as follows: The number of specimens is plotted

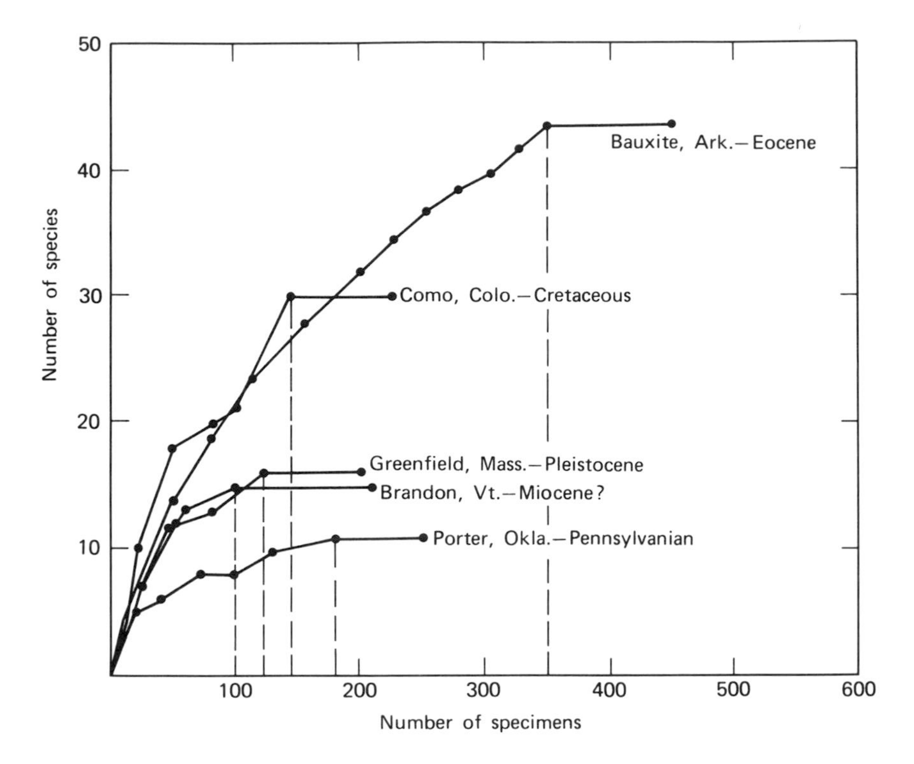

Figure 7-1. Species-stratum curves of five palynological assemblages. (From Wilson, 1959; reprinted with the permission of Oklahoma Geology Notes, Oklahoma Geological Survey.)

against the number of species until the curve flattens. This point Wilson refers to as the break-off point. After this point is reached, few additional species may be expected regardless of the number of species counted. In the example the number of specimens necessary to be counted ranges from 100 in the sample of lignite from Brandon, Vermont, to more than 350 in the sample of bauxite from Arkansas.

In the bauxite sample fewer than 50 species were recorded. Some samples may yield many more species; a single Upper Cretaceous sample from Kentucky has yielded more than 150 species.

A procedure that is often employed in making a relatively complete qualitative, as well as quantitative, estimation is to make the quantitative count first (200 to 500 specimens), then to scan the remainder of the slide or slides for species not included in the quantitative tally. Any additional species would contribute to a knowledge of the complete assemblage, and to the known vertical ranges of the scarce species, but would not be included in the percentage figures.

Presentation of Data

Information derived from the microscopic examination of samples should be made easily understandable to others. Commonly this presentation is in the form of tables or, preferably, graphs. The graphs may be bar graphs, or line or sawtooth diagrams, with one axis representing the stratigraphic position or depth and the other axis representing the abundance or percentage. Many graphic methods of presenting data are described by Erdtman (1943), Pokrovskaya (1958), and Sittler (1954). One example will suffice here.

A correlation diagram that incorporates relative frequencies, absolute frequencies, maxima, minima, increases and decreases in abundance, and ranges of groups of palynomorphs is presented by Jekhowsky and Varma (1959). This procedure makes maximum use of the data collected and presents it in a concise form.

Figure 7-2 shows the density of correlation horizons established by Jekhowsky and Varma (1959) between two wells 28 kilometers apart.

These horizons are in general parallel to and in agreement with the stratigraphic information derived from other sources.

Interpretation of Data

After data have been assembled it is necessary to interpret these data. Correlation lines can be drawn as illustrated in the chart by Jekhowsky and Varma (1959), but not all such correlations will have the same weight. Some will be more reliable than others. Numbers of specimens for each taxon or group of taxa will differ. Ranges of taxa that are numerically few in the samples will be less reliable than ranges based on more abun-

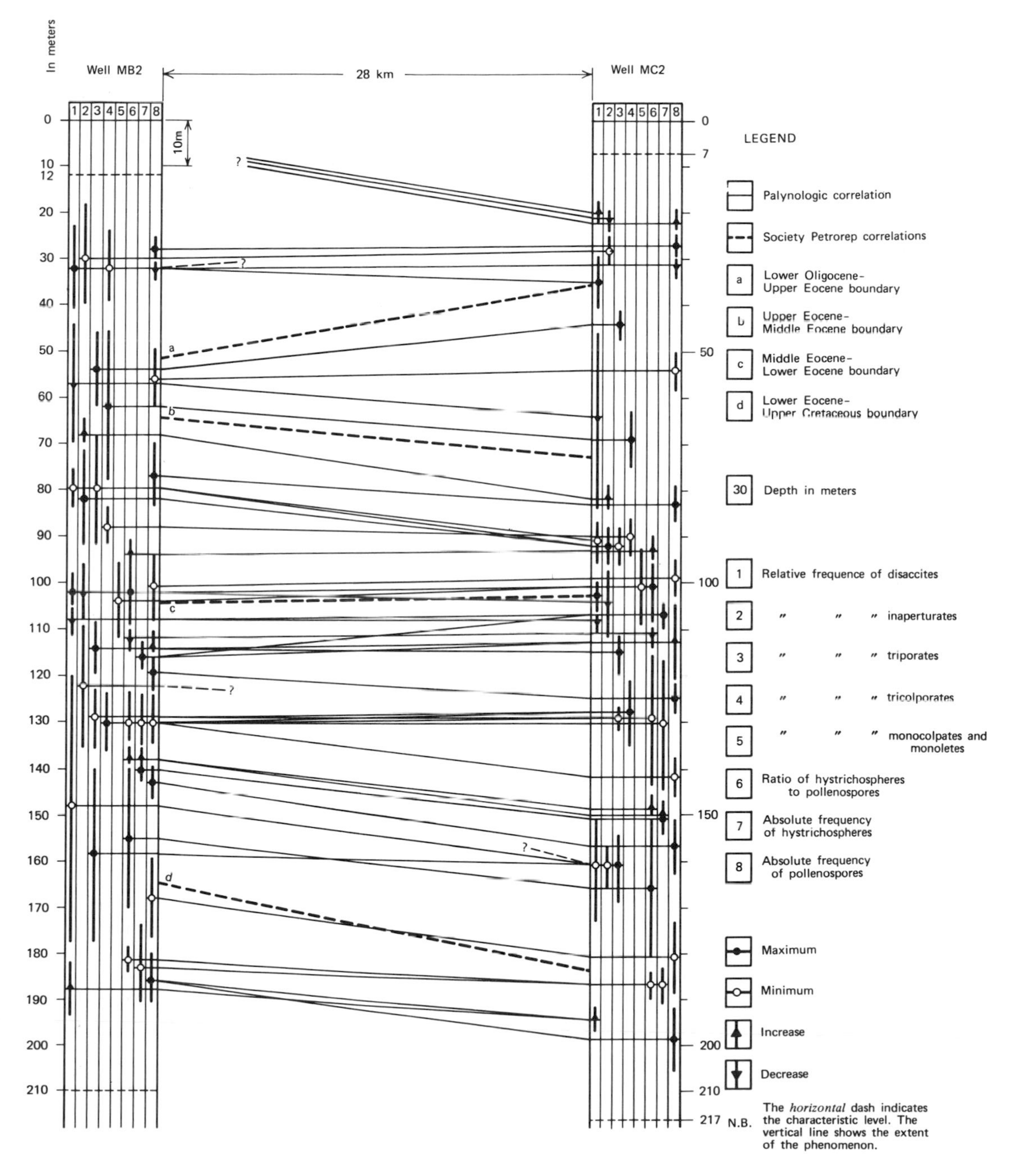

Figure 7-2. Palynological correlation of wells MB2 and MC2 in the Meaux area, France. From Jekhowsky and Varma (1959).

dant taxa. The palynologist's responsibility is to call attention to the weak or questionable correlations, as well as to point out the lines of evidence indicating strong probability of stratigraphic or time equivalence.

A detailed exposition of a statistical method of determining correlation between paleontologic sections, using primarily the upper and lower extremes of the ranges of taxa recovered, has been presented by Shaw in "Time in Stratigraphy" (1964). Basically the method derives an expression of time equivalence on a two-axis graph. One axis represents the standard section and the other the section being compared. In order to utilize the procedures outlined by Shaw certain basic requirements for effective paleontologic work must be met; these, paraphrased from Shaw (1964), are as follows:

1. Basic paleontologic taxa must be accurately defined; imprecise definition makes them potentially useless. The taxa must be recognized with consistency and must be used consistently within finite and determinable ranges.

2. Accurate records of stratigraphic position of species must be maintained. Because the limits of range of *some* species cannot be located at every stratigraphic level, separate collections must be made as precisely as possible at all times. A historic controversy known as the "Laramie problem" had its origin in the failure of certain early paleobotanists to keep separate the leaf collections from various different stratigraphic positions (Dorf, 1940).

3. Some sort of designation of each recognizable morphologic entity should be used. Morphologic units (taxa) must be defined and described with sufficient precision and consistency so that they will mean the same thing to all who use them. Undoubtedly taxa will eventually be coded for insertion into data-processing machines.

Another criterion for measuring floral similarities is that employed by Couper (1958). A percentage figure is obtained that is based on the number of localities from a given rock unit in which a species is present. The relative abundance of the species at the localities is unimportant.

Table 7-1 shows how treatment of data in this manner may serve to isolate different parts of a stratigraphic section (numbers are percentages).

In order to limit or minimize subjective bias in making comparisons of fossil assemblages attempts have been made to subject the data to statistical analysis. This method provides a numerical correlation coefficient or index of resemblance or lack of resemblance. One such method, the chi-square test, was used by Gray and Guennel (1961) to provide objective evaluation of the significance of relative-abundance values of spores in coal beds. In applying the chi-square test a probability value is obtained. This value tells whether the probability is high (e.g., 95 percent) that sample A and sample B consist of identical populations, or conversely, that the probability of their being identical is below acceptable limits.

Another statistical method of measuring faunal (or floral) resemblance, based on numbers of taxa and abundance of taxa, was suggested by Simpson (1960). He provided 13 formulas for obtaining indices of faunal resemblance. These formulas were based on manipulations of the following symbols:

E_1, Number of taxa (at a specified level) in the first, smaller (or equal) of the two faunas or samples compared but absent in the second.

E_2, Number of taxa in the second, larger (or equal) fauna or sample but absent in the first.

C, Number of taxa common to both.

$N_1 = E_1 + C$, Total taxa in first.

$N_2 = E_2 + C$, Total taxa in second.

$N_t = E_1 + E_2 + C = N_1 + N_2 - C$, Total taxa in both.

I_{E_1}, I_{E_2}, Number of individuals of taxa not in common in first and second samples, respectively.

I_{C_1}, I_{C_2}, Number of individuals of taxa in common in first and second samples, respectively.

$I_1 = I_{E_1} + I_{C_1}$, $I_2 = I_{E_2} + I_{C_2}$, Total number of individuals in first and second samples, respectively.

Index 2 and index 11 of Simpson (1960) were applied by Wilson (1964) in assessing palynological assemblage resemblances of a series of nine Croweburg coal collections. Index 2 is

$$\frac{C}{N_1} \times 100$$

This index tends to stress the most nearly similar parts of the two assemblages in question

Table 7-1. **Known Stratigraphic Ranges of "Key Forms" in British Jurassic and Lower Cretaceous Sediments**[1]

Species	*Lias (5 localities)*	*Lower Deltaic (8 localities)*	*Sycarham Beds (8 localities)*	*Gristhorpe Beds (20 localities)*	*Upper Deltaic (28 localities)*	*Oxford Clay (2 localities)*	*Kimeridge Clay (3 localities)*	*Purbeck Beds (5 localities)*	*Wealden Beds (16 localities)*	*Lower Greensand (5 localities)*
Abietineaepollenites dunrobinensis	40									
Monosulcites subgranulosus	40									
Cingulatisporites scabratus	20									
C. rigidus	20	25								
Calamospora mesozoica	40	75	88	70						
Todisporites major	20	50	75	80	60					
Concavisporites subgranulosus	20	12	25	20	18					
Matonisporites phlebopteroides	20	63	75	40	32					
Todisporites minor		63	63	65	18					
Klukisporites variegatus		75	100	70	11					
Trilites bossus		12	12	10	4					
Monosulcites carpentieri		12	25	35	20		33	20	12	
Eucommiidites troedssonii		100	75	65	64	100	33	80	50	
Spheripollenites scabratus		100	100	100	100	100	100	100	19	60
Leptolepidites major			37	15	25					
Cingulatisporites problematicus			20	30	18		33	20	31	40
Gleicheniidites senonicus			25	40	36		100	60	100	100
Perotrilites rugulatus				15						
Trilites equatibossus				10						
Lycopodiumsporites gristhorpensis				45	22					
Pilosisporites brevipapillosus				5	14					
Lygodioisporites perverrucatus				?	14					
Cingulatisporites dubius				15	28	50	33			
Concavisporites variverrucatus				10	4			80	75	
Cingulatisporites pseudoalveolatus				10	32			20	19	40
Foveotriletes microreticulatus					11					
Parvisaccites enigmatus					53					
Pteruchipollenites microsaccus					75	50	100			
Foveotriletes irregulatus						50	100			
Klukisporites pseudoreticulatus								60	50	
Trilobosporites bernissartensis								80	69	
Pilosisporites trichopapillosus								40	82	?
Cingulatisporites complexus								20	12	
Cicatricosisporites dorogensis								100	100	100
Trilobosporites apiverrucatus									63	
Peromonolites asplenioides									38	
Appendicisporites tricornitatus									56	60
Cicatricosisporites brevilaesuratus									63	20
Concavisporites punctatus									82	40
Cingulatisporites valdensis									63	40
Parvisaccites radiatus									94	100
Clavatipollenites hughesii									44	100
Spheripollenites psilatus									19	100
Cingulatisporites foveolatus										100
Trilites distalgranulatus										100

[1]From Couper (1958); reprinted with the permission of Palaeontographica, Stuttgart, Germany.

and minimizes the differences in size between N_1 and N_2. Index 11 is

$$\tfrac{1}{2}\left(\frac{I_{C_1}}{I_1} + \frac{I_{C_2}}{I_2}\right) \times 100$$

or

$$\left(\frac{I_{C_1}}{I_1} + \frac{I_{C_2}}{I_2}\right) \times 50.$$

This index is the mean of percentages of individuals in common taxa figured separately for each sample.

Wilson obtained values that ranged from 75 to 100 for index 2 and from 91.5 to 99.5 for index 11. In general the values were remarkably high, but such high values were to be expected if the samples came from the same seam.

For a close comparison the following conclusions were reached by Wilson (1964):

1. The samples must be processed essentially alike, for different or excessive treatment at one or another stage of sample preparation will destroy certain species.

2. The lithology of the samples should be as nearly identical as possible to eliminate paleoecological differences.

3. The taxonomy of the fossils in the assemblages must be known because confusion of closely related species will markedly affect the results.

Problems. Correlation problems may arise from several causes, including barren or poorly productive samples, differential preservation (some species being destroyed, whereas others are preserved), facies changes that mask the characteristic flora of the unit, poor or incomplete sampling, and failure to sample the correct lithotypes. These factors must be considered, as any one of them could prevent effective correlation. The most serious correlation problems arise from the failure to consider adequately the rocks in relation to the samples or to recognize either diastems or unconformities or faunal and lithologic breaks.

Diastems and Unconformities. Diastems and unconformities are the records of cessations of deposition. A *diastem* is defined as an intraformational depositional hiatus of minor duration unaccompanied by faunal or floral change. For the purposes of this discussion it may be referred to as a temporary surface representing an interval during which no permanent sediment is deposited owing to a change in current or a change in size or quantity of particles being discharged into the sedimentation area. Diastems also represent time intervals during which sediments have been removed by scour (Eaton, 1929).

An *unconformity* is a surface that represents a period of erosion or nondeposition that separates rocks of different ages. An unconformity represents an interval of time that is unrecorded at a particular locality (Blackwelder, 1909). A *disconformity* is a type of unconformity in which the beds on opposite sides are parallel.

The magnitude of the time lapse between successive waves of deposition is the critical factor. Diastems usually are of little consequence in palynological work. They occur between every bed. They may be recognizable in the rock, but they may not be indicated by the fossils from the rock. Unconformities representing appreciable intervals of time may, however, give rise to several problems.

A hiatus of very short duration may be represented in the rocks by a prominent lithologic change; for example, a bed of conglomerate may indicate either a diastem or an unconformity, and without the aid of fossils the distinction may prove difficult. Such a prominent lithologic change makes imperative the sampling of horizons immediately above and below. This must be done in order to determine in the absence of other evidence (such as may be derived from similar nearby sections) the significance of the time gap represented.

Conversely an unconformity may be obscure or not recognizable owing to similar conditions of deposition before and immediately after the time hiatus or lapse. Such a situation may be recognized from an abrupt change in the fossil assemblages. The simultaneous termination of several species from unrelated families should always raise the suspicion of a significant time lapse or an abrupt facies change. Entire floras can scarcely be expected to appear or vanish at the same point in time. Individual assemblages must be considered as consisting of a continuum of units that appear and disappear. A pronounced change in the rate at which taxa come and go should be examined critically to determine whether or not an unconformity is involved.

Apparently abrupt floral changes may reflect a hiatus of appreciable duration rather than reflecting climatic, edaphic, or evolutionary changes.

Correlation of Palynological Horizons with Faunal and Lithological "Breaks." That formation boundaries determined from the study of one group of animals or plants should not be expected to coincide precisely with boundaries determined from another group is the expected result of evolutionary theory. Different groups of plants and animals seldom evolve at the same rate, because such groups seldom are subjected to the same evolutionary pressures. A major change in a fauna of marine ammonites might not correspond in time with a change in the flora of gymnosperms. Some plant and animal groups have been shown to have evolved rather slowly; others have evolved comparatively rapidly. Principal zones of correlation have failed to coincide with major faunal changes. This phenomenon is to be expected on theoretical grounds.

Land plants are obviously subjected to different environmental factors than are aquatic organisms. Although increased rainfall might cause a floral change over a period, such a change would not necessarily affect the indigenous biota of the depositional site unless it simultaneously affected ecological conditions such as salinity or turbidity, which in turn might affect the biota.

Major developments in the evolution of plants do not necessarily correspond to the period boundaries of the geologic time system (Arnold, 1947; chap. 8). Historically, many of the systemic boundaries were based on certain faunal changes in marine organisms. Palynological changes, however, inevitably parallel the changes in plant megafossils, since pollen and spores are parts of plants. Major events in plant evolution have preceded, by approximately half a period, changes of corresponding magnitude in animals. Since plants are the basic link in the nutrition chain of all animal groups, such a situation might be expected. As a consequence of this discrepancy in time between major plant and animal evolutionary changes, at the presently defined boundaries of periods, both megafossils and microfossils commonly represent transition floras (Tschudy, 1964).

In a like manner paleontological zones need not, and commonly do not, coincide with the limits of a formation or rock unit. Formations are by their nature diachronous, and strict synchrony can only be expected along the strike. Since most clastic formations are laid down under conditions of transgression or regression, samples taken at right angles to the strike of a formation necessarily differ in age, and lithologic "breaks" commonly do not coincide with floral "breaks."

Horizons Within Formations. The concept that a zone that is recognized on the basis of fossils can be traced laterally is implicit in paleontologic work. Such a zone is a representation of biologic time. The zone thus defined may be one of many that are recognizable within a single rock unit. Extensions of these zones may occur in different formations at different localities. Formations transgress time, and, conversely, time horizons intersect formations.

Rock units are mappable lithic entities. The time required for their deposition at a specific locality may be short and may encompass only one or a few biozones (e.g., Cretaceous Fox Hills Sandstone from Western United States) or it may be relatively long and encompass several biozones (the Cretaceous Pierre Shale from Western United States). Some formations or parts thereof may be barren of palynomorphs, and consequently the number of recognizable zones within a given formation cannot be estimated a priori.

Age Determination

Palynological age determinations must depend for verification on dating methods other than the use of pollen and spores. The science of palynology is too young, and palynologists as yet have examined too little of the stratigraphic column, for age determinations to be made very often by means of palynomorphs alone. This limitation is shared to a lesser degree, perhaps, by paleontologists working with other fossil types. Through the years certain reliable age criteria have been established; for example, some vertebrate fossil species have been determined to be confined to a limited stratigraphic interval. Evolutionary stages preceding and succeeding these fossil species are well known. With the complete background knowledge available the presence of fossil bones of one of the species in a rock stratum serves to date that stratum. If a palynomorph assemblage is obtainable from rocks that are well dated by vertebrate or other well-established fossils, then these other fossils serve to date the plant-microfossil assemblage. Later this palynomorph assemblage can serve as a basis for other palynologic dating.

Even though every geologic system contains certain genera or species with short stratigraphic ranges, the stratigraphic value of the various biologic groups that occur in the several systems differs greatly. A group that produced short-ranging genera or species in one system may produce only long-ranging forms in another system. This principle is illustrated by Teichert (1958, p. 105) with a chart showing the relative biochronologic significance of the major groups of marine invertebrates during geologic time. He shows that, although graptolites persisted from Cambrian through Pennsylvanian time, their principal usefulness for worldwide correlation and zonation is limited to the Ordovician and Silurian Periods. Similarly trilete spores have persisted from the Middle Silurian to the present. Their principal usefulness, however, is essentially limited to pre-Upper Cretaceous strata.

Age Determination by Stratigraphic Position. In the absence of well-dated comparative material certain beds or stratigraphic intervals can be dated with some measure of precision if the overlying and underlying strata are well dated. A rock unit above a Late Devonian formation, and below a well-dated Early Pennsylvanian formation, and separated from these formations of known age by unconformities, may be of any age from Late Devonian to and including Early Pennsylvanian. On the other hand, a rock unit bounded below by rocks of Late Cretaceous age and above by rocks of latest Cretaceous age must have a Late Cretaceous age designation whether or not it is separated from the other formations by unconformities. A palynologic assemblage from such a rock unit can thereby be dated even though no exact stratum of equivalent date is available for comparison. Furthermore, a rock unit thus indirectly dated can be used as a control for dating of rocks of unknown stratigraphic position or age if these unknowns yield the same palynologic assemblages.

Comparison with Well-Dated Assemblages. Close similarity of fossil content, other things being equal, indicates age identity (redeposition may negate this generalization). Conversely, however, dissimilarity of assemblages is not always indicative of differing ages; for example, middle Eocene plant-microfossil assemblages from northern South America have very few species in common with palynologic assemblages of the same age from the Rocky Mountain region of the United States. The plants of northern South America had different origins or occupied different ecological habitats from those of the North American Rocky Mountain region. As a consequence a different pollen and spore suite was produced and deposited in each of the two regions. (See "Provincialism," p. 117.)

Type- or Standard-Section Concept. A type or control section is often required in palynological work for one or more of the following reasons:

1. Control or dating from stratigraphic position or from other fossils in the section often is not adequate or is questionable.

2. The palynologist commonly deals with rocks that are unfossiliferous except for pollen and spores. Established bases for correlations are therefore nonexistent.

3. Palynologists have not yet accumulated enough basic information concerning assemblages, ranges of species, evolutionary trends, and the affinities of fossil taxa. As a consequence the available data, except for limited floral provinces, permit only gross correlations and age estimates. A control section provides a known reference with which other samples or sections may be compared.

4. Floral provinces have existed in the past as at present. A control section from one part of the world may not yield a flora comparable to a synchronous flora from another part of the world. Therefore type sections from each distinct floral province must be examined.

5. Some formation type sections may not yield plant microfossils owing to factors such as composition of rock (conglomerates or offshore marine rocks), metamorphism, and oxidation; for example, the type Danian section from Denmark is made up of marine rocks that would not be expected to yield a complete pollen and spore flora of Danian age; the type Silurian is metamorphosed and would not yield recognizable palynomorphs; the Fountain Formation at its type locality in Colorado is a reddish conglomerate

with lenses of red or purple siltstone. Such rock types either never contained a characteristic or representative pollen and spore flora or the original flora may have been partly or completely destroyed. Other well-dated standard or reference sections must therefore be chosen.

Interval of Interest. The paleontologist is commonly concerned with only a part of the stratigraphic column at any one time or at any one locality. This part may be termed the interval of interest. This interval may be a single rock unit, a system or period, or even an era. If we are analyzing a formation or stratigraphic section of a known age or are trying to correlate samples from several wells, it may be necessary to know not only the floral changes that take place within the interval of interest but also any changes that may occur above and below that interval.

Floral changes do not necessarily coincide with either rock units or established time boundaries. If we are concerned, for example, with the floral changes that occur within the Paleocene of an individual basin, a knowledge of the local composition of the flora of the lowermost Eocene and the uppermost Cretaceous is absolutely necessary in order to understand these changes. The upper and lower limits of the flora that is characteristic of the interval of interest must be established, because the observed floral zone may extend above, below, or both above and below the interval of interest.

If a pronounced floral change is observed within an interval to be studied, then the top of the preceding (older) floral unit will have been established. The upper limits of the floral unit to be studied must then also be established, this identification requires examination of samples from above the critical interval.

In examining well sections, particularly those of exploratory wells, from areas where the subsurface stratigraphy is poorly known, disconformities representing a considerable time lapse may be confused with diastems. Should the interval of interest be bounded by two disconformities or diastems, then the sampling interval should include horizons above and below the interval of interest. The distance above and below will differ from section to section. Successive samples should be examined until an easily recognizable floral change is encountered. Only if the above procedure is followed can the investigator be sure of the stratigraphic position of the interval with which he is concerned.

Sampling Interval. The sampling interval is controlled by the problem to be solved. If we are trying to establish the floral changes that took place at or near a known time or rock boundary, a few samples on either side of the known boundary usually are sufficient. On the other hand, if we are examining a well section embracing 10,000 feet of Eocene sediments in an attempt to establish correlation zones within such a section, the number of samples may be in the hundreds.

If we are examining a succession of Pennsylvanian coal beds, then a channel sample must be taken from each distinct lithological unit. Such a sample may be a vertical channel representing a shale parting less than 1 inch thick or it may be a composite channel sample representing several feet of coal. Complete sampling is necessary in order that vertical changes in spore and pollen content will not be overlooked and in order that the characteristics of the individual beds and partings may be recorded.

Outcrop Samples. Outcrop samples should be taken from each recognizable rock unit that might produce palynomorphs. In a sequence consisting of sands, shales, and lignite streaks each lignite streak and shale bed should be sampled. Most sand, particularly if clean and porous, does not yield enough palynomorphs to make sampling worthwhile.

Cores. Cores provide the best samples for establishing a control section. If a well has been completely cored, we may then choose the most promising lithotypes for examination.

Cuttings. Well cuttings can be used, but their inherent limitations must be recognized. Some of the limitations include the following:

1. Uncertainty as to the level from which cuttings came. The degree of uncertainty is controlled by the size of the cuttings, by the circulation time, and by the mixing that inevitably takes place. (With special care and utilization of specific-gravity separation coal cuttings have been used in the Illinois basin.) Much of the palynologic work that is being done by oil companies is based on the examination of well cuttings.

2. The phenomenon of caving. This is particularly troublesome in some lithotypes. Lignites and subbituminous coals, as well as some swelling shales, are particularly prone to caving, thereby adding fragments of rock from up-hole horizons to the cuttings derived from the bottom of the hole.

If cuttings are used, the tops of floral zones may be determined with some reliability, provided no attempt is made to tie down such tops to absolute levels. The basal limit of a floral zone is much more difficult, and sometimes impossible to determine, owing to mixing of well cuttings in the circulating drilling mud.

Ranges of species are commonly unreliable because of mixing of cuttings. Vertical range of a species determined on the basis of cuttings will commonly appear to be greater than if the range were obtained from cores.

Correlative Zones. Palynological zones can be no closer than the sampling interval. In a rock section 500-feet-thick, if there are five equally spaced recognizable zones, then at a sampling interval of 250 feet a maximum of two zones may be recognized. If a sampling interval of 100 feet is used, a *maximum* of five zones may possibly be recognized. If a sampling interval of 50 feet is used, all five zones will be easily discerned.

The number of recognizable zones depends on the history of the vegetation within and surrounding a basin and on the rate of sedimentation. The number will of course be different in different regions. If sedimentation and consequent filling of the basin was slow, many horizons may be recognizable in a few hundred feet of rock. If sedimentation was rapid, then there may be no or only slight change in the plant microfossils obtained from thousands of feet of rock. The Venezuelan Tertiary rocks are a good example of thick sediments with few well-marked horizons. In these strata Kuyl, Muller, and Waterbolk (1955) recognized only eight major zones, yet the Eocene alone in some parts of Venezuela exceeds 10,000 feet in thickness.

Concept of Ranges of Genera and Species. Index or guide fossils may be useful, but commonly only within limited areas. The known stratigraphic range of a fossil genus or species, in areas in which intensive work has been carried out, can usually be relied on. It is, however, a hazardous procedure to rely on an index species of one area for stratigraphic evaluation in another.

The concept of index fossils is tenuous and in fact probably is not valid. If a particular index fossil is found above its known range, this occurrence then requires explanation. Either the range of the fossil must be extended or redeposition of the fossil in younger beds must be postulated.

Assemblages or groups of fossils are always more reliable indicators of zones or of time than are individual index fossils. Climatic changes that affect a significant segment of the total plant flora are more likely to produce recognizable and widespread time zones than are other factors that may affect only one element of a flora.

Range Charts. The known stratigraphic distribution of some groups of palynomorphs is shown in figure 7-3. This information is generalized for the reason that it has been derived from world literature. Absolute ranges are not precisely known, and those that are given can be expected to be extended slightly as knowledge increases. Age ranges of any given group may differ from one region to another, particularly in the Cenozoic.

Algal and fungal spores or sporelike structures are known from all periods from the Precambrian to the Quaternary. Other groups of plant microfossils have much more limited stratigraphic ranges. The Devonian-Lower Mississippian group of genera, which possess prominent grapnellike hooked appendages, and the group of bisaccate striate pollen genera, which are limited principally to the Permian and Triassic, are good examples.

Range charts representing only one or two periods are much more difficult to construct on a worldwide basis because of the insufficiency of published information. Such charts have been prepared for Cretaceous rocks by Couper (1964); a range chart of spore distribution for the Upper Carboniferous Namurian and Westphalian stages has been constructed by Butterworth (1964).

Figure 7-4 is a generalized chart, on a somewhat more restricted basis, showing the stratigraphic distribution of some plant groups and genera in post-Triassic strata. This chart is based on known occurrences in North America. In the younger parts of the stratigraphic column particularly, pollen and spore ranges at the species level must be resorted to.

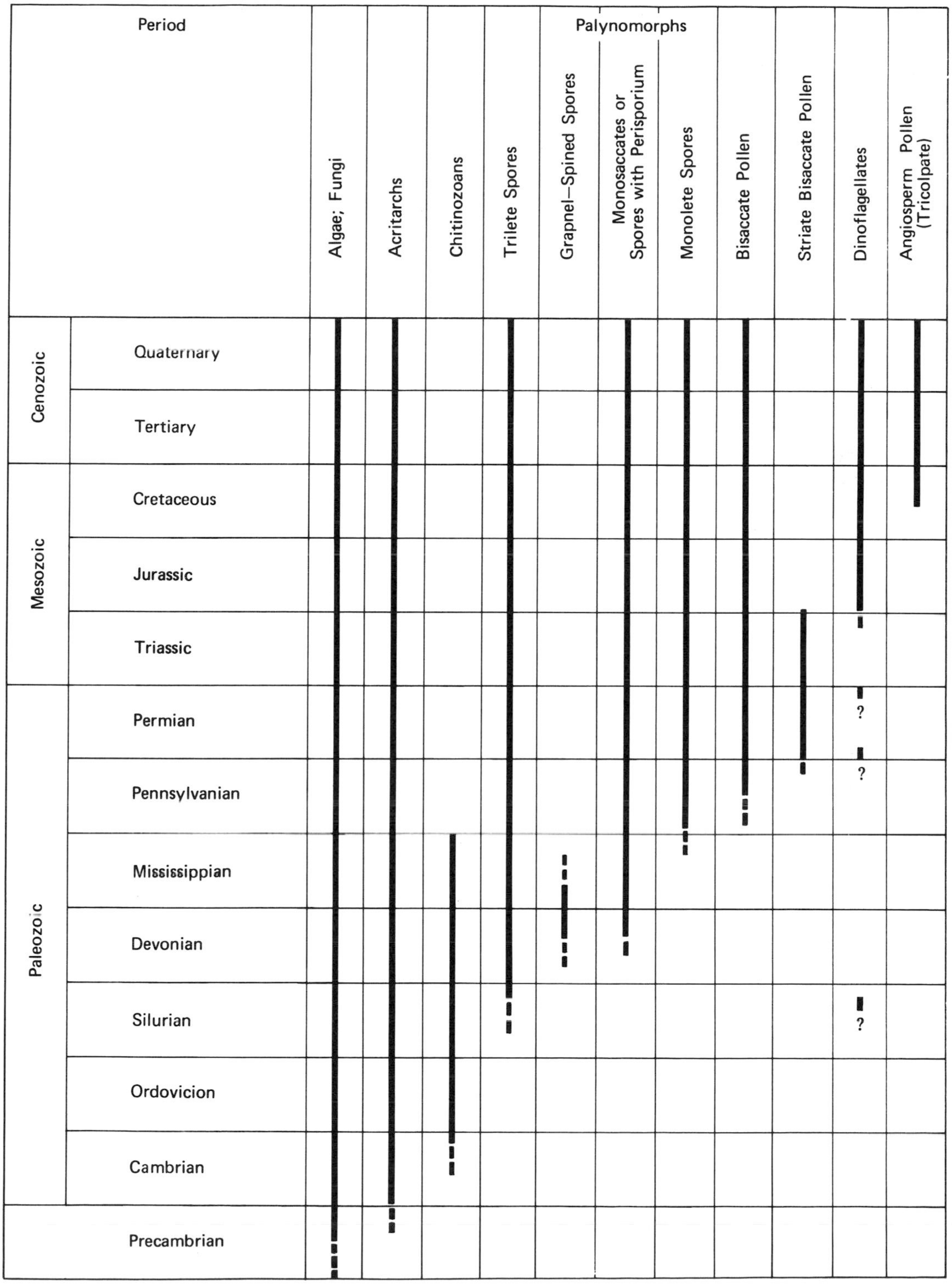

Period		Palynomorphs										
		Algae; Fungi	Acritarchs	Chitinozoans	Trilete Spores	Grapnel–Spined Spores	Monosaccates or Spores with Perisporium	Monolete Spores	Bisaccate Pollen	Striate Bisaccate Pollen	Dinoflagellates	Angiosperm Pollen (Tricolpate)
Cenozoic	Quaternary											
	Tertiary											
Mesozoic	Cretaceous											
	Jurassic											
	Triassic											
Paleozoic	Permian										?	
	Pennsylvanian										?	
	Mississippian											
	Devonian											
	Silurian										?	
	Ordovicion											
	Cambrian											
Precambrian												

Figure 7-3. Stratigraphic distributions of some palynomorph groups.

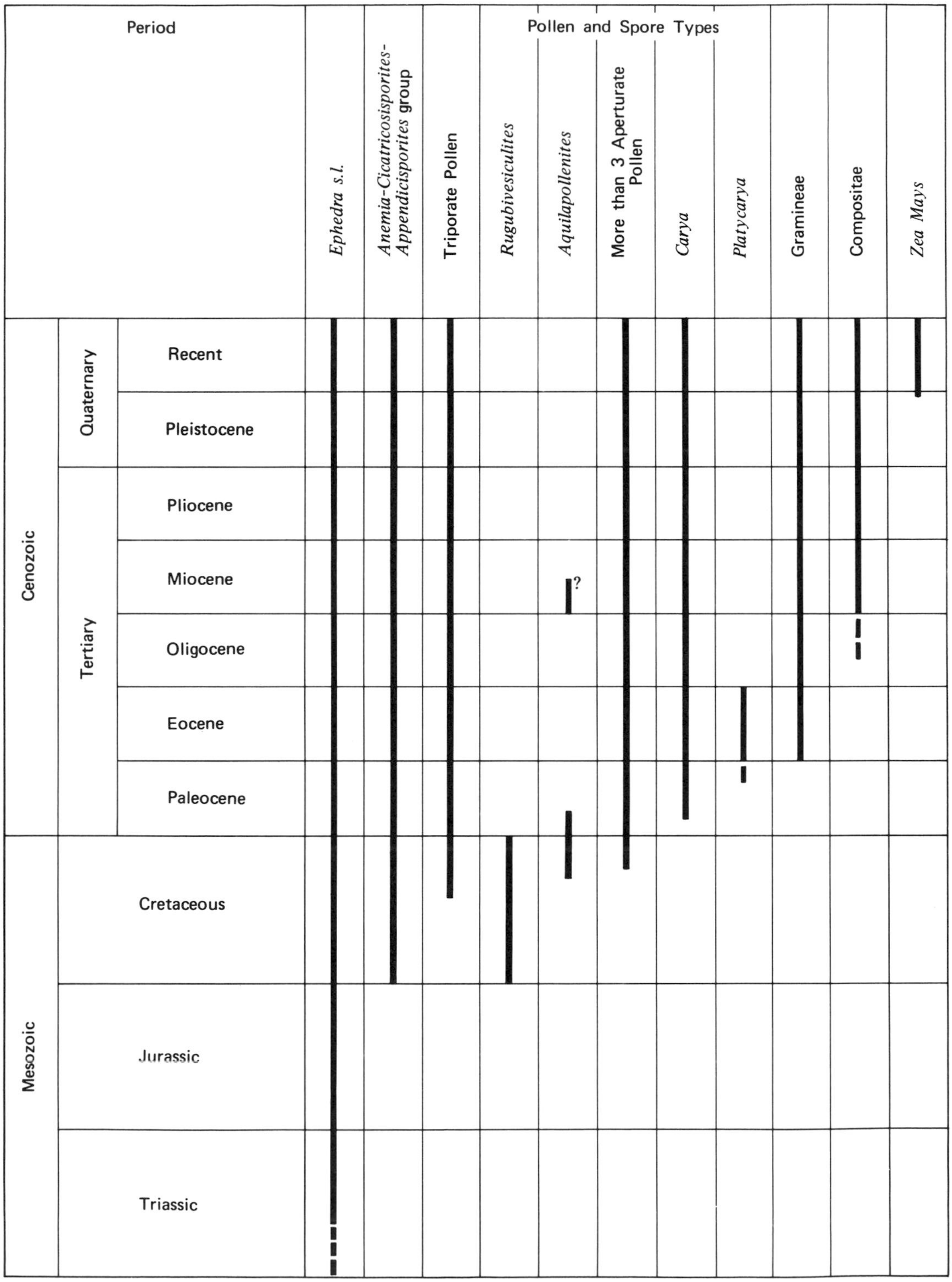

Period			Pollen and Spore Types										
			Ephedra s.l.	*Anemia-Cicatricosisporites-Appendicisporites* group	Triporate Pollen	*Rugubivesiculites*	*Aquilapollenites*	More than 3 Aperturate Pollen	*Carya*	*Platycarya*	Gramineae	Compositae	*Zea Mays*
Cenozoic	Quaternary	Recent											
		Pleistocene											
	Tertiary	Pliocene											
		Miocene					?						
		Oligocene											
		Eocene											
		Paleocene											
Mesozoic	Cretaceous												
	Jurassic												
	Triassic												

Figure 7-4. Stratigraphic ranges of some post-Triassic pollen and spores in North America.

Range charts have been made for parts of the stratigraphic column, but for the most part these are limited to individual studies or individual basins. They are of course extremely useful, but this usefulness is limited to application in a provincial rather than a worldwide sense. *Ginkgo* and *Platycarya* are genera living today in Asia; however, they became extinct on the North American continent in early Tertiary time. As more information is accumulated, more worldwide ranges of specific taxa can be determined.

Provincialism

During Early Cretaceous and pre-Cretaceous time world floras were more uniform than at the present time. Late Devonian floras from such diverse geographical localities as the Russian platform, the Canning basin of Australia, northern Canada, Central United States, Bolivia, and North Africa exhibit remarkable similarities in composition both on the generic and on the specific levels. Such similarities in general persisted from the time plants first invaded the land, presumably during the Silurian to approximately the end of the Early Cretaceous.

The first angiosperm pollen floras of late Early Cretaceous time exhibit great similarities, possibly owing at least in part to their primitive nature. Beginning with the Late Cretaceous floras, provincialism became more noticeable with time. Late Cretaceous floras of Eastern United States and Europe are distinctly different from floras of the same age from the Rocky Mountain region. With increasing floral diversification, by early Tertiary time, floral provincialism was progressively more marked; for example, middle Eocene floras from northern Venezuela are almost entirely different from middle Eocene floras of either the Midcontinent region or the Rocky Mountains in the United States. This diversification and the establishment of floral provinces has continued until the present. Today in North America, north of Mexico, we can recognize a number of floral provinces—namely, Tundra, Northern Conifer, Eastern Deciduous Forest, Coastal Plain, West Indian, Grassland, Cordilleran Forest, Great Basin, Californian, and Sonoran (Gleason and Cronquist, 1964).

Floral provincialism makes necessary, particularly in the younger part of the stratigraphic column, the establishment of control sections for each province or basin concerned. A Miocene control section established in Washington State probably would be nearly worthless as a basis for correlation of Miocene rocks from Florida. The palynologist should therefore obtain as much detailed control as practical before making determinations in an unfamiliar area.

Application to Other Geological Problems

Palynology may be applied to all problems amenable to solution by the use of fossils, assuming that palynomorphs can be obtained from the rocks in question. Synchrony of parts of some interfingering formations may be demonstrated. The fact that pollen and spores are deposited at the same time in contiguous or even separate continental and marine beds may permit time lines to be drawn across boundaries between marine and continental facies. Few other fossils can be utilized in this manner. Pollen and spores carried by wind and water can be simultaneously deposited in continental swamp or deltaic sites and in both brackish and wholly marine depositional basins, thus providing time markers across extremely varied depositional facies.

The diachronous nature of a channel sand and its matrix can be recognized in many places by the fact that the plant fossils present in the sand are younger than those in the adjacent rock. This recognition is particularly useful in subsurface investigations in which the channels are not visible as such and can be recognized only by their dissimilar fossil content.

Many disconformities can be distinguished from diastems by the pollen assemblages above and below the zone of interrupted deposition. Also, in some places disconformities can be recognized in rocks that do not clearly show physical evidence of an erosion surface. In southern Wyoming and in the Mississippi Embayment region it is difficult to separate by lithology Paleocene rocks from underlying Cretaceous rocks. In some parts of these areas no conglomerate or other evidence of an erosion surface is recognizable, yet the fossil pollen floras will distinguish Paleocene rocks from Cretaceous ones.

Studies related to the origin and development of some natural resources are particularly amenable to the palynological approach. The correlation of coal seams presents some special problems. Coal and associated strata may, but commonly do not, contain types of fossils other than pollen and spores. Therefore correlation of coal seams devolves almost entirely on the results obtained from the examination of the spores and pollen that they contain. The areal extent of an individual coal seam, whether it divides laterally into two or more seams, or pinches out and is replaced by another apparently identical seam—the knowledge of all of these aspects is necessary to the evaluation of the commercial possibilities of the bed in question and to the calculation of reserves. Such problems are being studied in many of the world's more important coal provinces.

In the exploration for oil, palynology is also playing a part. In exploratory wells the successive strata often can be dated palynologically and thereby can provide information about the depositional history of the basin. After a control well has been examined correlative horizons can be established on the basis of similarities in pollen-and-spore assemblages. This information, integrated with other data, provides knowledge to guide further drilling. Correlation diagrams involving several wells may indicate the direction in which to look for stratigraphic traps or other types of oil reservoirs. This tool has become so important in the oil industry that virtually all major companies have added palynology to their paleontological arsenal. The contributions that palynology can make in this field alone are limited only by time, cost, and a lack of trained personnel.

PALEOECOLOGY

Paleoecology is the part of paleontology that is concerned with the relationships of plants and animals of the geologic past with their habitats and environments. That the environments of the past have been varied and different from those of the present is beyond question. Paleoecology deals with the interpretation of these past environments by means of evidence from the fossil record.

Environmental interpretations based on palynomorphs comprise only a small part of the field of paleoecology. Pollen and spores, microscopic but vital elements in the life histories of the plants they represent, do no more than suggest the life form of the parent plant and are not necessarily found at the locality at which the parent plant grew.

Limitations

The general limitations common to the occurrence and interpretation of biological evidence in paleoecology have been summarized by Ager (1963) and are repeated below.

Limitations of Fossilization

1. The remains of most organisms are removed after death by scavengers, saprophytes, and bacterial decay.

2. The parts of an organism which do escape these processes are still very unlikely to be preserved as fossils.

3. Those that are preserved are likely to be removed by later erosion.

4. Only a minute proportion of those that survive are likely to be exposed at the surface or come to the notice of the paleontologist.

5. Only certain environments are at all likely to be represented in the rocks.

6. Many, if not most, fossils are preserved in environments other than those in which they lived.

7. It is impossible to make provable deductions about the ecology of organisms which are now extinct.

8. The uniformitarian approach of interpreting past environments in the light of modern ones may not be valid, since geographical factors may have changed.

9. Similarly, the organisms themselves may have changed their habits and habitats with time.

Most of these limiting factors apply to paleoecology based on pollen and spores; for example, almost all, if not all, of the Normapolles group of pollen grains (see chap. 2) prevalent in the Late Cretaceous are extinct. Not only are the ecological conditions under which the parent plants grew unknown to us but also we have no clear idea about the physical appearance or phylogenetic relationships of these plants. Ecological inferences based on this group are at present impossible.

That the plants themselves may have changed

in their ecological requirements with time must be seriously considered. The genus *Ephedra* today is confined to dry, semidesert localities. Pollen indistinguishable from modern pollen of this genus is known in the fossil record at least as early as the Triassic (Scott, 1960). Is it possible to assume that the genus *Ephedra* has occupied the same ecological niche since the Triassic? If we make such an assumption, then the discovery of *Ephedra* in some Tertiary rocks poses problems. To explain these apparent discrepancies additional speculation is often utilized; for example, Gray (1960) stated: "To account for the presence of the Claiborne ephedras in an environment of high atmospheric humidity it is possible to speculate that local conditions of edaphic aridity may have existed in highly isolated sites, on shifting sands, beach dunes, or sand flats adjacent to the coast, without surmising that this species had environmental requirements entirely different from modern ones. However, it seems as plausible to question the assumption that early Tertiary ephedras were physiologically as fully differentiated as the living species. Thus the tentative suggestion may be put forward that the Claiborne species did not have the limiting adaptive relations of living ephedras, although it may have been confined to drier, better drained sites where it survived largely because of limited competition with associates."

Although Gray mentioned two alternative hypotheses, the last clause of the above quotation is pure speculation, based on no information obtained from the fossil *Ephedra* or the pollen assemblage associated with it. If difficulty of interpretation is encountered with *Ephedra* in the Eocene (Claiborne group), how much more likely is it to encounter difficulty with the same genus in the Triassic!

Extreme caution in drawing ecological conclusions on plant microfossils from Tertiary and older rocks was advised by Kraüsel (1961). He cited the fact that to date studies have almost entirely been confined to applications to stratigraphy. He further emphasized "chaotic nomenclature." To a certain degree he is correct in his interpretation of nomenclature. The placing of a fossil pollen grain in a form genus with a name such as *Eucommiidites* is no reason to assume relationship with the modern *Eucommia*. For this reason form genera and most organ genera are of no value for ecological interpretation. In modern species of pollen and spores there is notable evidence of convergence. Morphological features may be common to several different families. When we find a fossil pollen grain of simple form with few distinguishing features the likelihood of determining to which natural genus it belongs becomes remote.

When dealing with assemblages of dispersed spores and pollen of Recent or near-Recent age palynologists have been able to do a remarkable job of reconstructing past climates and past plant communities. As we proceed back into the early Tertiary, where it is almost impossible to equate fossil plants at the species level with modern plants, the problems are multiplied; and when we are dealing with the Paleozoic very few conclusions of an ecological nature will stand critical evaluation.

Paleoecological Inferences

In spite of the problems referred to in the preceding cautionary words, paleoecological interpretations based on palynomorphs have been made and will continue to be made. Obviously many of the interpretations present in the literature are invalid, others are little more than suggestions, and still others are in all probability as valid as paleoecological conclusions based on any other group in paleontology. Furthermore, as knowledge increases in this comparatively young and rapidly growing field, the number of valid inferences will continue to grow.

Inferences from Kinds of Fossils. Families and genera known to be limited to restricted ecological conditions are rare; nevertheless, when such fossils are found careful inferences or conclusions based on them may be sound. All members of the Nymphaeaceae are water plants, for example, as are the ferns belonging to the Marsiliaceae and Salvineaceae. The Rhizophoraceae (mangroves) occupy a limited ecological niche at or just above the tidal zone in the tropics and subtropics. The Typhaceae, Naiadaceae, and Pontederiaceae occupy lacustrine, fluviatile, or paludal environments. Fossils belonging to all these families occur in Cretaceous rocks (Tschudy, 1961). Dinoflagellates, acritarchs, microforaminifera, silicoflagellates, and radiolarians for the most part signify a marine or, at least, a brackish environment of deposition. In using individual families or genera for interpretations the problem of exceptions must constantly be borne in mind. There are species of cactus that are adapted to nondesert localities and orchids that grow only in near-desert localities.

Inferences from Association of Fossils. Inferences based on fossil associations, especially in Tertiary and older rocks, are much more reliable than those derived from single species.

A study of recent sediments in the Orinoco delta exemplifies the utilization of groups of palynomorphs to elucidate problems of sedimentation as well as of ecology (Muller, 1959). Such work can now be directly applied to older sediments in the Orinoco Basin.

Assemblages from older rocks have provided abundant examples from which paleoecological inferences have been drawn. The nagging question often remains whether the conclusion is being drawn from the assemblage itself or whether the assemblage is being manipulated to strengthen or support a previous conclusion.

Inferences from Combination of Lithology and Fossils. An example of this kind of paleoecological inference may be seen in the work of Hacquébard, Cameron, and Donaldson (1965) and Smith (1962) in relation to some Paleozoic coals of Canada and England. Spores of the *Lycospora* type are known to have been produced by arborescent lycopods, and spores of the *Densosporites* and *Punctatisporites* types to have been produced by herbaceous lycopods. At least four types of plant associations were recognized. The coals derived from the different associations are distinct petrographically. These four types or successive phases may be described as follows:

1. An initial phase of peat formation, which probably took place in shallow stagnant water. The surrounding vegetation was a forest and gave rise to the *Lycospora* phase.

2. A gradual change in water level, possibly accompanied by a climatic change, which resulted in the gradual replacement of the forest by a more open vegetation. This Smith refers to as the transition phase.

3. A phase characterized by few species, with *Densosporites* dominant, that may be attributable to high precipitation and humidity. Decomposition proceeded further than in other environments, which suggests aerobic, shallow water or absence of surface water.

4. A return to climatic conditions of previous phases, which could have been brought about by subsidence, creating shallow stagnant water, decreased decomposition, and more anaerobic conditions.

These successions may have been responsible for two general results: (a) a long-lasting change due to change in ground-water level caused by either differential subsidence, or by long-range climatic changes, or both and (b) a superimposed change of shorter duration, which produced either flooding or desiccation but affected the vegetation in a minor manner, not drastically altering it.

Inferences To Be Derived from the Characteristics of the Fossils. These inferences are based on the morphology of the fossil spores or pollen grains and include such features as the presence of thick or thin walls and the distributive mechanisms inherent in the fossils themselves. On the basis of comparative data on the morphology of contemporary and fossil spores Naumova (1953) drew the following tentative conclusions:

1. Small spores possessing a thin exine and simple sculpture pertain to a hydrophytic, aquatic, and littoral aquatic flora.

2. Spores with a thick exine, with a perisporium, or with complex sculpture are characteristic of a land flora.

Float mechanisms, such as those on fossil *Azolla* spores, point to an aquatic habitat like that occupied by modern *Azolla* species.

Inferences from adaptive mechanisms (so-called) such as the "wings," or sacs, on conifers have been made. The "wings" of modern pines may be adaptive mechanisms developed primarily to protect the furrow, or sulcus, against excessive desiccation. The wings or bladders of the Pennsylvanian *Endosporites* or *Alatisporites* spores, however, had no such function. It is more probable that these structures were developed as "organs of flight," making the spores more readily transportable by wind. This information is of little ecological assistance; wings on spores do not contribute to the establishment of the ecological requirements of the parent plant.

Size and Sculpture. Modern pollen grains distributed by wind are commonly small and light and have a minimum of sculptural modifications (e.g., *Corylus*). Many modern pollen grains distributed by insects are highly sculptured, large grains, some even produced in masses (e.g., Oenotheraceae and some orchids). On the basis of size and sculpture we may conclude that a fossil pollen species was probably adapted to distribution either by wind or by insects. Entomophily, pollination by insects, is more common in tropical humid or rainy climatic conditions than is anemophily, pollination by wind. Airborne pollen is constantly washed out of the humid tropical air by rain; insect pollination under such conditions is a more effective fertilizing mechanism. Distribution of pollen by wind on a large scale is chiefly confined to temperate and cool climates.

Form and Organ Genera. In mid-Tertiary and older rocks it may be possible to assign fossil pollen to extant families, but only occasionally does the evidence support their assignment to extant genera. Some palynologists regard it as a dictum that palynomorphs older than late Tertiary should not be placed in modern genera. Consequently most systematic assignments of fossil pollen are made to form genera or organ genera.

This fact places a severe, if not commonly observed, limitation on paleoecological interpretations. In many palynological studies not a single form is assigned to a modern genus, yet in the paleoecological interpretations from these studies inferences are drawn from the ecology of the purported modern equivalents as though the identifications were established. This practice erroneously assumes that physiology is a more persistent and reliable attribute of plants than morphology. Names such as *Quercoidites*, *Nyssoidites*, *Faguspollenites*, *Ulmipollenites*, and *Juglanspollenites* seem to carry with them an implication of affinity to *Quercus*, *Nyssa*, *Fagus*, *Ulmus*, and *Juglans* that leads even careful authors to disregard the limitations that they acknowledged in originally choosing these terms. This pitfall in logic should be avoided in making paleoecological deductions.

Inferences To Be Drawn from Identity with Living Genera or Species.

Identity of fossil pollen with pollen from extant genera and species of plants is the most reliable basis for paleoecological interpretation. We know where modern plants grow. Inferences derived from identity of fossil with modern plants, however, are limited in value by the fact that they can be applied only to relatively young fossils.

A few examples of paleoecological inferences from modern plants or plant groups may be pertinent. Members of the Gramineae, if identified with certainty and if they occur in abundance, signify nearby grasslands. Juncaceae is a family that is composed almost entirely of aquatic or semiaquatic members. Members of the Droseraceae are limited to boggy or swampy regions. Many members of the Chenopodiaceae are common inhabitants of dry, open localities; however, exceptions such as *Beta vulgaris* are known. On the species level *Sarcobatus vermiculatus* pollen is indicative of saline, arid conditions.

The distribution of families, genera, and species of plants and the climatic and edaphic factors governing their distribution are discussed by Good (1953) in his "Geography of the Flowering Plants." This book is an excellent starting point for the interpretation of fossil-pollen floras in the light of what is known about the distribution of contemporary floras. An example of such interpretation is the Palmae, a family whose members are at present all limited to the tropical and subtropical zones. The discovery of fossil palms in southeastern Alaska (Hollick, 1936) seems to suggest a much warmer climate in Alaska during Tertiary time than at present. A second example is the Proteaceae, which at present is exclusively a tropical and south temperate family. However, the fossil-pollen genus *Proteacidites*—commonly associated with the Proteaceae on morphological grounds—has been reported from the Cretaceous of the United States, Canada, and Siberia, which suggests that the distribution of the family has been more widespread in the past than at present. On the basis of these facts what paleoecological conclusions, if any, can be drawn from the presence of fossil *Proteacidites* pollen?

Nothofagus is at present a genus confined to the south temperate zone. However, Zaklinskaya (1964, p. 85) has found Cretaceous records of it from the south of Eurasia, Kazakhstan, and Siberia, and I have found pollen grains that are morphologically similar to those of *Nothofagus* in both the Cretaceous of the Rocky Mountain region and the Tertiary of the Mississippi Embayment region. Until more is known about the past distribution of *Nothofagus*, to assume that this genus always has lived in the ecological niche it now occupies would be premature and might lead to erroneous conclusions.

A strong plea for meticulous care in the identification of fossil pollen with modern taxa was made by Kraüsel (1961), who said: "It is not enough to show that a fossil-pollen form is similar to a certain modern form, one must also know that it is identifiable with one *only*." This requirement can in many cases be adequately met by careful and extensive comparisons of grains of fossil pollen with those of modern pollen. An example is the study of pollen from the Brandon lignite by Traverse (1955), who concluded that the Brandon flora has no exact counterpart in North America today but that it is closely similar to the modern swamp vegetation of Florida. A careful comparison of modern pollen from the Florida swamp plants with the Brandon material provided a firm basis for significant paleoecological determinations.

Paleoecological Studies

Recent and Near-Recent. Many studies have been made dealing with the course of forest history across large areas and therefore with the fluctuations of climate to which such changes in forested areas are attributed. On the rather firm basis that the present is a key to the past, particularly the near past, Pleistocene climatic changes have been documented. The glacial epochs in Europe have been correlated with supposed synchronous events in North America. Particularly in the last few decades the dating of such events in forest history has been refined by numerous carbon-14 dates.

One of the many studies that have dealt with paleoecological implications was by Sears and Clisby (1955). It had as an objective the obtaining of the record of climate changes by means of an analysis of pollen from sediments of the Mexico City basin. The authors stated that in order to interpret the climatic, tectonic, volcanic, and biotic factors that influence the sedimentary complex, evidence from every possible source is required. In this study the chief palynologic determinants were as follows:

1. Climatic changes, controlled by available moisture, are reflected in the percentage composition of upland forest pollen.
2. Changes in sedimentary zones and in water content of the sediments are reflected by the abundance of pollen recovered.
3. Changes in sedimentation rate are reflected by pollen abundance; low pollen recovery is associated with rapid sedimentation and vice versa.
4. Habitat disturbances caused by volcanism, tectonic activity, and human occupation are reflected by changes in pollen from lowland-swamp plant communities.

The conclusions reached are that data from the Mexico City cores indicate a series of oscillations between moist and dry reaching as far back as early Wisconsin time. These oscillations were correlated with the advance and retreat of the ice fronts during the Wisconsin glaciation; as a result this forest history embraces about 35,000 years.

Details of near-Recent history are also described in studies by Faegri and Iversen (1950). In addition some of the problems in interpreting the significance of data obtained by analysis of Recent and near-Recent pollen assemblages are discussed by Davis (1963).

Cenozoic. Fossil-pollen flora obtained from Eocene rocks in Alabama was described by Gray (1960). The pollen genera she reported amplify the list of genera known from plant megafossils. On the basis of the combined floras she was able to postulate that the flora was mixed, consisting of a plexus of genera representing a tropical lowland flora and several genera confined to temperate latitudes. This led her to suggest that the pollen of temperate plants came from an upland community, much like the present forested area of the same region, and that a subtropical flora grew in the lowlands and along the coast. However, she did not rule out the possibility that some of the genera may have changed their climatic requirements through time.

Mesozoic. Kraüsel (1961) said: " . . . with the earliest Tertiary the study of paleoclimatology based on plant taxonomy ends, with but few exceptions." The flora extant during preangiosperm times gives us few clues concerning its climatic requirements. Although we may know the climatic implications of the living *Ginkgo*, the fossil Ginkgoales of the Mesozoic are too far removed from the present to allow us to make dependable inferences based on their pollen alone.

Although it is easy to prove that climatic conditions in Cretaceous time, for example, were warmer in Greenland than they are at present, such conclusions come from the presence of coal and the identification of megafossils rather than from the identification of spores and pollen. The plant microfossils from a coal on Disko Island, off the west coast of Greenland, showed no angiosperm fossils other than questionable angiosperm cuticles (Miner, 1932, 1935). The presence of megaspore genera, *Arcellites* and *Ariadnaesporites*, originally identified as *Selaginellites*, was reported. These megaspores probably belong to the Hydropteridae, or water ferns. If so—and their morphology supports this hypothesis—then on the basis of the ecological conditions required by their modern relatives some ecological inferences can be postulated. These genera, however, do not tell us much about the climate of Greenland at the time of their deposition, other than that conditions were probably much more temperate than they are today.

Paleozoic. If it is difficult to derive paleoecological inferences from the late Mesozoic, it is much more difficult to derive them from the Paleozoic. Nevertheless, a few conclusions can be drawn, as indicated by results such as those of Smith (1962) and Hacquébard, Cameron, and Donaldson (1965) mentioned in the section "Inferences from Combination of Lithology and Fossils."

Spores of *Tasmanites* are probably of marine origin. An association of *Tasmanites* with Paleozoic marine acritarchs and abundant Paleozoic spores would permit the inference that deposition took place near shore in a marine basin, but nothing could be deduced concerning the temperature of the sea or the climate on the adjacent land.

According to Chaloner (1959), most of the large Carboniferous megaspores are known to have been produced by arborescent lycopods, and at least some of the smaller Carboniferous megaspores to have been borne by herbaceous lycopods. Chaloner noted a fourfold decline in the mean size of megaspores from the Carboniferous to the Late Cretaceous, and he postulated that this decline may have been correlated with the decline in the arborescent lycopods, whereas the herbaceous lycopods persisted into the Mesozoic. Such an inference tells little about climate or ecology, but it does suggest a change in the appearance of the landscape.

Applications

An examination of the palynomorphs in a rock may be resorted to in order to assist in determining the facies of deposition—whether marine, continental, paludal, or deltaic. As more background information is collected more data of this type will be forthcoming. Data by Hoffmeister (in Woods, 1955) showed the abundance of pollen and spores per gram of sediment. The values ranged from above 25,000 per gram near shore to less than 5,000 per gram 40 miles offshore and less than half that number at a distance of 80 miles. With such data as a basis the character of near-shore and offshore facies can be postulated.

In many Recent and near-Recent deposits reasonable postulates concerning causes of floral succession and extinction of elements of the floras have been made. As more information is developed acceptable paleoecological postulates probably will be developed for the Tertiary, and perhaps even for the Mesozoic and Paleozoic.

References

Ager, D. V., 1963, Principles of paleoecology—an introduction to the study of how and where animals and plants lived in the past: New York, McGraw-Hill Book Co., Inc., 371 p.

Arnold, C. A., 1947, An introduction to paleobotany: New York, McGraw-Hill Book Co., Inc., 433 p.

Blackwelder, Eliot, 1909, The valuation of unconformities: Jour. Geology, v. 17, p. 289-299.

Butterworth, M. A., 1964, Miospore distribution in Namurian and Westphalian: Internat. Cong. Stratigraphie et Géologie du Carbonifère, 5th, Comptes rendus, p. 1115-1118.

Chaloner, W. G., 1959, Devonian megaspores from Arctic Canada [Northwest Territories]: Palaeontology, v. 1, pt. 4, p. 321-332.

Couper, R. A., 1958, British Mesozoic microspores and pollen grains; a systematic and stratigraphic study: Palaeontographica, v. 103, sec. B, no. 4-6, p. 75-174.

______1964, Spore-pollen correlation of the Cretaceous rocks of the Northern and Southern Hemispheres, p. 131-142 *in* Cross, A. T., ed., Palynology in oil exploration—a symposium, San Francisco, 1962: Soc. Econ. Paleontologists and Mineralogists Spec. Pub. 11.

Davis, M. B., 1963, On the theory of pollen analysis: Am. Jour. Sci., v. 261, no. 10, p. 897-912.

Dorf, Erling, 1940, Relationship between floras of type Lance and Fort Union formations: Geol. Soc. America Bull., v. 51, no. 2, p. 213-235.

Eaton, J. E., 1929, The by-passing and discontinuous deposition of sedimentary materials: Am. Assoc. Petroleum Geologists Bull., v. 13, no. 7, p. 713-761.

Erdtman, Gunnar, 1943, An introduction to pollen analysis: Waltham, Mass., Chronica Botanica Co., 239 p.

Faegri, Knut, and Iversen, Johannes, 1950, Text-book of modern pollen analysis, 2d ed., revised: Copenhagen, Munksgaard, 237 p.

Gleason, H. A., and Cronquist, Arthur, 1964, The natural geography of plants: New York, Columbia Univ. Press, 420 p.

Good, R. D., 1953, Geography of the flowering plants, 2d ed.: New York, Longmans, Green & Co., Inc., 452 p.

Gray, H. H., and Guennel, G. K., 1961, Elementary statistics applied to palynologic identification of coal beds: Micropaleontology, v. 7, no. 1, p. 101-106.

Gray, Jane, 1960, Temperate pollen genera in the Eocene (Claiborne) flora, Alabama: Science, v. 132, no. 3430, p. 808-810.

Grayson, J. F., 1960, Palynology as a working tool: Oil and Gas Jour., v. 58, no. 17, p. 136-140.

Guennel, G. K., 1958, Miospore analysis of the Pottsville coals of Indiana: Indiana Geol. Survey Bull. 13, 101 p.

Gutjahr, C. C. M., 1960, Palynology and its application in petroleum exploration: Gulf Coast Assoc. Geol. Socs. Trans., v. 10, p. 175-187.

Hacquébard, P. A., Cameron, A. R., and Donaldson, J. R., 1965, A depositional study of the Harbour seam, Sydney Coalfield, Nova Scotia: Canada Geol. Survey Paper 65-15, 31 p.

Hoffmeister, W. S., 1960, Palynology has important role in oil exploration: World Oil, v. 150, no. 5, p. 101-104.

Hollick, C. A., 1936, The Tertiary floras of Alaska, with a chapter on the geology of the Tertiary deposits, by P. S. Smith: U.S. Geol. Survey Prof. Paper 182, 185 p.

Jekhowsky, B. de, and Varma, C. P., 1959, Essai de corrélation d'après cuttings par voie palynologique simplifiée dans le Tertiare de MB.2 et MC.2: Inst. Français Pétrole Rev., v. 14, no. 6, p. 827-838.

Kosanke, R. M., 1950, Pennsylvanian spores of Illinois and their use in correlation: Illinois Geol. Survey Bull. 74, 128 p.

Kraüsel, R., 1961, Palaeobotanical evidence of climate, chap. 10 *in* Nairn, A. E. M., ed., Descriptive paleoclimatology: New York, Interscience Publishers, Inc., p. 227-254.

Kuyl, O. S., Muller, J., and Waterbolk, H. T., 1955, The application of palynology to oil geology with reference to western Venezuela: Geologie en Mijnbouw, no. 3, v. 17, p. 49-75.

Lang, W. H., 1925, Contributions to the study of the Old Red Sandstone flora of Scotland; I. On plant-remains from the fish-beds of Cromarty: Royal Soc. Edinburgh Trans., v. 54, pt. 2, p. 253-272 [1926].

Miner, E. L., 1932, Megaspores ascribed to *Selaginellites* from the Upper Cretaceous coals of western Greenland: Washington Acad. Sci. Jour., v. 22, nos. 18, 19, p. 497-506.

———1935, Paleobotanical examinations of Cretaceous and Tertiary coals; I. Cretaceous coals from Greenland: Am. Midland Naturalist, v. 16, no. 4, p. 585-625.

Muller, Jan, 1959, Palynology of Recent Orinoco delta and shelf sediments: Micropaleontology, v. 5, no. 1, p. 1-32.

Naumova, S. N., 1953, Sporovo-pyltsevye kompleksy verkhnego devona Russkoi platformy i ikh zanachenie dlya stratigrafii [Spore-pollen complexes of Upper Devonian of the Russian Platform and their meaning for stratigraphy]: Akad. Nauk SSSR Geol. Inst. Trudy [Akad. Sci. USSR Geol. Sci. Inst. Trans.], v. 143, Geol. Ser. 60, 203 p.

Pokrovskaya, I. M., 1958, Analyse Pollenique:[France] Bur. Recherches Géol., Géophys. et Minières, Serv. Inf. Geol., v. 24, 435 p. (French translation).

Scott, R. A., 1960, Pollen of *Ephedra* from the Chinle formation (Upper Triassic) and the genus *Equisetosporites*: Micropaleontology, v. 6, no. 3, p. 271-276.

Sears, P. B., and Clisby, K. H., 1955, Pleistocene climate in Mexico, pt. 4 *of* Sears, P. B., Palynology in southern North America: Geol. Soc. America Bull., v. 66, no. 5, p. 521-530.

Shaw, A. B., 1964, Time in stratigraphy: New York, McGraw-Hill Book Co., Inc., 365 p.

Simpson, G. G., 1960, Notes on measurement of faunal resemblances (Bradley volume): Am. Jour. Sci., v. 258-A, p. 300-311.

Sittler, C., 1954, Palynologie et stratigraphie, principe et application de l'analyse des pollens aux études de recherches du pétrole: Inst. Français Pétrole Rev., v. 9, no. 7, p. 367-375.

Smith, A. H. V., 1962, The paleoecology of Carboniferous peats based on the miospores and petrography of bituminous coals: Yorkshire Geol. Soc. Proc., v. 33, pt. 4, no. 19, p. 423-474.

Teichert, Curt, 1958, Some biostratigraphical concepts: Geol. Soc. America Bull., v. 69, no. 1, p. 99-119.

Thiessen, R., and Stand, J. N., 1923, Correlation of the coal beds in the Monongahela formation of western Pennsylvania and eastern Ohio: Carnegie Inst. Technology Bull. 9, 64 p.

Thiessen, R., and Voorhees, A. W., 1922, A microscopic study of the Freeport coal bed, Pennsylvania: Carnegie Inst. Technology Bull. 2, 75 p.

Traverse, A. F., Jr., 1955, Pollen analysis of the Brandon lignite of Vermont: U.S. Bur. Mines Rept. Inv. 5151, 107 p.

Tschudy, R. H., 1961, Palynomorphs as indicators of facies environments in Upper Cretaceous and lower Tertiary strata, Colorado and Wyoming, *in* Symposium on Late Cretaceous rocks, Wyoming and adjacent areas, Wyoming Geol. Assoc. Ann. Field Conf., 16th, 1961: p. 53-59.

______1964, Palynology and time-stratigraphic determinations, p. 18-28 *in* Cross, A. T., ed., Palynology in oil exploration—a symposium, San Francisco, 1962: Soc. Econ. Paleontologists and Mineralogists Spec. Pub. 11.

Wilson, L. R., 1946, The correlation of sedimentary rocks by fossil spores and pollen: Jour. Sed. Petrology, v. 16, no. 3, p. 110-120.

______1959, A method of determining a useful microfossil assemblage for correlation: Oklahoma Geology Notes, v. 19, no. 4, p. 91-93.

______1964, Palynological assemblage resemblance in the Croweburg coal of Oklahoma: Oklahoma Geology Notes, v. 24, no. 6, p. 138-143.

Woods, R. D., 1955, Spores and pollen—a new stratigraphic tool for the oil industry: Micropaleontology, v. 1, no. 4, p. 368-375.

Zaklinskaya, E. D., 1964, On the relationships between Upper Cretaceous and Paleogene floras of Australia, New Zealand, and Eurasia, according to data from spore and pollen analyses, *in* Cranwell, L. M., ed., Ancient Pacific floras—The pollen story: Hawaii Univ. Press, 114 p.

8

THE FOSSIL-PLANT RECORD

Chester A. Arnold

Fossil plants occur mostly in sedimentary rocks. Marine deposits may contain algae and other forms of sea life, but terrestrial vegetation is preserved in greatest abundance in sediments laid down under nonmarine conditions. Wherever coal seams occur fossil plants are likely to be found. Volcanic activity provides ideal conditions for preservation of plants in large numbers. Lava flows dam streams and form fresh-water lakes that quickly become filled with erosion products of loosely consolidated ash deposits. Many of the best known Tertiary floras, such as that at Florissant in Colorado, were preserved under such circumstances.

All parts of the plant body may be preserved as fossils, but they are usually disconnected from each other. Thus the bulk of our knowledge of paleobotany has been derived from detached organs—leaves, pollen, seeds, or stems. However, the organs that are preserved in the greatest quantities are made up of tissues with the greatest resistance to decay or abrasion. Thus fossil flowers are rare though they are occasionally found, and the same is true of soft fruits and succulent leaves. On the other hand, woody tissues, hard nuts, seeds, cutinized parts such as spores, pollen grains, and leaves of coriaceous texture are often preserved in countless numbers.

Plants are fossilized in several ways. The most familiar types are *impressions*, which are merely imprints left in soft sediments; *compressions*, or *compactions*, in which the plant parts are squeezed flat between layers of compacted sediments but under conditions that arrest decay; *casts*, in which a cavity left by decay of a plant part is secondarily filled; and *petrifactions*, in which some or all of the tissue structure is retained by infiltration with various minerals. Spores and pollen, as they are usually recovered from sedimentary rocks, belong to the compression category. Some plant fossils do not appear to belong to any of these classes, examples being flowers preserved in amber, wood pickled in asphalt or tar, and unfilled cavities left by the decay of stems and roots. Rarely, undecomposed and uncompressed plant parts are found embedded in rock in a mummified condition.

The process of petrifaction is responsible for the preservation of countless tree trunks found in many parts of the world, ranging in age from the Devonian to the Recent. Coal balls, carbonate, and pyritic nodular masses sometimes found in coal seams or roof shales, are other examples of petrifactions. These often contain well-preserved plant parts. Coal balls have been studied for more than a century in Europe, but they remained essentially unknown in North America until the 1920's. Since then they have been collected in large quantities in Kansas, Iowa, Illinois, and Indiana.

Petrifactions are of special value in paleobotanical research because they supply information not revealed in other types of fossils on the internal structure of extinct plants. How petrifaction occurs is not well understood, but it is essentially a process of infiltration of cells and tissues by dissolved mineral matter. It is not one of "molecule-by-molecule replacement" of plant substance, as is often stated in textbooks. Changes do take place in the chemical composition of plants during petrifaction. Volatile substances are expelled and often a residue of free carbon accumulates. Analyses of petrified wood have revealed the persistence of cellulose and lignin, though in proportions that are somewhat different from those found in living woods.

A TIME SCALE BASED ON PLANT FOSSILS

The conventional eras and periods of geological time are based principally on major changes in faunas revealed in the rock succession. *Proterozoic* means the age of earlier animal life. *Paleozoic* in turn means the age of ancient animal life, and *Mesozoic* and *Cenozoic* mean middle and new or recent life, respectively. It has been realized for some time that a chronology based on changes of comparable scope in the plant kingdom would be somewhat different. Such a system was actually devised a few years ago by some German geologists. It establishes five eras but retains the periods of the conventional scheme. The oldest era, the *Archeophytic*, embraces the oldest known rocks up through the early Precambrian. It would include the oldest living things and the simple organs that evolved from them. (See chap. 9, "Precambrian Microfossils."). The succeeding era is the *Eophytic*, which extends from the later Precambrian into the Silurian. This could be called the algal age. Vascular plants, which might have been in existence during the latter part of the Eophytic era, first become recognizable as floras at about the middle of the Silurian, which marks the beginning of the *Paleophytic* era. This begins with the Upper Silurian and continues through the Lower Permian. Within it appeared the early land floras of the Devonian and the Mississippian, Pennsylvanian, and Early Permian floras that followed. By Late Permian time the spread of colder climates and the disappearance of the lush coal swamp forests is everywhere manifest, and this marks the beginning of the *Mesophytic* era, which extends to about the middle of the Cretaceous Period. Then floras marked by the dominance of angiosperms characterize the upper half of the Cretaceous Period, which represents the earliest phase of the *Cenophytic* era, or the era of modern flowering plants. The Cenophytic embraces the Upper Cretaceous and the Cenozoic of the standard sequence. The Mesophytic and Cenophytic thus each began about half a period earlier than the conventional Mesozoic and Cenozoic, evidently due to the fact that plant evolution had preceded changes of corresponding magnitude in animals by approximately half a period. The stimulus to dinosaur evolution might have been major changes in floras during the Permian, just as mammalian evolution received a boost from the Late Cretaceous angiosperms. For obvious reasons the conventional system will be followed in this chapter.

THE DIM PLANT RECORD OF THE ARCHEOZOIC ERA

The fossil record fails to enlighten us as to when, where, or how life came into existence. However, recent biochemical studies, and the fact that the oldest rocks of the earth's crust appear to be composed of sediments laid down under a reducing rather than an oxidizing atmosphere, have led to the belief that the first life was anaerobic and that the dependence of plants and animals on free oxygen is a derived condition. However, plants capable of photosynthesis and the consequent release of free oxygen into the air had certainly come into existence by middle Precambrian time roughly 2.3 billion years ago. At about this time the oldest fossilized organisms were alive. From the middle Huronian Gunflint chert, in southern Ontario north of Lake Superior, Barghoorn and Tyler (1965) found minute objects that resemble colonies of blue-green algae and filamentous objects with attached spores that seem to represent fungi. If these organisms were there when the chert was formed and are not contaminants that found their way into the rock at a later time, which seems unlikely, there is no conceivable limit to the length of time cells with delicate walls may be preserved.

In 1911 the Swedish geologist Sederholm gave the name *Corycium enigmaticum* to some small saclike bodies composed of carbon found in Precambrian phyllites in Finland. Spectroscopic analysis of this carbon showed a high concentration of carbon-12, the isotope prevalent in plants (Rankama, 1948). These small bodies are believed to be ancient plant residues formed from organisms that grew in colonies similar to the living *Nostoc*.

Most of the evidence of life during the Archeozoic is indirect, in the form of precipitates of calcium, iron, or sulfur. In the Belt series of Montana large and distinctly formed reeflike structures show a close resemblance to similar ones formed by blue-green algae of the present day. First studied years ago by Walcott (1914), they have lately been reexamined by Rezak (1957), who has recognized four genera on the basis of form and probable manner of growth.

Extensive graphite deposits in the Adirondack Mountains may be metamorphosed coal beds, and minute objects resembling chains of the iron bacterium *Chlamydothrix* have been seen in the upper Huronian iron ores of northern Michigan and Minnesota. Whether they are remains of organisms or merely flocculent precipitates of inorganic origin has been a matter of considerable dispute.

DEVELOPMENT OF LAND FLORAS IN THE PALEOZOIC ERA

Cambrian, Ordovician, and Silurian

Remains of higher plants are scarce, almost to the point of nonexistence, in the predominantly marine rocks of the earlier half of the Paleozoic Era. There is ample evidence, however, of both calcareous and noncalcareous algae in the Cambrian seas, and the former were directly and indirectly responsible for many of the marine limestones and dolomites of the Cambrian and succeeding periods. The well known *Cryptozoon* reefs near Saratoga Springs, N. Y., are typical manifestations of the activity of these organisms. Charles D. Walcott (1919) described an algal flora associated with a large invertebrate fauna from the Burgess shale of British Columbia. These remains are shiny black films on the surfaces of slabs of hard dark shale. Of these *Marpolia* and *Morania* resemble certain living blue-green algae. *Yuknessia* may be a green alga, and several others resemble some of the living red algae.

An axis bearing small, simple, leaflike appendages from the Middle Cambrian of Siberia was named *Aldanophyton antiquissimum* by Kryshtofovitch (1953). Externally the plant resembles a herbaceous lycopod and has been cited as evidence of the great antiquity of the lycopod line. It is indeed the oldest known vascular plant if it has been correctly interpreted, but there has been some question concerning its identity since no tissue structure is visible in it.

Naumova (1949) described 12 types of cutinized spores belonging to the *Triletes* category from the Lower Cambrian of the Soviet Union. These spores have warty exines and triradiate tetrad scars. Some of the largest may be megaspores and the smallest, microspores. They resemble some of the vascular plant spores found in Devonian and Carboniferous rocks, but no Cambrian plants are known that could have borne them.

The Ordovician seas supported rich algal floras that supplied ample food for the many forms of invertebrates and primitive fishes that appeared during that time. An algal genus once regarded as an index fossil of the Ordovician is *Girvanella*, which, however, is now known to range from Late Cambrian to Jurassic. Forms showing characteristic features of the Codiaceae, Dasycladaceae, and Solenoporaceae appeared during the Ordovician.

The algal floras of the Ordovician seas persisted into the Silurian. The latter was a time of rapid development of calcareous green algae. An enigmatic plant that appeared in the Silurian was *Prototaxites*. It is preserved as silicified and occasionally calcified trunklike organs, sometimes as much as 3 feet across, that show evidence of irregular branching. The structure is unique, consisting of a ground mass of small, thin-walled, septate, branched, interwoven hyphae. Among these smaller strands, and for the most part extending vertically among them, are conspicuously larger, thick-walled, nonseptate tubes. Then scattered throughout are spaces filled with irregular thin-walled elements that are usually poorly preserved. In *Prototaxites logani*, the type species, this tissue is distributed as radially elongated masses that bear a crude resemblance to medullary rays.

Whether *Prototaxites* is an algal organism related to the brown seaweeds or a plant of some higher order is a matter of considerable difference of opinion. It is not a vascular plant, though it might have lived on land.

Aside from the rather strong possibilities that the Middle Cambrian *Aldanophyton* is a vascular plant, the oldest plants of this category come from the Middle Silurian. In 1935 Lang and Cookson described *Baragwanathia longifolia* and two species of *Yarravia* from the lower Ludlow beds of Australia. *Baragwanathia* bears considerable resemblance to an herbaceous lycopod, more so, in fact, than most of the lycopods of the succeeding Devonian floras. *Yarravia* is a leafless plant that has been placed in the Psilophytales. The Australian plants are the oldest with vascular tissues preserved. A few more vascular plants have been found in the Upper Silurian of Australia.

Devonian

In the Devonian emphasis shifts from the predominantly marine algal floras to land floras composed of vascular plants. This transformation was already under way during Silurian time, but too few Silurian land plants are known to justify the use of the term "floras" for them. It is not implied that algae are unimportant in the Devonian biota, but the significant changes of that period were the spread and diversification of land plants. For a long time the Devonian was regarded as the period in which land floras appeared on the earth.

Since a full treatment of Devonian land vascular floras would require a lengthy document, only some features are referred to here. The floras of the Lower and Middle Devonian were formerly referred to as the *Psilophyton* flora; and that of the Upper Devonian, as the *Archaeopteris* flora. Not only have the two floras been shown to overlap but *Psilophyton* is a rather rare plant that is often difficult to identify.

The Lower Devonian rocks have yielded several *Psilophyton*-like plants that include *Cooksonia, Zosterophyllum,* and *Hostimella*; lycopods referable to *Drepanophycus* and *Protolepidodendron*; the early sphenopsid *Protohyenia*; and stems bearing fan-shaped leaves known as *Psygmophyllum. Prototaxites*, and *Nematothallus*, though not in the vascular-plant category, were in existence.

A more diversified flora is found in the Middle Devonian. *Psilophyton*, described by Dawson 100 years ago, from the Gaspé sandstone, makes its appearance then. One of the top-ranking paleobotanical discoveries of the century was that of the silicified plants in the Rhynie chert in the Old Red Sandstone in Aberdeenshire in Scotland. A noteworthy feature of these plants, which were thoroughly described by Kidston and Lang (1917-1920) in a series of papers, is the completeness with which they were preserved. This is one of the rare instances in which whole fossil plants could be described. Associated with the three genera of vascular plants—*Rhynia, Horneophyton,* and *Asteroxylon*—was a considerable assortment of algal and fungal remains.

The lycopods are especially well represented in the Middle Devonian by several genera, some of which are *Archaeosigillaria, Colpodexylon, Leptophloeum,* and *Haplostigma. Calamophyton* and *Hyenia* are two early sphenopsids, the latter probably having evolved from the Lower Devonian *Protohyenia.* Fernlike plants are forecast by *Protopteridium*, known from compressions, and by *Arachnoxylon, Reimannia*, and *Iridopteris*, which are preserved as small pyritized axes. These four forms may be considered "psilophytic ferns" because they show evidence of connections with both groups.

Aneurophyton, from the Middle Devonian, is of questionable position among vascular plants. It reproduced by spores, and the small, veinless leaves were borne on the ultimate ramifications of much branched stems. Its habit was treelike, and the remains occur rather widely in Europe and North America. The large stump casts, described several years ago from the Catskill Mountains under the name of *Eospermatopteris*, are often cited as remains of the oldest known forest. These trees were similar to, if not identical with, *Aneurophyton*.

Upper Devonian floras contain a variety of lycopods similar to those from the preceding series. *Lepidosigillaria* was a fairly large treelike lycopod that foreshadowed the arborescent lepidodendrids and sigillarias of the coal-swamp forests of a later period. The Sphenopsida are represented by the distinctive *Pseudobornia*, and the ferns by *Rhacophyton. Archaeopteris*, formerly identified as a fern, was recently found by Beck (1960) to be connected with *Callixylon*, a plant whose internal structure is similar to that of a gymnosperm of the cordaitean alliance. The unique combination of characters revealed by this connection—that is, the pinnate foliage, the free-sporing heterosporous condition, the thick secondary wood made up of pitted tracheids and narrow rays and revealing rather well-marked growth rings, and the lofty habit—seem to place the plant in an intermediate position between ferns and seed plants. It probably represents some extinct class of spore-bearing woody plants that had evolved far along vegetative lines while retaining reproductive methods of vascular cryptogams.

Extending over a large area in Tennessee, Kentucky, Indiana, Ohio, Michigan, and southern Ontario is an extensive series of black-shale formations that range in age from late Middle Devonian to Early Mississippian. The greater portion of these shales belong to the Upper Devonian. Scattered widely through these shales at certain horizons are numerous trunks of *Callixylon* and an occasional one of *Prototaxites*. Then there are minute sporelike bodies called *Tasmanites* (or *Sporangites*) and thalloid spore-producing organisms called *Protosalvinia* (or *Foerstia*). *Callixylon newberryi* is the dominant species of *Callixylon* in the black shales. It was a large plant. Trunks 3 feet in diameter are on record. Three species of *Protosalvinia* have also been described. These plants were small bodies only a few millimeters high, without vascular tissue, but with waxy outer surfaces. Cutinized spores were borne in tetrads within specialized internal cavities. *Protosalvinia* apparently represents a class of advanced thallophytes or some extinct group of higher plants above thallophytes but below the vascular-plant level.

No objects definitely identified as seeds have been found in the Devonian. The bodies described as seeds that were attributed to *Eospermatopteris* were found on subsequent observation to be spore cases. *Callixylon* was originally assumed to be a seed plant because of its stem structure, but it is now known to have borne spores. Whether true seeds had evolved by Devonian time is unknown.

Mississippian

Floras evolved rapidly during the transition from the Devonian to the Mississippian Period, and the plants existed in the latter period in greater variety and abundance than in the rocks of the Devonian System.

Several new lycopods appear in the Lower Mississippian. These include *Lepidodendropsis*, which is characterized by low rhombic or fusiform leaf cushions that are spirally arranged but aligned in vertical rows so as to appear in close verticils. There are also forms that closely resemble the true *Lepidodendron* but with longer, more slender, and slightly sinuous cushions with tapering ends. *Sigillaria* was rare or probably nonexistent during this time, but *Archaeosigillaria* persisted from the previous period. *Archaeocalamites*, distinguished from the true *Calamites* of the Pennsylvanian Period by nonalternating ribs at the nodes, was a precursor of *Calamites*. Fernlike foliage, quite similar to the Devonian *Archaeopteris*, is abundant. The older Mississippian floras of eastern North America are characterized by *Triphyllopteris*, a form with narrowly triangular pinnules. In younger Mississippian beds this is replaced by forms of the *Cardiopteris* type, with broader pinnules. That strange group of fernlike plants, the Coenopteridales, had become well established. Species of *Botryopteris*, *Metaclepsydropsis*, and *Stauropteris* appear in the Lower Mississippian.

The oldest seed plants, the pteridosperms, are found in rocks of the earliest Mississippian age. Compressed, cupulate seeds occur in the Pocono and Price Formations of Pennsylvania and Virginia, respectively. In the calciferous sandstone of Scotland a number of seeds and seedlike fructifications have been found; two of those are *Calathospermum* (Walton, 1949) and *Genomosperma* (Long, 1960).

Related to the *Archaeopteris-Callixylon* complex of the Devonian is *Pitus* (usually spelled *Pitys*), which was a large tree with a structure much like that of *Callixylon*. The short, fleshy organs described by Gordon (1935) as the leaves of *Pitus dayi* are probably the dried bases of young fronds. *Pitus*, along with *Callixylon*, had been placed in the Cordaitales, but the recent discovery of a connection between *Archaeopteris* and *Callixylon* requires removal from that order.

The Mississippian phase of the New Albany black shale contains a rather large flora represented mostly by small stems and petiole fragments preserved in small phosphatic concretions. More than 30 genera have been identified, some of which are lycopods, some are sphenopsids, and others are early fernlike forms. Many represent unknown plants. No compressed foliage is associated with them.

Pennsylvanian

The uppermost Mississippian and the lowermost Pennsylvanian floras are rather difficult to distinguish because of the orderly transition from one period to the other. Most of the major plant groups of the Pennsylvanian floras were already in existence during the previous period.

The most important changes were increased adaptations to swamp environments that became wide-spread over the earth, especially in the Northern Hemisphere, during the Pennsylvanian Period.

Plant remains are unusually abundant in the Pennsylvanian rocks that represent deposition in swamps where coal was formed. In some places large quantities of plant material is preserved in coal balls, and these have yielded valuable information on the internal anatomy of the plants of that period. Casts and compressions, however, are much more abundant, and it is from these that the bulk of our knowledge of Pennsylvanian plants has been obtained.

Pennsylvanian floras, early and late, are set apart from those of other periods by an abundance of arborescent lycopods such as *Lepidodendron* and *Sigillaria*, giant-sized members of the scouring-rush group typified by *Calamites*, the low growing *Sphenophyllum*, true ferns and the fernlike Coenopteridales, seed ferns of the *Lyginopteris* and *Medullosa* types, and early forerunners of the conifer class, the Cordaitales. Members of these groups are often preserved in profusion in the shales that overlie coal beds. Coal itself rarely contains identifiable plant fragments except spores and fragments of leaf and stem cuticles. Coal balls constitute an exception, but their distribution is limited.

The two most prevalent genera of the arborescent lycopods, *Lepidodendron* and *Sigillaria*, are distinguished by the arrangement of the persistent leaf cushions on the trunk surfaces. In the latter they are in vertical series, one directly above the other; in the former and in the closely related *Lepidophloios* each cushion lies slightly oblique to the one below it. The foliar cushions remained on the trunks long after the leaves had fallen. Eggert (1961) has presented evidence that these cushions did not remain on the trunk surface indefinitely but were cast off in large sheets after considerable secondary cortex had developed beneath. This would account for the frequent occurrence in plant-bearing rocks of imprints of large areas covered with leaf cushions completely unassociated with remains of the inner tissues of the trunks.

The two most commonly preserved fructifications of *Lepidodendron* and its relatives are *Lepidostrobus* and *Lepidocarpon*. These are cones of variable size, ordinarily an inch or less wide and up to a foot long. The former released its microspores and megaspores for germination, whereas the other retained its single large megaspore and shed the whole sporophyll, which functioned much as a seed. Seedlike organs were also produced by *Sigillaria*, but the sporangium held several megaspores.

Stigmaria, the rootlike organ of the arborescent lycopods, is the most abundant plant fossil in rocks of Pennsylvanian age. It is especially prevalent in the underclay beneath coal seams, where it has often lain quite undisturbed in the fossil soil in which the coal-forming plants were rooted. Though rootlike in function and in position with respect to other parts of the plant, *Stigmaria* shows a structure that is more like that of a stem, and its so-called lateral rootlets are arranged in a definite pattern that is more leaflike than rootlike. *Stigmaria* is probably a modified underground branch system that developed in place of a true root system.

The arborescent lycopods of the Pennsylvanian coal swamps found a parallel in *Calamites*, a tree-sized member of the Sphenopsida, the scouring-rush class. Its trunks and branches were conspicuously jointed, like those of a modern *Equisetum*, but the leaves were larger and evidently contained chlorophyll. Fragments of these plants are extremely common in the plant-bearing rocks associated with coal seams. The parts that are most frequently encountered are compressed twigs bearing whorls of leaves (*Annularia* and *Asterophyllites*) and the casts of the large pith cavities that are distinctly marked by narrow longitudinal ribs extending the length of the internodes. At the nodes, which are marked by shallow constrictions that encircle the casts, most of the ribs alternate with those of the adjacent internodes.

Calamites bore its spores in strobili that were similar in size and shape to those of the arborescent lycopods, but they are readily distinguishable by the segmented structure that is particularly evident in compressed specimens. Sporangiophores that usually bore four sporangia were attached in whorls of six or more to the central axis. In *Palaeostachya* the sporangiophores were set at a 45-degree angle in the axils of whorled bracts. In *Calamostachys* the sporangiophores were somewhere along the internode between the bract whorls. *Cingularia* and *Mazostachys* bore their sporangiophores beneath the bracts. Heterospory prevailed but the difference in spore size was somewhat less than in the heterosporous lycopods. Calamitean spores lacked the prominent surface ornamentations that characterized many lycopod spores.

Sphenophyllum was a small trailing plant that probably grew as an undercover on the forest floor. The distinctly noded stems with whorled leaves bear some resemblance to leafy *Calamites* twigs but the leaves differ in being wedge shaped and in possessing forked veins. Those of *Calamites* are tapered or rounded at the apex and have but one unforked vein.

Detached strobili of *Sphenophyllum* are referred to *Bowmanites*. Some were rather large for the small stems that bore them, sometimes having been half an inch wide and several inches long. They were segmented like the cones of *Calamites* and bore their sporangiophores on axillary sporangiophores which, however, in some forms were elaborately branched. The spores were all of one kind as far as known, and were produced in simple sporangia on recurved sporangiophore tips.

One of the very few plants in the Pennsylvanian coal-swamp forests that would have looked at home in a forest of the present era was *Equisetites*, which closely resembles and may have been virtually indistinguishable from a modern *Equisetum*. It seems to have borne its strobili, however, on short lateral branches instead of at the apex of the main stem. If *Equisetites* was a true *Equisetum* the latter would be the oldest surviving genus of the present-day flora.

The Mississippian and Pennsylvanian Periods were for a long time referred to collectively as the Carboniferous, and this name is still widely used. Because of the abundance of fernlike foliage in rocks of the two periods, they were formerly referred to as the age of ferns. This term fell into disuse at the beginning of the present century when it was discovered that some of this fernlike foliage belonged to plants that bore seeds. These plants, long extinct, became known as seed ferns, or *pteridosperms*. Some of this fernlike foliage, however, shows remains of attached fructifications, and does represent true ferns, but in the absence of these remains no absolutely reliable criteria exist for separating vegetative foliage into the fern and seed-fern categories.

Many years before this problem developed a number of form genera had been created for the various kinds of fossil fernlike foliage. The most widely occurring ones are *Pecopteris*, *Sphenopteris*, *Neuropteris*, *Mariopteris*, and *Alethopteris*, which are distinguished from each other mainly by the form and venation pattern of the pinnules. There are now good reasons for believing that the most typical species of *Neuropteris* and *Alethopteris* are foliage of the Medullosaceae, whereas some species of *Pecopteris* and *Sphenopteris* belong to the Lyginopteridaceae, and others belong to ferns.

Some of the fructifications found on foliage of the *Pecopteris* and *Sphenopteris* types show general resemblances to the sori of certain modern ferns. Thus *Oligocarpia* is believed to represent an ancient member of the family Gleicheniaceae. *Senftenbergia* has been compared with the modern *Schizaea*; and *Asterotheca*, *Ptychocarpus*, *Scolecopteris*, and *Cyathotrachus* are quite typical of the fructifications of present day Marattiaceae.

Probably the largest and most diversified group of fernlike plants in the late Paleozoic floras was the Coenopteridales, which appear to have originated in the Devonian, became widely spread during the Pennsylvanian, and probably became extinct during the latter part of the Permian Period. However, during this long time interval some of the later ferns probably sprang from them. The Coenopteridales are preserved mainly in coal balls, and some have been studied in great detail. These plants had small and relatively simple stems. In *Botryopteris* and several others the stem had a small round protostele, and in *Asterochlaena* the xylem strand was deeply lobed. In the leaf stalks of the various genera the vascular strand presents a variety of forms, which provides the basis for generic distinctions. It is C-shaped in *Tubicaulis*, trident shaped in several species of *Botryopteris*, or like a double anchor in *Ankyropteris*. In some species of *Botryopteris* and *Anachoropteris* axes with petiolar structure bore adventitious buds from which stems developed.

Because the order is preserved mostly as petrifactions, little is known of the foliage of the

Coenopteridales. In a few instances pinnules of the *Pecopteris* type have been found on the frond ramifications. Some fronds have been seen to terminate as slender forked filaments that resemble the deeply cut pinnules of *Rhodea*. The sporangia of some of the coenopterids were clustered like bunches of grapes.

The largest of the known late Paleozoic ferns was *Psaronius*, which appears in the Early Pennsylvanian and extends into the Early Permian. The tall trunk bore a terminal crown of fronds believed to resemble *Pecopteris*. Adventitious roots emerged along the side of the trunk and surrounded it in a mantle of increasing thickness toward the base. The fructifications are not known, but the plant is classified with the ferns because of the numerous separate concentric vascular strands without secondary growth that occur in the relatively small stem.

The discovery of the pteridosperms at the beginning of the present century showed that ferns by no means monopolized the late Paleozoic coal-swamp forests. The pteridosperms are an extinct group of plants that taxonomically and morphologically stand between ferns and the more primitive gymnosperms such as cycads. However, their relation to either of these groups is not well understood.

Several families are recognized among the Paleozoic pteridosperms, but the best established ones are the Lyginopteridaceae and the Medullosaceae. The Lyginopteridaceae, typified by *Lyginopteris oldhamia*, have stems with single steles, foliage of the *Sphenopteris* and *Pecopteris* types, seeds with integument adherent to the nucellus, and microsporangia known in the detached condition as *Crossotheca* and *Telangium*. The stems, leaf stalks, and seed cupules are studded with rather coarse glandular hairs, which were the means by which the separate organs were recognized as belonging to one plant. The Medullosaceae have polystelic stems that in cross section show two or more (usually three) separate vascular units enclosed by a common cortex. Each has its own complete layer of secondary wood that is usually thicker on the side toward the center of the stem. The foliage belongs to *Neuropteris* and *Alethopteris*. The seeds, which were not cupule borne, appear to have replaced the pinnules of ordinary pinnae. They were rather large and known in the detached state as *Trigonocarpus* and *Pachytesta*. The microsporangiate organs were large synangia, called *Whittleseya*, *Aulacotheca*, and *Dolerotheca* when detached. The family is typified by *Medullosa anglica* from the British Coal Measures, and *M. noei* from the Upper Pennsylvanian of Illinois.

Detached seeds attributable to pteridosperms are often found in great numbers, but they are rarely attached to the parts that bore them. They are numerous in the petrified condition in coal balls. A significant fact is that none have ever been found to contain embryos; indeed, embryos have not been found in any Paleozoic seed belonging to any group. Seeds with embryos probably did not form until Mesozoic time.

The Cordaitales constituted another group of seed-bearing plants of the late Paleozoic coal-swamp forests. These plants were tall trees with massive woody trunks. The branched crown was shrouded in large but simple, parallel-veined leaves that were sometimes 4 inches wide and a yard long. Their flat, often heart-shaped seeds and the pollen were borne in loosely constructed conelike inflorescences that show significant morphological similarities to the seed cones of modern conifers.

The Coniferales apparently date from the Pennsylvanian Period. *Walchia*, a Pennsylvanian and Permian tree that has much the appearance of the living *Araucaria excelsa*, is rather sparsely preserved in the fossil series by leafy branches and cones. The scarcity of its remains is probably due to an upland habit remote from the swamps in which most of the Pennsylvanian vegetation was preserved.

Permian

Vast changes took place in the plant world during the Permian Period. In the Northern Hemisphere, particularly in North America, the oldest Permian floras differ in no essential respects from those of the Late Pennsylvanian, but soon afterward the cold climate that had spread over much of the Southern Hemisphere began to extend its influence over the rest of the earth. The lowered temperatures were accompanied by aridity. The swamps dried up, and the lush vegetation that they supported disappeared. It was replaced by newly evolved forms with smaller, thick, heavily cutinized leaves. Only the groups that were able to modify themselves to the adverse conditions were able to survive. The flora of the early Mesozoic had its inception at that time.

Gigantopteris, a plant with large, bifurcated, ribbonlike leaves, which was probably a pteridosperm, is found in the Lower Permian red beds of Texas and occurs also in Korea and China. Other typically Lower Permian plants are *Callipteris, Tingia,* certain species of *Odontopteris,* and *Calamites gigas. Trichopitys heteromorpha,* believed to be the oldest member of the Ginkgoales, was found in the Lower Permian of southern France, and *Sphenobaiera,* from the Upper Permian, is a more typical member of this order. Several genera of ferns attributable to the Osmundaceae come from the Upper Permian of the Ural Mountains.

The youngest Permian flora found in North America was described more than 30 years ago by White (1929) from the Hermit Shale of the Grand Canyon. It contains species of *Sphenophyllum, Taeniopteris, Callipteris,* and *Walchia* —which are genera that occur in the Early Permian elsewhere—but conspicuous by their absence are *Calamites, Cordaites, Odontopteris, Callipteridium, Neuropteris,* and *Alethopteris,* which are frequently encountered in the Permian elsewhere. The flora is distinguished by genera, previously unrecorded from typical Early Permian floras, that are similar to or identical with species from the Late Permian Kusnezk flora of Angaraland and the Gondwanaland flora of the Southern Hemisphere. The most important of the latter element is *Supaia,* probably a pteridosperm, of which eight species were named. *Supaia* resembles *Danaeopsis hughesi* of the lower Gondwana flora and *Thinnfeldia odontopterioides* of the early late Gondwana flora in India, Australia, and South Africa. The Hermit Shale flora therefore constitutes a link between typical Eur-American Permian floras and those of Gondwanaland.

Glossopteris Flora. This is the flora that spread throughout the Southern Hemisphere during the latter part of the Paleozoic Era, occupying ancient Gondwanaland, and remnants of it are found in southern Africa, India, Australia, and South America. It is doubtful, however, that throughout all or even the greater part of its existence Gondwanaland was one solid, uninterrupted land mass. Teichert (1958) believes that Australia was a separate continent as far into the past as our knowledge of it extends. The flora appeared during Namurian time in South America (Barbosa, 1958), though generally its range in time is given as late Carboniferous to Late Triassic. Altogether about 40 species of *Glossopteris* have been named, but studies by some Indian paleobotanists show that many of them intergrade and that some so-called species contain forms with more than one type of epidermal pattern (Surange and Srivastava, 1956). *Gangamopteris* is a smaller genus that occurs early in the time range of the flora.

The *Glossopteris* flora characterizes the lower of the two divisions of the Gondwana group; it has a total thickness of 30,000 feet in India and other places in the Southern Hemisphere. The upper Gondwana flora is quite different from that of the lower series.

No actual traces of the *Glossopteris* flora have been found in North America, though some of the constituents of the Hermit Shale flora resemble certain species in it. The specimens from Siberia that many years ago were classified as *Glossopteris* were later found to have been misidentified.

THE MESOZOIC FLORA

As previously pointed out, the Mesozoic flora was initiated during the latter part of the Paleozoic Era with the disappearance of the coal-swamp vegetation and the rise of new types better adapted to cope with the more severe climate. By the beginning of the Triassic Period the change to the new flora had become complete in the Northern Hemisphere, and by the end of the period (Rhaetic) the northern flora had again merged with the Gondwanaland flora, which was in turn succeeded in the Jurassic by a flora that was remarkably uniform the world over.

Triassic

We know less of Early Triassic floras than of those that appeared later in the period. In the Bunter (the earliest Triassic) the scouring-rush order is represented by *Equisetites* and *Schizoneura*. The latter is also a component of the *Glossopteris* flora. The principal lycopod of the Bunter is *Pleuromeia*, a plant more than a meter high that resembled a dwarf *Sigillaria* in some respects but was different in others. *Neuropteridium* is the most characteristic fern genus, and a few fronds referred to *Zamites* and *Pterophyllum* resemble some members of the cycadophyte group of the Paleozoic as well as those of succeeding periods. *Voltzia* is the best known of the Early Triassic Coniferales.

The most thoroughly studied Middle Triassic flora is the Ipswich flora of Queensland. It contains the probable pteridosperm *Stenopteris*, a few ferns identified as *Cladophlebis* and *Dictyophyllum*, several cycadophyte fronds, and leaves hardly distinguishable from the modern *Ginkgo*. Most of the ginkgophyte leaves of the Ipswich flora are the deeply cut types referable to *Ginkgoites* and *Baiera*.

The much richer Late Triassic flora contains *Neocalamites*, which is intermediate in size between *Calamites* and *Equisetum*, numerous pteridosperms (*Pteruchus*, *Umkomasia*, *Pilophorosperma*, *Zuberia*, *Lepidopteris*), an abundance of cycadophytic foliage types, and conifers resembling *Voltzia*. Outstanding among Late Triassic floras are the Keuper flora of Germany, the Chinle flora of the southwestern United States, and the flora of the Richmond coal basin in Virginia. The large silicified logs of *Araucarioxylon*, *Woodworthia*, and *Schilderia* so magnificently laid bare by erosion in the Petrified Forest National Monument in Arizona were preserved in the Chinle shale. The Chinle flora contains a number of ferns and cycadophytes in addition to conifers.

The Dolores Formation of western Colorado, which is of approximately the same age as the Chinle, has yielded what may be the oldest angiosperm, *Sanmiguelia*. The specimens are imprints of palmlike leaves that were up to 25 centimeters wide and 40 centimeters long. If found in the Upper Cretaceous or any horizon of later age, *Sanmiguelia* would be accepted as a palm without doubt.

Rhaetic. The Rhaetic is sometimes regarded as uppermost Triassic. From the Rhaetic of Sweden comes *Bjuvia simplex*, a member of the Cycadales allied to the living *Cycas*, and *Wielandiella angustifolia*, an early member of the Cycadeoidales. Harris (1931-37) has described a large flora of lycopods, ferns, cycadophytes, ginkgophytes, and conifers from the Rhaetic of Scoresby Sound in eastern Greenland. This flora is remarkable for the large quantities of cuticles in the compressions which have aided greatly in identification and classification. Of special interest in this flora are two Mesozoic pteridosperms, *Lepidopteris ottonis* and *Caytonia thomasi*, and a leaf, *Furcula granulifera*, that bears considerable resemblance to an angiosperm leaf.

Jurassic

Jurassic plants for the most part belong to the same classes and orders as those of the preceding period, though many new genera and families appear. Jurassic plants range from the Arctic to the Antarctic and are especially abundant in eastern Asia, Siberia, Argentina, South Africa, India, Australia, Great Britain, and central Europe. North American Jurassic floras are of lesser significance, though some well preserved forms are known from Mexico, northern California, Oregon, and Alaska. Almost all Jurassic floras, regardless of where they are found, consist of ferns, cycads and cycadeoids, ginkgophytes, and conifers. The cycadopsids and some of the conifers seem to have reached their maximum development during this period.

The one Jurassic flora that has contributed more to our knowledge of mid-Mesozoic floras than any of the others occurs in Yorkshire. A series of deltaic deposits known as the Oölite contain exceptionally well preserved foliage and fructifications of almost all of the plant groups known at that time, and some of the most prolific localities are at Whitby, Cayton Bay, and Roseberry Topping. From Whitby and elsewhere have come well-preserved compressions of foliage and fructifications of *Williamsonia* and *Williamsoniella*. Cayton Bay has yielded an ample crop of cycadophytic remains, including *Beania*, a seed-bearing cone that shows possible affinity with *Zamia*. One of the most significant of the discoveries at Cayton Bay was reported by H. H. Thomas (1925) when he described the fruits, pollen-bearing organs, and foliage of the Caytoniales. This group of plants had developed fruit-like bodies that when first discovered were compared with angiosperm fruits.

In Bihar in eastern India, the Rajmahal upper Gondwana series, which is believed to be of Late Jurassic age, contains plants similar to those found in Jurassic rocks elsewhere. They replace the *Glossopteris* flora of the older rocks. A group of plants peculiar to this region is the Pentoxylales. *Pentoxylon* had a slender stem with five or more separate steles that is somewhat reminiscent of *Medullosa*, but the secondary wood is coniferous in character though the foliage was of the *Taeniopteris* type. The seeds were borne in compact cones in which there were no cone scales or sporophylls.

Several modern fern families are recognizable in the Jurassic. Among these are the Matoniaceae, Marattiaceae, Cyatheaceae, Osmundaceae, and Schizaeaceae. Some of these are known in older rocks, but they show considerable development in the Jurassic. Of special interest are the petrified stems *Osmundites gibbeana* and *O. dunlopi* from New Zealand, which appear as intermediate types between the Permian *Thamnopteris* and the modern *Osmunda*.

Florin (1936) has described a number of Jurassic Ginkgoales from Franz Josef Land. These are forms with slender, almost linear leaves borne in clusters on short shoots. The largest of these is *Stephanophyllum* with leaves nearly 30 centimeters long and about 1 centimeter wide. *Windwardia* has smaller but proportionally wider leaves. *Arctobaiera* has the smallest leaves.

The Jurassic rocks are rich in remains of conifers, though most of the genera are extinct. However, *Sequoia* first appears in rocks of this age in China. Rather typical Jurassic genera are *Araucarites*, *Brachyphyllum*, *Pagiophyllum*, and *Podozamites*.

Silicified trunks of *Cycadeoidea* occur in Jurassic beds on the Isle of Portland off the southern coast of Great Britain and in the Morrison Formation in Utah, Wyoming, and other states.

The presence of angiosperms in Jurassic rocks is controversial; several have been reported, but proof is lacking (Scott, Barghoorn, and Leopold, 1960).

Cretaceous

A few plants are not known to occur outside the Cretaceous. Among these is *Tempskya*, a fern with a trunk consisting of a mass of small individual stems held together by a matrix of adventitious roots. *Tempskya* ranges from the Wealden to the Senonian, but in North America, where it is most abundant, it seems to be confined to the middle part of the Cretaceous System. *Weichselia*, another fern, is characteristic of the Wealden, but possibly it ranges into the Late Cretaceous. Two other ferns are *Knowltonella* and *Schizaeopsis* from the early and late Early Cretaceous, respectively. *Frenelopsis* is an extinct genus of conifers of the Early Cretaceous and early Late Cretaceous. *Sapindopsis*, so named because of its resemblance to the living soapberry, is one of the most common angiosperms of the Early Cretaceous, but it occurs sparingly in the Late Cretaceous. *Dewalquea* is another extinct Cretaceous angiosperm.

The Cretaceous was an important period in the history of the plant kingdom. It was during this time that the ferns and gymnosperms, which dominated Mesozoic floras up to that moment, surrendered to the flowering plants. The question whether angiosperms originated then or earlier is overshadowed in importance by the mere fact that during this period they became dominant.

Differences between Early and Late Cretaceous floras are much more pronounced than differences between Late Jurassic and Early Cretaceous or between Late Cretaceous and Paleocene floras. It is often difficult to tell the latter apart. In fact, were the eras delimited on the facts of plant evolution, the Late Cretaceous would be placed in the Cenozoic era.

Although plants that bear exact resemblances to angiosperms occur as far down in the stratigraphic sequence as the Upper Triassic, it is not until Lower Cretaceous rocks are reached that they become consistently established in the fossil record. There are some differences of opinion concerning their presence in the lower and middle parts of the Lower Cretaceous. In studying the pollen contents of the rocks of the Potomac group Brenner (1963) found no angiosperm pollen in the Patuxent and Arundel Formations. It was found to occur in the Patapsco Formation, which is equivalent to the Albian or the upper Lower Cretaceous. However, in spite of this a few angiosperm-type leaves have long been known to occur in the Patuxent and Arundel, and these have often been cited as evidence of Early Cretaceous angiosperms. It has been questioned whether these rather fragmentary leaves in the Patuxent and Arundel Formations are angiosperms or something else. Some investigators, including the late Professor E. W. Berry, have maintained that they could belong to the Gnetales because the leaves of some Recent species of *Gnetum* are much like those of angiosperms. This, however, seems improbable because no remains of this genus, or for that matter of any of the Gnetales except *Ephedra*, have been recognized anywhere in the Upper Cretaceous or Tertiary. It seems unlikely that this group could have flourished in relative abundance during Early Cretaceous times and could have been abruptly replaced by angiosperms, with no noticeable break in the sequence. No one questions the presence of angiosperms in the Patapsco Formation, and Albian age rocks often yield a few representatives of this group, always fewer in number than gymnosperms.

Overlying the Lower Cretaceous Potomac group with its early angiosperms are the Raritan and Magothy Formations, which are assigned to the lower Upper Cretaceous. These have large floras that contain up to 60 percent angiosperms. That the strong and steady increase in angiosperms revealed in the sequence from the Patuxent to the Magothy Formations was caused by an evolutionary outburst and was not merely a local phenomenon due to migration is shown by the fact that once angiosperms entered the flora they remained. The older Mesozoic elements never reappeared in large numbers. Once the Mesozoic elements disappeared, they were gone forever.

The flora of the Dakota Sandstone contains 460 named species; 99 percent of these are angiosperms. It is far more modern in aspect than any of the preceding ones, regardless of the fact that a long time range may be represented in the Dakota sequence.

All Late Cretaceous floras are dominated by angiosperms, and they consist largely of families in existence today. Even the ferns, of which several are known, are surprisingly modern. Most of the cycadophytes had disappeared, as had also the majority of the older ginkgophytes and conifers. However, a few, such as *Geinitzia* and *Brachyphyllum*, persisted longer than the others. *Metasequoia* and *Taxodium* make their earliest known appearances, though the latter is known only from pollen. *Sequoia*, having first appeared in the Jurassic, continued through the Cretaceous Period, into the Tertiary, and on with *Metasequoia* and *Taxodium* into the Recent.

The plant fossils that are most commonly encountered in Upper Cretaceous rocks are leaves that resemble those of laurels, figs, oaks, and other broadleaved trees that we see today in forests of moderately warm and well-watered regions.

MODERN FLORAS OF THE CENOZOIC ERA

Paleocene and Eocene Epochs

Warm climates extended into far northern latitudes during the Paleocene and Eocene Epochs. Palms thrived in southern Canada, and pines, birches, and willows grew in land areas now only 8° from the North Pole. One of the largest of the early Tertiary floras is the Wilcox flora that is preserved in a thick sequence of sediments laid down when the Gulf of Mexico extended north of its present borders. This flora ranges from Alabama to Texas and consists of several hundred species that represent 180 genera and 82 families. It bears a close resemblance to the Recent flora of the Antilles and Central America. Legumes are the dominating elements in this flora, but there are numerous members of the Lauraceae, Araliaceae, Meliaceae, Moraceae, and Palmaceae. The Green River flora of Wyoming, Colorado, and Utah contains abundant algal remains that must have originated in warm, shallow water. In places these algae formed thick limy crusts around silicified tree trunks. The Green River flora also contains cycads, conifers, palms, figs, sweet gums, laurels, and oaks.

For three centuries casts of seeds and dry indehiscent fruits have been collected in large numbers where they weather out of the Eocene London Clay along the Thames below London and on the Island of Sheppey. Reid and Chandler (1933) have described and figured more than 70 genera of these fruits. Some of the most copiously represented families are the Menispermaceae, Annonaceae, Lauraceae, Euphorbiaceae, Anacardiaceae, Icacinaceae, Vitaceae, and

Cornaceae. Of special interest are fruits of *Nipa*, a stemless palm that at present grows along tidal flats bordering the Indian Ocean.

An American counterpart of the London Clay seeds and fruits has been found in the Clarno tuffs of north central Oregon. Though not yet fully described, this flora contains representatives of the Menispermaceae, Juglandaceae, Icacinaceae, Vitaceae, and others. *Nipa* has not been found there.

Oligocene

Commencing in the Late Cretaceous but reaching its maximum intensity during the Miocene, an interval of widespread volcanic activity was responsible for the spread of thick deposits of lava and tuff over large areas of western North America. The blocking of streams by flowing lava produced numerous fresh-water lakes, which were in turn filled with falling ash and material freshly eroded from ash deposits. These lake beds contain the most extensive records of Tertiary floras known anywhere. The material consists predominantly of impressions of angiosperm leaves, but there are in addition numerous seeds and fruits, and occasional flowers. There are also ample representations of conifers, a few ferns, occasional cycads, and *Ginkgo* leaves. Silicified tree trunks, sometimes quite large, are of widespread occurrence and sometimes abundant.

Remains of the floras are the best indicators of Tertiary climates. They show the increase in warmth over northern latitudes that reached its maximum in the early Oligocene, and a reverse of that trend during late Oligocene and Miocene times. They also show the effect of proximity to ocean basins by revealing marked differences between inland and coastal floras at similar latitudes.

Western American floras of the early and middle Tertiary contain several genera that are restricted at present to Eastern Asia but which existed in North America up to Miocene or Pliocene time. Some of these are *Ginkgo*, *Pseudolarix*, *Metasequoia*, *Ailanthus*, *Koelreuteria*, *Cercidiphyllum*, *Trapa*, and *Zelkova*.

Large floras of early to middle Oligocene age are preserved in the lake beds at Florissant in Colorado and in the Ruby valley in southwestern Montana. These represent inland floras at about the time of maximum warmth and contain such genera as *Metasequoia*, *Salix*, *Morus*, *Populus*, *Quercus*, *Mahonia*, *Carya*, *Zelkova*, *Sassafras*, *Persea*, *Cercis*, and *Sapindus*, to mention only a few.

The effect of proximity to the sea is shown by the Weaverville flora in California, which is of approximately the same age though possibly slightly younger than the Florissant flora. It is quite different, being a subtropical or warm-temperate assemblage, as indicated by such genera as *Taxodium*, *Nyssa*, *Tetracera*, and *Ficus*.

The late Oligocene or early Miocene Bridge Creek flora of the John Day valley reflects the return of slightly lower temperatures after the peak of the warmth. This flora has been designated as a *Metasequoia*-oak-birch association that contains in addition species of *Alnus*, *Ostrya*, *Acer*, and *Umbellularia* that at present are associates of the coastal redwood belt of northern California.

Miocene

During the Miocene Epoch temperate mesophytic species had largely replaced the warm-temperate and subtropical types that had previously decorated the landscapes of western North America. Miocene floras are rich in such genera as *Acer*, *Alnus*, *Quercus*, *Populus*, *Salix*, *Pinus*, *Picea*, *Platanus*, *Fagus*, and *Mahonia*. The summers became drier, though rainfall in general was not deficient, and seasonal changes became more pronounced. Several genera such as *Carpinus*, *Ulmus*, *Tilia*, and *Fagus* made their last stand in the western part of the continent during this time. They persisted, however, in the eastern half.

Pliocene

The cooling trend that culminated in the Pleistocene ice age continued to develop during the Pliocene. It was then that the Arctic tundras probably originated. Elevation of the Cascade range during late Miocene time reduced the rainfall to the eastward, thus initiating the desert environments of the Great Basin and adjoining areas. During much of Pliocene time conditions not greatly unlike those existing today in northern latitudes probably prevailed.

The last remaining link between Tertiary floras and those of the present are revealed to some extent by pollen and other plant remains preserved in peat bogs of Pleistocene and post-Pleistocene times. Since most of the species are still in existence, this phase of the subject belongs to the realm of modern botany.

A SUMMARY OF ANGIOSPERM HISTORY

As explained earlier, angiosperms have dominated the land flora of the earth since mid-Cretaceous time. The angiosperm-fossil record, which consists mostly of leaves, is the most extensive from the standpoint of numbers of specimens of any vascular-plant group. When, where, and from what angiosperms originated are questions to which answers have been sought for a century. The problem, however, is not essentially different from that pertaining to any major group. The oldest conifers, for example, show up in the Middle Pennsylvanian, but by that time they had already developed strongly differentiated male and female cones and small needlelike leaves that present difficulties in postulating a direct cordaitalean ancestry for the group. Seeds attributed to pteridosperms occur in the oldest Mississippian age floras, but such organs are virtually absent from the Devonian rocks that are immediately beneath.

The oldest known plants that can reasonably be called angiosperms are *Sanmiguelia*, the palmlike plant from the Late Triassic, and *Furcula*, from the Rhaetic. The remains consist only of leaf impressions and a few fragmentary stem casts, though some cuticle is retained in *Furcula*. *Angiospermophyton americanum*, the alleged monocotyledon found several years ago in an Illinois coal ball, ultimately turned out to be a *Medullosa* petiole.

The difficulties in recognizing the oldest angiosperms, should they be found either by accident or design, are due to the lack of distinctive features that could be expected to be preserved in the fossils. Practically the only characters that set angiosperms completely apart from other plant groups equal in rank are the structure of the embryo sac and the double fertilization process. It is quite safe to predict that these will never be observed in fossils. Moreover, the embryo sac varies considerably in different angiosperm groups, and much remains to be learned about double fertilization because the actual process is rarely observed.

Whatever the causal factors might have been, there is little basis for disputing the fact that angiosperms did undergo remarkable spread and diversity during Cretaceous time. If we assume that they started at or not far above the zero point at the onset of the period, we are a little startled at the appearance of about 40 families in the Dakota Sandstone flora of the early Late Cretaceous. Even after making liberal allowances for errors in identification and for incomplete data the fact that flowering plants had evolved rapidly during the Early Cretaceous interval is indisputable. Moreover, many of the Dakota Sandstone types either became extinct during the remainder of Cretaceous time or served as the ancestors of later families and genera.

At least 80 percent (and maybe more) of the living angiosperm families have fossil records of sorts. A considerable number are limited to remains in Pleistocene peat deposits, but more than half of the extant families have Tertiary records, and a considerable number can be traced into the Cretaceous. (See list at end of this chapter.) The recently developed and rapidly expanding science of palynology and studies of leaf cuticles have added to the list of families with recognized fossil members. However, an attempt to utilize fully the data from these sources is unfeasible at present.

Of the families of flowering plants represented in the Cretaceous, a selected few are commented on below.

Leaf impressions and silicified trunks of palms occur in a number of Upper Cretaceous localities. If the Late Triassic *Sanmiguelia* should ultimately prove to be a palm, which it resembles more than anything else, the Palmae would become the oldest known family of flowering plants. There are no wholly convincing instances of grasses having been found in the Cretaceous, but the discovery of such would not come as a surprise.

Seeds believed to belong to an extinct genus of the Annonaceae have been reported from the Late Cretaceous of Nigeria. Leaves attributed to several genera are found in Europe and North America.

If the Early Cretaceous *Dewalquea* belongs to the Araliaceae, as commonly believed, this family becomes one of the oldest. Silicified wood from the Cretaceous of Egypt probably constitutes the oldest remains of the Celastraceae.

Certain Cretaceous leaves originally identified as *Populus* have been transferred to *Cercidiphyllum* (Cercidiphyllaceae). This is of special significance with respect to certain leaves described many years ago from Greenland, Alaska, and other places in the Arctic. The oldest probable record of the Ericaceae consists of leaves of *Andromeda* found in the Late Cretaceous of Alaska.

The Fagaceae was probably a large family during Late Cretaceous time, and several genera appear to have become extinct. The Late Cretaceous *Dryophyllum* is thought possibly to represent a genus that is intermediate between the Fagaceae and the Betulaceae, this indicating a common origin of these families.

Existence of the Guttiferae during the Cretaceous Period is indicated by *Guttiferoxylon* from Colombia, Egypt, and India, and by seeds from France. Leaves unquestionably of *Liquidambar* (Hamamelidaceae) were found in the Upper Cretaceous Aspen Shale of Wyoming. The Icacinaceae are represented by characteristic pitted endocarps in the lower Upper Cretaceous of New York State. Small emarginate leaflets from the Upper Cretaceous of Greenland have been identified as *Dalbergia* (Leguminoseae), and this family probably spread widely into warm regions during Late Cretaceous time.

The Lauraceae have long been regarded an ancient family, as is indicated by a considerable number of *Sassafras*-like leaves in the Dakota Sandstone. The same may be said of the Magnoliaceae. *Magnolia*-type leaves occur commonly in Late Cretaceous floras.

Figlike fruit casts from the Upper Cretaceous of Montana and Saskatchewan, first described as *Ficus ceratops*, have been found on reexamination to bear a greater resemblance to fruits of the genus *Guarea* of the Meliaceae. Leaves called *Menispermites* in the Dakota Sandstone and still older floras are often cited as evidence of the Menispermaceae in the Cretaceous. It may be an old family, but the oldest undoubted remains are fruits in the Eocene.

Ficus (Moraceae) has long been a favorite genus among paleobotanists for reception of Cretaceous leaves with apparent coriaceous texture. Two species of *Ficophyllum* are listed in the Patuxent flora, and in the early Late Cretaceous Magothy flora 13 species of *Ficus* are included. The Dakota Sandstone flora contains 22 so-called species. Few, if any, of these can be relied on as absolutely authentic occurrences of the genus in the past. The most reliable record of Cretaceous Moraceae is *Arithmicarpus hesperius* from South Dakota (Delevoryas, 1964).

Leaves showing the characteristic epidermal structure of *Myrica* (Myricaceae) have been reported from the Upper Cretaceous of the Harz Mountains. The Nymphaeaceae was once believed to be present in the Jurassic, but the pollen on which the determinations were based is now believed to be conifer pollen. Fossils bearing the characteristic markings of *Nymphaea* rhizomes occur in the Upper Cretaceous.

Winged fruits of *Fraxinus* (Oleaceae) have been reported from the Upper Cretaceous of Greenland. *Fraxinopsis*, once reported from the Triassic, is probably a conifer seed. *Credneria*, a large unlobed leaf from the Late Cretaceous, was originally placed in the Platanaceae, but its familial affinities are uncertain. Shallowly lobed leaves of the *Platanus* type are common in Late Cretaceous floras.

The Salicaceae have often been regarded on morphological grounds as a primitive (hence ancient) family, but the fossil record is not clear on this point. *Salix* and *Populus* were probably both in existence during Late Cretaceous time. As explained, some supposedly Cretaceous *Populus* leaves are now believed to belong to *Cercidiphyllum*.

If the Early Cretaceous *Sapindopsis* is a member of the Sapindaceae, this family, with the Araliaceae, is one of the oldest angiosperm taxa. Leaves closely resembling *Sapindus* do occur in the Upper Cretaceous, but they could easily be confused with other genera. Several conspicuously lobed Upper Cretaceous leaves have been assigned to the Sterculiaceae, but the evidence is not conclusive. Certain large leaves from the Upper Cretaceous have been identified as *Vitiphyllum*.

No attempt is made here to review these angiosperm families that appear first in the Tertiary. The list is continually expanding as investigations on cuticles and pollen are completed and published, and as old collections are reexamined and analyzed by modern techniques. For additional information Engler (1964) or Gothan and Weyland (1964) should be consulted.

Partial List of Cretaceous Angiosperm Families

Palmae	Ericaceae	Meliaceae	Rosaceae
Aceraceae	Fagaceae	Menispermaceae	Salicaceae
Annonaceae	Guttiferae	Moraceae	Sapindaceae
Araliaceae	Hamamelidaceae	Myricaceae	Sterculiaceae
Betulaceae	Icacinaceae	Nymphaeaceae	Tiliaceae
Celastraceae	Lauraceae	Oleaceae	Ulmaceae
Cercidiphyllaceae	Leguminosae	Platanaceae	Vitaceae
Cornaceae	Magnoliaceae	Proteaceae	

References

Barbosa, O., 1958, On the age of the lower Gondwana floras in Brazil and abroad: Comision para la Correlacion del Sistema Karroo, Congr. Geol. Intern., 20th, Mexico, 1956, p. 205-236.

Barghoorn, E. S., and Tyler, S. A., 1965, Microorganisms from the Gunflint chert: Science, v. 147, p. 563-577.

Beck, C. B., 1960, Connection between *Archaeopteris* and *Callixylon:* Science, v. 131, p. 1424-1525.

Brenner, G. J., 1963, The spores and pollen of the Potomac group of Maryland: Dept. Geol. Mines and Water Res., State of Maryland, Bull. 27.

Delevoryas, T., 1964, Two petrified angiosperms from the Upper Cretaceous of South Dakota: Jour. Paleont., v. 38, p. 584-586.

Eggert, D. A., 1961, The ontogeny of Carboniferous arborescent Lycopsida: Palaeontographica, v. 108B, p. 43-92.

Engler, A., 1964, Syllabus der pflanzenfamilien, 12th ed., v. II, Angiospermen. Berlin.

Florin, R., 1936, Die fossilen Ginkgophyten von Franz-Joseph-Land, nebst Erörtungen über vermeintlich Cordaitales mesoischen Altern: Palaeontographica, v. 81B, p. 1-173; ii. ibid., v. 82B, p. 1-72.

Gordon, W. T., 1935, The genus Pitys, Witham, emend: Royal Soc. Edinburgh Trans., v. 58, p. 279-311.

Gothan, W., and Weyland, H., 1954, Lehrbuch der Paläobotanik: Berlin.

Harris, T. M., 1931-1937, The fossil flora of Scoresby Sound, east Greenland: Medd. om Grønland, v. 85, 2, 3, and 5; ibid., v. 112, 1 and 2.

Krishtofovich, A. N., 1953, Discovery of lycopodiaceous plants in the Cambrian of eastern Siberia (in Russian): Doklady Akad. Nauk SSSR, v. 91, p. 1377-1379.

Kidston, R., and Lang, W. H., 1917-1920, On Old Red Sandstone plants showing structure from the Rhynie chert bed, Aberdeenshire: Royal Soc. Edinburgh Trans., v. 51, p. 761-784; ibid., v. 52, p. 603-627, 643-688, and 855-902.

Lang, W. H., and Cookson, I. C., 1935, On a flora, including *Monograptus*, in rocks of Silurian age, from Victoria, Australia: Royal Soc. [London] Phil. Trans., v. 224B, p. 421-449.

Long, A. G., 1960, On the structure of *Calymmatotheca kidstoni* Calder (emend.) and *Genomosperma latens* gen. et sp. nov., from the Calciferous sandstone series of Berwickshire: Royal Soc. Edinburgh Trans., v. 64, p. 29-44.

Naumova, S. N., 1949, Spores in the Lower Cambrian (in Russian): Bull. Geol. Acad. Sci. USSR, v. 4, p. 49-56.

Rankama, K., 1948, New evidence of the origin of Pre-Cambrian carbon: Geol. Soc. America Bull., v. 59, p. 389-416.

Reid, E. M., and Chandler, M. E. J., 1933, The London clay flora: London, British Mus. Nat. Hist., 561 p.

Rezak, R., 1957, Stromatolites of the Belt series in Glacial National Park and vicinity, Montana: U.S. Geol. Survey Prof. Paper 294-D, p. 127-154.

Scott, R. A., Barghoorn, E. S., and Leopold, E. B., 1960, How old are the angiosperms?: Amer. Jour. Sci., v. 258B, p. 284-299.

Surange, K. R. and P. N. Srivastava, 1956, Studies of the *Glossopteris* flora of India. 5. Generic status of *Glossopteris, Gangamopteris,* and *Palaeovittaria:* The Palaeobotanist v. 5, p. 46-49.

Teichert, C., 1958, Australia and Gondwanaland: Comision para la Correlacion del Sistema Karroo, Congr. Geol. Intern., 20th, Mexico, 1956, p. 115-138.

Thomas, H. H., 1925, The Caytoniales, a new group of angiospermous plants from the Jurassic rocks of Yorkshire: Royal Soc. [London] Phil. Trans., v. 213B, p. 299-363.

Walcott, C. D., 1914, Pre-Cambrian Algonkian algal flora: Smithsonian Misc. Coll., v. 64, p. 77-157.

———1919, Middle Cambrian algae: Smithsonian Misc. Coll., v. 67, p. 217-261.

Walton, J., 1949, *Calathospermum scoticum*—an ovuliferous fructification of Lower Carboniferous age from Dumbartonshire: Royal Soc. Edinburgh Trans., v. 61, p. 719-736.

White, D., 1929, The flora of the Hermit shale, Grand Canyon, Arizona: Carnegie Inst. Washington Publ. 405, 221 p.

9

PRECAMBRIAN MICROFOSSILS

James M. Schopf

This chapter presents information about the earliest vestiges of the palynologic record through the long period of aquatic evolution before there was any life on land. As living creatures multiplied, they became more diversified. Somewhere early in the process of occupancy of the sea a stable, functional division of heterotrophic as well as autotrophic protobionts occurred. Sequentially the various types of functional activity were developed as part of the early evolutionary progression. This implies the capacity to synthesize virtually all the various food and structural products of plants. The chemistry of plant products by the end of the Precambrian must have been comparable with that of plant products today. This chapter, then, is concerned with the palynological record produced while biotic processes and primary taxa were being established on earth, prior to occupancy of dry land.

Form, function, and composition have complex but essential interrelations. Organic functions and products are characteristic of different levels of organization, and, to some degree, biologic innovations are implied by evidence derived from palynology. Whenever possible the points in time at which these innovations are determinable are pointed out in this chapter because of their possible application to stratigraphy. In general an accurate knowledge of the tempo of phyletic progression should constitute the best guide for paleontologic interpretation of historical geology.

Publication authorized by the Director of the U.S. Geological Survey.

Phyletic Progression

The phyletic-progression chart given in figure 9-1 represents a schematic interpretation of the historical relationships in organic evolution. The progression in time, indicated by the column on the right, is scaled logarithmically to allow representation on a single page. Also, specific information about early organismal history is relatively scant and based for the most part on inference from determinations of later developments. The advent of ancient biologic innovations cannot be dated accurately, and therefore the progressive condensation that is characteristic of a logarithmic scale is no disadvantage in presenting this type of information. Accordingly the lower portions of the chart must be interpreted as representing progressively greater degrees of generalization.

Major Groups of Ancient Organisms

Some of the major distinctions within the organic world are suggested by figure 9-1. Groups included here have been assigned by various authors to as many as five separate organic kingdoms. However, when the various major groups of organisms are considered in relation to their supposed origins sharp lines of definition and distinction are notably lacking. Lines are shown in the diagram for purposes of presentation, but it is not possible to identify many of these as yet in terms of available fossils, and probably very few sharp lines of real distinction ever did exist. Later results from organic geochemistry may at least serve to define the stages based on derivatives of a few metabolic products. Probably geochemical distinctions actually are the most im-

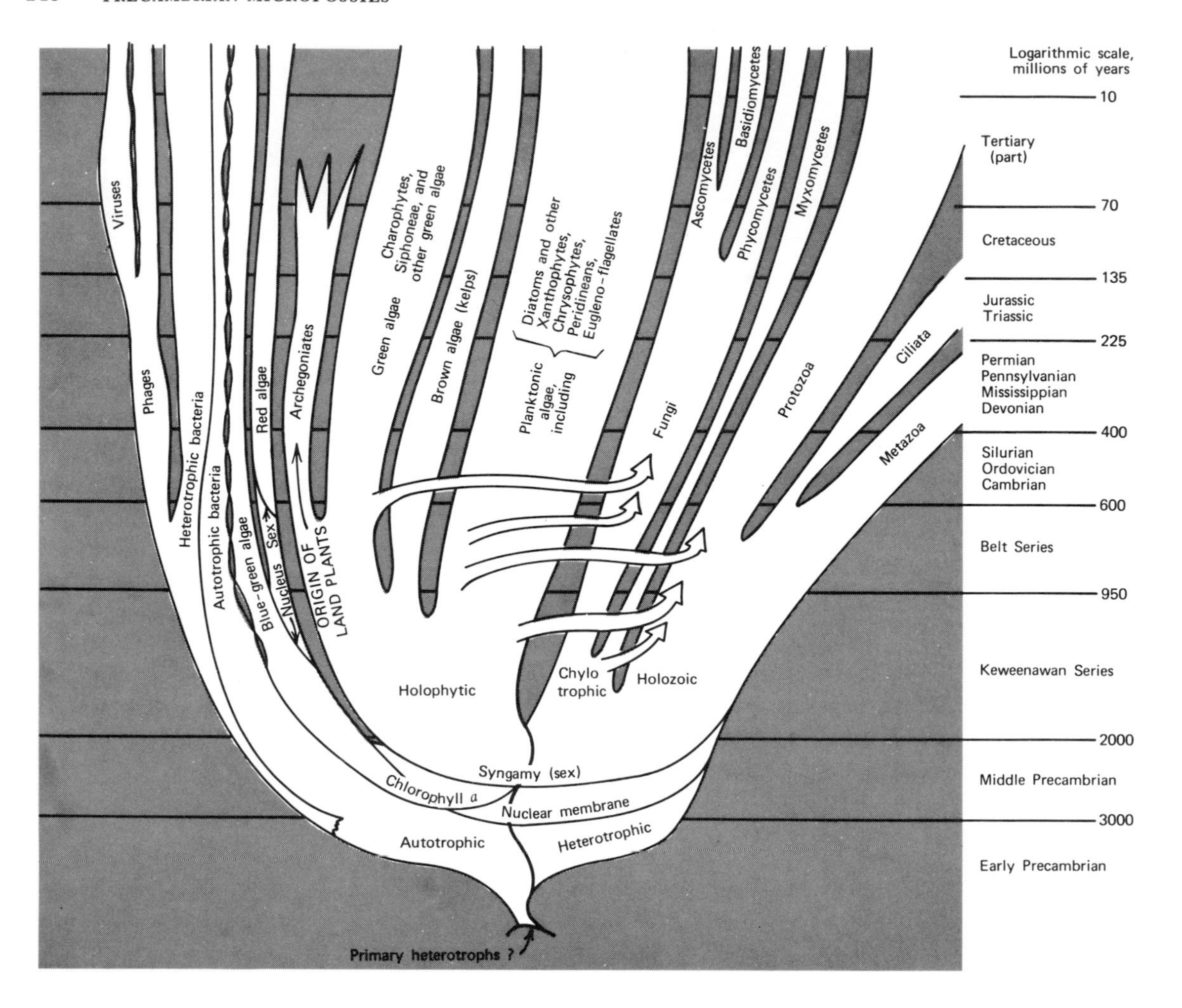

Figure 9-1. Phyletic progression in organic evolution.

portant, because, at or near the points of dichotomy, there may have been very little morphologic difference between the biotic elements that later gave rise to major groups of organisms. Shaded areas inserted between the labeled developmental "pathways" signify phyletic isolation.

When we consider the taxonomic subdivision of very ancient biotic assemblages in which morphologic evidence is at a minimum we inevitably encounter uncertainties. The major taxa have necessarily been defined in relation to their modern diversification, and, if phylogeny is projected to a point of origin, even major distinctions become matters of technicality. The modern groups of microorganisms that are presumed on various grounds to have a long Precambrian ancestry are the aquatic organisms whose habitats also have been maintained down to the present. Uniformity of environment evidently exerts a strong conservative influence.

Most of the major groups of organisms may contribute microfossil remains of one kind or another, but these remains may be so nearly alike that assured recognition is a problem. Recent studies show that several kinds are present even in very ancient deposits. These fossils are highly significant from a theoretical as well as a more practical stratigraphic standpoint. They present a practical problem in classification.

Classification of ancient microfossils must be based largely on inferences derived from morphologic evidence. Consequently their classification must be arbitrary to a considerable degree. It is desirable, however, that any arbitrary and conventional system of classification be designed to avoid conflict with fundamental evolutionary conceptions. A conventional taxonomic classification of the ancient microfossils should be judged according to criteria of convenience, with appropriate regard for criteria that appear to have a fundamental significance.

Attributes of Early Organisms

According to present concepts a primitive heterotrophic type of protobiont was evolved initially from an abiotic, coacervate "soup." Some authors (Oparin, 1938; Clark and Synge, 1959; Gaffron, 1960) have speculated that "organic" compounds were formed abiotically, by irradiation or electric discharge, and may have been assimilated by initially heterotrophic organisms before the mechanisms for autotrophic synthesis existed. Probably heterotrophic organisms (primary heterotrophs) would have exhausted the supply of abiotic organic nutrients fairly soon. In any event such a demand would be destined to exceed the supply, and, until autotrophic synthesis was possible, the biotic cycle that is essential to modern organisms would not have existed. However, structures resembling bacteria existed as long as 3 billion years ago (Barghoorn and Schopf, 1966). Whether they were heterotrophic or autotrophic remains uncertain. Once a system of organic autotrophic synthesis was established, heterotrophic organisms also would have had a continuing food supply. Later heterotrophs might have been derived from the primary heterotrophs, or they might have been derived later from the autotrophic population. Probably the heterotrophic mode of existence originated separately several times. In either event the available microfossils must be derived from the autotrophic stock.

The initial autotrophic mechanism may have been anaerobic, and it may have catalyzed reactions that are relatively inefficient in terms of energy conversion. A more efficient synthesis was possible with the origin of a photosynthetic pigment such as chlorophyll *a*. Since this pigment is present in virtually all green plants, an ancient origin is indicated. Detailed recent knowledge of the mechanism of photosynthesis suggests that the process is too complicated and consistent to have been independently and separately duplicated in different groups of organisms.

Another feature of general import in relating organic groups is the nuclear membrane. In bacteria and in blue-green algae a true nuclear membrane is lacking. The nuclear membrane differs from the external membrane of microorganisms. The external membrane must be more involved with organismal definition, and it probably reflects essential environmental relationships. The external membrane must have been present in all discrete organisms. The nuclear membrane is a more specialized structure, very ancient, but not as primitive in origin. There is reason to believe that as the red algae differentiated from the blue-green algae (in which a nuclear membrane is lacking) a nucleus with its nuclear membrane was differentiated de novo. Aside from this example, probably the organization of a specific nucleus was not independently duplicated in other groups.

Fusion of sex cells, or gametes, is another innovation of ancient standing that must be fundamentally considered whether organisms are classed as plants or animals. Presumably much of the cause of evolutionary progression, at least as far as morphologic diversification is concerned, must be attributed to development of an effective means of genetic recombination, and segregation, when sex cells are initiated. Organisms that have not developed a specific sexual mechanism, such as bacteria and blue-green algae, seem to have remained amazingly conservative in their morphology in spite of very ancient origin. Evidence (pigments and cytologic features) links the origin of red algae to blue-green algae, but this relationship necessitates an independent origin of sex mechanism in red algae. The red algae also show several unusual features of sex adaptation, and they are relatively more complex anatomically than the blue-green forms. The bacteria, with their derivative phage and virus forms, blue-green algae, and red algae must all trace back, separate from other groups, to an early protobiotic form of life that existed in the early and middle Precambrian.

Emphasizing the indistinctness of some relatively ancient major groups of organisms is the evidence that fungi and protozoa, in particular, have a polyphyletic derivation. Arrows connecting separate phyletic paths indicate probable points of polyphyletic infusion; those indicated are ancient, but more recent "cross connection" is not excluded. Not only green algae but also some of the planktonic chrysophytes, and possibly others, have probably entered the ancestry of the algalike fungi (Phycomycetes). The autotrophic (holophytic) mode of life probably has been lost in some algae that concurrently had developed a means of utilizing organic nutrients supplied directly from their environment. Fungi obtain this food in solution, commonly by enzymatic action external to the cells—an adaptation that Cronquist (1960) has called chylotrophic. Life processes generally depend on the coordinated catalytic reactions of many enzymes. Duplication of function involving similar, but not necessarily identical, enzymes may have a greater possibility of independent origins.

Other microorganisms similar to planktonic flagellated algae not only possess chlorophyll but also are able to engulf particles of food in an apparently holozoic fashion. As a consequence opinion commonly differs as to whether such organisms are properly classed as animals or as plants. Detailed studies of their life histories are required to disclose true relationships. A sufficient number of these studies have been made to show that the protozoa, as generally conceived, must be of polyphyletic derivation. For palynology, however, groups of microorganisms of equivocal alliance are conventionally treated for nomenclatural purposes according to the Botanical Code (Lanjouw, 1966). Such a practice is justified when there is legitimate uncertainty of alliance. Possibly even the resistant membranes of ova of some groups of invertebrate animals might be so classified since their generalized morphology or association may not provide sufficient evidence. However, such extreme examples are probably not numerous. Most of the Precambrian palynomorphs probably do represent phytoplankton. However, generalized structure will always remain a source of uncertainty and difficulty for any phyletic or stratigraphic purpose. Simple structures may be reduplicated in groups of divergent alliance beyond our capacity to distinguish them.

A point of phyletic significance linking all the sexually differentiated organisms, animal and plant (except for red algae, ascomycetes, and basidiomycetes), is that all of their life cycles include the flagellum, an organelle of amazingly uniform, yet highly complex, organization. The mode of energy exchange that activates flagella is still not fully understood. Among the flagellated microorganisms the consistency of flagellar organization assumes particular importance. Electron micrographs show that two central fibrils and nine peripheral fibrils compose the various flagella (axoneme) of all the sexual organisms. A strong contrast is evident between these structures and the much simpler (anaxoneme) flagella found in some types of bacteria.

It is difficult to conceive of an organelle that is more likely to be environmentally influenced than the flagellum. The fact that an amazing conservatism in this structure actually exists in all "higher" organisms is striking evidence of the unity of derivation of the axoneme organisms (archegoniates; green, brown, and planktonic algae; algalike fungi; myxomycetes; protozoans; and higher animals). This evidence also must be evaluated for fossils in spite of the fact that fossilized flagella have never been observed. All sex-differentiated organisms (except for some red algae that seem to have independently attained a sexual distinction) probably have a common ancestry that can be estimated to date from sometime during the middle Precambrian.

Clearly enough, the distinction between autotrophs and heterotrophic organisms ranges back to great antiquity, and through the greater part of the time span under consideration these organisms have for the most part been phyletically isolated from each other. We do not *know*, of course, whether the myxomycetes (or "mycetozoa") have existed since almost the middle Precambrian, but their organization is so different from that of any other organisms that this estimate of their antiquity seems reasonable. Kelp-like algae (fucoids), presumably similar to brown algae (though there is as yet no means of detecting in fossils the specific brown pigment that distinguishes modern forms), apparently extend through the late Precambrian. Even the archegoniates that include bryophytes and all higher plants must have an ancestry touching the Precambrian.

The first archegoniates were surely advanced types of algae, although modern algae seem not to include any representation very close to an ancestral archegoniate type. However, the archegonium is a special organ that provided a basis for important evolutionary advancement. The really primitive arrangements of archegoniates may have become extinct as they gave rise to progressive types of plants with a more diversified morphologic organization. Certainly precursors to land plants existed before the advent of a land flora during the Silurian.

Many of the Precambrian carbonaceous microfossils of sporelike form have been classed as phytoplankton. Certainly the probabilities favor this assignment, but we must keep in mind that chylotrophic and even holozoic microorganisms of this antiquity probably did not differ greatly in form. All of these organisms probably produced some type of resistant disseminule that could be preserved under appropriate conditions of rapid burial and mild tectonism. Reeflike calcareous deposits of Precambrian age suggest the presence of algae or bacteria that stimulated calcareous precipitation, but such deposits do not commonly include actual organic remains. They do indicate that the Precambrian microbiota was abundant. Future discoveries will no doubt provide further evidence of this vast assemblage that existed and became diversified during a period of time more than four times as long as the whole Phanerozoic.

CHANGES IN EARLY ENVIRONMENTS

Actualistic or uniformitarian principles apply in the interpretation of the Phanerozoic environments. During the Phanerozoic the sea and the atmosphere were relatively constant in composition and the land was affected by weathering and erosion at relatively constant rates. All these elements of course are interrelated, and a substantial change in one affects the others (Rubey, 1955).

Geology has made a remarkable progress in the interpretation of ancient environments according to the actualistic concept, yet at some point in the ancient past considerable variation must have occurred (Rutten, 1962; Rubey, 1955). Vital processes have evidently existed during more than half of the earth's past history. Photosynthesis by green plants is presumed to account for the present concentration of free oxygen in the atmosphere, but Rubey believes that at least a small amount of free oxygen must have been present very early in earth history. Further study of plant microfossils in Precambrian rocks should contribute important evidence about the early history of the earth before the atmosphere had acquired its present composition.

A current concept of the evolution of the atmosphere during geologic time is illustrated in figure 9-2. It illustrates the general differences and rate of change between the ancient and modern composition of the atmosphere. Time is plotted according to a logarithmic scale, and a generalized projection is given of the evolution of bacteria, plants, and animals. Although this illustration was drawn up entirely independently from the chart presented in figure 9-1, there appears to be general agreement. Unfortunately the progressive reduction in free chlorine, ammonia, and methane, the waxing and waning of carbon dioxide and hydrogen sulfide, and the enormous increase of free nitrogen and, to a lesser degree, free oxygen that prevails in the atmosphere today, as presented in figure 9-2, do not entirely agree with Rubey's (1955) interpretation. Any quantitative estimate is difficult, and some of Vologdin's (1962) data may be poorly founded. The reader should refer to Rubey's work for more detailed discussion. Differences in the atmospheric composition must have affected the Precambrian biota. A valid stratigraphic succession of microfossils should confirm the interpretation of ancient environments. Cloud (1965) already has suggested a microbiologic control for the occurrence of widespread Precambrian iron formations.

THE TIME SCALE

Traditionally paleontologists have been little concerned with absolute time and have concentrated on relative ages and sequence of fossils. An emphasis on relative age continues to be accepted paleontologic practice for provincial correlation, but for more extended use dating by determination of certain isotopes produced by radioactive decay of unstable elements may be generally more reliable. This is particularly the case in ancient rocks of the Precambrian in which, until very recently, fossils were thought to be essentially lacking. Nevertheless, the samples and techniques used for radiometric dating commonly involve a wide range of error. Such time determinations are "absolute" only in theory.

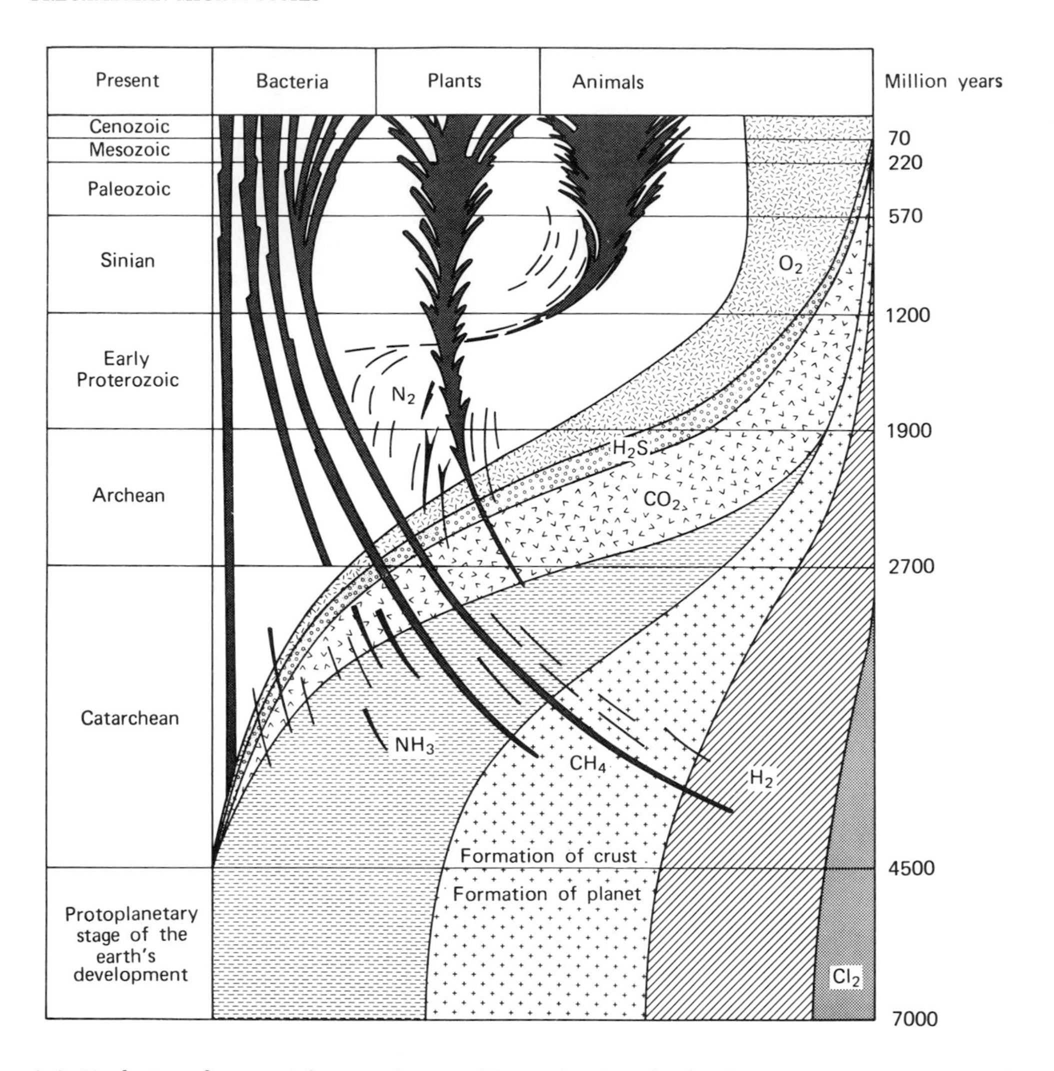

Figure 9-2. Evolution of terrestrial atmosphere and biota. (A. G. Vologdin from Kuznetsov, Ivanov, and Lyalikova, 1963.)

Isotopic decay proceeds at a constant rate independent of temperature or pressure, and a numerical age determination is possible providing daughter products are all in situ. Some daughter products are gaseous and may be lost, essentially by "leakage," and some may be lost by leaching if exposed to permeating water. Moreover, the radioactive elements must be present in sufficient quantity to permit an accurate quantitative determination. Ordinarily such amounts are present only in igneous rocks. The concentrations of mother and daughter isotopes are altered each time a rock is melted. As a rule isotopic determinations indicate the age of the last period of melting or of igneous intrusion, but less profound tectonic effects also are sometimes suspected of changing the composition. The precision of isotopic age determination varies according to the nature of the sample and other errors, and this is generally suggested by a plus-or-minus value that signifies the standard deviation of mass-spectrometer readings. Under favorable conditions for analytical determination relatively good agreement between duplicate samples may be achieved. The significance of these determinations always must be interpreted geologically, and we are forced finally to settle for a value showing relative consistency. Nonetheless, important progress has been made by this means in dating the ancient rocks.

The fossils that have been found in Precambrian rocks so far do not constitute an established independent basis for dating. Thus, even if isotopic dating provides only an inexact correlation, it commonly is the best information available. The older occurrences of Precambrian microfossils are difficult to interpret stratigraphically. Timofeyev (1959) has attempted stratigraphic correlation on the basis of palynomorphs from upper Precambrian (Sinian) deposits. These, however, appear very similar to forms that occur in the Lower Cambrian. Precambrian microfossils are still very incompletely known, and their stratigraphic use is just beginning. In some deposits microfossils are abundant, but not enough occurrences have been studied, and more refined study methods may be needed.

The chart presented in figure 9-3 illustrates in the left-hand column the now conventional isotopically determined dates used in most of the Western World for the Precambrian and Early Paleozoic. Note that, in order to accommodate an arithmetic scale of progression, the Precambrian section of this chart is represented at one-tenth the scale of the Paleozoic divisions. In the right-hand column similar dates are given according to a recent committee report of the Soviet Academy of Science (Akademii Nauk SSSR, 1960). Considering that the two series of age determinations have been based on different samples obtained in different parts of the world, the general agreement is highly significant. Some differences in stratigraphic terminology are evident. The apparently shorter interval determined by Soviet scientists for the Silurian (Upper Silurian as originally named by Murchison) may signify some discrepancy of correlation rather than simply a divergence in age determination. The Precambrian stratigraphic terminology of the left-hand column is in part that proposed by Goldich and others (1961). Terms designating subdivisions of the Precambrian generally have not been used as consistently as those in the Paleozoic, although greater stability as based on recent isotopic determinations now seems likely.

Figure 9-3 also shows the relative ages of noteworthy occurrences of fossils in the Precambrian, all but one of which are reasonably well controlled by isotopic dating. The Fig Tree occurrence is evidently oldest at about 3 billion years; that from the Gunflint Iron Formation is best known. The fossils from the "Molar Tooth" Limestone of the Belt Series (Pflug, 1965) are more diversified and, since they were isolated by a novel electrostatic and mechanical dispersion technique well adapted to general palynologic use, are of considerable interest. The Bitter Springs, Gunflint, and Fig Tree fossils are all preserved in petrified condition in primary deposits of chert. These occurrences are discussed in a subsequent section of this chapter.

THE FIRST VESTIGES OF LIFE

Search for Organized Extraterrestrial Objects

The advent of rocket exploration of space has spurred interest in the possible existence of extraterrestrial life. The most promising investigative techniques have been palynological. A greater appreciation of the value of morphological evidence provided by microscopic physical objects has developed as a result of this interest. Students who have become engaged in the quest for extraterrestrial evidence of life without adequate preparation or warning of the pitfalls peculiar to palynology are apparently most fallible; in spite of premature reports, the problem is not solved. Whether such reports are true or not, they at least sustain interest. The evidence favoring the existence of extraterrestrial life is unconvincing, but the question is extremely important, and the obstacles to positive proof remain impressive. The search for evidence of extraterrestrial life has contributed an additional understanding of the technical problems associated with the palynology of "unique" structures.

Search for evidence of extraterrestrial life has been concentrated on the carbonaceous chondrites, a type of meteorite with a large proportion of carbonaceous material. Such meteorites are relatively fragile, but the meteoritic origin of those studied is relatively well authenticated because all were observed to fall. Nevertheless, possibilities of local contamination with terrestrial material may not always be excluded (Fitch and Anders, 1963; Anders and others, 1964). The general status of these studies has been recently reviewed by Urey (1966).

Although Urey is a chemist, he believes that the palynologic evidence is "not of negligible importance." He gives good illustrations of objects described by Timofeyev (1963) as derived from the Migei meteorite and shows an object

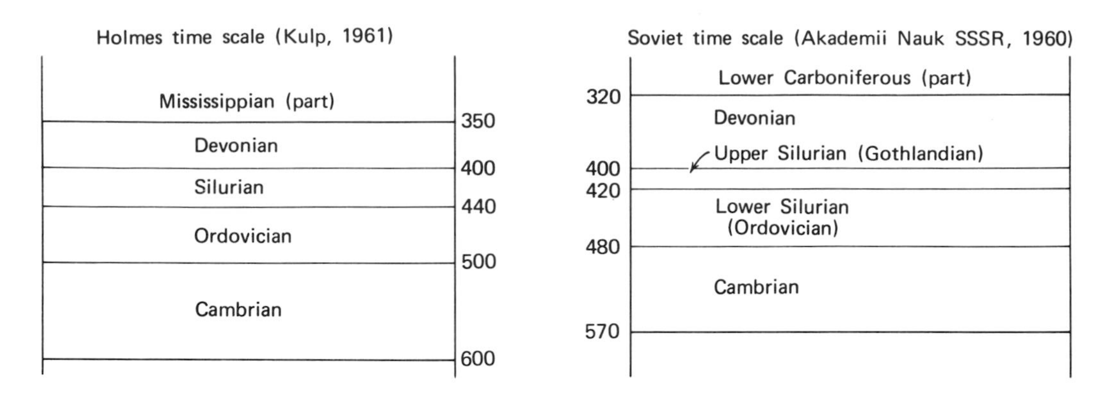

Precambrian units below are shown at one-tenth of the vertical scale of the Paleozoic units above

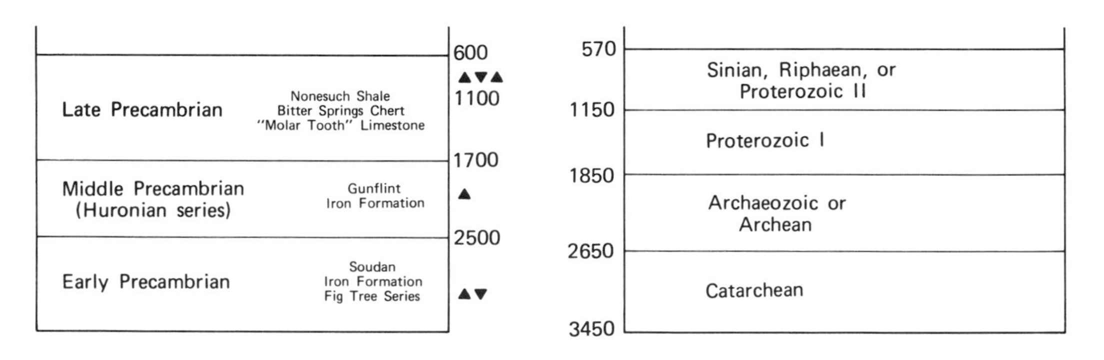

Figure 9-3. Comparison of "absolute" age scales and stratigraphic subdivision of ancient strata. Relative positions of Precambrian petrified microfossil material are indicated. See Goldich and others (1961) for discussion of Precambrian subdivisions; for fossils of Bitter Springs Chert see Barghoorn and Schopf (1965), and Schopf (1968); for Nonesuch Shale, Meinschein, Barghoorn, and Schopf (1964), and Barghoorn and others (1965); for "Molar Tooth" limestone. Pflug (1964, 1965, 1966); for Gunflint Iron Formation, see Tyler and Barghoorn (1954), Barghoorn and Tyler (1965), Cloud (1965), Schopf and others (1965); for Soudan Iron Formation, Cloud, Gruner, and Hagen (1965); for Fig Tree Series, Barghoorn and Schopf (1966), and Schopf and Barghoorn (1967).

previously reported by Nagy and others (1963) photographed in a thin section from the Orgueil meteorite that is very difficult to explain. Some other unusual objects derived from solution residues and given "extraterrestrial" names (Claus and Nagy, 1962; Staplin, 1962a) have since been shown to correspond with terrestrial materials (Fitch and Anders, 1963; Staplin, 1962b). Not all, however, have been explained. The organic microchemical evidence also is important, though it also, to a greater or lesser extent, is controversial. As Urey emphasizes, few questions are more fundamentally exciting today than the question of the existence of extraterrestrial life.

Although an early positive solution to the primary problem of demonstrating the existence of life extraneous to our planet is still lacking, the methods applied to investigate this problem will continue to be of interest. Microscopists now are more aware of the great inherent difficulty of establishing the authentic source of any *particular* particle observed under the microscope. Apparently palynologists most accustomed to microscopical search procedures for common objects habitually disregard many of the kinds of particles that are not pertinent to their specific studies. However, no type of particle can justifiably be ignored in the critical search for extraterrestrial biotic organization, and specific isolated particles of unusual form are difficult to identify or explain conclusively. The fact that so many of the "unique" microscopic objects have come to be identified and reasonably explained as terrestrial suggests a certain stage of advancement in palynology. Many more bizarre types of terrestrial particles are known now than ever before.

One great principle is reemphasized. Different biophysical relationships exist within the "world" of microsizes, as Wodehouse (1935) pointed out many years ago. In this world the forces of surface attraction, also called van der Waals forces, are generally more significant than the force of gravity. A minuscule electric charge will influence attraction of small particles. Particles may adhere to an exposed surface long after the original force of attraction has been dissipated. Attraction varies for different types of material in different situations, and, if not physically disturbed, some of the small particles may remain indefinitely. Unless surfaces have been scrupulously monitored recently, it is rash to suggest that any surface is truly free from vagrant particles. As any kind of particle may seem significant if we are searching for an extraterrestrial organized object, the severity of the requirements is extreme. If the vagrant appears to be unique (and uniqueness becomes more likely the more vigorously precautions are observed), problems of identification also are more difficult. Probably in preparations designed for microscopic optical observation there can never be an *absolute* safeguard against foreign small-particle contamination.

Petrified Microfossils in Precambrian Rocks

Three microfossil occurrences have been described from Precambrian deposits of primary chert. The chert matrix is apparently best suited to the preservation of biologic structures and biogenic substances of great antiquity. In such a matrix the fossils are not replaced; carbonaceous substances remain and are permineralized. The oldest microfossils yet known occur in cherty bands of the Fig Tree Series, Swaziland System, near Barberton in South Africa, and have been regarded, on the basis of isotopic dating, as about 3 billion years old (Nicolaysen, 1962). The Swaziland deposits are of economic as well as theoretical interest and are now generally regarded as the oldest sediments known on earth. It is of great interest, then, to discover that these rocks also contain definite remains of fossil microorganisms. One species of bacilliform fossils has been described recently by Barghoorn and J. W. Schopf (1966) and named *Eobacterium isolatum*. Spheroidal algalike fossil objects are also present (Schopf and Barghoorn, 1967), and residues of fibrillar carbonaceous matter occur; these have been interpreted as a residue of dispersed organic matter.

Tracings from Barghoorn and Schopf's electron micrographs are given in figure 9-4, reduced to approximately 25,000× magnification. The organisms are cylindrical and short, 0.45 to 0.75 microns by 0.18 to 0.32 microns. The wall appears to be two layered and granulose on the external surface. Figure 9-4*b* is a tracing of the opaque internal cast of the replicated body (an external mold) as shown in figure 9-4*a*, that was associated in the same micrograph. Figures 9-4*c* and 9-4*d* represent transverse sections of other individuals with the internal cast in place. Figure 9-4*e* shows a flattened and distorted specimen in which the polishing and etching apparently has laid bare an external surface with characteristic granulosity. A fissure (black) encircles much of this specimen. This extremely ancient isolated occurrence of a cellular organism is of interest, but its significance will be difficult to evaluate without additional studies. It apparently demonstrates the extreme conservatism of the bacilliform morphology. The occurrence must at least be interpretable in placing broad

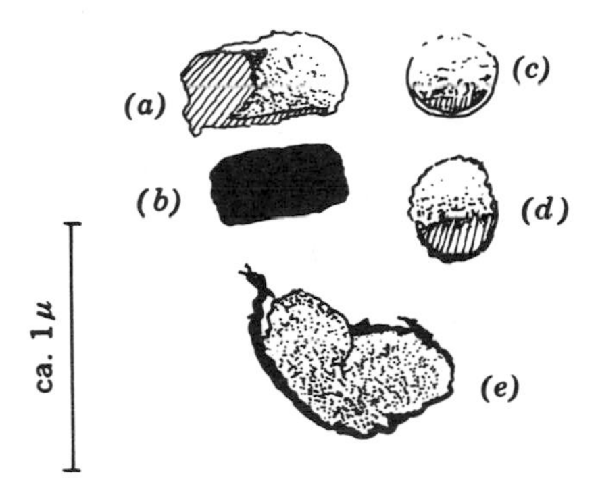

Figure 9-4. Bacilliform organisms (× ca. 25,000) drawn from electron micrographs of chert from the Fig Tree Series: (*a*) external mold of bacilliform organism, *Eobacterium isolatum* Barghoorn and Schopf, side view; (*b*) electron-opaque cast of (*a*); (*c*) and (*d*) transverse sections of *Eobacterium* specimens; (*e*) distorted specimen of *Eobacterium*, showing apparent surface granulation.

biotic limitations (temperature, media, etc.) on the original environment.

The best known of the Precambrian microfossils are those from the chert deposits of the Gunflint Iron Formation in Ontario (Tyler and Barghoorn, 1954; Barghoorn and Tyler, 1965; Cloud, 1965; Oró and others, 1965; Barghoorn, Meinschein, and Schopf, 1965). A variety of filamentous, sporelike, and anomalous microorganisms are present in this black, stromatolitic chert of late Huronian age in the upper third of the middle Precambrian. Isotopic dating suggests an age of at least 1.7 billion years, probably about 2 billion years, for this deposit.

The fossils consist of carbonaceous remains embedded in chalcedony. One of the more distinctive forms is a tiny (12 to 30 μ) umbrella-shaped object with a bulbous appendage, named *Kakabekia*. (See fig. 9-5, *a-d*.) Sporelike bodies have been assigned to the genus *Huroniospora*, examples of which are given in figure 9-6, *a-d*. Figure 9-6*e* illustrates a stellate thallus with branching filamentous appendages. Examples of filamentous thalli are given in figure 9-7, *a-i*. Perhaps most distinctive are the septate filaments shown in figure 9-7, *b* and *c*, which are assigned to *Gunflintia grandis*. The septa are variably spaced and tend to be obscured by residual contents of the cells. Occasionally the cell contents seem to show a spiral pattern of uncertain significance. Similar residues may be observed in dead cells of some types of modern filamentous algae. The authors of *Gunflintia* do not speculate about the contents of the cells, but if the organization in the protoplasmic residue can be interpreted it might imply that these forms had progressed beyond the level of blue-green algae. Similar, but much smaller, filaments are assigned to a different species, *Gunflintia minuta* (fig. 9-7, *d* and *e*).

The unbranched filaments described as having closely spaced septa are assigned to *Animikiea septata*. The septa shown in figure 9-7*a* appear so closely spaced and tenuous that some question remains whether these forms are truly septate or coenocytic with transversely striate walls. The structure of the filaments is distinctive.

Other unbranched filaments contain internal sporelike bodies of somewhat vague preservation. Similar bodies also occur external to the filaments and associated with them, but the suggestion may be made that a fertile phase is represented (fig. 9-7*f*, *Entosphaeroides amplus*). A much less regular type of filamentous organism is shown in figure 9-7*g*, *h*, and *i* (*Archaeorestis schreiberensis*). No septation is noted, but the filaments are irregular, branched, and intergrown. Interpretation is quite uncertain, but the irregular outlines suggest comparison with haustorial growths of some heterotrophic organisms.

In addition electron-micrographic studies of the Gunflint Iron Formation have disclosed numerous bacteria resembling the modern iron bacillus, *Sphaerotilus natans* (Schopf and others, 1965). Other forms resembling modern coccoid types of iron bacteria are also present. The ancient forms are smaller than modern analogs. The association of the fossil *Sphaerotilus* type in chains, sometimes with false branching and

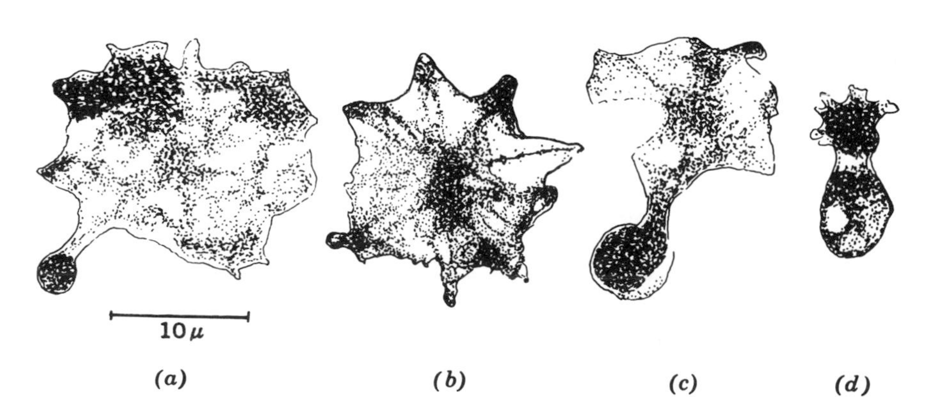

Figure 9-5. *Kakabekia umbellata* Barghoorn (× ca. 1,763), see Barghoorn and Tyler, 1965, (*a*) (type specimen); (*b*); (*c*); (*d*).

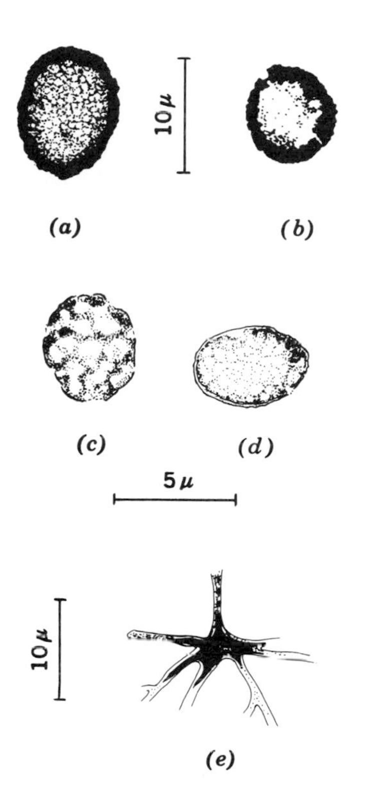

Figure 9-6. *Huroniospora* and *Eoastrion*: see Barghoorn and Tyler, 1965, (*a*) *Huroniospora microreticulata* Barghoorn (× 1,566); (*b*) *H. macroreticulata* Barghoorn (× 1,620); (*c*) *H. macroreticulata* Barghoorn (× 3,132) (type specimen); (*d*) *H. psilata* Barghoorn (× 3,132) (type specimen), (*e*) *Eoastrion bifurcatum* Barghoorn (× 1,312).

remnants of a sheath, distinguishes the Gunflint bacteria from the much older bacilliform organisms from the Fig Tree Series.

The possible importance of the Gunflint microorganisms in modifying (oxygenating) the sea and atmosphere and in iron deposition has been discussed by Cloud (1965). It is likely that many of the organisms were autotrophic, but it is not clear whether they were aerobic.

One similarly preserved assemblage from upper Precambrian rocks at Bitter Springs, Australia, has now been studied (Barghoorn and Schopf, 1965; Schopf, 1968). Spheroidal bodies, small septate filaments, and large septate filaments have been illustrated and described, including 14 species as members of the Oscillatoriaceae and Chroococcaceae. Possibly the Rivulariaceae and Nostocaceae also may be represented. Published figures suggest excellent preservation, with residues of protoplasmic contents more or less intact in many cells. The occurrence of green (eucaryotic) as well as blue-green (procaryotic) algae in this assemblage is noteworthy. Certainly the forms seem to have a surprisingly modern aspect in comparison with the Gunflint fossils. The late Precambrian age of the deposit is attested by reference to the overlying base of the Cambrian in the Bitter Springs area. Isotopic age determinations that have recently become available (Milton, 1966) suggest an age well in excess of 760 million years. Thus it may be safe to suggest that a modern type of filamentous algal assemblage had appeared 800 to 1,000 million years ago. Like the Gunflint and Fig Tree occurrences the delicate fossils have been preserved in primary chalcedonic deposits in an area that escaped subsequent metamorphic alteration.

Microfossils in Detrital Sediments

Most plant microfossils (palynomorphs) are not preserved in a petrified condition like those previously discussed. They are incorporated into sediment at the time of deposition and undergo much the same diagenetic and eometamorphic changes that affect the deposit generally. In preservation they may be compared with common types of plant megafossils that show the coalified type of compression preservation. Because they consist of carbonaceous rather than mineral matter, they have commonly been thought to be very susceptible to metamorphic destruction, but, evidently, this is not necessarily true. However, as a consequence of this belief older sediments that are somewhat altered by pressure have not been searched as diligently as younger deposits. Recent discoveries of plant microfossils in these older deposits suggest that they may not be barren of fossils, merely because of changes associated with incipient metamorphism. Fossils that are damaged to some extent may still be present and sufficiently distinctive to provide facies or stratigraphic indications of significance.

Late Precambrian and Cambrian palynomorphs obtained by usual maceration methods have been described chiefly by Soviet authors. Naumova (1951) described sporelike microfossils from Precambrian deposits of the southern Urals and compared them with forms of the Baltic Blue Clay (Naumova, 1949), discussed in the

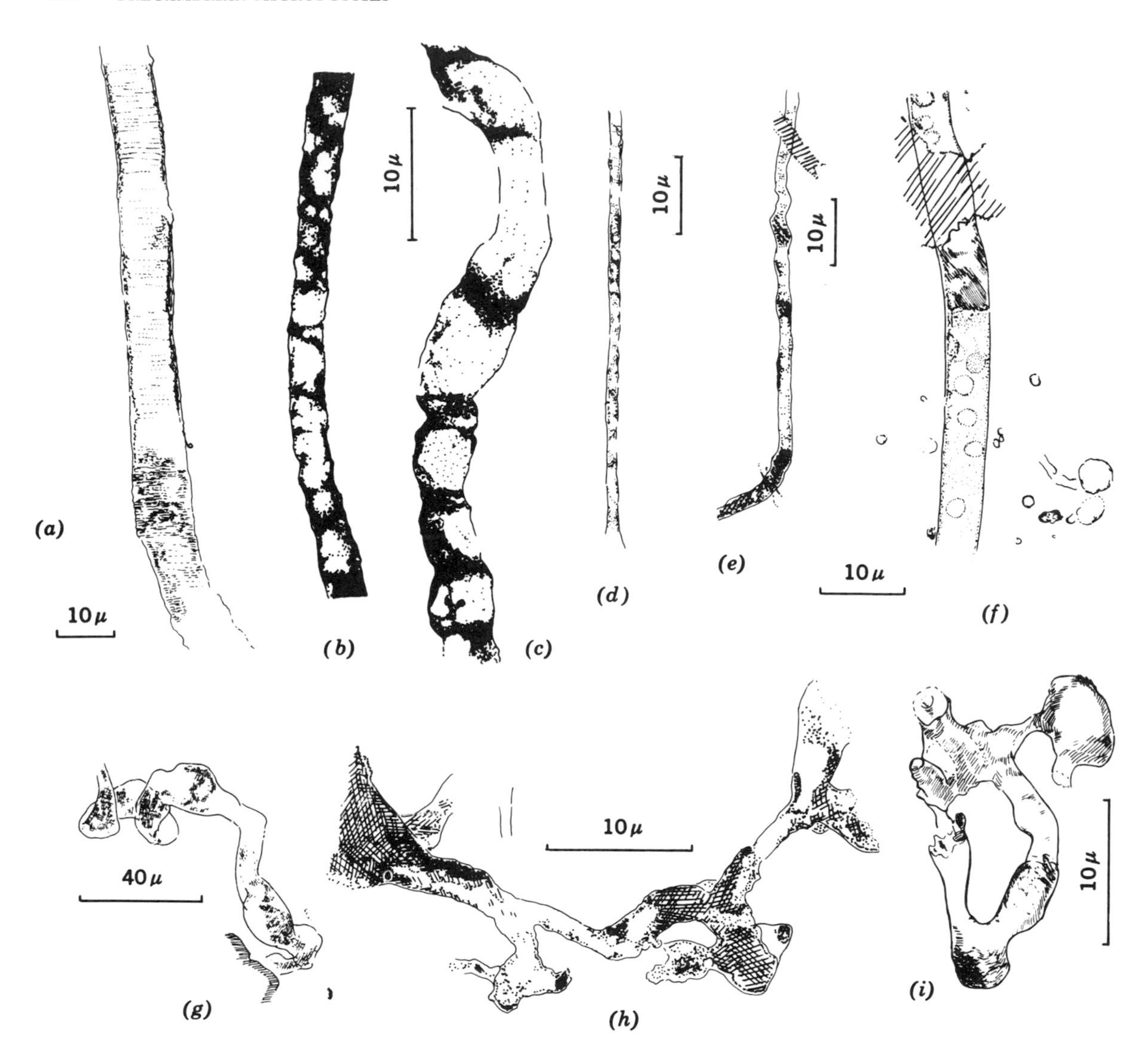

Figure 9-7. *Animikiea, Gunflintia, Entosphaeroides, Archaerestis,* see Barghoorn and Tyler, 1965: (*a*) *Animikiea septata* Barghoorn (× 830); (*b*) *Gunflintia grandis* Barghoorn (× 2,135); (*c*) *G. grandis* Barghoorn (× 2,135) (type specimen); (*d*) *G. minuta* Barghoorn (× 1,440) (type specimen); (*e*) *G. minuta* Barghoorn (× 935); (*f*) *Entosphaeroides amplus* Barghoorn (× 1,235) (type specimen); (*g*) probably *Archaeorestis schreiberensis* Barghoorn (× 435); (*h*) *A. schreiberensis* Barghoorn (× ca. 2,088) (type specimen); (*i*) *A. schreiberensis* Barghoorn (× ca. 2,088).

next chapter of this book. Timofeyev (1955, 1957, 1958) has described a more extensive similar suite of sporelike forms from Riphaean deposits in Saxony, Siberia, and Karelia. The sporelike, or cystose, structures represent fossils that definitely indicate the existence of abundant plant life in the Precambrian. Membranous forms are more common than thick-walled forms.

Many of the fossils are small and lack very definitive features. Most distinctions are based on differences of surface ornamentation and texture. Many of them seem to be poorly illustrated, either by obscure photographs at too low magnification or by drawings that imply a degree of generalization. Commonly the artists have shown a restricted, closed, three-rayed feature on the surface of the wall, but its significance is uncertain. Merely the obscure juncture of folds or striae may be represented. It is even more questionable that these cystose bodies show consistent evidence of orientation, or polarity, that would be implied by a functional three-rayed suture. Soviet authors have tended to interpret fossils having a tetrad imprint as land-plant spores, but remains of land plants are not likely to occur in Precambrian and Cambrian deposits. Probably the microfossils represent algae, and several authors prefer to class these forms noncommittally as "microplankton." A number of microfossils of this type are illustrated in figures 10-5 and 10-7.

Many of the Precambrian and Cambrian microfossils described by Timofeyev and his colleagues are illustrated by indistinct drawings or photographs that suggest difficulty in preparing this material for observation. The Precambrian forms have a simple morphology, and even the fact that these forms have been designated by lengthy names, intended to suggest morphology or relationship, does not conceal the real paucity of definable evidence. Forms such as those called *Prototrachysphaeridium nevelense* Tim., *P. conglutinatum* Tim.; *Protomycterosphaeridium marmoratum* Tim.; *Trachyoligotriletum obsoletum* (Naum.) Tim.; *Protoleiosphaeridium*; *Protolophosphaeridium crispum* Tim.; *Protozonosphaeridium patelliforme* Tim.; *Stictosphaeridium sinapticulaeferum* Tim., *S. implexum* Tim.; *Leioligotriletum nitidum* Tim.; and *Prototrematosphaeridium* are not very distinctive microfossils. Nevertheless, sufficient distributional evidence has been obtained for Timofeyev to base geologic correlations on it.

Studies by Marie-Madeleine Roblot (1963, 1964a,b,c) disclosed microfossils in authigenic cherts (phthanites) of the middle Brioverian Series, said to be from below the widespread tillite of the late Precambrian. Plutonic rocks emplaced in the late Brioverian have been isotopically dated at 550 and 580 million years, but the Precambrian tillite has been recognized in many parts of the world. The deposits must be somewhat older. The middle Brioverian has been assigned to the later stages of the Riphaean. (See fig. 9-3.) Correlation of absolute age and the paleontological base of the Cambrian increasingly may be subject to divergence of opinion in the light of new information.

Roblot's fossils are well illustrated, and most of them are similar to those shown in figure 9-8 discussed on a later page. Many of them appear to be delicate, discoidal objects that probably were compressed from an originally spheroidal form. Some show a patterned surface, but others are rather ambiguous with a texture suggestive of corrosion. Whether these forms should be regarded as hystrichospheres is uncertain. Spiny or hirsute appendages are lacking, but knoblike projections occur on at least two examples. Roblot has designated types of objects by a system of numbers and has not identified her material with that of Timofeyev or Naumova, although some of the objects seem very similar.

Other Precambrian microfossil assemblages have been reported by Pflug (1964, 1965, 1966) from the Belt Series in Montana. Material reported by Timofeyev and Naumova is more sporelike and seems to have been obtained by normal palynological solution and preparation procedures. Pflug, however, has used physical methods of disaggregation and an electrostatic means of concentrating the organic microfossils from the associated detrital mineral fragments. Perhaps partly as a result of avoiding chemical treatment, an assemblage of different aspect has been obtained. Most of the fossil remains appear to be collapsed by desiccation rather than crushed by pressure of overburden. Some cells contain evident bubbles, probably introduced during material preparation. Intaglios of mineral grains, fairly common in larger types of microfossils, are relatively rare. Some of the organized cells in Pflug's material are so minute that they may have been preserved within interstitial spaces that withstood compression of the rock. Many fossils consist of tiny, short, multicellular filaments. Many of the cells appear to contain a darkened residue, possibly consisting of altered protoplasmic material. Other examples, observed in situ within the rock by means of high-magnification, dark-field optics, seem to be organized as simple tissue. Isotopic determinations from related beds within the Belt Series suggest an age of about 1.1 billion years for this material. Some of the fossil types illustrated by Pflug are shown in figure 9-8.

These fossils are different from Precambrian fossils observed elsewhere by other workers. The fossils might represent the remains of anaerobic heterotrophic microorganisms that were buried in situ under accumulating sediment. If the fossils can be so interpreted, the technique employed by Pflug might open an additional extensive field for paleomicrobiologic investigation. The determination of botanical relationships of this material has significance with re-

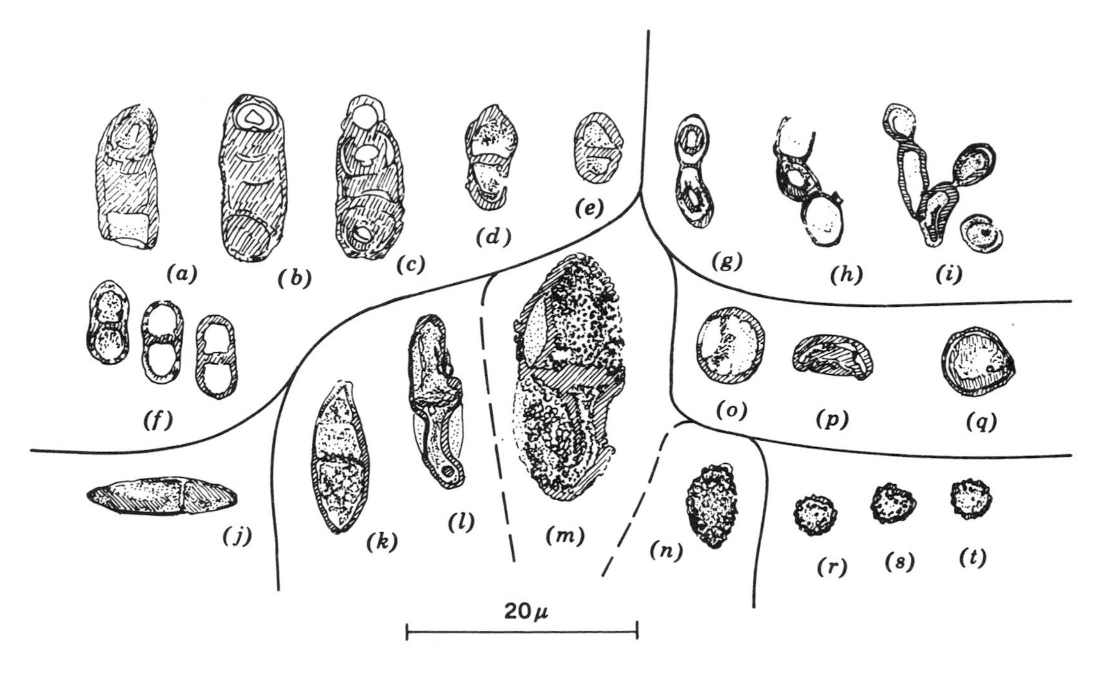

Figure 9-8. *Millaria, Fibularix, Catinella, Scintilla* (see Pflug, 1966; all × ca. 1,500); (*a*), (*b*), and (*c*) *Millaria implexa* f. *multiplex* Pflug—short, thick-walled, uniseriate filaments, with dark contents occupying some of the cells; (*d*), (*e*), and (*f*) *M. implexa*, f. *bipartita* Pflug—two cell filaments; (*f*) illustrates the same filament at different planes of focus; (*g*), (*h*), and (*i*) *Fibularix funicula* Pflug—laxly branched segments of multicellular filaments or mycelium; note constricted neck at points of cellular attachment; (*g*) and (*h*), form II; (*i*) form III; (*j*) Fusiform bicell with darkened contents; Unassigned for lack of information. (*k*) and (*l*) *F. funicula* Pflug—lenticular and fusiform terminal cells with characteristic collapse pattern and smooth or shagreen surface texture. Vesicular contents are present in (*k*), which may represent a juvenile stage of development. These forms are connected by various intergradational stages with filaments shown in (*g*), (*h*), and (*i*). (*k*), form IV; (*l*) form V. (*m*) *F. porulosa* Pflug—isolated terminal cell with punctate ornamentation, form V, with collapse configuration similar to (*l*), but larger. (*n*) *F. spinosa* Pflug—sporelike cell, form IV. (*o*), (*p*), and (*q*) *Catinella polymorpha* Pflug—sporelike microfossils, surface smooth, shagreen, or flecked, with darkened cell contents; possibly represents unicells of blue-green or green algae. (*o*), form E; (*p*) (collapsed), form G; (*q*), form G. (*r*), (*s*), and (*t*) *Scintilla perforata* Pflug—sporelike body with ornamentation, similar to some minute acritarchs.

spect to evolutionary history of ancient plants and for its biostratigraphic use in the vast range of Precambrian deposits.

SUMMARY

In only a few favored instances is the paleontologist's search for evidence of Precambrian life much less exacting than the quest for extraterrestrial organized objects. The search involves repetitive search for fossil structures that provide some convincing indication of biologic variability because virtually no value can be attached to objects that really seem unique. Unique objects provide no satisfactory basis for biologic comparison or geologic evaluation. Very simple objects similarly are reduced in value as simplicity obviates critical comparison and increases the possibility of incidental duplication, not necessarily from sources that may be easily anticipated. Great paleontologic virtue is to be attached to the ability to reproduce observations, a requirement that always involves the presence of fossils in convincing numbers.

Literature on these and other Precambrian fossils described prior to 1962 has been reviewed by Glaessner (1962). He indicates the abundance of biogenic structures (not necessarily identifiable fossils) in Precambrian deposits and suggests that future studies will prove most rewarding. An extensive bibliography on Precambrian fossils has more recently been compiled by Murray (1965).

References

Akademii Nauk SSSR, 1960, Geokhronologicheskaya skhala v absolyutnom letoischislenii po dannym laboratorii SSSR na 1960 g [The absolute geochronologic age scale from 1960 data of USSR laboratories]: Commission on the absolute-age determination of the section of geologic and geographic sciences, Akad. Nauk SSSR Izv., Ser. Geol., no. 10, p. 10-13, [translation Am. Geol. Inst., Nov. 1961].

Anders, Edward, DuFresne, E. R., Hayatsu, Ryoichi, Cavaillé, Albert, DuFresne, Anna, and Fitch, F. W., 1964, Contaminated meteorite: Science, v. 146, no. 3648, p. 1157-1161, 6 figs.

Barghoorn, E. S., Meinschein, W. G., and Schopf, J. W., 1965, Paleobiology of a Precambrian shale: Science, v. 148, no. 3669, p. 461-472.

Barghoorn, E. S., and Schopf, J. W., 1965, Microorganisms from the late Precambrian of central Australia: Science, v. 150, no. 3694, p. 337-339.

———1966, Microorganisms 3 billion years old from the Precambrian of South Africa: Science, v. 152, no. 3723, p. 758-763.

Barghoorn, E. S., and Tyler, S. A., 1965, Microorganisms from the Gunflint Chert: Science, v. 147, no. 3658, p. 563-577.

Clark, F., and Synge, R. L. M., eds., 1959, The origin of life on the earth: New York, Pergamon Press, International Symposium, 1st, Proc., Moscow, Aug. 19-24, 1957, 691 p.

Claus, George, and Nagy, Bartholomew, 1962, Taxonomical consideration of certain incerta sedis: Phycological Soc. America News Bull., v. 15, no. 1, p. 15-19.

Claus, George, Nagy, Bartholomew, and Europa, D. L., 1963, Further observations on the properties of the "organized elements" in carbonaceous chondrites, *in* Nagy, Bartholomew, consulting ed., Lifelike forms in meteorites and the problems of environmental control on the morphology of fossil and Recent protobionta, Conf., New York, April 30-May 1, 1962: N.Y. Acad. Sci. Annals, v. 108, art. 2, p. 580-605.

Cloud, P. E., Jr., 1965, Significance of the Gunflint (Precambrian) microflora: Science, v. 148, no. 3666, p. 27-35.

Cloud, P. E., Jr., Gruner, J. W., and Hagen, Hannelore, 1965, Carbonaceous rocks of the Soudan Iron Formation (early Precambrian): Science, v. 148, No. 3678, p. 1713-1716.

Cronquist, Arthur, 1960, The divisions and classes of plants: Bot. Rev., v. 26, no. 4, p. 425-482.

Fitch, F. W., and Anders, Edward, 1963, Organized element; possible identification in Orgueil meteorite: Science, v. 140, no. 3571, p. 1097-1100.

Gaffron, Hans, 1960, The origin of life, p. 39-84 *in* Tax, S., ed., Evolution after Darwin: Chicago Univ. Press.

Glaessner, M. F., 1962, Precambrian fossils: Biol. Rev., v. 37, p. 467-494.

Goldich, S. S., Nier, A. O., Baadsgaard, Halfdan, Hoffman, J. H., and Krueger, H. W., 1961, The Precambrian geology and geochronology of Minnesota: Minnesota Geol. Survey Bull. 41, 193 p., 5 pls., 37 figs., 32 tables.

Kulp, J. L., 1961, Geologic time scale: Science, v. 133, no. 3459, p. 1105-1114.

Kuznetsov, S. I., Ivanov, M. V., and Lyalikova, N. N., 1963, Introduction to geological microbiology [translation P. T. Broneer]: New York, McGraw-Hill Book Co., Inc., 242 p.

Lanjouw, J., and others, eds., 1966, International code of botanical nomenclature: Utrecht, Netherlands, Internat. Bur. Plant Taxonomy and Nomenclature, 402 p.

Meinschein, W. G., Barghoorn, E. S., and Schopf, J. W., 1964, Biological remnants in a Precambrian sediment: Science, v. 145, no. 3629, p. 262-263.

Milton, D. J., 1966, Drifting organisms in the Precambrian sea: Science, v. 153, no. 3733, p. 293-294.

Murray, G. E., 1965, Indigenous Precambrian petroleum?: Am. Assoc. Petroleum Geologists Bull., v. 49, no. 1, p. 3-21.

Nagy, Bartholomew, Fredriksson, Kurt, Urey, H. C., Claus, George, Andersen, C. A., and Percy, Joan, 1963, Electron probe microanalysis of organized elements in the Orgueil meteorite: Nature, v. 198, p. 121-125.

Naumova, S. N., 1949, Spori nizhnego kembriya [Spores of the Lower Cambrian]: Akad. Nauk SSSR Izv., Ser. Geol., no. 4, p. 49-56.

———1951, Spory drevnikh svit zapadnogo sklona yuzhnogo Urala [Spores of ancient formations of the slopes of the southern Urals]: Moskov. Obshch. Ispytateley Prirody Byull., Otdel. Geol. [Moscow Soc. Naturalists Bull., Geol. Div.], v. 1, p. 183-187.

Nicolaysen, L. O., 1962, Stratigraphic interpretation of age measurements in southern Africa, p. 569-598 *in* Engel, A. E. J., James, H. L., and Leonard, B. F., eds., Petrologic studies—A volume in honor of A. F. Buddington: Geol. Soc. America.

Oparin, A. I., 1938, Origin of life [translation Sergius Morgulis]: New York, The Macmillan Co., 270 p.; 2d ed., 1953, New York, Dover Pubs.; 3d ed., The Origin of Life on the Earth, 1957, New York, Academic Press.

Oró, J., Nooner, D. W., Zlatkis, A., Wikström, S. A., and Barghoorn, E. S., 1965, Hydrocarbons of biological origin in sediments about two billion years old: Science, v. 148, no. 3666, p. 77-79.

Pflug, H. D., 1964, Nieder Algen und ähnliche Kleinformen aus dem Algonkium der Belt-Serie: Oberhessischen Gesell. Natur- u. Heilkunde Giessen Ber., Neue Folge, Naturw. Abt., v. 33, no. 4, p. 403-411.

———1965, Organische Reste aus der Belt-Serie (Algonkium) von Nordamerika: Paläont. Zeitschr., v. 39, no. 1-2, p. 10-25.

———1966, Einige Reste niederer Pflanzen aus dem Algonkium: Palaeontographica, v. 117, Abt. B, no. 4-6, p. 59-74, pls. 25-29.

Roblot, Marie-Madeleine, 1963, Découverte de sporomorphes dans des sédiments antérieurs a 550 M.A. (Briovérien): Acad. sci. [Paris] Comptes rendus, v. 256, p. 1557-1559.

———1964a, Valeur stratigraphique des sporomorphes du Précambrien Armoricain: Acad. sci. [Paris] Comptes rendus, v. 259, pt. 3, p. 4090-4091.

———1964b, Sporomorphes du Précambrien Normand: Rev. Micropaléontologie, v. 7, no. 2, p. 153-156.

———1964c, Sporomorphes du Précambrien Armoricain: Annales de Paléontologie (Invertèbres), v. 50, no. 2, p. 105-110.

Rubey, W. W., 1955, Development of the hydrosphere and atmosphere, with special reference to probable composition of the early atmosphere, p. 631-650 *in* Poldervaart, A., ed., Crust of the earth—a symposium: Geol. Soc. America Spec. Paper 62.

Rutten, M. G., 1962, The geological aspects of the origin of life on earth: Amsterdam and New York, Elsevier Publishing Co., 146 p.

Schopf, J. W., 1968, Microflora of the Bitter Springs Formation, late Precambrian, Central Australia: Jour. Paleontology, v. 42, no. 3, pt. 1, p. 651-688, pls. 77-86.

Schopf, J. W., and Barghoorn, E. S., 1967, Algalike fossils from the early Precambrian of South Africa: Science, v. 156, no. 3774, p. 508-512.

Schopf, J. W., Barghoorn, E. S., Maser, M. D., and Gordon, R. O., 1965, Electron microscopy of fossil bacteria two billion years old: Science, v. 149, no. 3690, p. 1365-1367.

Staplin, F. L., 1962a, Microfossils from the Orgueil meteorite: Micropaleontology, v. 8, no. 3, p. 342-347, pl. 1.

———1962b, Organic remains in meteorites—a review problem: Alberta Soc. Petroleum Geologists Jour., v. 10, no. 10, p. 575-580.

Timofeyev, B. V., 1955, Nakhodki spor v kembriyskikh i dokembriyskikh otlozheniyakh vostochnoy Sibiri [Spore discoveries in Cambrian and Precambrian deposits of eastern Siberia]: Akad. Nauk SSSR Doklady, v. 105, no. 3, p. 547-550.

———1957, Spory onezhskoy svity Karelii [Spores of the Onega Formation in Karelia]: Geol. i geokhimii (Doklady i stat'i) Vses. Neft. Nauchno-Issled. Geol.-Razved. Inst. Trudy, v. 1, no. 7, p. 153-155.

———1958, Über das Alter sächsischer Grauwacken; mikropaläophytologische Untersuchungen von Proben aus der Weesensteiner und Lausitzer Grauwacke: Geologie, v. 7, no. 3-6, p. 826-839.

———1959, Drevneishaya flora Pribaltiki i ee stratigraficheskoe znachenie [Ancient flora of the Baltic region and its stratigraphic significance]: Vses. Neft. Nauchno-Issled. Geol.-Razved. Inst. Trudy, v. 129, 320 p., 25 pls., 4 figs., 14 charts.

———1963, Lebenspuren in Meteoriten, Resultate einer microphytologischen Analyse: Grana Palynologica, v. 1, p. 92-99.

Tyler, S. A., and Barghoorn, E. S., 1954, Occurrence of structurally preserved plants in pre-Cambrian rocks of the Canadian Shield [Ontario]: Science, v. 119, no. 3096, p. 606-608.

Urey, H. C., 1966, Biological material in meteorites; a review: Science, v. 151, no. 3707, p. 157-166.

Vologdin, A. G., 1962, Zhizn' Zemli [Life on the Earth]: Izdvo Akad. Nauk SSSR.

Wodehouse, R. P., 1935, Pollen grains and worlds of different sizes: Sci. Monthly, v. 40, p. 58-62.

10

EARLY PALEOZOIC PALYNOMORPHS

James M. Schopf

This chapter is concerned with the palynomorphs that reflect diversification of aquatic plants already in existence in the Precambrian and subsequent events leading to the most primitive land flora. During the early Paleozoic many planktonic forms continued their aquatic existence with minimal modification. The more complex types of algae that were also represented suggest stages of evolutionary progression. Eventually some of the more specialized types of algae were able to develop means of moisture control independent of their aquatic environment. Some of these plants evidently were enabled to move onto and eventually above the strand and thus to initiate population of the naked terrestrial surfaces. The definition and identification of the group or groups of plants able to accomplish transmigration from sea to land have long been prime objectives of paleobotanical study.

Palynological materials and methods are likely to contribute important information about the diversification of algae and the origin of terrestrial plants. They offer the possibility of outlining to an extent formerly undreamed of locations and environments in which land plants could originate. The early Paleozoic aquatic plants have contributed abundant palynomorphic material, and this literature has recently been extensively reviewed by Obrhel (1958). Pre-Devonian microfossils have also been reviewed by Chaloner (1960). He concludes that a firm basis for interpretation of the origin of land plants from algal ancestors is not completely evident. Further work is needed to identify semiaquatic and land-plant microfossils. In the course of such work much new knowledge will be gained about floristic diversification, paleoecology, and the palynologic applications of biostratigraphy.

Tappan (1968) recently has noted that the autotrophic planktonic algae greatly increased and diversified during the early Paleozoic. The importance of these organisms is emphasized by the fact that she considers phytoplankton to control the oxygen-carbon dioxide balance of the atmosphere. No one can question that planktonic algae had this function during the late Precambrian and early Paleozoic. Since the advent of a land flora their importance may be only slightly less. Tappan has presented a graphic illustration of the oxygen concentration in the atmosphere, based on estimated abundance of microfossils, that is here reproduced with modifications in figure 10-1. The oxygen curve represents a subjective evaluation based on the known abundance of autotrophic plants. This diagram differs from other charts that show primarily the diversification of taxa. The abundance of individuals, rather than their taxonomic diversification, must be paramount with reference to geochemistry.

If we accept the estimated oxygen-concentration curve as a reasonable estimate, the relations to abundance of algae and terrestrial life also may be sketched in. The changes in the number of algae should anticipate and be subject to more extreme fluctuations than the atmospheric oxygen reservoir. Presumably atmospheric concen-

Publication authorized by the Director of the U.S. Geological Survey.

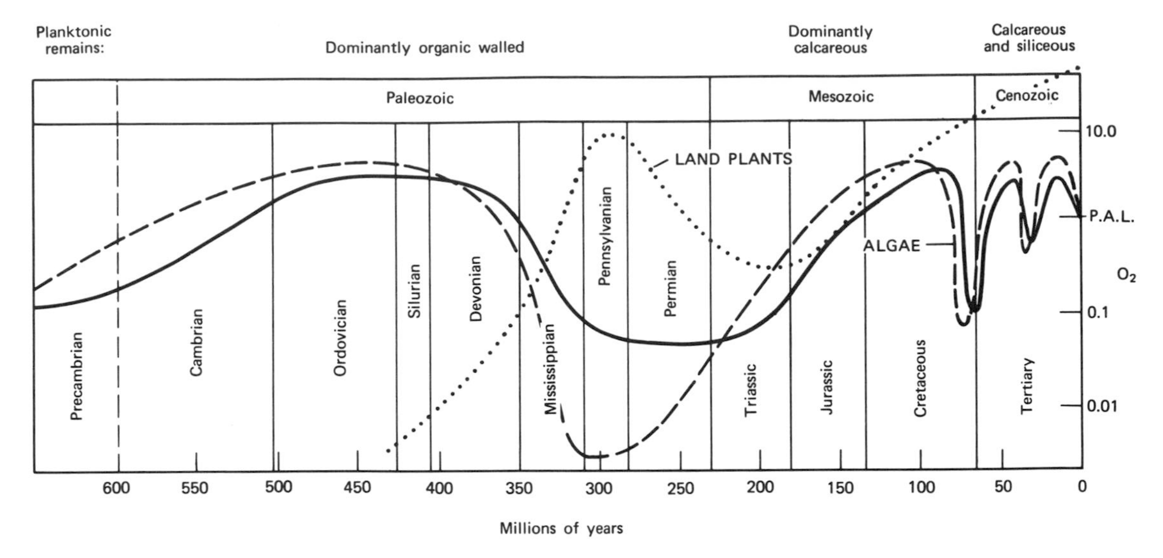

Figure 10-1. Past O_2 content of the atmosphere (solid line) compared with present atmospheric level (P. A. L.) of O_2 (after Tappan, 1968), and relative abundances of algae and vascular land plants. Numerical values (ordinate) apply only to O_2.

tration of carbon dioxide is largely reciprocal to the concentration of oxygen. Larger concentrations of carbon dioxide are known to stimulate growth of higher plants and may be responsible for the increased land-plant populations during the Carboniferous that are so evident in the fossil record and are indicated in figure 10-1.

A moderate decline in land-plant populations may be postulated during the Permian and Triassic owing, perhaps, to greater aridity. It seems at any rate that there is a less ample record of land fossils during these periods, and it is not as likely that this decline in terrestrial plants was due to a change in atmospheric composition.

An expansion of terrestrial plants in the Early Cretaceous, that continues to the present, must be postulated as the result of increased numbers of angiospermous plants. These plants have greater adaptability to varied environments than all but some blue-green algae and bacteria, owing largely to efficiencies of their seeds and an increased sensitivity to photoperiodicity. The present ecologic range of the higher land plants is very extensive because of rapid metabolic response and productivity when growth conditions are favorable. Modern planktonic algae, on the other hand, have not greatly expanded their range beyond that prevailing during the early Paleozoic.

Tappan believes, on the basis of the fossil record, that minor crises in planktonic algae also occurred in the Late Cretaceous and mid-Tertiary. Virtually coincident evolutionary outbursts of the angiosperms also occurred during these intervals, but one hesitates to suggest that they might be correlated with temporarily increased concentrations of carbon dioxide. Nevertheless, a considerable degree of coincidence is apparent. Evidently many modern families of angiosperms originated during the Cretaceous, and there was a great diversification of herbaceous angiosperms about the middle of the Tertiary.

Figure 10-1 presents a diagrammatic idea of interrelationships that accord with our evaluation of the paleontologic record. It expresses ideas that are difficult to quantify, and, since it is itself based on a subjective evaluation of the abundance of fossils, it cannot be used, by itself, in explanation of the reasons for such variations without circularity in reasoning. A more accurate means of estimating the abundance of populations through time is much to be desired. Nevertheless, the generalizations that are reflected by the abundance curves in this diagram are important.

RELATION OF EARLY CAMBRIAN AND LATE PRECAMBRIAN PALYNOMORPHS

From a palynological standpoint the Cambrian lower boundary has virtually ceased to exist. Consequently treatment in this chapter must overlap considerably and consider microfossils that are present in both. Timofeyev, (1958b) the noted Soviet palynologist, who may have studied more microfossils of this range of age than any one else, commonly finds it necessary to treat late Precambrian and Early Cambrian assemblages together. There are at least two reasons why distinction is difficult. First, such ancient sediments are likely to be so highly indurated that the fossils are damaged either in situ or subsequently and cannot easily be recovered intact. Second, none of the late Precambrian or Early Cambrian fossils has proved to be very distinctive. Micro-fossils of the types illustrated by Mlle. Roblot (1964) from Normandy are reasonably representative of fossils from Early Cambrian as well as the late Precambrian. Such fossils have been reported from Karelia, Bohemia, the southern Urals, Siberia, Scotland, and the Antarctic. Walton (1962) has reported occurrences in the Cambrian of western Canada, and the writer has observed similar forms from erratics within the Permian tillite of the Falkland Islands (see fig. 10-2), thus suggesting that upper Precambrian or Lower Cambrian beds were eroded and incorporated into the tillite. There is very little evidence to show what types of plants are represented by these fossils, but they clearly were wide ranging.

EARLY PALEOZOIC STRATIGRAPHIC UNITS

The relationships of some of the named rock units, time units, and fossil zones of the early Paleozoic are shown in figure 10-3. The sequences of European and American units are compared according to present conceptions of approximate time equivalence. European stage names that refer to the sequence of biostratigraphic zones in the areas from which the names of systems have been taken are generally regarded as standards of reference for expressing geologic time. Other stage names may be somewhat less authoritative geographically but, because of proximity to the material under investigation, are more meaningful and pertinent for a particular purpose.

The evolution of rock stratigraphic, biostratigraphic, and time stratigraphic nomenclature has a history that is too long and involved to be discussed here. Almost any given area has its own history of stratigraphic study, which has resulted in subdivision of the geologic section into meaningful local units. The palynologist will generally be familiar with the local terms. Such specifics are not shown in figure 10-3, which is intended as an aid in recalling the relative positions of the larger units and some of those of common reference in the literature.

Not all of the units named have equal validity because in different areas stratigraphic studies have not received equal treatment or emphasis; for example, only recently has it been understood that the Eden Stage of the Cincinnati region is equivalent to the Barneveld Stage of New York (Thomas J. M. Schopf, 1966), and the effects of correlation of the Cincinnati and New York sections on series classification have not yet been resolved. It is hoped that the stratigraphic names given in figure 10-3 can serve as a concise

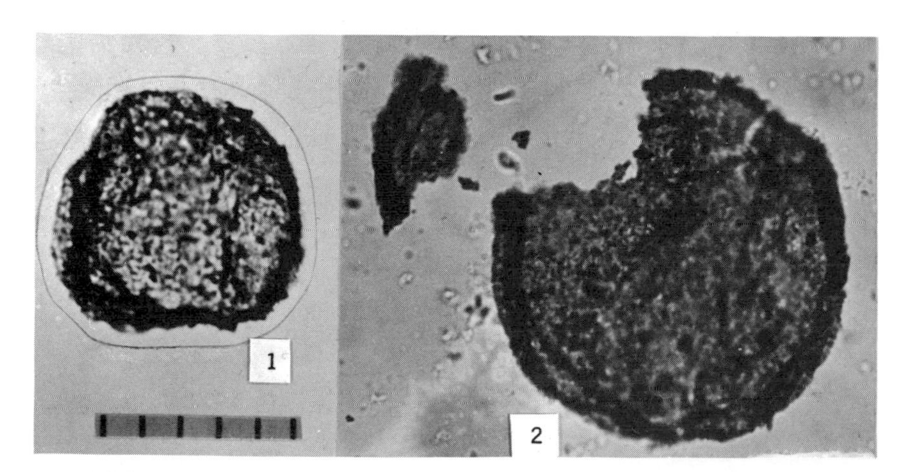

Figure 10-2. Examples of late Precambrian or Early Cambrian microfossils from erratic gravels incorporated in Permian tillite of the Falkland Islands, × 500; attached scale shows 10-micron divisions. The example on the right was broken in mounting and adjacent fragments convey some indication of the fragile nature of this material.

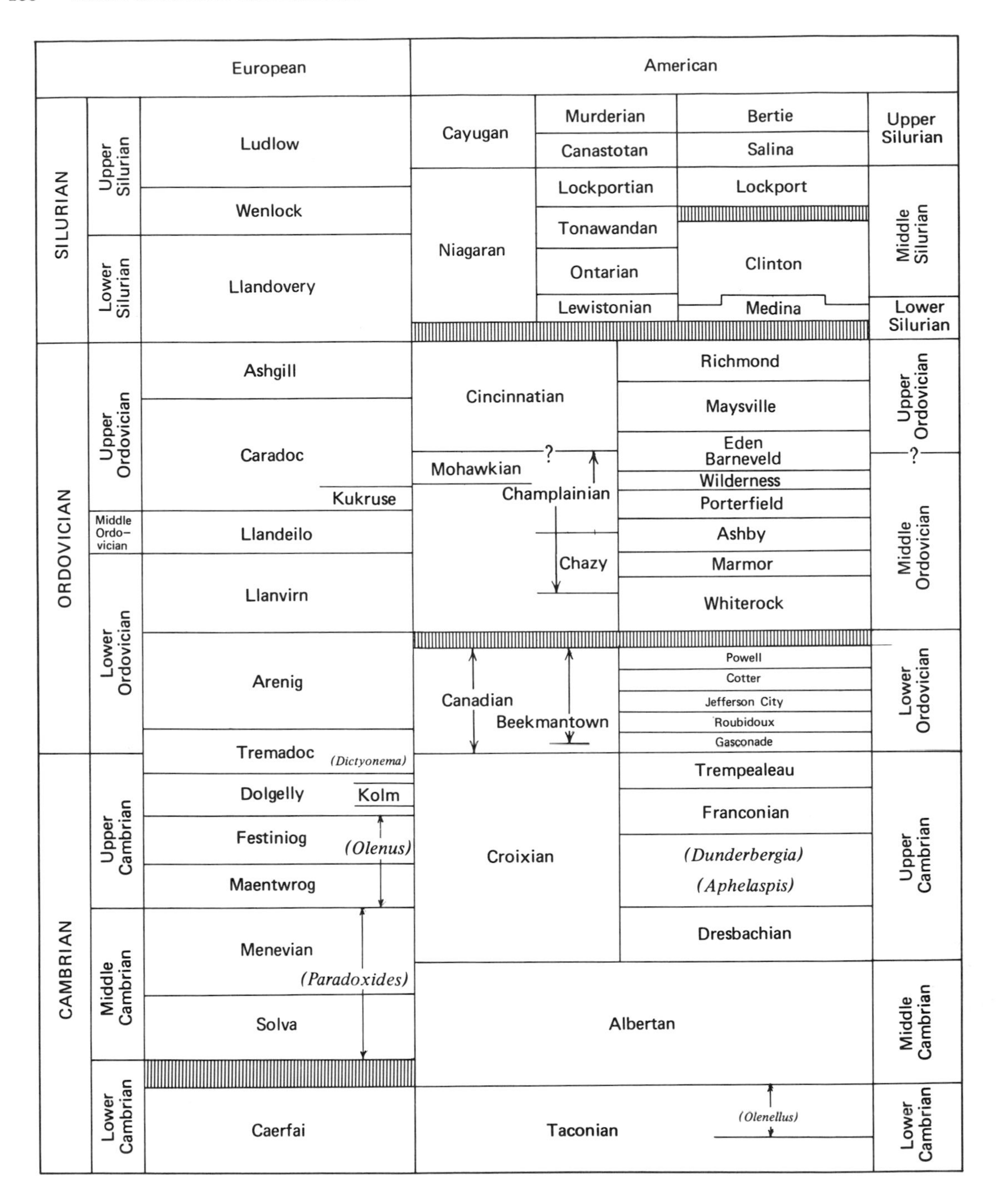

Figure 10-3. Stratigraphic units used for lower Paleozoic systems in Europe and North America.[1] Compiled from the following sources: Fisher (1959; 1962a,b); Howell and others (1944); Swartz and others (1942); Twenhofel and others (1954); Kummel (1961); Brinkmann (1960); Gignoux (1950); Wilmarth (1925, 1938); Keroher and others (1966); and American Commission on Stratigraphic Nomenclature (1961).

[1]The stratigraphic nomenclature used in this figure has been compiled from many sources and may not necessarily follow the nomenclature of the U.S. Geological Survey.

reference for relating the terms used in this text with other terms. The stratigraphic table is not intended to be complete, and many related stratigraphic terms can easily be added from reference books available to any student.

Timofeyev			Naumova
Ordovician	Middle	Kukruse (oil shale) *Echinosphaerites* beds	Lower Silurian
Ordovician	Lower	*Orthoceratites* beds Glauconitic Sandstone	Lower Silurian
Cambrian	Upper	*Dictyonema* beds *Obolus* Sandstone	Lower Silurian
Cambrian	Middle	Ischorian Suite	Lower Cambrian
Cambrian	Lower	*Eophyton* Sandstone Baltic Blue Clay Supra *Laminarites* beds	Lower Cambrian
Sinian System		*Laminarites* beds Gdovian Suite Priozersk Suite	
Proterozoic	Middle	Jotnian Formation Ladogan Formation	
Proterozoic	Lower	Karelian Formation	

Figure 10-4. Stratigraphic classifications used by Timofeyev (1959) and by Naumova (1949, 1950) in reporting plant microfossils of the Baltic region.

CAMBRIAN TO SILURIAN PALYNOMORPHS

As in the case of Precambrian sporomorphs, we are much indebted to Soviet scientists for knowledge of plant microfossils of Cambrian and Early Ordovician age. During much of Cambrian and Ordovician time the Russian platform was occupied by a shallow epicontinental sea. The deposits that accumulated there are relatively thin and the section is incomplete, but the beds have been little altered by any subsequent orogeny. In 1949 Naumova reported abundant sporelike microfossils from the Lower Cambrian Blue Clay of the Baltic region. She later (1950) reported on palynomorphs of her Lower Silurian (= Ordovician of fig. 10-3) ranging from Tremadoc (*Dictyonema*) to lower Caradoc (Kukruse) in the same region (see figs. 10-3 and 10-4.) Timofeyev (1959) has enlarged on Naumova's observations. Apparently most of the Cambrian and Ordovician argillaceous deposits in this region provide abundant microfossils in a favorable state of preservation.

Timofeyev and Naumova differ somewhat in their stratigraphic treatment, as shown in figure 10-4. Timofeyev assigns the *Obolus* Sandstone and *Dictyonema* beds to Upper Cambrian, whereas Naumova regards them both as "Lower Silurian" (= Ordovician), substantially in conformity with the normal Western classification. Thus there is a considerable range of stratigraphic overlap in the material covered by these reports. Timofeyev places the Ischorian suite in the Middle Cambrian, but, if the Ischorian is the equivalent of the Fucoidal Sandstone as I suppose, this unit is more conventionally (but not necessarily correctly) assigned to the Lower Cambrian along with the underlying *Eophyton* Sandstone and Baltic Blue Clay. Timofeyev's observations also extend to lower strata now classed as Sinian and to underlying sedimentary beds of Proterozoic age in Karelia.

According to Timofeyev, diversification of the fossil assemblage starts in the Middle Cambrian Ischorian at the top of, or just above, the Baltic Blue Clay. Hystrichosphaerids and diacrodioids are well represented and continue through to the top of the section at Vologda, about 400 kilometers north-northwest of Moscow. The same sort of diversification occurs near the top of the Ischorian at Valday, about midway between Moscow and Leningrad. The contrast seems to confirm Timofeyev's assignment of Ischorian to the Middle Cambrian instead of to the Early Cambrian. In the *Obolus* Sandstone and *Dictyonema* beds the diversified assemblage is continued. One of the greatest contrasts in the invertebrate faunas occurs between richly fossiliferous Ordovician deposits and the generally much less fossiliferous beds of Cambrian age. It is of interest that diversification of the plant-microfossil assemblages anticipates this faunal change.

In his monograph on the ancient flora of the Baltic region Timofeyev (1959) assigned the fossils within 34 genera, distributed in 5 families (2 with subfamilies), as follows:

LEIOSPHAERIDACEAE

Subfamily Protoleiosphaerideae

Protoleiosphaeridium
Symplassosphaeridium
Trematosphaeridium

Subfamily Leiosphaerideae

Leiosphaeridium
Trachysphaeridium
Lophosphaeridium
Orycmatosphaeridium
Vavosphaeridium
Zonosphaeridium

HYSTRICHOSPHAERIDACEAE

Archaeohystrichosphaeridium (79 species)
Hystrichosphaeridium (26 species)

DIACRODIACEAE

Subfamily Homodiacrodeae

Trachydiacrodium
Trachyrytidodiacrodium
Trachyzonodiacrodium
Lophodiacrodium
Lophorytidodiacrodium
Lophozonodiacrodium
Acanthodiacrodium
Acanthorytidodiacrodium
Acanthozonodiacrodium

Subfamily Heterodiacrodeae

Dasydiacrodium
Dasyrytidodiacrodium

SPHAEROLIGOTRILETACEAE

Leioligotriletum
Mycteroligotriletum
Bothroligotriletum
Trachyoligotriletum
Ocridoligotriletum
Tyloligotriletum
Lopholigotriletum
Acantholigotriletum
Trematoligotriletum
Stenozonoligotriletum

OOIDACEAE

Ooidium
Zonooidium

Deflandre and Deflandre-Rigaud (1962a) questioned whether all the genera are justified. By an apparent minor emendation of diagnoses they have combined *Trachyrytidodiacrodium* Tim. 1959 and *Trachyzonodiacrodium* Tim. 1959 with ***TRACHYDIACRODIUM;*** *Lophorytidodiacrodium* Tim. 1958a, *Lophozonodiacrodium* Tim. 1958a, and *Diornatosphaera* Downie 1958 with ***LOPHODIACRODIUM;*** *Acanthorytidodiacrodium* Tim. 1959 and *Acanthozonodiacrodium* Tim. 1959 with ***ACANTHODIACRODIUM;*** and *Dasyrytidodiacrodium* Tim. 1959 with ***DASYDIACRODIUM.***

It seems that the Deflandres' simplified generic arrangement is more appropriate at present, but

Figure 10-5. *Acritarchs from upper Precambrian and lower Paleozoic deposits* (All × ca. 300; redrawn from Timofeyev (1959). Dimensions in microns.) 1—*Protoleiosphaeridium conglutinatum*, 25-35 to 40-45, Proterozoic to Middle Ordovician inclusive, common. 2 and 3—*Bothroligotriletum exasperatum*, 30 to 110, Proterozoic to Lower Ordovician inclusive; 4—*B. exasperatum*, 30 to 110, Proterozoic to Lower Ordovician inclusive. 5—*Trachyoligotriletum planum*, 30 to 70, Proterozoic to Glauconitic Sandstone, inclusive (except Upper Laminarites); most common in the Ischorian. 6—*Leioligotriletum minutissimum*, 15 to 40, Proterozoic to Lower Cambrian inclusive. 7—*T. nevelense*, 40 to 80, Proterozoic, Sinian, Cambrian (except Dictyonema), occasionally Lower Ordovician, common; 8—*T. minutum*, 10 to 40, Proterozoic, Sinian, Cambrian (most common), Lower Ordovician?; 9—*T. incrassatum*, 15 to 40, Proterozoic, Sinian, Cambrian, occasionally Lower Ordovician, common; 10—*T. obsoletum*, 40 to 130, Proterozoic to Cambrian inclusive, Lower Ordovician?. 11—*L. bistrovi*, 30 to 33, lower beds of the Baltic Blue Clay; rare. 12—*Stenozonoligotriletum sokolovi*, 10 to 50, Proterozoic, Sinian, Cambrian, and Lower Ordovician (rare) (except Obolus and Echinospherites), most common in Lower Cambrian. 13—*L. compactum*, 40 to 100, upper Proterozoic to Eophyton inclusive. 14—*Mycteroligotriletum marmoratum*, 30 to 80, Proterozoic and Sinian; Middle Cambrian?; 15 and 16—*M. marmoratum*, 30 to 80, Proterozoic and Sinian; Ischorian?. 17 and 22—*L. glumaceum*, 35 to 105, Proterozoic, Laminarites, Baltic Blue Clay, Eophyton, lower Ischorian beds; occurrence discontinuous. 18—*T. asperatum*, 30 to 90, Proterozoic (some deposits) to Lower Ordovician inclusive (except Laminarites and Supra Laminarites). 19 and 20—*L. minutissimum*, 15 to 40, Proterozoic to Lower Cambrian inclusive; common. 21—*B. plicatile*, 75 to 135, Baltic Blue Clay; rare. 23—*L. ochroleucum*, 150 to 190, Eophyton; rare. 24—*B. auctum*, 110 to 170, upper laminarites, Baltic Blue Clay, and Ischorian; rare.

questions regarding species will have to be answered as additional studies are conducted; for example, Downie (1958) has emphasized the small size and intergrading characters of Tremadoc microplankton forms in Britain. In the meantime Timofeyev's method of meticulous discrimination will aid in providing a survey of the whole floral population.

Selected illustrations of Timofeyev have been carefully redrawn and reproduced at about ×300 to illustrate the morphology of these interesting microfossils. Figure 10-5 shows the more common and simple types that occur in both the Proterozoic and Cambrian. Simple sporelike forms, like those in figure 10-5-1, commonly adpressed in irregular groups, occur in deposits as young as

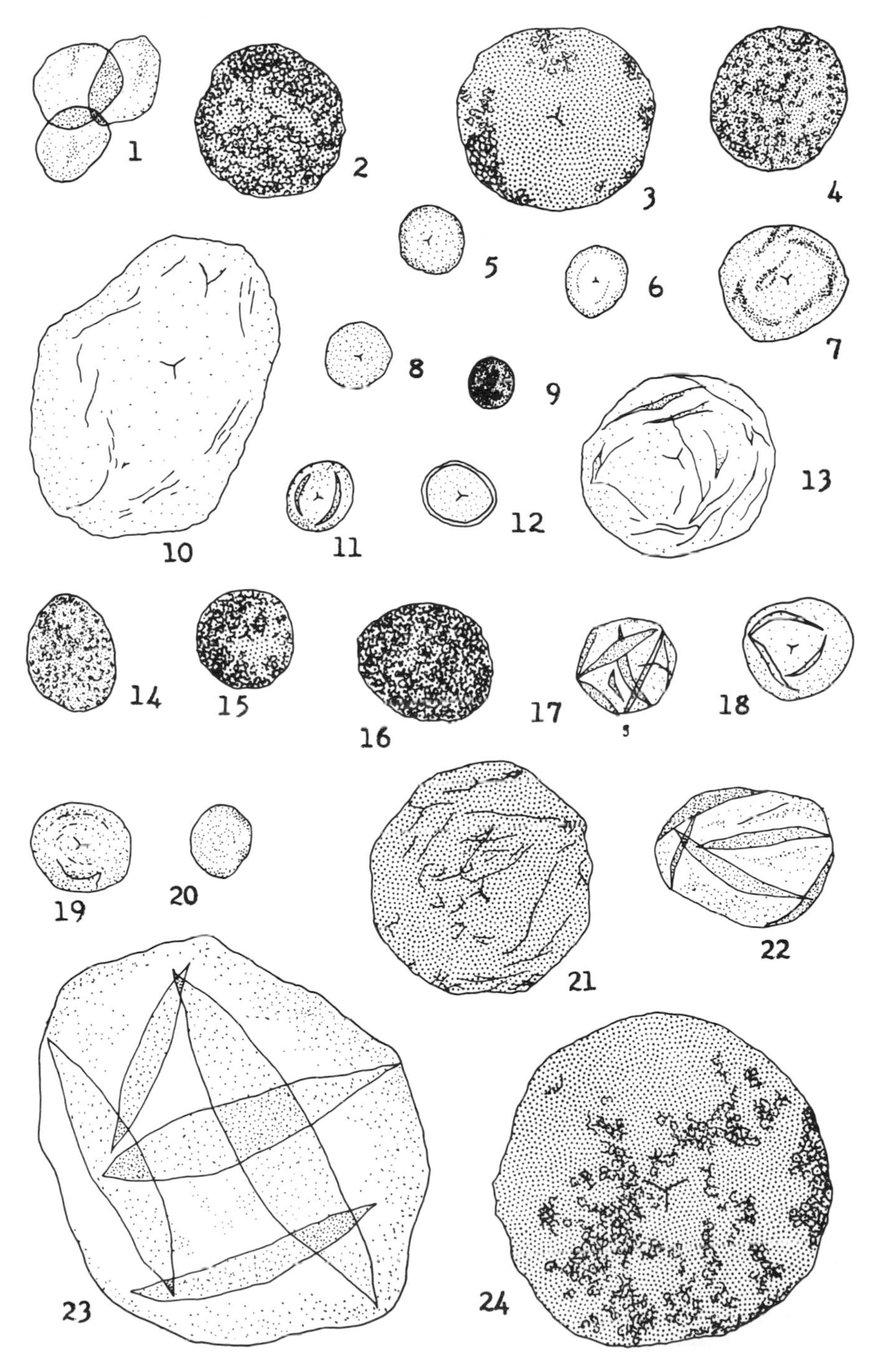

Ordovician. In spite of the fact that Timofeyev's drawings commonly show a small triradiate mark, it is doubtful that the mark indicates polarity. Its possible significance is discussed later. The form that Timofeyev calls *Stenozonoligotriletum soklovi* (fig. 10-5-12) shows the oldest expression of an "otorochka," or differentiation of the equatorial zone. Other examples are illustrated on figures 10-7-1 and 2. We must wonder whether the "otorochka" of this sporelike form consists of more than a rather uniformly thickened wall, the two surfaces of which can easily be observed.

The names that Timofeyev has applied to these examples are given in the caption to figure 10-5, along with listed stratigraphic ranges.

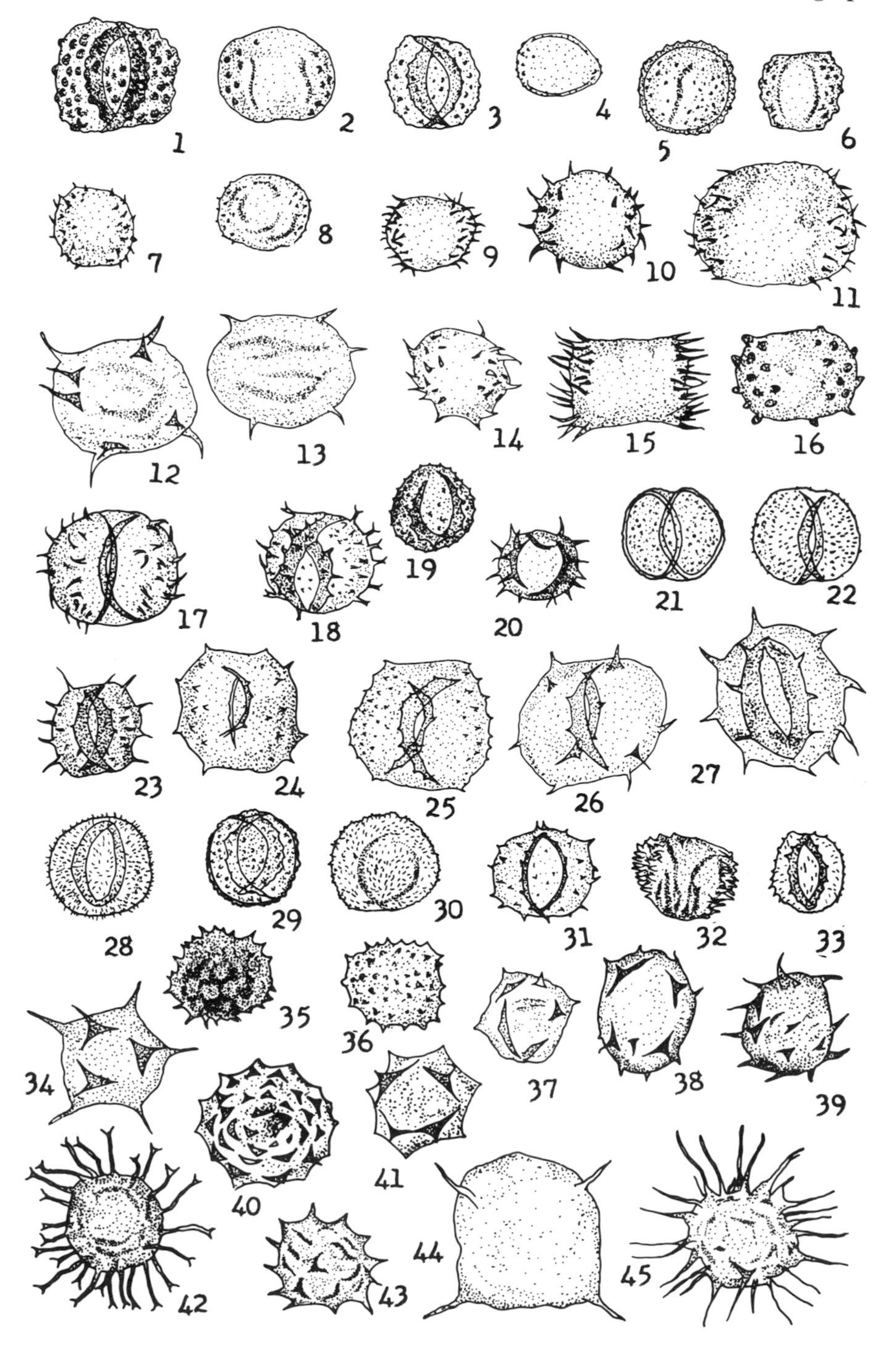

Longer ranging forms are placed nearer the top. We may readily note the difficulty in making a meaningful comparison of these generalized types of microfossils.

Microfossils of quite a different and more distinctive type, many of which are bilaterally symmetrical, occur in the Ischorian (Middle Cambrian) beds and above, as illustrated on figure 10-6. Those largely restricted to the Ischorian, shown in the upper three rows on the figure (1 to 16), show a variety of ornamentation that appears readily distinguishable. Some of the examples, although they have been given different specific names, may represent minor differences in growth or preservation of resting cells belonging to the same species. The spheroidal types, shown at the bottom of the figure (35 to 45), no doubt are acritarchs or "hystrichs" that are widely distributed in the middle as well as lower Paleozoic. The stratigraphic ranges given by Timofeyev are indicated in the caption to the illustrations.

Resistant "Spores" and the Three-Rayed Marking

Misleading inferences have been based on the assumption that only aerially exposed sporangia can form spores with resistant coats. The assumption evidently is based on the observation that *some* algal spores, swarmers, etc., do not have a resistant coat and that all land-plant spores exposed to desiccation are equipped with a durable outer membrane. The fact is, however, that many aquatic plants also have resistant coats on cysts and similar cells, and a broad generalization cannot be justified. Whether or not ancient palynomorphs are "truly" cutinized can be argued because the fossil membranes are difficult to analyze and they probably never retain full evidence of their original chemical composition. The changes are comparable to the alteration that takes place in spore coats during coal formation. Waksman (1938) has shown that marine plants form humus much as land plants do,

Figure 10-6. *Diacrodioid palynomorphs and acritarchs.* (All × ca. 300; dimensions given are in microns.)

Diacrodioid palynomorphs from Middle Cambrian deposits at Vologda, Soviet Union: 1—*Lophorytidodiacrodium primarium*, 40 × 50, Ischorian, rare; 2—*Lophory. salebrosum*, 35 × 50, Ischorian, common; 3—*Lophory. tuberculatum*, 35 × 45, Ischorian, common; 4—*Lophozonodiacrodium tumidum*, 27 × 39, base of Ischorian, monotypic; 5—*Lophozo. spectabile*, 35 × 45, Ischorian, common; 6—*Lophory. turulosum*, 30 × 40, Ischorian, rare; 7—*Acanthodiacrodium capsulare*, 35 × 40, Ischorian, rare; 8—*Ac-di. duplicativum*, 30 × 40, Upper Ischorian, common; 9—*Ac-di. adelphicum*, 30 × 40, Ischorian, rare; 10—*Ac-di. abortivum*, 45 × 55, with spines 50 × 60, Upper Ischorian, very rare; 11—*Ac-di. enodum*, 55 × 80, Ischorian, rare; 12—*Ac-di. cornutum*, 55 × 62, with spines 60 × 75, Upper Ischorian, monotypic; 13—*Ac-di. immarginatum*, 40 × 60, with spines 45 × 70, Ischorian, rare; 14—*Ac-di. crenatum*, 35 × 45, with spines up to 50, Ischorian, rare; 15—*Ac-di. barbullatum*, 30 × 50, with spines 45 × 75, Upper Ischorian, rare; 16—*Ac-di. endemicum*, 36 × 48, Lower Ischorian, monotypic.

Diacrodioid palynomorphs from Middle and Upper Cambrian: 17—*Acanthorytidodiacrodium diffusum*, 51 × 59, Upper Ischorian (Vologda) and Dictyonema (Valday), rare; 18—*Ac-ry. dichotomum*, 40 × 45, with spines 51 × 57, top of Ischorian (Vologda), very rare; 19—*Ac-ry. decipiens*, 34 × 36, Obolus (Vologda), Lower Dictyonema (Valday), common; 20—*Ac-ry. scaberrimum*, 28 × 33, with spines 33 × 45, Lower Obolus (Vologda), common; 21—*Acanthozonodiacrodium duplex*, 44 × 45, Lower Obolus (Vologda), rare; 22—*Ac-ry. echinatum*, 41 × 46, Dictyonema (Vologda), rare; 23—*Ac-ry. aciferum*, 40 × 44, with spines 45 × 55, Upper Ischorian and Obolus (Vologda), rare; 24—*Ac-ry. vulgare*, 51 × 54, Obolus (Vologda) and Dictyonema (Valday), rare; 25—*Ac-ry. unigeminum*, 58 × 59, Upper Ischorian (Vologda), rare; 26—*Ac-ry. patulum*, 55 × 69, with spines 62 × 75, top of Ischorian (Vologda), Obolus and base of Dictyonema (Valday), rare; 27—*Ac-ry. typicum*, 55 × 60, with spines 57 × 75, Obolus (Valday), rare; 28—*Ac-ry. lanatum*, 40 × 47, Lower Dictyonema (Valday), common; 29—*Ac-zo. hypocrateriforme*, 40 × 42, Lower Obolus (Vologda), rare; 30—*Ac-ry. hippocrepiforme*, 40 × 43, top of Ischorian (Vologda), common; 31—*Ac-ry. lentiforme*, 36 × 40, with spines 42 × 48, Upper Ischorian and Lower Obolus (Vologda), rare; 32—*Ac-ry. flabellatum*, 33 × 43, base of Obolus (Vologda), monotypic; 33—*Ac-ry. varium*, 30 × 33, Obolus and Ischorian (Vologda), common.

Acritarchs from Middle and Upper Cambrian: 34—*Archaeohystrichosphaeridium nalivkini*, 40 × 44, with spines 75, Lower Obolus (Vologda), rare; 35—*Ar. semireticulatum*, 37 × 42, Ischorian (Vologda), very rare; 36—*Ar. angulosum*, 36 × 44, Ischorian (Vologda), common; 37—*Ar. minimum*, 36 × 40, Dictyonema (Vologda), rare; 38—*Ar. protensum*, with spines 44 × 55, Obolus (Vologda), rare; 39—*Ar. scandeum*, 40 × 44, with spines 60, Lower Obolus (Vologda), common; 40—*Ar. mickwitzi*, 50 × 54, Lower Dictyonema (shales, Valday), monotypic; 41—*Ar. septangulum*, 47 × 48, Lower Obolus (Vologda), rare; 42—*Ar. waltzi*, 39 × 41, with spines 84, Obolus (Vologda), rare; 43—*Ar. genuinum*, 39 × 42, with spines 55 × 57, Lower Obolus (Vologda), common; 44—*Ar. scrotiforme*, 65 × 69, with spines 105, Obolus (Valday), monotypic; 45—*Ar. lüberi*, 48 × 51, with spines 95, Obolus (Vologda), rare. (The atypic abbreviations are used to distinguish the similar generic names.)

so the association of humic matter in the *Dictyonema* beds is not indicative of the presence of land vegetation, though this has been suggested. The cell walls and membranes of ancient plants, like those of modern plants, probably lacked chemical homogeneity. To add to analytic difficulty, much of the fossil substances are not soluble without decomposition, so there are many reasons why the original specific composition of fossil membranes remains in doubt. The morphology and optical properties of the fossil palynomorphs are evidence of physical stability and "resistance."

The term "spore" is not restricted, botanically, to mean "spores of terrigenous vascular plants," but some authors seem to presume that it has this implication. In the literature it is commonly stated that particular microfossils should *not* be called spores, presumably because the examples do not unquestionably represent cryptogamic vascular plants. All the higher plants, nearly all the fungi, many algae, and some bacteria produce some kind of spores during their life cycle, and therefore the term "spore" should not be used to imply identification of any group of plants. Types of microfossils that are ambiguous as to function and affinity probably should be referred to as *sporelike*, in order to avoid misinterpretation, but the term "spore" can properly apply to disseminules of algae *if* the life-cycle functions of the disseminules justify the usage.

The problematic and dubious nature of the triradiate aperture on Precambrian and early Paleozoic plant microfossils has recently been discussed by Volkova (1965), who questions the general occurrence of this structure in sporelike bodies of older deposits and particularly doubts the validity of identifying Proterozoic and early Phanerozoic vascular terrigenous plants on this basis. In more recent Soviet literature, cited by Volkova, many students of these ancient microfossils have come to regard the sporelike fossils as remains of microscopic algae. In any event more evidence is required than the presence of a three-rayed imprint (or the lack of it) as evidence of affinity.

Timofeyev (1959) suggested that disseminules of the Leiosphaeridaceae, Hystrichosphaeridaceae, and Diacrodiaceae are spores of algae and planktonic unicellular organisms. Direct proof seems to be lacking, and the sporelike bodies may, in part, represent cysts rather than spores, but probabilities based on association and sedimentary facies generally favor his suggestion. Indeed, it would seem most reasonable generally to presume that all represent *marine* organisms, though Timofeyev does not so indicate.

The sporelike fossils assigned to the Sphaeroligotriletaceae and the Ooidaceae, however, are presumed by Timofeyev to represent spores of terrestrial and semiterrestrial plants. The evidence for this interpretation is extremely vague. Timofeyev, Naumova, and perhaps others assume that the short-rayed, triradial imprint exhibited by some of these sporelike forms is an indication of bryophytic or pteridophytic alliance. This assumption is probably invalid for the following reasons:

1. A tetrad mark, by itself, can signify only that spores formed in tetrads. Thus a tetrad mark could identify any of the plant groups in which meiotic division occurs in a spore precursor cell. Some such plants are algae.
2. A distinction should be made between a tetrad mark and a functional germinal suture. The latter implies an early determination of polarity and organization in the gametophyte. Sutures are lacking in the Cambrian palynomorphs, just as they are in disseminules of *Tasmanites* in later deposits. Probably a few algae, such as *Foerstia* of the Devonian, have not only a tetrad mark but also structural sutures, even though the occurrence of these plants may signify a pelagic, sargassumlike mode of existence. However, other resistant algal (?) spores, as in *Parka*, lack such features.
3. Spores of many terrigenous bryophytes are apolar and lack the tetrad mark on mature spores (Udar, 1964; Erdtman, 1957, 1965). Probably the apolar condition in bryophytes is a result of post-tetrad spore enlargement and modification. Gametophytic polarity may not be established at the time of meiotic division, but the spores of most free-sporing pteridophytes have a well-defined polarity and germinal sutures. Functional sutures are lacking on the Cambrian and Ordovician microfossils, and their alliance with pteridophytes seems unlikely.

The presence or absence of the tetrad mark, a simple morphologic character, probably should not be taken arbitrarily as proof of affiliation with any major taxon. More convincing vegetative remains, as well as spores, should be in evidence before we assume the presence of land plants. Sporelike bodies may be relatively abundant, but credible vegetative remains of possible land plants are very scarce in the early Paleozoic, and none has been conclusively identified. Interpretation of "spores," such as those reported from the Lower Cambrian by Naumova (1949) and from Middle Ordovician limestone in Oklahoma by Taugourdeau (1965, p. 478) is very uncertain. The uncertainties of early Paleozoic origin of vascular plants are emphasized by Stewart (1960) and by Chaloner (1960).

Some algae are known to form their oospores in tetrads; therefore some relict of a tetrad mark or imprint could be preserved if the coat is durable and not much modified after tetrad separation. Although the presence of spores with a weak tetrad mark, irregularly developed and inconsistent in lower Paleozoic deposits, cannot be regarded as evidence of land plants, it may signify a mode of origin. Such spores probably should be regarded as representatives of algae, some of which, such as fucoids and laminarians, of which there is additional evidence (White, 1901, 1903; Fry and Banks, 1955), may have been fairly advanced. The evolutionary progression of these advanced types of algae, however, probably was not in the direction of land plants.

The Radially Asymmetrical Palynomorphs

The bilateral symmetry shown by some of the sporelike microfossils of early Paleozoic assemblages is a most striking indication of floristic differentiation. This symmetry is shown by the Diacrodiaceae and by forms assigned to the Ooidaceae by Timofeyev (1959). Representative examples of radially asymmetrical palynomorphs are illustrated in figures 10-6, 1 to 33, and figures 10-7, 6 to 11, all redrawn from examples illustrated by Timofeyev (1959). The fossil names and known stratigraphic range are given in cap-

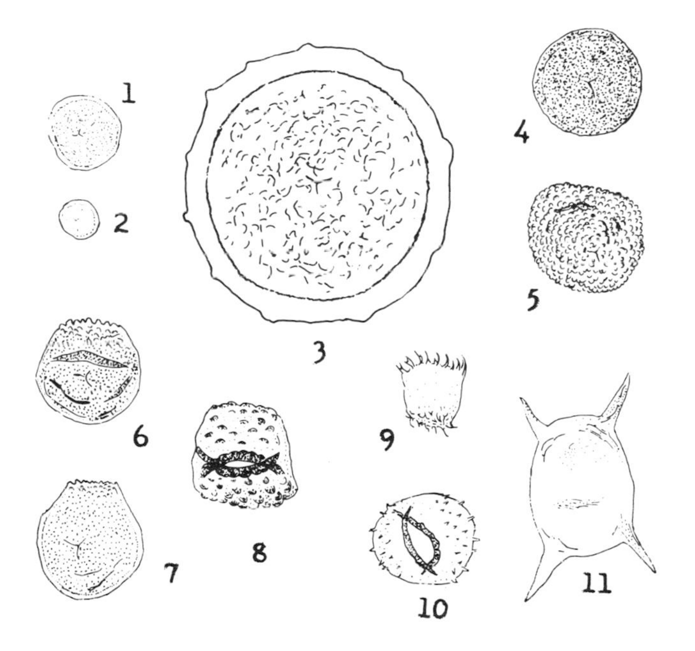

Figure 10-7. Sphaeroligoid, ooid, and diacrodioid palynomorphs from Precambrian and Cambrian deposits (All illustrations about × 300; all dimensions given are in microns.) 1 and 2—*Stenozonoligotriletum sokolovi*, 10 to 50, note otorochka, Proterozoic, Sinian and Cambrian (except Obolus and Echinosphaerites), Lower Ordovician (rare), abundant in Lower Cambrian; 3—*S. dedaleum*, 87 × 96, Obolus, monotypic; 4—*S. validum* (holotype), 44 × 47, Gdovian, Laminarites, Supra Laminarites, Baltic Blue Clay, Eophyton, and Glauconitic Sandstone, always sparse; 5—*Lopholigotriletum crispum* (holotype), 40 × 48, Obolus, common; 6—*Ooidium invisum* (holotype), 40 × 42, Ischorian, common; 7—*Zonooidium piriforme* (holotype), 45 × 50, Ischorian, rare; 8—*Lophorhytidodiacrodium lutkevischi*, 35 × 45, Ischorian, common; 9—*Dasydiacrodium trapezoideum*, 30 × 36, Upper Ischorian, monotypic; 10—*Acanthorhytidodiacrodium unigeminum*, 25 × 35, with spines 50 × 60, Upper Ischorian, rare; 11—*Acanthodiacrodium schmidti*, 35 × 55, with spines 50 × 90, Ischorian, rare.

tions of the figures. Since similar fossils are reported from Upper Cambrian and Lower Ordovician deposits in Russia by Timofeyev (1959) and Naumova (1950), and from Britain by Downie (1958), their general morphology is well established. Probably similar fossils occur elsewhere but have not yet been investigated.

The diacrodioid fossils commonly are compressed to form two arcuate folds, which must be prominent microscopic features. Nonetheless, the specific appearance of these folds is due to chance occurrences in compression, and probably they do not deserve the taxonomic importance that has been placed on them by Soviet authors. The variation that would justify species separation must be evaluated according to vital morphology and not according to accidents in preservation. No noncompressed reconstruction drawings have been presented.

Deflandre and Deflandre-Rigaud (1962a) have suggested comparison of some diacrodioid forms with *Palaeostomocystis* Def. 1937 which is presumed to be a planktonic organism represented in Cretaceous age flints. They also suggest comparison with the little-known modern form, *Echinum* A. Meunier 1910, originally discovered in colored snows and among Arctic plankton. The asymmetry of bilateral sporelike bodies could also be a reflection of the confining wall of a sporangium. Zygospores of conjugate green algae commonly are ovoid, with the long axis of the resistant spore parallel to the elongate outline of the cell that serves as the sporangium. Diacrodial spores are generally drawn as "alete," as they should be if zygospores actually are represented.

Ooid forms described by Timofeyev (1959) are pearshaped or unilaterally asymmetrical (figs. 10-6-6 and 10-6-7). They are supposed to show a tetrad mark, and, if this mark is substantiated, they could not represent zygospores. The tetrad mark is by no means evident on all members of the Ooidaceae, however, and wherever shown in illustrations it appears faint and perhaps questionable. There is no evidence that these forms have any relationship with land plants or incipient land plants. Of the forms illustrated by Timofeyev, possibly some may be interpreted as teratologic examples of diacrodioid types and others may be related to a different type of algae.

The same reasoning also applies to forms assigned to the Sphaeroligotriletaceae, all of which show a very short rayed tetrad imprint. None of the forms illustrated shows a suture along the rays, nor is there any indication that the imprint had any relation to germination or polarity of vital structures enclosed by the resistant wall. For the present these types may be regarded as additional evidence of differentiation in the prolific algal flora that existed in early Paleozoic seas. A definite gametophytic polarity with a functional trilete suture would be a more reliable indication of the existence of land plants. However, this suggestion is based only on a coincidence of empirical observation. Habitat and disseminule morphology probably were not precisely coordinated in incipient land plants. More critical evidence is needed.

ACRITARCHS

The palynomorphs of the early and middle Paleozoic have usually been reported as hystrichosphaerids (hystrichs or hystrix) or, more recently, as acritarchs. (See also Chapter 18). The term "acritarch" (Gr. *akritos*, unchosen, unsorted (confused), + *arche*, beginning, first cause, origin) was proposed by Evitt (1963) after hc realized that younger members of the "hystrich" genera (*Hystrichosphaeridium* and *Hystrichosphaera = Spiniferites*) probably were assignable to the Dinophyceae. Acritarchs have been discussed as if they constituted some kind of taxonomic group, but, because the cardinal element of their definition is the "quality" of nonassignment, there is a degree of unreality about this point of view. Nevertheless, "acritarchs" offer perplexing taxonomic problems and much latitude for a reasonable divergence of opinion. A great deal of work is in progress, and therefore it is difficult to offer a summary discussion that will not appear to be badly dated.

The most fundamental question concerning these microfossils relates in evaluation of the degree of polyphyleticism within the group. Some authors consider that objects as divergent as the ova of chordates and asexual spores of fungi, as well as spores and cysts of several algal and invertebrate phyla, may be included. Others (see Eisenack, 1963a) believe that most of the microfossils encountered represent planktonic algae that are reasonably interrelated. These two opinions are not squarely in opposition. Advocates of the latter philosophy probably would not deny that a few of the multifarious objects that some authors have attempted to classify with this group have been badly misassigned because of a purely superficial resemblance. When such mistakes are recognized they can be corrected simply by reclassifying the specimens that have been misinterpreted.

If we adopt the latter viewpoint, then we are required to consider the elements that, according to form resemblance, are presumed to be closely related and are in fact so treated in form classification. One of the most pertinent elements that these fossils have in common is in the eye of the beholder and *his* microscopical equipment. Some of the same descriptive terms are used simply because all are attempting to view them in the same way. Regardless of this, a variety of material is being regarded by persons who inevitably express their own personalized viewpoint. Individual bias always is possible as we attempt to apply subjective classification based on "form." The "allies" in a form classification always seem to be more numerous than those that show clear evidence of phyletic affinity. As a consequence very reasonable doubts can be generated about the biologic validity of any classification that lacks some means of functional orientation.

No doubt a great deal of the similarity of appearance reflects a universal biologic application of principles governing size and form and dissemination. It would not be surprising if over a long period of time divergent groups of unicellular (or paucicellular) organisms did not find a rather similar common pattern that served these functions best. From this standpoint the more obvious functional criteria are less likely to be useful indicators of phyletic alliance than incidental features of ornamentation or composition that do not interfere with the common functional objective and may have no apparent survival value. In ambiguous and difficult material incidental features are more likely to be atavistic and to serve better as indicators of affinity. For purposes of taxonomic assignment it may be necessary to emphasize and attach more significance to incidental features and minor resemblances than seems, on casual inspection, reasonable. Only retrospect will actually show whether the practices were justified.

Morphologic terminology also is a source of difficulty. It may be used either in the sense of functional analogy or in the sense of homology. Determination of which of these senses is implied is most difficult to establish from study of the literature. These potential divergences in the common use of terms add another dimension of semantic confusion to the problem.

Under the circumstances a pragmatic point of view is needed. This should be based on biologic principles most likely to provide a correct evaluation of the material, because, if analogies rather than homologies are given weight of emphasis, we lose the preeminent paleontologic value of comparison that is essentially based on hereditary relationship. The value of homologous comparison must be regarded as of overriding importance, but the proof of homologous interpretation is difficult and is likely to be convincingly demonstrated only by many records amassed over a considerable period of study. Such a study would reflect happenings in phyletic history over a long span of geologic time. In the meantime, and while this record is being accumulated, students still are confronted with a practical problem of classification.

Artificial Groups

Downie, Evitt, and Sarjeant (1963) have proposed a series of groups ("subgroups") that they equate with families proposed earlier under the Zoological Code of Nomenclature for microfossils of uncertain biological interpretation that share certain arbitrary features. The "Index of Genera of Fossil Dinophyceae and Acritarchs" by Norris and Sarjeant (1965) already referred to on page 73 has been arranged partly according to this system. With modifications in detail, these groups are similar to the form categories previously listed by Staplin (1961, p. 402).

It seems doubtful that these suprageneric taxa can be regarded as having formal status in taxonomy because an all-important functional biologic justification appears to be lacking for each of them. They represent arbitrary groups of genera. Nonetheless, the groups proposed can probably provide a useful artificial basis for identification.

A strenuous effort is being made to describe and identify valid species. These species are very numerous and are widely dispersed geographically, and therefore at least an artificial means of sorting the material is essential. The artificial subdivisions probably should not, however, be treated as if they were formal groups because in some instances we shall later be capable of assigning some of the species to taxa with phyletic implications. This process of reassignment, supported by accumulated evidence, should be encouraged. However, a useful artificial key for practical purposes of arrangement and identification needs no special justification. Some of the possible groups that may be used are the following:

Vesicles Simple, with Single Wall

Spheroidal, surfaces *smooth to papillate:* SPHAEROMORPHS; example: *Symplassosphaeridium* Tim. 1959.

Spheroidal, surfaces *apiculate to spinose*: ACANTHOMORPHS; examples: *Baltisphaeridium* Eis. 1958, *Micrhystridium* Defl. 1937.

Spheroidal, with uniform system of *external crests:* HERKOMORPHS; example: *Cymatiosphaera* O. Wetzel 1933.

Spheroidal, surface variable, with *membranous equatorial flange*: PTEROMORPHS; example: *Pterospermopsis* W. Wetzel 1952.

Polyhedral, surface with *crests varying* in prominence: PRISMATOMORPHS; example: *Polyedrixium* Deunff 1954.

Polyhedral, surface *apiculate to spinose*: POLYGONOMORPHS; examples: *Estiastra* Eis. 1959, *Veryhachium* Deunff, 1954.

Unipolar, globular, surface *smooth to papillate*: OOMORPHS; example: *Zonooidium* Tim. 1957 (*Fide* Timofeyev, 1959).

Bipolar, *elongate or fusiform,* surface variable: NETROMORPHS; examples: *Leiofusa* Eis. 1938, *Anthatractus* Deunff 1954.

Vesicles Duplicate

Concentrically spheroidal, surface *smooth or papillate*: DISPHAEROMORPHS; example: *Pterocystidiopsis* Defl. 1937.

Flattened, outline *circular or angular*: PLATYMORPHS; example: *Platycystidia* Cookson and Eis. 1960.

Bipolar, elongate or fusiform, with *spheroidal* inner membrane: DINETROMORPHS; example: *Diplofusa* Cookson and Eis. 1960.

Illustrations of specimens assigned to various genera that exemplify these morphologic groups have been redrawn from published illustrations and are shown in figure 10-8, 1 to 12. The "fichier" of Deflandre (1962b, 1964, 1965) provide further specific illustrations of these microfossils.

In addition more than a dozen genera have been named that are less easily categorized by simple elements of morphology. These might be designated "variomorphs" or, more simply, "others." The infinity of possible biologic variants will from time to time embarrass any attempts to formulate a comprehensive artificial scheme. An arbitrary arrangement that provides for most of the practical requirements should be adequate. A scheme that is exhaustive almost certainly would be so unwieldy that its practical purpose would be defeated.

The authors who proposed these groupings of acritarchs agree that many of the rather similar microfossils in Mesozoic and Tertiary deposits are referable to the Dinophyceae. Eisenack and Klement (1964), Downie and Sarjeant (1964), and Norris and Sarjeant (1965) have included dinophyte taxa in their nomenclatural indexes. A more complete listing of dinophyte genera was subsequently prepared by Loeblich and Loeblich (1966), and to these compendia a reader is referred. It is likely that late Paleozoic ancestors of the dinophytes also will be recognized, but it will doubtless be a long time before many of the forms that occur in lower Paleozoic deposits, which have been arranged in the informal groups listed above, can be assigned to major taxonomic groups with confidence.

Occurrences

Characteristic types of acritarchs are illustrated in figure 10-8, and others, from Timofeyev's extensive studies of the older Paleozoic, are shown on figure 10-6, 35 to 45. However, an adequate sampling of the diversity of form and preservation is beyond the scope of this treatment. A useful discussion of the definition of some common genera of acritarchs is given by Downie and Sarjeant (1963) with check lists of species they regard as included in them. Staplin (1961) had objected to the arbitrary definition of *Micrhystridium*, based solely on size, and with this conclusion Downie and Sarjeant seem largely to agree. However, they insist on retaining the 20-

micron distinction in the diagnosis because it reflects a break in the size range of species they assign to *Micrhystridium* and to *Baltisphaeridium*. In both reports taxonomic emendations are proposed and both are valid efforts to improve our means of taxonomic expression. Each person is entitled to his opinion about problems of classification. It may be a long time before enough of the abundant material has been surveyed to show which method of subdivision is practical and advisable to provide an adequate basis for a stable classification.

The abundant bilateral types of acritarchs that characterize Middle and Upper Cambrian and Tremadoc appear to be much diminished or lacking in younger deposits. Simple, thin-walled, spheroidal types, known as leiospheres, become abundant in the Upper Ordovician, Silurian, and

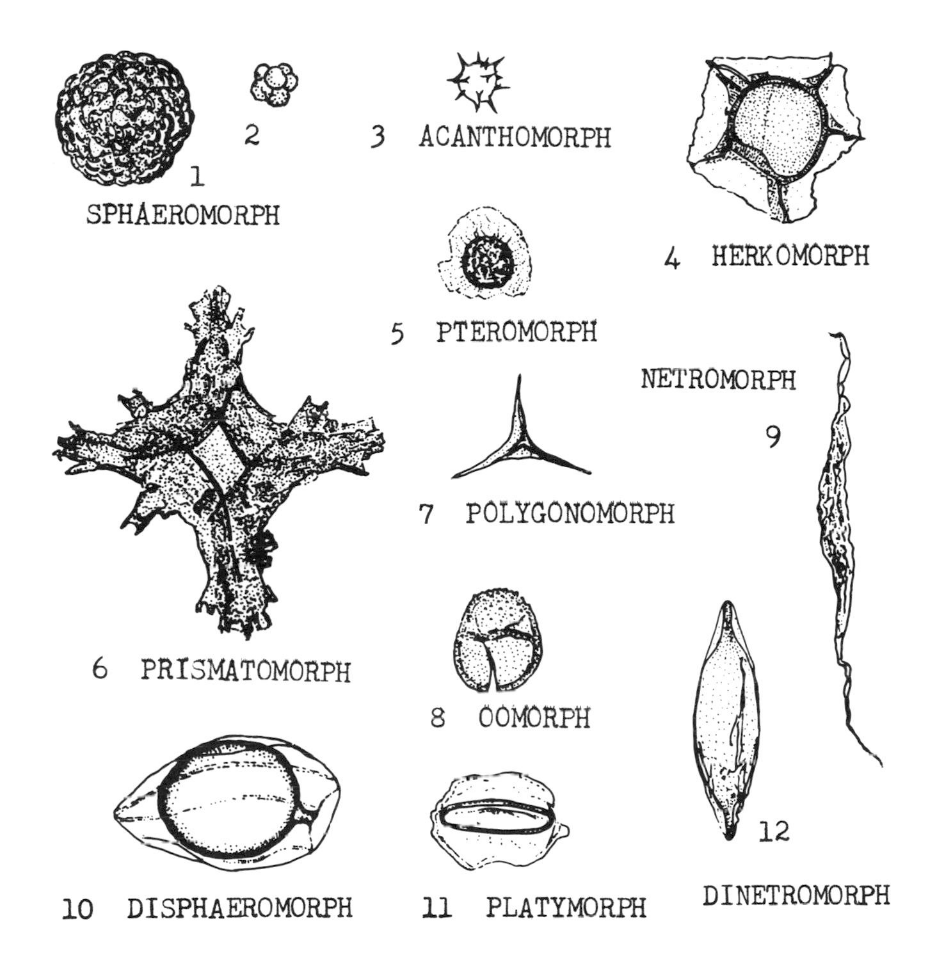

Figure 10-8. Acritarch microfossils, representative of morphologic groups (all dimensions given are in microns). 1—*Symplassosphaeridium tumidulum* Tim. (× 300), 40 to 55, Obolus beds, Upper Cambrian; from Timofeyev (1959, pl. 1, fig. 11). 2—*S. subcoalitum* Tim. (× 300), 18 to 20, Obolus beds, Upper Cambrian; from Timofeyev (1959, pl. 1, fig. 12). 3—*Micrhystridium inconspicuum* Def. (× 1,075), 8½, Silex de Paris, Cretaceous; from Deflandre (1937, pl. 12, fig. 12). 4—*Cymatiosphaera* sp., cf. *C. fagoni* Def. (× 425), 34, Onondaga Limestone, Middle Devonian; from Deunff (1957, fig. 12). 5—*Pterospermopsis* sp., cf. *P. onondagensis* Deunff (× 250), 12, Wenlock, Middle Silurian; from Downie (1959, pl. 12, fig. 8). 6—*Polyedryxium deflandreyi* var. *major* Deunff (× 340), 130, Onondaga Limestone, Middle Devonian; from Deunff (1955, pl. 1, fig. 3). 7—*Veryhachium centrigerum* Deunff (× 425), 45, Onondaga Limestone, Middle Devonian; from Deunff (1957, fig. 4). 8—*Zonooidium guttiforme* Tim. (× 300), 37 × 40, Ischorian and Obolus beds, Middle and Upper Cambrian; from Timofeyev (1959, pl. 13, fig. 6). 9—*Leiofusa filifera* Downie, (× 250), length 240, Wenlock Shale, Middle Silurian; from Downie (1959, pl. 11, fig. 7). 10—*Pteridocystidiopsis stephaniana* Def. (× 500), 36 × 60, Silex de Paris, Cretaceous; from Deflandre (1937, pl. 17, fig. 7). 11—*Platycystidia diptera* Cookson and Eis., (× 400), 48 × 30, Cretaceous; from Cookson and Eisenack (1960, pl. 3, fig. 22). 12—*Diplofusa gearlensis* Cookson and Eis. (× 225), 34 × 119, Cretaceous; from Cookson and Eisenack (1960, pl. 3, fig. 10).

Devonian (Deunff, 1954a,b). Occasional examples may possess a pylome, but the great majority do not (Staplin, 1961, p. 404; Downie, 1963, p. 629). Leiospheres are distinguished from *Tasmanites* (discussed later) by their lack of punctation. Other common types of acritarchs, with bizarre as well as simple appendages and many with hollow appendages, are associated with them. Ornamentation serves as a convenient means of species differentiation, but, as Downie (1958) emphasized in his study of the Tremadoc Shineton Shale, apparent intergradation in several directions may occur. Staplin (1961), in particular, has attempted to indicate the range of variation that he treated as characterizing taxa.

The smooth-walled acritarchs with a few hollow, elongate appendages—classed as species of *Leiofusa, Veryhachium, Domasia*, and *Deunffia*—have been studied by Downie (1963) from an extensive series of samples representing a thickness of about 1000 feet of section through the Wenlock Shale, as shown in figure 10-9. Such studies that follow closely the evolution of gradual changes would seem to offer advantages for stratigraphy. However, it will be necessary to confirm the general occurrence of features of the Wenlock progression elsewhere before it is relied on.

Downie reported a more or less progressive change in the acritarch populations throughout the Wenlock sequence. Some of the forms that Staplin has reported from the Upper Devonian reef area in Alberta are also present. Evidently many of the early Paleozoic acritarchs persist through a long span of time. Staplin (1961) has attempted to relate the abundance of acritarchs to distances from reefs in Alberta, but in general the factors determining abundance of these microfossils are poorly understood.

One means of testing the advisability of using a system of classification, when dealing with fossils that provide meager evidence of phyletic affinity, is with reference to their stratigraphic distribution. A first attempt at defining stratigraphic distribution of acritarch genera was provided by Eisenack (1963a), who simply listed ranges. Genera he listed for the older Paleozoic are tabulated in figure 10-10. The papers in which the listed genera are defined are listed in the References. A generalized picture is presented that reflects a stage of progress, but it is surely only a rough approximation of reality.

Downie and Sarjeant (1967) have recently presented range charts including many acritarchs as organic-shelled cysts of Dinophyceae. Many of the ranges given (cf. their fig. 2.6A) are excep-

Figure 10-9. Silurian and Devonian acritarchs. 1—Distribution and possible relationships of some polygonomorphic and netromorphic acritarchs in the Wenlock Shale of Britain: *a, Deunffia monospinosa; b, Deunffia brevispinosa; c, Leiofusa* cf. *tumida; d, Domasia bispinosa; e, Domasia trispinosa; f, Domasia elongata; g, Veryhachium elongatum; h, V. trispinosum; i, V. europaeum* var. *wenlockium; j, V. formosum; k, Leiofusa tumida.* The arrows indicate where phyletic transition seems possible. After Downie (1963, p. 634). 2—*Deunffia brevispinosa* Downie (× 950), Wenlock, common; (Downie, 1960, p. 198; pl. 1, fig. 4). 3—*Domasia bispinosa* Downie (holotype) (× 950), Wenlock, very rare (Downie, 1960, p. 200; pl. 1, fig. 3). 4—*Deunffia monospinosa* Downie (holotype) (× 500), Wenlock, common (Downie, 1960, p. 198; pl. 1, fig. 8). 5—*Deunffia furcata* Downie (holotype) (× 500), Wenlock, common; (Downie, 1960, p. 199; pl. 1, fig. 9). 6—*Veryhachium rhomboidium* Downie (holotype) (× 500), Wenlock, very rare; (Downie, 1959, p. 62; pl. 12, fig. 10). 7—*Veryhachium trisphaeridium* Downie (holotype) (× 500), Wenlock, very rare (Downie, 1963, p. 637; pl. 92, fig. 7). 8—*Veryhachium elongatum* Downie (holotype) (× 500), Wenlock, rare (Downie, 1963, p. 637; pl. 92, fig. 10). 9—*Domasia trispinosa* Downie (holotype) (× 500), Wenlock, very rare (Downie, 1960, p. 199; pl. 1, fig. 7). 10—*Leiofusa tumida* Downie (holotype) (× 500), Wenlock, very rare (Downie, 1959, p. 65; pl. 11, fig. 5). 11—*Multiplicisphaeridium spicatum* Staplin (holotype) (× 500), Devonian (Staplin, 1961, p. 411; pl. 49, fig. 21). 12—*Protoleiosphaeridium diaphanium* Staplin (holotype) (× 500), Upper Devonian (Staplin, 1961, p. 406; pl. 48, fig. 8). 13—*Baltisphaeridium arbusculiferum* Downie (holotype) (× 500), Wenlock, rare (Downie, 1963, p. 644; pl. 91, fig. 5). 14—*Baltisphaeridium robustispinosum* Downie (holotype) (× 500), Wenlock, very rare (Downie, 1959, p. 61; pl. 10, fig. 7). 15—*Baltisphaeridium granulatispinosum* Downie (× 500), Wenlock, common (Downie, 1963, p. 640; pl. 10, fig. 7). 16—*Baltisphaeridium granulatispinosum* Downie (holotype) (× 500), Wenlock, common (Downie, 1963, p. 640; pl. 91, fig. 1). 17—*Micrhystridium octospinosum* Staplin (holotype) (× 500), Devonian (Staplin, 1961, p. 410; pl. 48, fig. 18). 18—*Baltisphaeridium longispinosum* (Eis.) Eis. (× 500), Wenlock, common (Downie, 1959, p. 48; pl. 10, fig. 1). 19—*Lophosphaeridium* sp. cf. *Protoleiosphaeridium papillatum* Staplin; (× 500), Wenlock, common; (Downie, 1963, p. 631; pl. 92, fig. 12). 20—*Baltisphaeridium microspinosum* (Eis.) Downie (× 500), Upper Llandovery, rare (Downie, 1959, p. 60; pl. 10, fig. 10).

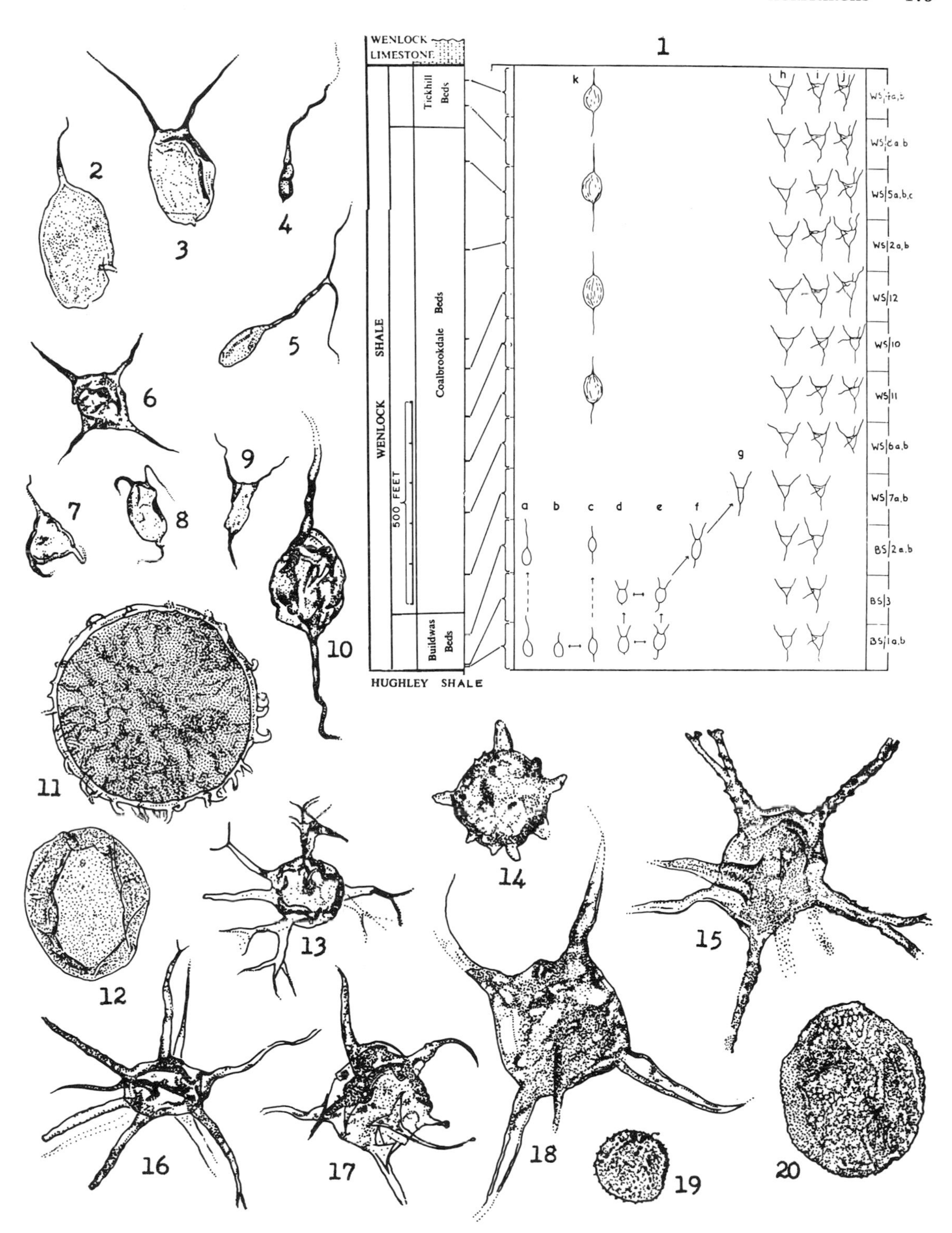

tionally long. Such a long stratigraphic range may signify that general, probably polyphyletic, features of morphology have assumed disproportionate emphasis in classification. On the other hand, it may be argued that algae are inherently conservative in many features and that a very long period of existence for many taxa can be reasonably anticipated. From both practical and theoretical viewpoints a system of classification that presents a more detailed view of the histori-

Figure 10-10. **Stratigraphic range of acritarch genera in the lower Paleozoic (From Eisenack, 1963a).**

Cambrian	*Ordovician*	*Silurian*	*Devonian*	
100 my	*60 my*	*40 my*	*50 my*	
xxxxxxxxxxxxxxxxxxxxxxxxxxxxxxx				*Acanthodiacrodium* Timofeyev 1958b, em., Deflandre & Deflandre-Rigaud 1962a
xxxxxxxxxxxxxxxxxxxxxxxxxxxxxxx				*Dasydiacrodium* Timofeyev 1959, em., Deflandre & Deflandre-Rigaud 1962a
xxxxxxxxxxxxxxxxxxxxxxxxxxxxxxx				*Lophodiacrodium* Timofeyev 1958b, em., Deflandre & Deflandre-Rigaud 1962a
xxxxxxxxxxxxxxxxxxxxxxxxxxxxxxx				*Trachydiacrodium* Timofeyev 1959, em., Deflandre & Deflandre-Rigaud 1962a
xxxxxxxxxxxxxxxxxxxxxxxxxxxxxxx				*Protoleiosphaeridium* Timofeyev 1956
xxxxxxxxxxxxxxxxxxxxxxxxxxxxxxx				*Symplassosphaeridium* Timofeyev 1956
xxxxxxxxxxxxxxxxxxxxxxxxxxxxxxx				*Trematosphaeridium* Timofeyev 1956
	xxxxxxxxxxxxxxxxxx			*Lophosphaeridium* Timofeyev 1956
	xxxxxxxxxxxxxxxxxx			*Orygmatosphaeridium* Timofeyev 1956
	xxxxxxxxxxxxxxxxxx			*Trachysphaeridium* Timofeyev 1956
	xxxxxxxxxxxxxxxxxx			*Vavosphaeridium* Timofeyev 1956
	xxxxxxxxxxxxxxxxxx			*Zonosphaeridium* Timofeyev 1959
	xxxxxxxxxxxxxxxxxx			*Aremoricanium* Deunff 1955 (Py.)[1]
	xxxxxxxxxxxxxxxxxx			*Lunulidia* Eisenack 1958
	xxxxxxxxxxxxxxxxxx			*Vulcanisphaera* Deunff 1961
	xxxxxxxxxxxxxxxxxx			*Priscotheca* Deunff 1961
	xxxxxxxxxxxxxxxxxx	xxxxxxxxxxxx	xxxxxxxxxxxxxxx	*Leiosphaeridia* Eisenack 1959 (Py.)
	xxxxxxxxxxxxxxxxxx	xxxxxxxxxxxx	xxxxxxxxxxxxxxx	*Leiofusa* Eisenack 1938 (Py.)
	xxxxxxxxxxxxxxxxxx	x x x x x x	xxxxxxxxxxxxxxx	*Cymatiosphaera* O. Wetzel 1933, em., Deflandre 1954
		xxxxxxxxxxxx		*Pulvinosphaeridium* Eisenack 1955 (Py.)
		xxxxxxxxxxxx		*Deunffia* Downie 1960
		xxxxxxxxxxxx		*Estiastra* Eisenack 1959
		xxxxxxxxxxxx		*Domasia* Downie 1960
		xxxxxxxxxxxx		*Cannosphaeropsis* (?) O. Wetzel 1933 (Py.)
		xxxxxxxxxxxx	xxxxxxxxxxxxxxx	*Dictyotidium* Eisenack 1955
			xxxxxxxxxxxxxxx	*Pterospermopsis* W. Wetzel 1952
			xxxxxxxxxxxxxxx	*Anthatractus* (?) Deunff 1954c
			xxxxxxxxxxxxxxx	*Duvernaysphaera* Staplin 1961
			xxxxxxxxxxxxxxx	*Polyedryxium* Deunff 1954c
			xxxxxxxxxxxxxxx	*Paleopedicystus* Staplin 1961
xxxxxxxxxxxxxxxxxxxxxxxxxxxxxxx	xxxxxxxxxxxxxxxxxx	xxxxxxxxxxxx	xxxxxxxxxxxxxxx	*Veryhachium* Deunff 1954d
xxxxxxxxxxxxxxxxxxxxxxxxxxxxxxx	xxxxxxxxxxxxxxxxxx	xxxxxxxxxxxx	xxxxxxxxxxxxxxx	*Baltisphaeridium* Eisenack 1958 (Py.)
	xxxxxxxxxxxxxxxxxx	x x x x x x	x x x x x x x x x	*Micrhystridium* Deflandre 1937
	xxxxxxxxxxxxxxxxxx	xxxxxxxxxxxx	xxxxxxxxxxxxxxx	*Tasmanites* Newton 1875 (Py.)

[1]Py. = with pylome according to Eisenack.

cal progression of these groups is much to be desired.

TASMANITES AND THE PRASINOPHYCEAE[1]

For over 100 years punctate sporelike bodies have been noted in Devonian black shales of North America and occasionally in other types of deposits, most notably in the sediments reworked from glacial till in the northern midwest United States where the till includes erratics of the Devonian shale. These bodies have often been called "Sporangites," a name of ambiguous application that reflects several of the misconceptions that have arisen concerning the spore-

[1]See chapter 18.

like bodies. Available information concerning these forms was reviewed in 1944 by Schopf, Wilson, and Bentall, who concluded that the scientific name *Tasmanites* was applicable.

Tasmanites was defined and named by Newton in 1875. Characteristic disseminules of the type species, *T. punctatus*, make up a large proportion of the marine black shale of Permian age in the Mersey Valley of Tasmania. Solid deposits of the *Tasmanites* disseminules from Alaska have been reported recently (Tourtelot, Tailleur, and Donnell, 1966). Recognition of *Tasmanites* as a planktonic alga suggests that such pure tasmanite deposits accumulated from algal blooms. For this and other reasons such deposits are likely to be relatively local. Dispersed occurrences, as in the Ohio Shale (Upper Devonian) and Chattanooga Shale (Devonian and Mississippian), are of course remarkably persistent. Formerly shale oil was obtained from the Tasmanian shale by subjecting it to destructive distillation. The product that was obtained resembled the oil obtained from "kerosene shale" or "white coal" (a torbanite rich in *Botryococcus*) in New South Wales. The Alaskan material recently studied by Tourtelot also will yield a large amount of shale oil. *Tasmanites* was not the first of the sporelike microfossils to be recognized, but it is one of the first to be identified with a valid scientific name. Thus it has been important to determine the nature of the material to establish application and status of the name because of its seniority over other plant-microfossil nomenclature. Further information about *Tasmanites*, through 1956, has been discussed by Winslow (1962, p. 77-79).

Until 1962 the affinity and interpretation of *Tasmanites* disseminules were obscure, although most of the later authors believed that the sporelike bodies were derived from algae. At that time Wall (1962) compared *Tasmanites* disseminules with the cysts of modern marine planktonic algae *Pachysphaera* and *Halosphaera*, specimens of which had been made available to him by Dr. Mary Parke of the Marine Biological Station at Plymouth, England. Parke (1966) has continued her studies of the modern material and succeeded in culturing *Pachysphaera pelagica* and in determining the major features of its life cycle. The likelihood of alliance between *Tasmanites* and *Pachysphaera* now seems very plausible. As a result of the exemption of algae from the provisions of article 58 of the "International Code of Botanical Nomenclature" (Lanjouw and others, 1966), it appears that if the names *Pachysphaera* and *Tasmanites* are accepted as synonymous, *Pachysphaera* must become a synonym. There is of course a real question whether features of the life cycle of *Pachysphaera* can be projected on the basis of available evidence into the distant Paleozoic where *Tasmanites* is identified. For this reason it seems desirable that the present taxonomic separation be maintained. However, there is no doubt that the relationship is close. As a result of the studies by Wall and by Parke, it seems most reasonable to interpret the fossil disseminules of *Tasmanites* as cysts of members of the class Prasinophyceae.

The modern planktonic algae assigned to Prasinophyceae seem to have the normal complement of chlorophyll *a* and *b*, xanthophyll, and carotene that characterizes green algae; they have a biflagellar or quadriflagellar motile stage somewhat comparable to aplanospores of members of the suborder Chlamydomonidineae (Volvocales), but relations of the cyst and the scaly flagella are different. Other differences may be discovered as further details are determined. The significance of puncta in the cyst wall is still unknown, but a respiratory function may be suggested. The occurrences of puncta that end blindly in the wall may be a result of growth. These suggestions need to be checked by detailed observations by use of the electron microscope on fossil material (see Manton, Oates, and Parke, 1963).

The Life Cycle of *Pachysphaera*

Pachysphaera cysts develop from motile swarmers and may be as small as 10 microns in diameter. The young cysts contain one nucleus, one chloroplast, and one pyrenoid, but as the cyst grows in diameter the chloroplast and pyrenoid continue to divide. When the cyst becomes mature (100 to 175 μ in diameter), the nucleus begins its process of repeated division until the number of pyrenoids, chloroplasts, and nuclei are equal. A mucilaginous layer develops within the wall of the cyst, and the protoplast separates into small units, each containing a full organelle complement. Each unit develops flagella and the contents are expelled, as a result of mucilaginous swelling, through a linear fissure in the outer resistant wall of the cyst. The discarded walls evidently may be preserved in sediment. The motile swarmers free themselves from mucilage and repeat the vagile phase of the life history. Whether sexual differentiation and conjugation occurs has not been determined. Presumably the cysts that develop later are not sexually differentiated.

Comparisons with *Tasmanites*

The capacity of the cyst of *Pachysphaera* to enlarge during growth and maturation contrasts strongly with the more stable size characteristics of cryptogamic spores. Winslow (1962) studied the measurable biocharacters in *Tasmanites* disseminules and demonstrated an extreme range of size as well as intergradation and overlap between types usually recognized as separate species. She wrote (1962, p. 82): "It seems evident that species of *Tasmanites* are frequently indistinctly separated and must be regarded a little differently than those of other genera in which specific differentiation is more distinct." It would seem most reasonable now to interpret the great range of size in *Tasmanites*, as in the cysts of *Pachysphaera*, as a result of ontogeny and normal growth.

In optical appearance the cysts of *Tasmanites* generally resemble spore coats, but the resemblance may be only superficial. Both apparently are of lipid composition and are highly durable and resistant to oxidation. However, they probably form in an entirely different manner, and the wall of the *Tasmanites* cyst is evidently much more adapted to processes of appositional growth. Differences in micellar organization may make the cysts of *Tasmanites* more anisotropic than spores of higher plants when viewed by means of polarized light. Cysts of *Tasmanites* also accept biologic stains much less readily. Chemical studies tend to illustrate the general similarities of all the disseminule lipids, and therefore an electron-micrographic study might show more of the differences.

Recently, Kjellström (1968) has shown by means of infrared spectra that specimens of *Tasmanites punctatus* Newton and *Tasmanites erraticus* Eisenack from the original Mersey River Tasmanite deposit in Tasmania do not differ in composition. The wall material is remarkably stable and seems to include a larger amount of the CH_2 and CH_3 than COOH groups, suggesting structural similarity with a long chain, saturated carbohydrate. Ultrathin sections (300 to 800 Å) examined by electron microscopy show two kinds of pores in *Tasmanites punctatus* (fig. 10-11) the smaller of which are ultramicroscopic. The wall shows definite concentric banding and a radially oriented ultrastructure that tends to explain the good optical anisotropy of the *Tasmanites* membrane. A different and less oriented type of ultrastructure is visible in the membrane of Silurian and Ordovician specimens of *Leiosphaeridia*. In some sections about the same type of ultrathin "chatter marks" are visible that McCartney and others (1966; 1967, fig. 35) have obtained from fusain in high volatile coal. Both infrared and EM studies are likely to produce additional significant information in the future.

Pylome?

Palynologists disagree concerning the nature of exit openings in the walls of *Tasmanites* cysts; for example, Eisenack (1963b, fig. 4) illustrated a circular perforation that he called a pylome, but it might also be the impression of a crystal of authigenic pyrite. Generally circular pylome openings are lacking. A few of the many macerations containing examples of *Tasmanites* prepared in the U.S. Geological Survey Laboratory in Columbus, Ohio, show specimens with many holes, but this is exceptional. The holes appear to have formed during diagenesis or to have been produced as artifacts during maceration. Some phytoplankton are known to have a naturally constituted exit hole, or pylome; and a mixed assemblage of fossil cysts, some of which have a pylome and some which lack it, might be present at any locality to explain the discrepancy among reports about cysts of *Tasmanites*. However, we would expect that other differences—at least in fine texture of the surfaces, anisotropism, or other features—would be apparent in these instances, but such a mixture has not been detected. In spite of evidence to the contrary, the occurrence of a pylome in *Tasmanites* seems very questionable.

Some microfossil assemblages include an abundance of *Tasmanites* cysts that are split into two lenticular segments. These examples suggest comparison with the split outer walls of cysts of *Pachysphaera* illustrated by Parke

Figure 10-11. Electron micrograph of the wall of *Tasmanites punctatus*. 16,000×. After Kjellström (1968).

(1966). The smooth tear of the wall may reflect merely the physical composition of the wall membrane when subjected to tension by imbibition of water, rather than a detectable morphologic differentiation. Such an extrusion mechanism contrasts greatly with that of a differentiated pylome. Very closely related groups of organisms probably would not show two very different modes of evacuating the living contents from the protective wall.

Tasmanites from Kettle Point, Ontario

Common types of *Tasmanites* occurring in the Huron Shale at Kettle Point, Ontario, are shown in figure 10-12, 1 to 10. This is the locality most commonly cited in reference to Dawson's early

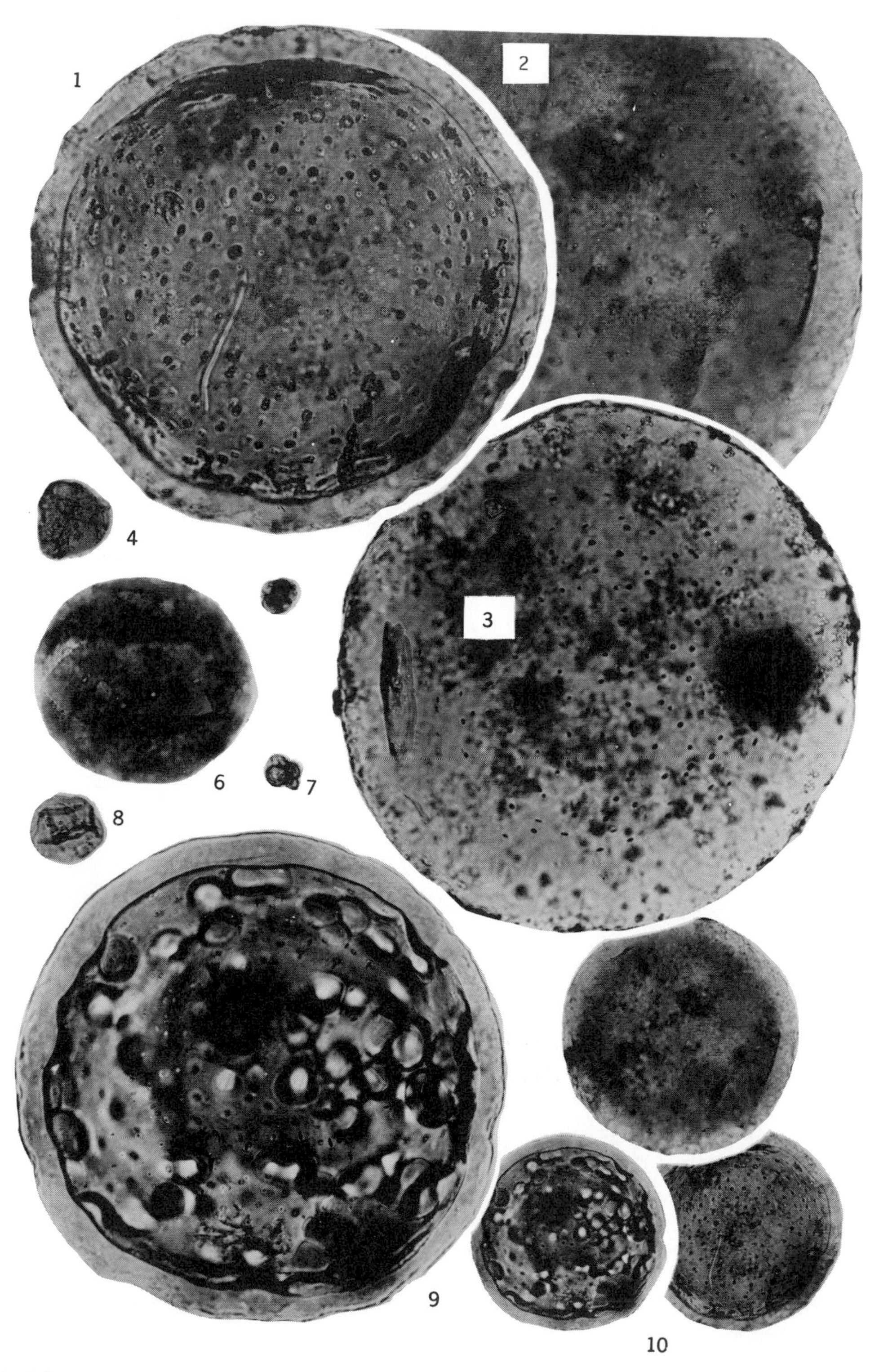

Figure 10-12. Palynomorphs from the Huron Shale at Kettle Point, Ontario. 1 – *Tasmanites huronensis* (Dawson) Winslow (1962); thick-walled disseminule with puncta internally flared (topotype) × 250; 2 – *T. huronensis*; somewhat larger example, puncta not as evidently flared, × 250; 3 – *T. huronensis*; wall thickness evident only in the area of an internal inclusion at the left side; × 250; scale at left. 4 – Small cryptogamic spore with trilete suture – an infrequent associate with *Tasmanites*; × 250; 5 and 7 – Groups of small sporelike bodies, possibly juvenile cysts

work on "Sporangites," and is the type source of *Tasmanites huronensis*. Figures 10-12-1 to 10-12-3 illustrate specimens 300 to 400 microns in diameter generally assigned to *T. huronensis* (Dawson) Winslow. Wall (1962) has interpreted linear markings such as that shown in number 1 as a preformed suture or line of dehiscence in *Tasmanites*. The inconsistencies of the occurrence of such a line leads the writer to suppose that such structures are a result of minor folds induced by compression. Various smaller forms are also present in the Kettle Point deposit. Occasionally very small bodies (20 to 30 μ) occur in groups as in figures 10-12-5 and 10-12-7. Originally the writer supposed that these might represent tetrads but, in view of Parke's work on *Pachysphaera*, they might be interpreted with equal justification as fortuitous aggregations of juvenile cysts. Members of the groups are not quite the same size, and most fossil tetrads show a more regular arrangement. Comparison of figures 10-12-5 with figures 10-12, 1-3 shows the magnitude of size range that must be considered as part of the *Tasmanites* ontogeny if alliance with *Pachysphaera* is admitted.

A peculiar type of punctate cyst with internal tubercles is also shown (10-12-9). One of the common types of cysts, about 300 microns in diameter, with evident fissures is shown at lower magnification in 10-12-6. The example in 10-12-8 resembles *Tasmanites sinuosus* Winslow, except for its smaller size (about 45 μ). A small cryptogamic spore with definite trilete suture, of about the same size, is shown in figure 10-12-4 for comparison. Such forms are infrequent, but they probably signify that some debris from land plants also contributed to the Kettle Point deposit.

Part of the difficulty in reporting adequately on the cysts of *Tasmanites* is again a reflection of the range of size. Figure 10-12-10 represents the same cysts shown in 10-12-1, 2, and 9, reproduced at lower magnification (× 100). A greater magnification is desirable even though at × 100 the image of these fossils is of a size that might be considered adequate for illustrating other palynomorphs.

Scanning electron micrographs showing general form and punctation morphology of *Tasmanites punctatus* Newton, from Tasmania, are shown in plate 10-1 through the courtesy of Dr. Göran Kjellström of Stockholm.

Occurrences of *Tasmanites*

The time range of the Tasmanaceae is Ordovician and younger according to Downie and Sarjeant (1967). The taxonomic problem of assigning Ordovician specimens that have a pylome (see Eisenack, 1958, 1963b; Martinsson, 1956) still remains unsolved, but many other forms of Ordovician and Silurian age lack a pylome and show the features characteristic of *Tasmanites*. Devonian occurrences in Brazil (Sommer, 1956) and in North America (see Winslow, 1962) are most widely known. Records of *Tasmanites* in the Mississippian are scarce, and, strangely, the genus seems generally to be lacking in Pennsylvanian age deposits. It is hard to understand how these algae, if they were present, could have been excluded from Pennsylvanian epicontinental seas. Tappan (1968) concludes that the Pennsylvanian represents a period of drastic reduction in the phytoplankton population, and the lack of *Tasmanites* in the Pennsylvanian supports her conclusion. Such ambiguous bodies might occasionally be overlooked, but the writer has been alert to this possibility for over 20 years and has yet to identify *Tasmanites* from Pennsylvanian deposits. *Tasmanites* of Permian age has been reported in abundance (Newton, 1875) from marine beds in Tasmania, correlative with Greta Coal Measures of Australia, and *Tasmanites* of Early Jurassic age has been reported (Wall, 1965) from Britain. White (1929), Winslow (1962), and Tourtelot, Tailleur, and Donnell (1966) have described *Tasmanites* of Jurassic, or possibly Lower Cretaceous, rocks from Alaska. Felix (1965) has reported *Tasmanites* from Neogene deposits in southern

---

of *Tasmanites*. Arrangement differs from that of a tetrad; × 250; 6—*Tasmanites huronensis*. Disseminule with arcuate fissures and folds, a common type of preservation; × 100; 8—*Tasmanites* cf. *T. sinuosus* Winslow (1962); sporelike body lacking haptotypic features and similar to *T. sinuosus* except for its smaller diameter; × 250; 9—*Tasmanites* sp.; note wall canals and internal "blisters." A unique example; × 250; 10—*Tasmanites* spp.; specimens shown in numbers 1, 2, and 9 illustrated at lower magnification, × 100.

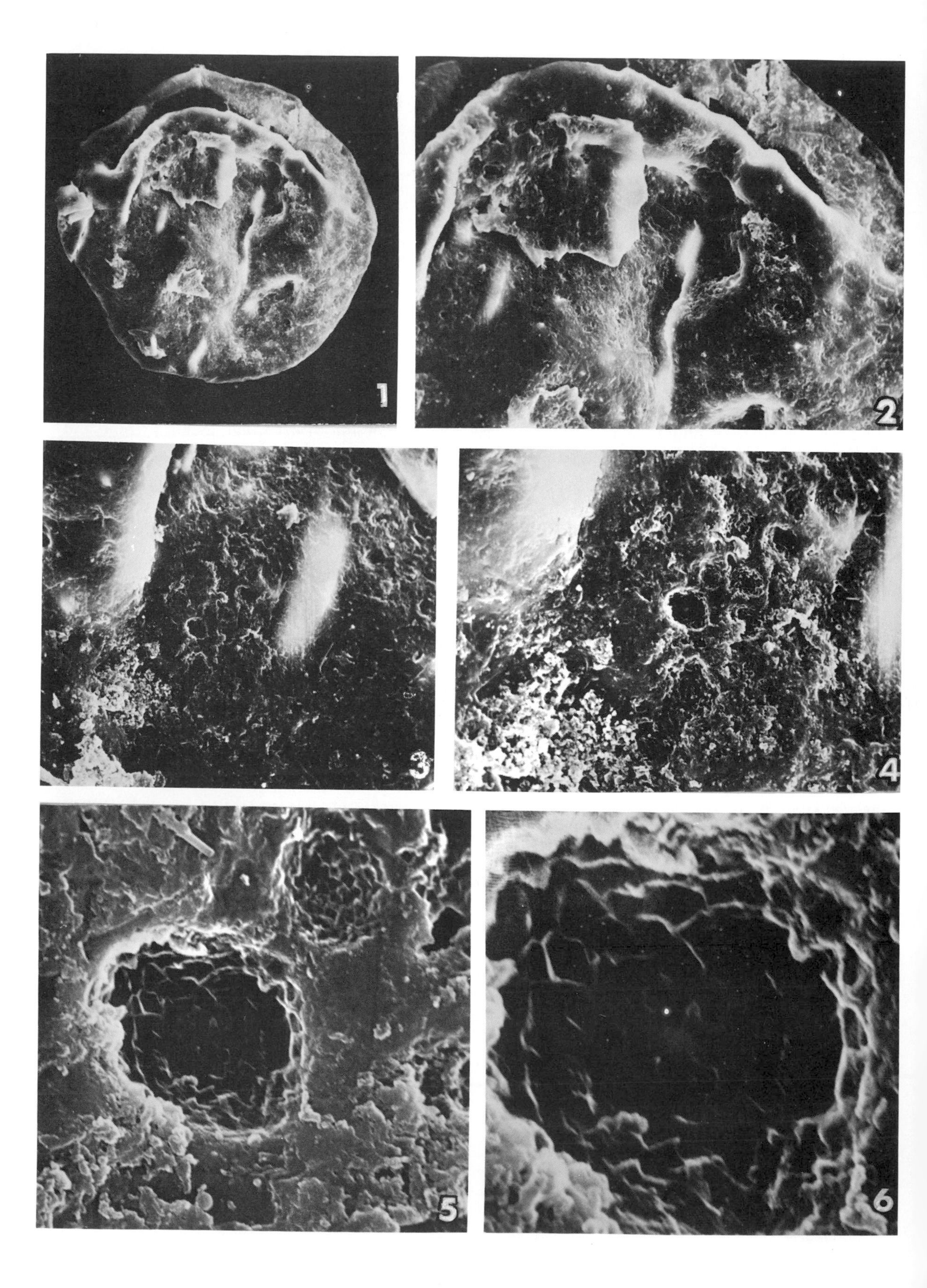
1
2
3
4
5
6

Louisiana, many of which show fissure openings like those of abandoned cysts of *Pachysphaera*. These reports document a long geologic record that will be given in much greater detail in the years ahead. It is sufficient, however, to show the ancient origin of the group and to link its habitat decisively with the marine domain.

Phyletic antiquity implies a corresponding genetic isolation. If we accept the present indication that modern *Pachysphaera* is equivalent to or allied in the same isolated phylogeny with *Tasmanites*, and possibly with *Tytthodiscus* and a few other genera, the fossil disseminules become identified as cysts of planktonic green algae with an alternation of cystose and motile life-cycle phases. Variations discernible from the modern examples alone (Parke and den Hartog-Adams, 1965) have suggested that these forms are taxonomically separated by important differences from other extant groups of algae. Fossil evidence, indicating that the history of this phyletic isolation extends at least as far as the early Paleozoic, seems independently to support the identification of these plants as a separate class of green algae, the class that is called the Prasinophyceae.

PORTENTIOUS SPORES WITH FUNCTIONAL SUTURES

The sporelike microfossils in Ordovician and older deposits evidently may show a suggestion of triradiate marking, but there is no indication that the marking had any relation to polarity or exit of the living contents. The oldest examples of spores with sutures which appear to be functional and which may denote a polarity gradient for vital elements originally enclosed were described by W. S. Hoffmeister in 1959. The material was derived from a deep oil test hole in Libya where associated evidence based on graptolites indicated Early Silurian age. Downie (1963) has reported similar spores from the Wenlock Shale of Middle Silurian age in Britain. Hoffmeister describes the fossils as "spores of vascular plants," but, whether or not the plants were vascular, the spores seem truly to be harbingers of land vegetation. (See chap. 11.)

The spores are subtriangular to suborbicular in proximo-distal plane and are generally 40 to 50 microns in diameter. They were evidently originally oblate, as judged by preferential compression normal to the axis. The trilete rays extend to the margin, and a commissural line of separation between the opposed labra commonly is visible. Some of the examples show a marginal area of denser coloration that Hoffmeister regards as a special zone of equatorial thickening or crassitude. The spore coats are smooth, and their thickness decreases from about 4 microns at the margin to about $1\frac{1}{2}$ microns in the contact areas. Drawings have been prepared based on Hoffmeister's published illustrations and are shown in figure 10-13, 1 to 5.

Two species based on the Libyan material are distinguished, largely on the basis of the prominence of the equatorial crassitude. Those with a denser equatorial zone were assigned to a new genus *Ambitisporites*, but those in which the thickening is less conspicuous were assigned, with a question mark, to *Punctatisporites*, a genus with its most common occurrence in the Pennsylvanian. The size range and preservation of both forms are very similar.

In view of the occurrence of these spores in Silurian deposits, isolated stratigraphically from land plants that are better documented, generic separation from younger forms seems justified. The plants represented by the spores evidently were pioneers that probably lacked even an ordinal relationship with plants of the Carboniferous. On the other hand, the similarities between the two Silurian spore types are striking, and, in view of their origin, a relationship would seem quite reasonable. The writer believes both species are better treated taxonomically as species of *Ambitisporites* rather than under separate genera. The species *Punctatisporites*? *dilutus* is therefore transferred as *Ambitisporites dilutus* (Hoffmeister) n. comb.

The spores of *Ambitisporites* constitute an

Plate 10-1. 1-6. *Tasmanites punctatus* Newton. Scanning electron (Stereo-Scan) micrographs of disseminule isolated from tasmanite, Mersey River, Tasmania. Large pores, barely visible at low magnification, seem to show an internal reticulate surface. Fig. 1, × 130; fig. 2, × 260; fig. 3, × 650; fig. 4, × 1,300; fig. 5, × 6,500; fig. 6, × 13,000. Photographs provided through the courtesy of Dr. Göran Kjellström, Geological Survey of Sweden, Stockholm.

important paleobotanical landmark. Such evidence is scanty, of course, but in basic configuration these spores correspond with the simpler types of spores known from Lower Devonian psilophytes. Tentatively the writer would assign *Ambitisporites* to the Psilophytaceae. Later discoveries will doubtless prove whether a different assignment is desirable, but it would seem plausible that these spores represent the oldest record of this family of plants.

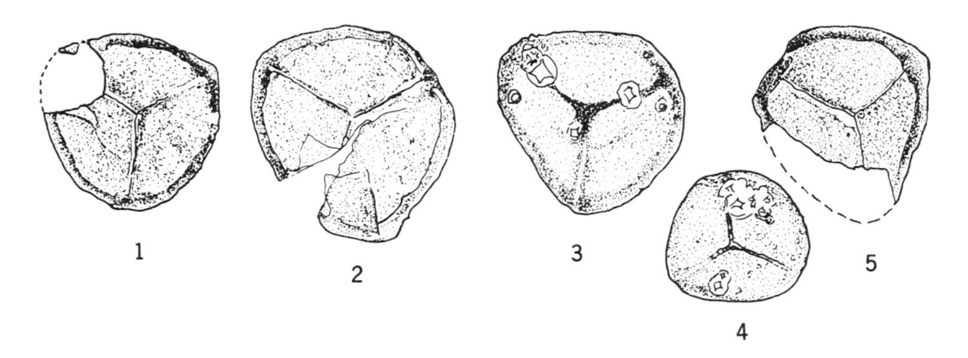

Figure 10-13. Land-plant spores from the Lower Silurian of Libya (× 500; redrawn). 1 and 2—***Ambitisporites avitus*** Hoffmeister: 1 = holotype, figure 1; 2 = figure 3 from Hoffmeister (1959). 3-5—***Ambitisporites dilutus*** (Hoffmeister) n. comb.: 3 = holotype, figure 10; 4 = figure 9; 5 = figure 8, from Hoffmeister (1959).

References

American Commission on Stratigraphic Nomenclature, 1961, Code of stratigraphic nomenclature: Am. Assoc. Petroleum Geologists Bull., v. 45, no. 5, p. 645-665.

Brinkmann, Roland, 1960, Geologic evolution of Europe: New York, Hafner Publ. Co., 161 p., 19 pls., 46 figs., 18 charts.

Chaloner, W. G., 1960, The origin of vascular plants: Science Progress, v. 48, no. 191, p. 524-534.

Cookson, I. C., and Eisenack, Alfred, 1960, Microplankton from Australian Cretaceous sediments: Micropaleontology (Am. Mus. Natl. History), v. 6, no. 1, p. 1-18, pls. 1-3, figs. 1-6.

Deflandre, Georges, 1937, Flagellés incertae sedis, hystrichosphaeridés, sarcodinés, organismes divers, pt. 2 *of* Microfossiles des silex crétacés: Annales Paleont., v. 26, pts. 1-2, p. 51-172.

———1954, Systématique des hystrichosphaeridés; sur l'acception du genre *Cymatiosphaera* O. Wetzel: Soc. Géol. France Comptes rendus, no. 11, p. 257-259.

Deflandre, Georges, and Deflandre-Rigaud, Marthe, 1962a, Nomenclature et systématique des hystrichosphères (sens. lat.) observations et rectifications: Rev. Micropaléontologie, v. 4, no. 4, p. 190-196.

———1962b, Dinoflagellés III—Peridinida à tabulation conservée, *in* Fichier Micropaléontol., sér. 11: Centre Natl. Recherche Sci., no. 383, pts. 1-5, p. 1751-1947.

———1964, Acritarches I—Polygonomorphitae-Netromorphitae *pro parte*—Appendice: Genres *Deflandrastrum* Combaz et *Wilsonastrum* Jansonius, *in* Fichier Micropaléont., sér. 12: Centre Natl. Recherche Sci., no. 392, pts. 1-10, p. 1948-2172.

———1965, Acritarches II—Acanthomorphitae 1, Genre *Micrhystridium* Deflandre (sens. lat.), *in* Fichier Micropaléont., sér. 13: Centre Natl. Recherche Sci., no. 402, pts. 1-5, p. 2176-2521.

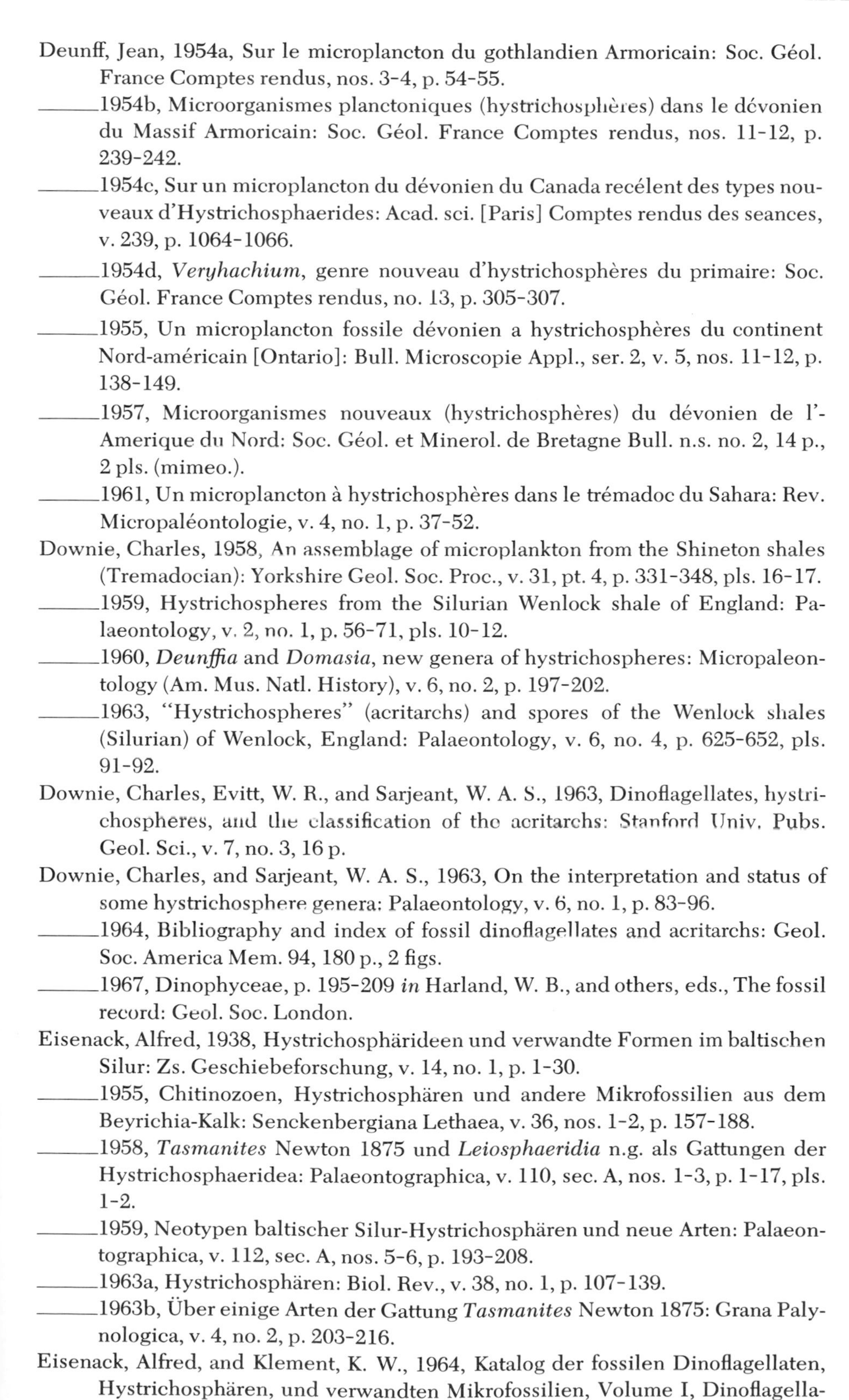

Deunff, Jean, 1954a, Sur le microplancton du gothlandien Armoricain: Soc. Géol. France Comptes rendus, nos. 3-4, p. 54-55.

———1954b, Microorganismes planctoniques (hystrichosphères) dans le dévonien du Massif Armoricain: Soc. Géol. France Comptes rendus, nos. 11-12, p. 239-242.

———1954c, Sur un microplancton du dévonien du Canada recélent des types nouveaux d'Hystrichosphaerides: Acad. sci. [Paris] Comptes rendus des seances, v. 239, p. 1064-1066.

———1954d, *Veryhachium*, genre nouveau d'hystrichosphères du primaire: Soc. Géol. France Comptes rendus, no. 13, p. 305-307.

———1955, Un microplancton fossile dévonien a hystrichosphères du continent Nord-américain [Ontario]: Bull. Microscopie Appl., ser. 2, v. 5, nos. 11-12, p. 138-149.

———1957, Microorganismes nouveaux (hystrichosphères) du dévonien de l'-Amerique du Nord: Soc. Géol. et Minerol. de Bretagne Bull. n.s. no. 2, 14 p., 2 pls. (mimeo.).

———1961, Un microplancton à hystrichosphères dans le trémadoc du Sahara: Rev. Micropaléontologie, v. 4, no. 1, p. 37-52.

Downie, Charles, 1958, An assemblage of microplankton from the Shineton shales (Tremadocian): Yorkshire Geol. Soc. Proc., v. 31, pt. 4, p. 331-348, pls. 16-17.

———1959, Hystrichospheres from the Silurian Wenlock shale of England: Palaeontology, v. 2, no. 1, p. 56-71, pls. 10-12.

———1960, *Deunffia* and *Domasia*, new genera of hystrichospheres: Micropaleontology (Am. Mus. Natl. History), v. 6, no. 2, p. 197-202.

———1963, "Hystrichospheres" (acritarchs) and spores of the Wenlock shales (Silurian) of Wenlock, England: Palaeontology, v. 6, no. 4, p. 625-652, pls. 91-92.

Downie, Charles, Evitt, W. R., and Sarjeant, W. A. S., 1963, Dinoflagellates, hystrichospheres, and the classification of the acritarchs: Stanford Univ. Pubs. Geol. Sci., v. 7, no. 3, 16 p.

Downie, Charles, and Sarjeant, W. A. S., 1963, On the interpretation and status of some hystrichosphere genera: Palaeontology, v. 6, no. 1, p. 83-96.

———1964, Bibliography and index of fossil dinoflagellates and acritarchs: Geol. Soc. America Mem. 94, 180 p., 2 figs.

———1967, Dinophyceae, p. 195-209 *in* Harland, W. B., and others, eds., The fossil record: Geol. Soc. London.

Eisenack, Alfred, 1938, Hystrichosphärideen und verwandte Formen im baltischen Silur: Zs. Geschiebeforschung, v. 14, no. 1, p. 1-30.

———1955, Chitinozoen, Hystrichosphären und andere Mikrofossilien aus dem Beyrichia-Kalk: Senckenbergiana Lethaea, v. 36, nos. 1-2, p. 157-188.

———1958, *Tasmanites* Newton 1875 und *Leiosphaeridia* n.g. als Gattungen der Hystrichosphaeridea: Palaeontographica, v. 110, sec. A, nos. 1-3, p. 1-17, pls. 1-2.

———1959, Neotypen baltischer Silur-Hystrichosphären und neue Arten: Palaeontographica, v. 112, sec. A, nos. 5-6, p. 193-208.

———1963a, Hystrichosphären: Biol. Rev., v. 38, no. 1, p. 107-139.

———1963b, Über einige Arten der Gattung *Tasmanites* Newton 1875: Grana Palynologica, v. 4, no. 2, p. 203-216.

Eisenack, Alfred, and Klement, K. W., 1964, Katalog der fossilen Dinoflagellaten, Hystrichosphären, und verwandten Mikrofossilien, Volume I, Dinoflagellaten: Stuttgart, E. Schweizterbart'sche Verlagsbuchhandlung (Nägele u. Obermiller), 895 p., 9 pls., 420 figs.

Erdtman, Gunnar, 1957, Pollen and spore morphology—Plant taxonomy; Gymnospermae, Pteridophyta, Bryophyta (An introduction to palynology, 2): Stockholm, Almqvist & Wiksell, 151 p., 5 pls., 265 figs.

——1965, Pollen and spore morphology—Plant taxonomy; Gymnospermae, Bryophyta (An introduction to palynology, 3): Stockholm, Almqvist & Wiksell, 191 p., 24 pls., 28 figs.

Evitt, W. R., 1963, A discussion and proposals concerning fossil dinoflagellates, hystrichospheres, and acritarchs: Natl. Acad. Sci. U.S. Proc., v. 49, no. 2, p. 158-164; ibid., no. 3, p. 298-302.

Felix, C. J., 1965, Neogene *Tasmanites* and leiospheres from southern Louisiana, U.S.A.: Palaeontology, v. 8, pt. 1, p. 16-26, pls. 5-8.

Fisher, D. W., 1959, Correlation of the Silurian rocks in New York State: New York State Mus. and Sci. Service, Geol. Survey Map and Chart Ser., no. 1 [1960].

——1962a, Correlation of the Cambrian rocks in New York State: New York State Mus. and Sci. Service, Geol. Survey Map and Chart Ser., no. 2.

——1962b, Correlation of the Ordovician rocks in New York State: New York State Mus. and Sci. Service, Geol. Survey Map and Chart Ser., no. 3.

Fry, W. F., and Banks, H. P., 1955, Three new genera of algae from the Upper Devonian of New York: Jour. Paleontology, v. 29, p. 37-44.

Gignoux, Maurice, 1950, Géologie stratigraphique, 4th ed.: Paris, Masson et Cie, 735 p., 155 figs.

Hoffmeister, W. S., 1959, Lower Silurian plant spores from Libya: Micropaleontology (Am. Mus. Natl. History), v. 5, no. 3, p. 331-334.

Howell, B. F., chm. and others, 1944, Correlation of the Cambrian formations of North America [chart 1]: Geol. Soc. America Bull., v. 55, p. 993-1004.

Keroher, G. C., and others, 1966, Lexicon of geologic names of the United States for 1936-1960: U.S. Geol. Survey Bull. 1200, 3 v.

Kjellström, Göran, 1968, Remarks on the chemistry and ultrastructure of the cell wall of some Palaeozoic Leiospheres: Geol. Föreningens i Stockholm Förhandlingar, v. 90, pt. 2, no. 533, p. 221-228, 8 figs. June 28, 1968.

Kummel, Bernhard, 1961, History of the earth—An introduction to historical geology: San Francisco, W. H. Freeman & Co., 610 p., illus.

Lanjouw, J., and others, eds., 1966, International code of botanical nomenclature: Utrecht, Netherlands, Internat. Bur. Plant Taxonomy and Nomenclature, 402 p.

Loeblich, A. R., Jr., and Loeblich, A. R., III, 1966, Index to the genera, subgenera, and sections of the Pyrrophyta: Miami Univ. Inst. Marine Sci., Studies in tropical oceanography, no. 3, 94 p., 1 pl.

McCartney, J. T., O'Donnell, H. J., and Ergun, Sabri, 1966, Ultrafine structures in coal components as revealed by electron microscopy, *in* Gould, R. F., ed., Coal Science: Washington, D.C., American Chem. Soc., Adv. in Chem. ser., no. 55, p. 261-273.

McCartney, J. T., and Ergun, Sabri, 1967, Optical properties of coals and graphite: U.S. Bur. Mines Bull. 641, 49 p., 36 figs., 11 tables.

Manton, I., Oates, K., and Parke, M., 1963, Observations on the fine structure of the *Pyramimonas* stage of *Halosphaera* and preliminary observations on three species of *Pyramimonas*: Marine Biol. Assn. U. K. Jour., v. 43, p. 225-238.

Martinsson, Anders, 1956, Neue Funde kambrischer Gänge und ordovizischer Geschiebe im südwestlichen Finnland: Uppsala Univ. Geol. Inst. Bull., v. 36, pt. 1, p. 79-105.

Naumova, S. N., 1949, Spori nizhnego kembriya [Spores of the Lower Cambrian]: Akad. Nauk SSSR Izv. Ser. Geol., no. 4, p. 49-56.

———1950, Spory nizhnego silura [Spores of the Lower Silurian], *in* Trudy konferentsii po sporovo-pyltsevomu analizu 1948 goda [All-Union spore and pollen conf., 1st, Moscow, 1948, Proc.]: Moskov Univ., Izd., p. 165-190.

Newton, E. T., 1875, On "Tasmanite" and Australian "White Coal": Geol. Mag., ser. 2, v. 2, no. 8, p. 337-342.

Norris, Geoffrey, and Sarjeant, W. A. S., 1965, A descriptive index of genera of fossil Dinophyceae and Acritarcha: New Zealand Geol. Survey Paleont. Bull. 40, 72 p.

Obrhel, Jiri, 1958, Über Funde von Sporen und Pollen (Sporae dispersae) in altpaläozoischen und vorpaläozoischen Formationen: Geologie, v. 7, no. 7, p. 969-983.

Parke, Mary, 1966, The genus *Pachysphaera* (Prasinophyceae), p. 555-563 *in* Barnes, Harold, ed., Some contemporary studies in marine science: London, George Allen & Unwin, Ltd.

Parke, Mary, and I. den Hartog-Adams, 1965, Three species of *Halosphaera*: Marine Biol. Assn. U. K. Jour., v. 45, p. 537-557.

Roblot, Marie-Madeleine, 1964, Sporomorphes du Précambrien Armoricain: Annales Paléont. (Invertèbres), v. 50, no. 2, p. 105-110.

Schopf, J. M., Wilson, L. R., and Bentall, Ray, 1944, An annotated synopsis of Paleozoic fossil spores and the definition of generic groups: Illinois Geol. Survey Rept. Inv. 91, 73 p., 3 pls., 5 figs.

Schopf, Thomas J. M., 1966, Conodonts of the Trenton Group (Ordovician) in New York, Southern Ontario, and Quebec: New York State Mus. and Sci. Service Bull. 405, 105 p., 6 pls., 7 figs., 9 tables.

Sommer, F. W., 1956, South American Paleozoic sporomorphae without haptotypic structures: Micropaleontology (Am. Mus. Natl. History), v. 2, no. 2, p. 175-181.

Staplin, F. L., 1961, Reef-controlled distribution of Devonian microplankton in Alberta: Palaeontology, v. 4, pt. 3, p. 392-424, pls. 48-51.

Stewart, W. N., 1960, More about the origin of vascular plants: Plant Sci. Bull. v. 5, no. 2, p. 1-5.

Swartz, C. K., chm. and others, 1942, Correlation of the Silurian formations of North America [chart 3]: Geol. Soc. America Bull., v. 53, no. 4, p. 533-538.

Tappan, Helen, 1968, Primary production, isotopes, extinctions and the atmosphere: Palaeogeography, Palaeoclimatology, Palaeoecology, v. 4, p. 187-210.

Taugourdeau, Philippe, 1965, Chitinozoaires de l'Ordovicien des U.S.A.; Comparison avec les faunes de l'Ancien monde: Inst. Francais Pètrole Rev., v. 20, no. 3, p. 463-485.

Timofeyev, B. V., 1956, Hystrichosphaeridae kembriya: Akad. Nauk SSSR Doklady, v. 106, no. 1, p. 130-132.

———1958a, Über das Alter sächsischer Grauwacken; Micropaläophytologische Untersuchungen von Proben aus der Weesensteiner und Lausitzer Grauwacke: Geologie, v. 7, nos. 3-6, p. 826-839.

———1958b, Spory proterozoyskikh i rannepaleozoyskikh otlozheniy vostochnoy Sibiri i ikh stratigraficheskoe znachenie [Spores of Proterozoic and Early Paleozoic deposits of eastern Siberia and their stratigraphic significance], p. 226-230 *in* Study of unified stratigraphic charts of Siberia (Reports on stratigraphy of Precambrian deposits): Interdept. conf., 1956, Trans., Akad. Nauk SSSR.

———1959, Drevneishaya flora Pribaltiki i ee stratigraficheskoe znachenie [Ancient flora of the Baltic region and its stratigraphic significance]: Vses. Neft. Naucho-Issled. Geol.-Razved. Inst. Trudy, v. 129, 320 p., 25 pls., 4 figs., 14 charts.

Tourtelot, H. A., Tailleur, I. L., and Donnell, J. R., 1966, Tasmanite and associated organic-rich rocks, Brooks Range, northern Alaska, *in* Abstracts for 1966: Geol. Soc. America Spec. Paper 101.

Twenhofel, W. H., chm. and others, 1954, Correlation of the Ordovician formations of North America: Geol. Soc. America Bull., v. 65, no. 3, p. 247-298.

Udar, Ram, 1964, Palynology of bryophytes, p. 79-100 *in* Nair, P. K. K., ed., Advances in palynology: Lucknow, India, Natl. Botanic Gardens.

Volkova, N. A., 1965, O prirode i klassifikatsii mikrofossiliy pastitel'nogo proiskhozhdeniya iz Dokembriya i nizhnego Paleozoya [Nature and classification of microfossils of plant origin from Precambrian and lower Paleozoic]: Paleont. zhur., no. 1, p. 13-25 [translation Internat. Geology Rev. Dec. 1965, v. 7, no. 12, p. 2121-2129].

Waksman, S. A., 1938, Humus; origin, chemical composition, and importance in nature, 2d ed., revised: Baltimore, Williams & Wilkins Co.

Wall, David, 1962, Evidence from Recent plankton regarding the biological affinities of *Tasmanites* Newton 1875 and *Leiosphaeridia* Eisenack 1958: Geol. Mag., v. 99, p. 353-362, pl. 17.

———1965, Microplankton, pollen, and spores from the Lower Jurassic in Britain: Micropaleontology (Am. Mus. Natl. History), v. 11, no. 2, p. 151-190, pls. 1-9.

Walton, H. S., 1962, Cambrian hystrichospheres from Western Canada (abs.): Pollen et Spores, v. 4, no. 2, p. 387.

Wetzel, Otto, 1933, Die in organischer Substanz erhaltenen Mikrofossilien des baltischen Kreide-Feuersteins: Palaeontographica, v. 75-77, sec. A, p. 141-186; ibid., v. 78-79, sec. A, p. 1-110.

Wetzel, W., 1952, Beitrag zur Kenntnis des dan-zeitlichen Meeresplanktons: Geol. Jahrb., v. 66, p. 391-419.

White, C. D., 1901, Two new species of algae of the genus *Buthotrephis*, from the Upper Silurian of Indiana: U.S. Natl. Mus. Proc., v. 24, p. 265-270.

———1903, Description of a fossil alga [*Thamnocladus*] from the Chemung of New York with remarks on the genus *Haliserites* Sternb.: New York State Mus. Ann. Rept. for 1901, p. 593-605.

———1929, Description of the fossil plants, *in* Some "mother rocks" of petroleum from Northern Alaska: Am. Assoc. Petroleum Geologists Bull., v. 13, no. 7, p. 841-848.

Wilmarth, M. G., 1925, The geologic time classification of the United States Geological Survey compared with other classifications, accompanied by the original definitions of era, period, and epoch terms, a compilation: U.S. Geol. Survey Bull. 769, 138 p., 1 pl.

———1938, Lexicon of geologic names of the United States (including Alaska): U.S. Geol. Survey Bull. 896, 2 v., 2396 p.

Winslow, M. R., 1962, Plant spores and other microfossils from Upper Devonian and Lower Mississippian rocks of Ohio: U.S. Geol. Survey Prof. Paper 364, 93 p., 22 pls., 12 figs.

11

DEVONIAN SPORES

J. B. Richardson

The wealth of data on Devonian spore assemblages published over the last decade has inevitably led to an explosion in numbers of Devonian spore taxa. In addition there is little uniformity in spore nomenclature and often lack of understanding of spore structure and preservational features; this has resulted in chaos in Devonian spore taxonomy. Consequently this combination of factors has obscured the stratigraphic and ecologic value of spores, especially to the nonspecialist. In spite of this, many interesting facts are available on the stratigraphic and geographic distribution of Devonian spore types; a broad picture of their "evolution" is emerging which is not readily apparent from studies of lists of spore species. It is intended here to describe in general terms the appearance and development of spore types throughout the Devonian mainly in morphologic terms and then to discuss their biologic significance.

STRATIGRAPHY

Figure 11-1 is a summary of the correlations, suggested by various authors, for European marine, British continental, and North American strata. It is intended as a quick source of reference for the stratigraphical terms used in the text. However, in using it two points should be understood. Firstly, although the chart represents the opinions of several authors based on the stratigraphical information available at the time of writing, many of these correlations, and hence the lines on the chart, will be subject to alteration as further stratigraphic work is completed. A second and related point concerns specifically correlations between British continental strata and the European stages. These two sets of stratigraphic divisions are based on two entirely different suites of fossils—fish and plants in the continental strata and brachiopods, corals, and molluscs in the marine strata. Consequently correlation between these two facies is notoriously difficult, and a precise correlation is still a distant goal awaiting further stratigraphic work. Work on fossil spores is making significant contributions to these correlation problems since spores occur in both marine and continental strata. However, it should be remembered that much of the work is in a reconnaissance stage—and unfortunately some of it is based on one or two samples, which occasionally are not even tied down accurately in a sedimentary sequence! So in studying palynological literature we must constantly ask the question, What is the nature of the stratigraphic control for the beds from which the spores were obtained?

SPORE SUCCESSION

Pre-Devonian Records[1]

No indisputable trilete spores have been recorded from pre-Silurian strata, although acritarchs (microfossils with organic tests, some of which may be algal spores) are known from the Precambrian and occur abundantly in Lower Paleozoic marine strata. However, it should be pointed out that continental strata are practically absent from the geological column before Late Silurian-Early Devonian time, and trilete spores

[1]See also chapter 10.

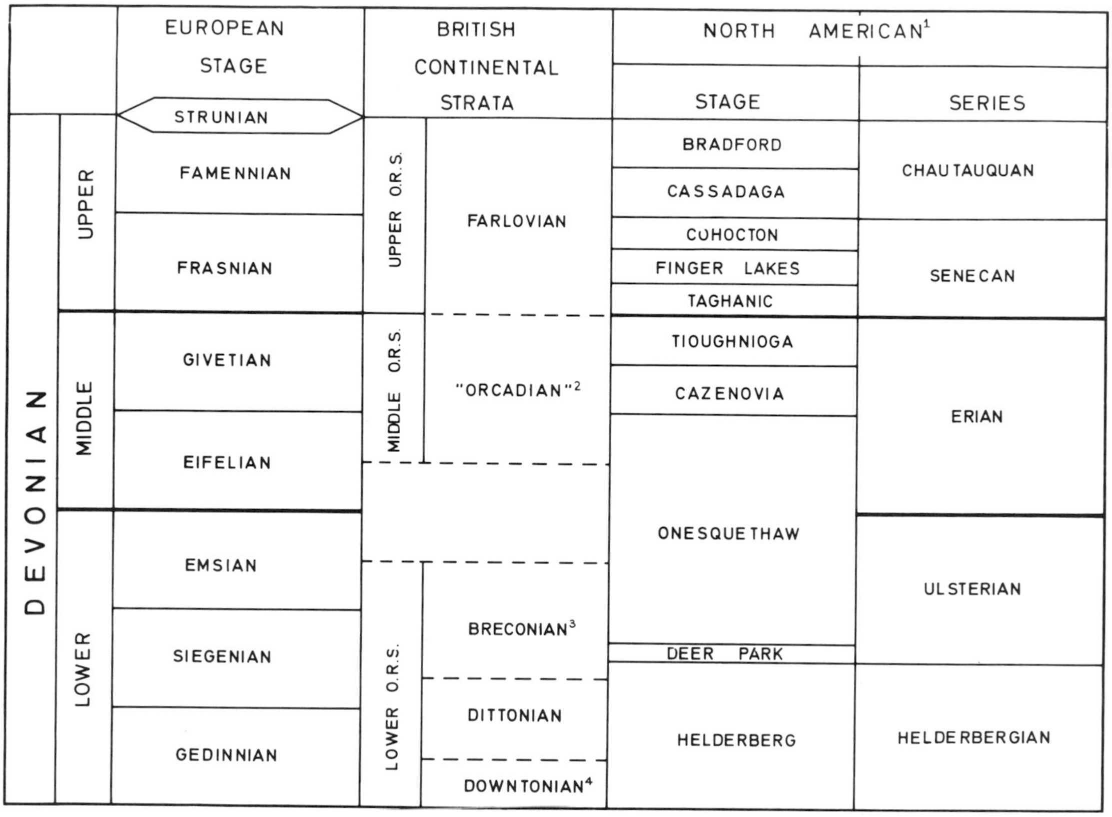

		European Stage	British Continental Strata		North American[1] Stage	North American[1] Series
		Strunian				
Devonian	Upper	Famennian	Upper O.R.S.	Farlovian	Bradford	Chautauquan
					Cassadaga	
		Frasnian			Cohocton	Senecan
					Finger Lakes	
					Taghanic	
	Middle	Givetian	Middle O.R.S.	"Orcadian"[2]	Tioughnioga	Erian
					Cazenovia	
		Eifelian			Onesquethaw	
	Lower	Emsian				Ulsterian
		Siegenian	Lower O.R.S.	Breconian[3]		
					Deer Park	
		Gedinnian		Dittonian	Helderberg	Helderbergian
				Downtonian[4]		

Figure 11-1. Equivalent Devonian stratigraphic units used in Europe and North America.

[1]After Rickard 1964.
[2]Strata of the Orcadian basin represent the greatest known development of continental Middle Devonian in the British Isles but also include probable Lower Devonian and Upper Devonian continental deposits.
[3]After White, 1956
[4]The Downtonian is now regarded as pre-Gedinnian and, therefore, pre-Devonian, by many stratigraphers.

are most abundant in continental and marginal marine strata.

Descriptions of trilete spores from well-dated Silurian strata are few in number. Hoffmeister (1959) records an assemblage consisting of two smooth trilete forms from the Lower Silurian; Downie (1963) also records two spore forms, one smooth-walled and one apiculate, from the Wenlock. More recently Richardson and Lister, (1969) have increased these records. In addition to forms that are closely comparable to Hoffmeister's spores (referred to the genus *Ambitisporites*) smooth-walled patinate spores occur (*Archaeozonotriletes*), and also spores with verrucate sculpture. In the succeeding Ludlovian most of these forms are still present, but in addition there are rare specimens with apiculate, granulate, and murinate sculpture. These records show a gradual increase in the number and variety of spore forms from their earliest record in the upper part of the Lower Silurian through the Wenlock and Ludlow. (See table 11-1.)

***Table 11-1.* Number of Spore Forms Recorded Through the Silurian and Lower Devonian[1]**

Stage	Total Number of Spore Forms	Sculptured Forms
Dittonian	29	21
Downtonian	24	13
Ludlovian	14	6
Wenlockian	8	1
Upper Llandoverian (Hoffmeister, 1959)	2	0
Pre-Silurian strata	No records of bona fide trilete spores	

[1]All records are from the Welsh Borderland (Richardson and Lister, 1969) except for the Upper Llandoverian material.

The spores found so far are mainly azonate, smooth, and retusoid (*Retusotriletes*), although some of them have thin proximal walls that thicken at the equator and over to the distal hemisphere (*Archaeozonotriletes*), and others have a narrow equatorial thickening (*Ambitisporites*). Sculptured forms are much more rare, but they occur in an increasing variety from the Wenlock to the Ludlovian. However, none of the spores has been found in sufficient numbers for new taxa to be described, in spite of intensive search. Except for the apiculate retusoid spores (*Apiculiretusispora*), they cannot be referred to existing genera and are consequently not included in the range chart.

The apparent paucity of spores and the lack of variety they show in the Silurian are in marked contrast to some rich assemblages in continental and marginal marine lower Gedinnian (Devonian) strata; in these latter assemblages the spores are much more abundant and show greater variety.

Devonian Records

Gedinnian. Relatively little has been published on Lower Devonian assemblages, and many of the studies available are based on poorly preserved material with little stratigraphic control. The few descriptions there are for the Gedinnian suggest very simple spore types, which are dominantly laevigate and azonate. Work in progress (Cramer, 1966; Richardson, 1967b; and Richardson and Lister, 1969), however, shows that a surprising variety of types already is present in the lowermost Gedinnian strata. (See pls. 11-1 and 11-2.) Admittedly this variety is based on only a few basic themes, but nevertheless these assemblages, and those from succeeding Gedinnian strata, are very distinctive. Compared with records of bona fide trilete spores from Silurian rocks, the lower Gedinnian assemblages represent a considerable increase in the number of spore types; for example, in the Welsh Borderland region of the British Isles Richardson and Lister (1969) have recorded eight spore forms (seven smooth-walled and one sculptured) from the Wenlock and 14 spore forms (six of which are sculptured) from the succeeding Ludlovian. These spores are all from marine strata deposited in open sea environments frequently far from the shore, and consequently the spores are usually rare (especially the sculptured forms) compared with palynomorphs such as acritarchs and Chitinozoa. In contrast, overlying basal Gedinnian strata (Downtonian marine—nearshore deposits) contain 24 distinct spore types, and 29 occur in upper Gedinnian (Dittonian, fluviatile flood-plain deposits); sculptured forms are much more abundant. How much of this increase in the number of spore forms is due to the clear-cut facies changes in this succession is difficult to assess, but the possibility of some facies control cannot be ignored.

Considering Gedinnian assemblages as a whole, we find several distinct features that separate them from succeeding Devonian assemblages. Firstly, the spores are very small. The size mode for measurements of 500 specimens from Downtonian (lower Gedinnian) and Dittonian (upper Gedinnian) from the Welsh Borderland is 17.5 microns, and the overall size range is 8 to 62 microns. (See fig. 11-2.) Secondly, well-developed contact areas and curvaturae perfectae (frequently forming thickened arcuate ridges) are a constant feature. The contact areas in many of the spores are thin, and the distal walls are comparatively thick (pl. 11-1, figs. 5-6). Thickened curvaturae frequently coincide with the equatorial margin so that spores appear cingulate in polar compression, and the generally thin nature of the contact faces enhances this appearance.

Sculpture is surprisingly varied compared with Silurian forms; granulate, apiculate, spinose, individually biform, verrucate, murinate, and reticulate patterns are all developed, but many of these sculptured spores have smooth proximal faces. Some show proximal sculpture, but this is clearly differentiated from that on the distal face. One such proximal development that is relatively common is the presence of prominent interradial papillae (pl. 11-1, fig. 10; pl. 11-2, figs. 6 and 7).

Another proximal development of importance is the presence of proximal radial ribs (pl. 11-2, figs. 10 and 11). This is a relatively rare, sporadically developed feature in the lowermost Gedinnian studied by the writer. In contrast, in the Dittonian, proximal ribs are abundantly represented; they become a constant character within a species and occur on spores that are distally smooth, apiculate, or verrucate.

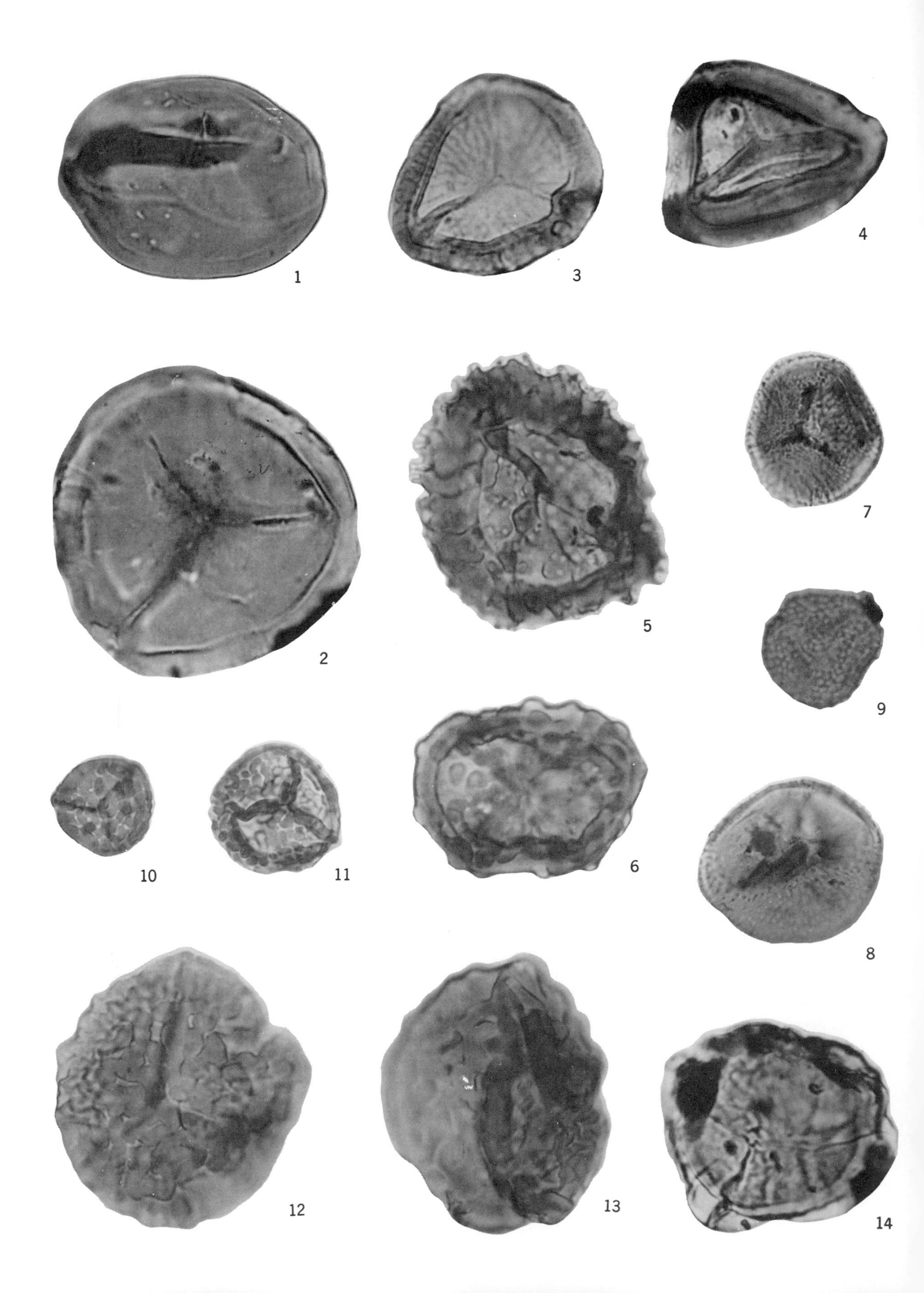

1
3
4
2
5
7
9
10
11
6
8
12
13
14

Siegenian and Emsian. Descriptions of well-dated Siegenian and Emsian assemblages are rare, but the few that are available indicate a similar pattern of development. A striking feature is the early appearance of important Devonian genera, many of which had appeared by the upper Emsian. (See fig. 11-3.)

Spore assemblages from these strata tend to be larger in size and continue to have proximal differentiation although not of the extreme type shown in the Gedinnian. Forms with proximal radial ribs are frequently present, are clearly differentiated, and are more diversified than in the Gedinnian. The proximal ribs are often more thickened, and robust, and these forms are clearly differentiated as a distinct group of spores (the genus *Emphanisporites* McGregor, 1961). There is variation in the type and arrangement of the proximal ribs. *Emphanisporites* with well-developed annulate distal thickenings (E. *annulatus-erraticus* types, pl. 11-3, fig. 4) occurs in possible Siegenian strata and continues into Middle Devonian and lower Upper Devonian strata. The incoming of annulate species will probably prove to be of considerable stratigraphic importance. So far as well-dated strata are concerned the species *E. annulatus* has not been reported from pre-Emsian strata.

The genus *Emphanisporites* ranges from lowermost Gedinnian (rare) to Lower Carboniferous, but has its maximum number of species (so far recorded) in the Emsian and lower Eifelian and then diminishes gradually through the remainder of the Devonian. Strong proximal sculpture is an unusual feature, and consequently *Emphanisporites* forms a very distinctive spore grouping. However, although the genus is most common in Lower Devonian rocks, it is not apparently invariably present; for example, Moreau-Benoît (1966) has not recorded any specimens of *Emphanisporites* in the Siegenian and Emsian (Anjou, France), the period of maximum diversification according to the number of species. On the other hand, Doubinger (1963) has recorded the genus in the middle Siegenian (Cotentin, France); Allen (1965) finds four species in middle and upper Siegenian and Emsian of Spitzbergen; and the writer has found two species in the lower Gedinnian and three species from the upper Gedinnian of the Welsh Borderland, two species in ?Siegenian Scotland, and two species in lower Emsian assemblages. In addition there are several records probably belonging to this genus from Lower and Middle Devonian in the Soviet Union (e.g., *Stenozonotriletes ornatissimus* Chibrickova, 1959, 1962; Naumova, 1953; Venozhinskene, 1964). The writer has seen forms of the *E. rotatus* and *E. annulatus-erraticus* complex in the Frasnian of New York State and the species *E. rotatus* in Famennian and Lower Carboniferous strata of New York and Pennsylvania. Winslow (1962) was the first to describe these forms (under the name *Radforthia*) from the Upper Devonian and Lower Carboniferous.

Pseudosaccate and zonate smooth and sculptured types are also present, and the incoming of forms of both these important types apparently occurs at least as far back as the Siegenian (Allen, 1965; Moreau Benoît, 1966; Doubinger, 1963). The writer has also seen well-developed zonato-pseudosaccate spores in strata of possible Siegenian age. (See pl. 11-3, fig. 5.) Elaboration of these types continues through the Siegenian and Emsian, and they become prominent in Eifelian and Givetian assemblages.

Records of pre-Middle Devonian types with anchor-shaped spines are rare, but Allen (1965) records three spore types from the Emsian with this distinctive type of sculpture. Although it is unlikely that a stratigraphy based on spores will be afforded the luxury of finding anchor-shaped spines to be an exclusive Middle and Upper Devonian feature (with waning importance in

---

Plate 11-1. Upper Silurian and Lower Devonian spores—Ludlovian and Downtonian (lower Gedinnian), Welsh Borderland. (All figures × 1,000 except for figure 9, which is × 2,000.) 1-2 *Retusotriletes* spp., 1—oblique compression showing curvaturae perfectae; 2—polar view, curvaturae coincide with the equatorial margin. 3-4 ?*Archaeozonotriletes* spp., 3—specimen from the upper Ludlow; 4—oblique compression showing thin proximal wall and thick equatorial and distal wall. 5-6 *Cymbosporites* spp., 5—showing biform sculptural elements; 6—verrucate form. Both figures show the thin nature of the contact areas. 7-8 *Apiculiretusispora* spp., 7—polar compression; 8—lateral compression. 9—?*Dictyotriletes* sp. 10-11 ?New genus spp., 10—showing interradial papillae; 11—slightly tipped spore showing thickened curvaturae perfectae. 12-13 ?New genus sp., 12—distal view showing convolute muri; 13—oblique compression showing thickened curvaturae perfectae. 14—?*Perotrilites* sp.

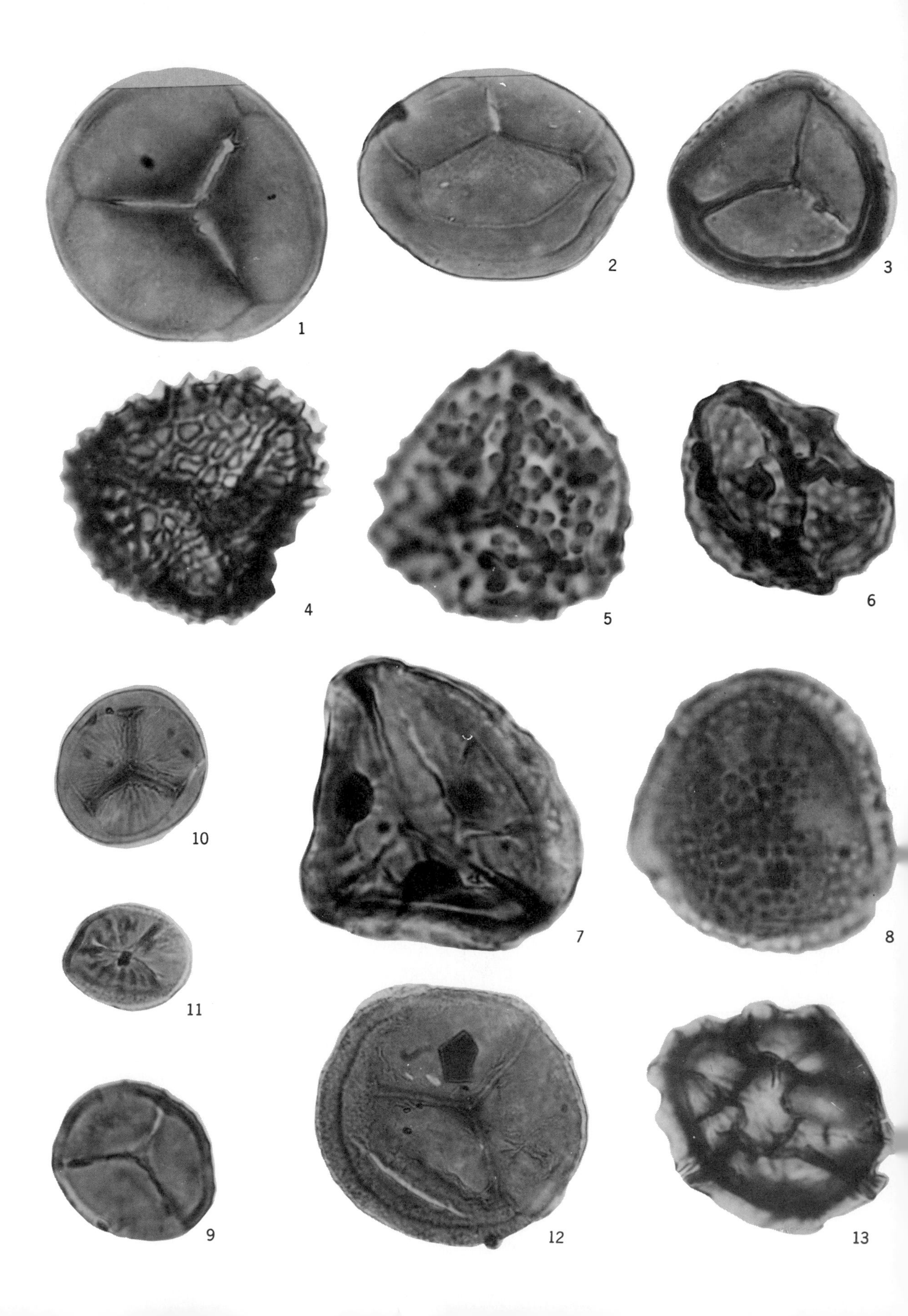

1
2
3
4
5
6
10
7
8
11
9
12
13

the Lower Carboniferous), it should be pointed out that there is still no record of such forms from Emsian strata accurately dated by marine faunas.

Spores of the megaspore size range (i.e., over 200 μ) occur in the Siegenian (Allen, 1965; Richardson, 1967b). Allen records laevigate spores (*Trileites oxfordiensis* Chaloner) with a size range from 186 to 530 microns (mean 290 μ), and with a range from Siegenian to Givetian. (The largest size of his Siegenian specimens is not recorded.) Chaloner (1963) described this species from probable Lower Devonian strata; his specimens range in size from 354 to 435 microns (mean 390 μ).

At least these records suggest possible heterospory or incipient heterospory as far back as the Siegenian. (See fig. 11-2.) However, it should be mentioned that microspores over 200 microns are not unknown (e.g., Carboniferous *Monoletes ovatus* Schopf 1938, 330 to 495 μ). Regardless of these considerations, these assemblages form a contrast in size alone with the Gedinnian where spores over 80 microns have so far not been recorded.

Eifelian-Givetian-Frasnian. Middle Devonian strata contain large pseudosaccate and zonate forms frequently with prominent sculpture which is of various types although often spinose. (See pls. 11-4 and 11-5.) Spores with well-developed bifurcate spines are varied and frequently abundant. (See pl. 11-5, figs. 1-3.) The lower limit of this distinctive spore association, so characteristic of Middle Devonian continental strata, is not known. Many spores are over 200 microns, and spores around 300 to 500 microns are relatively common, although there does not appear to be any distinct size differentiation. Middle Devonian strata from some parts of the world are characterized by fewer of these types and an abundance of spores of the genus *Geminospora* (pl. 11-5, figs. 5 and 6); these forms have a thick outer wall and a thin inner body separated by a cavity.

Upper Givetian and lower Frasnian spore assemblages are quite similar. *Geminospora* is frequently abundant. Spores with anchor-shaped spines are widely distributed. Furthermore, there appears to be a greater degree of size differentiation than is seen in the Middle or Lower Devonian. Few descriptions have been made of Frasnian megaspores, but Chaloner (1959) records four genera and six species from probable Frasnian strata of arctic Canada. He records species with a size range up to 1,610 microns, considerably larger than anything so far described from the Middle Devonian. In variety, however, they do not exceed that already present in the Middle Devonian (Allen, 1965; Chibrickova, 1959; Richardson, 1965a).

Frasnian assemblages from parts of the Soviet Union and northwest Canada are frequently characterized by monolete spores of the genus *Archaeoperisaccus*. (See pl. 11-5, figs. 9 and 10.) These spores sometimes occur in great abundance, and relatively heavily sculptured, finely sculptured, and smooth-walled species have been described from Frasnian deposits. From the published work this genus appears to be confined to the Frasnian, and McGregor (1969) records the same range in his comprehensive review of *Archaeoperisaccus*. Pashkevitch (1964) records that the sculptured varieties (early to middle Frasnian) precede the smooth-walled varieties (middle to late Frasnian) in the Russian succession.

Spores with multifurcate spinose appendages first described by Winslow (1962) from Ohio also appear in the Frasnian of New York State. Such types have so far not been recorded from the Givetian.

Plate 11-2. Lower Devonian spores—Dittonian (upper Gedinnian, Welsh Borderland). 1-2 *Retusotriletes* spp. 1, *R.* cf. *Phyllothecotriletes triangulatus* Streel 1964 (× 1,000), showing thickened apical area; 2—*Retusotriletes* sp. (× 1,000), lateral view showing contact areas and curvaturae perfectae. 3—?*Archaeozonotriletes* sp. (× 1,000), specimen showing thin proximal area and equatorial thickening. 4—*Dictyotriletes* sp. (× 2,000). 5-9 ?*Aneurospora* spp. (× 2000), 5-6, ?*Aneurospora* sp. A; 5—polar view, showing distal and equatorial ornament of cones; 6—lateral view, showing thickened curvatural ridges coinciding with the equator and prominent proximal papillae; 7-8, ?*Aneurospora* sp. B; 7—proximal view, showing proximal papillae; 8—distal view, showing ornament of grana; 9—?*Aneurospora* sp. C, specimen with sparse fine cones. 10-11 *Emphanisporites* spp. × 1000; 10—specimen showing fine proximal ribs and smooth exine outside the contact areas; 11—another species with fine distal sculpture (× 1,000). 12—*Perotrilites* sp. (× 1,000), obliquely compressed specimen, showing variable thickness of the exine and thin closely adhering perisporal membrane with granulose ornamentation. 13—*Chelinospora* sp. (× 1,000), distal view, showing reticulate ornament.

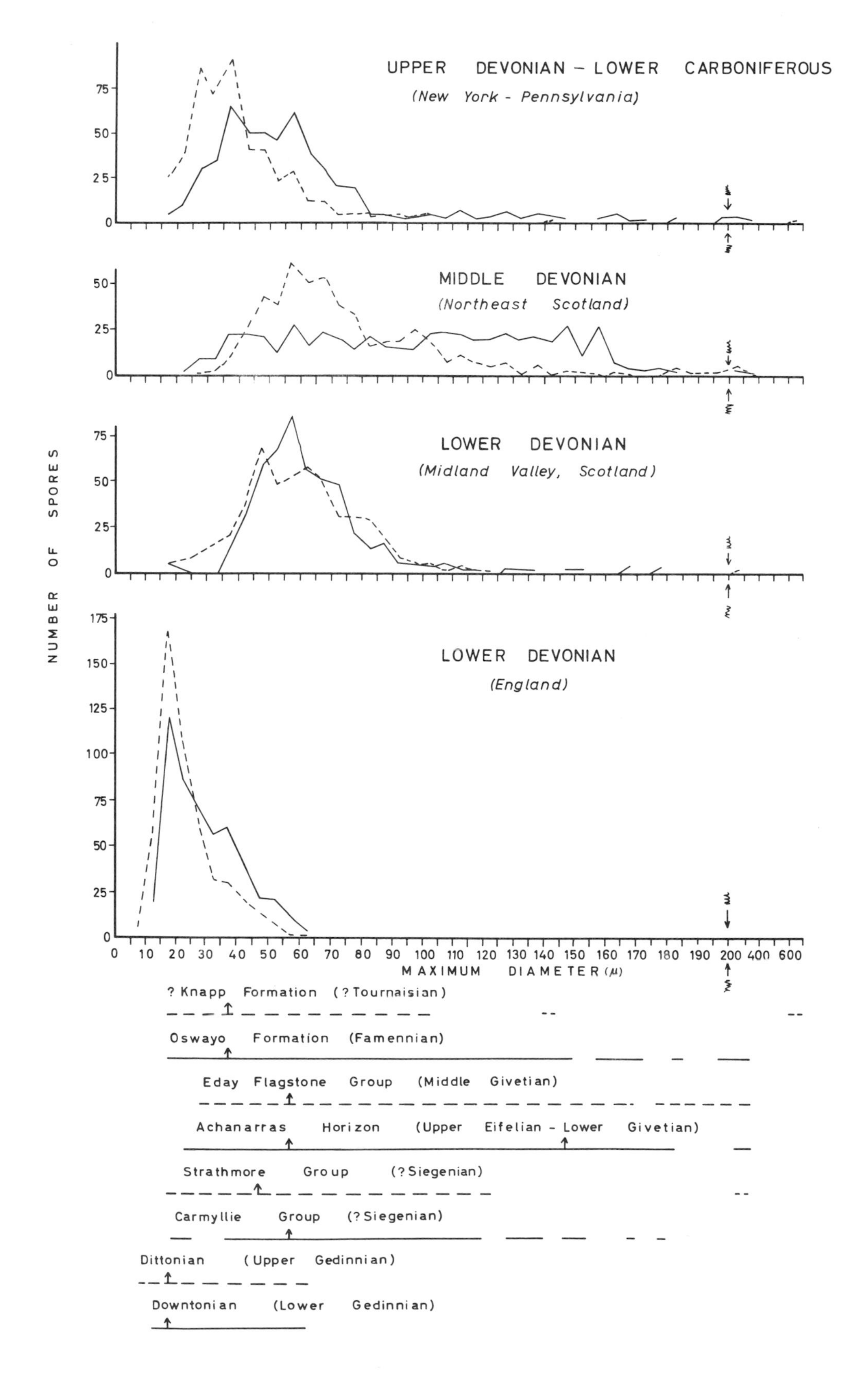

UPPER DEVONIAN – LOWER CARBONIFEROUS
(New York - Pennsylvania)
MIDDLE DEVONIAN
(Northeast Scotland)
LOWER DEVONIAN
(Midland Valley, Scotland)
LOWER DEVONIAN
(England)
NUMBER OF SPORES
MAXIMUM DIAMETER (μ)
? Knapp Formation (? Tournaisian)
Oswayo Formation (Famennian)
Eday Flagstone Group (Middle Givetian)
Achanarras Horizon (Upper Eifelian - Lower Givetian)
Strathmore Group (? Siegenian)
Carmyllie Group (? Siegenian)
Dittonian (Upper Gedinnian)
Downtonian (Lower Gedinnian)

Also present are forms with large distal thickenings of the genus *Archaeozonotriletes* (Naumova) Allen. (See pl. 11-5, figs. 7 and 8.) Many of these forms corrode in rather a distinctive way, giving the appearance of lophozonate or verrucate types. Some, if not all, of these corrosion patterns can be related to a distinctive wall structure. This was beautifully illustrated by Allen (1965, pl. 101, figs. 1-7). Some of the forms placed by the writer (Richardson, 1965, pl. 90, figs. 1-3) in the genera *Raistrickia* and *Verrucosisporites* have rather an uneven sculptural development, which gives them an asymmetrical appearance; these may well represent advanced stages in the corrosion of similar types. These patinate forms are found in Middle and Upper Devonian deposits but appear to be particularly characteristic of the Frasnian, especially in parts of the Soviet Union.

Famennian-Tournaisian. Famennian and Lower Carboniferous assemblages are now receiving a good deal of attention. Much of the work is not yet fully published, and the only published work based on long stratigraphic sections outside the Soviet Union is that by Winslow (1962). However, published and unpublished data illustrate the widespread occurrence of some of the spore species; for example, spores of the species "*Hymenozonotriletes*" *lepidophytus* Kedo 1957 (pl. 11-6, fig. 17) have been described under various disguises from many Famennian and lowermost Carboniferous deposits: Soviet Union (Kedo, 1957, 1962; Chibrickova, personal communication), Australia (Balme and Hassell, 1962), North Africa (Wray, 1964), Belgium (Streel, 1965), Ohio (Winslow, 1962), New York State and Pennsylvania (Richardson, 1965b), and Canada (McGregor and Owens, 1966).

Other elements besides "*Hymenozonotriletes*" *lepidophytus* also appear to be widely distributed. Forms with distinct distal sculpture of irregular or reticulate pattern (e.g., "*Archaeozonotriletes*" *literatus*—see pl. 11-6, figs. 10 and 11) also characterize these assemblages, as do certain species of *Vallatisporites* (e.g., *Hymenozonotriletes pusillites* Kedo 1957—see pl. 11-6, fig. 13). The latter species is used by Kedo to define the base of the Carboniferous, and similar forms are described from the Bedford shale and Berea sandstone by Winslow (these strata are placed near the Carboniferous boundary). The Russian and North American strata have not been correlated in detail with the type European succession, but it would seem that forms similar to *Vallatisporites pusillites* are restricted to a narrow time zone wherever the boundary may eventually be placed. So far in several parts of the world assemblages with "*Hymenozonotriletes*" *lepidophytus* are recorded in what are regarded as Famennian strata and are succeeded by and overlap with assemblages containing *Vallatisporites* cf. *pusillites* occurring in strata considered to be basal Mississippian (Tournaisian).[2]

Forms with bifurcate spines are still present and may be prominent in certain lithofacies but appear to die out rapidly in the Lower Carboniferous.

Pseudosaccate spores with prominent pointed spines (*Grandispora-Spinozonotriletes*) are also commonly present.

With regard to size, small and large spore species have distinct size ranges. Furthermore, megaspore species show greater diversification than in preceding strata and are much larger in size. Winslow (1962) records spores over 3,000 microns in diameter.

On the whole, Famennian and lowermost Carboniferous spore assemblages are closely similar and tend to differ considerably from Frasnian assemblages. In detail, of course, certain basic morphologies persist, but there appear to be greater numbers of newer types in the Famennian and Lower Carboniferous strata. Frasnian assemblages, on the other hand, are much more comparable with the Givetian.

GEOGRAPHIC DISTRIBUTION AND FACIES RELATIONSHIPS

As more spore assemblages are described it can

[2]Assemblages referred to as basal Tournasian or near the Carboniferous boundary may actually be pre-Carboniferous although post-Fammenian.

Figure 11-2. Comparison of size distribution in some Devonian and Lower Carboniferous spore assemblages. 500 spores were measured for each assemblage. This illustrates (a) the distinct size range and mode of the Downtonian-Dittonian spores, and (b) possibly reflects the inception and perfection of heterospory.

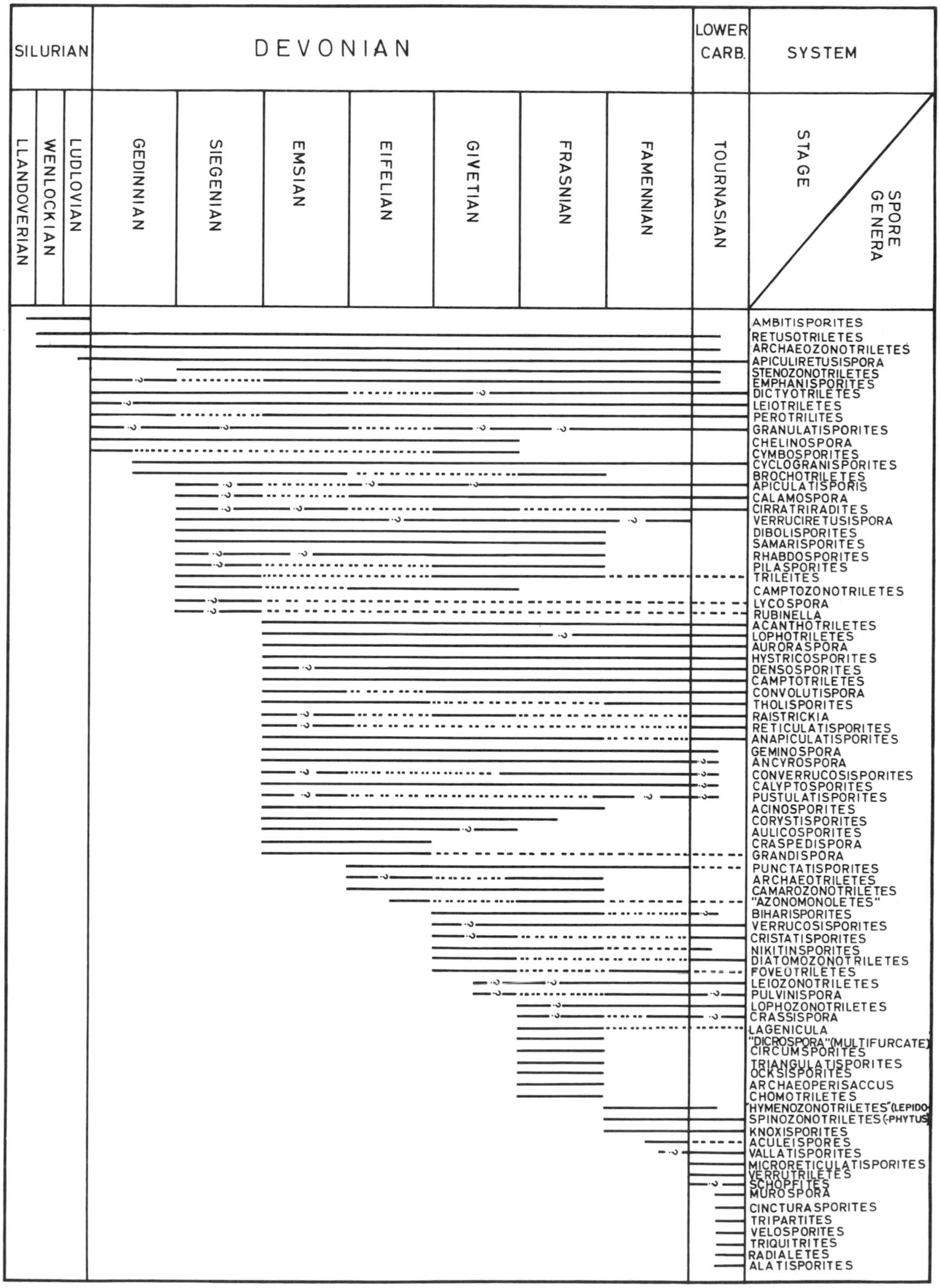

SYSTEM: SILURIAN | DEVONIAN | LOWER CARB.

STAGE: LLANDOVERIAN, WENLOCKIAN, LUDLOVIAN | GEDINNIAN, SIEGENIAN, EMSIAN, EIFELIAN, GIVETIAN, FRASNIAN, FAMENNIAN | TOURNASIAN

SPORE GENERA:

AMBITISPORITES
RETUSOTRILETES
ARCHAEOZONOTRILETES
APICULIRETUSISPORA
STENOZONOTRILETES
EMPHANISPORITES
DICTYOTRILETES
LEIOTRILETES
PEROTRILITES
GRANULATISPORITES
CHELINOSPORA
CYMBOSPORITES
CYCLOGRANISPORITES
BROCHOTRILETES
APICULATISPORIS
CALAMOSPORA
CIRRATRIRADITES
VERRUCIRETUSISPORA
DIBOLISPORITES
SAMARISPORITES
RHABDOSPORITES
PILASPORITES
TRILEITES
CAMPTOZONOTRILETES
LYCOSPORA
RUBINELLA
ACANTHOTRILETES
LOPHOTRILETES
AURORASPORA
HYSTRICOSPORITES
DENSOSPORITES
CAMPTOTRILETES
CONVOLUTISPORA
THOLISPORITES
RAISTRICKIA
RETICULATISPORITES
ANAPICULATISPORITES
GEMINOSPORA
ANCYROSPORA
CONVERRUCOSISPORITES
CALYPTOSPORITES
PUSTULATISPORITES
ACINOSPORITES
CORYSTISPORITES
AULICOSPORITES
CRASPEDISPORA
GRANDISPORA
PUNCTATISPORITES
ARCHAEOTRILETES
CAMAROZONOTRILETES
"AZONOMONOLETES"
BIHARISPORITES
VERRUCOSISPORITES
CRISTATISPORITES
NIKITINSPORITES
DIATOMOZONOTRILETES
FOVEOTRILETES
LEIOZONOTRILETES
PULVINISPORA
LOPHOZONOTRILETES
CRASSISPORA
LAGENICULA
"DICROSPORA" (MULTIFURCATE)
CIRCUMSPORITES
TRIANGULATISPORITES
OCKSISPORITES
ARCHAEOPERISACCUS
CHOMOTRILETES
"HYMENOZONOTRILETES" (LEPIDO-
SPINOZONOTRILETES (-PHYTUS)
KNOXISPORITES
ACULEISPORES
VALLATISPORITES
MICRORETICULATISPORITES
VERRUTRILETES
SCHOPFITES
MUROSPORA
CINCTURASPORITES
TRIPARTITES
VELOSPORITES
TRIQUITRITES
RADIALETES
ALATISPORITES

Figure 11-3. Range chart to show the known distribution of Devonian spore genera. Dubious generic assignments are either not included or indicated by a question mark.

be seen that many of them from strata of comparable age are closely similar from various parts of the world (Richardson, 1965a). However, differences are also apparent; some of these differences may be related to broad geographic control (floral provinces), whereas others appear to be more closely linked to lithofacies and depositional environment. The latter may partly be due to proximity to the site of growth of parent plants as well as mechanical sorting and preservation factors. Unfortunately too often spore assemblages are described from strata with little or no stratigraphic control, and therefore it is necessary to be cautious about comparisons. If we accept these limitations for the moment, certain interesting spore-distribution patterns appear to emerge.

Strata from the Middle Old Red Sandstone of the Orcadian basin, Scotland, are believed to have formed in a relatively large body of fresh water. Whether this was an enclosed lake as commonly supposed is not known with certainty, but no marine fossils occur, and the sediments compare closely with those forming at present in large fresh-water lakes (Rayner, 1963). Spore assemblages from the Orcadian strata are frequently dominated by two spore types; namely, spores with prominent bifurcate spines belonging to the genus *Ancyrospora* (pl.11-5, figs. 1-3) and the pseudosaccate species *Rhabdosporites langi* (pl. 11-4, fig. 5). These two genera frequently constitute as much as 50 percent of the assemblage. Specimens of *Ancyrospora* are especially abundant and in some beds make up 20 to 50 percent of the total spore content. Spore assemblages from comparable horizons in the Soviet Union (Kedo, 1955; Naumova, 1953; Chibrickova, 1959, 1962) have many species in common with the Scottish assemblages but differ from them in two main respects. Firstly, *Ancyrospora* is present but not abundant; on the other hand, thick-walled spores referable to the genus *Geminospora* (pl. 11-5, figs. 5 and 6) are relatively common (Kedo, 1955). Kedo's assemblages occur in sediments from a region where continental-marine strata interdigitate. Furthermore, the sequence of strata indicates an upward passage from lagoonal saline environments, to fresh-water lagoons intercalated with some littoral and marine deposits, and finally into marginal marine facies. It is interesting to note that it is the spore assemblages from the second group of strata, representing fresh-water lagoonal environments, that are most closely comparable to the Scottish material. The upper assemblages, from marginal marine facies, contain abundant *Geminospora* (Kedo, 1955, fig. 2) and are comparable with assemblages from marine or mixed facies from other parts of the Russian platform and the southwest Ural region (cf. Frasnian assemblages, Naumova, 1953; and Givetian assemblages, Chibrickova, 1959, 1963). It is possible therefore that the differences between the Scottish and Russian material are related to differences in plant ecology, with *Geminospora*-producing plants living in marginal deltaic or coastal floodplain areas, whereas plants living in or around fresh-water lagoons were shedding spores of *Ancyrospora* and *Rhabdosporites* (see also Streel, 1967).

In New York State a similar relationship exists in the Frasnian. Here fresh-water massive grey sandstones – siltstones (?upper floodplain environments) in the eastern part of the State frequently contain abundant *Ancyrospora, Hystricosporites,* and *Rhabdosporites langi* (the latter have been found in sporangia of *Tetraxylopteris* – Bonamo and Banks, 1967). In contrast, *Geminospora* occurs abundantly in association with red, and variegated red and green, silts and sandstones of lower flood-plain-marginal deltaic facies. Spore assemblages from marine strata near the delta margin are much more mixed, which reflects the fact that the spores are derived from several sources. In the latter sediments factors such as mechanical sorting and preservation are probably responsible for such variation as occurs. This contrasts to the situation in the on-delta areas, where abundance of one or two types suggests proximity to the habitat of the parent plants.

Not far from New York State spore assemblages of Frasnian age have been described from the Escuminac Formation of eastern Canada (Brideaux, manuscript). These assemblages closely resemble those from the Scottish Middle Old Red Sandstone in gross morphological aspect (although many of the species are different). It is interesting to note that the sediments and fossil association of fish and plants from Escuminac closely resemble the Orcadian basin facies and are believed to have formed in a fresh-water

1
2
4
5
3
6
7

lake (Dineley and Williams, 1968). Spore assemblages from this formation contain abundant spores with bifurcate spines (*Ancyrospora* and *Hystricosporites*) and also spores similar to *Rhabdosporites langi*. Thus it would seem that in eastern North America we have a situation similar to that in Western Europe, with the abundance of certain spore forms occurring in association with distinctive types of facies. It may be that *Geminospora* was produced by a halophytic plant growing in marginal delta swamps or by plants living in backswamp and levee environments along water courses of the lower flood plain (the associated sediments in New York State suggest the latter.) On the other hand, plants producing *Ancyrospora*, *Hystricosporites*, and *Rhabdosporites langi* probably lived in regions of the upper flood plain and in or around fresh-water lakes.

Naumova (1953) also comments on the variability of spore assemblages of Frasnian age and attributes this variation to transgression and regression of Devonian seas; for example, she records that the lower assemblages of the Voronezh beds are characterized by a predominance of the genus *Archaeozonotriletes* (mainly spores of the *Geminospora* type) which make up 30 percent of the assemblage, whereas such forms constitute only 5 percent of the upper assemblage. Naumova attributes these differences to the development of a hydrophilous vegetation corresponding to marine transgression in the lower Voronezh beds and a less hydrophilous vegetation corresponding to regression of the sea in the upper Voronezh beds. At the latter horizon the genera *Lophozonotriletes* (25 percent) and "*Hymenozonotriletes*" (10 percent) are the dominants. However, elsewhere Naumova refers to an association of small thin-walled spores and large thick-walled spores of the *Azolla* type (i.e., spores with bifurcate spines) as typical of hydrophilous vegetation.

Comparison of spore assemblages in different facies suggests that there is some evidence for ecological differentiation of Devonian plants. However, the exact nature of this differentiation is still largely unknown, and consequently there is a need for detailed combined sedimentological, stratigraphical, and palynological studies to determine relationships between environmental factors and spore associations. From the indications outlined above such investigations would be extremely fruitful—firstly, by adding to our understanding of the habitat and distribution of the Devonian flora and, secondly, as a consequence of this, by leading to a more accurate use of spores as stratigraphic indices.

Another interesting example of apparently restricted spore distribution is that of the dispersed-spore genera *Archaeoperisaccus* (pl. 11-5, figs. 9 and 10) and *Nikitinsporites*. Naumova (1953), Ozolinya (1963), Pashkevitch (1964), and Tuzova (1956) all record *Archaeoperisaccus* from Frasnian deposits of the Soviet Union, whereas several other writers (Archangelskaya, 1963, and Chibrickova, 1962) do not record this form in deposits of similar age. McGregor (1969) has drawn attention to the clear similarity of some species of *Archaeoperisaccus* to the microspores of *Krystofovichia africani* (a species based on spores, megasporangia, and microsporangia only) (Nikitin, 1934). Naumova described assemblages full of *Archaeoperisaccus* from the same area as that from which Nikitin described his material but makes no mention of this similarity. Several horizons containing *Archaeoperisaccus* also contain large spores of the genus *Nikitinsporites* Chaloner 1959, which resemble the distinctive *Krystofovichia* megaspores. Outside the Soviet Union these two genera again appear to be restricted, since they have only been recorded together from arctic and northwestern Canada. Neither genus has so far been recorded in the Frasnian of

---

Plate 11-3. Lower Devonian spores—"Dittonian" (?Siegenian, Midland Valley, Scotland). (All figures × 1,000.) 1—*Retusotriletes* sp., polar view showing curvaturae perfectae and thin apical area. 2-4 *Emphanisporites* spp.; 2—*E.* cf. *robustus* McGregor 1961, showing coarse proximal ribs; 3—*E.* cf. *rotatus* McGregor 1961, lateral compression, showing proximal ribs and curvaturae perfectae; 4—*E.* cf. *erraticus* McGregor 1961, showing distal annular thickening. 5—?*Samarisporites* sp., specimen shows trilete folds and ornament of cones and biform elements. 6—?*Dictyotriletes* sp., clearly showing reticulate ornament. 7—?*Perotrilites* sp., proximal view, showing contact areas and curvaturae perfectae. Spores of this type frequently show separation of the outer ornamented layer; sometimes this outer layer comes off completely, leaving spores similar to *Retusotriletes* cf. (*Phyllothecotriletes*) *triangulatus* Streel 1964. These spores are similar to those discovered by Banks, Hueber, and Leclerq (1964) in sporangia of *Dawsonites*.

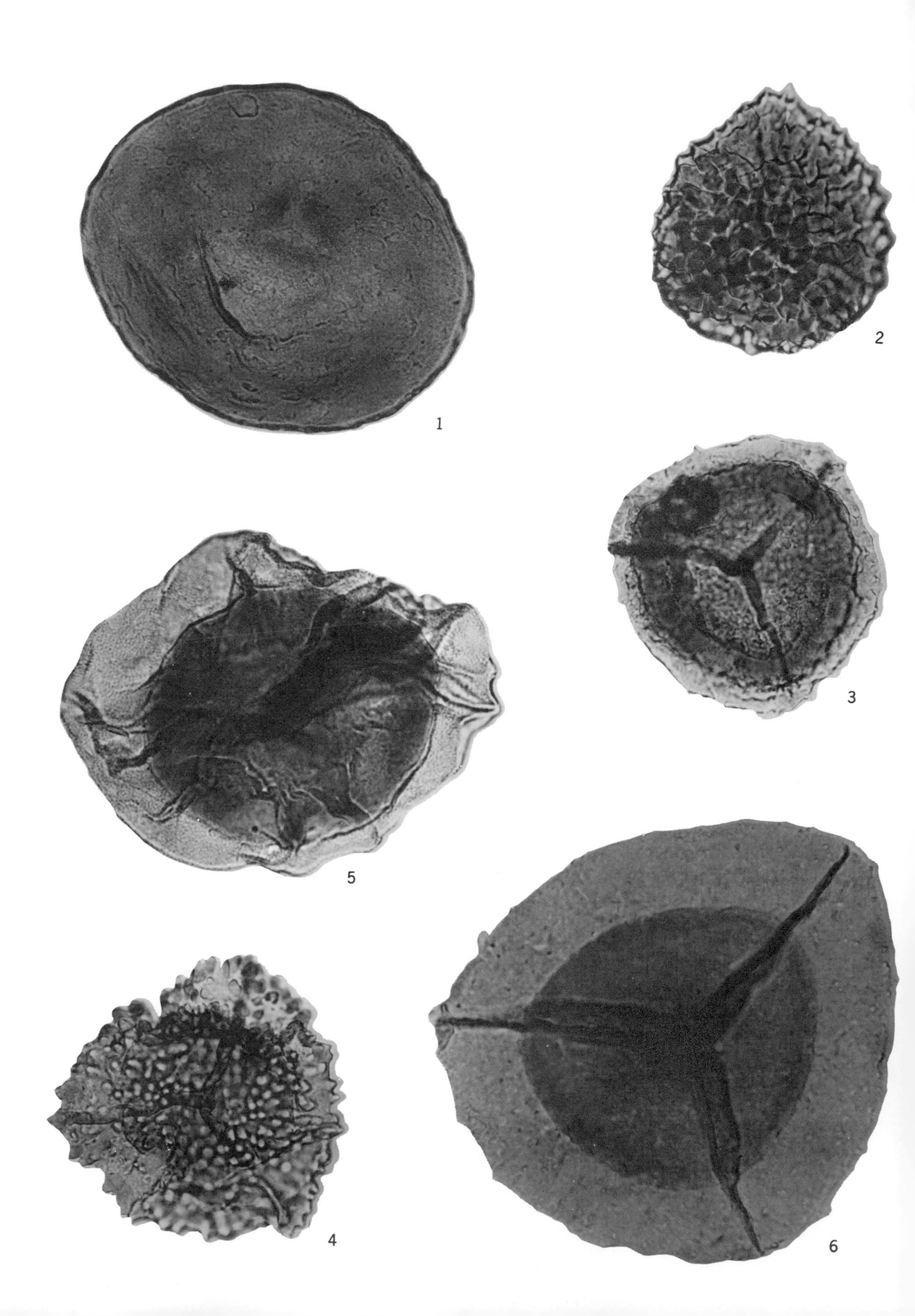
1
2
3
5
4
6

other parts of North America except for *Archaeoperisaccus* in probable Frasnian rocks of Alaska (Scott and Doher, 1967). It is interesting to note that Nikitin considered that *Krystofovichia* megasporangia and microsporangia may have had lycopod affinities, whereas Naumova classified *Archaeoperisaccus* as gymnospermous pollen, and other Soviet workers follow this (Tuzova, 1956).

Schopf (personal communication) has material from arctic Canada showing monolete grains intimately associated with the apical area of spores of *Nikitinsporites*, a similar relationship to that described by Nikitin for *Krystofovichia*. It would be particularly interesting to find the parent plant of these spores, which apparently have such a restricted stratigraphic and geographic distribution. It will also be interesting to see whether arctic and northwest Canadian spore assemblages differ in other ways from those in the southeastern parts of the North American continent and to study factors that may relate to these differences. An obvious possibility is paleoclimatic control; if North American data are plotted on a palaeolatitude map, the occurrence of these types seems to run parallel to the Devonian palaeoequator in the approximate position of the Temperate Belt. However, it will be necessary to obtain much more data to test whether or not this distribution is actually climatically controlled.

BIOLOGICAL SIGNIFICANCE

Spore Distribution

Whilst trilete spores are not unequivocal evidence for the existence of land vascular plants (since some red and brown algae and bryophytes produce spores in tetrads), most of the spores shown in the spore-genera range chart (fig. 11-3) were probably produced by a land flora. In support of this there is ample evidence that the early vascular land plants produced trilete spores similar to those found dispersed in sedimentary strata. The chart therefore reflects to some extent the evolution of the land flora, although it has considerable limitations, including the subjective nature of taxonomy and the fact that relatively little detailed work has been done on the Lower Devonian. Consequently the picture of floral evolution as represented by figure 11-3 is at best only a very general one.

Examination of the spore record, however, reveals several interesting features. Firstly, there are no bona fide records of trilete spores before the Silurian. The claims of some Russian workers of Precambrian, Cambrian, and Ordovician trilete spores have not been substantiated by clear photographs of their material or by other workers studying microfossil assemblages of comparable age. Many of the organic bodies described as spores are marine microplankton (acritarchs) of uncertain affinities, although some of them may be related to the algae (Wall, 1962).

Trilete spores are clearly present in the Silurian, but the few so far recorded are predominantly simple smooth types possibly belonging to six genera. If these spores were produced by land vascular plants,[3] then this suggests little pre-Devonian diversity of the land flora (but see p. 194). In contrast, through the Lower Devonian (Gedinnian, Siegenian, and Emsian) there is a rapid increase in the number and in the morphological diversity of the spore genera, which perhaps reflects rapid colonization by vascular plants of the newly formed Devonian landmasses. Almost two-thirds of the Devonian spore genera plotted on figure 11-3 were present by Emsian time. Placed in the context of the time

[3]Vascular plants have been recorded from the Ludlow (Obrhel, 1962) but not with certainty from earlier strata. The total of six genera includes representatives of possibly two new genera (Richardson and Lister, 1969) not shown on the range chart.

Plate 11-4. Middle Devonian spores—Eifelian and Givetian spores (northeast Scotland). (All figures × 500, except figure 1, which is × 200.) 1—*Trileites langi* Richardson 1965; spore of the megaspore size range, simple, smooth with a triangular area of thickening near the proximal pole. 2—*Acinosporites acanthomamillatus* Richardson 1965, distal view, showing ornament consisting of convolute ridges on which cones are superimposed. 3—*Densosporites devonicus* Richardson 1960; proximal polar view, showing trilete folds and equatorial thickening. 4—*Samarisporites orcadensis* Richardson 1965; proximal polar view, zonate spore with closely packed sculpture of cones. 5—*Rhabdosporites langi* (Eisenack) Richardson 1960; pseudosaccate spore, showing large folds in the pseudosaccus, which is ornamented with fine rods. 6—*Calyptosporites velatus* (Eisenack) Richardson 1962; pseudosaccate form in which the pseudosaccus is closely adpressed to the central body; pseudosaccus ornamented with sparse fine cones.

1
2
4
3
5
7
6
8
9
10

scale (Geological Society Phanerozoic time scale, 1964) this means that for the first 35 million years (Silurian) of the trilete spore record only six genera are known. In contrast, 43 new genera are recorded in the Lower Devonian (25 million years long), but for the next 25 million years only 26 new genera are recorded—12 in the Middle Devonian (11 million years), and 14 in the Upper Devonian (14 million years).

In considering figure 11-3 in more detail we must bear in mind its limitations. However, it does show several interesting features. Firstly the Gedinnian appears to have been a time of gradual change of the land flora, followed by more rapid diversification in the Siegenian and Emsian. How far this is affected by the lack of knowledge of Gedinnian spores is difficult to assess, but in morphology, genera present, and size they appear to be quite distinct from later Early Devonian spore floras, which suggests that they parallel macrofloral changes. A further period of change appears to have taken place at the beginning of the Famennian with, on the whole, very little difference between the late Famennian and early Tournaisian (Early Carboniferous), and finally a further floristic change is indicated by the introduction of several important Carboniferous spore genera in the late Tournaisian.

Thus the spore evidence does not give support to any hypothesis that assumes a long pre-Silurian history for the vascular plants. On the contrary, the evidence from fossil trilete spores indicates little pre-Devonian diversification of the land flora, then fairly rapid diversification and development in the Early Devonian, especially in the latter part of this time, followed by the gradual introduction of new types through the Middle and Late Devonian.

The evidence outlined above on spore distribution in the Devonian appears to contradict Ananiev's (1964) conception, based on macroplant evidence, of three distinct Devonian floras; namely, Lower Devonian—*Psilophyton*; Middle Devonian—*Hyenia*; and Upper Devonian—*Archaeopteris*. There is no such threefold breakdown of the plant microfossils. In fact the spore evidence gives support to a suggestion of Banks (1965) that land plants spread widely in the Lower Devonian and rapidly became diverse, that there is no simple threefold floristic division in the Devonian, and that there was probably "a single Devonian flora that evolved gradually throughout the period." Further the spore evidence points to the possibility of considerable ecological differentiation (see "Geographic Distribution and Facies Relationships"), which suggests much greater floral variety and more complex ecological relationships than those possible within a simple threefold floral division.

Evidence from Spore Morphology and Affinity

Certain features of spore morphology are also worthy of mention since they have possible implications in relation to the development of the Devonian flora. For instance, the small size of the Gedinnian spores so far described and the well-defined modal size peak would suggest that the plants producing them were dominantly, if not all, homosporous. Spores frequently have thin proximal walls, which suggests that most of the spore development took place in the tetrad and little or none after separation. Further the relatively uniform basic morphology of the spores—which are all trilete, small, and azonate—does not suggest great differentiation among their parent plants. How close this is to the true situation is difficult to assess, because the spore types far outnumber the description of plant macrofossils. Plants recorded from Downtonian and early Dittonian strata include *Cooksonia*, *Zosterophyllum*, "*Prototaxites*," and *Pachytheca*. Of these *Cooksonia* and *Zosterophyllum* are vascular plants, but so far the spores of nei-

Plate 11-5. Middle and Upper Devonian spores—Eifelian, Givetian, and Frasnian. Figures 1-4: Eifelian and Givetian spores (northeast Scotland). 1-3—*Ancyrospora* spp.: spores with bifurcate spines typical of the Middle Devonian, 1—*A. ancyrea* var. *ancyrea* Richardson 1962 (× 500); 2—*A. longispinosa* Richardson 1962 (× 250); 3—*A. grandispinosa* Richardson 1960 (× 300). 4—*Perotrilites conatus* Richardson 1965 (× 500), showing wrinkling of the perispore over the proximal face and ornament of cones. Figures 5-10: Frasnian spores (Naumova, Russian platform, USSR). 5-6 *Geminospora* sp. (*Archaeozonotriletes* sensu Naumova), showing thick outer layer and thin inner only slightly separated from it, ornament consists of small cones (× 500). 7-8 *Archaeozonotriletes variabilis* (Naumova) Allen 1965, 7—obliquely compressed, showing thick distal patina (× 500); 8—probably same type but with corroded patina (× 500). 9-10 *Archaeoperisaccus* sp., 9—polar compression, showing bilateral-symmetry, monolete mark and strong proximal fold of exoexine (× 500); 10—lateral compression, showing that the pseudosaccus is attached distally (× 500).

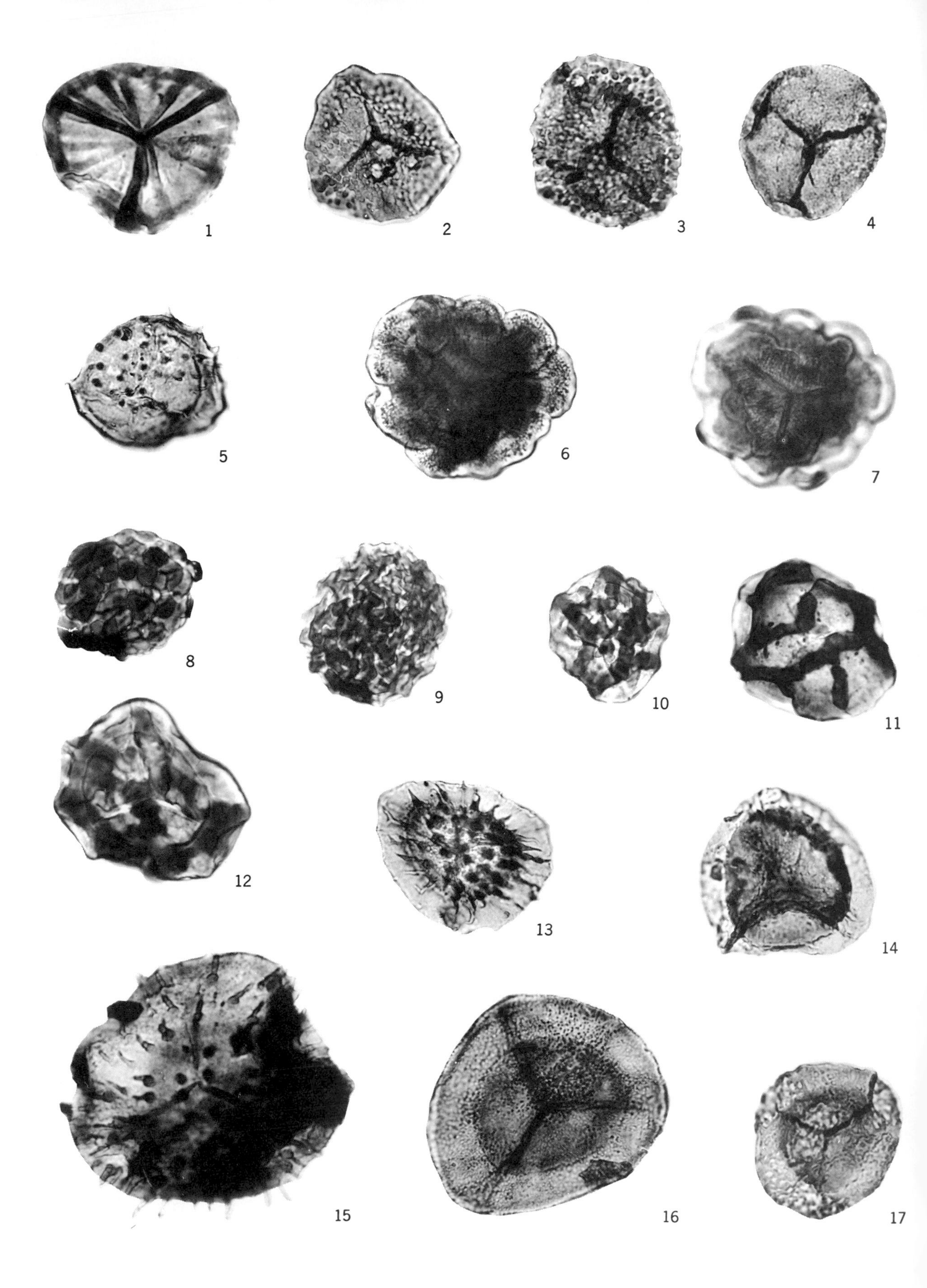

1
2
3
4
5
6
7
8
9
10
11
12
13
14
15
16
17

ther of these plants have been described in detail. Information on spores from the early psilophyte *Cooksonia* indicates only that they are small in size (25 to 35 μ) and simple (Lang, 1937). This does not contradict the dispersed spore evidence but adds nothing to it.

A further feature of interest in the Gedinnian assemblages studied by the writer is the presence of prominent proximal papillae in several of the spores. This is not a dominant feature in any other Devonian assemblages but is a character seen in spores of some Carboniferous lycopods. However, although the comparison is an interesting one, it could be simply due to a parallel development in spore morphology in unrelated groups of plants and should not be construed as evidence for lycopods in the early Gedinnian. Further it should be pointed out that the time gap between the Gedinnian and Carboniferous proximal-papillate spores is considerable.

The plant genus *Asteroxylon* from the Rhynie chert (probably Lower Devonian) is possibly a lycopod or close to the lycopod line of development; however, the spores of this plant have yet to be described. Spores and sporangia previously thought to belong to this genus have now been proved to represent a separate plant, the vegetative parts of which are still unknown (Lyon, 1964).

More is known about the spores of another Lower Devonian plant recently discussed by Banks, Hueber, and Leclerq (1964). This plant resembles *Dawsonites*, which is regarded as belonging to "probable coenopterid-type ferns" (Banks, 1965). The spores have what appears to be an ornamented perisporal membrane (Streel, 1967, pl. 2, figs. 18-20). Without the "perispore" the spores closely resemble those of the dispersed spore species *Retusotriletes* (*Phyllothecotriletes*) *triangulatus* (Streel, 1967) (pl. 11-2, fig. 1) and have a smooth exine that is thickened in the proximal polar region, giving a distinctive darkened triangular area. *Retusotriletes triangulatus* occurs in Lower and Middle Devonian strata, and the writer has found it especially abundant in the lower Dittonian (late Gedinnian) of the Welsh Borderland; in the latter assemblages only rare specimens show the "perisporal" covering. However, in some assemblages from the Scottish "Dittonian" (?Siegenian) many of the spores retain the outer sculptured layer (pl. 11-3, fig. 7), although frequently it is loosely attached.

The microspores of *Barinophyton richardsoni* (a heterosporous plant of uncertain affinities) are similar to those from *Dawsonites* in that they have a thickened apical area and a "perisporal" membrane. However, in the case of these *Barinophyton* microspores the "perispore" lacks sculpture (Pettit, 1965, p. 84). This is in contrast to the dispersed spores of this type seen by the writer in the Lower Devonian, which have a sculptured "perispore" and therefore most closely resemble spores from *Dawsonites*. However, dispersed spores that are similar to the megaspores and microspores of *Barinophyton* occur together in the Middle Devonian of Scotland—they are *Trileites langi* and "*Punctatisporites*" *confossus*, respectively. Plants placed in the genus *Barinophyton* have been reported from Lower, Middle, and Upper Devonian strata.

Similar spores to those recorded from *Dawsonites* have been seen by the writer in the upper Gedinnian (lower Dittonian, Welsh Borderland) and Siegenian to Emsian strata (Breconian, South Wales, and Scottish "Dittonian," Midland Valley), where they are abundant; and similar spores lacking the "perispore" occur

Plate 11-6. Upper Devonian and Lower Carboniferous spores—upper Famennian-lower Tournaisian (New York State and Pennsylvania). (All figures × 500.) 1—*Emphanisporites* cf. *rotatus* McGregor 1961 (al. *Radforthia radiata* Winslow 1962). 2-3 Cf. *Pustulatisporites pretiosus* Playford 1963; 3—specimen showing an inner body. 4—*Apiculiretusispora* sp., proximal view. 5—?*Apiculiretusispora* sp.; retusoid spore, with ornament of spines, proximal face thin. 6-7 *Secarisporites* sp., distal view, showing lobate ornament; 7—proximal view, showing short trilete mark. 8—*Verrucosisporites* cf. *nitidus* Playford 1963 (al. V. *gromosus* (Naumova) Sullivan 1964), distal view. 9—*Convolutispora* sp., distal view, showing convolute muri. 10-12 ?*Knoxisporites* cf. *Archaeozonotriletes literatus* Kedo 1963, 10—distal muri form irregular reticulum; 11—muri sparse, forming irregular coarse reticulum; 12—proximal view. 13—*Vallatisporites* cf. "*Hymenozonotriletes*" *pusillites* Kedo 1957, distal view, showing spinose ornament (form characteristic of the lowermost Tournaisian). 14—?*Auroraspora* sp., pseudosaccus corroded. 15—*Spinozonotriletes* sp., spinose pseudosaccus. 16—*Calyptosporites* cf. *velatus* (Eisenack) Richardson 1962; pseudosaccus triangular in outline, sculpture consists of cones. 17—"*Hymenozonotriletes*" cf. *H. lepidophytus* Kedo 1957; pseudosaccus, showing characteristic lumena or pits (form characteristic of the upper Famennian).

in the Middle Devonian. In contrast, spores similar to the megaspores and microspores from *Barinophyton* have not been seen in the Gedinnian but occur in the Eifelian and Givetian, ranging into the Upper Devonian (Frasnian).

Siegenian spore assemblages show a greater variety of spore types than those described from the Gedinnian, although the assemblages are still dominantly azonate. The spores tend to be larger in size than those from the Gedinnian. (See fig. 11-2.) The Emsian shows even greater morphological variety. However, none of the distinctive spore genera of this time has been related to its parent plant; examples are (a) the genus *Emphanisporites* McGregor (pl. 11-3, figs. 2-4), which is frequently abundant in the Lower Devonian, especially in the latter part of this time; (b) patinate forms such as *Archaeozonotriletes* (Naumova) Allen and *Chelinospora* Allen; and (c) zonate forms such as *Samarisporites* Richardson (Siegenian) (pl. 11-3, fig. 5) and pseudosaccate forms such as *Calyptosporites* Richardson (Emsian) (pl. 11-4, fig. 6), both of which become important in the Middle Devonian. All these types are striking structural types that are important in the evolution of Devonian spore assemblages.

The Middle Devonian saw an increase of spore size, with many spores grading into the megaspore size range and several spore species with a size mode of over 200 microns. It also witnessed the flourishing of many important structural spore types that first appeared in the Siegenian and Emsian, especially in the latter, continued on into the Late Devonian, and died out either in the Late Devonian or Early Carboniferous. These forms include genera with (a) distinctive biform sculpture, such as *Acinosporites* (pl. 11-4, fig. 2) and *Dibolisporites;* (b) forms with bifurcate processes, such as *Ancyrospora* (pl. 11-5, figs. 1-3), *Hystricosporites*, and *Nikitinsporites*; and (c) large pseudosaccate forms, such as *Auroraspora, Calyptosporites,* and *Rhabdosporites* (pl. 11-4, fig. 5). Until quite recently practically nothing was known of the affinities of any of these genera, but during the last few years two exciting finds have been made. Bonamo and Banks (1966, 1967) have described spores referable to *Dibolisporites* from *Calamophyton bicephalum* Leclerq and Andrews (Bonamo, 1965) and to *Rhabdosporites langi* from *Tetraxylopteris schmidtii* Beck (Bonamo, 1966). *Calamophyton* is regarded as a sphenopsid[4] and *Tetraxylopteris*, as a progymnosperm. Both of the spore genera associated with these plants are relatively common in Middle Devonian and Frasnian spore assemblages. The association of *R. langi* with *Tetraxylopteris* is particularly interesting since this distinctive pseudosaccate spore species is relatively widely distributed in strata of late Eifelian, Givetian, and Frasnian age and is frequently abundant—whereas *Tetraxylopteris* has not been reported outside North America. Similar spores have been recorded either in situ or associated with various species of *Protopteridium* (Lang 1925, 1926; Obrhel, 1961), which raises the interesting question of the possible relationships of these two plants. Bonamo (1966) gives a detailed discussion of these relationships.

Much less is known of the affinities of spores with bifurcate processes, which are also widely dispersed and frequently abundant in the Middle and Upper Devonian. However, Nikitin (1934) described megasporangia and microsporangia from the Frasnian of the Soviet Union, with megaspores resembling the dispersed-spore genus *Nikitinsporites* and microspores resembling the genus *Archaeoperisaccus*. (See pl. 11-5, figs. 9 and 10.) This development of a plant with trilete megaspores and monolete microspores is paralleled in the modern flora only by the aquatic lycopod *Isoetes*, and it is interesting to note that Nikitin considered that his megasporangia and microsporangia may have had lycopodiaceous affinities. Although we should not take the comparisons too far, some species of *Isoetes* have spiny megaspores and have double-layered monolete microspores. These comparisons at least make it unlikely that the genus *Archaeoperisaccus* represents gymnospermous pollen.

A further interesting spore-plant association involving a spore type occurring in the Middle and Upper Devonian is that between the dispersed-spore genus *Biharisporites* and the important Devonian plant genus *Archaeopteris* (Progymnospermopsida). Several *Archaeopteris* species have been shown to be heterosporous,

[4]Leclerq and Schweitzer (1965) consider that *Calamophyton* is not a sphenopsid but has cladoxylalean affinities.

and some of these—*A. hibernica, A.* cf. *halliana* (*A.* cf. *jacksoni*) (Pettit, 1965), and *A. macilenta* (Carluccio, 1966)—have sporangia that contain, or are associated with, megaspores referable to the dispersed-spore genus *Biharisporites*. Furthermore, *A.* cf. *halliana* and *A. macilenta* both have similar, small, apiculate microspores (?*Apiculatasporites*). Some of the latter appear to have a central body and resemble spores placed in the genus *Geminospora*, although the outer wall (exine) is not as thick as in typical *Geminospora* species. However, whatever the eventual generic assignment of the microspores, this does not alter the interesting fact that two species of *Archaeopteris* have similar megaspores and microspores, and three species have similar megaspores. Carluccio (1966, p. 60) also points out that the "strikingly similar megaspores" of three species of *Archaeopteris* coincide "with taxonomic evidence that the species are closely related and may intergrade." It is interesting to note that the species of *Biharisporites* resembling *Archaeopteris* megaspores so far have been reported from Givetian (Middle Devonian), Frasnian, and possibly lower Tournaisian strata, whereas the genus *Archaeopteris* has not been reported from the Middle Devonian or Lower Carboniferous. However, since there is a tendency to regard *Archaeopteris* as characteristic of the Upper Devonian, this is perhaps hardly surprising. In Middle and Upper Devonian strata there are a number of plants, in addition to *Archaeopteris* and *Tetraxylopteris*, regarded by several workers as progymnosperms, which have basically similar spores. The spores in all these plants (listed below) are two-layered and have an outer layer (exine) which bears small, closely-packed sculptural elements.

The spores of *Archaeopteris* (especially the microspores), *Aneurophyton*, and *Svalbardia* are also structurally similar in possessing a thick outer layer and a thin, frequently indistinct, inner layer. In spite of their present different generic assignments they could be placed together in a single genus *Geminospora*. The spore genera *Geminospora* and *Rhabdosporites* are widely distributed and often abundant in upper Eifelian, Givetian, and Frasnian assemblages. They appear to reach an acme in the upper Givetian and lower Frasnian and decline rapidly in the upper Frasnian and are rare or absent in the Famennian. The question of their significance in terms of plant evolution remains to be discovered, but at least the spore evidence is not as discouraging as it would appear to be at first glance in that a group of apparently related plants produced structurally similar spores. It is possible of course that other unrelated groups of plants also produced spores of this type, but against this is the evidence of the uniform stratigraphical distribution of *Geminospora* and *Rhabdosporites*, and their plant associations, although but few, are so far consistent.

The above review of some Devonian spore-plant associations reveals that spores have potentially a much greater palaeobotanical role as indicators of the distribution, evolution, and relationships of their parent plants because spores are much more abundant than identifiable large plant remains. This fact and the indications outlined above should give impetus to more careful descriptions of both dispersed and in situ spores.

Spore Evidence for the Development of Heterospory

The uniformly small size of Silurian and Ge-

Plant	Megaspores	Microspores or Isospores	
Tetraxylopteris		*Rhabdosporites* (Bonamo and Banks, 1967)	
Protopteridium		Probably *Rhabdosporites* (Lang, 1925, 1926; Obrhel, 1961)	
Aneurophyton		*Aneurospora* (Streel, 1964)	Possibly *Geminospora*
Svalbardia		*Lycospora* (Vigran, 1964)	Possibly *Geminospora*
Archaeopteris	*Biharisporites*	*Cyclogranisporites* (Pettit, 1965)	Possibly *Geminospora*

dinnian spores (fig. 11-2) suggests that the plants producing them were all homosporous. There is thus no spore evidence for heterospory in the Gedinnian. However, in the Siegenian and Emsian the size range of spore types is much greater. There is some indication that heterospory or at least incipient heterospory may have developed at this time, since spores of the megaspore size range have been recorded (fig. 11-2 and Allen, 1965). In the Middle Devonian the spore evidence for heterospory is stronger. Middle Devonian assemblages contain relatively large spores, but the modes of the spore assemblages (fig. 11-2) are poorly defined; relatively large spores, 200 to 400 microns, are fairly common, and some of the spores over 300 microns are almost certainly megaspores and are similar to megaspores from some Late Devonian plants (Richardson, 1965a, p. 600-601). By the Frasnian heterospory was clearly established, there are several dispersed spore species ranging in size from 200 to 1,600 microns (Chaloner, 1959), and several heterosporous plants are known. The Famennian and Early Carboniferous spore assemblages (fig. 11-2) show more clearly defined modes (37 μ), which are smaller than those from the late Early Devonian and the Middle Devonian (55 to 60 μ), which perhaps reflects to some extent the perfection of heterospory. Thus there is some spore evidence that the development of heterospory took place at least as far back as the Givetian, and it is possible that it may have occurred in late Siegenian-Emsian times, whereas the earliest heterosporous plant so far recorded is from the Frasnian.

Detailed studies of spore assemblages, especially those from well-controlled stratigraphic sequences, can be expected to throw a great deal of light in the next few years on the evolution and geographic distribution of Devonian spores. Not only are such studies a valuable tool for the determination of age, especially in continental sediments, but they also have an immense potential as indicators of plant relationships, the course of evolution, distribution, and habitat of the earliest land flora. A critical line of investigation, therefore, is the careful study and accurate description of spores from known plants. Only when many more of these studies are completed will the growing wealth of data on Devonian spores be fully utilized.

SOME IMPORTANT DEVONIAN SPORE GENERA

The short descriptions of the following genera reflect what the writer considers to be a usable concept for each genus. They may not correspond precisely with previous definitions of these genera and are intended merely as a general guide since detailed discussion of systematics is outside the scope of the present work.

Trilete Spores

Tetrad scar with three rays separating three contact areas, spores radially symmetrical.

Azonate. These are simple spores lacking equatorial structures. Nine genera are described below.

***Retusotriletes* (Naumova) Richardson 1965.** Type species—*R. pychovii* Naumova 1953. (See pl. 11-1, figs. 1 and 2; pl. 11-2, figs. 1 and 2, and Richardson (1965a).

Description. Trilete miospores with distinct curvaturae perfectae and well-developed contact areas, curvaturae perfectae often forming a wedge-shaped thickening when seen in lateral compression. Exine externally smooth or finely wrinkled, infrastructure varied.

Discussion. This genus is restricted to trilete, smooth, azonate miospores with distinct curvaturae perfectae. *Divisisporites* (Thomson) Potonié 1956 has curvaturae imperfectae. *Apiculiretusispora* Streel 1964 also has curvaturae perfectae but is sculptured with cones or small spines. Streel's emendation of *Retusotriletes* (1964) is not used here because it is based on a lectotype species with poorly defined contact areas and curvaturae imperfectae; as so defined the genus can be confused with *Divisisporites* (Thomson) Potonié 1956. The type species of *Phyllothecotriletes* Luber 1955 also has curvaturae perfectae and is thus synonymous with *Retusotriletes*.

***Trileites* (Erdtman 1945, 1947) ex Potonié 1956.** Type species—*T. spurius* (Dijkstra) Potonié 1956. (See pl. 11-4, fig. 1.)

Description. Trilete "megaspores" (i.e., spores with a mean diameter over 200 μ); subcircular to subtriangular equatorial outline, spores originally spherical or nearly so. Exine laevigate, curvaturae perfectae typically present but some species lack them.

Discussion. Laevigatisporites has shorter Y-rays and was originally concavo-convex in shape.

***Apiculiretusispora* Streel 1964.** Type species —*A. brandtii* Streel 1964. (See pl. 11-1, figs. 7 and 8).

Description. Trilete miospores with subcircular to subtriangular equatorial outline. Curvaturae perfectae distinct, contact areas well developed; sculpture consists of granules, cones, or spines, reduced or absent on the contact areas.

Discussion. Retusotriletes is similar but has a smooth exine. In the writer's experience species of *Apiculiretusispora* are especially common in Lower Devonian strata. *Anapiculatisporites* Potonié and Kremp 1954 also has no sculpture on the proximal contact surface but apparently lacks curvaturae perfectae. (See Potonié and Kremp, 1954 and 1955.)

***Lophotriletes* (Naumova) Potonié and Kremp 1954.** Type species—*L. gibbosus* (Ibrahim) Potonié and Kremp 1954.

Description. Trilete azonate miospores with clearly defined triangular to subtriangular equatorial outline. Exine covered on proximal and distal surfaces by coni.

Discussion. Apiculatisporis Potonié and Kremp 1956 has a circular to subcircular equatorial outline. *Planisporites* (Knox) as reemended by Potonié (1960) is similar to *Lophotriletes* but has coni of less than 1 μ.

***Dibolisporites* Richardson 1965.** Type species —*D. echinaceus* (Eisenack) Richardson 1965. (See Richardson, 1965a.)

Description. Trilete azonate miospores; equatorial outline subcircular to subtriangular. Sculptural elements dominantly biform (i.e., each element consists of two distinct parts, e.g., a tubercle surmounted by a cone) but otherwise very variable, consisting of cones, rodlike processes, pila, verrucae, and spines.

Discussion. The biform nature of the individual sculptural elements distinguishes this genus from most other miospores. *Acinosporites* Richardson 1965 has sculptural elements of various types superimposed on convolute and anastomosing ridges (Richardson, 1965a). *Bullatisporites* Allen 1965 is synonymous with *Dibolisporites*.

Botanical affinities. Spores referable to the genus *Dibolisporites* have been found in sporangia of *Calamophyton* (Bonamo and Banks, 1966); this plant may belong to the early articulates (Equisetinae), but see footnote 4.

***Biharisporites* Potonié 1956.** Type species—*B. spinosus* (Singh) Potonié 1956.

Description. Trilete azonate "megaspores"; equatorial outline subcircular to subtriangular, spores originally more or less spherical to hemispherical. Exine sculptured with small cones, grana, or biform elements. A thin intexine "mesosporium" is sometimes present.

Discussion. Verrutriletes Potonié 1956 has an ornament of verrucae, *Tuberculatisporites* has a concavo-convex shape (Chaloner, 1959).

Botanical affinities. Megaspores that can be assigned to the genus *Biharisporites* have been found by several workers in fructifications of heterosporous *Archaeopteris* (Pettit, 1965; Carluccio, 1966).

***Geminospora* Balme 1960.** Type species—*G. lemurata* Balme 1960. (See pl. 11-5, figs. 5 and 6.)

Description. Trilete miospores, equatorial outline subcircular to subtriangular. Exoexine thick especially distally, encloses a thin, frequently folded intexine variably separated from the exoexine; exoexine sculptured on the distal side with close-packed grana, cones, spines, baculae, and biform elements. Sculpture absent or reduced on the contact areas.

Discussion. According to Playford (1964) *Leiozonotriletes* Hacquebard 1957 has a similar construction but is laevigate. Most of the species placed by Naumova in the genus *Archaeozonotriletes* appear to be very similar to *Geminospora*. However, in emending the genus Potonié (1958) followed an arbitrary procedure and selected the first species described by Naumova as the lectotype; and in doing so he chose the species *A. variabilis*, which is patinate. Whilst some subdivision of Naumova's original genus is necessary, it would have been preferable to choose a more typical species since most forms placed in the genus by Naumova and other Soviet workers are spores of the *Geminospora* type.

Spores that probably belong to the genus *Geminospora* have been assigned to various other genera; for example, *Aneurospora goensis* Streel 1964, *Lycospora svalbardiae* Vigran 1964, and *Retusotriletes greggsii* McGregor 1964.

Botanical affinities. Microspores of *Svalbardia polymorpha* Hoeg 1942 appear to be identical to the spores of the genus *Geminospora*, and those of *Archaeopteris jacksoni* (Pettitt, 1965) also show some similarity but lack the typical thick exoexine.

***Dictyotriletes* (Naumova) Potonié and Kremp 1954.** Type species—*D. mediareticulatus* (Ibrahim) Potonié and Kremp 1954. (See pl. 11-2, fig. 4.)

Description. Trilete azonate miospores, exine with a network of muri (extrareticulum), which may be absent over the proximal surface.

Discussion. Reticulatisporites (Ibrahim) Neves 1964 has a differentially thickened cingulum (Neves, 1964). In the genus *Microreticulatisporites* (Knox) Potonié and Kremp 1954 the diameter of the lumina is less than 6 microns. In *Brochotriletes* Naumova 1953 the exine bears more or less circular pits.

***Emphanisporites* McGregor 1961.** Type species—*E. rotatus* McGregor 1961. (See pl. 11-2, figs. 10 and 11; pl. 11-3, figs. 2-4.)

Description. Trilete azonate miospores with well-developed radial ribs on the proximal surface. Spores distally laevigate or with sculpture of grana, cones, or verrucae, or a distal annulate thickening.

Discussion. The prominent proximal sculpture of radial ribs distinguishes this genus from other trilete spores. Some species of *Hystricosporites* also show this character, but such forms are easily distinguishable from *Emphanisporites* since they have in addition prominent bifurcate spines.

Patinate. These are trilete spores with a pronounced thickening of the distal exine.

***Archaeozonotriletes* (Naumova) Allen 1965.** Type species—*A. variabilis* Naumova 1953. (See pl. 11-5, figs. 7 and 8.)

Description Trilete miospores with circular, subcircular to subtriangular equatorial outline; exine laevigate or punctate; distally patinate—that is, with a pronounced thickening over the distal hemisphere.

Discussion. Allen distinguishes *Archaeozonotriletes* from *Tholisporites* on the basis of variation in the thickness of the patina. He states "in *Tholisporites* . . . the patina is thickened in the equatorial region ... whereas *Archaeozonotriletes* has a uniform or distal polar thickened patina." The writer doubts whether in practice this will be found to be a practical basis for subdivision.

Most spores show some thinning over the proximal surface, and some limit should be set to the differentiation between the distal and proximal surfaces. It is suggested here that distally the exine should have a minimum thickness of 5 microns and should be more than twice the thickness of the proximal exine.

***Cymbosporites* Allen 1965.** Type species—*C. cyathus* Allen 1965. (See pl. 11-1, figs. 5 and 6.)

Description. Miospores with circular to subtriangular equatorial outline. Exine thin proximally, equatorially and distally patinate, patina of even thickness or with its greatest thickness in the distal polar region. Patina variably sculptured with cones, spines, and granules.

Discussion. Distinguished from *Archaeozonotriletes* Allen 1965 by the ornament of granules, cones, or spines.

Zonate. These are trilete spores with an equatorial extension of the exoexine in the form of a flange or an equatorial thickening of the exoexine (cingulum).

***Ambitisporites* Hoffmeister 1959.** Type species—*A. avitus* Hoffmeister 1959.

Discussion. This genus was defined by Hoffmeister to cover smooth trilete spores with an equatorial thickening. As diagnosed the genus is synonymous with *Stenozonotriletes* Hacquebard 1957. The interest of Hoffmeister's spores is that they are from "high Lower Silurian" strata. However, very similar spores in Upper Silurian and lowermost Devonian samples from the Welsh Borderland have an equatorially and distally thickened exine, and consequently some of the Welsh Borderland spores are referred provisionally to the genus *Archaeozonotriletes* (Naumova) Allen 1965. Forms similar to *Ambitisporites* also occur, especially in the Wenlock.

***Samarisporites* Richardson 1965.** Type species—S. *orcadensis* (Richardson) Richardson 1965. (See pl. 11-4, fig. 4, and Richardson, 1965a.)

Description. Trilete zonate spores with smooth proximal surface, distally covered by prominent sculpture of cones, verrucae, or both, which often bear cones or short spines (i.e., biform elements); elements may be clearly separated, arranged in concentric patterns, fused together in regular rows or groups, or fused into irregular convolute groups.

Comparison. Cristatisporites has sculpture on the proximal and distal surfaces. *Cirratriradites* does not have such prominent distal sculpture.

Vallatisporites **Hacquebard 1957.** Type species—*V. vallatus* Hacquebard 1957. (See pl.11-6, fig. 13.)

Description. Trilete zonate miospores, equatorial outline circular to subtriangular. Exine two layered; intexine thin and laevigate. Equatorial border may be internally vacuolate. Adjacent to the junction with the spore cavity there is a narrow, lighter appearing internal "groove" to which the name cuniculus has been applied (Sullivan, 1964). Spores distally ornamented with granules, cones, verrucae, or spinose elements.

Discussion. Sullivan (1964) gives a different interpretation of the "groove" from that in Hacquebard's original description. Hacquebard described this feature as a "narrow 'ridge', characterized by single row of pits (visible in high focus); . . . area between body margin and 'ridge' may vary from a few microns, in which case it appears like a groove, to as much as 8 microns, becoming rampart-like in appearance." Some species are characterized by a row of pits adjacent to the cuniculus.

One species, *V.* (*Hymenozonotriletes*) *pusillites* Kedo 1957, has been used by Soviet workers to define the base of the Carboniferous and has been recorded from lowermost Carboniferous strata in several areas (e.g., Belgium, Soviet Union, United States).

Pseudosaccate.[5] In these spores a well-developed cavity separates any two layers of the spore coat. These spores lack the columellate structure typical of monosaccate pollen. Furthermore, in pseudosaccate genera the membranes are always clearly separated, whereas in forms such as *Geminospora* Balme separation is barely discernible in some forms and is never very great.

Auroraspora **(Hoffmeister, Staplin, and Malloy) Richardson 1960.** Type species—*A. solisortus* Hoffmeister, Staplin, and Malloy 1955.

Description. Trilete pseudosaccate spores; outline of pseudosaccus (exoexine) and central body (intexine) subtriangular to subcircular in proximal polar view. Pseudosaccus completely encloses the body; it is laevigate and without a limbus. Spores originally elliptical in cross section.

Discussion. Endosporites frequently has a limbus, and the central body has a finely reticulate external surface on the central wall. The latter genus is typical of Upper Carboniferous strata (see p. 249).

Calyptosporites **Richardson 1962.** Type species—*C. velatus* (Eisenack) Richardson 1962. (See pl. 11-4, fig. 6.)

Description. Trilete pseudosaccate miospores or megaspores; equatorial outline subtriangular to triangular, spores originally elliptical in cross section, usually preserved in polar compression with Y-folds only occasionally displaced from the center of the spore. Proximally laevigate, distally sculptured with cones or spines; the latter may bifurcate at their tips.

Discussion. The subtriangular outline, original elliptical cross section, strong trilete folds, and the presence of bifurcate spines in some species distinguish the genus from *Grandispora* Hoffmeister, Staplin, and Malloy 1955. *Spinozonotriletes* Hacquebard 1957 also appears to be pseudosaccate but has an intexine that is much thinner than the exoexine (Hacquebard, 1957; Dettman and Playford, 1963).

Rhabdosporites **Richardson 1960.** Type species—*R. langi* (Eisenack) Richardson 1960. (See pl.11-4, fig. 5.)

Description. Trilete pseudosaccate miospores; equatorial outline of both body and pseudosaccus subcircular, elliptical, or subtriangular. Spores originally more or less spherical, flattened at the proximal pole, with the body (intexine) attached on the proximal side. Pseudosaccus with an external ornament of evenly distributed, closely packed rods, which are parallel-sided elements with truncated tips. Intexine smooth.

Discussion. The pseudosaccus is typically collapsed into large folds and may also under certain preservational conditions separate into two distinct layers (Streel, 1964; Bonamo and Banks, 1966b).

Calyptosporites has a subtriangular outline, was probably originally elliptical in cross section, and has a sculpture of cones or spines.

Remysporites Butterworth and Williams 1958 is circular in shape, but the bladder is laevigate to microreticulate.

Botanical affinities. Bonamo and Banks (1967) have recently shown that spores identical with the type species occur in the fructifica-

[5] See Richardson, 1965a, p. 584.

tions of *Tetraxylopteris schmidtii*, a Devonian progymnosperm.

"*Hymenozonotriletes*" Naumova 1953

Remarks. Naumova's grouping includes zonate, cingulate, and pseudosaccate forms, most of which can be included under existing genera. It was emended by Potonié (1958), who chose *H. polyacanthus* as the type species; unfortunately the emended genus can be confused with various genera of the densospore type. For convenience in this work it is used for spores that have a distinctive pseudosaccus with a series of irregular to more or less circular pits in the pseudosaccus wall. Such spores were first recorded by Kedo (1957), who described them under the name *Hymenozonotriletes lepidophytus.* Since then similar spores have been described from Famennian-Lower Carboniferous strata by several authors; for example, *Leiozonotriletes naumovae* Balme and Hassell 1962, and *Endosporites lacunosus* Winslow 1962.

Forms with Prominent Anchor Spines.

***Ancyrospora* (Richardson) Richardson 1962.** Type species—*A. grandispinosa* Richardson 1960. (See pl. 11-5, figs. 1-3.)

Description. Trilete miospores or megaspores with a thick equatorial flange or pseudoflange. (The latter is formed from the confluence of spine bases, which in polar compression give the appearance of an equatorial flange.) Exoexine two-layered, frequently thick, and rather "spongy" in texture; intexine variable in thickness. Exoexine bears spinose processes that bifurcate at their tips. Trilete lips frequently elevated as an apical prominence.

Discussion. Archaeotriletes (Naumova) Potonié 1958 has a thin membranous zona, and club-shaped spines that arise from the central area. *Hystricosporites* McGregor 1960 is azonate; *Nikitinsporites* Chaloner 1959 has distinctive bifurcate spines that have thick, more or less parallel-sided stems, and each spine has a sharp constriction near the apex immediately behind the bifurcate tip.

***Hystricosporites* McGregor 1960.** Type species—*H. delectabilis* McGregor 1960.

Description. Trilete azonate miospores or megaspores that bear discrete bifurcate spinose processes; spines variable in size but uniformly tapered and without the constriction at the spine apex that is typical of *Nikitinsporites.* Trilete lips may be elevated as an apical prominence, and some species show proximal radial ribs.

Discussion. The nature of the spinose processes distinguishes *Nikitinsporites,* which is also azonate. *Dicrospora* Winslow 1962 includes several species referable to the genus *Hystricosporites;* Winslow's genus also includes forms with spine terminations that are multifurcate rather than bifurcate. In this chapter *Dicrospora* is used for multifurcate forms.

***Nikitinsporites* Chaloner 1959.** Type species—*N. canadensis* Chaloner 1959.

Description. Trilete azonate megaspores, triradiate lips greatly elevated to form an apical prominence. Exine covered with bifurcate processes with parallel-sided slightly tapered stems, sharply constricted near apices, and anchor-shaped tips.

Discussion. Similar megaspores have been reported in association with small spores of the genus *Archaeoperisaccus* (in the Voronezh region of the Soviet Union—Nikitin, 1934) and also in parts of arctic Canada, (McGregor, 1969). Nikitin described trilete megaspores that are comparable with *Nikitinsporites* and monolete microspores similar to *Archaeoperisaccus* from megasporangia and microsporangia that he thought may belong to a lycopod. If these spores are from the same plant, they form an interesting morphological parallel with spores of the modern aquatic lycopod *Isoetes,* which is probably unique among extant plants in having trilete megaspores and monolete microspores.

Monolete Spores

Tetrad scar with a single ray separating two contact areas, spores bilateral.

***Archaeoperisaccus* (Naumova) Potonié 1958.** Type species—*A. menneri* Naumova 1953. (See pl. 11-5, figs. 9 and 10.)

Description. Monolete, bilateral, pseudosaccate miospores. In polar compression elliptical in outline; central body usually thick walled. Exoexine frequently folded over the proximal surface, attached distally to the body but appears to be separate from the body at the equator and over the proximal surface, although it may be also attached in the area of the monolete mark. According to Naumova these forms may have small sculptural elements; however, the specimens examined by the writer appear to be externally smooth but infragranulate.

Discussion. Heavily sculptured forms described by Pashkevitch (1964) from north Timan (Soviet Union), such as *A. timanicus* and *A. verrucosus,* do not appear pseudosaccate and instead appear to have a solid flange or zona. In emending this genus Potonié (1958) excludes forms described by Naumova that have sculpture.

Botanical affinities. Similar spores have been described from microsporangia named *Krystofovichia africani* Nikitin 1934 (McGregor, 1969). Nikitin considered that the macrosporangia and microsporangia were probably lycopodiaceous.

Acknowledgements. The writer gratefully acknowledges the help of Professor S. N. Naumova in allowing him to publish photographs of some of her Frasnian material (pl. 11-5, figs. 5-10); Elsevier Publishing Co. for permission to reproduce figure 11-2 (Rev. Palaeobotan. Palynol., vol. 1, p. 111-129 [1967].); Drs. W. Brideaux, D. C. McGregor, and B. Owens for access to unpublished work; and The National Science Foundation for funds (grant G. P. 3606) to study Devonian and Lower Carboniferous palynology of N. Y. and Pa.

References

Allen, K. C., 1965, Lower and Middle Devonian spores of North and Central Vestspitsbergen: Palaeontology, v. 8, pt. 4, p. 687-748.

Ananiev, A. R., 1964, Recent studies on the Devonian floras of Siberia [abs.]: Internat. Bot. Cong., 10th, Edinburgh, p. 17-18.

Archangelskaya, A. D., 1963, New spore finds from Devonian deposits of the Russian platform: Ministerstvo geol. i ochroni Nedr SSSR, Moscow, v. 37, p. 18-30.

Balme, B. E., 1960, Upper Devonian (Frasnian) spores from the Carnarvon basin, Western Australia: The Palaeobotanist, v. 9, nos. 1, 2, p. 1-10.

Balme, B. E., and Hassell, C. W., 1962, Upper Devonian spores from the Canning Basin, Western Australia: Micropaleontology, v. 8, no. 1, p. 1-28.

Banks, H. P., 1965, Some recent additions to the knowledge of the early land flora: Phytomorphology, v. 15, no. 3, p. 235-245.

Banks, H. P., Hueber, F. M., and Leclercq, S., 1964, A probable fern in the Lower Devonian [abs.]: Internat. Bot. Cong., 10th, Edinburgh, p. 515.

Bonamo, P. M., 1965, *Calamophyton* in the Devonian of New York State: M.S. Thesis, Cornell Univ., Ithaca, N.Y.

______1966, *Tetraxylopteris schmidtii*: The fertile branching system: Ph.D. Thesis, Cornell Univ., Ithaca, N.Y.

Bonamo, P. M., and Banks, H. P., 1966, *Calamophyton* in the Middle Devonian of New York State: Am. Jour. Botany, v. 53, no. 8, p. 778-791.

______1967, *Tetraxylopteris schmidtii*: its fertile parts and its relationships within the Aneurophytales Am. Jour. Botany, v. 54, no. 6, p. 755-768.

Carluccio, L., 1966, Contributions to the morphology and anatomy of the Devonian progymnosperm *Archaeopteris*: Ph.D. Thesis, Cornell Univ., Ithaca, N.Y., 149 p.

Chaloner, W. G., 1959, Devonian megaspores from arctic Canada: Palaeontology, v. 1, pt. 4, p. 321-332.

______1963, Early Devonian spores from a borehole in southern England: Grana Palynologica, v. 4, p. 100-110.

Chibrickova, E. V., 1959, Spores from Devonian and earlier deposits in Bashkir: Izdatel. Akad. Nauk SSSR (Bashkir), p. 3-175.

______1962, Spores from Devonian terrigenous deposits of the Bashkir region and the southern slopes of the Urals: Izdatel. Akad. Nauk SSSR (Bashkir), p. 353-476.

Cramer, F. H., 1966, Palynology of Silurian and Devonian rocks in northwest Spain: Bol. Inst. geol. min. España, v. 77, p. 225-286.

Dettman, M. E., and Playford, G., 1963, Sections of some spores from the Lower Carboniferous of Spitzbergen: Palaeontology, v. 5, pt. 4, p. 679-681.

Dineley, D. L., and Williams, B. P. J., 1968, Sedimentation and Paleoecology of the Devonian Escuminac Formation and Related Strata, Escuminac Bay, Quebec: Spec. Pap. Geol. Soc. Am. 106, p. 241-264.

Doubinger, J., 1963, Étude palyno-planctologie de quelques échantillon du Dévonien inférieur (Siegénien) du Cotentin: Bull. Serv. Carte géol. Alsace Lorraine, v. 16, pt. 4, p. 261-273.

Downie, C., 1963, 'Hystrichospheres' (acritarchs) and spores of the Wenlock Shales (Silurian) of Wenlock, England: Palaeontology, v. 6, pt. 4, p. 625-652.

Hacquebard, P. A., 1957, Plant spores in coal from the Horton Group (Mississippian) of Nova Scotia: Micropaleontology, v. 3, no. 4, p. 301-324.

Harland, W. B., Smith, A. G., Wilcock, B. (eds.), 1964, Geological Society Phanerozoic time scale: Geol. Soc. [London] Quart. Jour., 120s, p. 260-262.

Hoffmeister, W. S., 1959, Lower Silurian plant spores from Libya: Micropaleontology, v. 5, no. 3, p. 331-334.

Kedo, G. I., 1955, Spores of the Middle Devonian of the northeastern Byelorussian SSR: Trudy Geol. Inst. Nauk., Akad. Nauk BSSR, Paleont. i stratig., v. 1, p. 5-59.

______1957, Spores from salt deposits of the Pripiat depression and their stratigraphic significance: ibid., Minsk (Akad. Nauk BSSR), v. 2, p. 3-43.

______1963, Tournaisian spores from the Pripiat depression and their stratigraphic significance: ibid., Minsk (Akad. Nauk BSSR), v. 4, p. 3-121.

Lang, W. H., 1925, Contributions to the study of the Old Red Sandstone flora of Scotland. I. On plant remains from the fish beds of Cromarty. II. On a sporangium-bearing branch system from the Stromness Beds: Royal Soc. Edinburgh Trans., v. 54, p. 253-280.

______1926, Contributions to the study of the Old Red Sandstone flora of Scotland. III. On *Hostimella* (*Ptilophyton*) *Thomasoni* and its inclusion in a new genus, *Milleria:* Royal Soc. Edinburgh Trans., v. 54, p. 785-790.

______1937, On the plant-remains from the Downtonian of England and Wales: Royal Soc. [London] Philos. Trans., ser. B, no. 544, v. 227, p. 245-291.

Leclerq, S., and Schweitzer, H. J., 1965, *Calamophyton* is not a Sphenopsid: Bull. Acad. Roy. Belg., v. 11, p. 1394-1402.

Lyon, A. G., 1964, Probable fertile region of *Asteroxylon mackiei* K. and L.: Nature, v. 203, no. 4949, p. 1082-1083.

McGregor, D. C., 1960, Devonian spores from Melville Island, Canadian Arctic Archipelago: Palaeontology, v. 3, pt. 1, p. 26-44.

______1961, Spores with proximal radial pattern from the Devonian of Canada: Geol. Survey, Canada Bull. 76, p. i-ix, 1-11.

______1964, Devonian miospores from the Ghost River formation, Alberta: Geol. Survey, Canada Bull. 109, p. 1-31.

______1969, Devonian plant fossils of the genera *Krystofovichia, Nikitinsporites,* and *Archaeoperisaccus*: Geol. Survey, Canada Bull. 182, p. 91-106.

McGregor, D. C., and Owens, B., 1966, Devonian spores of eastern and northern Canada: Canada Geol. Survey Paper 66-30, p. 1-65.

Moreau-Benoît, A., 1966, Étude des spores du Dévonien inférieur d'Avrillé (Le Fléchay), Anjou: Rev. Micropaléontol., v. 8, no. 4, p. 215-232.

Naumova, S. N., 1953, Spore-pollen complexes of the Upper Devonian of the Russian platform and their stratigraphic significance: Trudy Inst. Geol. Nauk., Akad. Nauk SSSR, v. 143, no. 60, p. 1-204.

Neves, R., 1964, *Knoxisporites* (Potonié and Kremp) Neves 1961: Congr. Int. Stratigraph. Géol. Carbonifère, 5th, Comptes rendus, 3, Paris, 1963, p. 1063-1070.

Nikitin, P., 1934, Fossil plants of the Petino horizon of the Devonian of the Voronezh region. 1. *Krystofovichia africani* n. gen. et. sp.: Bull. Acad. Sc. URSS, Cl. Sci. Math. Nat., 7, p. 1079-1092.

Obrhel, J., 1961, Die Flora der Srbsko-Schichten (Givet) des mittelböhmischen Devons: Sb. ústred. Ust. geol., Svazek, 26-1959, p. 1-40.

______1962, Die flora der Přídolí-Schichten (Budňany-Stufe) des mittelböhmischen Silurs: Geologie Jahrb. 11, Heft 1, p. 83-97.

Owens, B., (in press) Miospores from the Middle and early Upper Devonian rocks of the western Queen Elizabeth islands, Arctic Archipelago: Geol. Survey Canada Bull.

Owens, B., and Streel, M., 1967 *Hymenozonotriletes lepidophytus* Kedo, its distribution and significance in relation to the Devonian-Carboniferous boundary: Rev. Palaeobotan. Palynol., v. 1, p. 141-150.

Ozolinya, V. P., 1963, Spore-pollen spectrum of the Frasnian stage, Upper Devonian of Latvian SSR: Trans. Geol. Inst. Latvian SSR, v. 10, p. 299-310.

Pashkevitch, N. G., 1964, New Devonian species of *Archaeoperisaccus* (Gymnospermae) from North Timan: Akad. Nauk SSSR, Paleont. Zhur., no. 4, p. 126-129.

Pettit, J. M., 1965, Two heterosporous plants from the Upper Devonian of North America: British Mus. Nat. History Bull., v. 10, no. 3, p. 81-92.

Playford, G., 1963, Miospores from the Mississippian Horton Group, Eastern Canada: Geol. Survey, Canada Bull. 107, p. 1-47.

Potonié, R., 1956, Synopsis der Gattungen der Sporae dispersae. Teil 1. Sporites: Beih. Geol. Jahrb., v. 23, 103 p.

______1958, Teil 2. Sporites (Nachträge), Saccites, Alctes, Praecolpates, Polyplicates, Monocolpates: ibid., v. 31, 114 p.

Potonié, R., and Kremp, G., 1954, Die Gattungen der paläozischen Sporae dispersae und ihre Stratigraphie: Geol. Jahrb., v. 69, p. 111-194.

Rayner, D., 1963, The Achanarras Limestone of the Middle Old Red Sandstone, Caithness, Scotland: Yorkshire Geol. Soc. Proc., v. 34, pt. 2, no. 8, p. 117-138.

Richardson, J. B., 1960, Spores from the Middle Old Red Sandstone of Cromarty, Scotland: Palaeontology, v. 3, pt. 1, p. 45-63.

______1962, Spores with bifurcate processes from the Middle Old Red Sandstone of Scotland: ibid., v. 5, pt. 2, p. 171-194.

______1965a, Middle Old Red Sandstone spore assemblages from the Orcadian basin, northeast Scotland: ibid., v. 7, pt. 4, p. 559-605.

______1965b, Stratigraphical distribution of some Devonian-Lower Carboniferous spores. Provisional report: Subgroup 13B, CIMP, Sheffield, England, 3 p.

______1967a, A reconnaissance of some Upper Devonian and Lower Carboniferous spores from New York State and Pennsylvania [abs.]: Rev. Palaeobot. Palynol., v. 1, p. 63-64.

______1967b, Some British Lower Devonian spore assemblages and their stratigraphic significance: Rev. Palaeobot. Palynol., v. 1, p. 111-129.

Richardson, J. B., and Lister, R., 1969, Upper Silurian and Lower Devonian spore assemblages from the Welsh Borderland and South Wales. Palaeontology, vol. 12, pt. 2, p. 201-252.

Rickard, L. V., 1964, Correlation of the Devonian rocks in New York State: New York State Mus. and Sci. Service Geol. Survey, Map and Chart Ser. 1, no. 4.

Schopf, J. M., 1938, Spores from the Herrin (no. 6) coal bed in Illinois: Illinois Geol. Survey Rept. Inv. 50, 73 p.

Schopf, J. M., Wilson, L. R., and Bentall, R., 1944, An annotated synopsis of Paleozoic fossil spores and the definition of generic groups: Illinois Geol. Survey Rept. Inv. 91, 73 p.

Scott, R. A., and Doher, L. I., 1967, Palynological evidence for Devonian age of the Nation River Formation, East-Central Alaska: U. S. Geol. Survey Prof. Paper 575-B, p. B45-B49.

Streel, M., 1964, Une association de spores du Givétian inférieur de la Vesdre, à Goé (Belgique): Ann. Soc. Géol. Belg., v. 87 (7), p. 1-30.

——1966, Critères palynologiques pour une stratigraphic détaillée du *Tnla* dans les bassins Ardenno-Rhénans: Ann. Soc. Géol. Belg., v. 89 (3), p. 65-96.

——1967, Associations de spores du Dévonien inférieur Belge et leur signification stratigraphique: Ann. Soc. Géol. Belg., v. 90 (1), p. 11-54.

Sullivan, H. J., 1964, Miospores from the Drybrook Sandstone and associated measures in the Forest of Dean basin, Gloucestershire: Palaeontology, v. 7, no. 3, p. 351-392.

Tuzova, L. S., 1959, Stratigraphic significance of spores and pollen of the Devonian of Eastern Tartary: Izv. Kazan. Fil. Akad. Nauk SSSR, Ser. Geol. Nauk, v. 7, p. 97-154.

Venozhinskene, A. I., 1964, Spore assemblages from thc Stonishk'ay, Sh'ashives, and Viesite formations: Vopr. Stratigrafii Paleogeografii Devona Pribaltiki, Vilnius, Inst. Geol. Gosgeolkom. SSSR.

Vigran, J., 1964, Spores from Devonian Deposits, Mimerdalen, Spitsbergen: Norsk. Polarinst., Skr. 132, p. 1-32.

Wall, D., 1962, Evidence from Recent plankton regarding the biological affinities of *Tasmanites* Newton 1875 and *Leiosphaeridia* Eisenack 1958: Geol. Mag., v. 99, p. 353-362.

White, E. J., 1956, Preliminary note on the range of Pteraspids in Western Europe: Bull. Inst. Sci. Nat. Belg., v. 32, no. 10, p. 1-10.

Winslow, M. R., 1962, Plant spores and other microfossils from Upper Devonian and Lower Mississippian rocks of Ohio: U.S. Geol. Survey Prof. Paper 364, p. 1-90.

Wray, J. L., 1964, Paleozoic palynomorphs from Libya, *in* Palynology in Oil Exploration, Cross, A. T. ed.: Soc. Economic Paleontologists and Mineralogists Spec. Pub. 11, p. 90-96.

12

MISSISSIPPIAN AND PENNSYLVANIAN PALYNOLOGY

Robert M. Kosanke

INTRODUCTION TO SMALL SPORES

Many of the early investigations in palynology were concerned with spores and pollen from rocks of Pennsylvanian (Late Carboniferous) age. Because of this and the diversity of the floras, many genera have been described and systems of classification involving higher categories have been proposed. More recently, much activity has been concerned with palynological studies of Mississippian (Lower Carboniferous) rocks. More than 160 generic names are mentioned in this chapter, and a number of these are discussed in detail. The limited space available for this chapter does not permit unrestricted discussion of all of these genera, because such a task could constitute a topic for a book. With this limitation in mind certain genera were selected that are considered representative of the various morphologic types. These genera are discussed in detail under two main categories of small spores and megaspores. Classification, geologic and geographic distribution of genera and species, guide fossils, and morphology are the major topics considered with emphasis on morphologic and systematic aspects of palynology.

Most spores and pollen grains from Mississippian and Pennsylvanian rocks are believed to have been derived from vascular plants. These spores may be homospores, which are essentially the same size for a given species; male microspores and female megaspores of heterosporous plants; or the male spores (microspores, prepollen, or pollen) and female gametophytes of primitive seed plants. The homospores, microspores, prepollen, or pollen have been called small spores, denoting a size generally less than 200 microns. Guennel (1952) proposed the term "miospores" stating: "All fossil spores and spore-like bodies smaller than 0.20 mm, including homospores, true microspores, small megaspores, pollen grains, and prepollen arbitrarily are called miospores." Both the small spores, or miospores, and megaspores are considered in this chapter.

Most Paleozoic spores can be conveniently divided into two groups on the basis of symmetry: (a) bilateral and monolete and (b) radial and trilete. A third division, alete, would include spores and pollen grains that lack an aperture. Spores or pollen grains that lack an aperture are relatively few. This is fortunate, because the aperture is one of the main characters used in classification. Certain pollen grains—for example, the bisaccate pollen of the Pinaceae—usually lack a trilete aperture but are considered to possess radial symmetry. Within the genus *Abies*, a member of the Pinaceae, vestigial trilete apertures have been reported by Wodehouse (1935), who said: "The triradiate streak is a reliable diagnostic character when found, but it is generally very faint and difficult to see and often entirely absent. In all probability the triradiate

Publication authorized by the Director of the U.S. Geological Survey.

streak is homologous with the triradiate crest of the fern spores of which it represents a vanishing remnant." The presence of a vestigial trilete aperture in members of the Pinaceae is strong evidence that *Pityosporites* and other Paleozoic bisaccate genera have radial symmetry.

Bilateral, monolete spores assigned to the genus *Laevigatosporites* (pl. 12-1, figs. 1-4) are a conspicuous part of Pennsylvanian assemblages throughout the world. In proximal view these spores are elongate-oval and, in some species, roundly oval; in equatorial view these spores are reniform, or bean-shaped, and, in some species, broadly reniform.

Radial and trilete spores, the most abundant spore types, occur throughout the Mississippian and Pennsylvanian Periods. There are a number of basic morphologic types of radial and trilete spores. The simplest form is that expressed by a circular outline as viewed either proximally or equatorially. Such spores are spheroidal and bear a trilete aperture. Because of their shape, they were dispersed at random without preference to proximo-distal orientation and subsequently were flattened to varying degrees when incorporated in a rock matrix. As a result the spores show little preference for proximal or equatorial orientation, and therefore the trilete apertures may occur in almost any position when the spores were compressed. *Calamospora* (pl. 12-1, fig. 5) is a good example of this type of spore, although in many of these spores the thin spore coat was variously folded, obscuring the circular outline. Species of *Punctatisporites* (pl. 12-1, figs. 7 and 8), which are circular in outline, are another example of a random orientation of the aperture in compressed specimens.

Radial, trilete spores without equatorial structures, which are more or less triangular in overall proximo-distal view, tended to be preserved or flattened predominantly in good proximo-distal orientation. An example is *Granulatisporites* (pl. 12-2, fig. 13). This generalization is subject to limitations inasmuch as many small species of the genus can be found flattened in equatorial or oblique equatorial orientation.

Radial and trilete spores possessing continuous equatorial structures are commonly present in Mississippian and Pennsylvanian strata. In spores that are either circular or roundly triangular in proximo-distal view the equatorial structure may be a simple ridge, as in some species of *Lycospora* (pl. 12-1, fig. 17); a fimbriate flange, as in *Reinschospora* (pl. 12-2, fig. 6), or a distinct thickening, or cingulum, as in *Densosporites* (pl. 12-1, figs. 20-22). These spores and others possessing strong equatorial structures tended to be compressed and flattened in reasonably good proximo-distal orientation.

Radial and trilete spores which are roundly triangular to triangular in proximo-distal view and which possess a continuous equatorial thickening, such as *Murospora*, tended to be flattened in good proximo-distal orientation. Spores of similar shape that do not have continuous equatorial structures but have thickenings in the arcuate areas opposite the rays, such as *Triquitrites* (pl. 12-2, fig. 10) and *Tripartites* (pl. 12-3, fig. 5), were likewise compressed and flattened in good proximo-distal orientation. Spores with expanded radial extremities, such as *Waltzispora* (pl. 12-3, fig. 3), may tend to be flattened in good proximo-distal orientation, although folding of these extremities may have occurred.

Monosaccate spores, such as *Endosporites* (pl. 12-4, fig. 1), which have radial symmetry and are circular to roundly triangular in proximo-distal view, were flattened in good proximo-distal ori-

---

Plate 12-1. Mississippian-Pennsylvanian spore genera. 1, 2, 3, 4, and 6—Bilateral and monolete spores. 1 to 4, *Laevigatosporites* (Ibrahim) Schopf, Wilson, and Bentall 1944; spores of the type shown in figure 4 have been placed in *Thymospora* by Wilson and Venkatachala (1963); figure 6 represents *Renisporites* Winslow 1959 (maximum diameter of these spores is 58, 24, 25, 35, and 148 microns, respectively). 5—*Calamospora* Schopf, Wilson, and Bentall 1944; maximum diameter of 48 microns. 7 and 8—*Punctatisporites* Schopf, Wilson, and Bentall 1944; maximum diameters of 37 and 40 microns. 9—*Raistrickia* Schopf, Wilson, and Bentall 1944; maximum diameter of 66 microns. 10—*Convolutispora* Hoffmeister, Staplin, and Malloy 1955; maximum diameter of 43 microns. 11—*Schopfites* Kosanke 1950; maximum diameter of 106 microns. 12—*Dictyotriletes* (Naumova) Potonié and Kremp 1955; maximum diameter of 40 microns. 13—*Reticulatisporites* (Ibrahim) Schopf, Wilson, and Bentall 1944; maximum diameter of 78 microns. 14 to 17—*Lycospora* Schopf, Wilson, and Bentall 1944. The maximum diameter of these spores is 30, 30, 30, and 18, microns respectively. 18—*Cirratriradites* Wilson and Coe 1940; maximum diameter of 96 microns. 19—*Radiizonates* Staplin and Jansonius 1964; maximum diameter of 44 microns. 20 to 22—*Densosporites* (Berry) Butterworth, Jansonius, Smith, and Staplin 1964. Diameter of these spores is 28, 38, and 48 microns.

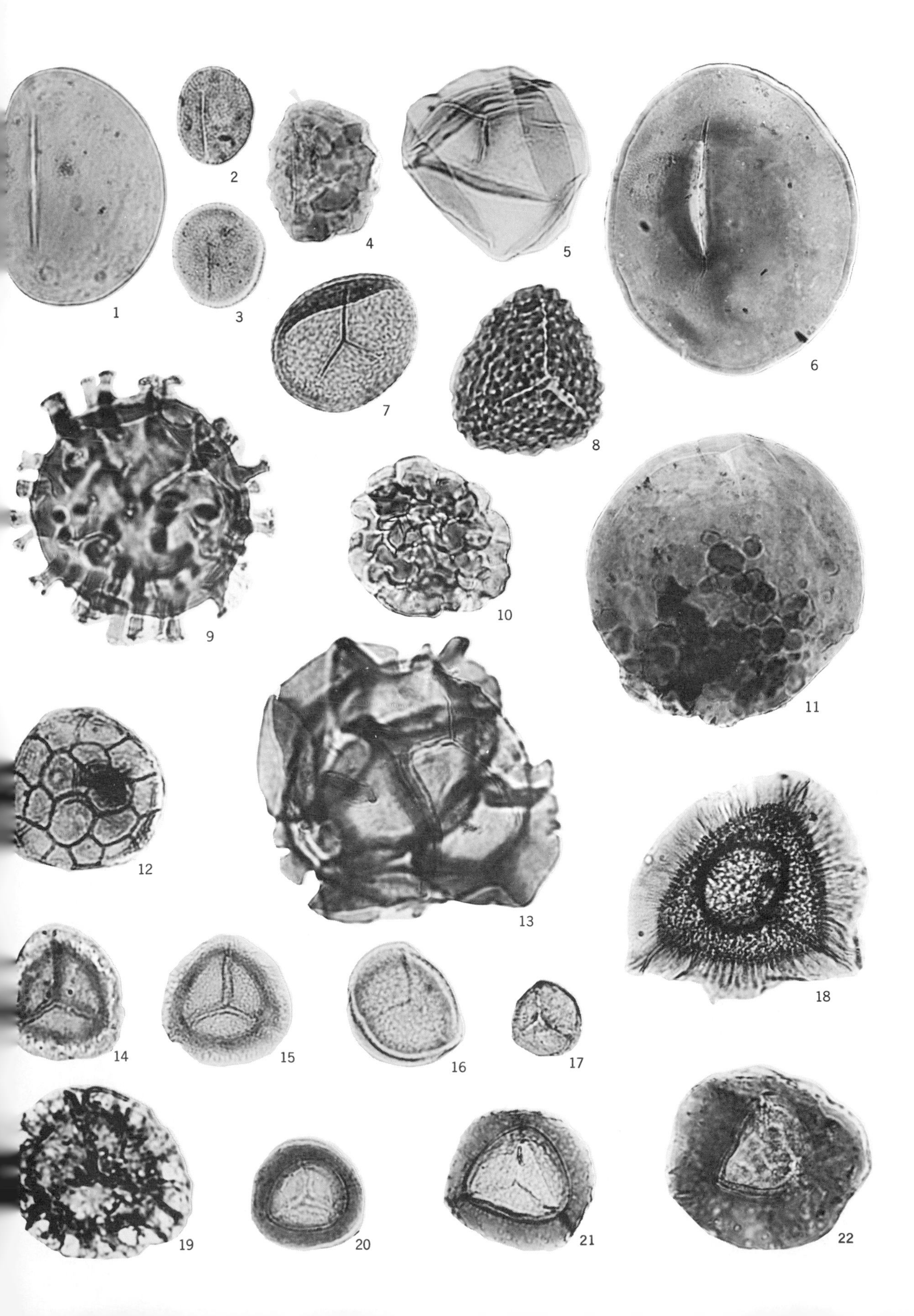

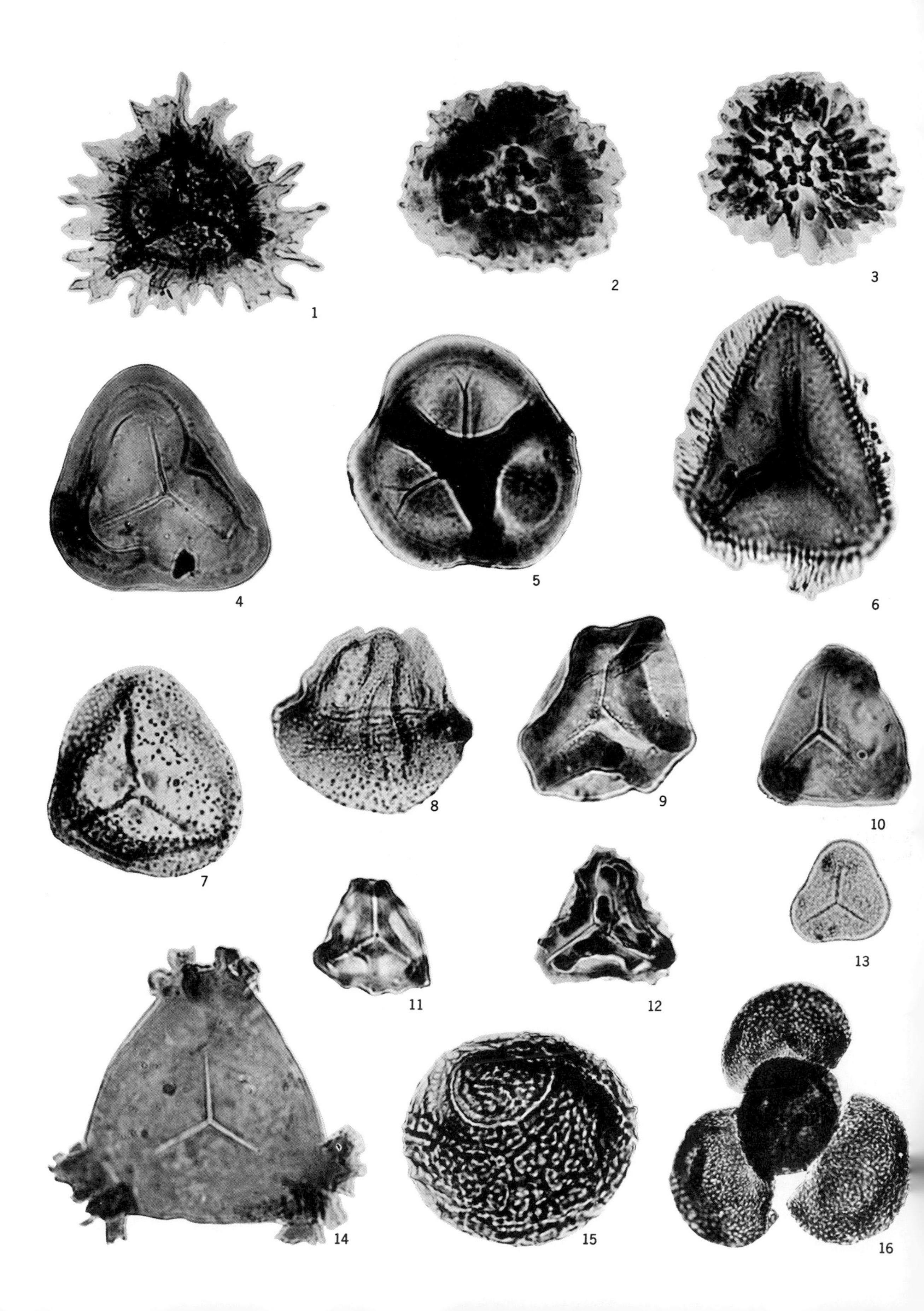

1
2
3
4
5
6
7
8
9
10
11
12
13
14
15
16

entation although the bladders may be variously folded. Virtually the same may be stated for *Alatisporites* (pl. 12-4, figs. 10 and 11), which is multisaccate. Bisaccate taxa such as *Pityosporites* and *Illinites*, which are of radial symmetry although usually elongate-elliptical in proximo-distal view, were usually flattened in either proximo-distal or oblique proximo-distal orientation. Equatorial compression and subsequent flattening does occur, but not commonly.

An ideal system of classification is one in which only morphologic features are required to classify fossil spores and pollen. Thus far basic morphologic features of selected spore types have been considered, and their ornamentation has been ignored. However, certain genera are recognized solely because they possess unusual or characteristic ornamentation. A few such genera are *Raistrickia* (pl. 12-1, fig. 9), *Schopfites* (pl. 12-1, fig. 11), *Dictyotriletes* (pl. 12-1, fig. 12), and *Reticulatisporites* (pl. 12-1, fig. 13). *Raistrickia* and *Reticulatisporites* are basically circular whether viewed in proximo-distal or equatorial view. The former is characterized by distinct spines, whereas the latter possesses a reticulate or netlike ornamentation. *Schopfites* and some species of *Dictyotriletes* possess dimorphic ornamentation; that is, different types of ornamentation on their proximal and distal surfaces. The proximal surfaces of both genera are nearly smooth or only finely ornamented. The distal surface of *Schopfites* is rough and composed of irregular projections; the distal surface of *Dictyotriletes* is reticulate.

Classification

The earliest attempts to classify spores and pollen in the United States were made by Thiessen and Staud (1923) by the simple designation of "Pittsburgh spore" or "Pittsburgh megaspore." In Britain Slater (1932) and Slater, Evans, and Eddy (1930) classified megaspores by means of symbols. This method was adopted by Raistrick and Simpson (1933) in Britain. The use of binomial nomenclature was adopted in Germany by Ibrahim (1932, 1933) and by Loose (1932, 1934).

Ibrahim's (1933) system for classifying Upper Carboniferous spores and pollen from some Ruhr coals of Germany included both small spores and megaspores. A type species was carefully designated for each genus, although holotypes for the various described species were not designated. Three main categories of spores were Triletes, Aletes, and Monoletes. In all, 18 genera were proposed and 79 species were described. The system of classification proposed by Ibrahim is shown in table 12-1.

All taxa with trilete apertures were classified under Triletes, spores without an aperture under Aletes, and those with a monolete aperture under Monoletes. The prefix of all spores with a trilete aperture ended in the letter "i," and this was separated from sporites by a hyphen. The hyphen following the prefix has been dropped in subsequent usage. Similarly the prefix of all alete spores ended in the letter "a," and all monolete spores in the letter "o." Under this system three genera could be separated only by the last letter of the prefix; for example, *Punctatisporites*, *Punctatasporites*, and *Punctatosporites*. In article 75 of the "1961 International Code of Botanical Nomenclature" such similar names are regarded as variants as follows: "When two or more generic names are so similar that they are likely to be confused, because they are applied to related taxa or for any other reason, they are to

Plate 12-2. Mississippian-Pennsylvanian spore genera. 1, 2, and 3—*Cristatisporites* (Potonié and Kremp) Butterworth, Jansonius, Smith, and Staplin 1964. The maximum diameter of these spores is 63, 57, and 48 microns. 4—*Simozonotriletes* (Naumova) Potonié and Kremp 1954. The maximum diameter is 59 microns. Staplin (1960) and Playford (1962) consider *Simozonotriletes* conspecific with *Murospora* Somers 1952. 5—*Knoxisporites* (Potonié and Kremp) Neves 1961; maximum diameter, 55 microns. 6—*Reinschospora* Schopf, Wilson, and Bentall 1944; maximum diameter of 73 microns. 7 and 8—*Crassispora* (Potonié and Kremp) Bhardwaj 1957; maximum diameters of 48 microns each; figure 7 represents a proximo-distal view and figure 8 an equatorial view. 9—*Ahrensisporites* (Potonié and Kremp) Horst 1955; maximum diameter of 48 microns. 10, 11, and 14—*Triquitrites* Wilson and Coe 1940. These spores have a maximum diameter of 37, 29, and 75 microns, respectively. Neves (1958) classified spores of the type shown in figure 14 under *Mooreisporites*. 12-*Savitrisporites* Bhardwaj 1955; maximum diameter, 33 microns. 13—*Granulatisporites* (Ibrahim) Schopf, Wilson, and Bentall 1944; maximum diameter, 24 microns. 15—*Vestispora* (Wilson and Hoffmeister) Wilson and Venkatachala 1963a; maximum diameter, 84 microns. 16—A tetrad of *Foveolatisporites* Bhardwaj 1955; maximum diameter, 135 microns. This genus is considered to be conspecific with *Vestispora* by Wilson and Venkatachala (1963a).

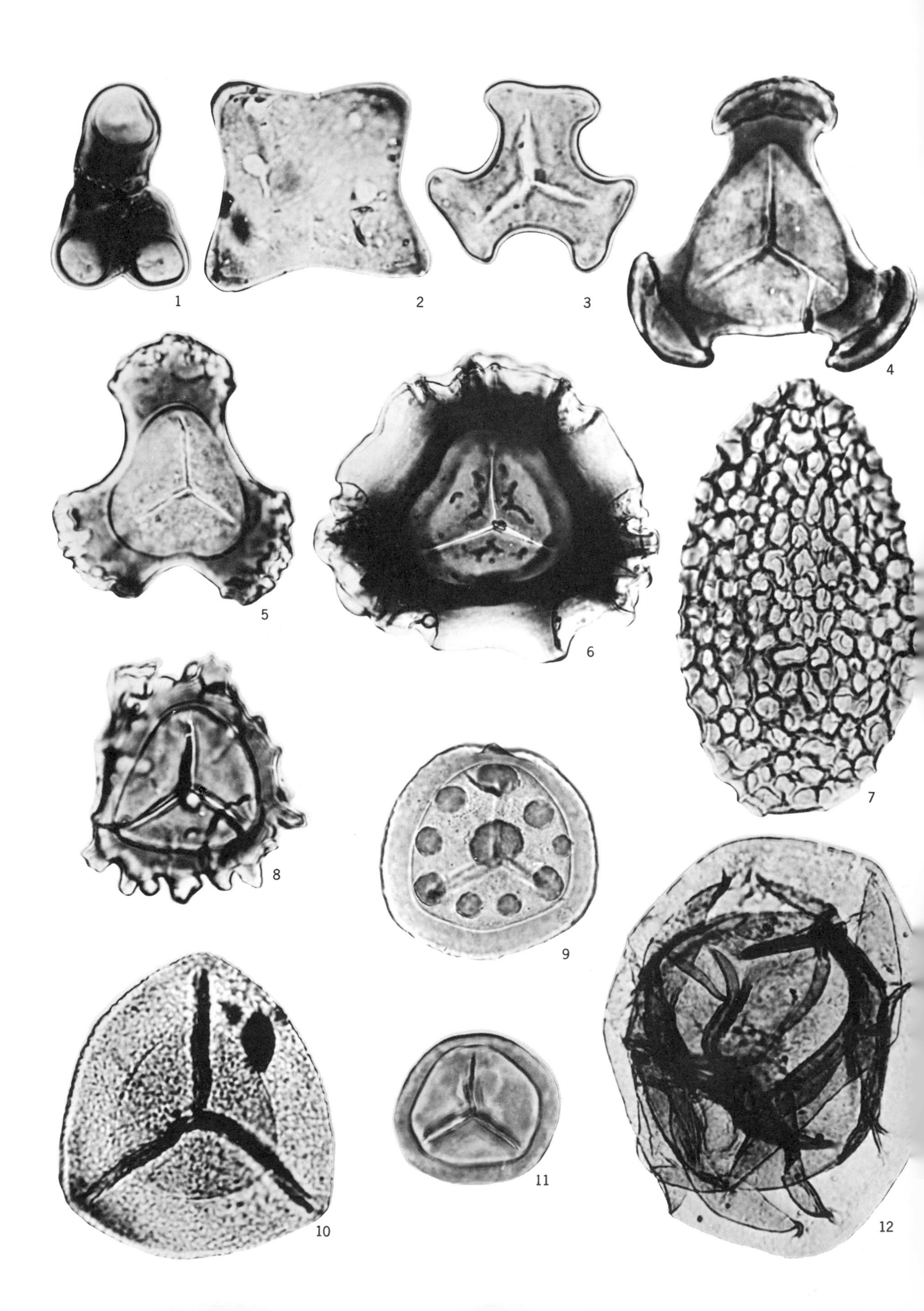
1
2
3
4
5
6
7
8
9
10
11
12

Table 12-1. System of Classification proposed by Ibrahim
Sporites H. Potonié 1893

Triletes Reinsch 1881	Aletes Ibrahim 1933	Monoletes Ibrahim 1933
Laevigatisporites Bennie and Kidston 1886		*Laevigatosporites* Ibrahim 1933
Punctatisporites Ibrahim 1933	*Punctatasporites* Ibrahim 1933	*Punctatosporites* Ibrahim 1933
Granulatisporites Ibrahim 1933		
Tuberculatisporites Ibrahim 1933		
Apiculatisporites Bennie and Kidston 1886	*Apiculatasporites* Ibrahim 1933	
Verrucosisporites Ibrahim 1933		
Setosisporites Ibrahim 1933		
Zonalesporites Bennie and Kidston 1886	*Zonalasporites* Ibrahim 1933	*Zonalosporites* Ibrahim 1933
Alatisporites Ibrahim 1933		
Valvisisporites Ibrahim 1933		
Reticulatisporites Ibrahim 1933	*Reticulatasporites* Ibrahim 1933	

be treated as variants, which are homonyms when they are based on different types."

Schopf, Wilson, and Bentall (1944) accepted *Punctatisporites, Granulatisporites, Alatisporites, Reticulatisporites, Laevigatosporites,* and *Zonalosporites* of Ibrahim (1933) and reinstituted *Triletes* and *Monoletes* as genera. In addition the following Mississippian and Pennsylvanian genera were accepted, revised, or proposed:

Calamospora Schopf, Wilson, and Bentall 1944
Cirratriradites Wilson and Coe 1940
Densosporites (Berry) Schopf, Wilson, and Bentall 1944
Endosporites Wilson and Coe 1940
Florinites Schopf, Wilson, and Bentall 1944
Lycospora Schopf, Wilson, and Bentall 1944
Parasporites Schopf 1938
Pityosporites Seward 1914
Raistrickia Schopf, Wilson, and Bentall 1944
Reinschospora Schopf, Wilson, and Bentall 1944
Tasmanites Newton 1875
Triquitrites Wilson and Coe 1940

Spores with known affinities were placed in appropriate genera, such as *Calamospora, Florinites, Lycospora,* and *Raistrickia.* Recognition was given to work of Soviet colleagues, but, because most of this palynological work was not available to the authors in 1943 and 1944, full consideration of it had to be postponed.

Kosanke (1950) used the taxa annotated by Schopf, Wilson, and Bentall (1944) and proposed five new genera: *Cadiospora, Illinites, Schopfites, Schulzospora,* and *Wilsonia.* The name *Wilsonia* subsequently was discovered to be illegitimate and was changed to *Wilsonites* by Kosanke (1950). Knox (1950) proposed four new genera: *Microreticulatisporites, Planisporites, Spinososporites,* and *Verrucososporites.* She further adopted most of the taxa discussed by Schopf, Wilson, and Bentall.

Potonié and Kremp (1954, 1955-1956) proposed a classification for Paleozoic spores and

Plate 12-3. Mississippian spore genera. (All illustrations are from Playford (1962 and 1963a through the courtesy of G. Playford, with the permission of "Palaeontology.") 1—*Chaetosphaerites pollenisimilis* (Horst) Butterworth and Williams 1958; about 45 microns; from Playford, 1962, plate 78, figure 2. 2—*Tetraporina incrassata* Naumova 1950; about 46 microns; from Playford 1963a, plate 95, figure 13. 3—*Waltzispora albertensis* Staplin 1960; about 40 microns; from Playford, 1962, plate 79, figure 10. 4—*Triquitrites trivalvis* (Waltz) Potonié and Kremp 1956; about 66 microns; from Playford, 1962, plate 85, figure 13. 5—*Tripartites incisotrilobus* (Naumova) Potonié and Kremp 1956; about 63 microns; from Playford, 1962, plate 85, figure 16. 6—*Monilospora triungensis* Playford 1963; about 107 microns; from Playford, 1963a, plate 92, figure 2. 7—*Retialetes radforthii* Staplin 1960; about 150 microns; from Playford, 1963a, plate 95, figure 3. 8—*Lophozonotriletes dentatus* Hughes and Playford 1961; about 53 microns; from Playford, 1963a, plate 91, figure 5. 9—*L. rarituberculatus* (Luber) Kedo 1957; about 75 microns; from Playford, 1963a, plate 91, figure 9. 10—*Endosporites micromanifestus* Hacquebard 1957; about 66 microns; from Playford, 1963a, plate 93, figure 17. 11—*Stenozonotriletes clarus* Ishchenko 1958; about 55 microns; from Playford, 1962, plate 86, figure 5. 12—*Remysporites albertensis* Staplin 1960; about 186 microns; from Playford, 1963a, plate 94, figure 3.

pollen in which recognition was accorded Soviet colleagues with validation of a number of their taxa. They also created a more elaborate system of categories above the rank of genus than was proposed by Ibrahim (1933).

The major divisions above the rank of genus are the following:

Oberabteilung (Superdivision) Sporonites (R. Potonié) Ibrahim 1933 (for fungi)

Oberabteilung (Superdivision) Sporites H. Potonié 1893 (for spores)

- Abteilung (Division) Triletes (Reinsch) Potonié and Kremp 1954 (for trilete spores with a number of subdivisions and a series of trilete spores)
- Abteilung (Division) Zonales (Bennie and Kidston) Potonié and Kremp 1954 (for trilete spores with equatorial structures)
- Abteilung (Division) Monoletes Ibrahim 1933 (for monolete spores)

Oberabteilung (Superdivision) Pollenites R. Potonié 1931 (for pollen grains)

- Abteilung (Division) Saccites Erdtman 1947 (for saccate pollen)
- Abteilung (Division) Napites Erdtman 1947 (for inaperturate pollen)
- Abteilung (Division) Praecolpates Potonié and Kremp 1954 (for *Schopfipollenites*)
- Abteilung (Division) Monocolpates Iverson and Troels-Smith 1950 (for monocolpate pollen)

All these divisions may be further subdivided into Subdivisions and Series, as indicated for Abteilung Triletes. The purpose of ranks above the genus level is to group together natural associations of genera into families, families into orders, etc. Because this is not accomplished by the categories above the generic level proposed by Potonié and Kremp, in my opinion such categories should not be used.

The classification of Potonié and Kremp accepted a number of taxa of Schopf, Wilson, and Bentall (1944) and all of the taxa proposed by Ibrahim (1933) except for *Apiculatasporites, Zonalasporites,* and *Zonalosporites*. Thus it reinstituted the "i," "a," and "o" single-letter differences between some genera. Potonié (1956) changed the names of his categories above the rank of the genus to Anteturma, Turma, Subturma, and Infraturma to correspond to Superdivision, Division, Subdivision, and Series.

Theoretically, according to the "International Code of Botanical Nomenclature," a single taxon may have but one valid name based on priority and other features of the code. Unfortunately at the present time some people use the system of classification annotated by Schopf, Wilson, and Bentall (1944), others use that of Potonié and Kremp (1954, 1955-1956), whereas Soviet workers for the most part retain their own classification. Thus a taxon may occur in the literature under more than one name. Regrettable as this may be, it is not necessarily a catastrophe. Good photomicrographs and descriptions ultimately should permit resolution of these problems of nomenclature.

Guide Fossils

Sufficient information has been published within the past 10 years to permit consideration of Mississippian spore and pollen taxa that are potentially useful for correlation purposes. Much of this information is available through the efforts of Hoffmeister, Staplin, and Malloy (1955), Luber (1955), Horst (1955), Ischenko (1956), Hacquebard (1957), Hacquebard and Barss (1957), Dybová and Jachowicz (1957), Butterworth and Williams (1958), Love (1960), Butterworth and Millott (1960), Staplin (1960), Neves (1961), Playford (1962; 1963a, b), and others.

Richardson (1964) and Butterworth (1964b) functioning as subgroup chairmen of Working Group 13 of the Commission Internationale de Microflore du Paléozoïque (C.I.M.P.) reported their findings relative to the stratigraphic distribution of genera. Richardson reported on spore occurrences in the Devonian through Namurian A; Butterworth reported on spore occurrences from Namurian A through Westphalian D. If we accept the Mississippian-Pennsylvanian boundary between Namurian A and B (fig. 12-1) and the Devonian-Mississippian boundary slightly below the Horton Group (Canada), certain statements may be made concerning genera that are restricted to Mississippian rocks. Such genera are *Waltzispora* Staplin 1960 (pl. 12-3, fig. 3), *Cincturasporites* Hacquebard and Barss 1957, *Rotaspora* Schemel 1950, *Tripartites* Schemel 1950 (pl. 12-3, fig. 5), *Remysporites* Butterworth and Williams 1958 (pl. 12-3, fig. 12), and *Chaetosphaerites* Felix 1894 (pl. 12-3, fig. 1). Genera that originate in Devonian rocks and appear to

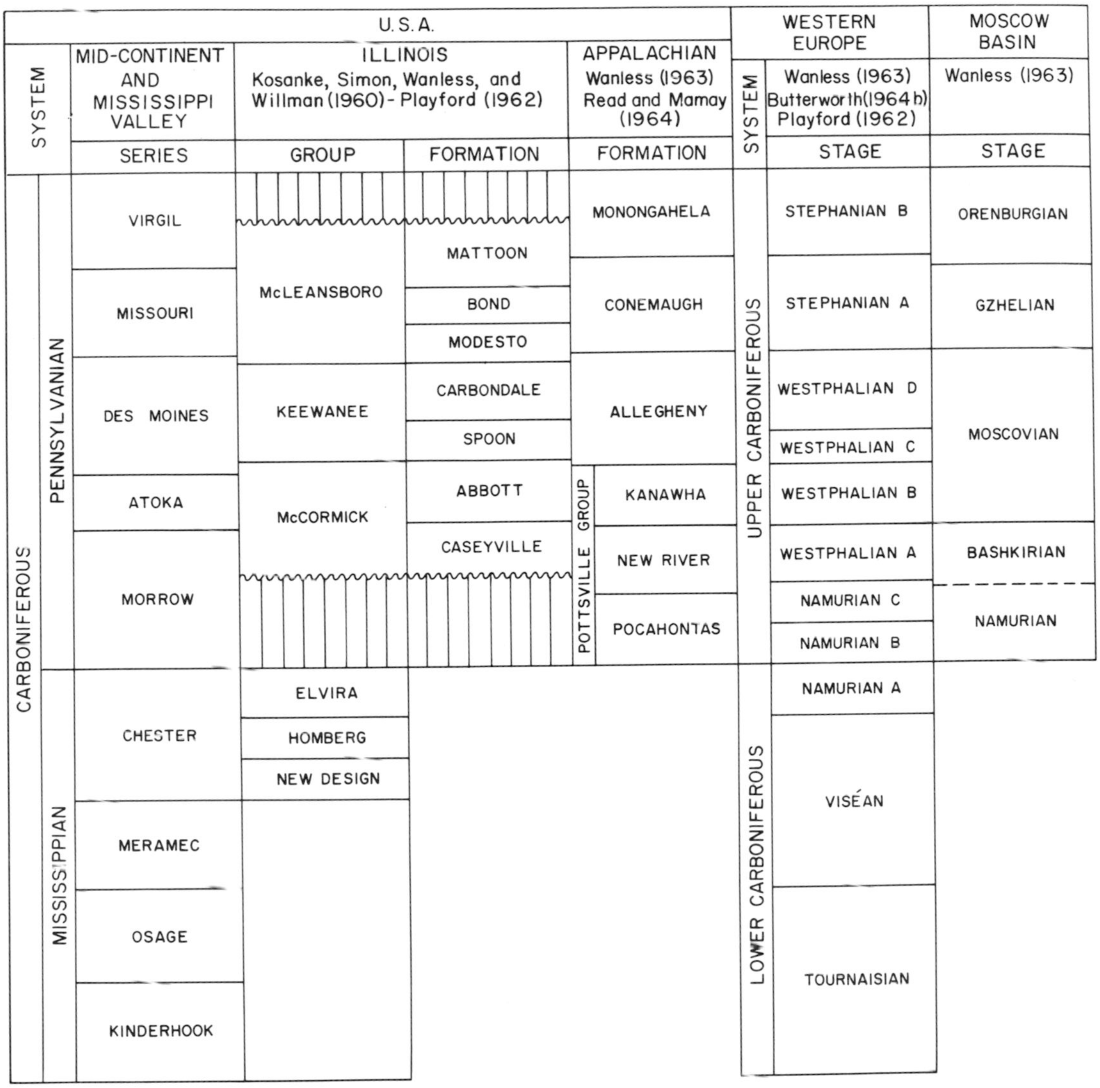

System		U.S.A. Mid-Continent and Mississippi Valley: Series	U.S.A. Illinois (Kosanke, Simon, Wanless, and Willman (1960) - Playford (1962)): Group	U.S.A. Illinois: Formation	U.S.A. Appalachian (Wanless (1963), Read and Mamay (1964)): Formation	Western Europe System	Western Europe (Wanless (1963), Butterworth (1964b), Playford (1962)): Stage	Moscow Basin (Wanless (1963)): Stage
CARBONIFEROUS	PENNSYLVANIAN	VIRGIL			MONONGAHELA	UPPER CARBONIFEROUS	STEPHANIAN B	ORENBURGIAN
		MISSOURI	McLEANSBORO	MATTOON	CONEMAUGH		STEPHANIAN A	GZHELIAN
				BOND				
				MODESTO				
		DES MOINES	KEEWANEE	CARBONDALE	ALLEGHENY		WESTPHALIAN D	MOSCOVIAN
				SPOON			WESTPHALIAN C	
		ATOKA	McCORMICK	ABBOTT	POTTSVILLE GROUP: KANAWHA		WESTPHALIAN B	
		MORROW		CASEYVILLE	NEW RIVER		WESTPHALIAN A	BASHKIRIAN
					POCAHONTAS		NAMURIAN C	NAMURIAN
							NAMURIAN B	
	MISSISSIPPIAN	CHESTER	ELVIRA			LOWER CARBONIFEROUS	NAMURIAN A	
			HOMBERG					
			NEW DESIGN					
		MERAMEC					VISÉAN	
		OSAGE					TOURNAISIAN	
		KINDERHOOK						

[1] Only Upper Carboniferous part of Namurian shown (Wanless, 1963, p. 35)

Figure 12.1. Selected stratigraphic divisions of the Carboniferous Systems of United States, Western Europe, and Russia.

terminate in the Mississippian are *Hystricosporites* McGregor 1960, *Spinozonotriletes* Hacquebard 1957, and *Diatomozonotriletes* (Naumova) Playford 1962. I regard the Mississippian species of *Diatomozonotriletes*, assigned to this genus by Playford (1963a), as belonging to *Reinschospora* Schopf, Wilson, and Bentall 1944. The basis for this decision is that the feature separating *Diatomozonotriletes* from *Reinschospora*, other than a stratigraphic hiatus, is the coarseness of the equatorial structure of most species assigned to *Diatomozonotriletes*. I believe that this is an excellent species character and, furthermore, that the equatorial structure of *D. trilinearis* Playford 1963 is similar to *Reinschospora triangularis* Kosanke 1950, a Pennsylvanian species, even to the splitting of the fimbrae. If we compare the reports of Richardson (1964) and Butterworth (1964b), it is possible to conclude, erroneously, that *Murospora* is restricted to Mississippian rocks. However, this is simply because the Mississippian occurrences were plotted by Richardson, whereas the Pennsylvanian distribution is not recorded by Butterworth.

Other genera that could indicate Mississippian age are *Costaspora* Staplin and Jansonius 1960 (in Staplin, 1960); *Endoculeospora* Staplin 1960; *Grandispora* Hoffmeister, Staplin, and Malloy 1955; *Labiadensites* Hacquebard and Barss 1957; *Leioaletes* Staplin 1960; *Lepidozonotriletes* Hacquebard 1957; *Monilospora* (Hacquebard and Barss) Staplin 1960 (pl. 12-3, fig. 6); *Perianthospora* Hacquebard and Barss 1957; *Pteroretis* Felix and Burbridge 1961; *Radialetes* Playford 1963; *Retialetes* Staplin 1960 (pl 12-3, fig. 7); *Retispora Staplin* 1960; *Tendosporites* Hacquebard and Barss 1957; *Tetraporina* (Naumova) Potonié 1960 (pl. 12-3, fig. 2), including *Azonotetraporina* Staplin 1960; *Tetrapterites* Sullivan and Hibbert 1964; *Tholisporites* Butterworth and Williams 1958; *Veliferaspora* Staplin 1960; *Velosporites* Hughes and Playford 1961; and *Zonaletes* Staplin 1960. In addition Butterworth (1964b) lists *Procoronaspora* Butterworth and Williams 1958 as ranging upward to the boundary of the Namurian A and B. The complete stratigraphic range of some of these genera has not been evaluated.

Butterworth (1964b) lists the following genera that occur in both Mississippian and Pennsylvanian rocks: *Schulzospora* Kosanke 1950; *Laevigatosporites* Ibrahim 1933; *Florinites* Schopf, Wilson, and Bentall 1944; *Ahrensisporites* (Potonié and Kremp) Horst 1955; *Reinschospora* Schopf, Wilson, and Bentall 1944; *Knoxisporites* (Potonié and Kremp) Neves 1964; *Mooreisporites* Neves 1958; *Crassispora* Bhardwaj 1957; and *Vestispora* Wilson and Hoffmeister 1956. For *Laevigatosporites, Florinites, Ahrensisporites, Mooreisporites, Crassispora,* and *Vestispora* these Mississippian occurrences are recorded only from the United Kingdom to date. The occurrence of *Schulzospora* and *Knoxisporites* in Mississippian and Pennsylvanian rocks is well documented on a worldwide basis. From my experience *Schulzospora* occurs rather consistently in Illinois coals of the Chester Series and becomes more abundant toward the Mississippian-Pennsylvanian boundary, declines in abundance in Early Pennsylvanian time, and is absent from coals of the Abbott Formation (Kosanke, 1964). Cropp (1963) reported *Schulzospora* from five coals of the Lee Formation (Early Pennsylvanian) of Tennessee. *Knoxisporites* has a longer range and occurs in both older and younger rocks than *Schulzospora*.

A comparison of the report of Butterworth (1964b) with a similar report by Alpern (1964), which covers the distribution of spores and pollen for the Stephanian and Permian, indicates that the following genera are restricted to the Pennsylvanian: *Dictyotriletes* (Naumova) Potonié and Kremp 1954, *Novisporites* Bhardwaj 1957, *Cadiospora* Kosanke 1950, *Foveolatisporites* Bhardwaj 1955 (pl. 12-2, fig. 16), and perhaps *Speciososporites* Potonié and Kremp 1954. Alpern (1964) suggests the possibility that *Foveolatisporites* might occur in Lower Permian rocks. Wilson and Venkatachala (1963a) have placed *Novisporites* and *Foveolatisporites* in synonymy with *Vestispora* (Wilson and Hoffmeister) Wilson and Venkatachala 1963. Other genera that could be added are *Gravisporites* Bhardwaj 1954, *Parasporites* Schopf 1938, *Radiizonates* Staplin and Jansonius 1964, *Renisporites* Winslow 1959, *Striatosporites* Bhardwaj 1954, *Trilobates* Somers 1952, and *Torispora* Balme 1952.

Additional genera that may be of value, when they are better known, in delineating Pennsylvanian rocks are *Bellispores* Artuz 1957, *Columinisporites* Peppers 1964, *Discernisporites* Neves 1958, *Latipulvinites* Peppers 1964, *Ibrahimispores* Artuz 1957, *Sinuspores* Artuz 1957, *Trinidulus* Felix and Paden 1964, and *Trivolites* Peppers 1964.

Densosporites (Berry) Butterworth, Jansonius, Smith, and Staplin 1964 and *Alatisporites* Ibrahim 1933 are examples of long-ranging genera that apparently do not extend to the Permian. An excellent example of a genus apparently restricted geographically and stratigraphically is *Pteroretis* Felix and Burbridge 1961. This genus is reported from two samples, 16 feet apart, in a single core from a well in Texas. The age is reported as Late Mississippian. *Tetrapterites* Sullivan and Hibbert 1964, an unusual spore-bearing structure from the Viséan of Britain, may be somewhat restricted in geographic occurrence. Lastly, *Trinidulus* Felix and Paden 1964 might represent a genus restricted in both stratigraphic occurrence and geographic distribution. It is reported from Lower Pennsylvanian rocks of Oklahoma and Texas. Whether or not *Pteroretis, Tetrapterites,* and *Trinidulus* remain restricted either stratigraphically or geographically or both, will be determined only by subsequent studies.

The range zones of a number of genera are characterized by consistency of occurrence throughout the entire zone. Other genera have distinctive range zones in which occurrence is not consistent throughout the entire zone. In other words, there are gaps in occurrence. Such gaps are relatively minor; that is, a genus is present in one coal, absent from the next coal above, and present in a still younger coal. Such interruptions may be the result of ecological factors or lack of complete information. *Schopfites* and *Laevigatosporites* are known to have range zones with significant gaps in occurrence. *Schopfites* was originally reported from a series of Pennsylvanian coals in the Carbondale Formation of Illinois. Subsequently Butterworth (1964b) reported that this genus was restricted to the entire Westphalian D section in Britain. Barss, Hacquebard, and Howie (1963) reported that *Schopfites* occurred in the *Ptycharpus unitus* zone or Westphalian D of Canada. However, Playford (1963b) reported this genus from the Horton Group of Early Mississippian age of eastern Canada. The hiatus or gap between the occurrence of *Schopfites* in the Horton Group and its occurrence in Middle Pennsylvanian rocks of the United States, Canada, and Europe is very large. The reason for the absence of this genus between Early Mississippian and Middle Pennsylvanian time is not known. A possible explanation might be that morphologically similar spore types arose, evolved, and became exinct at different geologic times in different plants. This is convergent evolution; that is, the evolution of similar spore types by unrelated parent plants.

The range zone of *Laevigatosporites* is also interrupted by gaps in occurrence, but these gaps are not as large as that discussed for *Schopfites*. Spores similar to *Laevigatosporites* have been reported from the Devonian, Lower and Upper Mississippian rocks, and the genus does not occur consistently until Lower Pennsylvanian rocks are encountered.

One of the primary factors used by palynologists in correlation studies is the range zone of individual species. Generally range zones of individual species are appreciably shorter than those for genera except where monotypic genera are involved or where relatively few species are assigned to a genus. Some species, like some genera, occur consistently through a given portion of the geologic column, whereas others have interrupted ranges.

Morphology and Affinities of Bilateral and Monolete Spores

Spores with bilateral symmetry and a monolete aperture occur very rarely in Mississippian rocks, but they are a prominent aspect of the spore and pollen assemblages of most Pennsylvanian sequences. The aperture is usually relatively straight, although in one genus, *Monoletes* (Ibrahim) Schopf, Wilson, and Bentall 1944 = *Schopfipollenites* Potonié and Kremp 1954, an angular deflection or median deviation of the aperture was reported by Schopf (1938). In some specimens this deflection is accompanied by a short, vestigial? ray or "crack," suggesting the possibility of a trilete aperture. Basically *Monoletes* is monolete and is here regarded as such.

Bilateral spores of this type occurring in Mississippian rocks have been classified as *Laevigatosporites* sp. by Schemel (1950) and Winslow (1962). Spores with a monolete aperture occurring in Mississippian rocks are recorded by Richardson (1964). *Schopfipollenites* is recorded in rocks of Namurian A in England by Neves (1961).

Bilateral, monolete spores that are oval in outline in proximo-distal view and reniform in equatorial view (long axis of spore) have been classified as *Laevigatosporites* (Ibrahim) Schopf, Wilson, and Bentall 1944 (pl. 12-1, figs. 1-4). *Phaseolites* Wilson and Coe 1940 (which is a synonym of *Laevigatosporites*), *Latosporites* Potonié and Kremp 1954, *Verrucososporites* (Knox) Potonié and Kremp 1954, *Punctatosporites* Ibrahim 1933, *Speciososporites* Potonié and Kremp 1954, and *Thymospora* Wilson and Venkatachala 1963 are other names applied to bilateral, monolete spores.

In the sense of Schopf, Wilson, and Bentall (1944) *Latosporites*, *Verrucososporites*, *Punctatosporites*, *Thymospora*, and even perhaps *Speciososporites* would be incorporated within *Laevigatosporites*. The separation of all of these genera except *Latosporites* is based on ornamentation. In the case of *Latosporites*, Potonié and Kremp (1954) believed a distal enlargement was present, so that as viewed equatorially in the long axis of the spore the width would be nearly eight-tenths of the length. The genus was based on *Laevigatosporites latus* Kosanke 1950. In this species the width viewed in proximo-distal plane is at least eight-tenths of the length. Most specimens viewed in equatorial plane are less than eight-tenths of the length and are very broadly reniform.

Knox (1950) proposed *Verrucososporites* for trilete spores possessing a warty or tuberculate exine. Under the system of classification proposed by Ibrahim (1933) the "o" ending of the prefix should give the connotation of a monolete spore. Because of this Potonié and Kremp (1954) emended Knox's description of *Verrucososporites* so that the genus was monolete and designated *Laevigatosporites obscurus* Kosanke 1950 as the type of the emended genus *Verrucososporites*. Wilson and Venkatachala (1963b) proposed *Thymospora* (pl. 12-1, fig. 4) and selected *Laevigatosporites pseudothiessenii* Kosanke 1950 as the type.

Apparently palynologists are reluctant to consider ornamented, bilateral, monolete spores as members of the genus *Laevigatosporites*. Whether or not these ornamented species should merit generic consideration does not seem appropriate to debate here. To the palynologist the segregation of these ornamented species into *Thymospora* or *Verrucososporites* has the advantage of appearing as a significant generic floral break because they appear in Middle Pennsylvanian rocks for the first time.

The use of *Verucososporites* by Knox (1950) was ill-advised because *Verrucosisporites* Ibrahim 1933 basically covered spores of a similar circumscription, and Schopf, Wilson, and Bentall (1944) had already transferred the type of *Verrucosisporites* to *Punctatisporites*. The transfer of the type of *Verrucosisporites*, *V. verrucosus* Ibrahim, to *Verrucososporites* by Knox (1950) accomplished very little. The alteration of the circumscription of *Verrucososporites* by Potonié and Kremp (1954) merely added to the confusion. It is perhaps for this reason that there has been some acceptance of *Thymospora*. In any event the question of whether or not coarse ornamentation is a generic character still remains.

Tuberculatosporites Imgrund 1960, *Pericutosporites* Imgrund 1960 and *Pectosporites* Imgrund 1960, were originally described from China from reportedly Permian rocks and have been arbitrarily deleted from consideration here. *Speciososporites* appears to be related most closely to the coarser types of *Laevigatosporites*.

Renisporites Winslow 1959 (pl. 12-1, fig. 6), was proposed for large spores between 125 and 225 microns that are bilateral, monolete, broadly oval in proximo-distal view, and reniform in equatorial long axis view. The spore coat is comparatively thick, up to 10 microns around the ends of the aperture. The spore coat is golden-yellow to brown, and it is levigate with scattered puncta or pits which tend to be concentrated into two areas equatorially on either side of the aperture. *R. confossus* Winslow is the type species. *Renisporites* was reported to be restricted to the Willis coal of southern Illinois and to its correlative in western Illinois, the Tarter coal of the Abbott Formation of the McCormick Group. I have observed a limited number of specimens of *Renisporites* from three coals in a diamond drill core from eastern Kentucky. These coals occur in the Breathitt Formation.

Possibly *Renisporites* has not been widely recognized by most palynologists because of its size. For some unexplained reason, *Renisporites* does not readily pass through the conventional 210-micron mesh screen used to separate large spores from small spores. Thus, its occurrence is frequently overlooked unless the plus 210-micron size fraction is examined. The affinities of *Renisporites* are unknown.

Monoletes (Ibrahim) Schopf, Wilson, and Bentall 1944 = *Schopfipollenites* Potonié and Kremp 1954 has not been reported from rocks older than Pennsylvanian in the United States. *Monoletes* ranges in maximum diameter from about 100 to 500 microns, but most are larger than 200 microns, and therefore most specimens are not observed by palynologists who concentrate on small spores or miospore assemblages. In proximo-distal view *Monoletes* is elliptical to oval to roundly oval, whereas it is more elongate when viewed equatorially in the long axis. Two prominent distal grooves are commonly present with a rounded umbo occurring between the distal grooves. Depending upon magnification, the spore coat may appear to be smooth to granular. *Monoletes* is basically monolete and in many specimens this aperture may display a median deflection. The parent plants of *Monoletes* are undoubtedly related to the Medullosaceae as reported by Schopf (1948). Delevoryas (1964) described a microsporangiate organ containing spores referrable to *Monoletes*. Based on the microsporangiate organ and the type of spores it contains, Delevoryas suggests an affinity with the pteridosperms.

The affinities of *Laevigatosporites* (Ibrahim) sensu Schopf, Wilson, and Bentall, 1944, apparently lie with the Pteropsida and Sphenopsida. Schopf (1938, pl. 6, fig. 8) illustrated spores now assignable to *Laevigatosporites pseudothiessenii* that were identified as derived from ferns based on studies of sporangial masses occurring in the plus 210-micron maceration residue. Mamay (1950) illustrated and described monolete and bilateral fern spores occurring in two species of *Scolecopteris* and in one species of *Cyathotrachus*. Reed (1938) observed bilateral and monolete spores in some fragmental fructifications which she thought might be related to *Calamites*. On the basis of new and more complete specimens, Baxter (1950), and more recently Leisman and Graves (1964), have established that these fructifications of *Peltastrobus* (Baxter) Leisman and Graves are allied with the Sphenophyllales.

Balme (1952) described an unusual spore from the Upper Carboniferous of Britain under the name of *Torispora* with *T. securis* as the type species. According to Balme *Torispora* is bilateral in proximal view, distinctly monolete, with one extremity of the exine expanded or thickened into a crescentic or broadly rectangular projection. *T. securis* ranged in overall length from 26 to 44 microns, with a mean diameter of 34 microns. Basically this would be *Laevigatosporites* with an unusual thickening at one end of the spore.

Horst (1957) found similar *Torispora*-like material in a coal from Saxony and referred it to Wandzellen (cell walls) of a microsporangium and proposed the genus *Bicoloria* for such sporangia.

Alpern (1958) added additional species to *Torispora* and proposed *Crassosporites* with *C. triletoides* as the type species. This species is described as possessing a "pseudotrilete" aperture. Staplin (1961) questionably assigned a new species to *Torispora* (*T.? tiara*) in which the spores were reported to be "monolete-triletoid." Staplin's illustrations of this species show that figures 1 and 2 are monolete, figures 3*a*-3*c* are trilete, and probably the specimen shown in figure 4 is trilete. The pronounced thickening in many specimens of *Torispora* obscures the nature of the aperture. Nevertheless, the generic diagnosis requires the monolete condition, and spores that are trilete should be assigned to a different genus. This has been done by Staplin (1961) for *Cornispora*.

Guennel and Neavel (1961) described an unusual occurrence of *Torispora securis* from the Indiana paper coal which they regard as equivalent to one of the Upper Block coals of late Pottsville age. They restrict the name *Bicoloria* to intact sporangia and *Torispora* to isolated spores. This approach seems realistic and in accord with the rules of botanical nomenclature. Guennel and Neavel illustrated the important fact that spores assignable to *Torispora* occurred along the periphery of the sporangia and that normal uniformly thin-coated spores occur within the interior of the sporangia. Further, gradations occurred from a minor thickening to a distinct and pronounced hemispherical cup. This presents an interesting problem because all of the specimens of a single sporangium must represent a single species and yet the dispersed spores from such a sporangium could be classified under two genera: *Torispora* and *Laevigatosporites*. The cause for the development of the thickenings of *Torispora* is unknown. Guennel and Neavel postulate that the exine of *Torispora* thickened in a manner similar to cuticularization. "Fatty acids presumably migrated to the outer, exposed parts of the spore exine where they condensed and oxidized. The 'cup' or hemispherical thickening of *Torispora* thus represents the part of the exine that was exposed to oxidation on the periphery of the sporangium." Because the unusual thickenings of *Torispora* often make the relationship to the proper species of *Laevigatosporites* obscure, use of the name *Torispora* in the sense of Balme seems reasonable.

Doubinger and Horst (1961) reported the range of *Torispora* as Tremadoc through Stephanian B. *Crassosporites punctatus* was questionably equated to *Torispora securis*. If subsequently established, this would increase the range to the Autunian. When I described the small spore flora of Illinois coals some years ago a very limited number of specimens of *Laevigatosporites* with unusual thickening were considered to be maverick specimens of that genus. The first occurrence of such specimens was in the Tarter coal in the middle of the Abbott Formation of the McCormick Group. More recently I have observed abundant specimens of similar construction from eastern Kentucky. Today these would be assigned to *Torispora*. The earliest occurrence of *Torispora* in eastern Kentucky is in the Richardson coal at its type locality. The Richardson coal occurs in the Breathitt Formation, equivalent to rocks near the top of the Kanawha Formation of the Appalachian area of eastern United States. (See fig. 12-1.)

Morphology and Affinities of Radial and Trilete Spores, Prepollen, and Pollen

Most Mississippian and Pennsylvanian spore and pollen genera are radial and trilete. We must subdivide these genera in some orderly manner for a convenient means of discussion. Accordingly the subject is treated under the following categories: spores lacking equatorial structures, spores with continuous equatorial structures, spores with discontinuous equatorial structures, and the saccate spores, prepollen, or pollen that can be subdivided into the monosaccate, bisaccate, and multisaccate groupings.

Spores Without Equatorial Structures. *Calamospora* Schopf, Wilson, and Bentall 1944 was proposed for radial, trilete spores that are essentially round in outline both in proximo-distal and equatorial views, are compressed without preferred orientation, and are characteristically smooth. Species assigned to this genus range in size from less than 30 microns to more than several hundred microns because the generic circumscription permits classification of homospores, microspores, and megaspores within the same genus. The reason for this is simply that calamarian plants produced similar types of spores, except for size, in homosporous and heterosporous species. Schopf, Wilson, and Bentall were aided greatly in the circumscription of the genus by the early and important contribution of Hartung (1933). The type species is *Calamospora hartungiana* Schopf 1944 (in Schopf, Wilson, and Bentall, 1944).

As previously indicated, the spore coats of species of this genus are levigate or smooth, but critical examination with an oil-immersion objective of some species may reveal a finely granulose condition. The walls are usually thin, although their thickness is variable as a result of the large size range. Many species in the smaller size ranges have walls less than 2 microns thick. The trilete aperture is distinct and in many species the rays of the aperture are short, about one-fourth to one-half the spore radius; however, some species do have longer rays. The lips may be lacking to well developed, and the commissure may be thin or wide. The *area contagionis*, or contact area, may be totally lacking, faintly discernible, or strikingly developed, depending on the species.

The genus was established to contain calamarian spores based, in part, on Hartung's (1933) investigations of the spores of a number of calamite fructifications. Further, Nêmejc (1935, 1937) showed that the spores isolated from *Noeggerathiostrobus* and *Discinites* are similar. Schopf, Wilson, and Bentall (1944) considered *Calamospora* allied to the entities mentioned. Fructification studies by Hoskins and Cross (1943), Arnold (1944), Kosanke (1955), and others have made contributions concerning spores of this type.

The stratigraphic range should and does parallel the range of plant megafossils of the sphenopsid alliance inasmuch as representatives of *Calamospora* are reported from the Devonian throughout the Paleozoic and into younger strata. Some species have extremely long stratigraphic ranges, whereas other species appear to have short stratigraphic ranges that are useful in correlation studies.

Spores of the calamarian type are difficult to segregate into species because morphologically they are rather simple. If the size variation of these spores is as great as reported, the problem of speciation is further complicated. Scott (1920) comments on the possible development of heterospory in the calamarians, and Schopf, Wilson, and Bentall (1944) discuss the evidence for incipient heterospory in the calamite alliance. The available evidence is fascinating and suggests a real potential for size variation with which we are concerned. Should some of the spores of a certain species of calamarians abort, with the result that spores of a single species were of two distinct sizes, they could be classified as two different species of *Calamospora*. This is not a cause for serious concern from the aspect of applied palynology because the ranges of both species would be identical.

The type species of *Raistrickia* is *R. grovensis* Schopf 1944 (in Schopf, Wilson, and Bentall, 1944). This species is roundly triangular in proximo-distal view, and in this aspect it is clearly differentiated from *Calamospora*. However, a number of other species of *Raistrickia* (pl. 12-1, fig. 9) are related to *Calamospora* in outline and are separated on the basis of the distinct spine type of ornamentation. Similarly there are species assigned to *Punctatisporites* (Ibrahim) Schopf, Wilson, and Bentall (pl. 12-1, figs. 7 and 8), *Reticulatisporites* (Ibrahim) Schopf, Wilson, and Bentall (pl. 12-1, fig. 13), and *Convolutispora* Hoffmeister, Staplin, and Malloy (pl. 12-1, fig. 10) that are in the same category as *Raistrickia*, but are separated on the basis of punctate, reticulate, and convolute types of ornamentation. *Cadiospora* Kosanke is readily identified on the basis of lip development and the presence of distinct arcuate ridges. *Vestispora* (Wilson and Hoffmeister) Wilson and Venkatachala 1963 (pl. 12-2, fig. 15), is also readily distinguished from the preceding genera by the presence of a two-layered spore coat and an operculum. *Schopfites* Kosanke (pl. 12-1, fig. 11) qualifies for discussion with genera of a round outline in both proximo-distal and equatorial view, but it has a dimorphic type of ornamentation that separates it from the genera already discussed. The proximal surface is smooth to granular, whereas the distal surface is strongly ornamented by variously shaped projections. In many specimens these distal projections overlap onto the proximal surface, and as a result the equator of the spore may appear to have some sort of a structure. *Dictyotriletes* (Naumova) Potonié and Kremp 1955 (pl. 12-1, fig. 12) is discussed here because the type species has a dimorphic type of spore coat and in proximo-distal view is round in outline. In equatorial view, however, at least some of the species have a pyramidal proximal surface. The proximal surface of *Dictyotriletes bireticulatus* is levigate, whereas the distal surface is reticulate. As in *Schopfites*, the distal ornamentation overlaps a portion of the proximal surface.

Punctatisporites (Ibrahim) Schopf, Wilson, and Bentall 1944 (pl. 12-1, figs. 7 and 8), are radial, trilete spores that range in proximo-distal view from circular to roundly triangular, with little preference for proximo-distal orientation when compressed. These spores are levigate, granulose, punctate, rugose, verrucose, apiculate, or even somewhat reticulate. The trilete rays are of variable length depending on the species, and the lips are only modestly developed if at all. The type species is *P. punctatus* (Ibrahim) Ibrahim 1933. The size ranges from about 30 to 90 microns.

Spores of this nature were classified by Potonié and Kremp (1954) as follows:

Punctatisporites (Ibrahim) Potonié and Kremp 1954 is restricted to spores of essentially circular outline, but, contrary to the circumscription of Schopf, Wilson, and Bentall (1944), only species without sculpture (punctation), although they may have "infrasculpture" (within the wall), are assigned to *Punctatisporites*. Ibrahim's (1933) original description stated that the surface appears finely sandy and that in the type species the spore surface is distinctly punctate and the surface is rough. A photomicrograph of the type specimen in Potonié and Kremp (1955, fig. 122) indicates a very finely punctate condition, but the spore margin is essentially smooth.

Punctatisporites (Ibrahim) Schopf, Wilson, and Bentall is divided by Potonié and Kremp as follows:

Punctatisporites (Ibrahim) Potonié and Kremp 1954 for spores without sculpture, but with infrasculpture. Type species—*P. punctatus* Ibrahim 1933.

Verrucosisporites (Ibrahim) Potonié and Kremp 1954, differentiated from *Punctatisporites* by the presence of warts. Type species—*V. verrucosus* Ibrahim 1933.

Converrucosisporites (Ibrahim) Potonié and Kremp 1954, for spores as in *Verrucosisporites* but with triangular outline. Type species—*C. triquetrus* (Ibrahim) Potonié and Kremp 1954. This species, because of its triangular outline, was assigned to *Granulatisporites* by Schopf, Wilson, and Bentall (1944).

Cyclogranisporites Potonié and Kremp 1954, for spores with a circular outline and ornamentation as in *Granulatisporites*. Type species—*C. leopoldi* (Kremp) Potonié and Kremp 1954.

Planisporites (Knox) Potonié and Kremp 1954, for spores with a circular outline and short spines with a broad base. However, the type species is more or less triangular. Type species—*P. granifer* (Ibrahim) Knox 1950.

Apiculatisporis Potonié and Kremp 1956, for spores basically circular in outline with broad based spines and spines longer than width of spine base. Type species—*A. aculeatus* (Ibrahim) Potonié and Kremp 1956.

Pustulatisporites Potonié and Kremp 1955, for roundly triangular spores with widely spaced pyramid-shaped bulges. Type species—*P. pustulatus* Potonié and Kremp 1955.

Reticulatisporites (Ibrahim) Schopf, Wilson, and Bentall 1944 (pl. 12-1, fig. 13) is a trilete, radial spore with reticulate ornamentation covering both proximal and distal spore surfaces. The basic outline is circular in either proximo-distal or equatorial view. The trilete aperture in many specimens is obscured by the ornamentation, which in many instances traverses across the rays indicating that at least the proximal ornamentation was formed after separation from the tetrad stage. Further, because species of the genus exhibit little preference for proximo-distal orientation when compressed, the trilete aperture may occur in almost any position. The reticulate ornamentation consists of walls called muri and the spaces or the lumen between the muri. The size ranges from about 25 to more than 90 microns, depending on the species. The genus is present, although rarely abundant, throughout the Mississippian and Pennsylvanian. Some species with a restricted stratigraphic range are useful in correlation studies. The type species is *R. reticulatus* Ibrahim 1933.

Dictyotriletes (Naumova) Potonié and Kremp 1955 (pl. 12-1, fig. 12), is sometimes confused with *Reticulatisporites* because both genera possess reticulate ornamentation. Validation of *Dictyotriletes* by Potonié and Kremp (1955-1956, pt. 1) and the selection of *D. bireticulatus* (Ibrahim) Potonié and Kremp as the type species seem to be proper because this species, in addition to having low muri, has dimorphic ornamentation in that the proximal surface is levigate. The confusion existing now arises from the generic description, which is not limited to spores with dimorphic ornamentation. The only difference between *Reticulatisporites* and *Dictyotriletes* is the height of the muri. The diagrammatic drawings in Potonié and Kremp (1955-1956, pt. 1, figs. 28 and 29) of *Dictyotriletes* clearly show dimorphic ornamentation, but this feature is absent from the description of the genus.

It seems to me that in the present state of our knowledge of affinities of fossil spores the condition of two types of ornamentation restricted to proximal and distal surfaces should merit consideration for segregation of separate genera. Examination of the species assigned to *Dictyotriletes* by Potonié and Kremp (1955-1956, pt. 1, figs. 296-305) shows a striking difference between *D. bireticulatus* with a dimorphic ornamentation and the other species that more clearly are assigned to *Reticulatisporites*. Butterworth (1964b) reports *Dictyotriletes bireticulatus* from Westphalian A and B, and I have observed this species in the Middle Pennsylvanian part of the Breathitt Formation of Kentucky.

Spores with a Continuous Equatorial Structure. Many radial, trilete spores are differentiated on the basis of continuous equatorial structures. These structures may be thickenings, ridges, flanges, or fimbriate flanges. The terms "cingulum," "crassitude," "zona," and "corona" have been used to denote these structures. A cingulum is a massive structure that may be wedge shaped in cross section. A crassitude is a smaller structure than a cingulum and may extend onto the body. A zona is a more or less membranous structure at the equator of the spore. A corona is a fimbriate structure that may or may not unite laterally.

One genus commonly present in the Mississippian and a large part of the Pennsylvanian is *Lycospora* Schopf, Wilson, and Bentall 1944; the type species is *L. micropapillata* (Wilson and Coe) Schopf, Wilson, and Bentall 1944. Representatives of *Lycospora* are allied with the arborescent lycopsids on the basis of the work of Felix (1954), Chaloner (1953a), and others.

All species of *Lycospora* (pl. 12-1, figs. 14-17) are characterized by a continuous equatorial structure. In some species, such as *L. granulata* (pl. 12-1, fig. 16), this structure is a strong ridge. In other species, such as *L. pseudoannulata* (pl. 12-1, fig. 14), the equatorial structure is extended and more flangelike, but not to the degree observed in most species of *Cirratriradites* (pl. 12-1, fig. 18). *Lycospora* in proximo-distal view may be round to roundly triangular, whereas in equatorial view it is somewhat flattened-lenticular in outline. Most species assigned to *Lycospora* are in the 20- to 50-micron size range. Many of the species have granulose spore coats, although the ornamentation ranges from almost levigate to rather rugose. The flangelike equatorial structures associated with some species of *Lycospora* may be smooth, somewhat granular, or even pitted. Apical papillae are characteristically associated with certain species.

Lycospora is known to occur in Devonian, Mississippian, and Pennsylvanian rocks. In the Eastern Interior Basin of the United States *Lycospora* occurs in coals rather consistently up to about the middle of the Modesto Formation near the position of the Des Moines-Missouri boundary and is absent from younger coals. It is present in limited quantities from other lithologies in younger strata, according to Peppers (1964). I have observed *Lycospora* in shales up to approximately the Virgil-Wolfcamp boundary in Texas. Alpern (1964) reports *Lycospora* from the Permian of Europe, but this genus has not been reported from the Permian of the United States.

Cirratriradites Wilson and Coe 1940 is somewhat similar in construction to *Lycospora*, but it is generally larger, and all species possess a definite and distinct flange. The type species is *C. maculatus* Wilson and Coe 1940. *Cirratriradites* is associated with the herbaceous lycopsids on the basis of the work of Chaloner (1954a), Hoskins and Abbott (1956), and others.

In proximo-distal view species of *Cirratriradites* (pl. 12-1, fig. 18) may be nearly round to almost triangular, but in equatorial view they are more or less oblate. For details of compressed specimens see Hughes, Dettmann, and Playford (1962), who have demonstrated that the flange of *C. elegans* (Waltz) Potonié and Kremp is distinctly wedge shaped in thin section. This is surprising in view of the fact that the peripheral portion of the flange of this species rather readily transmits light in proximo-distal view. Some species have flanges with distinct radially alined structures termed striations. The body ornamentation, depending on the species, ranges from almost levigate to distinctly punctate. Some species possess distal structures that have been called enclosing areas. Generally species assigned to *Cirratriradites* are in the 40- to 90-micron size range.

On the basis of isolated specimens we might conclude that there is some relationship between *Cirratriradites* and *Densosporites*. This relationship is clarified by examination of thin sections of both genera, as reported by Hughes, Dettmann, and Playford (1962). The relationship is with those species of *Densosporites* in which the peripheral portions of the cingulum thin somewhat and transmit light. Such a relationship really should not be unexpected in view of the fact that all of the genera under discussion are related to one or another portion of the lycopsid alliance.

Cirratriradites is known from the Devonian, Mississippian, and Pennsylvanian. It was reported from the Permian of Europe by Alpern (1964) but was not reported from the Permian of United States by either Wilson (1962) or Jizba (1962). *Cirratriradites* is absent from most coals of Late Pennsylvanian age, and in none is it as abundant as *Lycospora*.

Kosanke (1950) described two species of spores which he assigned to *Cirratriradites*. They are abundant and are restricted to two Pennsylvanian coals in what now is called the Abbott Formation in Illinois. These spores were assigned to *Cirratriradites* because both species possess equatorial flanges. These distinct flanges are composed of ridges or ribs or radially arranged processes which fork and anastomose. A similar spore was assigned to *Cirratriradites* by Knox (1950) which she named *C. aligerens*. It is from the Westphalian A. Dybová and Jachowicz (1957) proposed *Cingulizonates* for spores possessing a flange composed of ridges or processes. In doing so they instituted an illegitimate combination by transferring *Cirratriradites difformis* Kosanke to *Cingulizonates asteroides* because they did not retain the specific epithet *difformis* as required by article 45 of the "International Code of Botanical Nomenclature." Staplin and Jansonius (1964) proposed *Radiizonates* (pl. 12-1, fig. 19) for spores of this kind and selected *R. aligerens* (Knox) Staplin and Jansonius as the type. As Knox did not designate a type specimen, Staplin and Jansonius designated a neotype specimen.

Spores of the *Radiizonates* type are circular to subtriangular in overall outline, in proximo-distal view, with ridges or anastomosing equatorial processes that unite to form a large distinctive flange. The body ornamentation may be granulate, punctate, or reticulate. The overall size ranges from about 45 to 80 microns.

Radiizonates aligerens, *R. difformis*, and *R. rotatus* all have restricted ranges in the Pennsylvanian. *Cingulizonates radiatus* Dybová and Jachowicz, not treated by Staplin and Jansonius (1964), may or may not properly belong to *Radizonates.* It was reported from Namurian A through C of Poland.

Densosporites (pl. 12-1, figs. 21 and 22) occurs in Devonian, Mississippian, and Pennsylvanian rocks throughout the world. The type species is *D. covensis* Berry 1937, and it represents a dominant element of many coal samples. *Densosporites* is allied with the lycopsids by Chaloner (1958b) and Bhardwaj (1959). In coals of the Eastern Interior Basin *Densosporites* becomes extinct at about the boundary of the Spoon and Carbondale Formations. Cross and Schemel (1952) report that the genus occurs in probably slightly younger strata in the Appalachian region. Peppers (1964) reported *Densosporites* occurring in noncoal lithologies higher in the section in younger Pennsylvanian rocks. I have observed the genus in shale samples from Texas up to about the position of the Des Moines-Missouri boundary. Alpern (1964) reports *Densosporites* in the Stephanian of France.

Densosporites is characterized by a distinct equatorial thickening or cingulum. In some species this thickening causes the spore to have the appearance of a doughnut in cross section. Species selected by Hughes, Dettmann, and Playford (1962) for sectioning show the equatorial thickening to be wedge shaped. The body of *Densosporites* is usually thin and transparent, and the trilete aperture may or may not be readily apparent depending on the species. It is commonly only faintly visible in many species. In proximo-distal view the outline may be circular to triangular or somewhat irregular. The ornamentation in many specimens is rather fine and is levigate, granulate, punctate, or even spinose. The genus *Cristatisporites* (Potonié and Kremp) Butterworth, Jansonius, Smith, and Staplin 1964 (in Butterworth, 1964a) is similar in many respects to *Densosporites,* and the type species is *Cristatisporites indignabundus* (Loose) Potonié and Kremp 1954. *Cristatisporites* is related to those species of *Densosporites* in which the equatorial thickening is wedge shaped in cross section, permitting light to be transmitted through the peripheral portion of the cingulum. In addition, it has numerous distal spines. *Cristatisporites* (pl. 12-3, figs. 1-3) has a range similar to that of *Densosporites,* but it is not known to occur as high in the Pennsylvanian as *Densosporites.* Chaloner (1962) described *Sporangiostrobus ohioensis,* a lycopod cone containing both microspores and megaspores. The microspores are similar to *Cristatisporites solaris* (Balme) Butterworth 1964.

Vallatisporites Hacquebard 1957 is a radial, trilete spore that is roundly triangular in overall proximo-distal view. The spore possesses a perispore that extends beyond the body, at which point it thickens into a ridge with a single row of pits; it tapers to the spore margin forming a flange. The distal surface is ornamented with cones, grana, verrucae, or spinose elements. The proximal surface may be similarly ornamented in the region of the equatorial border. The thin body has a well-defined margin and is subcircular in proximo-distal view. The trilete aperture is usually faint, but the lips extend into the flange area. The overall size ranges from 48 to 84 microns. The type species, *V. vallatus* Hacquebard, is of Early Mississippian age. Staplin (1960) reported *Vallatisporites*? sp. A from the Golata Formation (Upper Mississippian) of Canada. Hacquebard (1957) remarked, "This genus resembles *Densosporites* in construction of equatorial portion; it differs in possessing a well-defined central body that is separated from the equatorial portion by a distinct groove or rampartlike area, from which the name *Vallatisporites* is derived."

Hacquebard (1957) called attention to the similarity of *Vallatisporites* to *"Hymenozonotriletes"* Naumova 1937, whereas Playford (1963b) discussed the similarities between *"H." pustillites* Kedo and *Vallatisporites vallatus* and concluded that *V. vallatus* had a distinct lip development that is lacking in *"Hymenozonotriletes" pustillites.*

Sullivan (1964) described two new species of *Vallatisporites* and transferred *Zonotriletes ciliaris* Luber to the genus. Sullivan stated, "*Densosporites* is distinguished from *Vallatisporites* by the more abrupt thickening of the cingulum, by the less pronounced cuniculus, the usual absence of vacuolation and folds accompanying the rays, and also by the thinner distal exoexine." The cuniculus is probably the space between the peripheral margin of the spore cavity and the inner surface of the equatorial portion of the exoexine according to Sullivan (1964).

Three Mississippian genera that have rather distinct equatorial structures are *Monilospora* (Hacquebard and Barss) Staplin 1960, *Labiadensites* Hacquebard and Barss 1957, and *Lophozonotriletes* (Naumova) Potonié 1958. (See pl. 12-3, figs. 8 and 9.) These genera have equatorial structures considered to be a cingulum by Playford (1962, 1963a). Staplin (1960) introduced the terms "patella" for an acorn structure and "capsula" for a layer that completely covers a spore. Both structures are prominent at the spore equator. The patella and capsula are reported to be transparent when covering the body so that their presence is difficult to determine. Butterworth and Williams (1958) had previously proposed the term "patina" for a structure in *Tholisporites* Butterworth and Williams 1958. The patina was described as an exinal thickening covering at least one hemisphere, and in *Tholisporites* this is distal. *Tholisporites* is reported to be related to the cingulate genera in that the patina is thickest at the equator of the spore and thins somewhat over the distal portion. Obviously patina and patella are related if indeed they are not synonymous. For practical purposes *Monilospora* (pl. 12-3, fig. 6), *Labiadensites*, and *Lophozonotriletes* (pl. 12-3, figs. 8 and 9) are here considered to be cingulate in that they have equatorial structures.

The interpretation by Neves (1961) of *Tholisporites? bianulatus* is somewhat different from that of *Tholisporites* Butterworth and Williams. Neves reported the presence of two raised bands encircling the spore subequatorially, with a large thickening centered over the distal pole.

Monilospora (pl. 12-3, fig. 6) in proximo-distal view may be nearly spherical to roundly triangular in outline, whereas the body is roundly triangular and is either slightly convex or concave interradially. The broad equatorial structure, or cingulum, has a peripheral portion that is fluted or scalloped, although this condition in *M. dignata* Playford is not pronounced. The body is levigate or at best only finely ornamented. The size ranges from about 45 to 120 microns.

Labiadensites, which is circular to subcircular in proximo-distal view has distinct rays, thickened lips, and a circular to subcircular body. The broad cingulum is marked by a peripheral zone that is somewhat translucent and ruffled. The body and cingulum may be levigate or only finely ornamented. *Labiadensites* is similar morphologically to *Monilospora* except for the strong lip development in *Labiadensites*, which also appears to have a more distinctly demarcated peripheral portion of the cingulum. *Labiadensites* is large, ranging from about 90 to 145 microns in overall diameter.

Lophozonotriletes is roundly triangular in proximo-distal view, and its body generally conforms to the outline of the cingulum. The cingulum is broad and rather uniform, although it tends to be irregular in *L. dentatus* (pl. 12-3, fig. 8) and *L. variverrucatus* because of the ornamentation. Species assigned to this genus have rounded distal projections. The size is extremely variable – it ranges from 40 to 170 microns.

Stenozonotriletes (Naumova) Hacquebard 1957 is reported by Richardson (1964) as occurring in Devonian and Mississippian rocks. *Stenozonotriletes* (pl. 12-3, fig. 11) is nearly circular to subtriangular in proximo-distal view. The rays of the trilete aperture are long, the lips may or may not be developed, and the cingulum may be thinly to modestly developed. The presence of a cingulum, in species in which this feature is only modestly developed, can be difficult to determine (Playford, 1962, p. 605). Species assigned to *Stenozonotriletes* occur in the isolated state in good proximo-distal orientation because of the cingulum. The ornamentation may be levigate, finely granulose, or punctate. The size ranges from about 40 to 120 microns.

Thus far for spores with a continuous equatorial structure we have discussed genera that are predominantly round in proximo-distal view, although some species are roundly triangular. Luber and Waltz (1938) described a large number of species of rather diverse morphology assigned to *Zonotriletes*. Naumova (1939) introduced the name *Simozonotriletes* (pl. 12-2, fig. 4) for a group of morphologically related spores included within *Zonotriletes* Waltz 1938. Brief descriptions and drawings were presented, although no type specimens were designated. Grace Somers, now Mrs. Grace Somers Brush, encountered specimens in the Lower Jubilee coal from Nova Scotia that obviously were related to *Z. intortus* Waltz 1938 (in Luber and Waltz, 1938). However, Somers was not certain, on the basis of available drawings and brief descriptions, that her specimens and those described as *Z. intortus* were conspecific. Consequently Somers (1952) proposed *Murospora* for her material, described and illustrated two species, and designated *M. kosankei* the type species.

Potonié and Kremp (1954), unaware of Somers' publication, transferred *Zonotriletes intortus, Z. sublobatus,* and *Z. auritus* of Luber and Waltz (1938) to *Simozonotriletes*; they presented a description and selected *S. intortus* as the type. Sullivan (1958) followed Potonié and Kremp and used *Simozonotriletes* in a paper describing the possible variation occurring in spore types assigned to *Simozonotriletes*. He proposed a number of varieties of *S. intortus* and reduced *S. sublobatus* to a varietal status.

Staplin (1960) considered *Simozonotriletes* (Naumova) Potonié and Kremp as well as *Westphalensisporites* Alpern 1958 as synonymous with *Murospora* Somers 1952. Playford likewise adopted *Murospora* over *Simozonotriletes*, on the basis of the rule of priority.

The cingulum of *Simozonotriletes* displays considerable variation in thickness and shape. These variations are not known to occur in specimens Somers assigned to *Murospora*. Further, the maximum overall diameter of *Murospora* as reported by Somers is 31 microns, whereas the maximum diameter of *Simozonotriletes* is 95 microns. Because of the differences in cingulum thickness, shape and overall spore size, I believe both genera serve a useful purpose and should be retained.

Murospora may be roundly triangular to nearly triangular in proximo-distal view, it may be concave or convex interradially, and it may have a distinct and in many specimens variable equatorial structure, or cingulum. The rays of the trilete aperture are long, and the lips may be poorly developed to strongly thickened. Somers (1952) reported the size range for *Murospora* to be 20 to 31 microns.

Simozonotriletes (pl. 12-2, fig. 4) is similar to *Murospora* except for a thicker and more angular cingulum. Also, *Simozonotriletes* is significantly larger than *Murospora*, with many specimens in the 50- to 60-micron size range and a maximum of about 95 microns.

Knoxisporites (Potonié and Kremp) Neves 1961 has been reported from rocks that range in age from Devonian through Mississippian to Middle Pennsylvanian. The type species is *K. hageni* Potonié and Kremp 1954.

Knoxisporites (pl. 12-2, fig. 5) is a radial, trilete spore with a prominent equatorial cingulum and distinctive distal bands. The outline of the spore is round to roundly triangular in proximo-distal view. The trilete aperture is distinct, and lip thickness varies with the species. The spore ranges in size from about 40 to at least 100 microns.

Reinschospora Schopf, Wilson, and Bentall 1944 is one of the more unusual genera of Paleozoic spores. The type is *R. bellitas* Bentall (in Schopf, Wilson, and Bentall, 1944). The genus is named in honor of Paulus F. Reinsch for his early contributions to palynology. In 1884 he illustrated specimens and gave diagnoses that are assignable to the genus. The genus has been reported from the Mississippian and Pennsylvanian in many parts of the world, and Balme (1964) reported the genus from the Early Permian of Australia.

The body of *Reinschospora* (pl. 12-2, fig. 6) may be triangular to roundly triangular in proximo-distal view, with convex, straight, slightly concave, or strongly concave interradial areas. The trilete aperture is usually long, lips are indistinct to modestly developed, and the spore equator is surrounded by a fimbriate flange that has a maximum length interradially. The overall spore outline in proximo-distal view tends to be nearly round. In many isolated specimens of *Reinschospora* the fimbriae are not united by a thin membrane, but this probably is a matter of preservation or the maceration process. The fimbriae are rather straight and are like setae. In some species the fimbriae bifurcate, and in other species they have small terminal knobs. In some species the fimbriae are deeply imbedded in the spore coat.

Although some of the species of *Reinschospora* assigned to *Diatomozonotriletes* by Playford (1963a) appear to be distinct, their basic morphology is that of *Reinschospora* and are here considered to belong to that genus. The overall maximum diameter of species assigned to *Reinschospora* ranges from about 40 to 80 microns.

Rotaspora Schemel 1950, a genus that is apparently restricted to Upper Mississippian rocks, has been reported from Russia, Europe, and United States. The type is *R. fracta* Schemel.

Rotaspora is characterized by a nearly triangular body and an equatorial flange structure that in proximo-distal view gives the entire spore a circular outline. The flange, as in *Reinschospora*, is widest interradially. The peripheral portion of the flange is thickened to form a rim. The nearly triangular body may be concave or convex interradially. The rays are long, without lip development. *Rotaspora knoxi* Butterworth and Williams is described as smooth and vitreous appearing, whereas *R. fracta* Schemel is reported to be variously ornamented from levigate to verrucose. The spores of *Rotaspora* are generally small, ranging from about 25 to 45 microns.

Crassispora Bhardwaj 1957 (pl. 12-2, figs. 7 and 8) was proposed for radial, trilete spores that are circular to subtriangular in proximo-distal view, with an equatorial structure in the form of a band or narrow thickening that generally appears as a slightly darker narrow peripheral ring. Such a thickening has been termed a crassitude by Potonié and Kremp (1955-1956). The ornamentation is typically composed of numerous blunt cones (coni), which Bhardwaj reported to be present on both proximal and distal surfaces. There are also exceedingly fine punctations (infrapunctat) between the blunt cones. The proximal surface is thin; in many specimens it is torn, and the trilete aperture is not apparent. The rays of the aperture are long and without significant lip development; the aperture may be elevated in well-preserved specimens. Apical papillae apparently are present in the type species, as pointed out by Peppers (1964) and Sullivan (1964). The type species is *C. ovalis* (Bhardwaj) Bhardwaj 1957, and the size range for this species is reported to be 46 to 55 microns. *Crassispora kosankei* (Potonié and Kremp) Bhardwaj 1957 ranges in size from 65 to 85 microns. *C. plicata* Peppers 1964 lacks the proximal conelike ornamentation, but it has distinct trilete rays and apical papillae. The size is 48 to 75 microns. Sullivan (1964) restricted the conelike ornamentation to the distal surface, as reported by Peppers (1964) for *C. plicata*, and accepted the presence of apical papillae without emending the genus. I have observed specimens of *C. ovalis* in equatorial view (pl. 12-2, fig. 8) that clearly show the nature of the equatorial structure (crassitude) and the conelike ornamentation on both proximal and distal surfaces. Tetrads of *Crassispora* clearly show the nature of the equatorial structure as sharply defined narrow bands.

Crassispora had been considered to be restricted to Pennsylvanian rocks, but recently species assigned to this genus have been reported from rocks of Tournaisian and Viséan age (Sullivan, 1964).

Triangular Spores with Equatorial Structures Opposite the Rays. *Triquitrites* Wilson and Coe 1940 (pl. 12-2, figs. 10-11, 14), *Tripartites* Schemel 1950, and *Ahrensisporites* (Potonié and Kremp) Horst 1955 (pl. 12-2, fig. 9) possess distinctive equatorial structures opposite the rays of the aperture.

The type species of *Triquitrites* is *T. arculatus* Wilson and Coe. The original descriptions of the genus and the type species indicate that the thickened corners were considered to be flangelike. A photomicrograph of this species by Wilson (1958) shows these corners to be more padlike, conforming somewhat to the generic description presented in Schopf, Wilson, and Bentall (1944). On the basis of the original drawing in Wilson and Coe (1940) it could not be clearly determined whether the corners are flange or merely a thickening. Sullivan and Neves (1964) commented that considerable variation exists in the degree of equatorial thickening opposite the rays. In *Triquitrites arculatus* the corners are only modestly thickened, whereas in *T. batillatus* Hughes and Playford 1961 they are very prominent features. Kosanke (1950) described and illustrated *T. inusitatus* with unusual thickenings opposite the rays. These thickenings are somewhat irregular processes. Neves (1958) proposed *Mooreisporites* (pl. 12-2, fig. 14) for spores with these irregular processes opposite the rays and transferred *T. inusitatus* to *Mooreisporites*. Regardless of the proper placement of spores assigned to *Mooreisporites* by Neves, they have a thickening at the corners like that in *Triquitrites*.

The body of *Triquitrites* (pl. 12-2, figs. 10 and 11) is basically triangular in outline in proximodistal view, and the corners are truncated in some species and more or less pointed in others. The interradial area may be concave, essentially straight, or convex. In species with strongly convex interradial areas the overall proximo-distal outline is roundly triangular. The ornamentation varies considerably. It may be levigate, granulose, punctate, verrucose, spinose, or reticulate; spores of the *T. inusitatus* type have irregular processes. The size also ranges from about 25 to 95 microns. Species of *Triquitrites* are reported by Sullivan and Neves (1964) to range throughout the Mississippian and Pennsylvanian into the Early Permian.

Ahrensisporites was proposed by Potonié and Kremp (1954). The type species is *A. guerickei*, which was described and illustrated in a doctoral dissertation by Ulrich Horst 11 years earlier under the binomial *Triletes guerickei*. Potonié and Kremp provided a generic description of *Ahrensisporites* and a drawing. Horst's doctoral dissertation is not a valid publication, although it is regarded as such by Potonié and Kremp. Horst (1955) described and illustrated *A. guerickei*.

Ahrensisporites (pl. 12-2, fig. 9) is basically triangular in proximo-distal view, with more or less truncated to rounded corners; the interradial area between the rays is concave, straight, or convex. The corners possess equatorial structures that may be thin and membranelike or thickened. They continue distally and are typically united interradially. These distal structures, called kyrtomes, differentiate *Ahrensisporites* from *Triquitrites*. There is no question that *Ahrensisporites* and *Triquitrites* are related morphologically; as a matter of fact, incomplete development of the kyrtomes can present problems in the proper generic assignment.

Ahrensisporites is common in Upper Mississippian and Lower to Middle Pennsylvanian rocks (Sullivan and Neves, 1964). The rays of the trilete aperture tend to be somewhat longer than those of *Triquitrites*, and the lip development in *Ahrensisporites* is, at best, only modest. *Ahrensisporites*, like *Triquitrites*, displays extremes in both size variation and ornamentation. The size ranges from about 23 to 70 microns. Peppers (1964) reports a distal projection in *Ahrensisporites exertus* Peppers.

Tripartites Schemel 1950 (pl. 12-3, fig. 5) was proposed for spores that are radial, trilete, and are of subtriangular outline in proximo-distal view; they have truncated or rounded corners and concave, straight, or convex interradial margins. Schemel (1950) reported the body to have an equatorial flange that is expanded at the corners opposite the rays. The truncated and fluted corners unquestionably have a significant development in the type species, *T. vetustus*, as shown in Schemel's photomicrograph, but the pronounced interradial flange shown in Schemel's figure 3 is not nearly so apparent in the photomicrograph. Staplin (1960) proposed two sections of the genus: Section A for species of the *T. vetustus* type in which the spores do not possess a continuous thickened structure (girdle) but may have a narrow interradial flange; and Section B for spores of the *T. inciso-trilobus* (Waltz) Potonié and Kremp type in which the corners and interradial thickening are continuous. The rays of the trilete aperture are generally long and have only modest lip development; the ornamentation may be levigate, granulate, coarsely punctate, or finely reticulate. The size ranges from about 30 to 73 microns. The type species was described from Upper Mississippian rocks of Utah. The genus has been reported from many areas of the world from rocks of Mississippian and, perhaps, Early Pennsylvanian age.

Hacquebard and Barss (1957) recognized an unusual spore occurring in a coal from the Nahanni River area of the Northwest Territories of Canada which they compared with *Azonotriletes lobophorus* Waltz 1938 (in Luber and Waltz, 1938). Differences between the Canadian and Russian specimens were noted by Hacquebard and Barss, but they considered these differences to be minor. Hacquebard and Barss recognized that *A. lobophorus* was not validly published, but, because the species was rare in their assemblage, a simple comparison with the Russian material was sufficient. Staplin (1960) proposed *Waltzispora* for spores of this general construction and designated *W. lobophorus* (Waltz) Staplin as the type. He also described *W. albertensis* from the Golata Formation and differentiated it from the type of the genus by the fact that the radial extremities are more expanded in *W. albertensis*.

Waltzispora (pl. 12-3, fig. 3) is basically triangular in proximo-distal view, with radial extremities or corners expanded so as to be mushroom shaped, saddlelike, or T-shaped. These expanded corners are not thickened or flangelike, but rather are continuous with the spore body cavity. This feature differentiates *Waltzispora* from *Triquitrites* and *Ahrensisporites*. Staplin (1960) reported the trilete aperture to be distinct and without lip development. Ornamentation is levigate to finely granulate, and the known size range is 24 to 58 microns. Playford (1962) published some unusually fine photomicrographs and described *Waltzispora sagittata*, which is finely granulate. He also reported that the spore coat of *W. lobophorus* is granulate and that the granules tend to be closely spaced in the region of the distal surface. *Waltzispora* is restricted to the Mississippian, so far as is now known.

Savitrisporites Bhardwaj 1955 (pl. 12-2, fig. 12) was proposed for radial, trilete spores that are triangular in proximo-distal view, with straight to slightly concave interradial areas. The proximal surface is levigate, and a distinct distal ornamentation is composed of ridges with blunt or rounded spines that frequently join and extend equatorially to form a cingulum-type structure. The rays of the trilete aperture are long and have no distinctive lips; the type species is *S. triangulus* from the Stephanian of Europe.

Sullivan (1964) proposed that the type species of *Callisporites* Butterworth and Williams 1958 is in fact morphologically inseparable from *Savitrisporites*. Hence *Callisporites nux* Butterworth and Williams was transferred to *Savitrisporites*. In so doing it was necessary to broaden somewhat the circumscription of *Savitrisporites* to accomodate the lip development of *S. nux*. The size ranges from about 30 to 60 microns. *Savitrisporites nux* is known to occur from Late Mississippian to Middle Pennsylvanian in the United States.

There are two additional genera that are basically triangular in proximo-distal view, with either straight or slightly convex interradial areas and equatorial structures. These are *Gravisporites* Bhardwaj 1954 and *Angulisporites* Bhardwaj 1954. Both have long rays, but *Gravisporites* has well-developed lips, which are lacking in *Angulisporites*.

Trilobates Somers 1952 was described from the Lower Jubilee coal of Nova Scotia, Canada, and *T. belli* was designated as the type species. *Trilobates* was thought by Somers to be related to *Tripartites* in that the body is triangular in proximo-distal view and is surrounded equatorially by a "flange or bladder" that becomes cushionlike at the corners and is lobed. The lips adjacent to the trilete aperture are prominent. Barss, Hacquebard, and Howie (1963) reported *Trilobates* occurring throughout the Morien Series of Canada (Westphalian C-D). In proximo-distal view *Trilobates* is similar to *Waltzispora*, but in *Waltzispora* the body forms the corners or cushions. The size reported for the genus is 31 microns.

Triangular Spores Without Equatorial Structures. A number of radial, trilete genera that are predominantly triangular in proximo-distal view will be discussed. This is not to say that species that are more or less triangular and assigned to predominantly circular spores do not occur; for example, *Raistrickia grovensis* Schopf, the type species of *Raistrickia* Schopf, Wilson, and Bentall 1944, is definitely roundly triangular, but probably more species of the genus are circular in outline than triangular. Other examples could be cited, but the above-mentioned case is sufficient.

Granulatisporites (Ibrahim) Schopf, Wilson, and Bentall 1944 includes spores that are roundly triangular in proximo-distal view, with concave to convex interradial areas. Compressed specimens are subtriangular, with rounded corners. The type species is *G. granulatus* Ibrahim 1933. Spores assigned to *Granulatisporites* (pl. 12-2, fig. 13) are generally small and range from about 20 to 50 microns, although a few species are larger. The ornamentation is greatly varied, according to the species, including levigate, granulate, punctate, verrucose, reticulate, and apiculate forms. The rays of the trilete aperture extend reasonably well into the corners, and the lips generally are not developed.

Spores of the same general construction were classified into the following genera by Potonié and Kremp (1954):

Granulatisporites (Ibrahim) Potonié and Kremp 1954 includes only species with granulose spore coat. Type species—*G. granulatus* Ibrahim 1933.

Leiotriletes (Naumova) Potonié and Kremp 1954 includes only levigate species. Type species—*L. sphaerotriangularis* (Loose) Potonié and Kremp 1955.

Lophotriletes (Naumova) Potonié and Kremp 1954 includes only species with apiculate ornamentation. Type species—*L. gibbosus* (Ibrahim) Potonié and Kremp 1954.

Acanthotriletes (Naumova) Potonié and Kremp 1954 includes only species with closely spaced spines which are longer than twice their basal width. Type species—*A. ciliatus* (Knox) Potonié and Kremp 1954.

Converrucosisporites Potonié and Kremp 1954 includes only species with verrucose ornamentation. Type species—*C. triquetrus* (Ibrahim) Potonié and Kremp 1954.

Anapiculatisporites Potonié and Kremp 1954 includes species with ornamentation similar to *Acanthotriletes* but more or less restricted to the distal surface. Type species—*A. isselburgensis* Potonié and Kremp 1955.

Obviously ornamentation is the sole controlling factor for the creation of the above-mentioned genera. A possible exception is *Anapiculatisporites*, which has a dimorphic type of ornamentation like *Schopfites* and *Dictyotriletes*. On this basis *Anapiculatisporites* should receive special consideration.

Affinities of species assigned to *Granulatisporites* should be at least in part, with the ferns. Some modern species of ferns possess the essential characters of *Granulatisporites*, as demonstrated by Selling (1946). Mamay (1950) has shown this to be true of certain fossil ferns. However, both these authors and Knox (1951) and others have conclusively demonstrated that a wide variety of spore types are associated with the ferns. These include bilateral and monolete as well as radial and trilete spores. In addition a great variation exists in both size and kind of ornamentation, depending on the species.

Microreticulatisporites (Knox) Bhardwaj 1955 originally included both radial and bilateral reticulate spores in which the width of the reticulum was less than 6 microns. Potonié and Kremp (1954) emended the genus to exclude bilateral, monolete spores. Bhardwaj (1955) emended the genus further, and as a result it is currently restricted to spores of a triangular outline in proximo-distal view, with the width of the reticulum less than 3 microns. Sullivan (1964, p. 366) stated, "Bhardwaj (1955) has described the ornament of the triangular spores of *Microreticulatisporites* as 'extrareticulate', but perhaps a better description would be punctate or foveolate." In any event the genus exists for spores of a triangular outline with more or less rounded corners, generally distinct and long rays, and in the 30- to 60-micron size range. Species of *Microreticulatisporites* are reported from both the Mississippian and Pennsylvanian, and *M. lacunosus* (Ibrahim) Knox is the type species.

Two additional radial trilete genera, which basically are triangular in proximo-distal view and which lack equatorial structures, are *Latipulvinitus* Peppers (1964) and *Trivolites* Peppers (1964). These two genera are similar in that each possesses a distinct proximal thickening.

Latipulvinites has rounded corners and convex, straight, or concave interradial areas, with straight rays extending nearly to the spore body margin, and slightly elevated lips. Developed on the proximal surface, adjacent to the lips and surrounding the aperture, are thickened triradiate ridges. The spore coat is levigate, the known size range is 38 to 48 microns, and the genus occurs in Upper Pennsylvanian rocks of Illinois and Kentucky. The type species is *L. kosankii*.

Trivolites has rounded corners and interradial areas that are straight to slightly concave. The rays of the aperture are distinct, and it has well-developed lips. The proximal thickening is interpreted as the contact area, or *area contagionis*, which extends at the radial points or corners beyond the margin of the spore body. The outer margin of the thickening is sharply delineated, but the inner portion of the thickening thins gradually toward the trilete rays. The spore coat is levigate, the known size range is 29 to 44 microns, and the type species is *T. laevigata*.

Spores assigned to *Procoronaspora* Butterworth and Williams 1958 are radial, trilete, roundly triangular in proximo-distal view, and are characterized by ornamentation restricted to the interradial areas. The ornamentation is composed of granules, verrucae, spines, or rods that are relatively low and do not reach the proportions exhibited by *Reinschospora*. The spore coat is relatively thick in proportion to spore size, with rays extending at least three-fourths the spore radius. Lip development varies with the species, but usually it is only slightly raised. The known size range is 25 to 50 microns. The type species is *P. ambigua*, which was described by Butterworth and Williams from the upper part of the Limestone Coal Group (Lower Carboniferous) of Scotland. Staplin (1960) described two additional species from the Golata Formation (Upper Mississippian) of Canada.

Saccate Spores, Prepollen, and Pollen. The saccate genera, those with bladders or sacci, can conveniently be divided into three groups: those with a single bladder (monosaccate), those with two bladders (bisaccate), which some people term disaccate, and those with three or more bladders (multisaccate). The air sacs, or bladders, have been termed wings, because it was believed that their function was primarily concerned with distribution by air currents. Wodehouse (1935) disputed the value of air sacs for this purpose, although he stated, "Unquestionably they do actually give pollen grains a greater range of flight, but whether or not this is of value to their possessor is doubtful, and that they were developed in these grains to meet such a need is still more doubtful." The placement and construction of the bladders in the Coniferales is such that as the grains dry, the bladders close, protecting the distal furrow. Wodehouse further stated, "So that if the bladders are organs of flight, pollen grains are possibly the only flying organisms of which it can be said that they fold up their wings and fly away." Felix and Burbridge (1961) in their description of the Mississippian spore genus *Pteroretis* suggest reserving the term "wing" for appendages of the type found in *Pteroretis*; these are of single thickness and winglike rather than inflated and saclike.

All saccate genera possess a body and one, two, or more bladders, or sacci. During Mississippian and Pennsylvanian time the monosaccate genera were more numerous than the bisaccate and multisaccate genera. The bisaccates increased in abundance in Late Pennsylvanian time and were very abundant by Early Permian time. Wilson (1962) reported that the bisaccates of the Flowerpot Formation (Permian) of Oklahoma represent 87 percent of the spore and pollen population. Such an abundance indicates a significant floral change between Pennsylvanian and Permian time.

About a dozen saccate genera will be discussed subsequently. Some of these are considered to be spores because the germinal exit is believed to have been proximal; others are considered to be pollen grains because the germinal exit is believed to have been distal. Other genera are called prepollen, a term used by Schopf (1938) and originally proposed by Renault (1896). The germinal exit in prepollen probably was proximal, but these taxa display advanced organization.

Monosaccate taxa. Among the 10 or so important monosaccate genera occurring in Mississippian and Pennsylvanian rocks *Spencerisporites* Chaloner 1951 (pl. 12-4, fig. 4) is perhaps the most unusual. Spores assigned to this genus are unusual because of their size and the presence of a marginal flange. These spores rarely are observed in the minus-210-micron residue. Felix and Parks (1959) reported the overall size range of *Spencerisporites* to be 245 to 370 microns, whereas Winslow (1959) found two species that range in overall size from 272 to 468 microns. The marginal flange, equatorial in position, may be narrow and inconspicuous or it may be striking and prominent. No other monosaccate genus possesses a marginal flange.

Winslow (1959) observed that the marginal flange represents about 10 percent of the spore diameter for one species and nearly 20 percent of the overall diameter for another species. In proximo-distal view the overall outline is triangular to subtriangular, although the body is circular to subcircular. The rays of the trilete aperture are prominent as are the lips, but they are narrow and high. They tend to fold and this folding extends beyond the margin of the spore body well into the bladder. Radial striations occur on the proximal surface between the rays of *Spencerisporites radiatus*. The bladder completely surrounds the body and has faint anastomosing lines that form a network. The marginal flange may also have anastomosing or intersecting lines, and the margin of the flange may be nearly smooth; in most specimens it is somewhat crenulate. Excellent photomicrographs of *Spencerisporites* are shown by Felix and Parks (1959) and Winslow (1959).

Chaloner (1951) investigated the fructification of *Spencerites insignis* Scott and reported that the spores were associated with the isolated species *Triletes karczewskii* Zerndt 1934. Horst (1955) indicated that *Zonales-sporites radiatus* Ibrahim 1933 was conspecific with *Triletes karczewskii*. Dijkstra proposed the name *Microsporites* for spores of this type but later rejected it by placing these spores in *Endosporites*. The nomenclatural problems associated with spores of this nature are discussed in detail by Felix and Parks (1959) and Winslow (1959). It is generally agreed today that *Spencerisporites* Chaloner 1951 is the proper generic name to be applied to spores of this nature and that *S. radiatus* (Ibrahim) Winslow 1959 is the type species.

The importance of Chaloner's discovery lies in the fact that these large monosaccate spores were

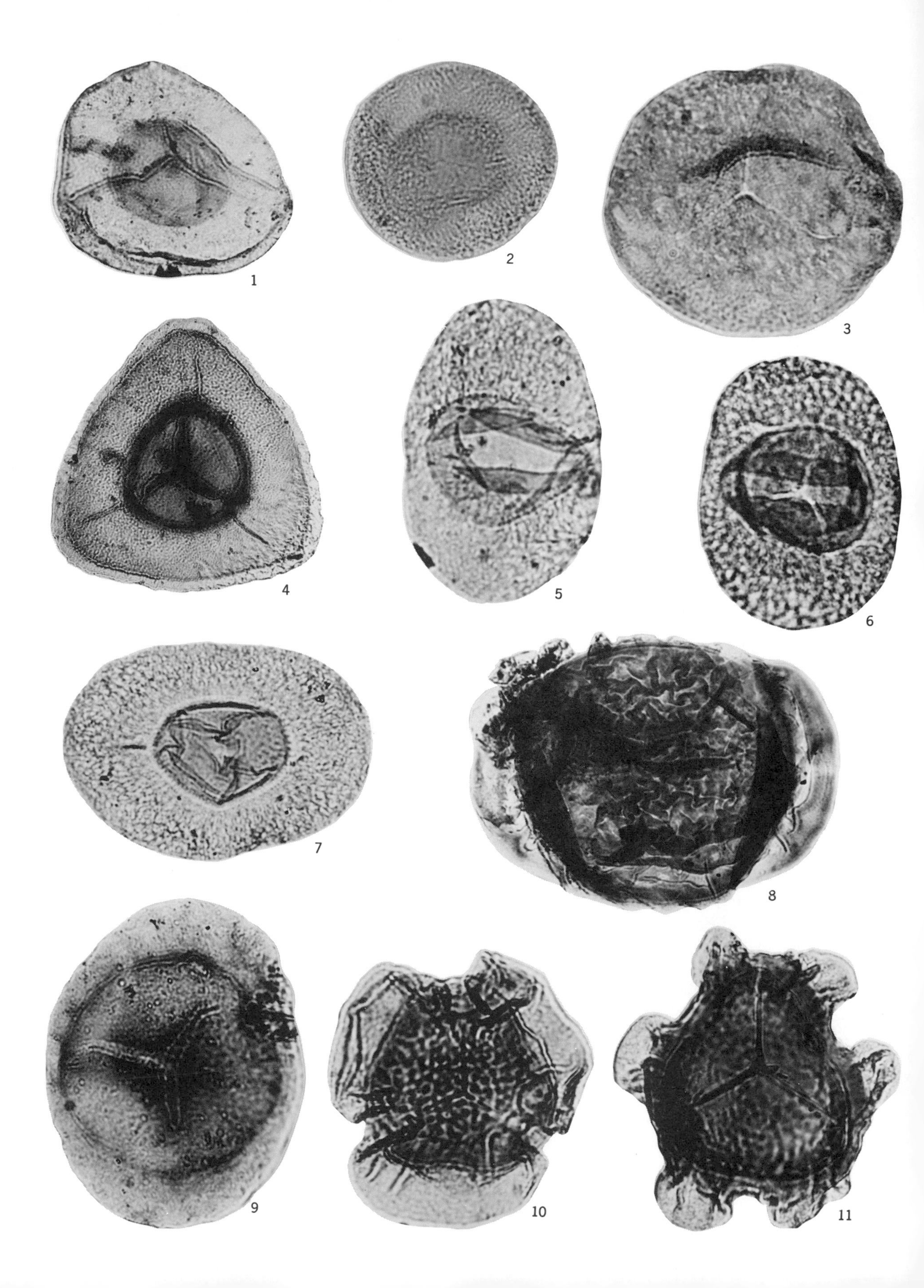

1
2
3
4
5
6
7
8
9
10
11

produced by the homosporous lycopsid *Spencerites*. This was the first proof of the occurrence of the monosaccate condition in the lycopsids.

Spencerisporites in Illinois, according to Winslow (1959), is restricted to the Pennsylvanian. Potonié and Kremp (1954) reported *Spencerisporites* (*Microsporites*) occurring as low as Namurian A in Asia and the Soviet Union.

Endosporites Wilson and Coe 1940 has the longest stratigraphic range of any Paleozoic monosaccate genus. *Endosporites* is present from Devonian through Permian rocks. Some years ago this genus was thought to be related to the cordaitaleans, as reported by Schopf, Wilson, and Bentall (1944): "*Endosporites* is related to some Pennsylvanian cordaitaleans. They correspond to spores observed by Wilson in male strobili . . . ." Chaloner (1953b, 1958a) has shown a relationship between *Endosporites* and the lycopsids. This is most interesting because *Spencerisporites*, a monosaccate genus, is related to the homosporous and herbaceous lycopods, whereas *Endosporites* is related to heterosporous and arborescent lycopods. *Lycospora* is also related to the heterosporous and arborescent lycopsids. Lycopsid plants producing *Endosporites* survived beyond those producing spores assignable to *Lycospora*.

Endosporites (pl. 12-4, fig. 1) is circular to roundly triangular in proximo-distal view, with the body spheroidal; in some species the body tends to be roundly triangular. The body of *Endosporites* is somewhat darker than the bladder, and it is generally distinct in all species. The original circumscription of the genus did not state whether the bladder enclosed the body on both proximal and distal surfaces, although the drawings might be taken to indicate this. Subsequently Schopf, Wilson, and Bentall (1944, fig. 14a) suggest the proximal surface to be free of the bladder. The interpretation of Potonié and Kremp (1954) shows the bladder covering both the proximal and distal surfaces. Frequently *Endosporites* displays a distinct preference for good proximo-distal orientation, indicating that the equatorial view is elongate-elliptical to somewhat lens shaped. The bladder is thin in proportion to its size, and folding of the bladder is not uncommon. The bladder of some species is marked by the limbus, which is a line paralleling the bladder outline. The limbus represents the position of adjacent bladders of the tetrad. The bladder is generally smooth externally and is granular to reticulate internally.

The body of *Endosporites* possesses a trilete aperture with thin but usually distinct lips, and apical papillae are present in some species. The rays of the trilete aperture are long, extending at least two-thirds of the body radius. The nature of the aperture suggests that it was functional. The body may be levigate or finely ornamented, granular to finely punctate.

The size of various species of *Endosporites* ranges from about 50 to at least 125 microns. The type species is *E. ornatus* Wilson and Coe 1940.

Wilsonites (Kosanke) Kosanke 1959, of somewhat similar structure to *Endosporites*, is known to occur in rocks of Pennsylvanian and Permian age. In overall size species of *Wilsonites* range from about 50 to more than 100 microns.

Wilsonites (pl. 12-4, figs. 2 and 3) differs from *Endosporites* in a number of respects. The body is consistently vague in appearance because it is covered by a thicker and more coarsely ornamented bladder. The bladder is attached to the body on the proximal surface, at least along the rays of the aperture, and is enclosed by, but free of, the bladder distally. The outline of *Wilsonites* in proximo-distal view is circular, whereas in equatorial view it is rather oval in outline. The limbus, as pointed out by Bhardwaj (1957), is absent in *Wilsonites* and may or may not be present in *Endosporites*, depending on the species. Apical papillae are not known to occur in *Wilson-*

Plate 12-4. Mississippian-Pennsylvanian saccate genera. 1—*Endosporites* Wilson and Coe 1940; maximum diameter is 91 microns. 2 and 3—*Wilsonites* (Kosanke) Kosanke 1959; maximum diameter is 52 and 71 microns, respectively. 4—*Spencerisporites* Chaloner 1951. This illustration is from Winslow (1959, pl. 13, fig. 4) through the courtesy of the Illinois Geological Survey. Total diameter of specimen is 345 microns. 5 to 7—*Florinites* Schopf, Wilson, and Bentall 1944; figure 6 is focused to illustrate the presence of a trilete aperture, which is not commonly associated with the genus: the maximum diameter of these specimens is 68, 66, and 73 microns, respectively. 8—*Parasporites* Schopf 1938, with a total length of 309 microns. This illustration is from Winslow (1959, pl. 14, fig. 12) through courtesy of the Illinois Geological Survey. 9—*Schulzospora* Kosanke 1950; maximum diameter, 84 microns. 10 and 11—*Alatisporites* Ibrahim 1933: figure 10 illustrates a specimen with three bladders and a maximum diameter of 68 microns; figure 11 illustrates a specimen with six bladders and a maximum diameter of 80 microns.

ites. The type species is *W. vesicatus* (Kosanke) Kosanke 1959.

Guthörlisporites Bhardwaj 1954 is similar to *Wilsonites* and *Endosporites* but differs in that the proximal body surface is free of the bladder and tends to be more oval than round in proximo-distal view, although it is elongate elliptical in equatorial view. Specimens of *Guthörlisporites* are usually preserved in good proximo-distal orientation. *Guthörlisporites* has not been reported as consistently as *Wilsonites*, possibly because recognition of a free proximal body surface is difficult in many specimens. The type species of *Guthörlisporites* is *G. magnificus* Bhardwaj 1954. The known size range of the genus is about 60 to 125 microns. The genus has been reported from the Westphalian D and the Permian.

Florinites Schopf, Wilson, and Bentall 1944 is a rather commonly occurring monosaccate genus. *Florinites* is fairly common throughout the Pennsylvanian and Permian, although through most of this span it is not abundant. Exceptions are in the Pennsylvanian, largely in shales as reported by Peppers (1964). Some coal samples may contain a modest abundance of *Florinites*, but these are not common.

The overall proximo-distal outline of *Florinites* (pl. 12-4, figs. 5-7) is broadly elliptical, whereas the body is spheroidal and is enclosed by the bladder except for a portion of the distal side. The distal surface of the body that is free of the bladder can be distinguished in most specimens, even when they are viewed from the proximal side. Folding of the body and bladder adjacent to the bladder attachment on the distal side is common. In many specimens the bladder is randomly folded. The body of *Florinites* may be levigate, granulose, or sparsely punctate, and the internal reticulate ornamentation of the bladder is its most outstanding feature. As originally described, no evidence of an aperture was reported other than the possibility of a vestigial proximal structure. Subsequently I observed and described a species (Kosanke, 1950) with a trilete proximal aperture, which has all of the other characteristics of *Florinites*; this I named *F. triletus* (pl. 12-4, fig. 6). Recently I have observed this same species from Middle Pennsylvanian coals from Kentucky. Other palynologists have recorded specimens of *Florinites* that have a trilete aperture, although most do not. The known size range of *Florinites* is about 50 to 200 microns. Neves, Sullivan, and Owens (in Butterworth, 1964b) reported the occurrence of *Florinites* in Britain in Namurian A.

Florinites was associated with the cordaitaleans by Schopf, Wilson, and Bentall (1944); Delevoryas (1953) among others has reported the pollen grains of *Cordianthus* to be related to *Florinites*.

Schulzospora Kosanke 1950 was described from a Lower Pennsylvanian coal from Illinois. This genus subsequently has been reported from Upper Mississippian and Lower Pennsylvanian rocks from many areas of the world. *Schulzospora* (pl. 12-4, fig. 9) is a trilete, radial spore in which the body is completely surrounded by a bladder; it is monosaccate. The overall outline in proximo-distal view is elliptical to broadly oval shaped; it is usually compressed in good proximo-distal orientation. The body is circular in proximo-distal view, and the trilete aperture may be faint to fairly distinct, depending on the species. The aperture generally does not possess any significant lip development. The bladder and body are finely ornamented, generally granulate or punctate, and the body may or may not be distinctly delimited from the bladder. However, the bladder is visible even when not distinct. The size ranges from about 50 to at least 110 microns. The type species is *S. rara* Kosanke 1950.

Schemel (1951) described *Vesicaspora* from the Middle Pennsylvanian Mystic and Marshall coals of Iowa. He also reported the genus to occur in the Weir-Pittsburg and younger Cherokee and Marmaton coals of Oklahoma, Kansas, Missouri, and Iowa. I observed and recorded (Kosanke, 1950) this genus from Middle Pennsylvanian coals from Illinois and simply reported it as a gen. nov. *Vesicaspora* is not abundant in its known Pennsylvanian occurrences, and it has an interrupted stratigraphic distribution. The genus was emended by Wilson and Venkatachala (1963c), who provided 30 photomicrographs of specimens of the type species *V. wilsonii* (Schemel) Wilson and Venkatachala. Schemel (1951) suggested that *Vesicaspora* is monosaccate, although he did not so state in his description. Wilson and Venkatachala (1963c, p. 148) clearly indicate that *Vesicaspora* is monosaccate and that distal inclination of the bladder ends is the result of oblique orientation. By the original circumscription of Schemel and the emendation of Wilson and Venkatachala, equatorial union of the bladder is required. However, in some species of *Vesicaspora* this union is not much more than a very thin connection, and distal inclination of the bladder ends is apparent. Such a distal inclination occurs only in bisaccate or multisaccate forms. Obviously, from a technical point of view, if oppositely placed bladders are united even by a thin equatorial connection, a monosaccate condition exists.

Vesicaspora lacks a proximal aperture and possesses a distal sulcus. The body in proximo-distal view is circular and may be levigate to granulate. The bladder is levigate externally and finely reticulate internally. *Vesicaspora wilsonii* is of relatively small size; the maximum overall length is less than 50 microns. Several species of the genus have been described from Permian rocks.

Auroraspora Hoffmeister, Staplin, and Malloy 1955 was proposed for radial, trilete, monosaccate spores in which the body is significantly darker than the bladder, which is reported to be transparent, thin, and colorless. The validity of the separation of *Auroraspora* from *Endosporites* was questioned by Playford (1963a). The type species is *Auroraspora solisortus* Hoffmeister, Staplin, and Malloy. The photomicrograph of the holotype specimen suggests that the bladder might have at least some internal ornamentation. The type is from the Chester Series of the Eastern Interior Basin. Richardson (1960) transferred *Endosporites micromanifestus* Hacquebard (pl. 12-3, fig. 10) to *Auroraspora*, but Playford (1963a) reassigned this species to *Endosporites*. The size range of *Auroraspora solisortus* is reported to be 61 to 78 microns.

Staplin (1960) reported the occurrence of *A. solisortus?* from the Golata Formation (Upper Mississippian) of Canada. The original description of the genus was based on specimens from the Hardinsburg Formation of the Chester Series of the Eastern Interior Basin of the United States. Sullivan (1964) described *A. balteola* from the Drybrook Sandstone of Viséan age from Gloucestershire, England.

Remysporites Butterworth and Williams 1958 was proposed for large, monosaccate, radial, trilete spores in which the bladder enclosed the body except on the proximal surface. Playford (1963a) concluded on re-examination of the type specimen that the bladder completely encloses the body. In overall proximo-distal view *Remysporites* (pl. 12-3, fig. 12) is circular to somewhat oval and the body, which is relatively large, essentially matches the contour of the bladder. The body is distinct from the bladder. The bladder may be levigate to finely reticulate externally, in contrast to the more common internal bladder ornamentation of *Endosporites*, *Wilsonites*, and *Florinites*. Folding of the bladder is common and may be associated in part with the attachment of the body to the bladder. The distinct trilete aperture is without significant lip development. The contact area may be prominently ornamented by vermiculate to verrucose structures. The size range for the type species was reported as 120 to 250 microns, whereas *Remysporites albertensis* Staplin 1960 was reported as 141 to 160 microns.

Remy (1953, 1954) described and illustrated a pteridosperm fructification under the name of *Paracalathiops stachei*, which contained spores that Butterworth and Williams believed to be conspecific with *Remysporites*. The type species is *R. magnificus* (Horst) Butterworth and Williams. *Remysporites* is restricted to the Mississippian, having been reported from Viséan through Namurian A.

Grandispora Hoffmeister, Staplin, and Malloy 1955 was proposed for radial, trilete, monosaccate spores covered with prominent spines scattered over the bladder. Overall outline in proximo-distal view is circular to subcircular, and the body outline essentially parallels the bladder. The body and bladder may be granulate or punctate; and the bladder spines of the type species, *G. spinosa*, are reported to be 2 to 8 microns in length and 8 to 25 microns apart, with the overall size ranging from 100 to 143 microns. The body of *Grandispora* is not known to be greatly thicker than the bladder, and the trilete aperture is thin, without significant lip development. The rays are long, extending possibly to the margin of the body. *Grandispora* was originally described from Upper Mississippian rocks.

Velosporites Hughes and Playford 1961 is unquestionably similar to *Remysporites* (see Playford, 1963a). When Hughes and Playford proposed *Velosporites* it was separated from *Remysporites* by the fact that the bladder completely enclosed the body of *Velosporites*. This distinction is no longer valid according to Playford (1963a), who decided to retain *Velosporites* on the basis of ornamentation and a thicker body wall in *Velosporites*. The type species of *Velosporites* is *V. echinatus*. *Velosporites* is known from Early to Late Mississippian.

Discernisporites Neves 1958 is a radial, trilete spore somewhat roundly triangular in proximo-distal view and convex lens shaped in equatorial view. The type species is *D. irregularis*, which has a size range of 50 to 100 microns. The region of the contact area of the proximal surface is ornamented with warts and elongate ridges. The overall appearance suggests the body of a monosaccate spore. The periphery of the central thickened area is thin and pale yellow. *Discernisporites concentricus* Neves 1958 lacks the heavy ornamentation of the type species; its proximal area is termed infragranulate and is distinctly darker than the rest of the spore. The trilete aperture of both species is more than one-half the spore radius, and the lips are raised. These two species were described from the *Gastrioceras subcrenatum* marine shales (Westphalian) of England.

Proprisporites Neves 1958 is a radial, trilete spore that is roundly triangular in outline as observed from a proximo-distal view. It possesses a thin perisporial membrane, characteristically folded in the form of linear strips, largely on the distal surface. These folded strips cross the spore equator and are present on a portion of the proximal surface adjacent to the equator. The type species is *P. rugosus*, which has a known size range of 65 to 85 microns. The trilete aperture is clearly visible, and the rays are two-thirds the spore radius. The lips are thin but well defined; the body is punctate. The type species was reported from the *Gastrioceras subcrenatum* marine shales (Westphalian) of England. Neves (1961) described *Proprisporites levigatus*, which is distinguished from *P. rugosus* by a levigate spore coat, as occurring in marine shales of the Namurian A and B of England. Neves (1958) classified *Proprisporites* within the saccate spore and pollen group but, to conform with the Potonié and Kremp system of supergeneric categories, he created a new series. This he termed Series Membranati to house spores possessing a membrane but lacking the characteristic internal bladder ornamentation of true saccate spores and pollen.

Hymenospora Neves 1961 was placed in his Series Membranati because the spore body is encircled by a light-yellow membrane that lacks the characteristic internal ornamentation of saccate spores and pollen. This is similar to the membrane of *Proprisporites*. In proximo-distal view the light-brown body is more or less circular, the trilete aperture has rays three-fourths the body radius, and the lips are thin. The surrounding membrane is attached to the body along a series of narrow linear furrows in the region of the trilete aperture. The type species *H. paliolata* has a size range of 70 to 105 microns. Neves reported this genus from marine shales of Namurian A and C in England.

Secarisporites Neves 1961 is a highly ornamented, radial, trilete spore that in proximo-distal view is subcircular to very roundly triangular. The outer portion of the exine (exoexine) has developed a number of large lobate structures, suggesting a flangelike development that is deeply dissected in the interlobed areas. The distal surface of the spore is covered with ridges and warts. The rays of the trilete aperture are long and thin. Two species described by Neves were distinguished on the basis of size and spacing of the lobed structures. The type species is *S. lobatus*, known from Namurian A through C. Another species, *S. remotus*, is from Namurian B and C from England. Peppers (1964) described a species of *Secarisporites* that occurred in Middle to Upper Pennsylvanian rocks in Illinois and Kentucky.

Spinozonotriletes Hacquebard 1957 was proposed for radial, trilete spores with a body surrounded by a thick membrane that has acutely pointed spines. It is not known whether the membrane is perisporal or whether the body represents a mesosporium. The outline in proximo-distal view is subcircular to roundly triangular; the rays of the trilete aperture are long; lips are faint to distinct. The known size range is 51 to 150 microns, and the genus is known to occur in Devonian and Lower Mississippian rocks (Richardson, 1964).

Bisaccate Taxa. Prepollen or pollen grains of radial symmetry represent a significant morphological advance in organization that occurred in the Middle to Late Pennsylvanian. They are generally not abundant until Permian time. The modern counterparts are found exclusively within the Podocarpaceae and Pinaceae of the Coniferales. To my knowledge bisaccate prepollen or pollen occurs only in the Gymnospermae in the geologic past.

The pollen of any one of several species of *Pinus* can serve to illustrate the morphology of a typical bisaccate grain. The body in proximodistal view may be circular to oval, with two oppositely placed bladders, or sacci. The bladders occupy approximately two-thirds of the maximum length. If such a grain were oriented in longitudinal equatorial view, the oppositely placed bladders would be observed to be distally inclined, inasmuch as the body tapers toward the distal sulcus. The proximal surface is covered with a thickened layer, termed the cap, which extends from bladder to bladder over the proximal surface and approximately to the equator of the body between the bladders. The bladders are internally reticulate. Germination of the pollen tube in true pollen grains of this type would be distally through the thin sulcus area. No functional proximal aperture is present, although, as mentioned earlier, a vestigial proximal aperture may occur, as in *Abies*.

In essence we have described *Pityosporites* (Seward) Manum 1960, and the type species is *P. antarcticus* Seward. The precise origin in time of *Pityosporites* is in doubt. Neves (1961) illustrated and compared a saccate grain with *Pityosporites* from the Namurian C of England. Williams (1955) described *P. westphalensis* from the Westphalian A-C Coal Measures of Great Britain. Williams reported that this species is a rare constituent of the assemblages and postulated that the parent plants formed a part of the upland flora rather than being present in the coal swamps proper. I have observed a few specimens of *Pityosporites* from coals of Ohio and Illinois, and in both places these occurrences were Late Pennsylvanian. Recently I have observed *Pityosporites* from coals and coal partings (Middle Pennsylvanian) from eastern Kentucky.

Parasporites Schopf 1938 (pl. 12-4, fig. 8), is a radial trilete bisaccate prepollen grain with an overall appearance of bilateral symmetry because of its oppositely placed bladders. In proximo-distal view the body is circular to oval, dense, thick, and appears rugose. The bladders are translucent, smooth externally, granulose to faintly reticulate internally. The bladders are distally inclined to some extent, a condition that varies somewhat with individual specimens. The trilete aperture displays two well-marked rays that extend toward the bladders and a third ray that may be nothing more than a deflection or a shortened third ray. This shortened third ray varies considerably in its distinctiveness from specimen to specimen. The type species is *P. maccabei* Schopf, which was reported from the Nos. 5 and 6 coals of the Carbondale Formation (Middle Pennsylvanian) from Illinois. One of the more unusual aspects of *Parasporites* is its size, which Schopf reported to be as much as 300 microns in maximum diameter. Winslow (1959) reported that the genus occurs in coals from Illinois from the Rock Island No. 1 coal at the base of the Spoon Formation, and in the Carbondale and Modesto Formations (Middle and Late Pennsylvanian). Winslow also reported that the maximum size range of *Parasporites* is 257 to 340 microns. She thought that the bladders joined proximally and that the folding of these caused the rugose proximal ornamentation. If this is true, then technically *Parasporites* would be monosaccate rather than bisaccate. *Parasporites* has not been reported in the literature as occurring outside of Illinois. In part this undoubtedly is because *Parasporites* occurs in the plus-210-microns residue and is not customarily examined by palynologists studying the small-spore, or miospore, assemblages. I have observed a single specimen of *Parasporites* from the Carbondale Formation of western Kentucky.

I have described (Kosanke, 1950) a prepollen grain under the name *Illinites* that is somewhat similar in construction to *Parasporites*. It is only a fraction of the size of *Parasporites*, being 56 to 70 microns in maximum diameter. Subsequently the genus was emended by Potonié and Klaus (in Potonié and Kremp, 1954); and a number of new species have been added, increasing the known size range to about 40 to 85 microns. The type species is *I. unicus*, and the genus occurs in five coals of Late Pennsylvanian age in the Bond and Mattoon Formations of Illinois (Kosanke, 1950). The genus was reported by Hoffmeister and Staplin (1954) from the Morehouse Formation of Louisiana; by Barss, Hacquebard, and Howie (1963) in the Pictou Group of Canada, and by

Peppers (1964) from Middle and Upper Pennsylvanian samples from Illinois and Kentucky. The genus has been reported by several authors from the Permian.

Illinites is not at the same stage of advancement as *Pityosporites*. We have every reason to believe that germination in *Pityosporites* was distal, as in modern bisaccate pollen of the Coniferales. This is not necessarily true for *Illinites* because a proximal aperture is present and was conceivably functional. Further, no strongly developed cap is present in *Illinites*, at least none that is comparable to the caps of many bisaccate grains described from Permian rocks or to the caps of modern conifer pollen. The distal sulcoid area is not a thin membranelike area as in modern conifer pollen. Germination in *Illinites* could easily have been proximal rather than distal. The affinity of *Illinites* is unknown, but on the basis of the bisaccate condition we can postulate that a gymnospermic origin is possible.

Jizba (1962) recorded the presence of five species of what she described as bisaccate pollen from the Pennsylvanian (Virgil Series) of Kansas. Throughout most of her Pennsylvanian samples these taxa are rare, although *Complexisporites polymorphus* Jizba 1962, *Striatiosaccites tractiferinus* (Samoilovich) Jizba 1962, and *Alisporites zapfei* (Potonié and Klaus) Jizba 1962, occur abundantly in at least one sample from the Pennsylvanian. These bisaccate taxa commonly occur in Permian strata.

Multisaccate Taxa. Spores with a definite and presumably functional proximal aperture and possessing three or more bladders are not common members of Paleozoic assemblages. Only one genus is known and that is *Alatisporites* Ibrahim 1933; the type species is *A. pustulatus*. Pollen grains of modern plants with three bladders occur only in the Podocarpaceae: *Microcachrys*, *Pherosphaera*, and *Dacrycarpus*. These three-bladdered, or trisaccate, grains of the Podocarpaceae are significantly different from the spores of *Alatisporites*. They possess three distally inclined bladders without any vestige of a proximal trilete aperture.

Alatisporites (pl. 12-4, figs. 10 and 11) is radial, trilete, with a body that is subtriangular in proximo-distal view. The interradial margins are often concave, although they may be almost straight to slightly convex. The rays of the trilete aperture are long, without lips or with poorly developed lips. The body may be levigate, granulose, or punctate to reticulate, depending on the species. The wall thickness of the body ranges from less than 2 to 5 microns. Many species have three bladders, whereas some species consistently have six. Other species apparently vary considerably in the number of bladders; for example, Morgan (1955) reported that *A. hoffmeisterii* Morgan has 7 to 11 bladders. The bladder ornamentation varies from levigate to granular to punctate, appearing almost finely reticulate. The overall size range of species of *Alatisporites* is about 50 to 150 microns.

Alatisporites has an interrupted stratigraphic range in both Mississippian and Pennsylvanian rocks. In some areas, such as the Eastern Interior Basin, the genus consistently occurs in the upper portions of coals; this suggests that the parent plant was not an early member of the plant community. Some recent studies in eastern Kentucky have shown that *Alatisporites* occurs almost anywhere in the coals and, in some instances, in the underclay. The reason for this discrepancy may be that different species are involved, some of which were present throughout a coal deposition cycle, whereas others were not present until later in the cycle.

INTRODUCTION TO MEGASPORES

A megaspore may be defined as a spore, produced by heterosporous plants, that gives rise to the female gametophyte or megagametophyte. Division of the individual spore mother cell (meiosis and mitosis) results in four megaspores. Within the free-sporing lycopods the tendency for three of the four spores to abort was reported by Arnold (1938) in *Lepidostrobus braidwoodensis*. Schopf (1938) proposed the genus *Cystosporites* to house the seed megaspores and abortive megaspores of the Lepidocarpaceae. In this genus one spore was fertile and three were abortive in each tetrad. Chaloner (1958c) found that in the tetrads of the fern *Stauropteris burntislandica* two fertile and two abortive spores were formed. Usually megaspores from Paleozoic plants are significantly larger than their corresponding microspores. However, Thompson (1934) demonstrated that some pollen grains of modern gymnosperms and angiosperms are as large as or larger than their corresponding megaspores.

Megaspore differentiation may be difficult to make on the basis of size of Devonian representatives because many species are as large as 200 microns (Richardson, 1960). The size is extremely variable, however, and representatives of these species would be found also in the small size residue (—65 mesh). In the lack of evidence to the contrary I believe these spores should be considered as small spores for the present.

The genus *Dicrospora* Winslow has a known size range, exclusive of appendages, of about 70 to 550 microns. Winslow (1962) reported, "The most evident relationships are with the genus *Kryshtofovichia* as described by Nikitin (1934), and with the *Archaeotriletes* 'subgroup' of Naumova . . . *Kryshtofovichia* Nikitin (1934) is regarded as heterosporous with megaspores and microspores entirely differing in form." *Dicrospora* is related morphologically to the megaspores of *Kryshtofovichia*. Certainly the size attained within the genus *Dicrospora* is well within that of known megaspores, but the lower limit of this size range is well within that of small spores. Winslow (1962) reported, "Whether these spores functioned as megaspores, microspores, or isospores is unknown." (See p. 218.)

We do not have to look far among Paleozoic spore genera—that is, nonmegaspore genera—to find large spores. Both well-known prepollen genera *Parasporites* and *Monoletes* attain sizes in excess of 300 and 500 microns, respectively. Chaloner (1951) demonstrated that the spores of the eligulate cone genus *Spencerites insignis* compare very closely with the isolated spores of *Spencerisporites karczewskii*. He reported discrepancies in the size ranges that were attributed to different modes of preservation. *Spencerites* is homosporous, and therefore the possibility exists that the isolated spores of *Spencerisporites karczewskii* are examples of homospores greater than 300 microns in diameter.

Size differentiation as a basis for distinction between megaspores and small spores must be viewed with caution. Evidence for heterospory as presented by Arnold (1939) for the Late Devonian genus *Archaeopteris* is unquestioned. Chaloner (1959) made some interesting comparisons of megaspore size on material from the Devonian through Upper Cretaceous. He noted that with one exception, Devonian megaspores are appreciably smaller in diameter than Carboniferous megaspores and that there is a continuous decrease in megaspore size from the Carboniferous to the Upper Cretaceous. The mean diameter of the six taxa of Devonian megaspores described by Chaloner range from 270 to 925 microns and averages 525 microns.

The stratigraphic occurrence of megaspores is inevitably linked with the occurrence of heterosporous plants. Andrews (1961) discussed heterospory and the evolution of the seed. The oldest known heterosporous megafossils, according to Andrews, would be Late Devonian or possibly late Middle Devonian. Andrews also considered evidence based on isolated spore studies.

Winslow (1962) reported the occurrence of megaspores assignable to the Lycopodineae, the genus *Triletes*, from Upper Devonian rocks of Ohio. She also reported eight species of *Triletes* that originate in Lower Mississippian rocks; this indicates a rapid development of the genus at that time. These are unquestioned megaspores based not only on size but also on morphology.

Classification

Schopf (1938) proposed that the megaspores of heterosporous free-sporing lycopods classified as members of the genus *Triletes* be divided into four sections: *Aphanozonati*, *Lagenicula*, *Auriculati*, and *Triangulati*. The isolated seed megaspores of *Lepidocarpon* were placed in the genus *Cystosporites*. Schopf, Wilson, and Bentall (1944) described *Calamospora* and remarked, "*Calamospora* is unique among genera typically represented by plant spores in that megaspores, microspores, and probably isospores, are included in it." Subsequently Dijkstra and Vierssen Trip (1946) proposed an additional section, which they called Zonales. Winslow (1959), using this system of classification, made no sectional assignment for four species and three varieties assigned to *Triletes*.

This system of classification would now appear as follows:

Genus *Triletes* (Bennie and Kidston) ex Zerndt 1930
- Sectio Lagenicula (Bennie and Kidston) Schopf 1938
- Sectio Aphanozonati Schopf 1938
- Sectio Auriculati Schopf 1938
- Sectio Zonales Dijkstra 1946
- Sectio Triangulati Schopf 1938
- Sectio Incertus

Genus *Cystosporites* Schopf 1938

Genus *Calamospora*, Schopf, Wilson, and Bentall 1944

Morphology and Affinities of *Triletes*

The genus *Triletes* (Bennie and Kidston) ex Zerndt 1930 has radial symmetry, is virtually circular in proximo-distal view and is either circular or somewhat elongate in longitudinal view. Species assigned to the genus range from about 180 to at least 4,000 microns in diameter. Species of *Triletes* may be smooth, punctate, spinate, or variously ornamented distally; the proximal surface is ornamented to a significantly lesser degree, if at all. Some species have an equatorial rim or flange. The trilete aperture is clearly defined, the lips are generally thick or prominent, and arcuate ridges and an apical prominence are commonly present.

Triletes definitely is associated with heterosporous, free-sporing lycopods. They have a stratigraphic range from Late Devonian to Holocene because isolated megaspores of *Selaginella* could be herein classified.

The type species of *Triletes* is *T. glabratus*, and a specimen is illustrated in plate 12-5, figure 5.

Sectio Lagenicula (Bennie and Kidston) Schopf 1938. Spores assigned to this section are characterized by a unique structural development of a portion of the pyramic surface resulting in an elongation, or beak, in the apical areas of the pyramic segments. Arcuate ridges are generally well developed, but zonal structures characteristic of the sections Zonales and Triangulati are lacking. Spores assigned to this section range in size from 250 microns in *Triletes papillaephorus* to about 2,300 microns in *T. crassiaculeatus* based on measurements of 21 Carboniferous species. The spores are circular in proximo-distal view, more or less prolate in longitudinal plane. The spore coat ranges in thickness from as little as 4 microns to as much as 50 microns. Ornamentation ranges from levigate to strikingly spinose.

Chaloner (1953a) has shown that *Triletes horridus* (Zerndt) Schopf, Wilson, and Bentall (pl. 12-5, fig. 6) is to be correlated with *Lepidostrobus dubius* Bennie; *Triletes crassiaculeatus* (Zerndt) Schopf, Wilson, and Bentall, with *Lepidostrobus allantonensis* Chaloner; and *Triletes rugosus* (Loose) Schopf, with *Lepidostrobus russelianus* Bennie and *L. olryi* Zeiller. Felix (1954) reported that the megaspores of *Lepidostrobus diversus* Felix are "referable to the Lagenicula section of *Triletes* and conform most closely to *Triletes rugosus* (Loose) Schopf." Dijkstra (1958) reported that spores belonging to the Lagenicula section occur in *Sigillariostrobus* cf. *S. major* (Germar) Zeiller.

Winslow (1959) reported that spores of the Lagenicula section are present in Illinois coals from the Chester Series through the entire Pennsylvanian sequence. She noted that the spinose species are prominent and range from the Chester Series through the Caseyville Formation to the basal Reynoldsburg coal of the Abbott Formation. Smooth-walled species first occur in the Spoon Formation in Illinois.

Arnold (1950) reported species of this section in the coals of Cycles D and F of Michigan, from a coal of a strip mine in Section 14 of Eaton County, Michigan, and from a shale at the Big Chief No. 8 mine, St. Charles, Michigan.

Cross (1947) reported species of this section in the Pottsville, Allegheny, Monongahela, and possibly from a portion of the Conemaugh Series of Pennsylvania, West Virginia, and Kentucky.

Dijkstra (1952a) reported species of this section in the Dutch Carboniferous throughout the Westphalian. He further reported a species from Turkey as low as Namurian A, and possibly as low as the Dinantian from Poland.

Potonié and Kremp (1954) reported species of this section from the Viséan throughout most of the Stephanian of Europe. Dijkstra and Piérart (1957) reported lageniculate megaspores from the Lower Carboniferous of the Moscow Basin.

Winslow (1962) reported four species of the Lagenicula section at the base of the Mississippian in Ohio, and two other taxa are also present that may be assigned to this section.

Plate 12-5. Megaspore genera. (All illustrations from Winslow (1959) through the courtesy of the Illinois Geological Survey.) 1—*Cystosporites breretonesis* Schopf 1938; total length 3,390 microns. 2—*Triletes brasserti* Stach and Zerndt 1931; total diameter, 1,795 microns. 3—*T. auritus* Zerndt 1930 (sensu Potonié and Kremp 1956); total diameter, 1,205 microns. 4—*T. triangulatus* Zerndt 1930 (sensu Dijkstra 1946); total diameter, 2,345 microns. 5—*T. glabratus* Zerndt 1936 (sensu Dijkstra 1946); total diameter, 2,345 microns. 6—*T. horridus* (Zerndt) Schopf, Wilson, and Bentall 1944 (sensu Dijkstra 1946); total diameter, 865 microns.

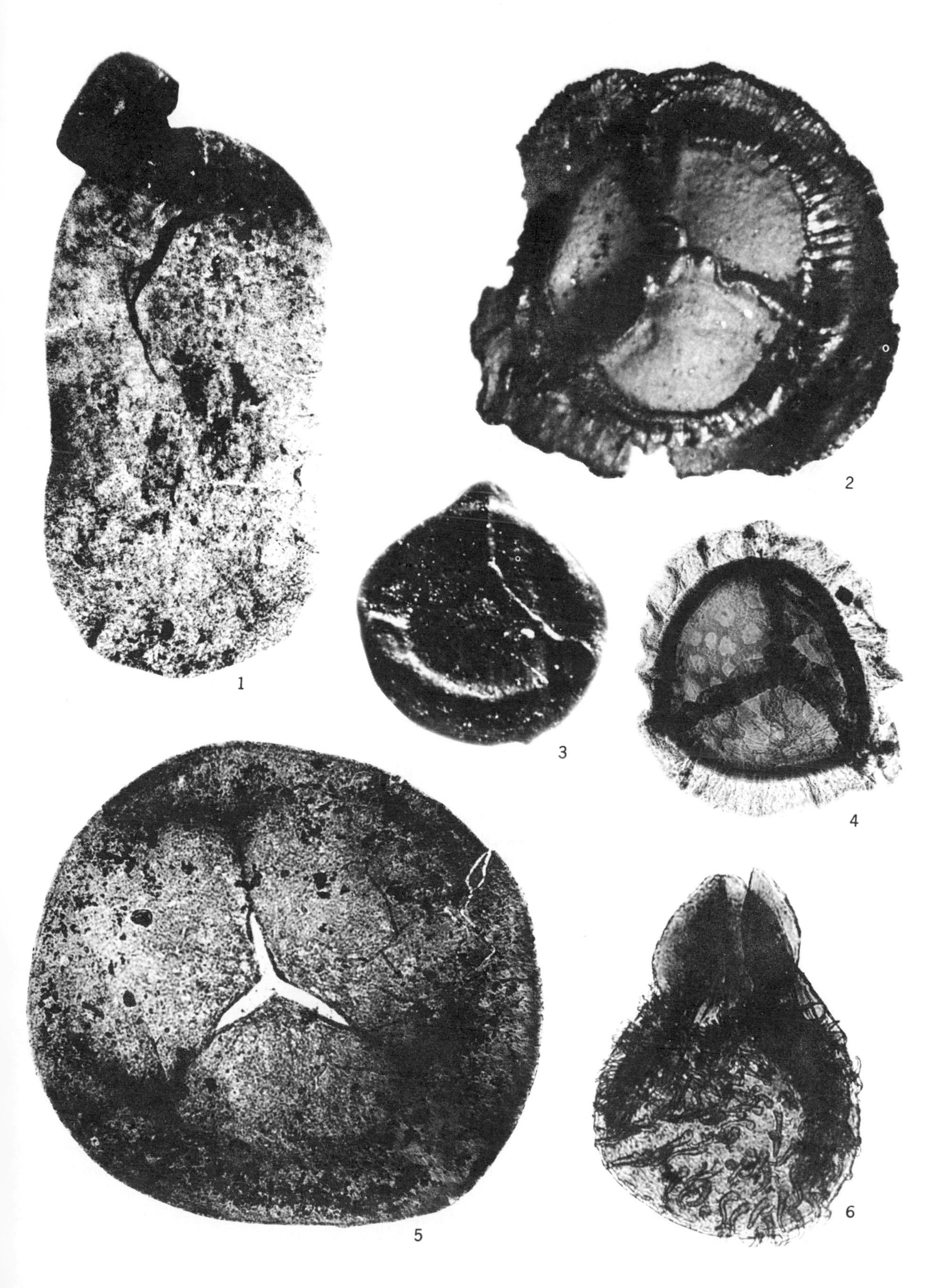
1
2
3
4
5
6

Sectio Aphanozonati Schopf 1938. Spores of this section are characterized by being originally more or less saucer shaped and appearing circular to oval shaped in proximo-distal view. These spores range in size from 180 microns for *Triletes microgranulatus* to 3,000 microns for *T. glabratus* (pl. 12-5, fig. 5) based on measurements of 17 Carboniferous species. The question may be raised as to whether or not *T. microgranulatus* is a valid member of the genus based on size, which according to Dijkstra and Piérart (1957) ranges from 180 to 230 microns and averages 215 microns. The surface in this taxon is covered with granulations 1 to 2 microns in diameter, and the wall thickness is reported to be 7 microns. Should this taxon ultimately be excluded from *Triletes*, the minimum size range would be raised to 270 microns in *T. ignobilis*. The trilete aperture is distinct in species assigned to this section, but lip development is usually modest. Arcuate ridges are present to sharply defined. The spore coat ranges in thickness from 7 to at least 40 microns and it may be smooth to spinate.

Bocheński (1936) found a variation in the size of the megaspores of *Sigillariostrobus czarnokii*, which are of the *Triletes glabratus* type. Schopf (1941a) related the megaspores of *Mazocarpon oedipternum* Schopf to the Aphanozonati section, specifically to *Triletes glabratus*. Chaloner (1953c) reported that the megaspores of the compression megafossil *Sigillariostrobus rhombibracteatus* (Kidston) Chaloner are related to *Triletes mamillarius* as well as to the megaspores of the petrifaction *Mazocarpon shorense* Benson.

Winslow (1959) reported that spores of the Aphanozonati section occur in all of the Pennsylvanian formations of Illinois. Cross (1947) reported a similar occurence in his study of the spore floras of Pennsylvania, West Virginia, and Kentucky. Arnold (1950) reported that taxa of this section are present in a shale of Cycle A, and in a shale from the Big Chief No. 8 mine at St. Charles, Mich. Dijkstra (1952a) reported that spores assignable to this section occur in Turkey throughout the Westphalian, in Poland from Namurian B throughout the Westphalian and possibly as low as the Namurian A, and that they occur in the Namurian B and all of the Westphalian of Holland. Potonié and Kremp (1954) indicated the occurrence of these spores in Europe from Namurian A to Stephanian B. Dijkstra and Piérart (1957) described spores assignable to this section from the Lower Carboniferous of the Moscow Basin. Winslow (1962) reported a taxon from the base of the Mississippian in Ohio, which she assigned to this section of *Triletes*.

Sectio Auriculati Schopf 1938. The spores of this section are characterized by the presence of arcuate thickenings that are bulbose projections or "ears" developed at the radial extremities. The spores are subtriangular to trilobate in proximo-distal view. They range in size on the basis of measurement of five Carboniferous species, from 650 microns in *Triletes auritus* Zerndt (sensu Dijkstra) to 1,600 microns in *T. grandiosus* Dijkstra and Piérart. If *T. hamatus* Dijkstra is assigned to this section, the maximum size range would be increased to 2,600 microns. The trilete rays are sharply defined and extend to or beyond the arcuate ridge, with a strong lip development. Dijkstra and Piérart (1957) reported that the spore coat of *T. grandiosus* possesses two layers; the outer layer is 120 to 180 microns thick, and the inner layer is about 10 microns thick. Winslow (1959) reported the presence of an inner membrane in *T. auritus* Zerndt (pl. 12-5, fig. 3) and a spore coat that is 20 to 52 microns thick, being thinnest in the contact areas.

Potonié and Kremp (1955-1956), Dijkstra (1955), and Dijkstra and Piérart (1957) have reported species of the Auriculati section as occurring from Westphalian A through the Stephanian of Europe. Cross (1947) reported their occurrence from the Pottsville through the Monongahela Series of the Eastern United States. Winslow (1959) reported their occurrence in Illinois from the Caseyville through the Mattoon Formations.

Chaloner (1953b) described *Lepidostrobus zea*. The megaspores of this species were identified with *Triletes auritus* Zerndt, whereas the microspores were of the *Endosporites globiformis* (Ibrahim) Schopf, Wilson, and Bentall type. Chaloner (1958b) demonstrated the occurrence of *Triletes auritus Zerndt* and *Endosporites globiformis* (Ibrahim) Schopf, Wilson, and Bêntall in the heterosporous lycopod cone *Polysporia mirabilis* Newberry. This suggests that the range of *Endosporites* should parallel that of the auriculate megaspores. Winslow (1959) reported that auriculate megaspores are generally present throughout the

Pennsylvanian of Illinois, and reported *Endosporites* present through much of the Pennsylvanian of Illinois (Kosanke, 1950). *Endosporites* is reported present in Devonian and Mississippian strata by Richardson (1964); however, auriculate megaspores are not known to occur in this portion of the Paleozoic.

Secto Zonales Dijkstra 1946. The species assigned to this section are characterized by the presence of an equatorial rim and a zone composed of anastomosing appendages that form a more or less solid flange, as in *Triletes brasserti* Stach and Zerndt 1931 (pl. 12-5, fig. 2), or open, as in *T. rotatus* Bartlett 1929. The flange is located just to the proximal side of the equator in many specimens of *T. superbus* Bartlett 1929. As indicated by Arnold (1961) this may be due to spore enlargement during later stages of development. In proximo-distal view the taxa of this section are subtriangular to subcircular in outline. These spores range in size from 250 microns in *T. annulatus* Dijkstra to 4,100 microns in *T. superbus* Bartlett (see Winslow, 1959). The trilete rays are prominent; the lips are either straight or sinuous, usually thick or high, and the upper edges of the lips may be plain or possess small spines. The spore coat is generally thinner proximally and thicker distally or at the juncture of the rim; it ranges from 12 to 83 microns, according to Winslow. An inner membrane is present in *T. ramosus* and *T. brasserti*, and in the latter Winslow (1959) reported the occasional presence of small tubercles. The spore coat may be smooth to tuberculate proximally or with spines originating on the tubercles. Distally the spore coat is smooth or may possess blunt processes.

Species of this section are known to occur in Europe from the Dinantian through Westphalian D (Zerndt, 1930, 1937, and 1938; Dijkstra, 1952b, 1955, 1957; Potonié and Kremp, 1954; and others). Cross (1947) reported their presence in the Pottsville Series of Eastern United States with two questionable occurrences in the Allegheny Series. Winslow (1959) reported that spores of this section occur in Illinois in the Caseyville, Abbott, and Spoon Formations. Chaloner (1956) reported the presence of *Triletes superbus* presumably from the Francis Creek Shale Member of the Carbondale Formation in Illinois. These megaspores of *T. superbus* were from a lycopsid cone *Sporangiostrobus langfordi* Chaloner.

Chaloner (1962) described *Sporangiostrobus ohioensis* from the Pennsylvanian of Ohio, which contained megaspores most similar to *Superbisporites superbus* (Bartlett) Potonié and Kremp. However, variation within the megaspore population was observed, and therefore the spores could be compared with three genera of dispersed megaspores in the most general terms. These three genera are the following:

Superbisporites (Bartlett) Potonié and Kremp 1954
Zonalesporites (Stach and Zerndt) Potonié and Kremp 1954
Rotatisporites (Bartlett) Potonié and Kremp 1954

Chaloner further commented, "In the structure of the equatorial flange the megaspores of *Sporangiostrobus ohioensis* might be said to be intermediate between *Superbisporites superbus* (with a more or less continuous flange of anastomosing processes) and *Rotatisporites ramosus* (with a more clearly defined 'rim with spokes' structure)."

Sectio Triangulati Schopf 1938. Spores assigned to this section are characterized by the presence of an equatorial, solid, membranous flange and are of small to medium size. They range in size from 350 microns in *Triletes tricollinus* Zerndt to at least 1,000 microns in the most commonly occurring species *T. triangulatus* Zerndt (pl. 12-5, fig. 4). The body of the spore is virtually circular in proximo-distal view; because the flange is widest opposite the rays, the overall outline is subtriangular. The trilete rays are distinct and extend to the flange margin. The lips are membranous and many are plicate. The sutures extend about one-half the body radius. The contact areas of *T. triangulatus* may be granulose or reticulate, or they may have radiating ridges. Distally a reticulate perisporial membrane is present. The spore coat is granulose in both *T. triangulatus* and *T. tricollinus*. The contact areas possess elliptical elevations, and the reticulate distal membrane is lacking in *T. tricollinus*. Winslow (1959) reported an inner membrane ornamented with small papillae in *T. triangulatus*.

Dijkstra and Vierssen Trip (1946) placed *Triletes gymnozonatus* Schopf in synonymy with *T. triangulatus* Zerndt. Guennel (1954) proved that the perisporial membrane and flange of *T. triangulatus* are readily detached and that the remaining spore body was in fact *T. gymnozonatus*. Guennel also summarized the stratigraphic distribution of *T. triangulatus*.

Schopf (1938) reported that the characteristics of the spores assigned to the Triangulati section were found in *Selaginellites* and modern heterosporous lycopods. For this reason he believed that the spores of this section should probably be correlated with *Selaginellites* or similar forms. Chaloner (1954a) reported that the megaspores of *S. suissei* Zeiller were to be identified with *Triletes triangulatus*. Hoskins and Abbott (1956) found that the megaspores of *Selaginellites crassicinctus* Hoskins and Abbott were of the *Triletes triangulatus* type.

Dijkstra (1952a, 1952b, and 1955) and Potonié and Kremp (1954) recorded species of this section occurring from Namurian A through Westphalian D in Europe and Turkey. Cross (1947) reported *Triletes triangulatus* occurring throughout most of the Pennsylvanian of the Eastern United States. Arnold (1950) reported this species occurring in Cycles A and E of Michigan as well as other shales and coals from mines and quarry locations. Winslow (1959) did not find *T. triangulatus* in the Mississippian of Illinois but reported it in all of the Pennsylvanian formations.

Sectio Incertus. Spores not readily assignable to one of the sections previously discussed are provisionally placed in Sectio Incertus until such time as their proper placement can be established, providing the author has reason to believe the spores in question are related to the free-sporing lycopods.

There are differences of opinion concerning the section assignment in some instances, but these differences are of minor importance in the overall picture. Winslow (1959) did not make sectional assignments for four species and three informal varieties. All possess spinate appendages and in one of these species, *Triletes praetextus*, the spines are somewhat restricted to an equatorial zone, although not to the degree found in species assigned to the Sectio Zonales.

Triletes globosus Arnold was not given section assignment by Winslow, although the presence of a small apical prominence suggested to Arnold (1950) that "The small apical prominence may indicate affinity with the *Lagenicula* group of spores ...." The apical prominence in both *Triletes globosus* and *T. praetextus* is formed by an increase in lip height at the spore apex, but in the spores of the section Lagenicula the apical prominence is a result of an elongation of the three contact areas. The apical prominence in some specimens of *T. globosus* is difficult to observe. Specimens treated with Clorox for 5 to 10 seconds are reduced in opacity so that the nature of the apical prominence can be observed.

Morphology and Affinities of *Cystosporites*

The genus *Cystosporites* Schopf 1938 (pl. 12-5, fig. 1) is radial and has a trilete aperture. Fertile spores are more or less oval in proximo-distal view and saclike in longitudinal view. Abortive spores are circular to oval in transverse and longitudinal planes. Fertile spores can attain a length greater than 1 centimeter according to Winslow (1959), whereas most abortive spores are one-third or less the diameter of fertile spores. Fertile spores are characterized by a fibrous spore coat in the median region. The wall may be ornamented with bumplike protrusions, which form an inverse reticulum, or it may be granulose or spinose. Abortive spores may be granulose or spinose, or they may have an inverse reticulation. The walls of abortive spores are comparatively thick for the spore size. An apical cushion, or tuft, that is spongy or granulose may occur on either fertile or abortive spores or both. The trilete apertures of *Cystosporites* are well developed in both the fertile and abortive spores. The apertures of fertile spores are not large in proportion to spore size. The apical cushion can mask the aperture of abortive spores. The sutures are distinct and extend to the arcuate ridges. The lips are of modest development and may be elongate toward the apex.

The genus was proposed by Schopf (1938) to include spores pertaining to lycopod seeds. Schopf (1941b) indicated the relationship to the Lepidocarpaceae. Chaloner (1952) demonstrated that the megaspores of *Lepidocarpon waltoni* Chaloner correlate with *Cystosporites giganteus* (Zerndt) Schopf. He discussed both fertile and abortive megaspores.

Dijkstra (1952a) reported *C. giganteus* from

Namurian A through D in Turkey, from possibly as low as the Dinantian through Westphalian D in Poland, and throughout the Westphalian in Holland. He reported *C. varius* (Wicher) Dijkstra in the Westphalian A, B, and D of Turkey, and throughout the Westphalian of Poland and Holland. Dijkstra reported *C. verrucosus* from the Westphalian B of Holland. Potonié and Kremp (1954) reported the genus in Europe from the Viséan and Namurian A, and Westphalian A through Stephanian A. Dijkstra (1955) reported *C. giganteus* and *C. varius* as ranging from Westphalian B into the Stephanian of Spain. Dijkstra (1957) reported *C. giganteus* and *C. barbatus* from the Lower Carboniferous of Egypt, and *C. giganteus* from the Lower Carboniferous Limestone Coal Group of Scotland and Ireland. Dijkstra and Piérart (1957) reported *C. zerndti* and *C. strictus* in the Lower Carboniferous of the Moscow Basin.

Chaloner and Pettitt (1963) reported the occurrence of a seed megaspore that they assigned to *Cystosporites* from an " . . . Upper Devonian sandstone bearing compressions of *Archaeopteris* from the Escuminac Formation, of Scaumenac Bay, Quebec, Canada . . . . Spores of the same type have also been obtained from the bedding planes of the Acanthodian Bed of the same locality."

Cross (1947) reported *Cystosporites* in the Pennsylvanian of the Eastern United States, specifically the Pottsville, Allegheny, and lower part of the Monongahela Series. Arnold (1950) reported the presence of *C. varius* from a Pennsylvanian coal of Michigan, Grand Ledge coal of Cycle B, and from the Williamston spore coal, Williamston, Mich. He reported *C. giganteus* from the coal of Cycle F at Grand Ledge, a strip mine in section 14, Eaton County, and a shale from the Big Chief No. 8 mine at St. Charles, Mich. Chaloner (1954b) reported *C. giganteus* from the Mississippian of Indiana. Winslow (1959) reported *C. giganteus* from the Mississippian Chester Series of Illinois as well as the Caseyville, Abbott, Spoon, Carbondale, and Modesto Formations of Pennsylvanian age of Illinois. She reported *C. verrucosus* from the Caseyville, Abbott, and Carbondale Formations, and *C. varius* from the Caseyville, Abbott, Spoon, and Carbondale Formations. A species of *Cystosporites* was recognized by Winslow as occurring within the Modesto Formation.

The Genus Calamospora

As indicated earlier in this chapter, *Calamospora* (pl. 12-1, fig. 5) is unique in that the generic circumscription includes homospores, microspores, and megaspores. The characteristics of the genus are discussed in this chapter under "Spores Without Equatorial Structures," p. 236. Horst (1955) reported megaspores of the genus larger than 1,000 microns in diameter, and Winslow (1959) indicated a wall thickness up to 18 microns.

Discussion of Classification

The system of megaspore classification proposed by Potonié and Kremp (1954) contrasts with that proposed by Schopf (1938) in that a number of genera are proposed for the free-sporing lycopods that are included in *Triletes* and sections of the genus by Schopf. From the standpoint of a natural system of classification it seems reasonable to classify all megaspores of the free-sporing lycopods within *Triletes*. One of the problems associated with the use of the section system of classification is simply that, inevitably, some taxa do not fit precisely within the available sections; for example, the apical thickenings of the spores of *T. globosus* Arnold are not equivalent to the apical thickenings of spores assigned to the Lagenicula section. Clearly such a difference should merit placing these spores in a new section.

Arnold (1950) believed that spores with an apical prominence deserved generic recognition, and he followed Zerndt by placing such spores in the genus *Lagenicula*. However, he did place spores with zonal structures, both flanges and restricted bulbous projections, as well as taxa without zonal appendages, in *Triletes*.

The system proposed by Potonié and Kremp (1954) meets one of Arnold's objections to the section system of classification in that more genera are available to house the diverse forms of megaspores associated with the lycopods. However, Arnold suggested the possible elevation of the sections to a generic rank. Potonié and Kremp created four genera—*Zonalesporites*, *Radiatisporites*, *Rotatisporites*, and *Superbisporites*—to house spores with a pronounced equatorial rim and flange zone. As a result of this, and other changes, the Potonié and Kremp system of classification would include 16 genera as contrasted to the 3 genera of Schopf.

I do not propose to evaluate the system of classification used in the Soviet Union without having access to all of the necessary literature and the opportunity of making comparisons at least by means of photomicrographs. This system was originally proposed by Naumova (1939). Arnold (1950) reviewed the system and indicated the lack of designated types. Types have subsequently been designated, and, regardless of any personal objections to the names, they should be adopted if they have priority and can qualify within the framework of the "International Code of Botanical Nomenclature."

The system of Potonié and Kremp (1954), together with the type species, is given below within the sectional groupings reviewed earlier in this chapter:

Triletes (Bennie and Kidston) ex Zerndt 1930
- Sectio Lagenicula
  - *Lagenicula* (Bennie and Kidston) Potonié and Kremp 1954
    - *L. horrida* Zerndt 1934
  - *Lagenoisporites* Potonié and Kremp 1954
    - *L. rugosus* (Loose) Potonié and Kremp 1954
- Sectio Aphanozonati
  - *Laevigatisporites* (Ibrahim non Bennie and Kidston) Potonié and Kremp 1954
    - *L. primus* (Wicher non Bennie and Kidston) Potonié and Kremp 1954
  - *Tuberculatisporites* (Ibrahim) Potonié and Kremp 1954
    - *T. tuberosus* Ibrahim 1933
- Sectio Auriculati
  - *Valvisisporites* (Ibrahim) Potonié and Kremp 1954
    - *V. trilobus* Ibrahim 1933
- Sectio Zonales
  - *Zonalesporites* (Ibrahim) Potonié and Kremp 1954
    - *Z. brasserti* (Stach and Zerndt) Potonié and Kremp 1954
  - *Radiatisporites* Potonié and Kremp 1954
    - *R. radiatus* (Zerndt) Potonié and Kremp 1954
  - *Rotatisporites* Potonié and Kremp 1954
    - *R. rotatus* (Bartlett) Potonié and Kremp 1954
  - *Superbisporites* Potonié and Kremp 1954
    - *S. superbus* (Bartlett) Potonié and Kremp 1954
- Sectio Triangulati
  - *Triangulatisporites* Potonié and Kremp 1954
    - *T. triangulatus* (Zerndt) Potonié and Kremp 1954
- Sectio Incertus
  - *Colisporites* Potonié and Kremp 1954
    - *C. bulbosus* (Horst) Potonié and Kremp 1954
  - *Triletisporites* (Potonié) Potonié and Kremp 1954
    - *T. tuberculatus* Zerndt 1930
  - *Setosisporites* (Ibrahim) Potonié and Kremp 1954
    - *S. hirsutus* (Loose) Ibrahim 1933
  - *Bentzisporites* Potonié and Kremp 1954
    - *B. bentzii* Potonié and Kremp 1954

Cystosporites Schopf 1938
- *C. breretonensis* Schopf 1938

Calamospora Schopf, Wilson, and Bentall 1944
- *C. hartungiana* Schopf 1944

References

Alpern, Boris, 1958, Description de quelques microspores du permo-carbonifère français: Rev. Micropaléontologie, v. 1, no. 2, p. 75-86.

——1964, La Stratigraphie Palynologique du Stephanien et du Permien: Avanc. Études Stratigraphie et Géologie Carbonifère Congr., 5th, p. 1119-1129.

Andrews, H. N., Jr., 1961, Studies in paleobotany: New York and London, John Wiley & Sons, Inc., 487 p.

Arnold, C. A., 1938, Note on a lepidophyte strobilus containing large spores, from Braidwood, Illinois: Am. Midland Naturalist, v. 20, no. 3, p. 709-712.

——1939, Observations on fossil plants from the Devonian of eastern North America; Pt. 4, Plant remains from the Catskill delta deposits of northern Pennsylvania and southern New York: Michigan Univ. Mus. Paleontology Contr., v. 5, no. 11, p. 271-314.

——1944, A heterosporous species of *Bowmanites* from the Michigan Coal Basin: Am. Jour. Botany, v. 31, no. 8, p. 466-469.

______1950, Megaspores from the Michigan Coal Basin: Michigan Univ. Mus. Paleontology Contr., v. 8, no. 5, p. 59-111.

______1961, Re-examination of *Triletes superbus, T. rotatus,* and *T. mamillarius* of Bartlett: Brittonia, v. 13, no. 3, p. 245-252.

Balme, B. E., 1952, On some spore specimens from British upper Carboniferous coals: Geol. Mag., v. 89, no. 3, p. 175-184.

______1964, The palynological record of Australian pre-Tertiary Floras, p. 49-80 *in* Ancient Pacific floras; the pollen story: Hawaii Univ. Press, 10th Pacific Sci. Cong. Ser., 114 p.

Barss, M. S., Hacquebard, P. A., and Howie, R. D., 1963, Palynology and stratigraphy of some Upper Pennsylvanian and Permian rocks of the Maritime Provinces: Canada Geol. Survey Paper 63-3, 13 p.

Baxter, R. W., 1950, *Peltastrobus reedae*, a new sphenopsid cone from the Pennsylvanian of Indiana: Bot. Gaz., v. 112, no. 2, p. 174-182.

Bhardwaj, D. C., 1955, The spore genera from the upper Carboniferous coals of the Saar and their value in stratigraphical studies: Palaeobotanist, v. 4, p. 119-149.

______1957, The palynological investigations of the Saar coals – Pt. 1, Morphography of Sporae dispersae: Palaeontographica, v. 101, sec. B, no. 5-6, p. 73-120.

______1959, On *Porostrobus zeilleri* Nathorst and its spores with remarks on the systematic position of *P. bennholdi* Bode and the phylogeny of *Densosporites* Berry: Palaeobotanist, v. 7, no. 1, p. 67-75.

Bocheński, Tadeusz, 1936, Über Sporophyllstände (Blüten) einiger Lepidophyten aus dem produktiven Karbon Polens: Polskie Towarz. Geol. Rocznik, v. 12, p. 193-240.

Butterworth, M. A., 1964a, *Densosporites* (Berry) Potonié and Kremp and related genera: Avanc. Études Stratigraphie et Géologie Carbonifère Cong., 5th, p. 1049-1057.

______1964b, Miospore distribution in Namurian and Westphalian: Avanc. Études Stratigraphie et Géologie Carbonifère Cong., 5th, p. 1115-1118.

Butterworth, M. A., and Millott, J. O'N., 1960, Microspore distribution in the coalfields of Britain: Internat. Comm. Coal Petrology Proc., no. 3, p. 157-163.

Butterworth, M. A., and Williams, R. W., 1958, The small spore floras of coals in the Limestone Coal Group and Upper Limestone Group of the Lower Carboniferous of Scotland: Royal Soc. Edinburgh Trans., v. 63, pt. 2, no. 17, p. 353-392.

Chaloner, W. G., 1951, On *Spencerisporites,* gen. nov., and *S. karczewskii* (Zerndt), the isolated spores of *Spencerites insignis* Scott: Annals and Mag. Nat. History, ser. 12, v. 4, no. 45, p. 861-873.

______1952, On *Lepidocarpon waltoni,* sp. n., from the lower Carboniferous of Scotland: Annals and Mag. Nat. History, ser. 12, v. 5, no. 54, p. 572-582.

______1953a, On the megaspores of four species of *Lepidostrobus*: Annals Botany, new ser., v. 17, no. 66, p. 263-292.

______1953b, A new species of *Lepidostrobus* containing unusual spores: Geol. Mag., v. 90, no. 2, p. 97-110.

______1953c, On the megaspores of *Sigillaria*: Annals and Mag. Nat. History, ser. 12, v. 6, no. 72, p. 881-897.

______1954a, Notes on the spores of two British Carboniferous lycopods: Annals and Mag. Nat. History, ser. 12, v. 7, no. 74, p. 81-91.

______1954b, Mississippian megaspores from Michigan and adjacent states: Michigan Univ. Mus. Paleontology Contr., v. 12, no. 3, p. 23-25.

______1956, On *Sporangiostrobus langfordi* sp. nov., a new fossil lycopod cone from Illinois: Am. Midland Naturalist, v. 55, no. 2, p. 437-442.

______1958a, *Polysporia mirabilis* Newberry, a fossil lycopod cone: Jour. Paleontology, v. 32, no. 1, p. 199-209.

______1958b, A Carboniferous *Selaginellites* with *Densosporites* microspores: Palaeontology, v. 1, pt. 3, p. 245-253.

______1958c, Isolated megaspore tetrads of *Stauropteris burntislandica*: Annals Botany, new ser., v. 22, no. 86, p. 199-204.

______1959, Devonian megaspores from Arctic Canada [Northwest Territories]: Palaeontology, v. 1, pt. 4, p. 321-332.

______1962, A *Sporangiostrobus* with *Densosporites* microspores: Palaeontology, v. 5, pt. 1, p. 73-85.

Chaloner, W. G., and Pettitt, J. M., 1963, A Devonian seed megaspore: Nature, v. 198, no. 4882, p. 808-809.

Cropp, F. W., 1963, Pennsylvanian spore succession in Tennessee: Jour. Paleontology, v. 37, no. 4, p. 900-916.

Cross, A. T., 1947, Spore floras of the Pennsylvanian of West Virginia and Kentucky, *in* Wanless, H. R., Symposium on Pennsylvanian problems: Jour. Geology, v. 55, no. 3, pt. 2, p. 285-308.

Cross, A. T., and Schemel, M. P., 1952, Representative microfossil floras of some Appalachian coals: Avanc. Études Stratigraphie et Géologie Carbonifère Cong., 3d, Heerlen, 1951, v. 1, p. 123-130.

Delevoryas, Theodore, 1953, A new male cordaitean fructification from the Kansas Carboniferous: Am. Jour. Botany, v. 40, no. 3, p. 144-150.

______1964, A probable pteridosperm microsporangiate fructification from the Pennsylvanian of Illinois: Palaeontology, v. 7, pt. 1, p. 60-63.

Dijkstra, S. J., 1952a, The stratigraphical value of megaspores: Avanc. Études Stratigraphie et Géologie Carbonifère Cong., 3d, Heerlen, 1951, v. 1, p. 163-168.

______1952b, Megaspores of the Turkish Carboniferous and their stratigraphical value [with discussion]: Internat. Geol. Cong., 18th, Great Britain, Rept., pt. 10, p. 11-17.

______1955, The megaspores of the Westphalian D and C: Geol. Stichting Med., new ser., no. 8, p. 5-11.

______1957, Lower Carboniferous megaspores: Geol. Stichting Med., new ser., no. 10 [1956], p. 5-18.

______1958, On a megaspore-bearing lycopod strobilus: Acta Botanica Neerlandica, v. 7, p. 217-222.

Dijkstra, S. J., and Piérart, P., 1957, Lower Carboniferous megaspores from the Moscow Basin: Geol. Stichting Med., new ser., no. 11, p. 5-19.

Dijkstra, S. J., and Vierssen Trip, P. H. van, 1946, Eine monographische Bearbeitung der Karbonischen Megasporen, mit besonderer Berücksichtigung von Südlimburg (Niederlande): Geol. Stichting Med., ser. C-3-1, no. 1, 101 p.

Doubinger, Jeanne, and Horst, Ulrich, 1961, Monographie de *Torispora, Crassosporites,* et *Bicoloria*: Internat. Comm. Microflore du Paleozoïque, 29 p. (mimeographed).

Dybová, Sona, and Jachowicz, Aleksander, 1957, Microspores of the Upper Silesian coal measures: Geol. Práce, v. 23, 328 p.

Felix, C. J., 1954, Some American arborescent lycopod fructifications: Missouri Bot. Garden Annals, v. 41, no. 4, p. 351-394.

Felix, C. J., and Burbridge, P. P., 1961, *Pteroretis,* a new Mississippian spore genus: Micropaleontology, v. 7, no. 4, p. 491-495.

Felix, C. J., and Parks, Patricia, 1959, An American [Okla.] occurrence of *Spencerisporites*: Micropaleontology, v. 5, no. 3, p. 359-364.

Guennel, G. K., 1952, Fossil spores of the Alleghenian coals in Indiana: Indiana Geol. Survey Rept. Prog. 4, 40 p.

______1954, An interesting megaspore species found in Indiana Block coal: Butler Univ. Bot. Studies, v. 11, Papers 8-17, p. 169-177.

Guennel, G. K., and Neavel, R. C., 1961, *Torispora securis* Balme—Spore or sporangial wall cell: Micropaleontology, v. 7, no. 2, p. 207-212.

Hacquebard, P. A., 1957, Plant spores in coal from the Horton group (Mississippian) of Nova Scotia: Micropaleontology, v. 3, no. 4, p. 301-324.

Hacquebard, P. A., and Barss, M. S., 1957, A Carboniferous spore assemblage in coal from the South Nahanni River area, Northwest Territories: Canada Geol. Survey Bull. 40, 63 p.

Hartung, Wolfgang, 1933, Die Sporenverhältnisse der Calamariaceen: Preuss. Geol. Landesenalt, Inst. Paläobotanik, Arb., v. 3, no. 1, p. 95-149.

Hoffmeister, W. S., and Staplin, F. L., 1954, Pennsylvanian age of Morehouse formation of northeastern Louisiana: Am. Assoc. Petroleum Geologists Bull., v. 38, no. 1, p. 158-159.

Hoffmeister, W. S., Staplin, F. L., and Malloy, R. E., 1955, Mississippian plant spores from the Hardinsburg formation of Illinois and Kentucky: Jour. Paleontology, v. 29, no. 3, p. 372-399.

Horst, Ulrich, 1955, Die Sporae dispersae des Namurs von Westoberschlesien und Mährisch-Ostrau; stratigraphischer Vergleich der beiden Gebiete an Hand der Sporendiagnose: Palaeontographica, v. 98, sec. B, no. 4-6, p. 137-232.

______1957, Ein Leitfossil der Lugau-Oetsnitzer Steinkohlenflöze: Geologie, v. 6, p. 698-721.

Hoskins, J. H., and Abbott, M. L., 1956, *Selaginellites crassicinctus*, a new species from the Desmoinesian series of Kansas: Am. Jour. Botany, v. 43, no. 1, p. 36-46.

Hoskins, J. H., and Cross, A. T., 1943, Monograph of the Paleozoic cone genus *Bowmanites* (Sphenophyllales): Am. Midland Naturalist, v. 30, no. 1, p. 113-163.

Hughes, N. F., Dettman, M. E., and Playford, G., 1962, Sections of some Carboniferous dispersed spores: Palaeontology, v. 5, pt. 2, p. 247-252.

Ibrahim, A. C., 1932, Beschreibung von Sporenformen aus Flöz Ägir, *in* Potonié, Robert, Sporenformen aus den Flözen Ägir und Bismarck des Ruhrgebietes: Neues Jahrb., v. 67, sec. B, p. 477-449.

______1933, Sporenformen des Aegirhorizonts des Ruhr-reviers: Würzburg, Konrad Triltsch, Tech. Hochschule Berlin dissert., 48 p.

Ischenko, A. M., 1956, Spore and pollen of the lower Carboniferous deposits of the western extension of the Donets basin and their stratigraphic significance: Akad. Nauk Ukrain. SSR Inst. Geol. Nauk Trudy, Ser. Strat. i Palaeont., v. 11, 143 p.

Jizba, K. M. M., 1962, Late Paleozoic bisaccate pollen from the United States Midcontinent area: Jour. Paleontology, v. 36, no. 5, p. 871-887.

Knox, E. M., 1950, The spores of *Lycopodium, Phylloglossum, Selaginella,* and *Isoetes* and their value in the study of microfossils of Palaeozoic age: Edinburgh, Bot. Soc. Trans., Proc., v. 35, pt. 3, p. 209-357.

______1951, Spore morphology in British ferns: Edinburgh, Bot. Soc. Trans., Proc., v. 35, pt. 4, p. 437-449.

Kosanke, R. M., 1950, Pennsylvanian spores of Illinois and their use in correlation: Illinois Geol. Survey Bull. 74, 128 p.; correction with title, *Wilsonites*, new name for *Wilsonia* Kosanke 1950, Jour. Paleontology, v. 33, no. 4, p. 700 [July 1959].

———1955, *Mazostachys*—a new calamite fructification: Illinois Geol. Survey Rept. Inv. 180, 24 p.

———1964, Applied Paleozoic palynology, p. 75-89 *in* Palynology in oil exploration—a symposium: Soc. Econ. Paleontologists and Mineralogists Spec. Pub. 11, 200 p.

Kosanke, R. M., Simon, J. A., Wanless, H. R., and Willman, H. B., 1960, Classification of the Pennsylvanian strata of Illinois: Illinois Geol. Survey Rept. Inv. 214, 84 p.

Leisman, G. A., and Graves, Charles, 1964, The structure of the fossil sphenopsid cone, *Peltastrobus reedae*: Am. Midland Naturalist, v. 72, no. 2, p. 426-437.

Loose, Friedrich, 1932, Beschreibung von Sporenformen aus Flöz Bismarck, *in* Potonié, Robert, Sporenformen aus den Flözen Ägir und Bismarck des Ruhrgebietes: Neues Jahrb., v. 67, sec. B, p. 449-452.

———1934, Sporenformen aus dem Flöz Bismarck des Ruhrgebietes: Preuss. Geol. Landesanst., Inst. Paleobot., Arb. v. 4, no. 3, p. 127-164.

Love, L. G., 1960, Assemblages of small spores from the Lower Oil-Shale Group of Scotland: Royal Soc. Edinburgh Proc., sec. B (Biology), v. 67, pt. 2, no. 7, p. 99-126.

Luber, A. A., 1955, Atlas of the spores and pollen grains of the Paleozoic deposits of Kazakhstan: Akad. Nauk Kazakh. SSR Vestnik, 126 p.

Luber, A. A., and Waltz, I. E., 1938, Classification and stratigraphic value of spores and some Carboniferous coal deposits in the USSR: Central Geol. Prosp. Inst. Trans., no. 105, 46 p. (Russian, English summ.).

Mamay, S. H., 1950, Some American Carboniferous fern fructifications: Missouri Bot. Garden Annals, v. 37, no. 3, p. 409-477.

Morgan, J. L., 1955, Spores of McAlester coal: Oklahoma Geol. Survey Circ. 36, 52 p.

Naumova, S. N., 1939, Spores and pollen of the coals of the USSR: Internat. Geol. Cong., 17th, USSR, 1937, Rept., v. 1, p. 355-366.

Nemějc, F., 1935, Note on the spores and leaf cuticle of *Noeggerathia foliosa* Stbg.: Internat. Acad. Tchècque Sci. Bull (Ceská Akad. věd a Umění), v. 36, p. 61-63.

———1937, On *Discinites* K. Feistm: Internat. Acad. Tchècque Sci. Bull. (Ceská Akad. vèd a Umění), v. 38, p. 3-9.

Neves, R., 1958, Upper Carboniferous plant spore assemblages from the *Gastrioceras subcrenatum* horizon, north Staffordshire: Geol. Mag., v. 95, no. 1, p. 1-18; discussion, by W. G. Chaloner, no. 3, p. 261-262.

———1961, Namurian plant spores from the southern Pennines, England: Palaeontology, v. 4, pt. 2, p. 247-279.

Nikitin, P. A., 1934, Fossil plants of the Petino horizon of the Devonian of the Voronezh region; 1, *Kryshtofovichia africani* nov. gen. et. sp. [pteridophyte]: Acad. Sci. USSR Bull. 7, p. 1079-1091; English summ., p. 1091-1092.

Peppers, R. A., 1964, Spores in strata of Late Pennsylvanian cyclothems in the Illinois Basin: Illinois Geol. Survey Bull. 90, 89 p.

Playford, Geoffrey, 1962, Lower Carboniferous microfloras of Spitsbergen: Palaeontology, v. 5, pt. 3, p. 550-618.

______1963a, Lower Carboniferous microfloras of Spitsbergen – Pt. 2: Palaeontology, v. 5, pt. 4, p. 619-678.

______1963b, Miospores from the Mississippian Horton Group, eastern Canada: Canada Geol. Survey Bull. 107, 47 p.

Potonié, Robert, 1956, Synopsis der Gattungen der Sporae dispersae; Pt. 1, Sporites: Geol. Jahrb. Beihefte, no. 23, 103 p.

Potonié, Robert, and Kremp, Gerhard, 1954, Die Gattungen der paläozoischen Sporae dispersae und ihre Stratigraphie: Geol. Jahrb., v. 69, p. 111-193 [1955].

Potonié, Robert, and Kremp, Gerhard, 1955-1956, Die Sporae dispersae des Ruhrkarbons, ihre Morphographie and Stratigraphie mit Ausblicken auf Arten anderer Gebiete und Zeitabschnitte [pts. 1, 2, and 3]: Palaeontographica, v. 98, pt. 1, sec. B, no. 1-3, 125 p. [1955]; v. 99, pt. 2, sec. B, no. 4-6, p. 85-186 [1956]; v. 100, pt. 3, sec. B, no. 4-6, p. 65-121 [1956].

Raistrick, A., and Simpson, J., 1933, The microspores of some Northumberland coals, and their use in the correlation of coal seams: Inst. Mining Eng. Trans., v. 85, pt. 4, p. 225-235.

Read, C. B., and Mamay, S. H., 1964, Upper Paleozoic floral zones and floral provinces of the United States, *with* a Glossary of stratigraphic terms, by G. C. Keroher: U.S. Geol. Survey Prof. Paper 454-K, p. K1-K35.

Reed, F. D., 1938, Notes on some plant remains from the Carboniferous of Illinois: Bot. Gaz., v. 100, no. 2, p. 324-335.

Reinsch, P. F., 1884, Micro-Palaeo Phytologia Formationis Carboniferae; v. 1, Continens Trileteas et Stelideas: Erlangae, Germania, Theo. Krische, 79 p.

Remy, Winfried, 1953, Untersuchungen über einige Fruktifikationen von Farnen und Pteridospermen aus dem mitteleuropäischen Karbon und Perm: Akad. Wiss. u. Literatur Abh. Math.-Naturw. Kl., Abh. Jg. 1952, no. 2, 38 p.

______1954, Die Systematik der Pteridospermen unter Berücksichtigung ihrer Pollen: Zeitschr. Angew. Geologie, v. 3, no. 3, p. 312-319.

Renault, B., 1896, Basin Houiller et Permien d'Autun et d'Epinac: Paris, Études des Gites mineraux de la France, pt. 4.

Richardson, J. B., 1960, Spores from the Middle Old Red Sandstone of Cromarty, Scotland: Palaeontology, v. 3, pt. 1, p. 45-63.

______1962, Spores with bifurcate processes from the Middle Old Red Sandstone of Scotland: Palaeontology, v. 5, pt. 2, p. 171-194.

______1964, Stratigraphical distribution of some Devonian and Lower Carboniferous spores: Avanc. Études Stratigraphie et Géologie Carbonifère Cong., 5th, v. 3, p. 1111-1114.

Schemel, M. P., 1950, Carboniferous plant spores from Daggett County, Utah: Jour. Paleontology, v. 24, no. 2, p. 232-244.

______1951, Small spores of the Mystic coal of Iowa: Am. Midland Naturalist, v. 46, no. 3, p. 743-750.

Schopf, J. M., 1938, Spores from the Herrin (no. 6) coal bed in Illinois: Illinois Geol. Survey Rept. Inv. 50, 73 p.

______1941a, Contributions to Pennsylvanian paleobotany; *Mazocarpon oedipternum*, sp. nov., and sigillarian relations: Illinois Geol. Survey Rept. Inv. 75, 53 p.

______1941b, Notes on the Lepidocarpaceae: Am. Midland Naturalist, v. 25, no. 3, p. 548-563; repr. 1941, Illinois Geol. Survey Circ. 73.

——1948, Pteridosperm male fructifications; American species of *Dolerotheca*, with notes regarding certain allied forms: Jour. Paleontology, v. 22, no. 6, p. 681-724; repr. 1949, Illinois Geol. Survey Rept. Inv. 142.

Schopf, J. M., Wilson, L. R., and Bentall, Ray, 1944, An annotated synopsis of Paleozoic fossil spores and the definition of generic groups: Illinois Geol. Survey Rept. Inv. 91, 73 p.

Scott, D. H., 1920, Studies in fossil botany, v. 1, 1st ed.: A. & C. Black, Ltd., 434 p.

Selling, O. H., 1946, Studies in Hawaiian pollen statistics—Pt. 1, The spores of the Hawaiian pteridophytes: Bernice P. Bishop Mus. Spec. Pub. 37, 87 p.

Slater, L., 1932, Microscopical study of the coal seams and their correlation: Inst. Mining Eng. Trans., v. 83, p. 191-206.

Slater, L., Evans, M. M., and Eddy, G. E., 1930, The significance of spores in the correlation of the coal seams—Pt. 1, The Parkgate seam, South Yorkshire area: D.S.I.R. Fuel Res., no. 17, 28 p.

Somers, Grace, 1952, Fossil spore content of the Lower Jubilee seam of the Sydney coalfield, Nova Scotia: Nova Scotia Research Found., 30 p.

Staplin, F. L., 1960, Upper Mississippian plant spores from the Golata formation, Alberta, Canada: Palaeontographica, v. 107, sec. B, no. 1-3, p. 1-40.

——1961, New plant spores similar to *Torispora* Balme: Jour. Paleontology, v. 35, no. 6, p. 1227-1231.

Staplin, F. L., and Jansonius, Jan, 1964, Elucidation of some Paleozoic densospores: Palaeontographica, v. 114, sec. B, no. 4-6, p. 95-117.

Sullivan, H. J., 1958, The microspore genus *Simozonotriletes*: Palaeontology, v. 1, pt. 2, p. 125-138.

——1964, Miospores from the Drybrook Sandstone and associated measures in the Forest of Dean basin, Gloucestershire: Palaeontology, v. 7, pt. 3, p. 351-392.

Sullivan, H. J., and Neves, R., 1964, *Triquitrites* and related genera: Avanc. Études Stratigraphie et Géologie Carbonifère Cong., 5th, p. 1079-1093.

Thiessen, Reinhardt, and Staud, J. N., 1923, Correlation of the coal beds in the Monogahela formation of Ohio, Pennsylvania, and West Virginia: Carnegie Inst. Technology Bull. 9, 64 p.

Thompson, R. B., 1934, Heterothally and the seed habit versus heterospory: New Phytologist, v. 33, p. 41-44.

Wanless, H. R., 1963, Pennsylvanian Period and System (or Epoch and Series), Art. Pennsylvanian *in* Termes stratigraphiques majeurs: Internat. Geol. Cong., 21st, Copenhagen 1960, Lexique Strat. Internat., v. 8, 64 p.

Williams, R. W., 1955, *Pityosporites westphalensis*, sp. nov., an abietineous type pollen grain from the Coal Measures of Britain: Annals and Mag. Nat. History, ser. 12, v. 8, no. 90, p. 465-473.

Wilson, L. R., 1958, Photographic illustrations of fossil spore types from Iowa: Oklahoma Geology Notes, v. 18, nos. 6-7, p. 99-101.

——1962, Permian plant microfossils from the Flowerpot Formation, Greer County, Oklahoma: Oklahoma Geol. Survey Circ. 49, 50 p.

Wilson, L. R., and Coe, E. A., 1940, Descriptions of some unassigned plant microfossils from the Des Moines series of Iowa: Am. Midland Naturalist, v. 23, no. 1, p. 182-186.

Wilson, L. R., and Venkatachala, B. S., 1963a, An emendation of *Vestispora* Wilson and Hoffmeister, 1956: Oklahoma Geology Notes, v. 23, no. 4, p. 94-100.

——1963b, Morphological variation of *Thymospora pseudothiessenii* (Kosanke) Wilson and Venkatachala: Oklahoma Geology Notes, v. 23, no. 5, p. 125-132.

______1963c, A morphologic study and emendation of *Vesicaspora* Schemel, 1951: Oklahoma Geology Notes, v. 23, no. 6, p. 142-149.

Winslow, M. R., 1959, Upper Mississippian and Pennsylvanian megaspores and other plant microfossils from Illinois: Illinois Geol. Survey Bull. 86, 135 p.

______1962, Plant spores and other microfossils from Upper Devonian and Lower Mississippian rocks of Ohio: U.S. Geol. Survey Prof. Paper 364, 93 p.

Wodehouse, R. P., 1935, Pollen grains, their structure, identification, and significance in science and medicine, 1st ed.: New York, McGraw-Hill Book Co., Inc., 574 p.

Zerndt, Jan, 1930, Megasporen aus einem Flöz in Libiaz (Stephanien): Acad. Polonaise Sci. Bull., Ser. Sci. Tech., p. 39-70.

______1937, Les mégaspores du bassin houiller polonais—Two parts: Acad. Polonaise Sci. Trans. Géol., no. 3, 78 p.

______1938, Vertikale Reichweite von Megasporentypen im Karbon des Bassin du Nord—Pionowy zasiag megaspor w karbonie Bassin du Nord: Polskie Towarz. Geol. Rocznik, v. 13, p. 21-30.

Appendix to References

Several important palynological contributions have been published since the manuscript for this chapter was submitted. Two of these publications are listed:

Felix, C. J., and Burbridge, P. P., 1967, Palynology of the Springer formation of southern Oklahoma, U.S.A.: Palaeontology, v. 10, pt. 3, p. 349-425.

Smith, A. H. V., and Butterworth, M. A., 1967, Miospores in the coal seams of Great Britain: London, Palaeont. Assoc., Spec. Papers in Palaeontology, no. 1, 324 p.

13

PALYNOLOGY OF THE PERMIAN PERIOD

George F. Hart

The flora of the Permian Period is distinctly different from that of the Carboniferous Period and from that of the succeeding Triassic Period. Because Cenozoic and Carboniferous palynological assemblages were studied earlier, the classification developed for them was utilized for the Permian. This approach was adequate for the study of local palynofloras and stratigraphic problems; however, it obscured the possibilities for broad comparisons among the various palynofloristic studies. The recent publication (Hart, 1965) of a uniform and comprehensive classification of Permian miospores has facilitated consideration of the Permian palynofloras on a worldwide basis, although, obviously, refinement of the classification scheme is necessary as more species are defined statistically and the limits of their variability are determined.

Temporal subdivisions of the Permian Period on the basis of palynofloristic characteristics are now slowly becoming discernible. These subdivisions reflect two striking influences on Permian plant life. One of these is the immense effect of the Gondwanian glacial epoch on the origin and development of the floras. The other is the existence of a definite areal separation of the earth into a northern Laurasian and a southern Gondwanian region; the classical paleobotanical provinces based on megafossils (fig. 13-1) can be delimited palynologically. Recognition in future studies of these two major influences is basic to the establishment of a worldwide Permian chronostratigraphy based on pollen and spores.

GENERAL CHARACTERISTICS OF PERMIAN PALYNOFLORAS

At the general level the Pollenite miospores form the major part of the Permian complexes in the Gondwanian and Euramerian provinces and are of considerably less significance in the Angarian and Cathaysian provinces. Permian Pollenites are represented by all the major subturmae of Potonié and in their generic composition form a sharp contrast to both the Carboniferous and Triassic complexes. Of greatest significance are the saccate genera, and particularly the Disaccites. Looked at on purely morphologic grounds, the Disaccites underwent a period of explosive evolution at the end of the Paleozoic Era, and we find numerous different lines evolving. As a consequence many of the earlier forms that occur in the late Paleozoic and early Mesozoic Eras are difficult to subdivide into infraturma, genera, or, often, species. However, the three infraturmae Striatiti, Disacciatrileti, and Disaccitrileti are easily differentiated and can be conveniently dealt with as distinct groups. Possible interrelationships of the saccate groups are given in figure 13-2.

The basis of saccate taxonomy is in general the degree of development of the saccus. In the Disaccites this is expressed by the nature of the distal and proximal roots, in particular the length and position on the central body of the sacci roots; the size and shape of the sacci with respect to the central body are apparently directly related to the nature of the distal and proximal

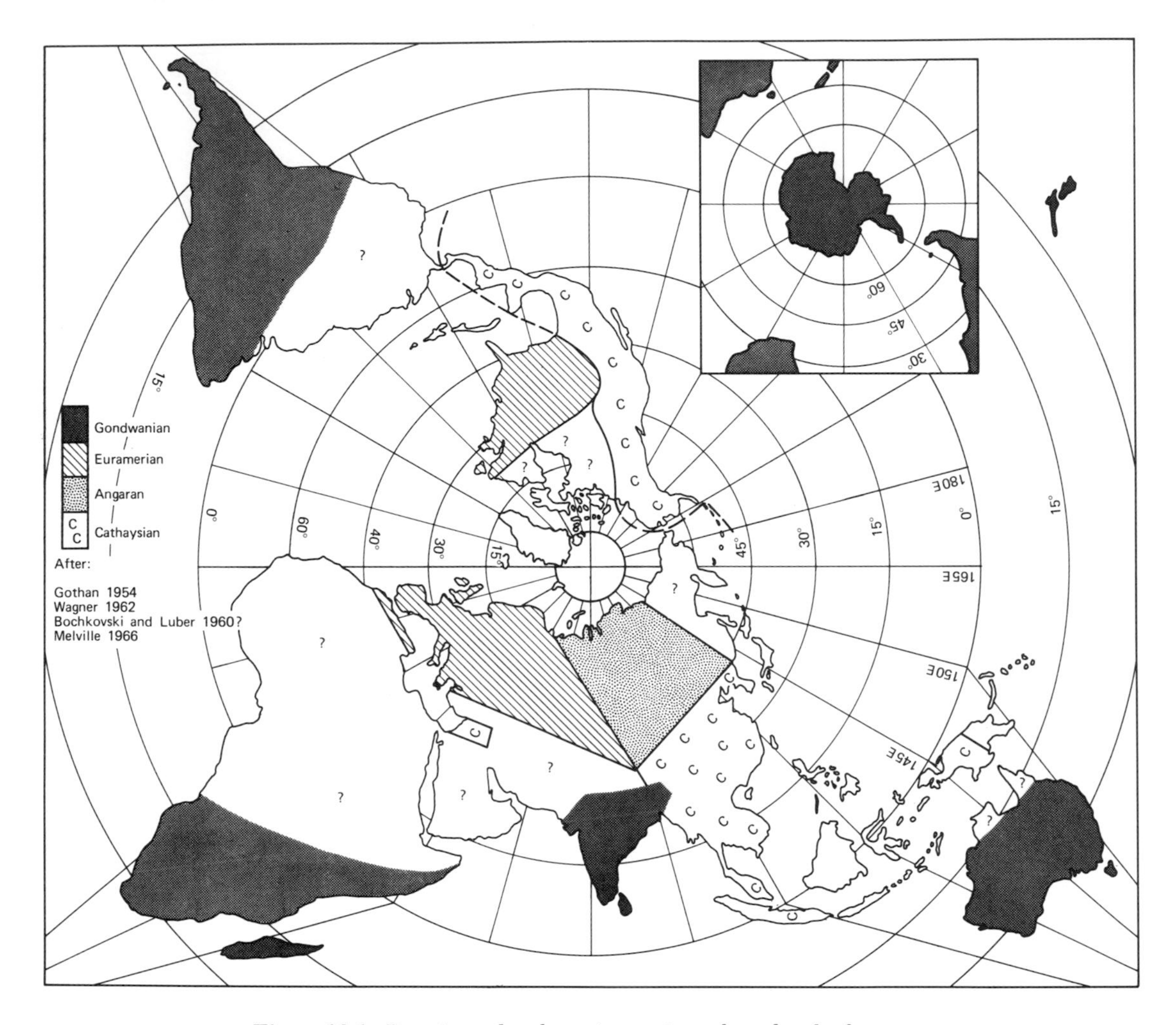

Figure 13-1. Permian palaeobotanic provinces based on leaf genera.

roots. In the Monosaccites the degree of saccus development is expressed by a proximal, distal, or proximal and distal attachment of the saccus to the central body and by the general symmetry of the saccus.

The disaccate Striatiti, in which the thickened proximal cap is sculptured by longitudinal ribs and striae, are typical and usually abundant in the Permian complexes, and in terms of its miospore components the Permian Period may be characterized as the *Striatiti complex*. The major genera of Permian disaccate Striatiti have been reviewed by Hart (1965) and may be described briefly as follows (fig. 13-3):

Protohaploxypinus. Specimens of this genus are generally haploxylonoid in outline and have four or more longitudinal ribs. (See fig. 13-3*a*.)

Striatopodocarpites. This genus is diploxylonoid in outline and has four or more longitudinal ribs. The central body is usually circular when observed in polar view. (See fig. 13-3*b*.)

Striatoabietites. Specimens of this genus are somewhat similar to *Striatopodocarpites* in their diploxylonoid outline, but their sacci are smaller when compared to the central body size and the distal zone between the sacci roots is wider. (See fig. 13-3*c*.)

Lueckisporites. This genus is clearly separated from other Striatiti by possessing only two longitudinal ribs. (See fig. 13-3d.)

Hamiapollenites. Specimens of *Hamiapollenites* have very small sacci attached almost terminally. Running transversely across the central part of the distal surface is one or more thickening. (See fig. 13-3g.)

Taeniaesporites (Leschik 1956) Jansonius 1962. This genus is diploxynoid to haploxynoid in outline. In polar view the proximal cap shows three to five longitudinal ribs. (See fig. 13-3f.)

Vittatina. Although assigned to the disaccate Striatiti, specimens of *Vittatina* usually lack sacci. They normally consist of a haploxylonoid body with a thickened and longitudinally striated, proximal cap. Terminally the longitudinal ribs of the proximal surface may pass onto the distal surface. If sacci are present, they are minute and almost unnoticeable bladderlike swellings. (See fig. 13-3e.)

The Disaccitrileti, which have a proximal mark on the central body, form only a small portion of the Permian miospore complexes; nevertheless, they are persistently present and may be considered as typically, but not exclusively, Permian forms. They are chiefly represented by the genera *Vestigisporites* (forms with a monolete mark and lateral bladderlike swellings connecting the two terminal sacci), *Limitisporites* (forms with a monolete mark and no lateral bladderlike swellings), and *Jugasporites* (forms with a bilete mark).

Finally, the Disacciatrileti are prominent representatives of the Permian complex; of all the Disaccites these are the most difficult to subdivide into genera and species, mainly through a lack of diagnostic features on the central body. This infraturma had its origin in the middle-upper Carboniferous, probably from a *Florinites*-like ancestor, along with the Striatiti. The major genera used to accommodate the Permian Disac-

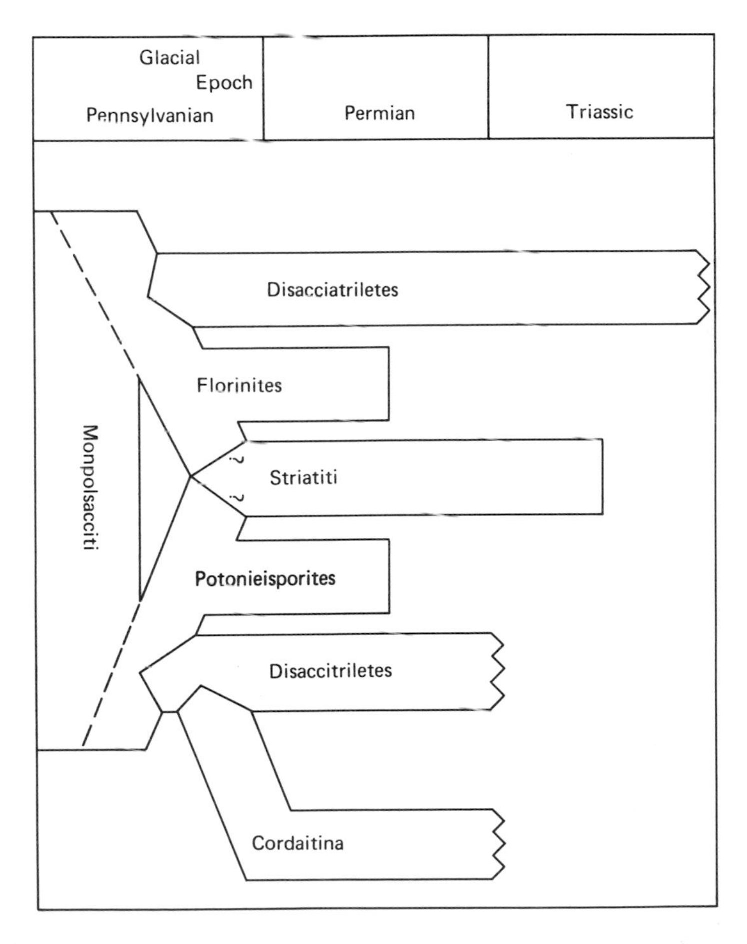

Figure 13-2. Relationships suggested by morphologic trends in upper Paleozoic Saccites. Morphographic transitions between the monosaccate *Florinites* and Disaccitrileti are common and presumably point to an ancestor-descendant relationship. Similarly the trend from the monosaccate Monpolsacciti *Potoniesporites* to the Disaccitrileti *Vestigisporites* to the monosaccate Dipolsacciti *Cordaitina* has been observed. The disaccate Striatiti has been observed trending from a Disaccitrileti (*Vesicaspora*) or Monpolsacciti (*Florinites*)-like ancestor, although the thickening of the proximal cap as observed in some *Potoniesporites* may point to their partial origin along that line.

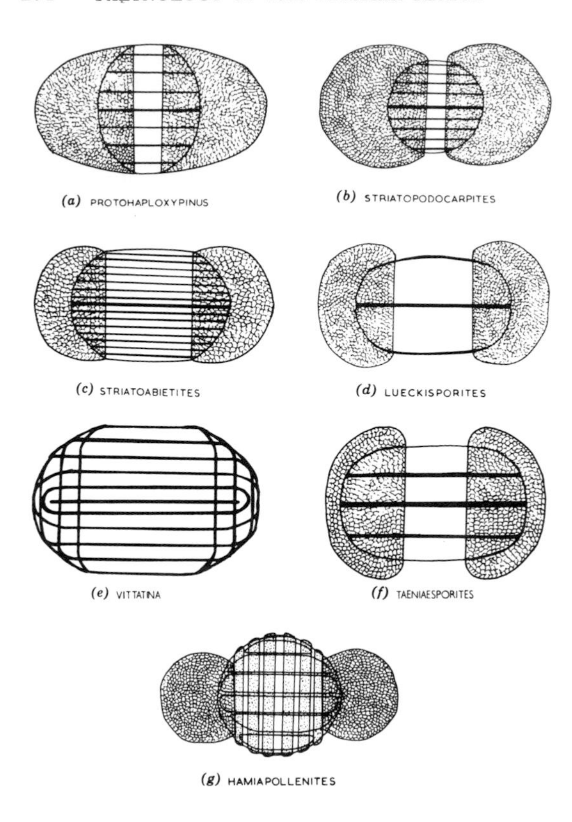

Figure 13-3. The genera of disaccate Striatiti: (*a*) *Protohaploxypinus;* (*b*) *Striatopodocarpites;* (*c*) *Striatoabietites;* (*d*) *Lueckisporites;* (*e*) *Vittatina;* (*f*) *Taeniaesporites;* (*g*) *Hamiapollenites.*

ciatrileti are *Pityosporites* (in which the sacci are greater than or equal to the central body in size and the complete outline is diploxylonoid), *Piceapollenites* (which contains usually haploxylonoid specimens with sacci equal to or less than the central body in size and less than or about semicircular in shape), and *Vesicaspora* (which is haploxylonoid in outline and has lateral bladderlike swellings connecting the two terminal sacci).

The Disacciatrileti, along with the Striatiti, underwent an explosive burst, presumably related to the Gondwanian glaciation, at the beginning of the Permian Period. However, unlike the Striatiti, the evolutionary burst of the Disacciatrileti did not reach its maximum until well into the Mesozoic Era. Perhaps because of this the Disacciatrileti are distinctly of less importance than the Striatiti during the Permian Period, but, unlike the Striatiti, they did not become extinct in the lower Mesozoic succession.

Monosaccate genera also figure prominently in the Permian complexes but are usually subsidiary to the disaccate genera in both diversity and abundance. Hart (1965) divided the monosaccates into two infraturmae according to the degree of development of the saccus. Monpolsacciti (including the Permian genera *Paleoconifera, Potoniésporites*, and *Sehoripollenites*), have the saccus attached only on one hemisphere, the other being free of the saccus. The Dipolsacciti have the saccus attached both proximally and distally, and include the genera *Cordaitina, Nuskoisporites*, and *Striomonosaccites*. Polysaccites are rare but may occasionally attain significance as zonal fossils (e.g., *Dulhuntyispora* in Australia).

Although the saccate miospores are usually the most important pollenite forms, they do not form the total complex. The turma Plicates, although not abundant, is nevertheless an important group. Permian Plicates have been assigned to the subturma Praecolpates, Polyplicates, and Monocolpates. The Praecolpates are interesting because they are "prepollen"; the Polyplicates because of their morphologic relationship to the disaccate Striatiti, and the existence of their living representatives, and the Monocolpates because the Permian types presumably represent the initial stock from which the later Mesozoic forms were derived.

Finally, in the Pollenites part of the Permian complex are miospores belonging to the Aletes, a group that is a persistent minor component in the majority of the successions studied palynologically.

The other major anteturma of miospores—the Sporites—in essential characteristics does not show such a great difference from the underlying Carboniferous complex as do the Pollenites. The majority of genera and all of the infraturmae are continuous from the Carboniferous palynoflora, and only at the species level are the temporal differences obvious. This may be largely a reflection of the simplicity in form of the Sporites; for example, the Zonales, particularly the Camerati, are the most complex of the Sporites and significantly are extremely useful for biostratigraphy. Triletes form the major part of the Permian Sporites palynoflora, occurring in all assemblages and locally becoming important zonal fossils. Monoletes are never abundant but may also have local biostratigraphic significance.

Representative genera of spores and pollen from Permian rocks are shown on plates 13-1 and 13-2.

TEMPORAL DISTRIBUTION OF PERMIAN PALYNOFLORAS

The generalized stratigraphic distribution of each miospore species for each continent has been plotted on range charts and analyzed in terms of the classical Permian geographic areas of Gondwanaland and Laurasia. (See Hart, 1965, figs. 411-420.) Each of these palaeogeographic areas shows distinct palynologic characteristics (see Hart, 1965, figs. 421-432) that allow them to be separated. However, it is now clear that a better understanding of Laurasian Permian palynofloras can be derived by considering each of the Laurasian palaeobotanic provinces of Euramerïa, Angara, and Cathaysia separately.

The Gondwanian Palynoflora

Typical Gondwanian palynofloras have been described by Balme and Hennelly (1955; 1956a,b), Bharadwaj (1962), Balme (1964), and Hart (in press, a and b). Essentially the complex is characterized by the presence of *Protohaploxypinus, Cordaitina, Striatopodocarpites, Vesicaspora*, and various species of Sporites such as *Punctatisporites gretensis, Cirratriradites splendens, C. africanensis, C. australensis, Granulatisporites trisinus, G. papillosus*, and *Acanthotriletes tereteangulatus*.

The palynologic similarities between Australia, Africa, India, South America, and Antarctica during the Permian Period are very strong. Work on the African palynoflora has shown a fourfold division of the Permian Period in terms of palynofloristic zones; namely,

STRIATITI FLORIZONE
ZONATI FLORIZONE
CINGULATI FLORIZONE
CAMERATI FLORIZONE

Details of these florizones are given by Hart (in press, a); and their equivalents in the other parts of Gondwanaland, by Hart (in press, b). Figures 13-4, 13-5, and 13-6 show some aspects of the palynofloristic zones as they occur in southern Africa.

The Euramerian Palynoflora

Typical Euramerian palynofloras have been studied in the European sector by Klaus (1953, 1960, 1963), Grebe (1957), Boulouard (1963), Leschik (1956), Samoilovich (1953), Shatkinskaya (1958), and Abramova and Marchenka (1960). In the North American sector they have been described by Wilson (1962), Jizba (1962), and Jansonius (1962). Characteristic is an abundance of saccate miospores with a smaller number of Sporites. *Cordaitina, Vittatina, Striatoabietites, Protohaploxypinus, Lueckisporites, Vesicaspora, Illinites, Limitisporites*, and *Piceapollenites* are common, but at the species level there is a great difference between the Gonwanian and Euramerian palynofloras.

In Soviet Europe, around the stratotype area, data are available from the Sakmarian, Artinskian, Kungurian, Kazanian, and Tatarian stages. The middle Permian Ufimian stage, which is not recognized by all Soviet Permian stratigraphers, does not have a particularly distinctive palynoflora but readily conforms to the Kungurian palynoflora. (see figure 13-7 for stage names of the Permian System.)

The Sakmarian complex, according to Shatkinski (1958), is dominated by saccate Pollenites, with Sporites forming usually less than 2 percent of the complex. The vast majority of the saccate forms belong to the genera *Cordaitina, Florinites, Protohaploxypinus, Striatoabietites*, and some Reticulonapiti.

The Artinskian complex, according to Samoilovich (1953) and Abramova and Marchenka (1960), has as its main elements *Tuberculatosporites marattiformis* (Samoilovich 1953) Hart 1965; *Lycospora subdola* (Luber and Valts 1941) Hart 1965; *Granulatasporites irregularisplicatus* (Samoilovich 1953) Hart 1965, *G. subreticulatus* (Samoilovich 1953) Hart 1965, and *G. pustillus* (Samoilovich 1953) Hart 1965; *Pilaspora bulbifera* (Luber and Valts 1953) Hart 1965; *Laricoidites levis* (Luber and Valts 1941) Hart 1965; *Cordaitina convallata* (Luber and Valts 1941) Samoilovich 1953; *Cycadopites erosus* (Luber and Valts 1941) Hart 1965, *C. caperatus* (Luber and Valts 1941) Hart 1965, and *C. glaber* (Luber and Valts 1941) Hart 1965; *Reticulatisporites compactus* (Luber and Valts 1941) Hart 1965; and some species of *Vittatina* and Disacciatrileti.

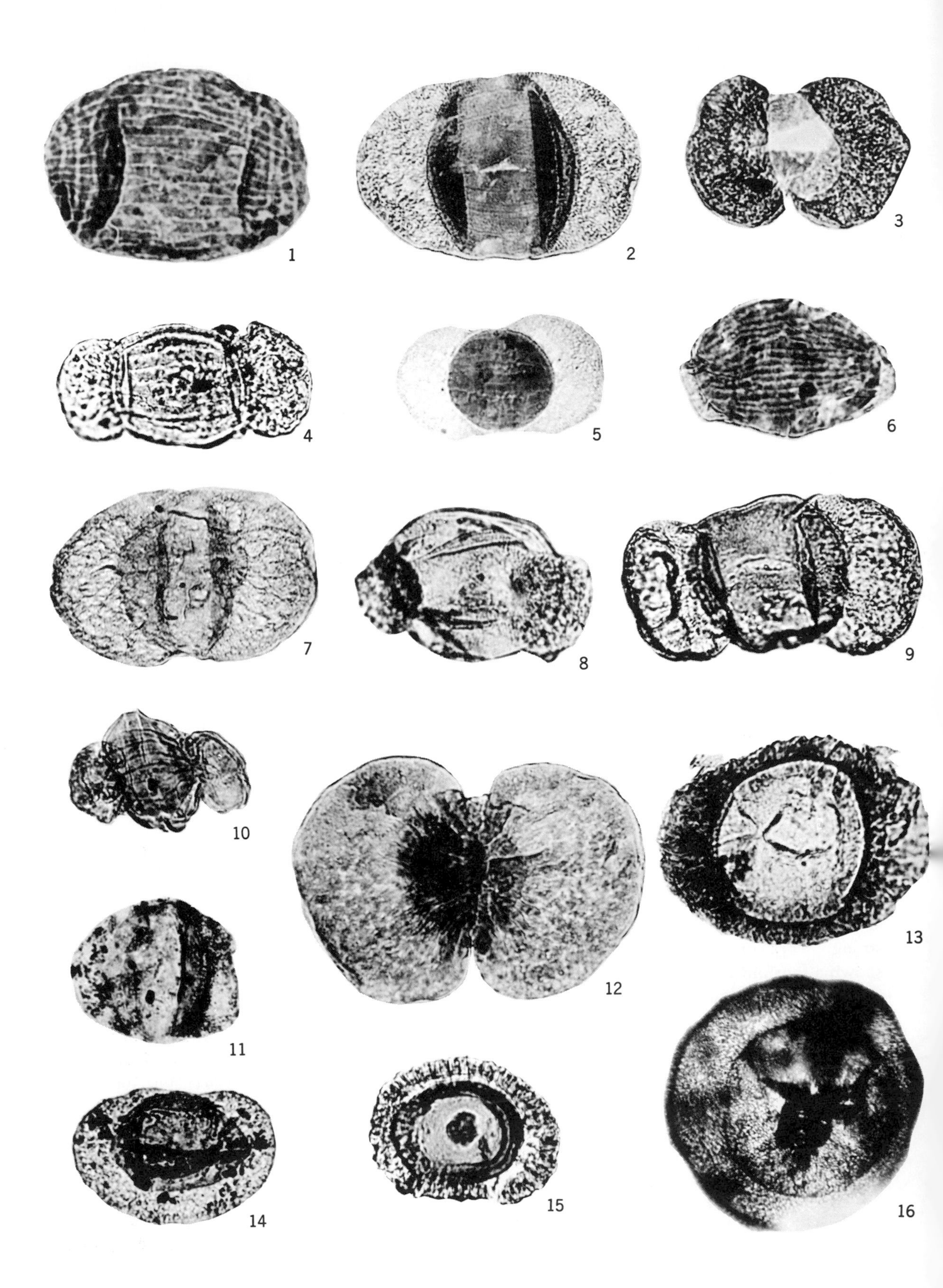
1
2
3
4
5
6
7
8
9
10
11
12
13
14
15
16

The Kungurian complex, according to Samoilovich (1953) and Abramova and Marchenka (1964), is characterized by the strong dominance of Pollenites over Sporites, by diversification amongst the genera *Cordaitina*, and *Vittatina*, by a significant number of *Cycadopites*, and particularly by the wholesale development of the disaccate Striatiti. Species occurring in this complex include *Cordaitina ornata* Samoilovich 1953, *C. uralensis* (Luber and Valts 1941) Samoilovich 1953, and *C. rotata* (Luber and Valts 1941) Samoilovich 1953; *Striatoabietites elongatus* (Luber and Valts 1941) Samoilovich 1953; *Vittatina vittifera* (Luber and Valts 1941) Samoilovich 1953, *V. striata* (Luber and Valts 1941) Samoilovich 1953, and *V. subsaccata* Samoilovich 1953; *Piceapollenites prolixus* (Luber and Valts 1941) Hart 1965 and *P. sublevis* (Luber and Valts 1941) Hart 1965; *Abiespollenites sylvestritypus* (Samoilovich 1953) Hart 1965; *Jugasporites auritus* (Luber and Valts 1941) Hart 1965; *Potoniéisporites turboreticulatus* (Samoilovich 1953) Hart 1965; *Cycadopites erosus* (Luber and Valts 1941) Hart 1965 and *C. tunguskensis* (Luber and Valts 1941) Hart 1965; *Monosulcites subrotatus* (Luber and Valts 1941) Hart 1965; *Laricoidites levis* (Luber and Valts 1941) Hart 1965; *Granulatasporites indefinitus* (Samoilovich 1953) Hart 1965 and *G. subreticulatus* (Samoilovich 1953) Hart 1965; and *Reticulatasporites fabuginus* (Samoilovich 1953) Hart 1965 and *R. microdictyus* (Luber and Valts 1941) Hart 1965.

The Kazanian complex, according to the work of Zoricheva and Sedova (1954), amongst others, consists of a great number of saccate Pollenites from the genera *Vittatina, Striatopodocarpites, Protohaploxypinus,* and *Cordaitina.* Various species of Disacciatrileti are also prominent and are associated with some types assumed by Zoricheva and Sedova to have a modern aspect (*Selaginella, Osmunda*). It is during this stage that the Disacciatrileti really commence their explosive burst, with *Pityosporites, Piceapollenites,* and *Vesicaspora* probably forming the initial stocks. An interesting aspect about the Perm region of Soviet Europe is the apparent absence of *Nuskoisporites* and *Lueckisporites*, and the species *N. dulhuntyi, L. virkkiae, Striatoabietites richteri*, and *Vesicaspora schaubergeri* in particular.

The Tatarian complex is once more characterized by disaccate Striatiti, particularly those belonging to the genera *Striatopodocarpites, Protohaploxypinus*, and *Vittatina.* Also prominent are Disacciatrileti and particularly *Limitisporites* and *Jugasporites.* Some characteristic species include *Striatoabietites bricki* Sedova 1956; *Protohaploxypinus latissimus* (Luber and Valts 1941) Samoilovich 1953 and *P. perfectus* (Naumova ex. Kara Murza 1952) Samoilovich 1953; *Hamiapollenites bullaeformis* (Samoilovich 1953) Hart 1965; *Vittatina striata* (Luber and Valts 1941) Samoilovich 1953, *V. cincinnata* (Luber ex. manuscript) Hart 1964, *V. subsaccata* Samoilovich 1953, and *V. vittifer* (Luber and Valts 1941) Samoilovich 1953. Additionally in western Europe *Nuskoisporites dulhuntyi* Potonié and Klaus 1954, *Lueckisporites virkkiae* Potonié and Klaus 1954, *Striatoabietites richteri* (Klaus 1953) Hart 1964, and *Taeniaesporites noviaulensis* Leschik 1956 are important.

The Cathaysian Palynoflora

Palynological studies of typical Cathaysian palynofloras are rare, but it appears that the main characteristics are those enumerated in the works of Ouyang (1962, 1964). Similar complexes have been studied by Singh (1964) from Iraq and by Hart and Wagner (unpublished manuscript) from Turkey. In general the characteristics of the palynoflora are an abundance of Carboniferous relict genera, such as *Torispora, Camptotriletes, Schopfites, Densosporites, Reinschospora,* and *Cyclogranulatisporites*; these forms are not exclusively Carboniferous genera but are common in those strata. Associated Pollenites are *Potoniéisporites, Cordaitina, Vesicaspora,* and *Hamiapollenites*, all of which are also known from the uppermost Carboniferous rocks,

Plate 13-1. Representative Permian saccate genera (Magnification all approximately × 500 except figure 14.) 1—*Vittatina;* 2—*Protohaploxypinus;* 3—*Lueckisporites;* 4—*Striatoabietites;* 5—*Striatopodocarpites;* 6—*Vittatina;* 7—*Vesicaspora;* 8—*Abiespollenites;* 9—*Limitisporites;* 10—*Hamiapollenites;* 11—*Piceapollenites;* 12—*Pityosporites;* 13—*Vestigisporites*; 14—*Potonieisporites* (×250); 15—*Cordaitina*; 16—*Nuskoisporites*.

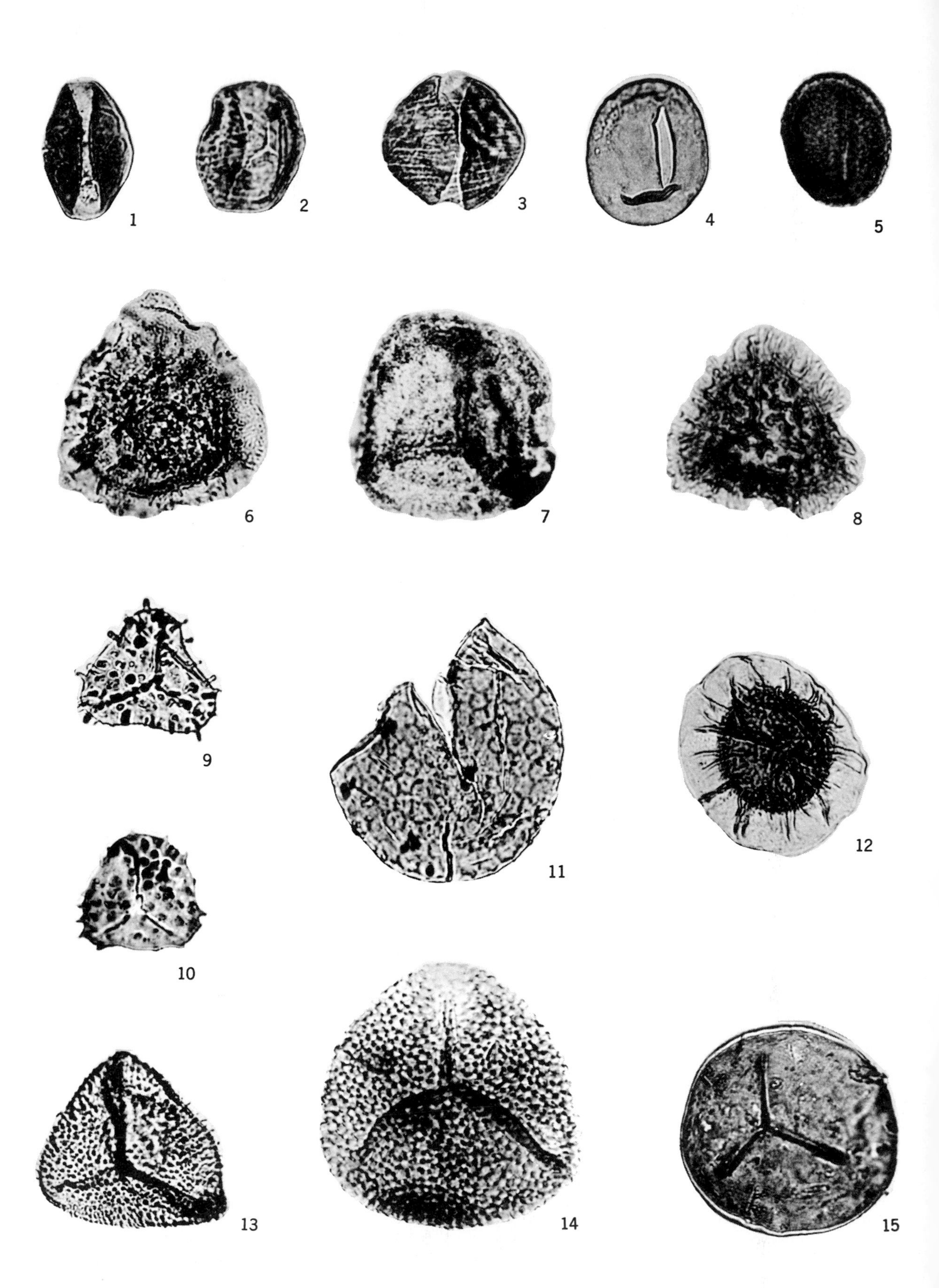
1
2
3
4
5
6
7
8
9
10
11
12
13
14
15

but which are more typical of the Permian Period.

Ouyang (1962, 1964) described miospore complexes from the Lungtan series of Chekiang and the lower Shihhotze series of Shansi. The Lungtan series of South China has been equated with the upper Shihhotze series of North China. Sze and others regard the age of the deposits as middle Permian or even Early Permian, although many Chinese geologists would prefer a Late Permian age. On the basis of the palynologic analysis of Ouyang, certainly the former age is indicated. The lower Shihhotze series, on the other hand, is regarded as definitely Late Permian in age and is thought to belong to the upper part of the Lower Permian System (presumably Kungurian Stage or uppermost Artinskian Stage).

The palynoflora described from the Lungtan series consisted of 70 morphotypes assigned to 26 genera; of these 19 species are closely comparable with forms known from the upper Carboniferous in Europe and North America, and only a few species, amongst which are *Cordaitina uralensis* and *Protohaploxypinus prolixus*, are comparable with the upper part of the Kuznets basin palynoflora described by Andreyeva (1956) from the Angaran palaeobotanic province. The miospores described from the lower Shihhotze series consisted of 64 species assigned to 31 genera, and of these the majority were new species.

The Angaran Palynoflora

The Angaran palynoflora, referred to as the Tunguskian flora by the Soviets, has been studied in greatest detail in the Kuznets basin, where the palynologic succession shows very marked differences from the Permian Euramerian palynoflora as seen in Soviet Europe. It has been studied by Andreyeva (1956) in the Kuznets basin, by Medvedeva (1960) in the Tungus Basin, and by Dibner (1967) from northern Siberia. Basically the complex consists of *Apiculatisporites, Acanthotriletes, Leiotriletes, Cordaitina, Vesicaspora,* and *Piceapollenites*. The genera are chiefly forms that occur in most geological formations, and during the Permian Period the Angaran province seems to be typified more by its negative characteristics than its positive ones.

The Lower Permian Balakhonian suite is characterized in its lower part by miospores belonging to the genera *Cordaitina* (*C. rotata, C. uralensis, C. psiloptera,* and *C. rugulifera*); and *Cycadopites* (*C. caperatus,* and *C. retroflexus*). Triletes are also important, particularly *Acanthotriletes rectispinus* and *A. obtusosetosus*, associated with the aletes *Laricoidites similis*. In the upper part of the Balakhonian suite large quantities of *Cycadopites* occur (*C. caperatus* and *C. glaber*), but the complex is dominated by triletes such as *Acanthotriletes obtusosetosus, A. rectispinus, A. stimulosus, Apiculatisporis asperatus,* and also the aletes *Laricoidites similis. Cordaitina* is still important and is represented chiefly by *C. rotata* and *C. rugulifera*.

The Strelkinskian complex studied by Medvedeva (1960) was equated with the lower Kungurian stage of the Euramerian palaeobotanic province. This complex shows a tremendous abundance of *Cycadopites* (36 percent) and of triletes such as *Granulatasporites pastillus* (Samoilovich 1953) Hart 1965; *Lophotriletes polypyrenus* (Luber and Valts 1941); and *Acanthotriletes heterodontus* (Andreyeva) Naumova, *A. mediaspinosus* (Andreyeva) Naumova, and *A. multisetus* (Luber) Naumova. In addition *Cordaitina* forms a significant percentage in the complex.

The Il'inian and Erynakovian complexes of the Kusnets basin have also been studied by Andreyeva (1956). The palynofloristic characteristics of the Il'inian suite are the presence of the miospores *Acanthotriletes heterochaetus* (Andreyeva 1956) Hart 1965 and *A. tenuispinosus* (Luber and Valts 1941) Hart 1965; *Baculatisporites lyctis* (Andreyeva 1956) Hart 1965; and *Apiculatisporis globulosus* (Andreyeva 1956) Hart 1965.

The species components of the Erynakovian

Plate 13-2. Representative Permian non-saccate genera (Magnification approximately × 500.) 1—*Cycadopites;* 2—*Marsupipollenites;* 3—*Pakhapites;* 4—*Laevigatosporites;* 5—*Thymospora;* 6—*Cirratriradites* (?); 7—*Grandispora* (?); 8—*Zinjispora;* 9—*Neoraistrickia;* 10—*Apiculatisporis;* 11—*Inderites;* 12—*Hymenozonotriletes* (?); 13—*Acanthotriletes;* 14—*Converrucosisporites;* 15—*Punctatisporites.*

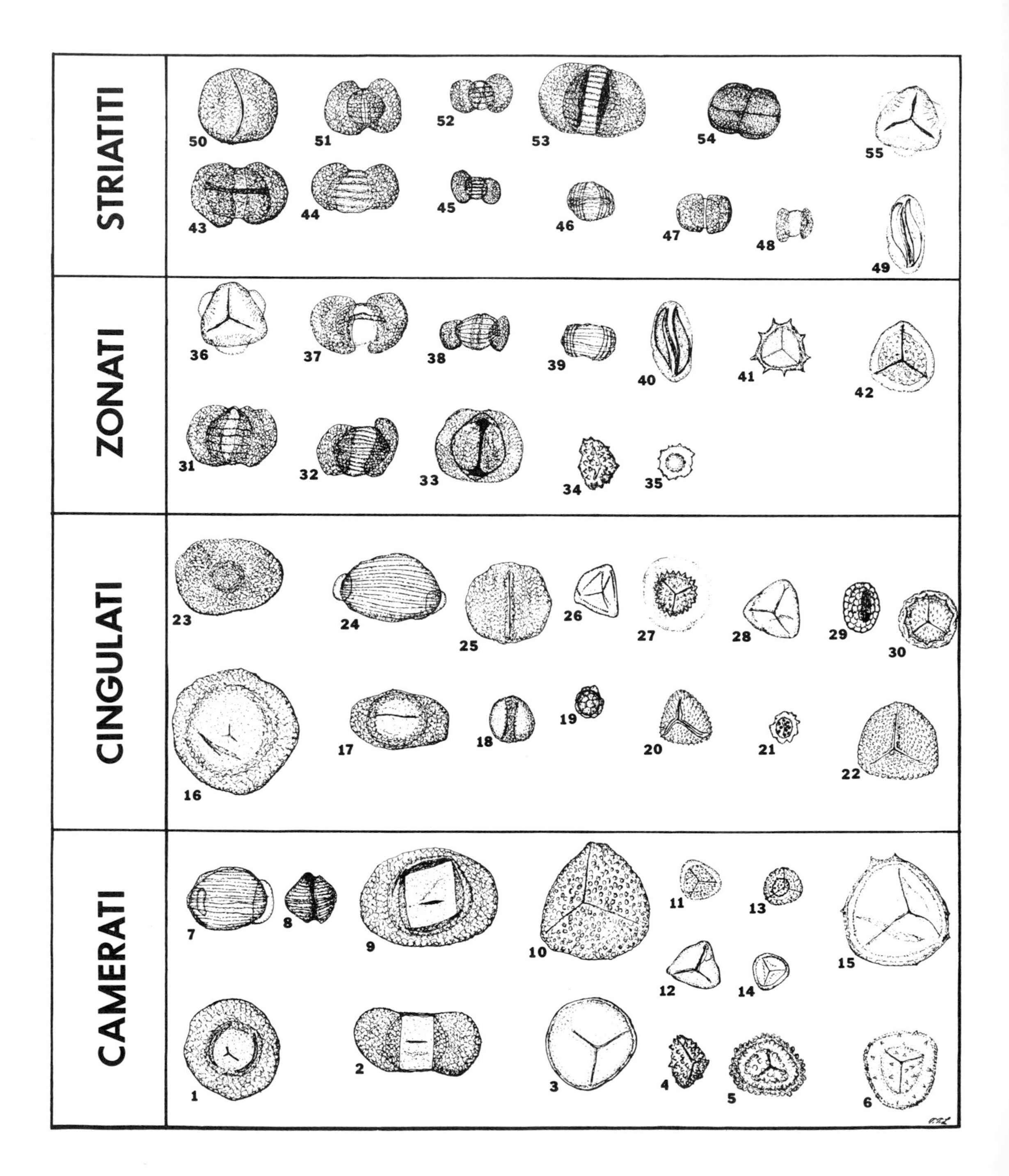

Figure 13-4. Major miospore characteristics of the Permian florizones from the Great Karroo basin, South Africa. 1—*Cordaitina triangulara* (Mehta 1944) Hart 1965. 2—*Vestigisporites rudis* Balme and Hennelly 1955. 3—*Punctatisporites gretensis* Balme and Hennelly 1956. 4—*Apiculatisporis cornutus* Balme and Hennelly 1956. 5—*Zinjispora zonalis* Hart 1965. 6—*Grandispora dwykensis* Hart 1963. 7—*Vittatina saccata* Hart 1960. 8—*Pakhapites fasciolatus* (Balme and Hennelly 1956) Hart 1965. 9—*Vestigisporites balmei* Hart 1960. 10—*Converrucosisporites naumovai* (Hart 1963) Smith and others 1965. 11—*Grandispora salisburyensis* Hart 1963. 12—*Granulatisporites papillosus* Hart 1965. 13—*Gondispora vrystaatensis* Hart 1963. 14—*Lycospora tritriangulara* Hart 1963. 15—*Grandispora punctata* Hart 1963. 16—*Cordaitina balmei* Hart 1965. 17—*Vestigisporites thomasi* (Pant 1955) Hart 1965. 18—*Cycadopites nevesi* Hart 1965. 19—*Reticulatasporites sp.* 20—*Granulatisporites micronodosus* Balme and Hennelly 1955. 21—Zonate form A. 22—*Converrucosisporites pseudoreticulatus* (Balme and Hennelly 1956). 23—*Florinites eremus* Balme and Hennelly 1955. 24—*Vittatina saccata* Hart 1960. 25—*Vesicaspora potoniei*

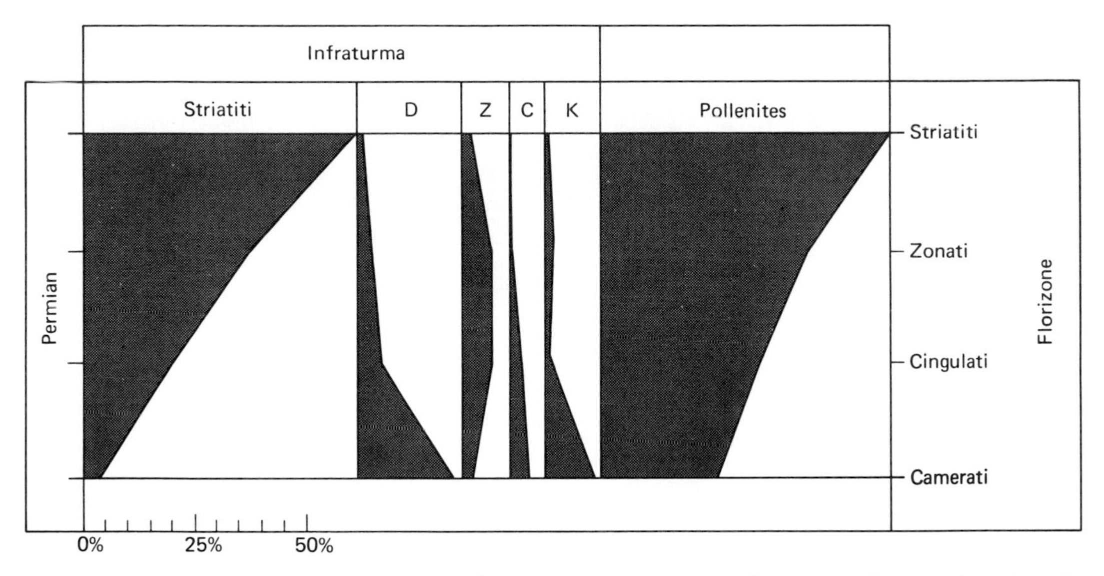

Figure 13-5. Arithmetic mean percent values of major miospore taxa in each Permian florizone from the African Karroo (calculated exclusive of Disacciatrileti). Abbreviations: D—Dipolsacciti; Z—Zonati; C—Cingulati; K—Camerati.

complex include *Cordaitina rotata* (Luber and Valts 1941) Samoilovich 1953, *C. uralensis* (Luber and Valts 1941) Samoilovich 1953; *C. gemina* (Andreyeva 1956) Hart 1965, *C. rugulifera* (Luber and Valts 1941) Samoilovich 1953, and *C. psiloptera* (Luber and Valts 1941) Hart 1965; *Cycadopites caperatus* (Luber and Valts 1941) Hart 1965 and *C. conjunctur* (Andreyeva 1956) Hart 1965; *Laricoidites similis* (Luber and Valts 1941) Hart 1965; *Granulatasporites pastillus* (Samoilovich 1953) Hart 1965; *Calamospora nigritella* (Luber and Valts 1941) Hart 1965; *Apiculatisporis asperatus* (Luber and Valts 1941) Hart 1965; *Raistrickia incurvispina* var. *hirtella* (Andreyeva 1956) Hart 1965, *R. heteromorpha* (Andreyeva 1956) Hart 1965, and *R. heteromorpha* var. *gongilocarpa*; *Acanthotriletes tenuispinosus* (Luber and Valts 1941) Hart 1965 and *A.* cf. *rectispinus* (Luber and Valts 1941) Hart 1965; and *"Lophotriletes" polypyrenus* Luber and Valts 1941. Although normally the Triletes are the most important, locally *Cycadopites* and *Cordaitina* become prominent.

AREAL DISTRIBUTION OF PERMIAN PALYNOFLORAS

So far the distribution of Permian miospores has been discussed in terms of the classical concept of Permian landmasses and palaeobotanic provinces. From the sum total of palynologic evidence it is valid to accept the existence of the two distinct Permian geographic areas of Laurasia and Gondwanaland, whether or not we ac-

(Lakhanpal, Sah, and Dube 1960) Hart 1965. 26—*Reinchospora plumsteadi* Hart 1963. 27—*Cirratriradites australensis* Hart 1963. 28—*Granulatisporites papillosus* Hart 1965. 29—*Thymospora leoparda* (Balme and Hennelly 1956) Hart 1965. 30—Camerate form A. 31—*Protohaploxypinus sewardi* (Virkki 1938) Hart 1964. 32—*Striatoabietites* cf. *multistriatus* (Balme and Hennelly 1956) Hart 1965. 33—*Vesicaspora potoniei* (Lakhanpal, Sah, and Dube 1960) Hart 1965. 34—*Apiculatisporis cornutus* Balme and Hennelly 1956. 35—New genus(?). 36—*Dulhuntyispora* cf. *D. dulhuntyi* Potonié 1956. 37—*Lueckisporites nyakapendensis* Hart 1960. 38—*Hamiapollenites karrooensis* Hart 1965. 39—*Vittatina africana* Hart 1966. 40—*Schopfipollenites sinuosus* (Balme and Hennelly 1956) Hart 1965. 41—New genus(?). 42—*Cirratriradites africanensis* Hart 1963. 43—*Lueckisporites nyakapendensis* Hart 1960. 44—*Protohaploxypinus sewardi* (Virkki 1938) Hart 1964. 45—*Striatopodocarpites sp.* 46—*Vittatina africana* Hart 1966. 47—*Pityosporites papilio* (Goubin 1965) nov. comb. 48—*Pityosporites tuscus* (Goubin 1965) nov. comb. 49—*Schopfipollenites sinuosus* (Balme and Hennelly 1956) Hart 1965. 50—*Vesicaspora potoniei* (Lakhanpal, Sah, and Dube 1960) Hart 1965. 51—*Striatopodocarpites octostriatus* Hart 1960. 52—*Striatopodocarpites cancellatus* (Balme and Hennelly 1955) Hart 1960. 53—*Protohaploxypinus amplus* (Balme and Hennelly 1955) Hart 1964. 54—*Guttulapollenites hannonicus* Goubin 1965 (Believed to be a variant of *L. nyakapendensis*). 55—*Dulhuntyispora* cf. *D. dulhuntyi* Potonié 1956.

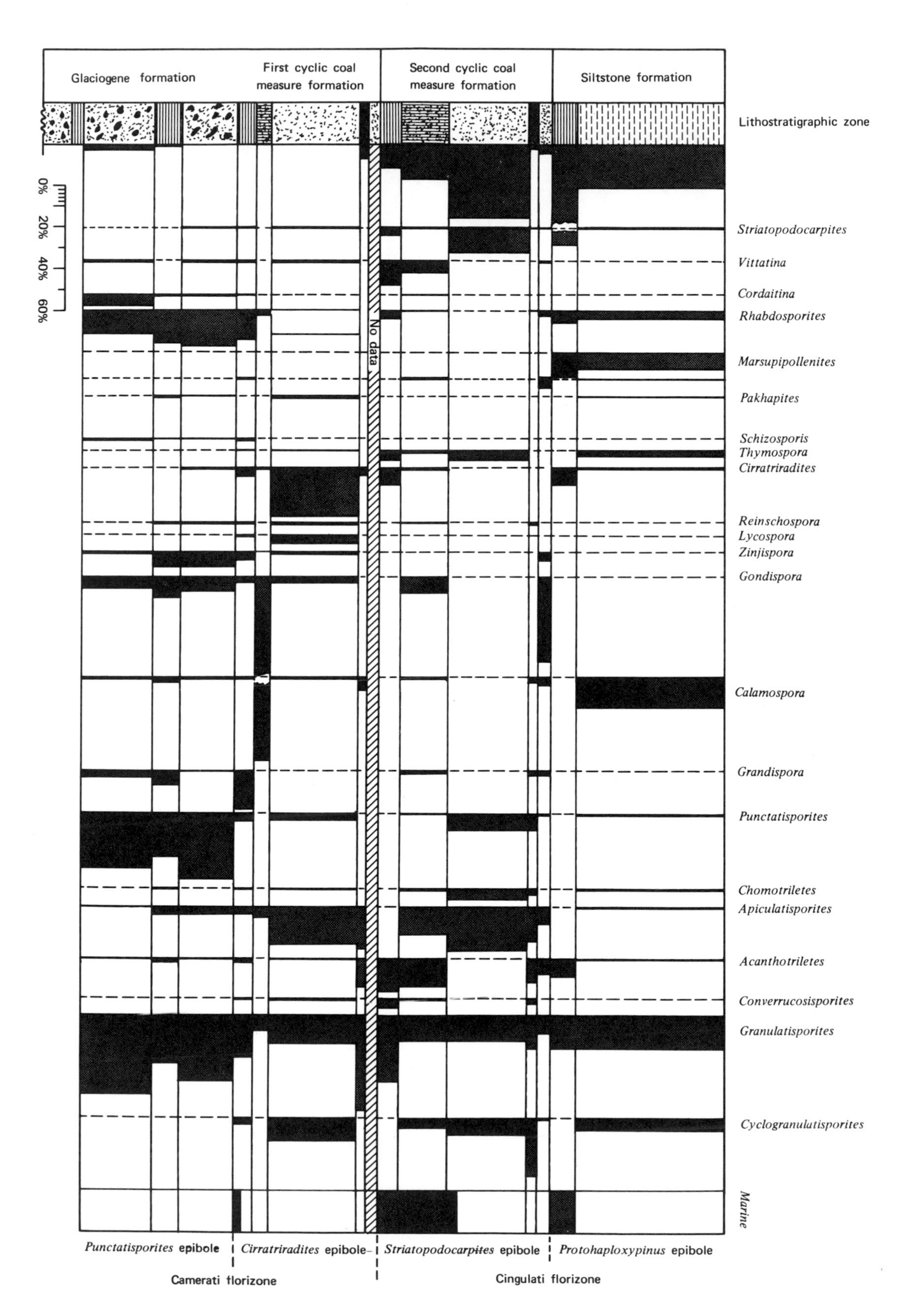
Glaciogene formation
First cyclic coal measure formation
Second cyclic coal measure formation
Siltstone formation
Lithostratigraphic zone
0%
20%
40%
60%
No data
Striatopodocarpites
Vittatina
Cordaitina
Rhabdosporites
Marsupipollenites
Pakhapites
Schizosporis
Thymospora
Cirratriradites
Reinschospora
Lycospora
Zinjispora
Gondispora
Calamospora
Grandispora
Punctatisporites
Chomotriletes
Apiculatisporites
Acanthotriletes
Converrucosisporites
Granulatisporites
Cyclogranulatisporites
Marine
Punctatisporites epibole
Cirratriradites epibole
Striatopodocarpites epibole
Protohaploxypinus epibole
Camerati florizone
Cingulati florizone

	United States	Soviet Union	Europe
Upper Permian	Ochoan	Tatarian	
			---?---
	Guadalupian	Kazanian	Thuringian
		?	
Lower Permian		Kungurian	

	Leonardian	Artinskian	Saxonian
			---?---
	Wolfcampian	Sakmarian	Autunian

Figure 13-7. Permian stratigraphic units in the United States, Soviet Union, and Europe.

cept continental drift. The palynologic separation of the two landmasses is one of the most striking features in studying any distribution map of Permian miospores at the global level. In all the major taxa there is an almost complete separation of the two landmasses, and it is clear that India belongs to the Gondwanian landmass.

In the more detailed analyses the discussion has centered around the temporal distribution of the Permian miospores within the confines of the classically defined palaeobotanic provinces as founded on megaplant and mainly leaf-genera studies. However, a more logical analysis is to plot world-distribution maps of all the Permian miospores and define palynofloristic palaeobotanic provinces with a purely empirical technique; that is, without reference at all to classically defined provinces. This can act as a check on the validity of both the palynofloristic and classical paleobotanic methods as paleobiogeographic tools.

Distribution maps were constructed for all of the Permian miospores, and figures 13-8, through 13-11 illustrate the distribution of some of the taxa (the infraturma Striatiti). The most striking feature of the complete analysis was the fact that few species occur beyond the limits of particular palaeobotanic provinces, as defined by classical theory. Moreover, as we would expect, the Saccites show the greatest number of species that transgress their normal endemic areas and become exotics in a foreign province. The main mixing is on the fringes of provinces; for example, *Cycadopites caperatus, C. retroflexus,* and *Acanthotriletes rectispinus* occur in the Eurameriam province in Kazakhstan, but they are presumed to be endemic to the Angaran province.

The Cathaysian palaeobotanic province is certainly the most interesting palynofloristically. As noted above it is characterized by an abundance of typically Carboniferous genera. As a consequence it is thought that the Cathaysian province was a refugee (relict flora) area during the Lower and middle Permian Period, where the Carboniferous flora temporarily survived into the Permian Period, due to some unknown palaeobiogeographic factor.

Figure 13-6. Percentage abundance of some important genera through the Lower Permian florizones of the northern margin of the Great Karroo basin, South Africa. The distribution chart shows clearly the influence of changing palaeogeography, in the position of the marine horizons and the detrital cycles, and the general climatic amelioration after the Dwyka glaciation, on the palynofloristic development of the area. Furthermore, the close relationship between palynofloral and formational boundaries is apparent.

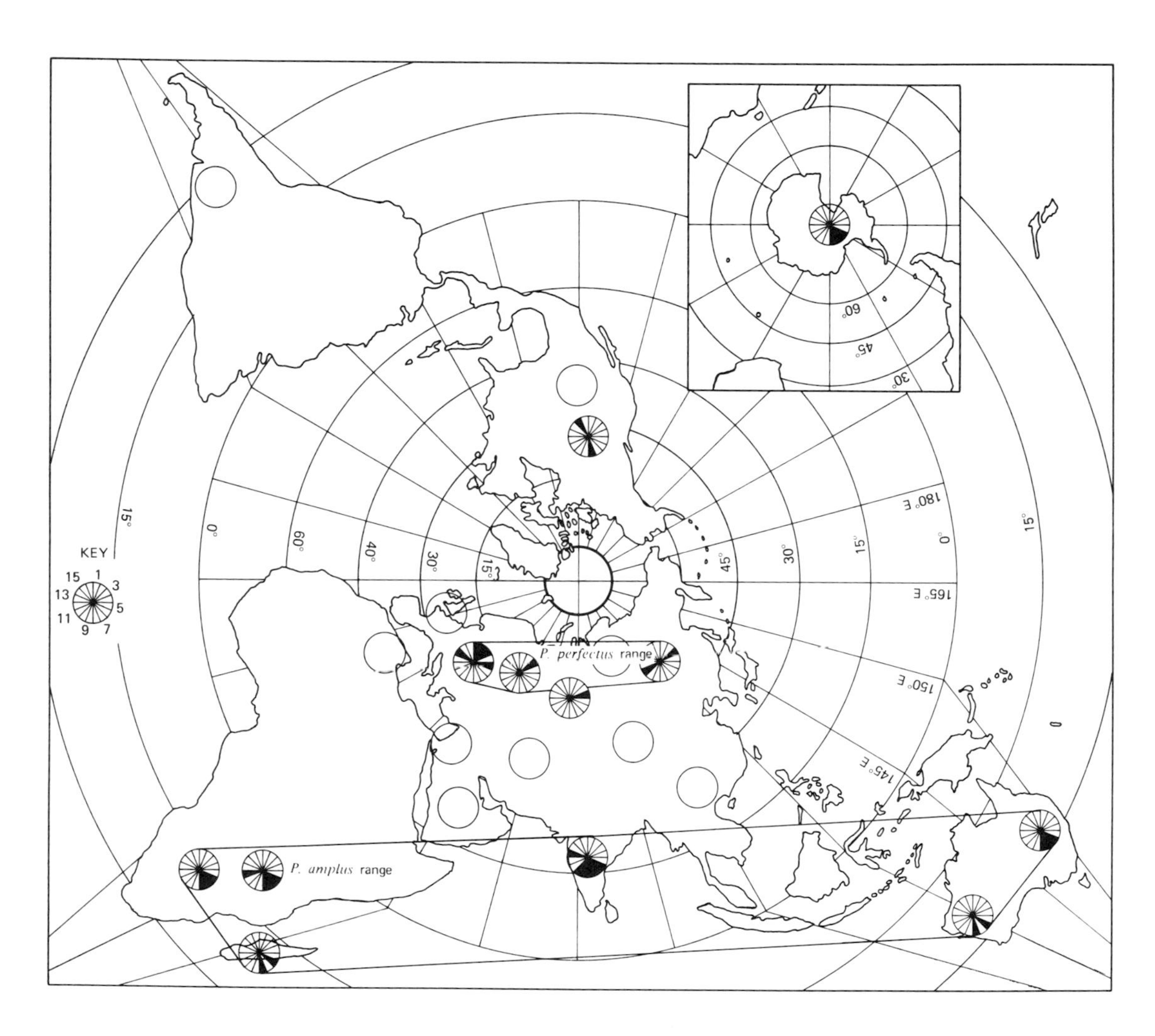

Figure 13-8. Global distribution of species of *Protohaploxpinus:* 1–*P. latissimus*; 2–*P. suchonensis*; 3–*P. perfectus*; 4–*P. pennatulus*; 5–*P. dvinensis*; 6–*P. amplus*; 7–*P. micros*; 8–*P. sewardi*; 9–*P. goriaiensis*; 10–*P. kumaonensis*; 11–*P. rhomboeformis*; 12–*P. globus*; 13–*P. jacobii*; 14–*P. volaticus*; 15–*P. samoilivichi*.

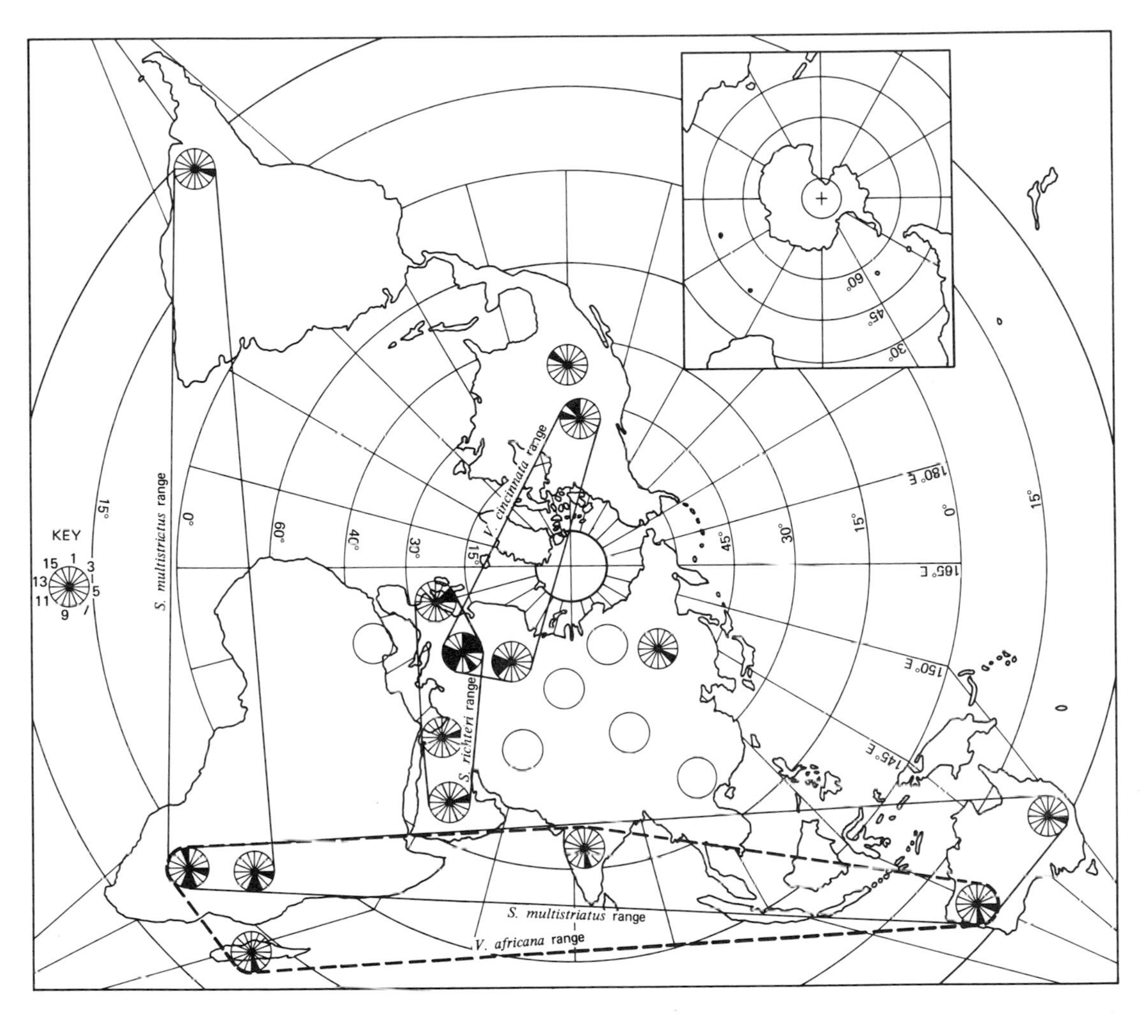

Figure 13-9. Gobal distribution of species of *Striatoabietites* and *Vittatina:* 1—*S. elongatus;* 2—*S. striatus;* 3—*S. giganteus;* 4—*S. richteri;* 5—*S. multistriatus;* 6—*S. sedovai;* 7—*S. bricki;* 8—*V. africana;* 9—*V. saccata;* 10—*V. subsaccata* and *V. persecta;* 11—*V. vittifera;* 12—*V. striata;* 13—*V. cincinnata;* 14—*V. costabilis* and *V. lata;* 15—*V. wodehousei, V. simplex,* and *V. saccifera;* 16—*V. minima.*

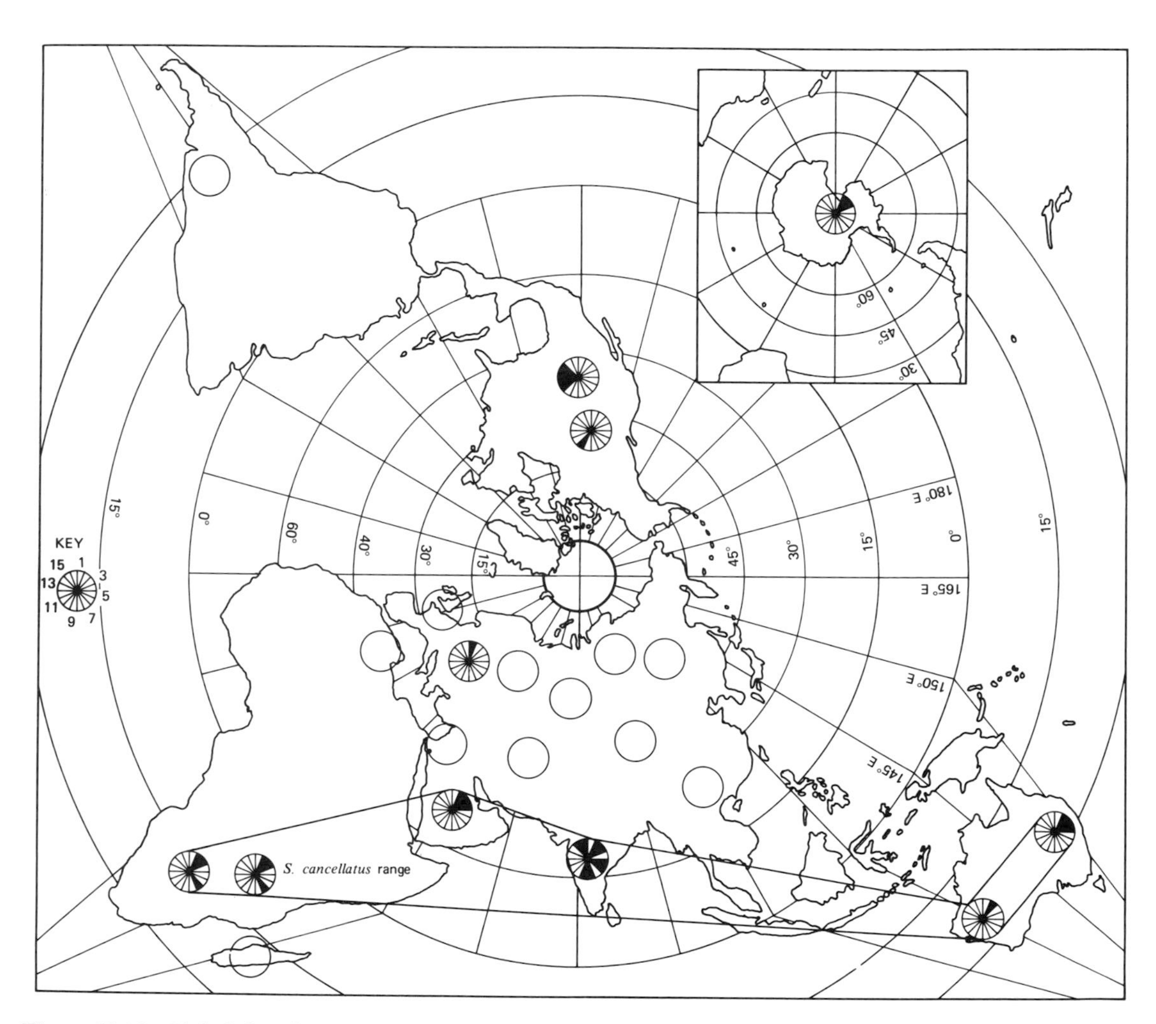

Figure 13-10. Global distribution of species of *Striatopodocarpites:* 1—*S. tojmensis;* 2—*S. fusus;* 3—*S. cancellatus;* 4—*S. phaleratus;* 5—*S. gondwanensis;* 6—*S. renisaccatus;* 7—*S. octostriatus;* 8—*S. balmei;* 9—*S. varius;* 10—*S. rugosus;* 11—*S. communis;* 12—*S. limnisaccatus;* 13—*S. radiatus;* 14—*S. divaricatus;* 15—*S. secretus;* 16—*S. raniganjensis.*

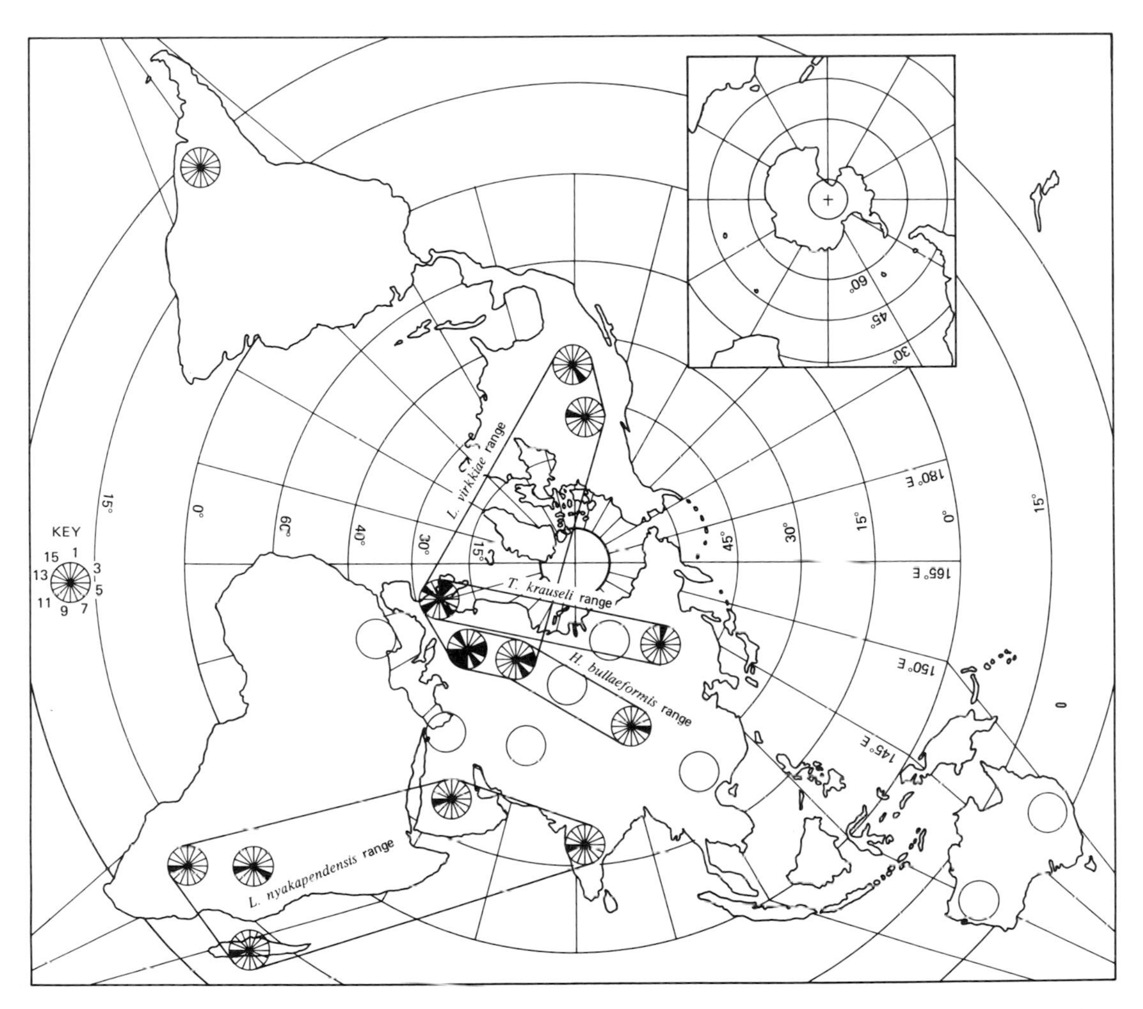

Figure 13.11. Global distribution of species of *Taeniaesporites*, *Hamiapollenites*, and *Lueckisporites*: 1–*T. krauseli*; 2–*T. noviaulensis*; 3–*H. saccata*; 4–*H. tractiferinus*; 5–*H. bullaeformis*; 6–*H. karrooensis*; 7–*L. virkkiae*; 8–*L. globosus*; 9–*L. ruttneri*; 10–*L. parvus*; 11–*L. junior*; 12–*L. nyakapendensis*; 13–*L. tattooensis*; 14–*L. microgranulatus*.

References

Abramova, S. A., and Marchenka, O. F., 1960, Material in the palynological study of the subsalt succession of the Kungurian deposits of the Kalin area: Trudy VNIIGAL, vyp. XL, Gosknimiizdat, p. 337-370.

Andreyeva, E. M., 1956, Spore-pollen characteristics of the Balakhonian and Erynakovian Suite of the Kuznets Basin *in* Atlas of the leading forms of fossil flora and fauna of the Permian System of the Kuznets Basin; ed. V. I. Yavorsk: Moskva, Trudy VSEGEI, p. 207-271.

Balme, B. E., 1964, The palynological record of Australian pre-Tertiary floras: Ancient Pacific floras, University of Hawaii Press, p. 48-80.

Balme, B. E., and Hennelly, J. P. V., 1955, Bisaccate sporomorphs from Australian Permian coals: Aust. J. Bot., v. 3, no. 1, p. 89-98.

———1956a, Monocolpate and Alete sporomorphs from Australian Permian sediments: Aust. J. Bot., v. 4, no. 1, p. 54-67.

———1956b, Trilete sporomorphs from Australian Permian sediments: Aust. J. Bot., v. 4, no. 3, p. 240-260.

Bharadwaj, D. C., 1962, The miospore genera of the coals of the Raniganj stage (Upper Permian), India: The Palaeobotanist, v. 9, nos. 1 and 2, p. 68-106.

Boulouard, C., 1963, Contribution a l'etude des Saccates, essai de classification morphographique et application stratigraphique: SNPA., Direction exploration et production, entre de recherches Pau, Dip. d'etudes superieures, Univ. Paris, vol. 1, p. 1-78.

Dibner, A. F., 1967, Permian Complexes of spore and pollen of the Noril'ian region and their significance for the purpose of comparing the deposits: Uchenye zapiski paleont. i. biostrat., NIIGA, Leningrad, vyp. 19, p. 51-80.

Grebe, H., 1957, Microflora des neiderrheinschen Zechsteins: Geol. Jb., v. 73, p. 51-74.

Hart, G. F., 1965, The systematics and distribution of Permian miospores: Univ. Witwatersrand Press, Johannesburg.

———(in press,a), Micropalaeontology of the Karroo deposits in South and Central Africa: Review section, Gondwana Symposium 1967, South America.

———(in press,b), The stratigraphic subdivision and equivalents of the Karroo sequence as suggested by palynology: Gondwana Symposium 1967, South America.

Hart, G. F., and Wagner, R. MS., Miospores from Permian coals in the Hazro district, southweast Anatolia (Turkey): in manuscript.

Imgrund, R., 1960, Sporae Dispersae des Kaiping-bekens, China: Geol. Jb., v. 77, p. 143-204.

Jansonius, J., 1962, Palynology of Permian and Triassic sediments, Peace River area, western Canada: Palaeontographica, v. 110, p. 35-98.

Jizba, K. M. M., 1962, Late Paleozoic bisaccate pollen from the United States midcontinent area: Jour. Paleontology, v. 36, no. 5, p. 871-887.

Klaus, W., 1953, Alpine Salzmikropaläontologie (Sporendiagnose): Paläont. Z., v. 27, p. 52-56.

———1960, Sporen der karnishen Stufe der ostalpinen Trias: Jb. Geol. B. A., Sonderbd. 5, p. 107-183.

———1963, Sporen aus dem südalpinen Perm: Jb. Geol. B. A., v. 106, p. 229-363.

Leschik, G., 1956, Sporen aus dem Saltzon des Zechsteins von Neuhof (bei Fulda): Palaeontographica B, v. 100, p. 122-142.

Luber, A. A., and Valts, I. E., 1941, Atlas of microspores and pollen of the Palaeozoic of USSR: Trudy VSEGEI, Fascicle 139, p: 1-107.

Medvedeva, A. M., 1960, The stratigraphic separation of the lower horizons of the Tungus series by the method of spore-pollen analysis: Inst. geol. i razrabot. goryuck. iskopayem., izdat. akad. nauk SSSR, p. 1-92.

Ouyang, S., 1962, The microspore assemblage from the Lungtan series of Changsing, Chekiang: Acta Palaeontologica Sinica, v. 10, no. 1, p. 76-119.

______1964, A preliminary report on sporae dispersae from the lower Shihhotze series of Hokii district, northwest Shansi: Act. Palaeontologica Sinica, v. 12, no. 3, p. 486-519.

Samoilovich, S. R., 1953, Pollen and spores from the Cherdynian and Aktybinian Pre-Urals: Leningrad-Moskva, Trudy VNIGRI, nov. ser., v. 75, p. 5-92.

Shatkinskaya, E. F., 1958, An atlas of spores and pollen from the upper Paleozoic deposits of the Aktubinian pre-Urals: Yn-iv. Saratov (unpublished report) Saratov.

Singh, H. P., 1964, Miospore assemblage from the Permian of Iraq: Palaeontology, v. 7, no. 2, p. 240-265.

Wilson, L. R., 1962, Permian plant microfossils from the Flowerpot Formation, Greer County, Oklahoma: Ok. Geol. Surv. Circ. 49, p. 1-50.

Zoricheva, A. I., and Sedova, M. A., 1954, Spore-pollen complexes of the Upper Permian deposits of some regions of the north European part of SSSR, p. 160-201 *in* Material po palinologii i stratigraphii: Moskva, Trudy VNIGRI, p. 160-201.

14

TRIASSIC SPORES AND POLLEN

W. G. Chaloner

GENERAL COMPOSITION OF TRIASSIC FLORAS

The principal floral changes that took place between the late Paleozoic and the Mesozoic seem to have occurred during the course of the Permian, rather than at the Permian-Triassic boundary. The major subdivisions of the stratigraphic column on the basis of plant evolution, proposed by Gothan (see Gothan and Remy, 1957), accordingly include a boundary between characteristically Mesozoic floras (Mesophytikum) and Paleozoic floras (Paleophytikum) somewhere in the Permian rather than at its top.

Of the several gymnospermous groups that are prominent in Mesozoic floras—namely, the conifers, cycadophytes, Ginkgoales, Caytoniales, and pteridosperms s.l.—all but the Caytoniales had appeared by the Late Permian. The pteridophytes, (i.e., the "spore-producing plants"—sphenopsids, lycopsids, and ferns), on the other hand, evidently played a less important role compared with the gymnosperms during the Permian and Early Triassic. However, by the close of the Triassic the ferns (represented by orders that were unimportant or not represented at all in the Carboniferous) had again become an important constituent of the flora.

It is difficult to assess the importance of various plant groups in successive floras simply on the basis of their frequency of occurrence as macrofossils—compressions and petrifactions of leaves, stems, and reproductive parts. Obviously the series of fortuitous events that result in any fossils being formed (and eventually found and studied) favors certain groups more than others. Plants growing in or close to an environment in which deposition is occurring will have a greater probability of representation than those in more remote habitats. Large plants with robust leaves and stems will be similarly favored over smaller and more perishable ones. The preservation of fossil spores involves a different series of events and thus of factors governing the "bias" of representation. The interpretation of the observed sequence of fossil floras from the Carboniferous to the Triassic is further complicated by the climatic changes during the period, and the effect that this may have had on the type of bias of the representation of different groups. This change, in the Eur-American province at least, appears broadly to have been from a more or less seasonless, humid subtropical climate to one of relative aridity, in the Permian and Early Triassic, with a return to relatively more humid conditions at the close of the Triassic. The relatively arid phase of the Permian and Early Triassic in the Northern Hemisphere corresponds with a general paucity of plant-bearing horizons compared with the preceding Carboniferous and succeeding Jurassic. It is hard to judge whether the observed floral changes (in both macrofossils and microfossils) are due to poor representation of the contemporaneous flora owing to lack of environments favorable to fossilization, and how far they represent genuine extinction of Paleozoic groups and their eventual replacement by Mesozoic ones.

The recognition of what parent plants are represented by successive fossil-spore assemblages is obviously of first importance in understanding their ecological and evolutionary meaning. In the Quaternary, and to a lesser extent in the Tertiary, similarity of fossil spores with living genera is significant enough that many can be assigned at least to living genera with some degree of confidence. In the Mesozoic such argument of affinity by analogy with the structure of spores of living plants is very much less secure and in some cases obviously misleading. Several Mesozoic groups of seed plants, for example, produced bisaccate pollen that is superficially similar to that of some living conifers, although unrelated to them. The only reliable basis for attempting to interpret the earlier spore assemblages in terms of the parent flora is a study of spores in situ in more or less contemporaneous fossil plants. Fortunately enough early Mesozoic fructifications have been investigated for some broad comments to be made concerning this aspect of Triassic spore assemblages, and these are attempted below. Detailed references will be found in Potonié (1962, 1967).

Lycopods

The Paleozoic miospore genera such as *Lycospora* and *Densosporites*, s.l., which have been shown to represent lycopod microspores, together with the many Carboniferous megaspore genera that collectively represent the same group all undergo a dramatic decline at the close of the Carboniferous (at least in the Northern Hemisphere). Rather few spore-bearing lycopod fructifications are known from the Triassic; these are the following:

Pleuromeia:
Megaspores: cf. *Trileites* (smooth, triradiate).
Microspores: cf. *Punctatisporites* (smooth, triradiate); also possibly monolete, cf. *Laevigatosporites*.

Lycostrobus:
Megaspores: *Nathorstisporites* (p. 305).
Microspores: *Aratrisporites* (p. 299).

Selaginellites (viz., *S. polaris*):
Megaspores: *Banksisporites* (p. 305).
Microspores: *Lundbladispora* (p. 299).

Cylostrobus Helby and Martin 1965:
Megaspores: *Banksisporites* (p. 305).
Microspores: *Aratrisporites* (p. 299).

The rather odd distribution of the megaspores *Nathorstisporites* and *Banksisporites*, and the microspores *Aratrisporites* and *Lundbladispora* among the cones of the three genera *Lycostrobus*, *Selaginellites*, and *Cylostrobus* suggests that there may be a closer relationship among these three parent genera than is generally believed.

Another interesting feature of these correlations is that in each case the spore records suggest a much wider geographic range than does the macrofossil record; for example, *Lycostrobus scotti*, a large bisexual cone, is known only from its holotype (Rhaetian of Sweden), whereas its megaspores have been found from equivalent horizons in Greenland, in many European localities from Sweden to Poland, and from Australia. Megaspores generically equivalent to those of *Selaginellites polaris* (known only from the Triassic of Greenland) and *Cylostrobus* (known only from Australia) have been found in the Triassic or Rhaeto-Liassic of several European localities, South America, Africa, and Tasmania (Dettman, 1961). In the intensively studied flora from the Upper Triassic of East Greenland many more species of megaspores have been recognized than species of *Selaginellites* (or other heterosporous plants). This suggests that with the Mesozoic lycopods, as with herbaceous plants in general, the spores are probably giving a fuller picture of the diversity of the group than the sparse macrofossil record. However, even if we accept that many of the unassigned miospore genera of the Mesozoic (especially those with prominent equatorial features) may represent lycopods, we must recognize that this group was probably a less prominent component of the land flora in the Mesozoic than in the Carboniferous.

Sphenopsida

More or less smooth-walled, circular, triradiate spores, comparable with the Paleozoic *Calamospora* have been recorded from Liassic and Triassic *Equisetites* (*Equisetum*, s.l.). (See Potonié, 1962.) These records support the evidence from the Paleozoic that the sphenopsids produced characteristically round, smooth, triradiate spores. In favorable lithologies in the Mesozoic (e.g., in deltaic facies in the Jurassic and Cretaceous) *Equisetites* is commonly one of the most abundant macrofossils; for this reason it is a little surprising that *Calamospora* is not more abundant. It may be that spore production was not very abundant in these plants (which may have relied predominantly on vegetative reproduction) or that the spores were relatively nonresistant compared with those of other groups.

Whatever the reason, it seems as though in this instance the situation is the reverse of that in the lycopods and that the spores may be underrepresenting the group.

Ferns

The very diverse triradiate spores of the Carboniferous represent collectively the ferns, the lycopods, the sphenopsids, the prepollen of some pteridosperms, together with, possibly, some bryophyte isospores. The great decrease in abundance and diversity of such spores seen in Upper Permian assemblages compared with those of the Upper Carboniferous presumably represent a major decline in the first three of these groups. This is of course equally apparent in their macrofossil record and has usually been correlated with (if not attributed to) the relative aridity of the climate in the Northern Hemisphere during the Permian. The various saccate gymnosperm pollen types accordingly appear in relatively greater proportion in the Early Triassic, as compared with the triradiate spores mainly of pteridophytic origin. Although relatively few accounts of Triassic assemblages give quantitative details, it appears that by the close of the Triassic, at least, triradiate spores were again typically equaling or even outnumbering gymnosperm pollen (as they had done in the Carboniferous). On the basis of the macrofossil evidence it may be supposed that this represents a resurgence mainly of the ferns rather than the lycopods or sphenopsids. This evidently involved fern groups that had already been present at the close of the Paleozoic—such as the Marattiales, Osmundaceae, Gleicheniaceae, and Schizaeaceae—together with more characteristic Mesozoic families—such as the Cyatheaceae, Dicksoniaceae, and Matoniaceae.

A number of fossil ferns from the Triassic and Jurassic are known in the fertile state, and study of the spores in their sporangia has yielded the following correlations:

Marattiales:

Asterotheca: *Latosporites* (smooth, monolete)
Danaeopsis: cf. *Punctatisporites* (smooth, triradiate)
Marattiopsis (or *Marattia*): *Marattisporites, Punctatosporites* (both monolete)

Filicales:

Osmundaceae:
Osmundopsis: *Osmundacidites* (triradiate, sculptured)
Todites: *Granulatisporites* (triradiate, sculptured)

Gleicheniaceae:
Gleichenites: Deltoidospora (p. 294)

Matoniaceae:
Phlebopteris: *Deltoidospora, ?Concavisporites.* (p. 294)

Dipteridaceae:
Dictyophyllum: *Triplanosporites, Dictyophyllidites*

Gymnosperms

The occurrence of bisaccate pollen in a number of very different gymnosperm groups (namely, Coniferales, Caytoniales, Pteridospermales, s.l.) suggests that this morphology arose independently as a direct adaptation to wind pollination, possibly involving a pollen-drop mechanism in an inverted ovule. The great diversity of bisaccate striate genera that dominates the Permian is still of unknown source, although the incomplete evidence at present available suggest that both *Glossopteris*-like plants and conifers (*Ullmannia, Rissikia*—see Potonié, 1962 and Townrow, 1967) may have produced pollen of this general type. The decline of bisaccate striate genera in the transition from Permian to Triassic in the Southern Hemisphere has been attributed to the decline of the *Glossopteris* flora and its replacement by one in which *Dicroidium* (represented by *Alisporites* or *Pteruchipollenites*) and other Mesozoic gymnosperms were dominant (Balme, 1963). At least some of the bisaccate nonstriate genera (e.g., *Pityosporites*, s.l.) which survive from the Permian through the Triassic presumably represent the conifers, which are well represented in both Permian and Triassic macrofloras. The rise of the Caytoniales appears to coincide with that of the bisaccate pollen *Vitreisporites* (see p. 302) in the course of the Triassic. Triassic monocolpate pollen—for example, *Cycadopites*, s.l. (probably a "primitive type" in the gymnosperms)—evidently represents several widely separated groups, including the Ginkgoales, Cycadales s.l., Bennettitales, and pteridosperms (in part). In view of the dominant role of these four groups in Triassic macrofloras it is a little surprising that monocolpate pollen is in general a relatively small constituent of Triassic assemblages compared with bisaccate pollen.

GENERA OF TRIASSIC SPORES AND POLLEN

This section contains a discussion of some of the genera of spores and pollen that occur in Triassic rocks. They have been selected either because of their common occurrence in the Triassic or because of their particular stratigraphic or biological significance. The ranges of these genera are shown in figure 14-1. Such a table involves many decisions regarding the generic identity of spores recorded in the literature and on the stratigraphic correlation of widely separated sections (see also figure 14-2); for these and other reasons such a compilation can only be a rather subjective and general record.

Figure 14-2 shows the approximate equivalence of the stratigraphic units from different parts of the world from which Triassic palynological data have been recorded and some of the more important sources of such data (the numbers on the table relating to works cited in the bibliography). The stratigraphic equivalence of any unit is normally that proposed or favored by the author of the relevant palynological work; in some cases this is based solely on the palynological data unsupported by any faunal evidence.

Azonate, Triradiate Spores

A considerable number of smooth, triradiate miospores have been reported from the Triassic; many of them agree superficially with those of a number of living and fossil leptosporangiate ferns. Triassic triradiate, smooth miospores of subtriangular amb, with concave to slightly convex interradial margins may be properly included in *Deltoidospora* Miner (1935). (See pl. 14-1, figs. 1-3.) Spores of this type, when flattened with the polar axis in the plane of compression, show that the proximal profile of the spore is more or less convex and the distal profile rather more strongly so. (See pl. 14-1, fig. 3.) The polar axis of the unflattened spore was evidently only slightly shorter than the longest measurement in the equatorial plane. Rather different appearances result from compression occurring more or less in the equatorial plane (as in pl. 14-1, fig. 1), obliquely with one lobe half overlapping an adjacent one (pl. 14-1, fig. 2) or with one lobe flattened symmetrically against the polar axis (pl. 14-1, fig. 3). These orientations were made the basis for supposedly distinct genera (*Triplanosporites* Pflug, as in pl. 14-1, fig. 2; *Poroplanites* Pflug, as in pl. 14-1, fig. 3) but it is questionable just how far this procedure is justified.

Cyathidites Couper (based on a Jurassic type) and *Leiotriletes* (Naumova) Potonié and Kremp (based on a Carboniferous type) are both very similar to *Deltoidospora*. Most authors have tended to use *Leiotriletes* for Paleozoic spores and *Deltoidospora* or *Cyathidites* for Mesozoic or Tertiary spores. It is doubtful whether these three genera can be separated consistently, and *Deltoidospora* has priority.

Spores agreeing with *Deltoidospora* were produced by various Mesozoic fern families—including Gleicheniaceae, Matoniaceae, and Cyatheaceae,—of which the two former are known from the Triassic. Spores substantially similar to *Deltoidospora* are known from the Devonian to the Recent.

Circular spores (originally more or less spherical), with a smooth, thin exine and short, triradiate sutures have been assigned to *Calamospora* Schopf, Wilson, and Bentall. Spores of this general type (some of which are known to have been produced by sphenopsids) are common in many Paleozoic assemblages but are rare in the Triassic (and subsequent Mesozoic periods—see chap. 15).

The name *Concavisporites* Pflug (pl. 14-1, fig. 4), of which *Paraconcavisporites* Klaus may be a synonym, is applied to Triassic and later spores resembling *Deltoidospora* but having proximal folds or thickenings encircling the triradiate mark. This encircling feature has been described as a torus, or margo (i.e., a difference in thickness of the proximal wall adjoining the aperture) and as a kyrtom (which as originally used for the Carboniferous genus *Ahrensisporites* referred to a distal feature). From photographs (and even in some cases from actual specimens) it is difficult to distinguish a proximal thickening from secondary folds occurring as a result of an original feature of the proximal face.

Paraconcavisporites Klaus has been separated from *Concavisporites* as having an asymmetrical displacement of the three laesurae where they meet at the apex, and spores showing this feature are reported to be restricted to the Triassic. It is

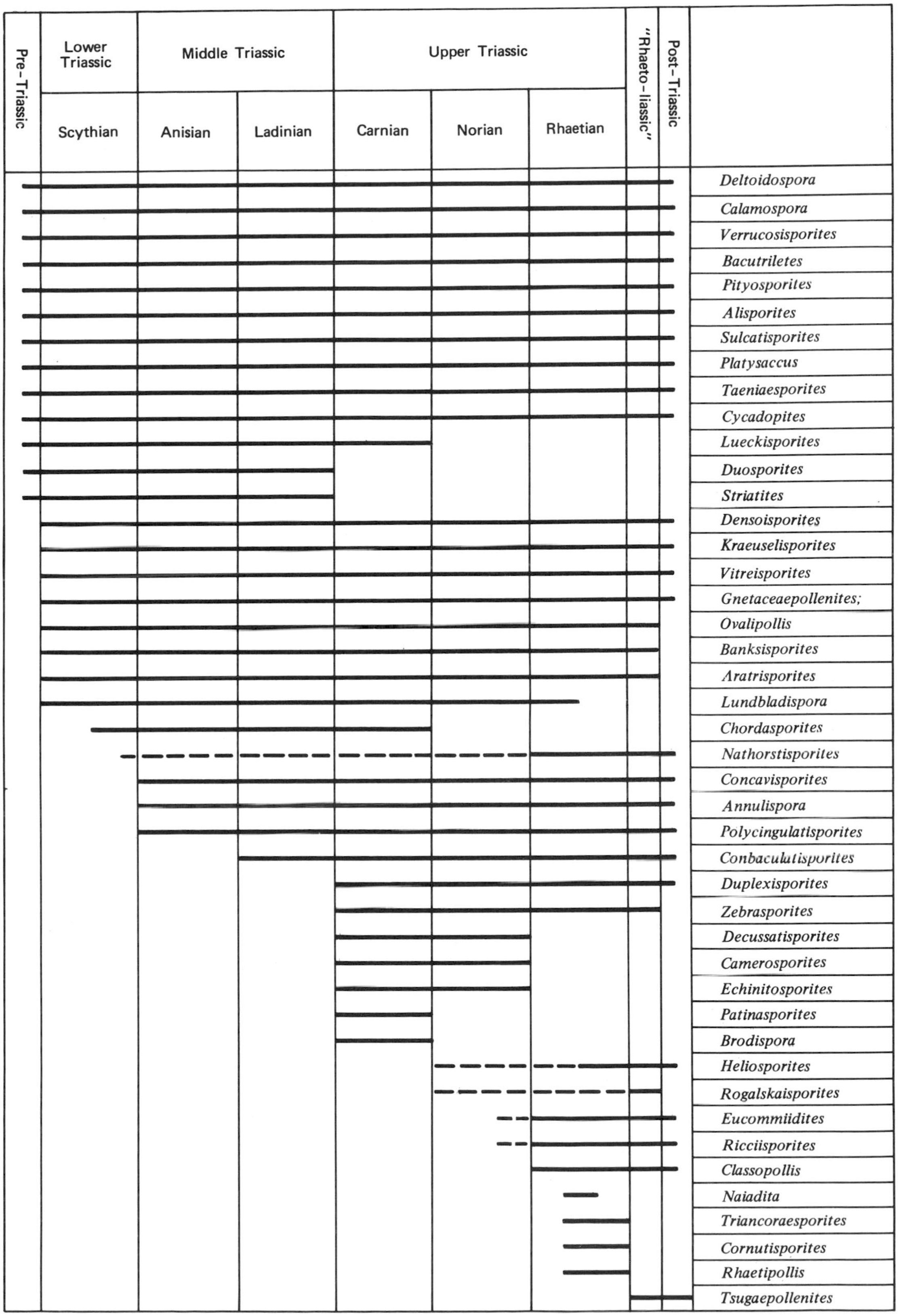

Pre-Triassic	Lower Triassic	Middle Triassic		Upper Triassic			"Rhaeto-liassic"	Post-Triassic
	Scythian	Anisian	Ladinian	Carnian	Norian	Rhaetian		

Deltoidospora
Calamospora
Verrucosisporites
Bacutriletes
Pityosporites
Alisporites
Sulcatisporites
Platysaccus
Taeniaesporites
Cycadopites
Lueckisporites
Duosporites
Striatites
Densoisporites
Kraeuselisporites
Vitreisporites
Gnetaceaepollenites;
Ovalipollis
Banksisporites
Aratrisporites
Lundbladispora
Chordasporites
Nathorstisporites
Concavisporites
Annulispora
Polycingulatisporites
Conbaculatisporites
Duplexisporites
Zebrasporites
Decussatisporites
Camerosporites
Echinitosporites
Patinasporites
Brodispora
Heliosporites
Rogalskaisporites
Eucommiidites
Ricciisporites
Classopollis
Naiadita
Triancoraesporites
Cornutisporites
Rhaetipollis
Tsugaepollenites

Figure 14-1. Stratigraphic ranges of selected Triassic genera.

	Alpine Facies	Germanic Facies	U.S.S.R.	U.S.A.	Canada	Ellesmere Island	Argentina	Western Australia	South and East Australia	Tasmania	Antarctica	Madagascar
Upper Triassic	Rhaetian	5, 8, 17, 24, 25, 31, Rhaetian ("Rhaetic System") 32, 37, 43, 47 / "Upper Keuper" of some authors		? Chinle		?			Leigh Creek Coal Measures (lower part) 42	? Brady	Beacon Group (in part) - Timber Peak Locality 38 ?	
	Norian	Keuper or Gipskeuper (Middle Keuper of some authors) 7, 30, 39, 44, 51, 55	Norian	Formation 9, 40, 52		Heiberg Formation 36 ?	? Los Rastros Formation 20		? Bundamba Group 11 ?	Formation 41 ?		Isalo Group
	Carnian 26 27		Carnian	?			?		Moorooka Formation 13	?		
Middle Triassic	Ladinian	Lettenkohle 2	Ladinian						Mount Crosby Formation 12 ?	Tiers		?
	Anisian	Muschelkalk	Anisian		Toad Formation 21				Tingalpa Formation 13 Hawkesbury Formation	Formation 41		Upper Sakamena Formation ?
Lower Triassic	Scythian (= Werfenian)	Bunter 15, 45, 46, 48, 57	Olenek		Grayling Formation 21	? Bjorne Formation 36		Kockatea Shale 1	? Wollar Sandstone 18 ? Narrabeen Group 19	? Ross Formation 41		Middle Sakamena Formation
			Ind									
		6, 28, 33, 49, 50, 53, 54, 58	3, 4, 29, 34, 35, 56						10, 14	14		16, 22, 23

questionable whether this feature can be consistently recognized. *Auritulinasporites* Nilsson 1958 and *Toroisporis* Krutzsch (see Reinhardt, 1961) and *Dictyophyllidites* of some authors are very similar to *Concavisporites* and may well be partial synonyms. Russian authors (e.g., Chalishev and Varyukhina, 1966) have attributed similar Triassic spores to living taxa as "Dipteridaceae," "*Matonia*," or to fossil (macrofossil) fern genera, as "*Dictyophyllum*" (Kopitova, 1963b).

A spore superficially resembling *Deltoidospora* but possessing a distal annular thickening (or "circumpolar crassitude") has been described as *Annulispora* by de Jersey. This genus is a characteristic Triassic spore, although it is also reported from some later horizons. Some spores of this type have a slight equatorial thickening, leading Playford (1965) to regard them as cingulate. The closely similar *Distalanulisporites* Klaus may be a synonym of *Annulispora*.

Numerous sculptured, triradiate, azonate spores occur in the Triassic, although generally they are less abundant (compared with associated gymnosperm pollen) than in the Carboniferous and Jurassic. A number of such genera have been based on Triassic spores (see Mädler, 1964) but the distinction between these and similar Paleozoic and Jurassic or Cretaceous forms is not easy to sustain at generic level. An example of one of these genera of long-ranging sculptured triradiate spores occurring in the Triassic is *Verrucosisporites* (Ibrahim) Smith and Butterworth, a genus based on a Carboniferous species, for triradiate, more or less circular spores with verrucose ornament. A large Late Triassic species of this genus that occurs in Britain and on the Continent is *V. morulae* (pl. 14-1, fig. 6).

A genus of triradiate, sculptured spores that appears to have a more limited range is *Conbaculatisporites* Klaus (pl. 14-1, fig. 5). This includes spores of subtriangular amb with baculate ornament and is a characteristic Triassic form. Similar genera of longer range include *Baculatisporites* Thomson and Pflug, with a circular amb, and *Neoraistrickia* Potonié, which has fewer and relatively larger sculptural elements.

Triradiate Spores with an Equatorial Feature

Kraeuselisporites Leschik (pl. 14-1, fig. 9) is a common and characteristic genus for triradiate miospores with a rounded-triangular amb, a broad zona, and a coarse ornament of coni, principally on the distal face. Some species show a gap between an inner and outer exine layer and accordingly have been regarded as cavate. The closely similar *Styxisporites* Cookson and Dettmann is probably synonymous.

Heliosporites Schulz is a monotypic genus of somewhat similar spores, with a rather narrower equatorial feature and longer, blunt, curved, and sometimes recurved spinose sculptural elements. These spores, which are restricted to the Rhaetian and basal Jurassic, are about 40 to 60 microns in diameter, commonly adhere in tetrads, and are strongly reminiscent of some living *Selaginella* microspores. There has been some confusion over the correct designation for this spore, which has been variously named *Lycospora reissingeri* (Harris, 1957), *Heliosporites altmarkensis* (Schulz, 1962), and *Styxisporites reissingeri* (Danze-Corsin and Laveine, 1963). If these species are accepted as synonymous, the

---

Figure 14-2. Selected stratigraphic divisions of the Triassic System. Numbers on the figure record the occurrences of some of the important palynological data. The numbers refer to the list of references as follows: 1—Balme, 1963. 2—Bharadwaj and Singh, 1964. 3—Bolkhovitina, 1959. 4—Chalishev and Vanyukina, 1966. 5—Chaloner, 1962a. 6—Chaloner, 1962b. 7—Clarke, 1965. 8—Danze-Corsin and Laveine, 1963. 9—Daugherty, 1941. 10—deJersey, 1962. 11—deJersey, 1964. 12—deJersey and Hamilton, 1965a. 13—deJersey and Hamilton, 1965b. 14—Dettmann, 1961. 15—Freudenthal, 1964. 16—Gothan and Remy, 1957. 17—Harris, 1957. 18—Helby, 1966. 19—Helby and Martin, 1965. 20—Herbst, 1965. 21—Jansonius, 1962. 22—Jekhowsky, de, and Goubin, 1964. 23—Jekhowsky, de, and Letullier, 1960. 24—Jung, 1958. 25—Jung, 1959. 26—Kavary, 1966. 27—Klaus, 1960. 28—Klaus, 1964. 29—Kopitova, 1963b. 30—Leschik, 1955. 31—Levet-Carette, 1964. 32—Lundblad, 1959. 33—Mädler, 1964. 34—Malyavkina, 1953. 35—Malyavkina, 1960. 36—McGregor, 1965. 37—Nilsson, 1958. 38—Norris, 1965. 39—Pautsch, 1958. 40—Peabody and Kremp, 1964. 41—Playford, 1965. 42—Playford and Dettmann, 1965. 43—Reinhardt, 1961. 44—Reinhardt, 1963. 45—Reinhardt, 1964. 46—Reinhardt and Schmitz, 1965. 47—Shulz, 1962. 48—Schulz, 1964. 49—Schulz, 1965. 50—Schulz, 1966. 51—Schulz and Krutzsch, 1961. 52—Scott, 1960. 53—Taugourdeau-Lanz, 1962. 54—Taugourdeau-Lanz, 1963. 55—Taugourdeau-Lanz and Jekhowsky, 1959. 56—Varyukhina, 1961. 57—Visscher, 1966. 58—Warrington, 1967.

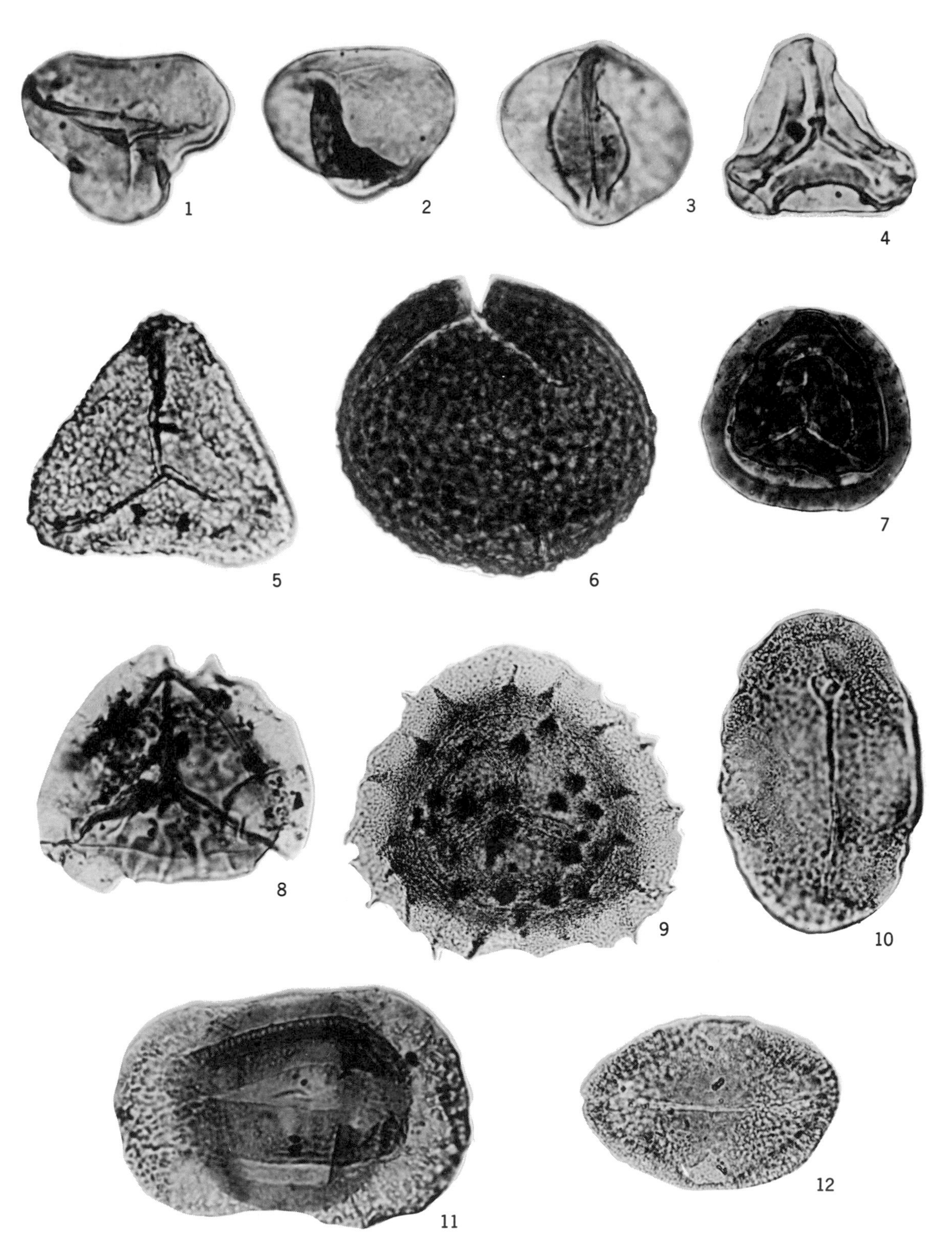

Plate 14-1. (All figures are × 750 unless otherwise stated.) 1-3—*Deltoidospora neddeni* (Potonié) *comb. nov.*: 1—flattened slightly obliquely down the polar axis; 47 microns; 2—flattened obliquely, one lobe partially folded on the remaining two; 43 microns; 3—flattened along one radius, showing the "*Poroplanites*" orientation (all from the Keuper, Lunz, Austria). 4—*Concavisporites lunzensis* (Klaus) *comb. nov.* Keuper, Lunz, Austria; 39 microns. 5—*Conbaculatisporites mesozoicus* Klaus. Keuper, Lunz, Austria; 56 microns. 6—*Verrucosisporites*

correct name for the combination is *Heliosporites reissingeri* (Harris) *comb. nov.*

Zebrasporites Klaus (pl. 14-1, fig. 8), a genus characteristic of the Upper Triassic, is based on triradiate spores with a thin membranous zona, which is widest interradially and is strengthened by a series of distal ribs extending on to the zona.

Two other Triassic genera that are restricted to the Rhaetian, *Cornutisporites* Schulz and *Trianchoraesporites* Schulz, have very pronounced radial features with expanded extremities, which are recurved in a grapnellike manner, especially in the former genus.

A spore (pl. 14-3, fig. 9) with a narrow equatorial feature, distinctive distal sculpture, and a proximal germinal exitus of irregular outline (described as "hilate" by Dettmann) is of interest as being one of the few dispersed spores that can be reliably attributed to a bryophyte. This is the spore described by Harris (1938) from the Rhaetian liverwort *Naiadita lanceolata*. It has never been given a name as a dispersed spore, but it occurs at the same horizon as the parent plant and appears on present records to be restricted to a very narrow vertical interval in the British Rhaetian. (The spores reported by McGregor, 1965, from the Ellesmere Lower Triassic appear to be specifically, and probably generically, distinct from the British material).

Duplexisporites Deak (pl. 14-1, fig. 7) includes spores with prominent round-topped muri on both proximal and distal surfaces, more or less paralleling the amb, and interrupted especially toward the distal pole. The genus appears in the Triassic and continues through the Mesozoic. The equatorial feature, which is no broader than the other muri, may be interpreted as a cingulum, although the spore is regarded by some (e.g., Playford and Dettmann, 1965) as azonate. Closely similar (?congeneric) spores have been described from the Russian Upper Triassic by Kopitova (1963a, b) as *Azonotriletes intertextus*. Other synonyms of this rather distinctive genus are discussed by Playford and Dettmann (1965).

Spores with a Saccus, Cavate Exine, or Mesospore

Two genera of triradiate spores with a gap between layers of the exine (a cavity, or saccus) occur commonly in the Triassic. These are *Densoisporites* Weyland and Krieger, with a smooth exine, and *Lundbladispora* Balme, with a sculptured outer surface. Some species of both these genera have a narrow equatorial thickening (rather like the limbus in *Endosporites*), and this has led Playford (1965) to refer to both as being cingulate. It is not clear from our present knowledge whether these spores—in which the double wall structure has been variously termed a mesospore, a cavate structure, or a saccus—had an air-filled space between the wall layers in life.

Aratrisporites Leschik (pl. 14-2, fig. 6) is a characteristic and widespread Triassic genus that resembles the last two genera in having a cavate, or saccate structure, with spines on the outer exine surface; unlike them, however, it has a monolete mark. The closely similar genus, *Saturnisporites* Klaus, is regarded by some as a later synonym of Leschik's genus. *Aratrisporites*, at one time thought to be zonate, is now widely acknowledged as being saccate, or cavate (Bharadwaj and Singh, 1964; Clarke, 1965; Playford and Dettmann, 1965). Apparently congeneric spores have been reported from the Russian Triassic under the name of *Zonomonoletes* Luber (e.g., *Z. tschalyschevii* Varyukhina 1961; *Z. spinosus* Kopitova, 1963a, b).

Saccate Pollen

A number of monosaccate pollen genera, some of which range upward from the Permian, occur in the Triassic. *Tsugaepollenites* Pot. and Ven., a genus characterized by a velum or irregularly attached saccus reminiscent of the pollen of the living *Tsuga*, first appears in the Upper Triassic and is a common constituent of later Mesozoic assemblages. In its broadest sense this genus includes not only late Mesozoic and Tertiary pol-

---

morulae Klaus. "Lower Keuper," Bromsgrove, England; 70 microns. 7—*Duplexisporites scanicus* (Nilsson) Playford and Dettman. Rhaetian, Foxholes No. 2 borehole, Owthorpe, Nottinghamshire, England; 45 microns. 8—*Zebrasporites fimbriatus* Klaus. Middle Keuper, Basle, Switzerland; 62 microns. 9—*Kraeuselisporites cuspidus* Balme. (holotype). (× 600). Lower Triassic, western Australia; 90 microns. (Photo by B. E. Balme.) 10 and 12—*Ovalipollis lunzensis* Klaus. 10—Middle Keuper, Basle, Switzerland, 72 microns; 12—Keuper, Longdon, Worcestershire, England, 56 microns. 11—*Taeniaesporites*, cf. *T. kraeuseli* Leschik. Rhaetian, Foxholes No. 2 borehole, Owthorpe, Nottinghamshire, England; 78 microns.

len of pinacean conifers but, within the Triassic at least, probably the pollen of other conifers also.

Of the large number of bisaccate striate pollen in the Permian (see chap. 13) relatively few forms survive into the Triassic. Among Triassic forms at least three genera may be clearly recognized. *Lueckisporites* Potonié and Klaus, with its proximal cap divided by a single striation into two broad, thick "muri," survives into the Upper Triassic. *Taeniaesporites* Leschik (synonymous with *Lunatisporites* Leschik), with from three to five proximal muri (pl. 14-2, fig. 1), ranges through the Triassic and possibly into the basal Jurassic. *Striatites* Pant, with numerous striae and muri, ranges at least into the Middle Triassic. There is no simple or obvious relationship between the fossil history of the bisaccate striate pollen and that of any single parent group. There is evidence (which is still not entirely convincing) linking bisaccate striate pollen to *Glossopteris* fructifications and to at least two distinct groups of conifers (see Potonié, 1962 and Townrow, 1967). If this heterogeneous origin of such seemingly distinctive pollen is borne out by further work, it will constitute a remarkable example of convergent evolution of a single plant organ.

Chordasporites Klaus (pl. 14-2, fig. 5) includes nonstriate bisaccate pollen rather reminiscent of *Lueckisporites* because of the division of the proximal face by a single thickening (or possibly a regularly occurring secondary fold) linking the two sacci. It ranges from the Lower Triassic to the Carnian. Kopitova (1963a, b) assigns *Chordasporites*-like pollen to *Florinites pseudostriatus*. Another bisaccate genus virtually restricted to the Triassic is *Ovalipollis Krutzsch* 1955 (pl. 14-1, figs. 10 and 12), of which *Unatextisporites* Leschik is a later synonym. This genus is characterized by very narrow sacci extending around the narrow ends of an almost rhomboidal body to produce a spore of elliptical amb. An aperture in the form of a narrow slit along the longest axis of the grain apparently represents a monolete suture (i.e., a proximal germinal aperture). In being both bisaccate and monolete *Ovalipollis* thus differs from all living bisaccate pollen but resembles some of the bisaccate striate genera.

Aside from *Chordasporites* and *Ovalipollis* there are a large number of less distinctive forms of nonstriate bisaccate genera in Triassic assemblages. On the average these probably exceed the number of bisaccate striates in both numbers and species within the Triassic, whereas in a typical upper Permian assemblage the reverse is probably true. Many genera have been recognized, based on relative size, configuration and extent of attachment of sacci, and on aperture and haptotypic features. A consistent separation between many of these genera is difficult to maintain.

The earliest described bisaccate genus was *Pityosporites* Seward; its holotype is a petrified specimen and can only be seen in a rock thin section cut parallel to the polar axis and including the sacci. As a result the character of the grain as seen down the polar axis is unknown. The genus has been widely used as a "portmanteau name" for nonstriate bisaccates lacking clearly defined characteristics that would place them in more narrowly defined genera. In this very broad sense *Pityosporites* ranges throughout the Triassic. *Alisporites* Daugherty (pl. 14-2, figs. 2 and 4), based on an Upper Triassic type, has sacci typically narrower than the body (measured perpendicular to the polar axis and in the plane of the sacci) and evidently not as deep (measured parallel to the polar axis) as in the hemispherical sacci of many living conifers. This latter feature may be seen in occasional specimens flattened so that one saccus is superimposed on another. (See pl. 14-2, fig. 2.) The genera *Pteruchipollenites* Couper and *Abietineaepollenites* Potonié as used by many authors includes species closely comparable to *Alisporites* (see also Rouse, 1959). Spores resembling *Alisporites* have been obtained from a disconcertingly wide group of fossil plant fructifications. Very similar forms have, for example, been extracted from *Pteruchus*

Plate 14-2. (All figures are × 750 unless otherwise stated.) 1—*Taeniaesporites* cf. *T. noviaulensis* Leschik. (× 600). Lower Triassic, western Australia; 75 microns. (Photo by B. E. Balme.) 2 and 4—*Alisporites* sp., cf *A. opii* Daugherty. 2—specimen flattened along its longest axis. Chinle Formation; Upper Triassic, Utah; 67 microns. 4—flattened obliquely down the polar axis, thus foreshortened central body and sacci. Rhaetian, Foxholes No. 2 borehole,

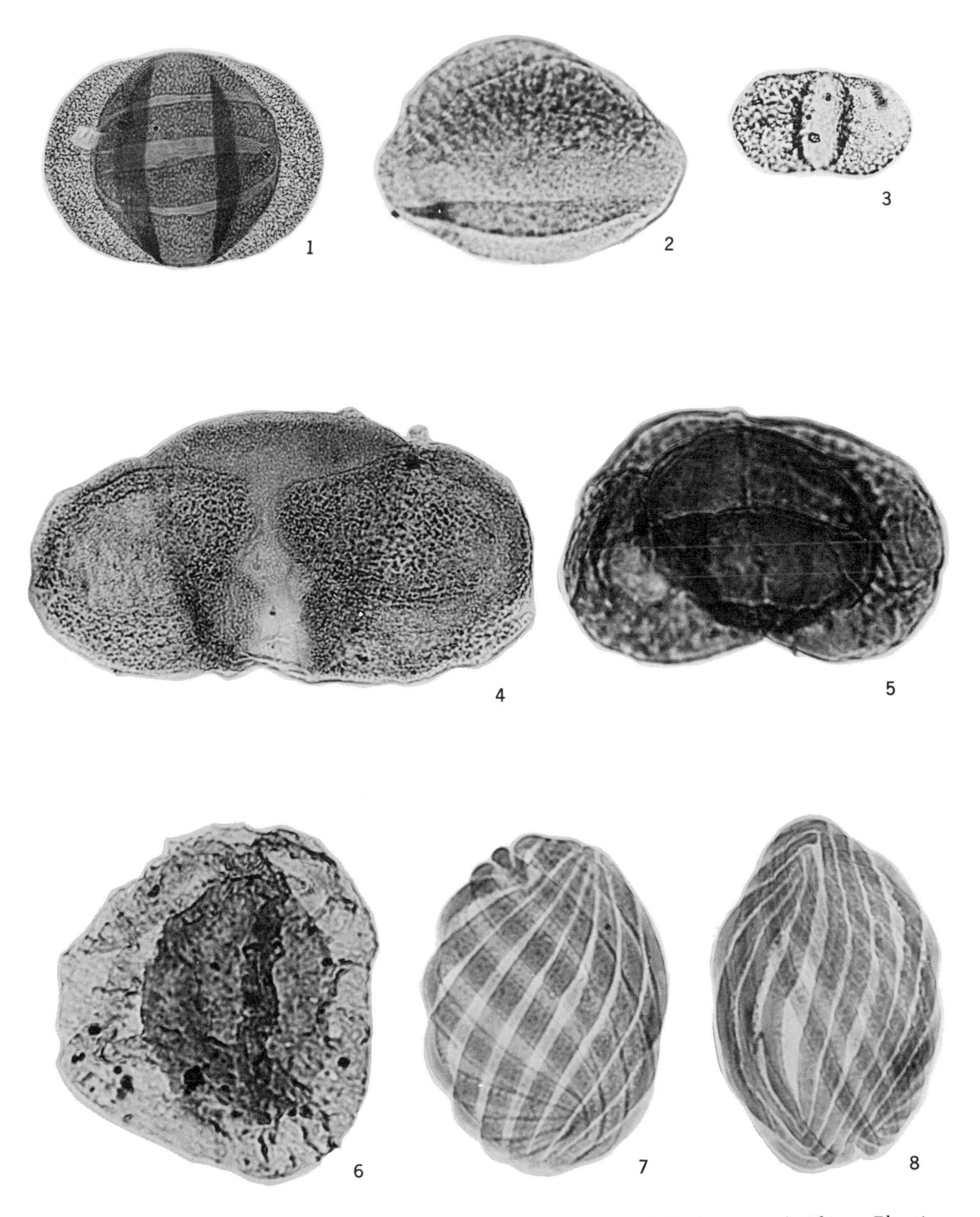

Owthorpe, Nottinghamshire, England; 165 microns. 3—*Vitreisporites pallidus* (Reissinger) Nilsson. Rhaetian, Foxholes No. 2 borehole, Owthorpe, Nottinghamshire, England; 38 microns. 5—*Chordasporites singulichorda* Klaus. "Lower Keuper," Bromsgrove, Worcestershire, England; 81 microns. 6—*Aratrisporites fischeri* (Klaus). Keuper, Lunz, Austria; 75 microns. 7 and 8—*Gnetaceaepollenites chinleana* (Daugherty) *comb. nov.*, Chinle Formation, Upper Triassic, Utah; 7—70 microns; 8—75 microns.

Thomas (a pteridosperm in the broad sense) and *Masculostrobus willsi*, a conifer; this is discussed fully by Townrow (1962).

Pollen that may be assigned to the genus *Platysaccus*, with its pronouncedly diploxylonoid sacci, occurs all through the Triassic, although perhaps less commonly than in the Permian.

Vitreisporites Leschik is a genus for small bisaccate pollen, typically not more than 40 microns in its greatest dimension. (See pl. 14-2, fig. 3.) The central body is subcircular to oval (elongation perpendicular to a line joining the sacci), and the attachment of the sacci is such as to give a more or less smooth oval outline to the whole grain. The sacci are minutely infrareticulate, and the central body may be smooth or show a similar but finer stippled pattern. The details of saccus attachment are obscure, but they were apparently somewhat offset distally (Townrow, 1962); the grain was sufficiently flattened in the equatorial plane in life to show strongly preferred proximo-distal orientation. As a result the fossil almost invariably presents a view down the polar axis, showing the two air sacs in apparently linear relationship with the central body. The small size of *Vitreisporites* distinguishes it immediately from the majority of bisaccate pollen genera.

Two rather distinct groups of parent plants seem to have produced pollen agreeing with *Vitreisporites*: the Caytoniales and *Harrisiothecium* Lundblad (a problematical pteridosperm, s.l.). *Vitreisporites* in the strict sense appears in the Triassic and ranges on into the Jurassic and Cretaceous. Occasional grains assigned to this genus have been reported by some authors from the Permian (Balme, 1963). In the Northern Hemisphere at least there seem to be no well-authenticated records before the Triassic.

Nonsaccate Pollen

Gnetaceaepollenites Thiergart (pl.14-2, figs. 7 and 8) is a genus representing originally prolate spheroidal grains, with a longitudinally striate (i.e., "polycolpate") exine. The striae presumably represent germinal areas (colpi) comparable to those of the very similar pollen of living *Ephedra*. Although *Gnetaceaepollenites* is based on a species showing zigzag striae (colpi), a broad interpretation of the genus may include forms such as the species illustrated here with more or less straight striae. This range, from forms with straight furrows to those with zigzag furrows, corresponds to that shown among the species of living *Ephedra*. The species illustrated is believed to be that originally described by Daugherty as *Equisetosporites*— a genus now to be regarded as a probable synonym of *Gnetaceaepollenites*. This same species was assigned by Scott (1960) to the living genus *Ephedra*; the more noncommittal course of placing this pollen in a genus based on fossil dispersed spores is favored here. It seems likely that most of these striate grains represent Ephedraceae; however, whereas *Gnetaceapollenites* ranges from the Triassic to the Recent, macrofossil evidence of the family extends back only as far as the Cretaceous. Permian, and even the earlier Triassic records (e.g., McGregor, 1965, Lower Triassic of Ellesmere Island) require confirmation.

One of the most ubiquitous of all Mesozoic forms is the monocolpate genus *Cycadopites* (Wodehouse) Wilson and Webster (pl. 14-3, fig. 8). This comprises grains of more or less elliptical amb with a smooth exine and a single presumably distal colpus. The genera *Monsulcites*, *Ginkgocycadophytus*, *Palmidites*, and *Entylissa* may be regarded as later synonyms. Several similar genera (and even suprageneric categories) have been recognized on the shape and length of the colpus. It seems impossible to recognize consistent taxa of generic or higher level on these characters, and *Cycadopites* is here treated in a broad sense and, as such, having priority. A number of widely separated groups of living plants produce pollen closely comparable to *Cycadopites*—namely, some monocotyledons, the Ginkgoales, and Cycadales. In addition at least two extinct Mesozoic groups, the Bennettitales and pteridosperms, s.l. (e.g., *Antevsia* Harris, Peltaspermaceae), produced similar pollen. Evidently this simple type of monocolpate pollen represents either a primitive form common to

---

Plate 14-3. 1-3—*Decussatisporites delineatus* Leschik. Keuper, Lunz, Austria: 1 and 2—42 microns, the same specimen viewed from opposite sides; 3—45 microns (all × 1,000). 4-6—*Classopollis torosus* (Reissinger) Balme. (× 750), Rhaetian: 4—flattened in the equatorial plane; Henfield borehole, Sussex, England, 38 microns; 5—obliquely flattened, 39 microns; 6—flattened in the plane of the polar axis, 34 microns; 5 and 6—from the **Fox-**

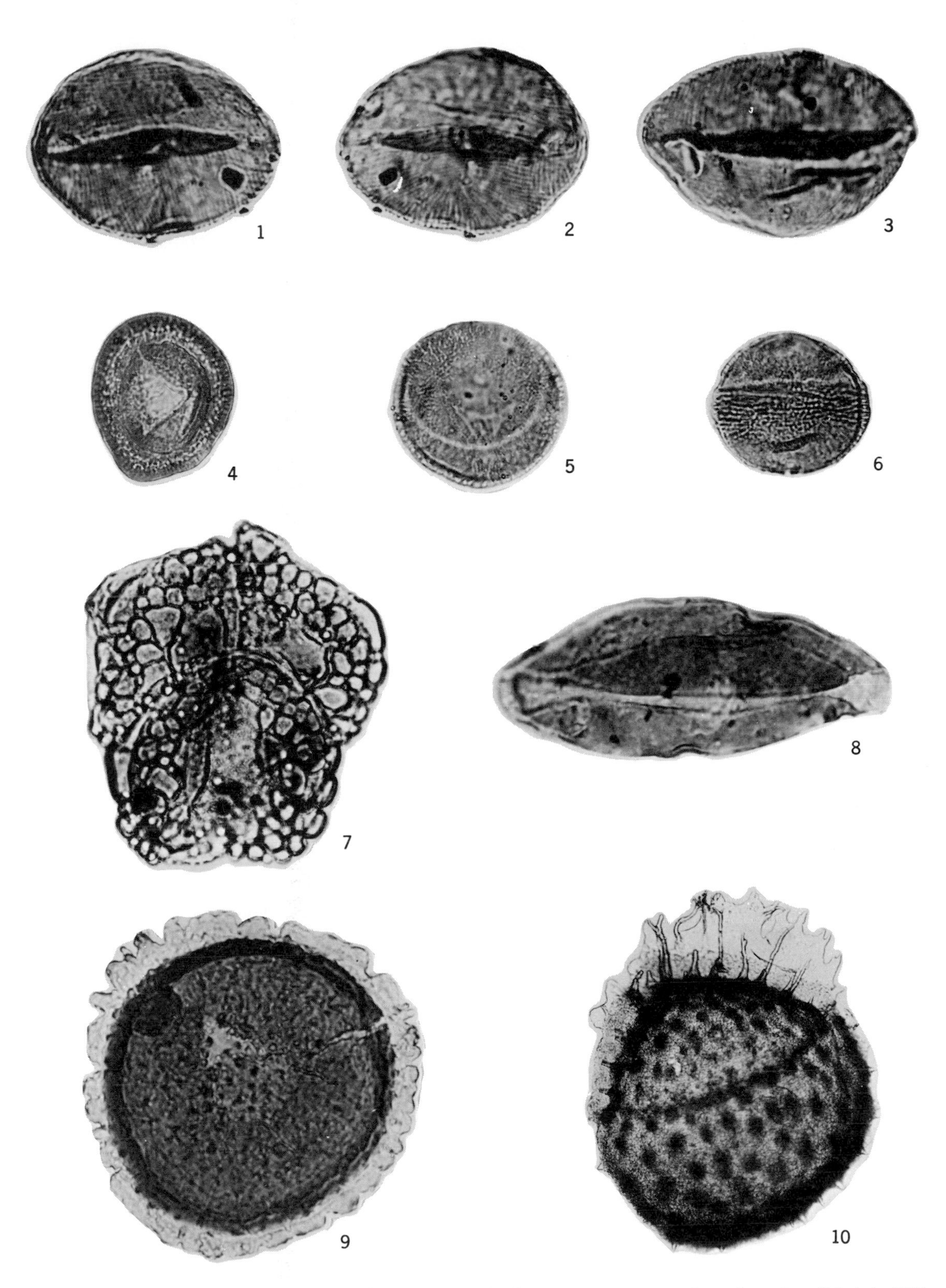

holes No. 2 borehole, Owthorpe, Nottinghamshire, England. 7–*Ricciisporites tuberculatus* Lundblad. (× 500). Rhaetian, Upton borehole, Burford, Oxfordshire, England. Tetrad; 126 microns. 8–*Cycadopites carpentieri* (Delc. and Sprum.) *ex* Couper (× 750). Keuper, Lunz, Austria; 81 microns. 9–Spore of *Naiadita lanceolata* Buckman. (× 500). Rhaetian, Hapsford Bridge, Somersetshire, England; 108 microns. 10–*Nathorstisporites hopliticus* Jung. (× 100). Rhaeto-Liassic of Australia; 500 microns. (Photo by M. E. Dettman.)

all these groups or possibly a morphologically simple type arrived at independently.

A genus similar to *Cycadopites* but readily distinguished from it by its sculpture is *Decussatisporites* Leschik. (See pl. 14-3, fig. 1-3.) (The name *Lagenella* (Malyavkina) Klaus 1969 pars. must be regarded as a later synonym.) This genus includes pollen of elliptical amb, of which the whole exine is finely striate, with the striae and intervening muri aligned transversely to the long axis on the proximal face and parallel to it on the distal face; there are two areas of convergence of the striations, near the equator at the center of each of the long sides of the flattened grain. The two margins of the colpus often overlap, as a result of compression. *Decussatisporites* is restricted to the Upper Triassic; it has been recorded in close association with *Stachyotaxus* by Potonié (1967). This conifer is placed in the Cheirolepidaceae, an attribution that gains no support from the vastly different morphology of the pollen of *Cheirolepidium* itself (see *Classopollis* below).

Eucommiidites Erdtman is comparable to *Cycadopites* in having one large colpus, but it has in addition two further colpoid apertures, which have led some authors to regard it as an angiospermlike tricolpate genus. (See chap. 15.) Spores evidently closely similar to Jurassic and Cretaceous *Eucommiidites* have been described from the Russian Triassic as *Paleoconiferus asaccatus* (Chalishev and Varyukhina, 1966). If the first record of this species by these authors is to be accepted as representing *Eucommiidites*, the earliest occurrence of the genus is pushed back to the Middle Triassic.

Classopollis Pflug (pl. 14-3, figs. 4-6) is a genus of remarkable and distinctive organization. The pollen grain is a slightly prolate spheroid with a proximal triradiate or triangular aperture and a distal more or less circular pore. A narrow striation or groove (a thin area of the exine—the "rimula" of Pflug) extends around the spore parallel to the equator, but rather nearer the distal pole. The exine is otherwise thickest in the equatorial region and thins toward either pole. The inner surface of the exine is covered with baculae, directed inward; in the equatorial region adjacent baculae are fused to form a series of beaded muri separated by internal striae, giving a set of about nine striations parallel to the equator. The grains commonly occur adhering in tetrads, although the exines of the four members are not joined as in *Ricciisporites* and separate individuals are equally common. There is no good evidence to suggest that the grain ever split around the subequatorial striation to detach an operculum, and it seems that it may have germinated either proximally through the triangular aperture or distally through the pore. *Corollina* (Malyavkina) Klaus is a synonym of *Classopollis*; this and other aspects of the somewhat involved synonymy of *Classopollis* are discussed by Pocock and Jansonius (1961), Klaus (1960), and Chaloner (1962a). *Classopollis* appears in the Rhaetian and ranges into the Cretaceous. Earlier records (middle Keuper of Mädler, 1964; Permian of Pocock and Jansonius, 1961) require confirmation.

Pollen indistinguishable from *Classopollis torosus* was produced by the Rhaetian conifer *Cheirolepidium muensteri* (Harris, 1957); but the long range and abundance of *Classopollis* make it unlikely that *Cheirolepidium muensteri* was its only source, and other (presumably closely related) conifers represent the most probable parent group.

Ricciisporites Lundblad 1959 (of which *Tetradosulcites* Erdtman is a synonym) is a genus (pl. 14-3, fig. 7) in which the four members of the tetrad are structurally united by the exine, much as in the pollen of living Ericaceae. Each member of the tetrad has a distal colpus, and the colpi of all four members are more or less parallel. The exposed exine surface, apart from the colpi, is densely covered with clavate processes of varied sizes; these are so closely packed as to produce the effect of a negative reticulum. The members of the tetrad are so firmly joined that even highly corroded tetrads have not separated into their constituent grains. The affinity of *Ricciisporites* is unknown, but a general comparison may be made with the several groups producing monocolpate pollen. *Ricciisporites* is monotypic; *R. tuberculatus* is known only from the Rhaetian and basal Liassic, from Greenland, Britain, Germany, and Poland.

Echinitosporites Schulz and Krutzsch (1961) from the Keuper of Germany is of interest in that it somewhat resembles the constituent grains of the *Ricciisporites* tetrad.

Anomalous Sporelike Bodies

Camerosporites Leschik is an Upper Triassic

spore of distinctive structure and unknown relationship. It is spheroidal in shape, with a series of equatorial, more or less hemispherical, exinous thickenings around the (presumed) equator, and most of one face (the ?distal) occupied by a germinal area or tenuitas. Leschik regarded the exinal thickenings as small cavities, or chambers—hence the name (see Clarke, 1965). *Camerosporites* is known from the Upper Triassic of Britain and Switzerland.

Brodispora Clarke is an Upper Triassic genus with an oblate, more or less spheroidal body marked by a series of concentric (? sometimes spiraled) striae centering on the two poles of the spore. Similar spores described from elsewhere in the Triassic (e.g., de Jersey, 1962; Norris, 1965; and Peabody and Kremp, 1964) have been assigned to *Circulisporites* De Jersey and *Chomotriletes* Naumova. It appears that a single group of spores of rather distinctive morphology and restricted, Triassic range may be involved here.

Rhaetipollis Schulz 1967 is a type of spore comparable to the striate forms just considered in being apparently formed of two symmetrical hemispheres each with a concentric ornament. This interesting spore has a very restricted range within the Rhaetian of Britain and Germany.

Megaspores

Although megaspores are not as abundant in the Triassic as a whole as they are in some Paleozoic coals and later Mesozoic autochthonous deposits, they are of considerable botanical interest, and some are of restricted stratigraphical range. *Banksisporites* Dettman is a genus for smooth or granular megaspores with mesosporium, or inner (? nexinous) body, that occurs throughout the Triassic. It is known to represent, at least in part, lycopod megaspores. *Nathorstisporites* Jung 1958 (pl. 14-3, fig. 10), also a lycopod megaspore, is characterized by tall, fimbriate lips and a sculptured exine. It is a widespread genus in the Rhaetian and Early Jurassic, but an earlier and rather isolated record (Lower Triassic) is reported from Australia by Helby (1966).

References

Balme, B. E., 1957, Spores and pollen grains from the Mesozoic of western Australia: CSIRO, Chatswood, western Australia, ref. T. C. 25, p. 1-48.

——1962, Some palynological evidence bearing on the development of the *Glossopteris* flora: Evolution of living organisms, Melbourne Univ. Press, p. 269-280.

——1963, Plant microfossils from the Lower Triassic of western Australia: Palaeontology, v. 6, no. 1, p. 12-40.

Bharadwaj, D. C., and Singh, H. P., 1964, An Upper Triassic miospore assemblage from the coals of Lunz, Austria: Palaeobotanist, v. 12, no. 1, p. 28-44.

Bolkhovitina, N. A., 1959, Spore-pollen assemblages of the Mesozoic of the Viljuj depression and their stratigraphic value [in Russian]: Trudy Geol. Inst. USSR, v. 24, p. 1–186.

Chalishev, V. I., and Varyukhina, L. M., 1966, Biostratigraphy of the Triassic of the Petschora district [in Russian]: Moscow-Leningrad, Akad. Nauk SSSR, Kom. Fil. Inst. Geol., 148 p.

Chaloner, W. G., 1962a, Rhaeto-Liassic plants from the Henfield borehole: Bull. Geol. Surv. Great Britain, v. 19, p. 16-28.

——1962b, British Rhaetic and Triassic spores: Pollen et Spores, v. 4, no. 2, p. 339.

Clarke, R. F. A., 1965, Keuper miospores from Worcestershire, England: Palaeontology, v. 8, no. 2, p. 294-321.

Couper, R. A., 1958, British Mesozoic microspores and pollen grains: Palaeontographica, v. 103, sec. B, p. 75-179.

Cookson, I., and Dettmann, M. E., 1958, Some trilete spores from Upper Mesozoic deposits in the eastern Australian region: Proc. Royal Soc. Victoria, v. 70, no. 2.

Danzé-Corsin, P., and Laveine, J. P., 1963, Flore infraliassique du Boulonnais. B. Microflore: Mem. Soc. Geol. Nord. v. 13, p. 57-143.

Daugherty, L. J., 1941, The Upper Triassic flora of Arizona: Carnegie Inst. Wash. Publ. 526, p. 1-108.

Dettmann, M. E., 1961, Lower Mesozoic megaspores from Tasmania and South Australia: Micropaleontology, v. 7, no. 1, p. 71-86.

Freudenthal, T., 1964, Palaeobotany of the Mesophytic. I. Palynology of Lower Triassic rock salt, Hengelo, The Netherlands: Acta Bot. Neerlandica, v. 13, p. 209-236.

Gothan, W., and Remy, W., 1957, Steinkohlenpflanzen: Essen, p. 1-248.

Goubin, N., 1965, Description et Repartition des Principaux Pollenites Permiens, Triassiques, et Jurassiques des Sondages du Bassin de Morondava (Madagascar): Rev. Inst. Français Petr., v. 20, no. 10, p. 1415-1461.

Harris, T. M., 1938, The British Rhaetic flora: Brit. Mus. [Nat. Hist.], p. 1-84.

______1957, A Liasso-Rhaetic flora in South Wales: Proc. Royal Soc. [London], v. 147B, p. 289-308.

Helby, R., 1966, Triassic plant microfossils from a shale within the Wollar sandstone [New South Wales]: Jour. Proc. Royal Soc. New South Wales, v. 100, p. 61-73.

Helby, R., and Martin, A. R. H., 1965, *Cylostrobus* gen. nov., cones of lycopsidean plants from the Narrabeen group (Triassic) of New South Wales: Austr. Jour. Bot., v. 13, p. 389-404.

Herbst, R., 1965, Algunos esporomorfos del Triasico de Argentina: Ameghiniana, v. 4, no. 5, p. 141-152.

Jansonius, J., 1962, Palynology of Permian and Triassic sediments, Peace River area, western Canada: Palaeontographica, v. 110, sec. B, p. 35-98.

Jekhowsky, B. de, and Goubin, N., 1964, Subsurface palynology in Madagascar: A stratigraphic sketch of the Permian, Triassic, and Jurassic of the Morondava basin, p. 116-130 *in* Palynology in oil exploration: Soc. Econ. Paleont. and Mineral. Spec. Pub. 11.

Jekhowsky, B. de, Sah, S. C. D., and Letullier, A., 1960, Reconnaissance palynologique du Permien, Trias, et Jurassique des Sondages affectués par la société des pétroles de Madagascar dans le bassin de Morondava: Comptes rendus Soc. Géol. Fr., v. 7, p. 166-167.

Jersey, N. J. de, 1959, Jurassic spores and pollen grains from the Rosewood coalfield: Queensland Govt. Mining Jour., v. 60, p. 346-66.

______1962, Triassic spores and pollen grains from the Ipswich coalfield: Geol. Survey, Queensland, Publ. 307, p. 1-18.

______1964, Triassic spores and pollen grains from the Bundamba group: Geol. Survey, Queensland, Publ. 321, p. 1-21.

Jersey, N. J. de, and Hamilton, M., 1965a, Triassic microfloras from the Mount Crosby formation: Queensland Govt. Min. Jour., July 1965, p. 1-4.

______1965b, Triassic microfloras of the Moorooka and Tingalpa formations: Queensland Govt. Min. Jour., July 1965, p. 5-10.

Jung, W., 1958, Zur Biologie und Morphologie einiger disperser Megasporen, vergleichbar mit solchen von *Lycostrobus scotti*, aus dem Rhät-Lias Frankens: Nordost-Bayern, Geol. Blätter, v. 8, no. 3, p. 114-130.

——1959, Die dispersen Megasporen der fränkischen Rhät-Lias Grenzschichten: Ph. D. Thesis, Ludwig Maximillians Univ., München, p. 1-97.

Jux, U., 1961, The palynological age of diapiric and bedded salt in the Gulf Coastal Province: Louisiana Geol. Survey Geol. Bull., v. 38, p. 1-46.

Kavary, E., 1966, A palynological study of the subdivision of the Cardita shales (Upper Triassic) of Bleiberg, Austria: Verhandl. Geol. Bundesanst. v. 1, no. 2, p. 178-189.

Klaus, W., 1958, Some lower Mesophytic microspores of Europe, with remarks on their relations to the Gondwana microflora: J. Palaeont. Soc. India, v. 3, p. 151-155.

——1960, Sporen der karnischen Stufe der ostalpinen Trias: Jahrb. Geol. B. A., v. 5, p. 107-183.

——1964, Zur sporenstratigraphischen Einstufung von gipsführenden Schichten in Bohrungen: Erdoel-Zeitschrift, v. 4, p. 119-132.

Kopitova, E. A., 1963a, New species of spores and pollen from Triassic beds of western Kazakhstan [in Russian]: Ministerstvo geol. i ochroni Nedr SSSR VNIGNI Trudy, v. 37, p. 65-69.

——1963b, Stratigraphy and spore-pollen complexes of the Triassic depression of the Ilek basin (Aktyubinsk pre-Urals) [in Russian]: Ibid., v. 37, p. 77-88.

Krutzsch, G., 1955, Uber einige liassische "angiospermide" Sporomorphen: Geologie, v. 4, p. 65-76.

Leschik, G., 1955, Die Keuperflora von Neuewelt bei Basel. **II.** Die Iso- und Mikrosporen: Schweiz. Paläont. Abh., v. 72, p. 1-70.

Levet-Carette, J., 1964, Étude de la microflore infraliassique d'un sondage effectué dans le sous-sol de Boulogne-sur-Mer (Pas-de-Calais): Ann. Soc. Geol. Nord, v. 83, p. 101-128.

Lundblad, B., 1956, On the Stratigraphical value of the megaspores of *Lycostrobus scottii*: Sveriges Geol. Undersök., C, v. 50, no. 3, p. 1-11.

——1959, On *Ricciisporites tuberculatus* and its occurrence in certain strata of the "Hollviken II" Boring in southwest Scania: Grana Palynologica, v. 2, no. 1, p. 77-86.

Mädler, K., 1964, Die geologische Verbreitung von Sporen und Pollen in der Deutschen Trias: Beihefte geol. Jahrb., v. 65, p. 1-147.

Malyavkina, V. S., 1953: Upper Triassic, Lower Jurassic, and Middle Jurassic spore-pollen assemblages from the eastern and western pre-Urals [in Russian]: Leningrad and Moscow, VNIGNI Trudy N. S., v. 75, p. 93-147.

——1960, Spore-pollen assemblages from the Triassic of the Russian platform [in Russian]: Leningrad and Moscow, VNIGNI Trudy, v. 29, p. 26-31.

Manum, S., 1960, On the genus *Pityosporites* Seward 1914, with a new description of *Pityosporites antarcticus* Seward: Nytt Mag. Bot., v. 8, p. 11-15.

McGregor, D. C., 1965, Illustrations of Canadian fossils: Triassic, Jurassic, and Lower Cretaceous spores and pollen of Arctic Canada: Geol. Survey Canada, Paper 64-155, p. 1-32.

Miner, E. L., 1935, Palaeobotanical examination of Cretaceous and Tertiary coals. II. Cretaceous and Tertiary coals from Montana: Amer. Midl. Nat., v. 16, p. 616-625.

Nathorst, A. G., 1908, Paläobotanische Mitteilungen. 3. *Lycostrobus scotti*, eine grosse Sporophyllähre aus den rätischen Ablagerungen Schonens: K. Svensk. Vet. Akad. handl., v. 43, no. 3, p. 1-12.

Nilsson, T., 1958, Über Das Vorkommen eines mesozoischen Sapropelgesteins in Schonen: Lunds Univ. Arsskr. N. F. Avd. 2, v. 54, no. 10, p. 1-112.

Norris, G., 1965, Triassic and Jurassic miospores and acritarchs from the Beacon and Ferrar groups, Victoria Land, Antarctica: New Zealand Jour. Geol. Geophys., v. 8, no. 2, p. 236-277.

Pautsch, M. E., 1958, Keuper sporomorphs from Swierczyna, Poland: Micropaleontology, v. 4, no. 3, p. 321-324.

Peabody, D. M., and Kremp, G. O. W., 1964, Preliminary studies of the palynology of the Chinle formation, Petrified Forest, p. 11-26 *in* Preliminary investigations of the microenvironment of the Chinle formation, Petrified Forest National Park, Arizona, Interim Res. Rept. No. 3, Geochronology Laboratories, Univ. of Arizona [unpublished report].

Pflug, H. D., 1953, Zur Entstehung und Entwicklung des Angiospermiden Pollens in der Erdgeschichte: Palaeontographica, v. 95, sec. B, p. 60-171.

Playford, G., 1965, Plant microfossils from Triassic sediments near Poatina, Tasmania: Jour. Geol. Soc. Australia, v. 12, no. 2, p. 173-210.

Playford, G., and Dettmann, M., 1965, Rhaeto-Liassic plant microfossils from the Leigh Creek coal measures, South Australia: Senckenbergiana leth., v. 46, p. 127-181.

Pocock, S. A. J., and Jansonius, J., 1961, The pollen genus *Classopollis* Pflug 1953: Micropaleontology, v. 4, p. 439-449.

Potonié, Robert, 1956, Synopsis der Gattungen der Sporae dispersae, Teil I: Beihefte Geol. Jahrb., v. 23, p. 1-103.

______1958, Ibid., Teil II: ibid., v. 31, p. 1-114.

______1960, Ibid., Teil III: ibid., v. 39, p. 1-189.

______1962, Synopsis der Sporae in situ: ibid., v. 52, p. 1-204.

______1966, Synopsis der Gattungen der Sporae dispersae, Teil IV: ibid., v. 72, p. 1-244.

______1967, Versuch der Einordnung der fossilen Sporae dispersae in das phylogenetische System der Pflanzen familien: Forschungsber. land. Nordrhein-Westf., v. 1761, p. 1-310.

Reinhardt, P., 1961, Sporae dispersae aus dem Rhät Thüringens: Monatsber. Deutsch. Akad. Wiss. Berlin, v. 3, no. 11/12, p. 704-711.

______1963, Megasporen aus dem Keuper Thüringens: Freiberger Forschungsh., C, v. 164, p. 115-128.

______1964, Einige Sporenarten aus dem Oberen Buntsandstein Thüringens: Monatsber, Deutsch. Akad. Wiss. Berlin, v. 6, no. 8, p. 609-614.

Reinhardt, P., and Schmitz, W., 1965, Zur Kenntnis der Sporae dispersae des mitteldeutschen Oberen Buntsandsteins: Freiberger Forschungsh., C, v. 182, p. 19-36.

Rouse, G. E., 1959, Plant microfossils from Kootenay coal-measures strata of British Columbia: Micropaleontology, v. 5, no. 3, p. 303-324.

Schulz, E., 1962, Sporenpaläontologische Untersuchungen zur Rhät-Lias-Grenze in Thüringen und der Altmark: Geologie, v. 11, no. 3, p. 308-319.

______1964, Sporen und Pollen aus dem mittleren Buntsandstein des germanischen Beckens: Monatsber. Deutsch. Akad. Wiss. Berlin, v. 6, no. 8, p. 597-606.

______1965, Sporae dispersae aus der Trias von Thüringen: Mitt. Z. G. I., v. 1, p. 257-287. (Seen only in preprint form.)

______1967, Sporenpaläontologische Untersuchungen rätoliassischer Schichten im Zentralteil des Germanischen Beckens: Palaont. Abh, B v. 2, no. 3, p. 541-633.

Schulz, E., and Krutzsch, W., 1961, *Echinitosporites iliacoides* nov. fgen. et fsp., eine neue Sporenform aus dem Keuper der Niederlausitz: Geologie, v. 10, Beih. 32, p. 122-127.

Scott, R. A., 1960, Pollen of *Ephedra* from the Chinle formation (Upper Triassic) and the genus *Equisetosporites*: Micropaleontology, v. 6, no. 3, p. 271-276.

Taugourdeau-Lanz, J., 1962, Contribution a la connaissance de la microflore du Trias: Pollen et Spores, v. 4, no. 2, p. 360.

———1963, Note préliminaire a une étude sur la microflore du Trias français: Mem. B. R. G. M., Paris, v. 15, p. 570-575.

Taugourdeau-Lanz, J., and Jekhowsky, G. de, 1959, Spores et pollens du Keuper, Jurassique, et Crétacé infèrieur d'Aquitaine: Comptes rendus Soc. Geol. Fr., v. 7, p. 167-168.

Thiergart, F., 1949, Der Stratigraphische Wert mesozoischer Pollen und Sporen: Palaeontographica, v. 89, sec. B, p. 1-34.

Townrow, J. A., 1962, On some disaccate pollen grains of Permian to Middle Jurassic age: Grana Palynologica, v. 3, no. 2, p. 13-44.

———1967, On *Rissikia* and *Mataia* podocarpaceous conifers from the lower Mesozoic of southern lands: Pap. Proc. Royal Soc. Tasmania, v. 101, p. 103-136.

Varyukhina, L. M., 1961, Spores and pollen of Triassic deposits of the southern Petschora basin [in Russian]: Dokl. Akad. Nauk SSSR, v. 138, no. 3, p. 631-634.

Visscher, H., 1966, Palaeobotany of the mesophytic. III. Plant microfossils from the Upper Bunter of Hengelo, The Netherlands: Acta Bot. Neerlandica, v. 15, p. 316-375.

Warrington, G., 1967, Correlation of the Keuper series of the Triassic by miospores: Nature, v. 214, p. 1323-1324.

15

JURASSIC AND EARLY CRETACEOUS POLLEN AND SPORES

N. F. Hughes

Paleobotanists have frequently suggested that macrofossil evidence from the Jurassic and Early Cretaceous Periods points to a lack of evolutionary activity in the land plants. Palynology shows clearly that this conclusion is ill-founded and that plant evolution was as eventful as in any other period of comparable duration. The marine stratigraphic succession is well correlated in Europe and in many other areas, and therefore dating of nonmarine successions by palynologic correlation can be particularly effective in these periods.

Stratigraphic Scale

For correlation of rock sections in other parts of the world a standard for the Jurassic Period and part of the Cretaceous Period is taken in Western Europe. In this standard section the divisions of time recorded in rock sequences have been arbitrarily selected, and their lower limits are in each case defined in terms of the marine (ammonite) faunas. The standard scale used is taken from Harland and others (1967), and the principles involved are discussed in George and others (1967). Where necessary relevant details of the Western European rock succession are given for reference. (See table 15-1.)

GENERAL FLORAL CHARACTERISTICS

Plant-microfossil assemblages of the Jurassic and Early Cretaceous reflect their provenance from a much more diverse group of gymnosperms than is known today; from pteridophytes, which were locally more important than in any single locality today but markedly less diverse; and from some bryophytes that have not yet been assessed. The pteridophyte spores form the most obvious element and have thus been used without reference to facies in coarse stratigraphical deductions. Of the gymnosperm representatives the bisaccates, when present, frequently predominate; little progress has been made with their taxonomy, which is still in an early stage of development. Other gymnosperm grains are often small or rather lacking in characters at the optical level and have received less detailed attention. Such bryophyte spores as are recognized have been recently distinguished from spores presumed to be of pteridophyte origin.

More than 100 genera have been used for the many organ species described from this period. Although these are all easily arranged in infraturmae of the supplementary suprageneric classification of Potonié (1956b, 1958, 1960) with additions by Dettmann (1963), they are described here around a few of the more prominent types that are encountered in the average preparation. These selected types are arranged as a convenience in the order of Potonié's morphographic classification. Although general botanical assignments are in many cases obvious, more detailed assignments to extant genera are unhelpful to any genuine study of the course of pre-Quaternary plant evolution (Hughes, 1963).

Table 15-1. **Stratigraphic Divisions of the Jurassic and Early Cretaceous in Western Europe**

Period	*Age (Stage)*	*Definition of Beginning of Division (Zone of)*	*Notes*
Early Cretaceous	Albian	*Leymeriella tardefurcata*	
	Aptian	*Prodeshayesites fissicostatus*	
	Barremian	*Paracrioceras strombecki*	
	Hauterivian	*Acanthodiscus radiatus*	
	Valanginian	*Kilianella roubaudiana*	
	Berriasian	*Berriasella boisseri* (approximately)	Including upper and middle Purbeck beds
Jurassic	"Tithonian"	*Gravesia* spp., *Taramelliceras lithographicum*	Including lower Purbeck beds, Portland beds, upper and middle Kimmeridge Clay
	Kimmeridgian	*Pictonia baylei*	Including lower Kimmeridge Clay
	Oxfordian	*Quenstedtoceras mariae*	
	Callovian	*Macrocephalites macrocephalus*	
	Bathonian	*Zigzagiceras zigzag*	
	Bajocian	*Leioceras opalinum*	Including French Aalenian Bajocian and Vesulian (sensu Arkell, 1956)
	Toarcian	*Dactylioceras tenuicostatum*	
	Pliensbachian	*Uptonia jamesoni*	
	Sinemurian	*Arietites bucklandi*	
	Hettangian	*Psiloceras planorbis*	

Smooth Azonotrilete Miospores (Infraturma Laevigati)

The most prominent type is classified as *Cyathidites* Couper 1953 (type species, *C. australis*), which has a concavely triangular amb and simple, long laesurae. (See pl. 15-1, figs. 1 and 2). It is thus sufficiently distinguished from *Deltoidospora* (cf. chap. 14, p. 294), and in any case little purpose is served by pursuing purely morphographic synonymy at the optical level when so few characters are available; fine-structure investigations should eventually provide a useful taxonomic basis. For general practical reasons it would also be undesirable for *Cyathidites* to be used in the Paleozoic in preference to *Leiotriletes*, over which it has priority of validation.

Some species of *Cyathidites* belong (Couper, 1958) to leaf species of the ferns *Coniopteris*, *Eboracia*, *"Dicksonia"* (Dicksoniaceae); some probably also represent the Cyatheaceae, but others are of unknown affinity and could well have come from different fern families. Spores with a circular amb are found (Couper, 1958) in compression ferns such as *Todites williamsoni* (Osmundaceae); dispersed specimens that have been placed in the genus *Todisporites* Couper 1958 occur widely, although they are only locally

Plate 15-1. (All figures × 500.) 1 and 2—*Cyathidites australis* Couper: 1—?Barremian; Kent, England; 2—Triplan view, Valanginian, Surrey, England. 3 and 4—*Cyathidites punctatus* (Delcourt and Sprumont): 3—?Hauterivian, Sussex, England; 4—Valanginian, Surrey, England. 5, 6, and 7—*Biretisporites potoniaei* Delcourt and Sprumont: 5 and 6—Valanginian, Surrey, England (5, low focus; 6, high focus); 7—?Valanginian, Belgium. 8 and 9—*Dictyophyllidites harrisii* Couper (holotype); Bajocian, East Yorkshire, England; high and mid focus. 10 and 11—*Osmundacidites wellmani* Couper; Bajocian, East Yorkshire, England. 12—*Matonisporites phlebopteroides* Couper; Barremian, Hampshire, England. 13 and 14—*Cyclosporites hughesi* (Cookson and Dettmann); Aptian, south Australia, low and high focus. 15, 16, and 17—*Pilosisporites* cf. *trichopapillosus* (Thiergart); Barremian, Hampshire, England. 18 and 19—*Marattisporites scabratus* Couper: 18—Bajocian, East Yorkshire, England; 19—Bathonian, Scotland. 20 and 21—*Coronatispora valdensis* (Couper); Valanginian Surrey, England; high and low focus. 22—*Kuylisporites* sp., Barremian, Surrey, England. 23—*Klukisporites variegatus* Couper; Bajocian, East Yorkshire, England. 24—*Dictyotriletes* sp., Barremian, Hampshire, England. 25 and 26—*Leptolepidites* sp., Valanginian, Sussex, England; high and low focus.

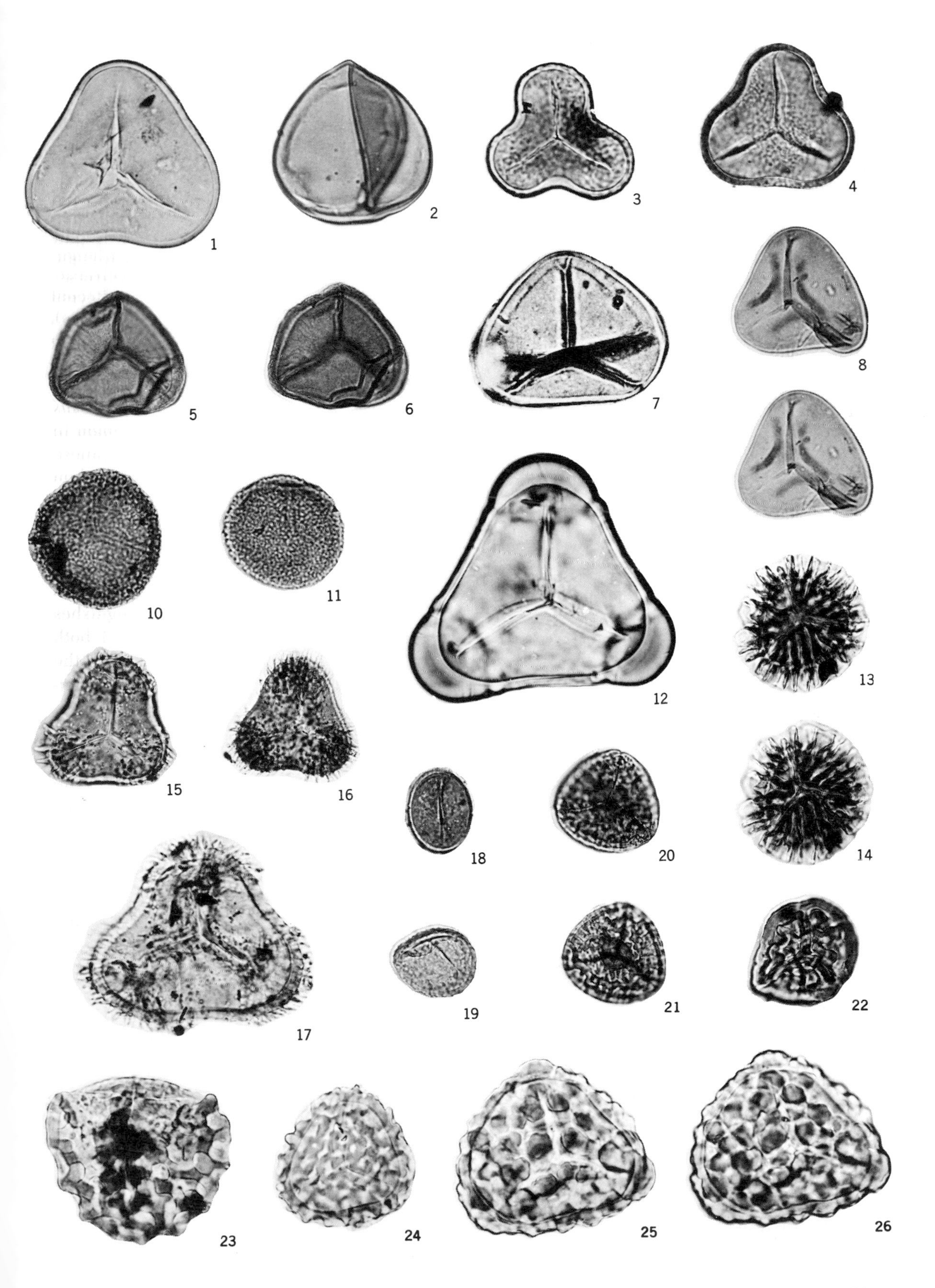
1
2
3
4
5
6
7
8
10
11
12
13
15
16
18
20
14
17
19
21
22
23
24
25
26

abundant (middle Bajocian, Yorkshire, England). *Cyathidites punctatus* (pl. 15-1, figs. 3 and 4) is a large Lower Cretaceous species of unknown affinity, with an undulose exine surface and apparent broad "lips," which are only a compression feature (Dettmann, 1963, p. 23).

Other smooth spores with the laesurae enclosed within elevated lips are classified (Dettmann, 1963) in *Biretisporites* (pl. 15-1, figs. 5-7), which has a uniform exine; and in *Dictyophyllidites* Couper 1958 (type species, *D. harrisii*), with thickening about the laesurate margins. There is a tendency to lateral compression in the thin-walled species of *Dictyophyllidites* such as *D. harrisii* (pl. 15-1, figs. 8 and 9), which is similar to spores found in the Jurassic *Dictyophyllum* (related to the Recent ferns *Dipteris, Cheiropleuria*, or to the Matoniaceae).

One of the most difficult spores to identify is *Calamospora mesozoica* Couper, 1958, and there are thus few records of it; the tetrad mark and laesurae are very short (not more than half the spore radius), and the spore wall is very thin and uniform (about 0.5 μ). Thus the laesurae tend to be concealed among the many irregular exinal folds of this originally spherical spore. *Neocalamites* is known as a stem compression from the Jurassic (Harris, 1961), although its spores are not known. Even less satisfactory is the failure as yet to identify the spores of the universal Mesozoic *Equisetites*. Recent *Equisetum* possesses small, thin-walled, apparently "alete" spores (with elaters); comparable Mesozoic spores dispersed (and thus without elaters) may well have been overlooked owing to their lack of characters (but see Batten, 1968).

Some small, thick-walled ($2\frac{1}{2}$ to 3 μ) spores of the genus *Stereisporites* Pflug 1953 are believed to represent the Sphagnales (Bryophyta—Musci). The attribution is not proved, although some species show a low distal polar thickening like that in some species of Recent *Sphagnum*.

Azonotrilete Miospores with "Unit" Sculpture (Infraturma Apiculati)

The sculpture consists only of discrete elements with approximately equidimensional bases. *Osmundacidites* Couper 1953 (type species, *O. wellmanii*) was erected for granulate spores (pl. 15-1, figs. 10 and 11) similar to those obtained from Bajocian *Osmundopsis* and *Todites* (Osmundaceae). *Leptolepidites* Couper 1953 (type species, *L verrucatus*) is densely verrucate equatorially and distally (pl. 15-1, figs. 25 and 26); *Rubinella* Maljavkina ex Potonié 1960 is verrucate proximally and distally, so that the tetrad mark is seldom clearly seen. *Concavissimisporites* Delcourt and Sprumont 1955 is now restricted (Dettmann, 1963) to concavely triangular spores covered uniformly with verrucae. Comparable spores are known both from Recent Cyatheaceae and from *Lygodium* (Schizaeaceae), but no Mesozoic species have been found in sporangia.

Species of *Pilosisporites* Delcourt and Sprumont 1955 (type species, *P. trichopapillosus* Thiergart) (pl. 15-1, figs. 15-17) are common in Lower Cretaceous rocks. The long spinose sculptural elements are variously distributed on the exine in the different species. Bolkhovitina (1961) and other Soviet authors attribute these spores to *Lygodium* (Schizaeaceae), but firm evidence is lacking. Distal clavate sculpture distinguishes *Ceratosporites* Cookson and Dettmann 1958, and overall clavate sculpture distinguishes *Neoraistrickia* Potonié 1956; affinities of both genera are uncertain but probably lie with the ancestors of the living *Lycopodium* and *Selaginella*.

Kuylisporites Potonié 1956 (type species, *K. waterbolki*) bears distally a number of crescentic pseudopores (pl. 15-1, fig. 22), in which it resembles spores of some species of the Recent tree fern *Hemitelia* (Cyatheaceae).

Azonotrilete Miospores with "Continuous" Sculpture (Infraturma Murornati)

The sculpture varies from somewhat irregular rugulae to the precise parallel muri of *Cicatricosisporites*. *Lycopodiacidites* Couper emend. Potonié 1956 was erected for spores with a circular amb and irregular rugulae distally, and some species are clearly not far from the verrucate *Leptolepidites* mentioned above. *Tripartina* Maljavkina ex Potonié 1960 includes triangular forms. The affinities of these genera are uncertain.

Cyclosporites Cookson and Dettmann 1959 (type species, *C. hughesi*) has a distal reticulum of high-crested muri with an unusual proximal radial arrangement of similar muri. (See pl. 15-1, figs. 13 and 14.) The laesurae are enclosed between high membraneous lips, as is the case in *Lycopodiumsporites* Thiergart ex Delcourt and Sprumont 1955, which has a distal reticulum only. Although this genus is poorly based (Delcourt, Dettmann, and Hughes 1963), it includes a number of species that are of constant occurrence, and the implied attribution to the *Lycopodium clavatum* group of Knox (1950) seems to be well founded. *Klukisporites* Couper 1958 (type species, *K. variegatus*) is distally foveoreticulate (pl. 15-1, fig. 23); it is also closely similar to spores found in situ in Bajocian species of *Klukia* and *Stachypteris* (Schizaeaceae). Pocock (1962) has placed some relatively similar but purely reticulate Upper Jurassic and Lower Cretaceous species in *Dictyotriletes* (Naumova) Potonié and Kremp 1955. (See pl. 15-1, fig. 24.)

Truly foveolate spores (example: pl. 15-3, fig. 3) are included in *Foveosporites* Balme 1957, which Balme believed to resemble spores of the *Lycopodium verticillatum* group of Knox (1950).

Staplinisporites Pocock 1962 (type species, *S. caminus* Balme) has radial and concentric distal muri and a distal polar thickening; Dettmann (1963) compares it with Recent *Encalypte* spores (Bryophyta—Musci).

Perhaps the most striking murornate spores fall in the genus *Cicatricosisporites* Potonié and Gelletich 1933 (type species, *C. dorogensis*), with distal and equatorial parallel muri. (See Cretaceous example in pl. 15-2, figs. 1-6.) Although the type species is from the Eocene of Hungary, there is a large number of Cretaceous species throughout the world; the spores were first common in the earliest Cretaceous, but Burger (1966) recorded them from highest Upper Jurassic of the Netherlands, and Norris (1963, unpublished) had a Late Jurassic record from just below the Portland beds in southern England. The affinity is well known from the existence of similar spores in Recent *Anemia* and Cretaceous *Pelletieria* and *Schizaeopsis* (Hughes and Moody-Stuart, 1966). It is interesting that in Early and mid-Jurassic time there were spores with rows of verrucae and some long, parallel, irregular ridges figured as *Cingulatisporites problematicus* by Couper (1958) and *Corrugatisporites* by Nilsson (1958) and reassembled by Playford and Dettmann (1965) under *Duplexisporites* Deak 1962; (type species, D. *generalis*). The precise parallel muri borne by these spores and by *Contignisporites* (Zonati) presumably had an as yet unidentified "insect" functional correlation; a harmomegathic explanation may apply if some of these spores prove to have come from any plant resembling *Ceratopteris* (Adiantaceae).

Zonotrilete Miospores

This group is here taken in the sense of Dettmann (1963) to include all acavate trilete spores with equatorial thickenings or extensions of any kind. The Mesozoic spores of this type are prominent in the record but particularly so in the Lower Cretaceous; the Mesozoic Auriculati and Cingulati differ from those of the Paleozoic in their much less pronounced thickening.

Zonotrilete with Radial Thickening Only (Infraturma Auriculati). The smooth valvate spores (Pl. 15-1, fig. 12) are included in *Matonisporites* Couper 1958; (type species, *M. phlebopteroides*) and are similar to spores from Mesozoic *Phlebopteris* (probably Matoniaceae). *Trilobosporites* Pant ex Potonié 1956 (type species, *T. hannonicus* Delcourt and Sprumont) has granulate to verrucate sculpture that is usually larger and sometimes coalescent on the valvae (pl. 15-2, figs. 10-12); the affinity of *Trilobosporites* is not known.

Plicatella Maljavkina 1949 (type species, *P. trichacantha*) may be included here as it has parallel regular equatorial and distal muri (cf. *Cicatricosisporites*) and also short radial equatorial appendages (pl. 15-2, figs. 7, 8, and 13); it enters early in the Cretaceous Period but after *Cicatricosisporites*. These spores have previously been placed by some authors in *Appendicisporites* Weyland and Krieger 1953, which is an Upper Cretaceous genus separated on the nature of the appendages.

Zonotrilete with Interradial Thickening Only (Infraturma Tricrassati of Dettmann 1963). Interradial crassitudes are clearly displayed by *Gleicheniidites* Ross emend. Skarby 1964 (type species, *G. senonicus*), which has a smooth exine. (See pl. 15-2, figs. 14 and 15.) Fairly certain affinity with the Gleicheniaceae is suggested by comparison with spores of several species of Recent *Gleichenia*. *Sestrosporites* and *Coronatispora* (pl. 15-1, figs. 20 and 21) were both erected by Dettmann (1963) for spores found in the Mesozoic of both England and Australia;

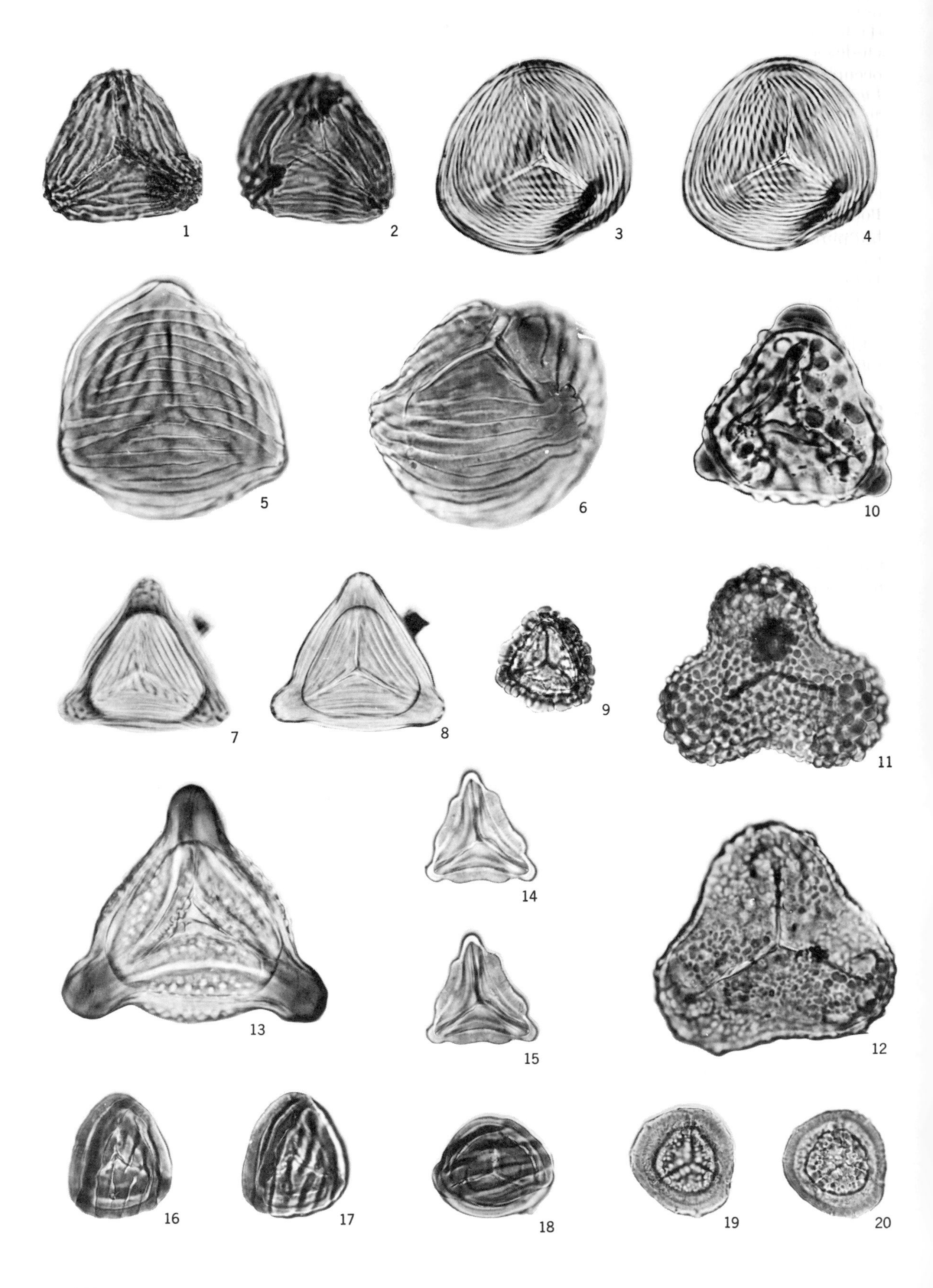
1
2
3
4
5
6
10
7
8
9
11
13
14
15
12
16
17
18
19
20

according to Dettmann the closest Recent parallel to *Sestrosporites* is the spore of a species of *Lycopodium*. Species of both were described originally under *Cingulatisporites* by Couper (1958), but that genus is now better left for the original Tertiary material.

Zonotrilete with Equatorial Thickening (Infraturma Cingulati). There are a number of spores of this general type in the Mesozoic, but they are less numerous than the Paleozoic representatives, and their provenance is unknown. *Cingutriletes* Pierce emend. Dettmann 1963 and *Taurocusporites* Stover are genera for spores with a circular amb, but others with a subtriangular amb have been placed in the Carboniferous genus *Murospora* Somers, 1952. These species (pl. 15-2, figs. 9, 19, and 20) are relatively similar in polar view and in thin section (Dettmann, 1963) to Carboniferous species of this genus, and in this author's opinion are not entirely cleared of the suspicion of reworking in spite of their records from North America, Australia, and Europe (unpublished).

Foraminisporis Krutzsch 1959 includes granulate to verrucate species with a very narrow, sculptured cingulum; these compare closely with spores of certain Recent Hepaticae (Dettmann, 1963). Species of *Krauselisporites* Leschik emend. Jansonius 1962 (described (p. 297) in the Triassic Chap.) occur throughout the Jurassic and Lower Cretaceous of Australia; they were recorded under the name *Styxisporites*, which is now in synonymy (Dettmann, 1963).

Contignisporites Dettmann 1963 (type species, *C. glebulentus*) shows a single distal set of parallel muri that coalesces with the cingulum, and thus there are two spores of each of two symmetry types from every tetrad. (See pl. 15-2, figs. 16-18.) The genus is represented by species in England, Soviet Union, and Australia from mid-Jurassic onward; it is now clearly distinguished from the schizaeaceous *Cicatricosisporites* (see above). From comparisons with Recent members of this group (Bolkhovitina, 1961) it seems unlikely that the affinities of *Contignisporites* lie with the fern family Schizaeaceae. This is more probably a case of parallel evolution of a character.

There do not appear to be any Mesozoic representatives of the infraturma Patinati Butterworth and Williams 1958.

Cavate Trilete Miospores (*Perinotrilites* Erdtman emend. Dettmann)

Spores with a cavate separation of exine layers are not common. They have in the past been described as perinate, but the true perine is only a thin, late-formed layer seen occasionally as a discontinuous preservation on the walls of numerous kinds of trilete spores; it is thus better to use the more precise cavate nomenclature in these cases. At least some of these spores probably belonged to waterplants.

Perotrilites Erdtman ex Couper 1953 (type species, *P. granulatus*) and *Crybelosporites* Dettmann 1963 (type species, *C. striatus*) include proximally cavate species that have been found directly associated with the megaspore *Pyrobolospora*. They are separated on the nature of the outer sculpture. (See pl. 15-3, figs. 5 and 6.) *Densoisporites* Weyland and Krieger 1953 emend. Dettmann 1963 has proximally attached, thick sculpture with equatorial thickening.

Monolete Miospores

Monolete fern spores are relatively rare in the Mesozoic, the most common being *Marattisporites* Couper 1958 (type species, *M. scabratus*), which is finely sculptured (pl. 15-1, figs. 18 and 19) and resembles spores taken from extant *Marattiopsis* species.

Plate 15-2. (All figures × 500.) 1 and 2—*Cicatricosisporites* sp. (cf. spores of *Ruffordia goepperti* Seward); dispersed, Berriasian; Hampshire, England. 3 and 4—*Schizaeopsis americana* (spores from fructification); ?Barremian, Virginia; low and high focus. 5 and 6—*Pelletieria valdensis* (spores from fructification); Berriasian; Sussex, England. 7 and 8—*Plicatella* sp.; Aptian; Hampshire, England; low and mid focus. 9—*Cingutriletes* sp.; Valanginian; Surrey, England. 10, 11, and 12—*Trilobosporites* Pant ex Potonié: 10—*T. bernissartensis* (Delcourt and Sprumont), Valanginian, Dorset, England; 11 and 12—Berriasian, Sussex, England (11, *T.* sp.; 12, *T. apiverricatus* Couper). 13—*Plicatella* sp.; Barremian; Surrey, England; 14 and 15—*Gleicheniidites* sp.; Valanginian; Sussex, England; mid and high focus. 16, 17, and 18—*Contignisporites* sp.: 16 and 17—Valanginian, Sussex, England, low and high focus; 18—?Hauterivian, Dorset, England. 19 and 20—*Taurocusporites* cf. *segmentatus* Stover; Valanginian; Surrey, England; high and low focus.

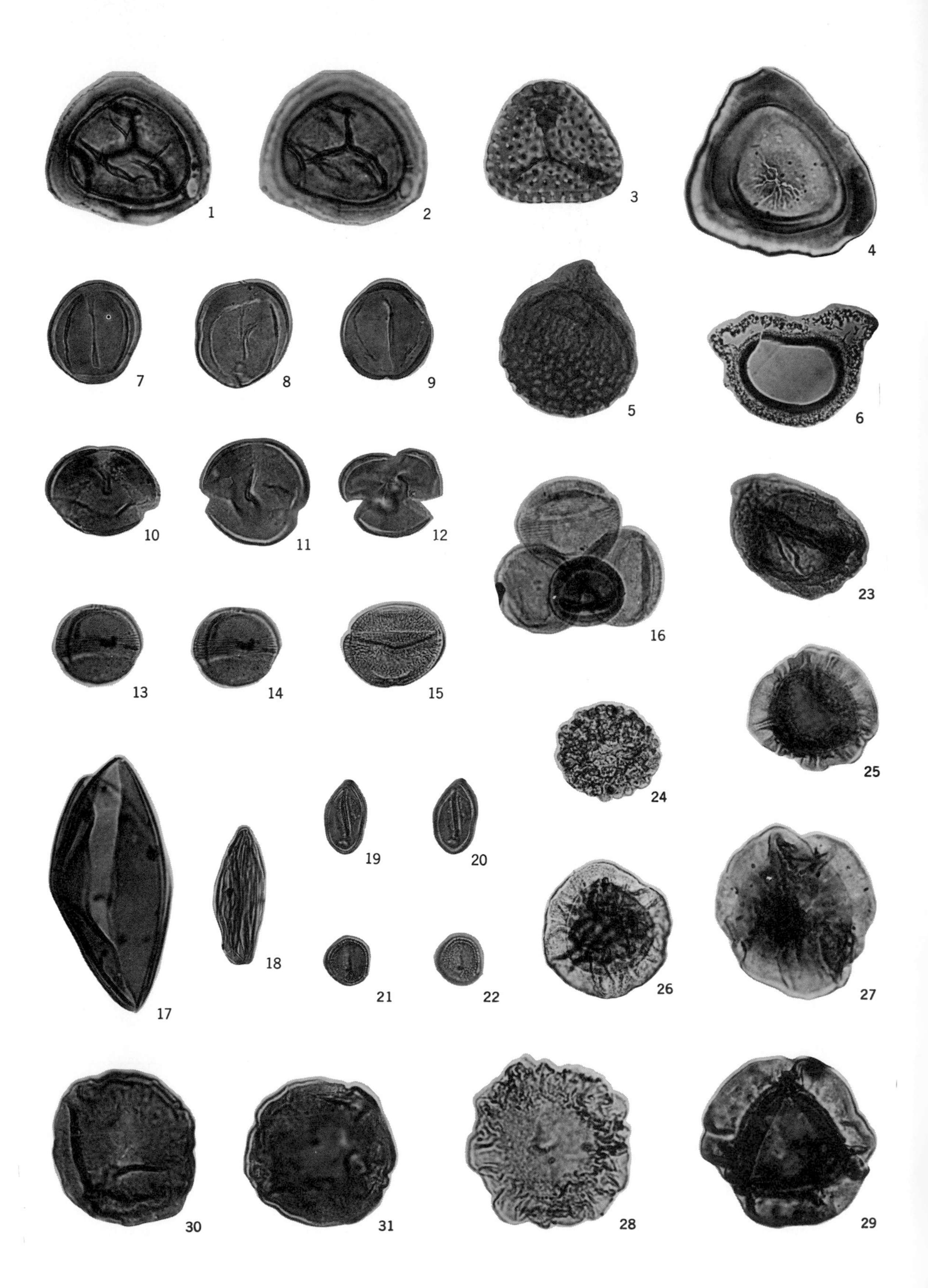
1
2
3
4
5
6
7
8
9
10
11
12
13
14
15
16
17
18
19
20
21
22
23
24
25
26
27
28
29
30
31

Hilate Miospores

Dettmann (1963) has brought together a number of interesting spores with two-layered exine, which, although they probably originated in tetrahedral tetrads, developed a distal (or sometimes proximal) polar hilum by natural sclerinous breakdown. The best known is *Aequitriradites* Delcourt and Sprumont 1955 emend. Delcourt and others 1953 (type species, *A. dubius*), which has a broad membraneous zona. (See pl. 15-4, figs. 6-10.) Other zonate genera are *Couperisporites* Pocock 1962 (type species, *C. complexus* Couper), with distinctive distal verrucae; and *Rouseisporites* Pocock 1962 (type species, *R. reticulatus*), with flask-shaped radial equatorial invaginations. (See pl. 15-3, figs. 1 and 2.) *Cooksonites* Pocock 1962 is cingulate. (See pl. 15-3, fig. 4.) Spores similar to all these are known from various living Hepaticae. They are more diverse in the Cretaceous than in the Jurassic.

Monosaccate Miospores

This type of pollen is to be seen in most Mesozoic assemblages and may well have belonged to the gymnospermous forerunners of *Tsuga*. Some of these grains have been interpreted and named as "acritarchs" by Döring (1961), and other authors have used four or five different generic names for them. Since a faint trilete tetrad mark is commonly present and since the main differences between species appear to be in the restriction of the saccus into separate vesicles, *Tsugaepollenites* Pontonié and Venitz 1934 emend. Potonié 1958 seems to be the most appropriate genus, although it has a Cenozoic type species (*T. igniculus*). Two examples are shown in plate 15-3, figures 25-29. *Cerebripollenites* also occurs throughout the Jurassic and Early Cretaceous. (See pl. 15-3, fig. 24.)

Bisaccate Pollen

These grains form a most important element of Mesozoic assemblages, from which they are seldom absent. Their nomenclature is unduly complicated for two reasons. First, many authors have tried to see representatives of well-known extant genera in this complex. Second, Soviet authors especially have produced large numbers of taxa named on the basis of "flat" drawings with no plane of focus indicated because the drawings are often composite.

It is peculiarly difficult to illustrate these grains because of the many possible aspects and the resulting many possibilities of compression. The compression characters will probably eventually be used in taxonomy. The best criteria for distinction of genera or species have not yet been established, nor have more than a very few species been adequately illustrated in view of the difficulties mentioned.

The generic name with priority is *Pityosporites* (Seward, 1904) but even careful attempts (Manum, 1960) to redefine the genus from reexamination of the type specimen can only satisfy nomenclatural rather than taxonomic doubts. It is probably wiser to avoid use of such a name.

Several of the most obvious types can at present be accommodated in *Alisporites* Daugherty 1941 (type species, *A. opii*), in which the length of the saccus (pl. 15-4, figs. 13-15) does not exceed the length of the corpus taken parallel to the length of the distal tenuitas (definitions in Dettmann, 1963, p. 17 and 18; or Couper, 1958);

Plate 15-3. (All figures × 500) 1 and 2—*Rouseisporites granospeciosus* (Delcourt and Sprumont); Valanginian, Belgium; mid and high focus. 3—*Foveosporites* sp.; Barremian; Hampshire, England. 4—*Cooksonites* sp.; Barremian; Hampshire, England. 5 and 6—*Crybelosporites* Dettmann: 5—*C. striatus* (Cookson and Dettmann); Aptian-Albian, south Australia; 6—*C. stylosus* Dettmann; Aptian, south Australia (section by M. E. Dettmann). 7-12—*Eucommiidites troedssonii* Erdtman; Bathonian, Scotland: 7—proximal view; 8 and 9—oblique views; 10-12—"end" views. 13-16—*Classopollis* sp.: 13 and 14—Berriasian; Surrey, England; low and high focus; 15—?Berriasian; Hampshire, England; 16—tetrad, Berriasian, Surrey, England. 17—*Schizosporis* cf. *parvus* Cookson and Dettmann, Valanginian; Surrey, England. 18—*Ephedripites* cf. *multicostatus* Brenner; Barremian; Surrey, England. 19-22—*Clavatipollenites hughesii* Couper; Barremian; Hampshire, England: 19 and 20—"front" view, low and high focus; 21 and 22—"end" view, low and high focus. 23—*Perinopollenites elatoides* Couper; Bajocian, Yorkshire, England. 24—*Cerebripollenites mesozoicus* (Couper); Hauterivian; Surrey, England. 25-29—*Tsugaepollenites*, Surrey, England: 25—*T. dampieri* (Balme), Valanginian; 26—*T. ?dampieri*, Valanginian; 27—*T. ?trilobatus* (Balme), Hauterivian; 28—*T.* cf. *segmentatus* (Balme), Hauterivian; 29—*T. trilobatus*, Barremian. 30 and 31—*Araucariacites australis* Cookson: 30—?Hauterivian; Hampshire, England; 31—Valanginian; Surrey England.

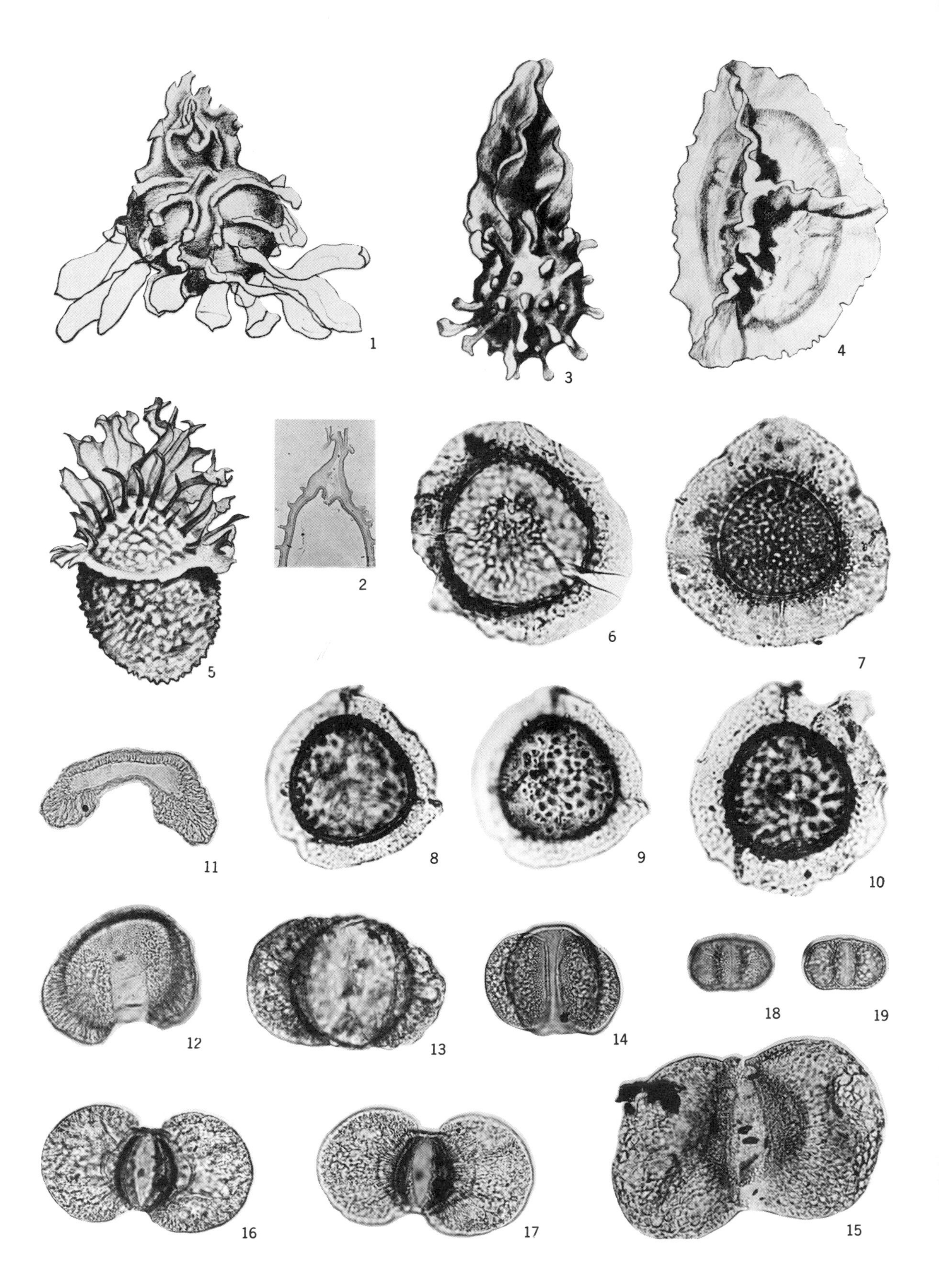
1
3
4
5
2
6
7
11
8
9
10
12
13
14
18
19
16
17
15

and *Podocarpidites* Cookson ex Couper 1953 (type species *P. ellipticus*), in which the length of saccus does exceed the length of corpus. For this character a polar view is necessary, as in the examples shown in plate 15-4, figures 16 and 17. *Podocarpidites* is of worldwide occurrence and is no way restricted to the Southern Hemisphere as its name may imply (perhaps quite wrongly). On the other hand, the trisaccate forms such as *Microcachyridites* Cookson ex Couper 1953 (type species, *M. antarcticus*) appear to have been mainly confined to the Southern Hemisphere in the Mesozoic, as they are now.

In the Jurassic there are some interesting unusual records such as that of the very large *Abietinaepollenites dunrobinensis* Couper 1958 (from the Lower Liassic), with a corpus length of about 100 microns; and *Parvisaccites enigmatus* Couper 1958 (from the Bajocian-Bathonian), with a thick proximal cap and small sacci. These and other taxa point to a Jurassic diversity about which little is yet known.

In the Early Cretaceous (Barremian) species of *Parvisaccites* Couper 1958 (type species, *P. radiatus*) became important stratigraphically. (See pl. 15-4, figs. 11 and 12.) Also, by the end of the Early Cretaceous it is perhaps justifiable to begin identifying bisaccate species to such extant groups as the family Pinaceae.

Believed to be of different origin is *Pteruchipollenites* Couper 1958 (type species, *P. thomasii*), with a clearly delimited colpus, corpus longer than wide, and no proximal thickenings; grains of this type in situ (in the so-called Mesozoic pteridosperms) mainly from the Jurassic and earlier are discussed by Townrow (1962). The very small grain *Vitreisporites* Leschik 1955 (type species, *V. signatus*), with a corpus length of only 20 microns, is of almost universal occurrence throughout the mid-Mesozoic and closely resembles grains taken from *Caytonanthus* (Caytoniales). (See pl. 15-4, figs. 18 and 19.)

Colpate Pollen

Monocolpate pollen is mostly unsculptured; thus, although it was presumably contributed by at least three separate groups of Mesozoic plants, few species of it have been distinguished on the restricted numbers of characters available at the optical scale; electron micrographs (Erdtman, 1965) of the surface of Recent grains show more characters.

A number of grains in fructifications have been described (see Potonié, 1962), but in some cases the affinity of the pollen-bearing organ itself is still in doubt. In general monocolpate pollen thought to have come from the Ginkgoales has a colpus running completely across the grain, resulting in some cases in a tendency to split in two. The colpus in the Cycadales is shorter than the width of a grain, and the grains are thus oval. At least some of the Cycadeoidales have a grain elongated parallel to the colpus to a spindle shape as much as 80 microns in "length." Although all the types occur widely in assemblages, only the Ginkgoales type usually occurs in any quantity. The generic names in current use are *Ginkgocycadophytus* Samoilovich 1953 (type species, *G. caperatus*), which has priority, and *Monosulcites* Cookson ex Couper 1953.

Grains like those of Recent *Ephedra* with several "longitudinal" furrows may be polyplicate (polycolpate of Steeves and Barghoorn, 1959) or monocolpate; even their orientation is uncertain (Erdtman, 1965), but the long axis is more probably polar. Such grains appear in quantity in Lower Cretaceous rocks of about Barremian-Aptian age (pl. 15-3, fig. 18), although Triassic exam-

Plate 15-4. (Figs. 1, 3, 4, and 5 × 100; figs. 6-19 × 500.) 1 and 2—*Pyrobolospora* (megaspores): 1—*P. vectis* Hughes, Barremian, Hampshire, England; 2—*P. reticulata* Cookson and Dettmann, Albian, south Australia (section, M. E. Dettmann, × 200). 3—*Arcellites hexapartitus* (Dijkstra) (megaspores); Barremian; Hampshire, England. 4—*Henrisporites undulatus* (Dijkstra) (megaspores); Barremian; Hampshire, England. 5—*Thomsonia reticulata* Mädler (megaspores); Barremian; Hampshire, England. 6-10—*Aequitriradites:* 6—*A.* cf. *spinulosus* (Cookson and Dettmann), Barremian-Aptian, Surrey, England; 7—*A. dubius* Delcourt and Sprumont, Barremian, Hampshire, England; 8-10—*A.* sp., Barremian-Aptian, Surrey, England (8 and 9, low and high focus). 11 and 12—*Parvisaccites radiatus* Couper; Barremian, Hampshire, England: 11—section, J. C. Moody-Stuart; 12—holotype. 13-15—*Alisporites:* 13—*A.* sp., Barremian-Aptian, Surrey, England; 14—*A.* cf. *similis,* Valanginian, Surrey, England; 15—*A.* cf. *grandis,* Barremian, Hampshire, England. 16 and 17—*Podocarpidites* sp.; Valanginian; Surrey, England; low and high focus. 18 and 19—*Vitreisporites pallidus* (Reissinger); Valanginian; Norfolk, England; mid and low focus.

ples were included in the preceding chapter under *Gnetaceaepollenites*.

Perhaps the most surprising colpate grain is *Eucommiidites* Erdtman emend. Hughes 1961 (type species, *E. troedssonii*), which appears to be monocolpate (colpus as in *Cycas*, presumably distal) and zonosulcate proximally; this gives the compressed grain an asymmetrical "tricolpate" appearance (Couper, 1958), from which Erdtman devised the generic name. (See pl. 15-3, figs. 7-12.) The type specimen is from the Early Jurassic of Sweden; there appear to be several other species in pre-Albian rocks, although the published nomenclature is not yet settled. Hughes (1961) found the grains of a small species in the pollen chambers and very large micropyles of some Early Cretaceous dispersed ovules; the length of the micropyle strongly suggests comparison with Recent *Ephedra*. There is a similar record from the Potomac series of Maryland by Brenner (1963); see also Reymanovna 1968.

Clavatipollenites Couper 1958 (type species, *C. hughesii*) is monocolpate, with a finely clavate exine that becomes tectate. (See pl. 15-3, figs. 19-22.) It occurs in Barremian and later rocks and may represent one of the very first true angiosperms (see Kemp 1968).

Porate Pollen

Throughout the Jurassic and most of the Early Cretaceous the small spherical monoporate *Classopollis* (discussed in the preceding chapter) occurs in a large proportion of assemblages and is the dominant form in certain facies. Possibly its record is exaggerated by reworking but samples showing its dominance still appear as late as the Valanginian. There are several different species in the Late Jurassic, as indicated by Burger (1965, 1966); two of these are echinate (pl. 15-3, figs. 13-16), but most of the variety is in equatorial structure rather than sculpture.

Perinopollenites Couper 1958 (type species, *P. elatoides*) has an exine of two distinct layers, the outer loose fitting; it folds easily, and the single pore is not always clearly visible. (See pl. 15-3, fig. 23.) Similar grains have been taken from the Jurassic conifer *Elatides williamsonii*, which is assigned to the Taxodiaceae.

Alete Pollen

Grains of this type may be better grouped with monoporates; Erdtman (1965) describes most of their Recent counterparts from the Araucariaceae and Cupressaceae as doubtfully monolept or atreme. The exine of any such grain is presumably thin walled in some area, however small.

Araucariacites Cookson ex Couper 1953 (type species, *A. australis*) is a large (60 to 70 μ) thin-walled scabrate grain common in the Early and mid-Jurassic, although the type specimen is from the Tertiary (pl. 15-3, figs. 30 and 31); grains of Recent *Araucaria* are very similar.

A number of species of very small grains, usually less than 30 microns in diameter (45 μ by formal definition) are conveniently grouped under *Spheripollenites* Couper 1958 (type species, *S. scabratus*).

Other Pollen

Cookson and Dettmann (1959) described several species of *Schizosporis* characterized by regular splitting into two halves (pl. 15-3, fig. 17); affinities are not known.

Dispersed Megaspores

These spores have a mean diameter of over 200 microns and could in many cases be accomodated on morphographic grounds in miospore taxa, but it seems unhelpful or even perverse to do so when their size distinction is clear. The number of Mesozoic species is greatly in excess of the number of known plants that could have provided them, and for convenience they are discussed below in four groups. Since derived megaspores from the richer Carboniferous flora have been found in Jurassic, Cretaceous, Tertiary, and Pleistocene strata in England, possibilities of reworking must be carefully eliminated.

Azonate megaspores. Smooth, echinate, verrucate, and reticulate spores are common although, because of the scale, the dimensions of such sculptural elements as verrucae do not fall within miospore definitions. Potonié (1956b) erected numerous generic names, some of which have been used widely; Harris (1961), however, prefers to place all his well-described and illustrated megaspores from the Yorkshire, England, mid-Jurassic under *Triletes*, and his policy gives no cause for difficulty or dispute. Comparable spores are found in certain Recent species of both *Selaginella* and *Isoetes*.

Although these spores could well be of stratigraphical value in the Jurassic and Cretaceous, very little has so far been done with them, perhaps because of the difficulty of extracting adequate numbers.

Zonate Megaspores. Spores with a narrow but distinct zona fall into *Minerisporites* Potonié 1956b (type species, *M. mirabilis*; Tertiary) and *Henrisporites* Potonié 1956b (type species, *H. affinis*; Cretaceous). They include as prominent examples *M. richardsoni* from the mid-Jurassic, *M. marginatus*, and *H. undulatus* from the Early Cretaceous (pl. 15-4, fig. 4). These in general resemble certain other Recent species of *Selaginella* that are zonate.

Barbate Megaspores. This group, which is marked by hairs or spines densely arranged on and around the laesurate lips, is characteristic of Mesozoic assemblages. *Lycostrobus* Nathorst 1908, although erected for a spore-bearing fructification of Rhaetian (Triassic) age and subsequently for another in the Upper Liassic (see Potonié, 1965), has been wrongly used for dispersed spores, which are better placed in *Thomsonia* Mädler 1954 (type species, *T. reticulata*) and allied genera. (See pl. 15-4, fig. 5.) These spores occur throughout the Jurassic and are most common in the Early Cretaceous. Mädler (1954) considered that they were related to *Azolla* (Salviniaceae), and they must surely have belonged to waterplants.

Pyrobolotrilete Megaspores. In the Early Cretaceous but particularly from Barremian times onward there occur numerous small megaspores (<400 μ and occasionally <200 μ) with elevated, laesurate, exoexinal lips that are often over 250 microns high. The lips form a "neck" structure that conceals the normal endexinal trilete.

Pyrobolospora Hughes 1955 (type species, *P. vectis*) has a coarse (murornate) reticulum and long exoexinal hollow appendages. (See pl. 15-4, figs. 1 and 2.) Other species with an apiculate or laevigate sculpture are now placed in *Arcellites* Miner 1938 (type species, *A. disciformis*), since Potter (1963) and Ellis and Tschudy (1963) have correctly shown that, although Miner misillustrated and completely misinterpreted his *Arcellites*, the material was similar to that subsequently described as *P. hexapartita* (Dijkstra). (See pl. 15-4, fig. 3.)

The smaller *Balmeisporites* Cookson and Dettmann 1958 (type species, *B. holodictyus*), which seldom exceeds 200 microns, is similarly organized, occurs in the same beds (at least in Australia), and is most conveniently placed here in spite of its size range.

It seems most probable that all the spores of this group belonged to aquatic plants of which the main organs are unlikely to have been fossilized; they may have belonged to the fern family Marsiliaceae, although it contains no precise Recent parallels.

DISTRIBUTION AND SEQUENCE OF FLORAS

Although there is a constancy about the major groups of plants represented, the detailed evolution at perhaps generic and specific level is as continuous as in any other geological period of comparable duration; it is frequently possible to date an assemblage to a geological stage on general grounds unless local facies restriction is very severe. Below the stage level better resolution is at present prevented by the insufficient standard of most published taxonomic work.

Geographical variation certainly existed, but it has not yet been of much use in elucidating Jurassic climates. In the middle of a long geologically normal (nonglacial) time, with presumably the atmospheric circulation belts slightly widened but not essentially different, the pattern has proved to be elusive.

Sequence of Floras in Europe

More published information relates to Europe than to other areas; there is a full rock succession, and, although much of it was of marine origin, there were always extensive islands and embayments with nonmarine facies.

The Lias α (Hettangian) of Poland was deposited in such an embayment, and the assemblages have been described by Rogalska (1956, 1962). Similar assemblages from southern Sweden (Nilsson, 1958) and other parts of Europe show a marked rise of *Osmundacidites* and the appearance of *Eucommiidites*. *Classopollis* becomes abundant and remains so for the rest of the Jurassic Period. The assemblages are not very different from those of the Rhaetian (Late Triassic) immediately below, although *Ovalipollis* and some other Triassic genera have disappeared.

European assemblages from the stages Sinemurian to Toarcian are less well known and thus less distinctive, because of the effect of fairly widespread marine transgression; this may be seen by study of work from England (Wall, 1965) and France (Danzé and Laveine, 1963). Very large bisaccate grains appeared at this time in Britain (Couper, 1958), but not in Europe (Döring and others, 1966).

Bajocian and Bathonian floras are well known from the classic area of Yorkshire, England, from which microfossils were described by Couper (1958) and more recently in greater detail by Muir (unpublished). It is probably the best documented Mesozoic area for correlation of macrofossils and miospore assemblages. Numbers of *Tsugaepollenites* and *Araucariacites* increase rapidly, as do several species of *Lycopodiumsporites*; among monosulcates the large benettitalean types become less common than the small oval species.

Callovian to Tithonian assemblages continue to be dominated by *Classopollis, Tsugaepollenites,* and *Araucariacites*; and there is less variety in bisaccates, although these include some grains with a short, wide corpus. *Contignisporites* appears, but other cicatricose and allied forms (*Cicatricosisporites*) are only found in the late English Kimmeridgian, which is perhaps equivalent to middle Tithonian. The greater part of the English Purbeck beds is now assigned to the Cretaceous (Berriasian).

The assemblages of the first four Early Cretaceous stages (Berriasian to Barremian) are marked by striking changes in the fern spores. *Cicatricosisporites* becomes universal, as do to a lesser degree *Trilobosporites, Pilosisporites*, and others. *Aequitriradites* becomes numerous among the hilates, and *Schizosporis reticulatus* is of regular occurrence. There are echinate species of *Classopollis* in the Berriasian, but the genus then declines in importance, as does *Tsugaepollenites*; it is difficult sometimes to decide how much their continued occurrence is attributable to reworking of Late Jurassic material. In Barremian times bisaccates became diversified to include numerous *Parvisaccites.*

Aptian and Albian assemblages are marked by a sharp increase in the importance of *Gleicheniidites* and a decrease in *Cicatricosisporites, Plicatella*, etc. Although there are of course new species, *Clavatipollenites* is common, and rare genuine tricolpate grains appear. *Ephedripites,* which appears to enter in the Barremian, becomes more common, and bisaccates appear with a clear resemblance to some Recent genera (*Pinus, Cedrus*). Among megaspores the sudden diversification of *Arcellites* and *Pyrobolospora* is striking.

Differences in Floras in Other Areas.

Although correlation by marine fossils is relatively good in most of the Jurassic Period, it is seldom yet possible to make long-distance palynological correlation below the level of stratigraphical series.

In northern temperate areas many assemblages have been described from Asia (Bolkhovitina, 1959 and earlier), but they are not very different from those in Europe. On general grounds it had been suggested that cycadophytes would be more common in lower latitudes and coniferophytes more abundant further north; Teslenko (1959), however, relates Asian Middle Jurassic cycadophyte abundance to proximity to extensive shelf seas rather than to control by latitude. Such Canadian evidence as is available from southern Alberta (Rouse, 1959; Pocock, 1962, 1964) and from the arctic islands (McGregor, 1965) may well support this.

Few assemblages from equatorial areas have been described, but Kuyl and others (1955) claim that in the "tropics" Chlamydospermae such as *Eucommiidites* and also *Classopollis* predominate over saccate and monoporate conifer grains. These interesting zonosulcate grains and their distribution are clearly of great importance to the problem of angiosperm origin.

The best information from the Southern Hemisphere is from Australia (Balme, 1957, 1964; Dettmann, 1963), where there are some clear-cut differences from Europe in species. *Cicatricosisporites* is much less diverse, although it may well enter at the same time, and *Plicatella* does not appear. Other spore genera have purely Australian species. *Exesipollenites* is an important element with *Classopollis* in the Early Jurassic, whereas its occurrences in Europe are more scattered. Polysaccate conifer grains (*Microcachryidites*) are suddenly important in the Early Cretaceous. *Eucommiidites* has not been reported, but the Albian in Australia is characterized by the unusual *Hoegisporis.*

EVOLUTIONARY TRENDS

There is no doubt that evolution in a broad sense

can be observed and described in a succession of miospores, although as they are only organ taxa the observations relate to something other than the evolution of genetic lineages of plants.

The direct Mesozoic evidence for the stratigraphical incongruity (Potonié, 1956a) of development of plant organs is meager, but the reality of such occurrences in the Mesozoic is indicated by the general failure of paleobotanists to identify Mesozoic plants satisfactorily in terms of Recent taxa. The only fixed points are the few "in situ" descriptions (Potonié, 1962, 1965) of spores from fructifications, and only a small number of these satisfy modern palynological requirements.

The Jurassic and Early Cretaceous were periods of very varied selection for new types of spore and pollen apertures, some of which originated in Late Triassic time; the pollen apertures seem to culminate in the tricolpate type just before the Cenomanian (Late Cretaceous) age. With the exceptions of waterplants and of *Classopollis*, elaboration of exine structure does not appear as a major trend. Spore exine sculpture, on the other hand, shows much greater variety than at any time since the Carboniferous, particularly in the Early Cretaceous. Although presence or absence of perine has produced much taxonomic discussion, true perine is an unimportant character that is to be expected only rarely in dispersed-spore preservation; fragments of perine have been seen adhering to numerous spore types.

Inadequate typification of species tends to conceal evolution in that many currently used species appear to have diagnoses based on a few scattered observations on material ranging sometimes over as long as two geological periods; naturally nothing concerning evolution can be "resolved" within this span of time. When authors begin to adhere rigidly to a principle of publishing formal descriptions only when adequate material from a single sample is available, knowledge of something approaching genuine taxa, and consequently of evolution, will accrue.

Although in general it is only the change of one whole assemblage to the next that may be stratigraphically useful, the following observations on certain groups of spores may well ultimately prove to be significant in the histories of the concerned plant groups and in broad first stratigraphical assignment to a geological period.

Saccate Pollen

Monosaccate pollen becomes much less important after the Triassic. *Cerebripollenites*, with a much divided saccus, is common throughout the Jurassic; the several species of *Tsugaepollenites* are very common in the Middle and Late Jurassic. It is not clear, however, which if any of these has any connection with Recent *Tsuga* or the Tertiary *Tsugaepollenites*.

The variety and size of the bisaccate conifer pollen decrease through the Jurassic, and the trend changes only with the sudden increase of *Parvisaccites* in the Barremian. Trisaccates are rare in the Northern Hemisphere and appear to be merely aberrant grains from "bisaccate plants"; natural selection of genuine trisaccites does, however, appear to have begun in the Cretaceous of the Australian region, where certain Recent conifer species bear these grains.

Podocarpidites is rare through the Jurassic and into the Early Cretaceous; since most of the records seem to be of very small numbers of specimens, it may at this time represent only aberrant grains of other bisaccates. By the Aptian and perhaps a little earlier, however, it may well represent genuine species even in the Northern Hemisphere.

Although there are Early Cretaceous macrofossils of *Pinites* leaves and probably cones, the precise provenance of Jurassic bisaccate pollen of the *Alisporites* type is not clear from existing macrofossil records. This is of course part of the general uncertainty about the role of such gymnosperms as dominants in the lowland forests of an angiosperm-free world. The tendency for bisaccate pollen to occur in a palynological facies of restricted variety has led to the suggestion that they came from upland floras. It seems more likely, however, from this frequently recurring pattern that the plants concerned occupied a restricted, perhaps coastal, facet of lowland forest. The fact that the surviving descendant species occur in temperate and/or upland areas is probably a post-Cretaceous phenomenon and therefore not relevant to Mesozoic interpretations.

NonSaccate Pollen

It is scarcely possible to find enough characters to discuss any evolution in the small-grained nonsaccate gymnosperms represented by *Spheripollenites* and *Inaperturopollenites*, which fairly closely parallel pollen of some living trees. It is more surprising that *Monosulcites* of the cycadophytic type shows no exine structure development or other special feature.

This type of grain does persist unaltered to the present day in living cycads, which are now only relict survivors; perhaps this lack of selection for change is connected with their relative failure as a group and with the assumption that they played no part in angiosperm ancestry. The *Ginkgo* type of monosulcate grain has few characters, and there is as yet no palynological indication of the Jurassic and Cretaceous macrofossil diversity of the Ginkgoales. The elongated cycadeoid monosulcates are never common, and possibly the plants concerned never occurred in riverine floras.

Zonosulcate Pollen

Classopollis appears to provide one extreme of the logical development of all-round germinal apertures by the zonosulcate method; the zonosulcus is just distal of the equator, and this position may be related to the well-known tendency in some species of this genus to continued tetrad attachment. The equatorial exine structural developments (Pettitt and Chaloner, 1964) are of unknown purpose, although perhaps they are simply necessary to provide sufficient rigidity to make such an aperture arrangement worthwhile. Although this kind of grain may just have survived into Late Cretaceous time, there appear to have been no further elaborations; its disappearance may signal extinction of an important conifer group that is not clear in the macrofossil record.

Eucomiidites was originally interpreted by Erdtman (1948) as asymmetrically tricolpate, but it seems unlikely that such an asymmetry would have developed first and persisted through several species spread over a whole geological period before genuine tricolpates appeared in the Aptian-Albian. The later interpretation (Hughes, 1961) as a distally monosulcate and proximally zonosulcate grain from a vertical tetrad involves a different grain orientation; such an innovation, however, does not appear to have led to any further development. Species of *Eucommiidites* are distinctly smaller in the Early Cretaceous than the type species (and almost certainly others) in the Jurassic. These smaller species eventually coexisted with the early tricolpates but then disappeared quite suddenly, unless any later occurrence has been overlooked.

Both these groups therefore appear to have been "experiments" that were not ultimately successful in attaining multiple germinal apertures in connection with approaches to angiospermy.

Strongly Sculptured Fern Spores

The scope of the relatively sudden Early Cretaceous development of schizaeaceous spores is well illustrated by Bolkhovitina (1961). The spores fall mainly into *Cicatricosisporites*, *Trilobosporites*, and *Lygodioisporites*, of which *Cicatricosisporites* persists in quantity beyond the Hauterivian and probably gave rise gradually by selection to *Plicatella* with radial equatorial crassitudes; the even longer radial extensions of *Appendicisporites* (Albian to Senonian) may well have had an elaterlike function in the sporangium.

Recent *Anemia* (with this kind of spore) and *Schizaea* (with rather different monolete spores) are both unimportant ground plants, and *Lygodium* is now specialized. It is not easy therefore to appreciate their Cretaceous ancestors as more widespread and dominant plants in a world free of angiosperm herbs.

The Early Cretaceous spore-sculpture development appears to be a culmination of a trend from the Devonian onward that is probably to be correlated with the steadily increasing general involvement of plants with land animals (particularly insects—Hughes and Smart, 1967). The Late Cretaceous angiosperm revolution, however, removed the importance of this trend in pteridophyte plants whose surviving successors began to fill only minor niches. The exine sculpture pattern of angiosperm pollen become subsequently modified in more subtle ways.

Hilate Spores

The importance of these spores in the Mesozoic was not fully realized until recently (Dettmann, 1963). Many of them almost certainly represent bryophytes of which macrofossil evidence is unlikely to be found for preservational and paleoecological reasons. No palynological evidence so far conflicts with the slowly accumulating (Lacey, 1967) macrofossil record of spread of these plants from a Devonian ancestor to an apparent Early Cretaceous maximum after which palynological records at least are submerged by angiosperm pollen.

Spores of Waterplants

By their very nature fresh-water vascular plants are unlikely to have sufficient cuticle to favor reasonable preservation of macrofossils; at the same time, however, their requirements for distribution lead to the development of elaborate structures for floating, for entangling, for water seal against premature growth in their usually thick-walled spores; possibly in the case of megaspores the neck structures may even protect the first stages of the gametophyte.

Throughout the Jurassic and Early Cretaceous megaspore "species" are much more numerous than the known heterosporous land plants, leading to the presumption that these megaspores must have mostly come from waterplants, such as the unknown forerunners of the existing Salviniaceae and Marsiliaceae.

Mesozoic evolution in such spores is from the development of a group of proximal processes in *Thomsonia* (Barbates) and *Hughesisporites* to the various elaborate neck structures of *Arcellites* and *Pyrobolospora* of the Early Cretaceous. At the same time these developments appear even in such small spores as *Balmeisporites* described from Australia (Cookson and Dettmann, 1958) and now known also from the Barremian of England (unpublished).

Conclusions

The period under consideration (Jurassic and Early Cretaceous) may be viewed palynologically and paleobotanically as the time of culmination of a very diverse gymnosperm and pteridophyte flora. The numerous probable extinctions at the end of it are masked in the record by forced classifications into extant higher taxa, and the changes involved in the survival of the existing "lines" in much reduced ecological circumstances appear to be underestimated.

References

Balme, B. E., 1957, Spores and pollen grains from the Mesozoic of western Australia: Commonw. Sc. Ind. Res. Org., Fuel Res., (TC-25) p. 1-48.

Balme, B. E., 1964, The palynological record of Australian pre-Tertiary floras: Ancient Pacific Floras, Univ Hawaii Press, p. 49-80.

Batten, D. J., 1968, Probable dispersed spores of Cretaceous *Equisetites*: Palaeontology, v. 11, p. 633-642.

Bolkhovitina, N. A., 1959, Spore-pollen complexes of the Mesozoic deposits in the Vilui Basin, and their stratigraphic importance (in Russian): Trudy Geol Inst Acad Nauk SSSR, v. 24, p. 1-185.

Bolkhovitina, N. A., 1961, Fossil and Recent spores of the family Schizaeaceae [in Russian]: Moscow, Trud. Geol. Inst. Nauk SSSR, v. 40, 1-176.

Brenner, G., 1963, The spores and pollen of the Potomac group of Maryland: Baltimore, Bull. Maryland Dept. Geol., Mines and Water Resources, v. 27, p. 1-215.

Burger, D., 1966, Palynology of uppermost Jurassic and lowermost Cretaceous strata in the eastern Netherlands: Leids. Geol. Meded., v. 35, p. 209-276.

Cookson, I. C., and Dettmann, M. E., 1958, Cretaceous "Megaspores" and a closely related microspore from the Australian region: Micropaleontology, v. 4, p. 39-49.

——, 1959, On *Schizosporis*, a new form genus from Australian Cretaceous deposits: Micropaleontology, v. 5, p. 213-216.

Couper, R. A., 1958, British Mesozoic spores and pollen: Palaeontographica, v. 103, sec. B, p. 75-179.

Danzé, J., and Laveine, J.-P., 1963, Etude palynologique d'une argile provenant de la limite Lias-Dogger, dans un sondage à Boulogne-sur-Mer: Annales Soc géol Nord, v. 83, p. 79-90.

Delcourt, A. F., Dettmann, M. E., and Hughes, N. F., 1963, Revision of some Lower Cretaceous microspores from Belgium: Palaeontology, v. 6, p. 282-292.

Dettmann, M. E., 1963, Upper Mesozoic microfloras from southeastern Australia: Royal Soc. Victoria Proc., v. 77, p. 1-148.

Döring, H., 1961, Planktonartige Fossilien des Jura-Kreide-Grenz-bereichs der Bohrungen Werle (Mecklenburg): Geologie Beihefte, v. 32, p. 110-121.

Döring, H., and others, 1966, Erlauterungen zu den Sporen—stratigraphischen Tabellen vom Zechstein bis Oligozän: Abh. Zentr. Geol. Inst., 8.

Ellis, C. H., and Tschudy, R. H., 1963, The Cretaceous megaspore genus *Arcellites* Miner: Micropalentology, v. 10, p. 73-79.

Erdtman, G., 1965, Pollen and spore morphology/Plant taxonomy, Gymnospermae, Bryophyta, Text: Almqvist and Wiksell, Stockholm, 191 p.

George, T. N., and others, 1967, Report of the Stratigraphical Code Subcommittee: Geol. Soc. [London] Proc., No. 1638.

Harland, W. B. *et al.* (Eds.), 1967, The Fossil Record: London (Geological Society), xii + 828 pp.

Harris, T. M., 1961, The Yorkshire Jurassic flora; I—Thallophyta-Pteridophyta: Brit. Mus. Nat. Hist., 212 p.

Hughes, N. F., 1955, Wealden plant microfossils: Geol. Mag., v. 92, p. 201-217.

——, 1961, Further interpretations of *Eucommiidites* Erdtman 1948: Palaeontology, v. 4, p. 292-299.

——, 1963, The assignment of species of fossils to genera: Taxon, v. 12, no. 9, p. 336-337.

Hughes, N. F., and Moody-Stuart, J. C., 1966, Descriptions of schizaeaceous spores taken from Early Cretaceous macrofossils: Palaentology, v. 9, p. 274-289.

Hughes, N. F., and Smart, J., 1967, Plant-insect relationships in Palaeozoic and later time: *In* Harland, W. B. *et al.* (Eds.), 1967, The Fossil Record: London (Geological Society), p. 107-117.

Kemp, E. M., 1968, Probable angiosperm pollen from British Barremian to Albian strata: Palaeontology, v. 11, p. 421-434.

Knox, E. M., 1950, The spores of *Lycopodium, Phylloglossum, Selaginella,* and *Isoetes* and their value in the study of microfossils of Palaeozoic age: Bot. Soc. Edinburgh Trans., v. 35, p. 207-257.

Kuyl, O. S., Muller, J. and Waterbolk, H. T., 1955, The application of palynology to oil geology with reference to Western Venezuela: Geologie Mijnb., v. 3, n.s. 17, p. 49-76.

Lacey, W. S., 1967, Bryophyta, *In* Harland, W. B. *et al.* (Eds.), 1967, The Fossil Record: London (Geological Society), p. 211-216.

Mädler, K., 1954, *Azolla,* aus dem Quartär und Tertiär sowie ihre Bedeutung fur die Taxonomie alterer Sporen: Geol. Jahrb., v. 70, p. 143-158.

Manum, S., 1960, On the genus *Pityosporites* Seward 1914, with a new description of *Pityosporites antarcticus* Seward: Nytt. Mag. Bot., v. 8, p. 11-16.

McGregor, D. C., 1965, Illustrations of Canadian fossils—Triassic, Jurassic, and Lower Cretaceous spores and pollen of Arctic Canada: Geol. Surv. Canada, Paper 64-55, p. 1-32.

Nilsson, T., 1958, Uber das vorkommen eines mesozoischen Sapropelgesteins in Schonen: Lunds. Univ. Arss., n.s. v. 54, p. 1-111.

Pettitt, J. M., and Chaloner, W. G., 1964, The ultrastructure of the Mesozoic pollen *Classopollis:* Pollen et Spores, v. 6, p. 611-620.

Playford, G., and Dettmann, M. E., 1965, Rhaeto-Liassic plant microfossils from the Leigh Creek coal measures, south Australia: Senck. leth., v. 46, p. 127-281.

Pocock, S. A., 1962, Microfloral analysis and age determination of strata at the Jurassic-Cretaceous boundary in the western Canada plains: Palaeontographica, v. 111, sec. B, p. 1-195.

Potonié, R., 1956a, Die stratigraphische Inkongruität der Organe des Pflanzenkörpers: Palaont. Zeitsch., v. 30, p. 88-94.

______, 1956b, Synopsis der Gattungen der Sporae dispersae, Teil I: Beihefte Geol. Jahrb., v. 23.

______, 1958, Synopsis der Gattungen der Sporae dispersae, Teil II: ibid., v. 31.

______, 1960, Synopsis der Gattungen der Sporae dispersae, Teil III: ibid., v. 39.

______, 1962, Synopsis der Sporae in situ: Beihefte Geol. Jahrb., v. 52, p. 1-204.

______, 1965, Fossile Sporae in situ: Köln, Forschber. Landes NRhein-Westf., v. 1483, 74 p.

Potter, D. E., 1963, An emendation of the sporomorph *Arcellites* Miner 1935: Oklahoma Geol. Notes, v. 23, p. 227-230.

Reymanovna, M., 1968, On seeds containing *Eucommiidites troedssonii* pollen from the Jurassic of Grojec, Poland: J. Linn. Soc. Lond. (Bot.), v. 61, p. 147-152.

Rogalska, M., 1962, Spore and pollen grain analysis of Jurassic sediments in the northern part of the Cracow-Wielun cuesta (in Polish): Inst. geol. Prace Polska, v. 30, p. 495-507. (Engl. summary).

Rouse, G. E., 1959, Plant microfossils from Kootenay coal-measures strata of British Columbia: Micropaleontology, v. 5, p. 303-324.

Steeves, M. W., and Barghoorn, E. S., 1959, The pollen of *Ephedra*: J. Arnold. Arbor., v. 40, p. 221-255.

Teslenko, Ju. V., 1959, Some features of the distribution of the cycadophytes in the South of Siberia (in Russian): Dokl. Akad. Nauk SSSR, v. 127(1), p. 191-193. (Transl. Amer. Geol. Inst. 1960, p. 790-791.)

Townrow, J. A., 1962, On some disaccate pollen grains of Permian to Middle Jurassic age: Grana Palynologica, v. 3, no. 2, p. 13-44.

Wall, D., 1965, Microplankton, pollen and spores from the Lower Jurassic in Britain: Micropaleontology, v. 11, p. 151-191.

16

LATE CRETACEOUS AND EARLY TERTIARY PALYNOLOGY

John S. Penny

On most standard sections (fig. 16-1) the Upper Cretaceous is separated from the Lower Cretaceous at the Albian-Cenomanian boundary. This division, originally established for Western Europe, is not supportable everywhere by reliable lithologic or paleontologic criteria. The European boundary between Upper and Lower Cretaceous and the major stratigraphic breaks in other continents were not necessarily concurrent (Loeblich and Tappan, 1961). For paleobotanical purposes the division just as plausibly might have been placed between the Aptian and the Albian (Seward, 1931, p. 383), and on the basis of palynological evidence alone the first clear evidence of angiosperm evolution begins with the Albian. By common assent, however, the Cenomanian is chosen as the base of the Upper Cretaceous. The approximately 80-million-year interval between the beginning of this epoch and the close of the Oligocene in mid-Tertiary time spans the relatively long episode in earth history that concerns the present chapter. (See table 1-1.)

PHYSICAL CHARACTERISTICS OF THE INTERVAL

The Late Cretaceous began under conditions of major worldwide marine transgressions that reached maxima during Cenomanian-Turonian and Maestrichtian times. After the lapse of some 47 million years the period closed with the onset of the greatest orogenic cycle since Precambrian time. The succeeding Paleogene, or older Tertiary, interval of some 38 million years was marked by less extensive transgressions but with considerable orogenic activity that began the unfolding and shaping of the modern world (Kummel, 1961). During this whole span of time the climates, ranging from temperate to tropical, were uniformly more genial than those prevailing in later Cenozoic time. The numerous floras of the whole interval reflect no evidence of the existence of extensive glacial conditions; and although, most certainly, there were steep latitudinal gradients in solar radiation, there is no floral evidence of extreme temperature gradients correlative with the varied geography of recovery sites.

PLANT WORLD OF THE LATE CRETACEOUS

The transition from Albian to Cenomanian time saw the continued increase in flowering plants, with some decline in pteridophytes. No striking changes were yet reflected in the abundances of cycadophytes and conifers, although both were to decline in post-Turonian time. Perhaps the most noteworthy feature of the Upper Cretaceous-Paleogene plant world was the increasingly widespread distribution of flowering plants, attesting to, whatever the geomorphological basis, extensive migration opportunities (Good, 1953). The transgressions, regressions, and orogenies occurring over the whole interval coincided with extraordinary evolutionary developments in insects, flowering plants, and pla-

<table>
<tr><th colspan="3">Series</th><th colspan="2">European Stages</th><th>U.S. Gulf Coastal Plain</th></tr>
<tr><td rowspan="9">Tertiary</td><td rowspan="9">Paleogene</td><td rowspan="3">Oligocene</td><td colspan="2">Chattian</td><td>(upper)</td></tr>
<tr><td colspan="2">Rupelian</td><td>(middle)</td></tr>
<tr><td colspan="2">Lattorfian</td><td>(lower)</td></tr>
<tr><td rowspan="3">Eocene</td><td colspan="2">Priabonian</td><td>Jackson Stage</td></tr>
<tr><td colspan="2">Lutetian</td><td>Claiborne Group</td></tr>
<tr><td colspan="2">Ypresian</td><td>Wilcox Group</td></tr>
<tr><td rowspan="3">Paleocene</td><td colspan="2">Sparnacian</td><td rowspan="3">Midway Group</td></tr>
<tr><td colspan="2">Thanetian</td></tr>
<tr><td colspan="2">Montian – Danian</td></tr>
<tr><td colspan="2" rowspan="6">Cretaceous</td><td rowspan="6">Upper Cretaceous</td><td rowspan="4">Senonian</td><td>Maestrichtian</td><td>Navarro Group</td></tr>
<tr><td>Campanian</td><td>Taylor Group</td></tr>
<tr><td>Santonian</td><td rowspan="2">Austin Chalk</td></tr>
<tr><td>Coniacian</td></tr>
<tr><td colspan="2">Turonian</td><td>Eagle Ford Shale</td></tr>
<tr><td colspan="2">Cenomanian</td><td>Woodbine Formation</td></tr>
</table>

Figure 16-1. Selected stratigraphic divisions of the Late Cretaceous and early Tertiary. Data on Gulf Coastal Plain from Popenoe, Imlay, and Murphy (1960); and Murray (1961).

cental mammals. Many biologists have detected herein significant cause-and-effect relationships. Seward (1931) speculating on the course of Cretaceous plant evolution wrote: "May we not see in this sinking and flooding a possible influence on the course of evolution in the organic world and almost worldwide interference with the physical environment which had its repercussion in the altered trend of plant development?"

Cenomanian Floras

Reconstruction and interpretation of the floral world of Late Cretaceous time rests primarily on the evidence afforded by the study of leaf impressions. Here floras from Greenland, Western Europe, Siberia, Japan, China, and North America have played preeminent roles. Information from Southern Hemisphere floras of similar age is accumulating slowly, but in general the data are incomplete, and few southern floras can compare yet in richness with those, for example, of the North American Dakota and Raritan or the Czechoslovakian Perutzer megafloras of mid-Cretaceous age. From these latter three floras alone the appearance of over 50 families of flowering plants by Cenomanian time can be postulated, attesting to the attainment of considerable modernity in the composition of the transitional forests of much of the mid-Cretaceous world. The pattern of evolutionary change that occurred within the major groups of vascular plants during the transition from Early into Late Cretaceous time seems remarkably similar wherever floral successions have been studied. (Compare figs. 16-2 and 16-3.) Berry (1916a) in reviewing the affinities of the then unknown Upper Cretaceous floras of the world cautiously suggested that they were indicative of "warm temperate rain-forest types, less tropical than succeeding Eocene and Oligocene floras."

Newer palynological analyses of mid-Cretaceous sediments—indeed even of beds that have produced some of the better known megafloras—have yielded preliminary evidence that is not

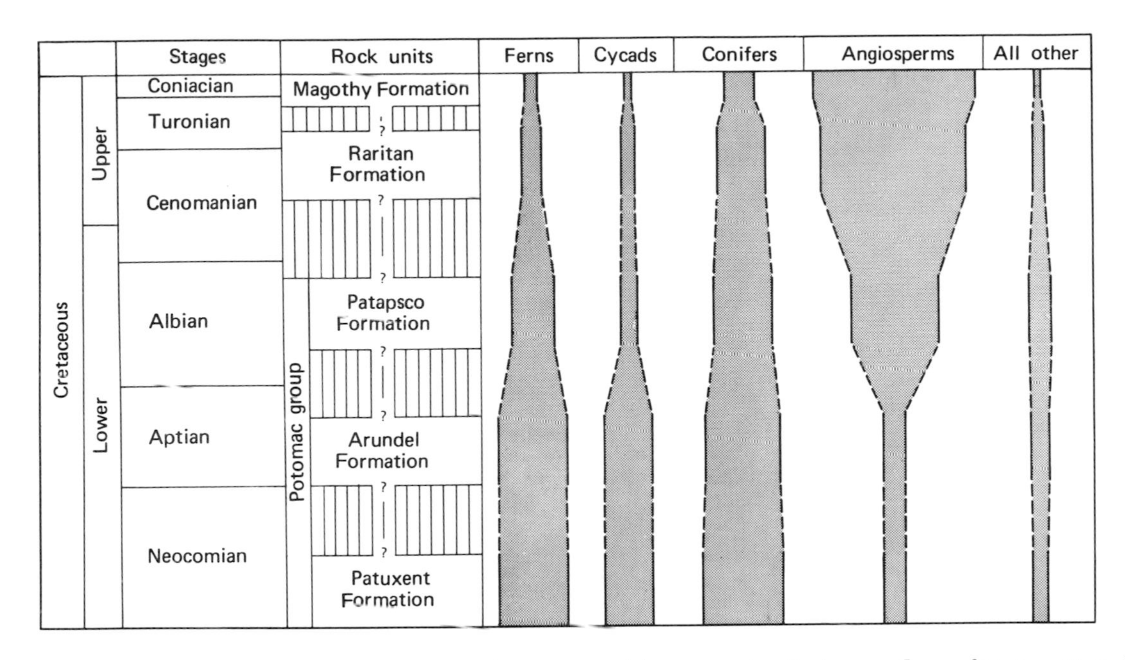

Figure 16-2. Relative proportions of genera in plant groups of the successive Cretaceous floras of Eastern United States (from Dorf, 1955).

always concordant with the paleofloristic and paleoecological interpretations based on leaf floras. Comparisons of the plant microfossils and megafossils of the Perutzer, Dakota, and Raritan Formations will illustrate this point. (See table 16-1.) The Perutzer flora of western (Bohemian) Czechoslovakia, generally regarded as lower Cenomanian, consists of more than 230 species of leaf, fruit, and seed remains, collected by a score of paleobotanists beginning with Sternberg in 1820 and culminating in the comprehensive "Flora Cretacea Bohemiae" of Velenovský and Viniklář (1926-1931). About half of the recorded species are angiosperms in company with relatively strong pteridophytic and coniferous components (each about 20 percent). Cycadophytes account for about 5 percent of the total flora, with the presence of Gnetales and Ginkgoales suggested by *Ephedropsis strobilifera* and *Pseudoginkgo bohemica*. More than 20 families of angiosperms are indicated, including the following:

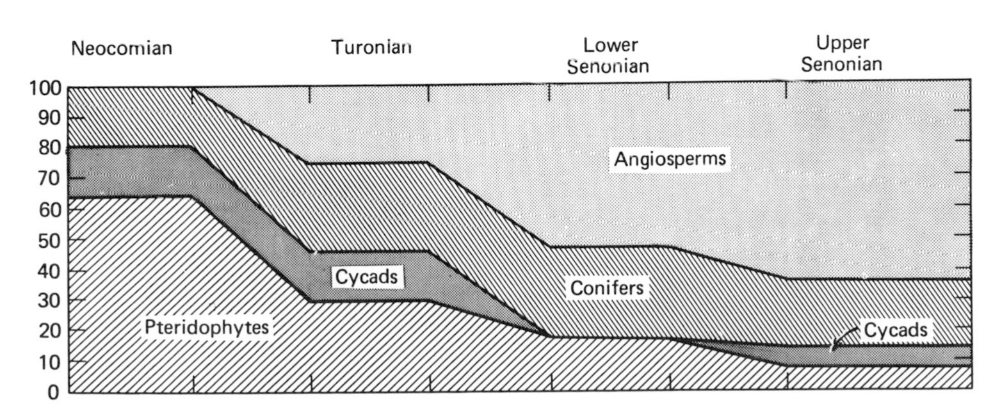

Figure 16-3. Changes in dominance of plant groups in New Zealand during the Cretaceous, based on percentages of species in collections from Waikato Heads (Neocomian), Seymour River coal measures (? Turonian), Paparoa coal measures (Lower Senonian), and from Shag Point and Pakawau (Upper Senonian). From McQueen (1956).

Araliaceae
Aristolochiaceae
Bignoniaceae
Bombacaceae
Butomaceae
Cyperaceae
Gramineae
Lauraceae
Magnoliaceae
Menispermaceae
Moraceae
Myricaceae
Myrtaceae
Platanaceae
Proteaceae
Rhizophoraceae
Salicaceae
Sapindaceae
Sparganiaceae
Sterculiaceae
Stachyuraceae
Theaceae
Vitaceae

Despite the difficulties in placing megafossil remains in modern taxa, and even assuming that 20 percent of the Perutzer flora are misidentified, the fact remains that the flora is a rich one and shows considerable angiosperm diversity. As shown in table 16-2, a palynological examination made of Perutzer sediments by Thiergart (1953) revealed, on the contrary, a flora strongly dominated by pteridophytes and gymnosperms, with only two species of angiosperm pollen, accounting for but 3 percent of the pollen and spore species.[1]

Apparent discordancy between the microfossil-megafossil evidence can be found within the large Dakota flora despite some dispute concerning stage correlation with European age equivalents. Lesquereux (1892), who monographed the Dakota megaflora from Kansas, Colorado, and Nebraska, favored a Cenomanian age. More recent stratigraphic assignments have suggested an age range from Aptian to Cenomanian for the Dakota Sandstone throughout the wide area of its development in western interior United States. Pierce (1961) examined the spores and pollen in certain premarine Dakota sediments from Minnesota classified by Cobban and Reeside (1952) as Cenomanian, and concluded the following: "A comparison of the plant microfossils from the Cretaceous of Minnesota with plant microfossils from the Cretaceous of Europe indicated that the premarine Cretaceous deposits of the State are Cenomanian equivalents."

[1]Firtion (1952) reported no angiosperm pollen from the lower Cenomanian sediments atop the Wealden beds at Nouvion-en-Thierache, France, although fern spores and gymnosperm pollen were recovered in good preservation. Thiergart (1954) reported only some "myrteoid" and "quercoid" pollen in Cenomanian coal from near Marseille. Durand and Ters (1958) recorded only three species of angiosperm pollen in sands and clays rich in Cenomanian fossils from Challons, France.

***Table 16-1.* Plant Family Affiliations Suggested by Leaf Impressions versus Pollen Recovery from Three Cenomian Floras (40°–50°N. Latitude)**

Family	*1*	*2*	*3*	*4*	*5*	*6*
Alismataceae			X			
Araceae			X			
Cyperaceae	X					
Gramineae	X		X			
Liliaceae			X			
Palmae			X			
Sparganiaceae	X					
Aceraceae		?	X			
Anacardiaceae			X			
Anonaceae			X			
Aquifoliaceae			X			?
Araliaceae	X		X			?
Aristolochiaceae	X		X			
Betulaceae			X			
Balanophoraceae			X			
Bignoniaceae	X					
Bombacaceae	X					
Butomaceae	X					
Caprifoliaceae			X			
Celastraceae			X		X	
Cornaceae			X		X	
Dioscoriaceae			X			
Ebenaceae			X		X	
Ericaceae			X		X	
Fagaceae		?	X	O		O
Hamamelidaceae			X	O		?
Juglandaceae			X			
Lauraceae	X		X		X	
Leguminosae			X		X	
Magnoliaceae	X		X	O	X	
Menispermaceae	X		X	O		
Monimiaceae			X			
Moraceae	X		X		X	
Myricaceae	X		X		X	
Myrsinaceae			X		X	
Myrtaceae	X		X			
Platanaceae	X		X	O		?
Proteaceae	X		X			
Rhamnaceae			X			
Rhizophoraceae	X					
Rosaceae			X			
Salicaceae	X		X		X	?
Sapindaceae	X		X			
Sapotaceae			X			
Stachyuraceae	X					
Sterculiaceae	X		X			
Theaceae	X					
Tiliaceae			X			
Ulmaceae			X			
Vitaceae	X		X			

1. Perutz Megaflora (Berry 1916; Velenovský and Viniklář 1926–1931).
2. Perutz Microfossil Flora (Thiergart 1953).
3. Dakota Megaflora (Lesquereux 1892; Pierce 1961).

4. Dakota Microfossil Flora (Pierce 1961).
5. Raritan Megaflora (Dorf 1952).
6. Raritan Microfossil Flora (Groot, Penny, and Groot 1961).

X = Leaf impressions; O = pollen; ? = questionable pollen.

Although the Minnesota segment of the Dakota megaflora is a rather small one (about 40 species), over 90 percent of the listed species are angiosperms. Of the 103 palynomorph species recorded by Pierce less than 25 percent were angiosperms; moreover, for 20 genera of dicotyledons represented by leaf impressions only 2 (*Magnolia* and *Platanus*) were indicated by fossil pollen.

A third well-known early Late Cretaceous assemblage, again with evidence of apparent discordancies, is the Raritan flora of Eastern United States. On strong megafossil evidence, supported by some faunal evidence, the formation in outcrop is of Cenomanian age. Dorf (1952), in a reassessment of the floral evidence, reaffirmed its Cenomanian age and suggested a possible extension into early Turonian. Groot, Penny, and Groot (1961) on the basis of the diversity of the angiosperm types in the assemblage obtained from the type locality, supported Dorf's interpretation, concluding that the lower part of the formation was probably Cenomanian and the upper part Turonian. As in the case of the Perutzer and Dakota floras, the evidence again from leaf fossils suggests a flora dominated by a variety of angiosperms. Even when the evidence is restricted to leaf impressions recovered from the "lower" part of the formation, the record of angiosperm diversity is compelling: of the 29 species whose identifications and stratigraphic position seem reliably ascertained 26 were angiosperms (Dorf, 1952). In contrast only nine species of angiosperm pollen were recorded by Groot, Penny, and Groot (1961) from the lower part (i.e., probable Cenomanian) of the New Jersey Raritan Formation, and of these, only four species were of common occurrence. Couper (1960b) reported angiosperm pollen as "still comparatively rare" into the late Turonian of New Zealand. Langenheim, Smiley, and Gray (1960) reported angiosperm pollen as "scarce or absent in all samples" of the Albian or Cenomanian Kaolak River sediments of the Alaskan arctic coastal plain.

Table 16-2. Abundances of Pollen and Spore Species in Perutzer Sediments[1]

Species	Abundance (percent)
cf. Zonales sp.	10
Schizaeaceae (total)	21
Concavisporites ? bohemiensis	47
Inaperturopollenites magnus	15
I. hiatus	1
cf. *Sciadopitys*	2
Pinus, haploxylon type	1
Pollenites laesus	2
Tricolpopollenites cf. *T. henrici*	1

[1] From Thiergart (1953).

In their study of eight Portuguese samples ranging in age from Aptian to Cenomanian Groot and Groot (1962b) recorded 46 species of spores and pollen of which some 31 species were of pteridophytic or gymnospermous affinities. Fewer than a dozen species of angiosperm pollen were recorded from the sample considered to be of Cenomanian age, and of these only three species (*Tricolpopollenites retiformis* and two species of *Latipollis*) were of common occurrence. The authors concluded that the older, *Classopollis*-dominated conifer forests of Portugal were replaced by angiosperm forests by Late Cenomanian time. Whether this underrepresentation of the angiosperm component will prove to be a general and worldwide characteristic of mid-Cretaceous microfossil floras cannot be affirmed categorically until considerably more palynological evidence is obtained, particularly from now tropical and high northern and southern latitudes. Not all palynological studies known at present, however, indicate such low percentages of angiosperm-pollen species. Steeves (1959) examined two deep well cores from Long Island, New York, that penetrated the total cover of unconsolidated sediments at the test site. Stratigraphic intervals designated as "upper Raritan" formation (interpreted as Upper Cretaceous) and "lower Raritan" formation (interpreted as Lower Cretaceous) were reported as having angiosperm percentages as high as 75 and 24 percent, respectively. However, the relationships between the units so designated by Steeves and the type Raritan remain to be clarified. Somewhat anomalous also are the relatively high angiosperm percentages reported by Bolkhovitina (1953) for Cretaceous deposits from western Kazakhstan in the southern Urals. About 33 palynomorph species of pteridophytes, conifers, and angiosperms were reported from marine sands, ostensibly of Cenomanian age, along the Ayat River. The suggested angiosperm component is about 47 percent, although the angiospermous affinity of the genus *Triporina* Naumova, under which Bolkhovitina listed several species, has been questioned (Scott, Barghoorn, and Leopold, 1960). The rather unusual feature of the Ayat River pollen flora is the precocious morphological evolution of the grains, judging from the close resemblances that Bolkhovitina saw between her fossil species and extant modern pollen.

Cenomanian Pollen	*Modern Pollen*
Morus pumila Bolk.	*Morus alba* L.: "completely corresponds to the fossil specimen"
Betula infucata Bolk.	*Betula pendula* Lodd., *B. verrucosa* Ehrh. ". . . the fossil pollen corresponds to these species"
Betula ajatensis Bolk.	*Betula raddeana* Trautv. "similar"
Rubus proximus Bolk.	*Rubus sanguineus* Friv. "very similar"
Eucalpytus colorata Bolk.	*Eucalptus pauciflora* Sieb. "analogous"
Ilex uralensis Bolk.	*Ilex crenata* Thunb. "similar"
Sambucus pseudo-canadensis Bolk.	*Sambucus canadensis* L. "identical"

In general palynologists have been somewhat less successful than Bolkhovitina in assigning Cretaceous pollen to infrafamily taxa, agreeing with Scott, Barghoorn, and Leopold (1960) that "in many cases Late Cretaceous pollen cannot be assigned to families; where families can be recognized, modern generic assignments may be difficult or impossible."

Several workers have considered possible causes for the apparent differences between the spore-pollen and leaf records of similar beds. As far as the plant microfossil record is concerned discussions have tended to emphasize the role of one or more of the following factors:

Fossilization: entombment and preservation opportunities.

Sampling: pattern and number of collections.

Treatment: laboratory and preparation techniques.

Phylogeny: evolution and radiation.

Pollination: entomophily versus anemophily.

Ecology: upland versus lowland site sources.

Taxonomy: identification and nomenclature.

In view of the need for considerably more data categorial statements here are best withheld. The evidence, such as it is, however, suggests that fossil-pollen floras are likely to contain mixed assemblages of grains from upland (or hinterland) vegetation, in company with those from littoral or riparian vegetation, but that neither the microfossil record alone nor the megafossil record alone is without potential bias. In this light the two sources of data should be regarded as not so much discordant as complementary. Langenheim, Smiley, and Gray (1960) have shown that separate evaluations of plant microfossil and megafossil evidence have led to equivocal interpretations. On the other hand, the use of a dual approach already has shown considerable promise for Late Cretaceous and Paleogene floras. Smiley (1961), using both microfossil and megafossil data, suggested a possible pattern of angiosperm migration into the Arctic, postulating that ". . . angiosperm forests of the Arctic may have developed in riparian habitats while gymnosperm forests still inhabited higher ground of the region."

Gray (1960) added several temperate genera to the known Claiborne (middle Eocene) flora of Alabama and suggested the survival of a temperate, essentially Late Cretaceous, upland community, coeval with the subtropical strand flora. Jones (1961) added evidence that the augmented Wilcox (lower Eocene) flora of Arkansas indicated ". . . a warm, humid coastal plain with adjacent highlands."

Palynological Characterization of the Cenomanian (plate 16-1)

Pteridophytes, especially ferns, tend to be well represented by trilete and monolete spores. Species of *Trilobosporites, Pilosisporites,* and the more bizarre schizaeaceous types known from Lower Cretaceous deposits are absent or rare. Families of Filicales probably represented include Polypodiaceae, Osmundaceae, Schizaeaceae, Gleicheniaceae, Matoniaceae, Cyatheaceae, and Hymenophyllaceae. Ferns of unknown taxonomic position are less common than in the Lower Cretaceous. Both vesiculate and nonvesiculate conifer pollen are abundant. Families represented in Northern Hemisphere fossil-pollen floras include Pinaceae, Taxodiaceae, Cupressaceae, and Podocarpaceae, Podocarpaceous pollen, together with fern spores, dominates the Southern Hemisphere mid-Cretaceous plant-microfossil record. Monosulcate pollen of bennettitalean, ginkgoalean, or cycadalean origin, although not uncommon, is rarely abundant. Similarly *Classopollis* pollen, although present, is less common than in Early Cretaceous time. Although undoubted tricolpate pollen appears for the first time in the lower Albian, monosulcate (angiospermous), triporate, and syncolpate pollen groups are first known only from the Cenomanian. Families probably represented include the Magnoliaceae, Hamamelidaceae, Aquifoliaceae, Myrtaceae, Menispermaceae, Fagaceae, and Platanaceae. Present palynologic evidence in general tends to support the Axelrod (1959) hypothesis of poleward migration of early angiosperms, since, to quote Couper (1964), ". . . angiospermous pollen grains are relatively much more abundant in the Upper Cretaceous sediments of present-day tropical and subtropical areas—than in sediments of the same age from present-day temperate regions." Megaspores seem to have been neglected in palynologic studies of Cenomanian and later sediments. Although recorded from the Cenomanian, Senonian, and even the Paleocene (Dijkstra, 1961; Hall, 1963), megaspore entities either have been overlooked or unrecovered in post-Albian sediments. Hall (1963) has suggested that perhaps "no large and distinctive [megaspore] floras occur in post-Cretaceous times." Much remains to be learned about the botanical affinities, distribution, and stratigraphic significance of Upper Cretaceous and Paleogene megaspores.

Turonian Floras

The widespread transgressions of the Cenomanian seas swelled to their maxima during the next, relatively short (4 million years), Turonian Epoch. Although only minor crustal and climatic changes are reflected by Turonian deposits, their fossil record indicates, if no major phylogenetic change, considerable evolutionary development. Turonian megafossil evidence, confirmed at least by Northern Hemisphere microfossil records, attests to the attainment of full dominance by the angiosperms and to a slow decline in the number of fern, cycadophyte, and conifer genera. (See fig. 16-4.)

The well-developed marine sequences of Western Europe have long served as stratigraphic standard sections for Cretaceous deposits. Where interfingering marine and nonmarine beds have occurred accurate age assignments for many Western European megafloras have been possible, and these in turn have served as reference floras for worldwide paleobotanical correlation. On the other hand, Cretaceous and Tertiary palynological studies from 1930 onward have tended to establish Central European sequences as standards for correlation purposes. Moreover, the nomenclatural priorities and practices initiated in large measure by German brown-coal palynologists have had considerable influence on palynologic studies throughout the world. From early Turonian upward the evolutionary development of many different palynomorph genera and species, particularly of dicotyledonous types, has enabled German palynologists to work out the stratigraphic sequences of European Mesozoic and Cenozoic deposits with considerable refinement (Krutzsch, 1957). How widely developed the horizontal distribution of the European stratigraphic marker species is, however, remains to be discovered. Preliminary evidence suggests that considerable generic similarity may exist in widely separated Turonian assemblages across much of Eurasia and North America, but perhaps not south of the Tethyan geosyncline. The small fern-*Podocarpidites* assemblage listed by Couper (1960b), for example, from the *Inoceramus*-dated Gridiron Formation of Turonian age in the Clarence valley of New Zealand lacks the Normapolles and associated sporomorphs so characteristic of Northern Hemisphere Turonian spore-pollen floras. Typical Normapolles forms were not reported by Brown and Pierce (1962) in their preliminary palynologic correlations within the late Cenomanian-Turonian Eagle Ford group of northeast Texas, although two triangular tricolporate forms were figured. Tropical spore-pollen floras of reliably dated Turonian age are not well known at this time, but petroleum company palynological records and other unpublished studies give some indications that these too lack the Normapolles palynomorphs of the northern latitude Turonian

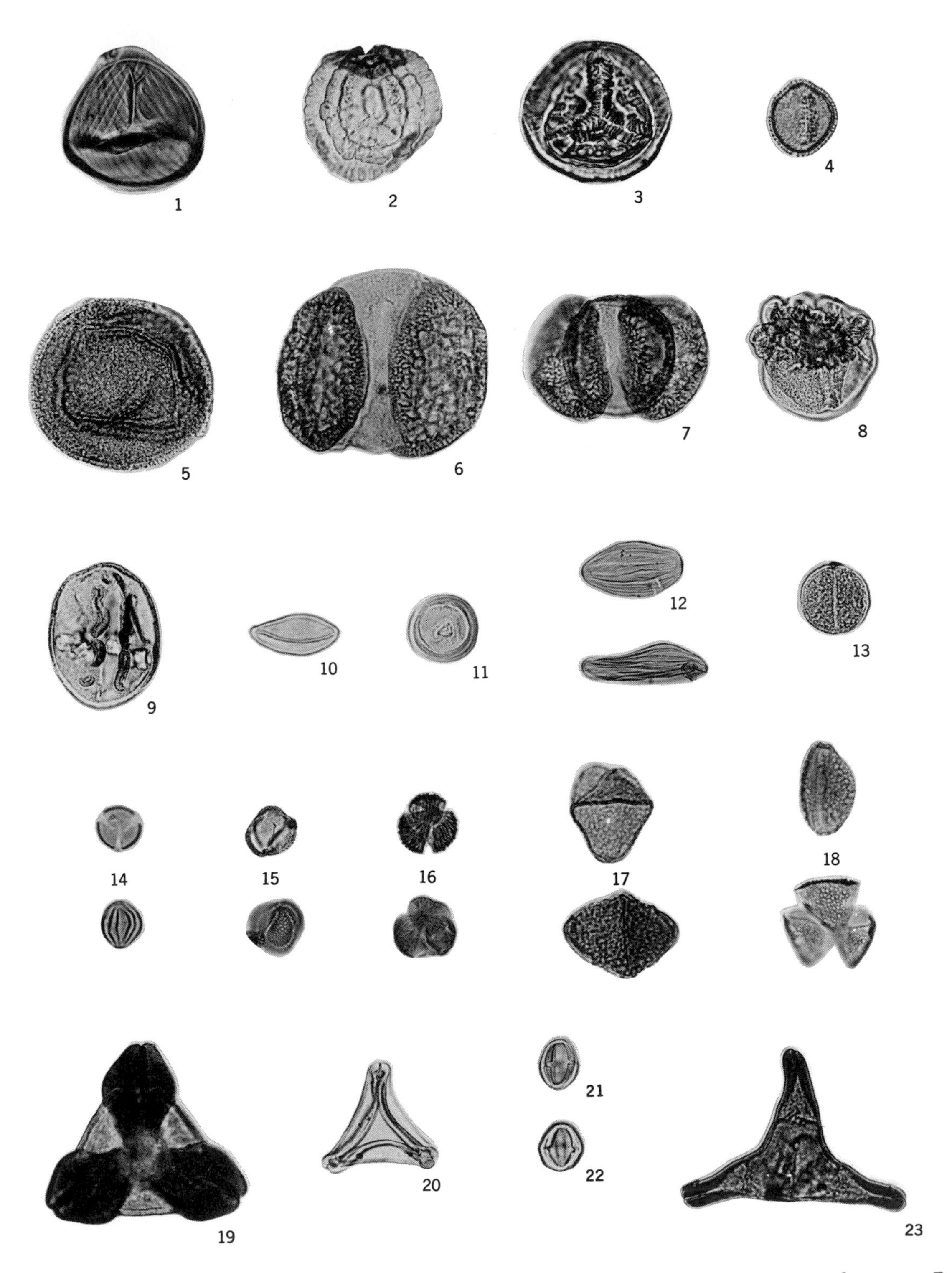

Plate 16-1. Cenomanian-lower Senonian pollen. (all figures × ca. 500.) 1—*Cicatricosisporites dorogensis* R. Potonié and Gelletich. 2—*Taurocusporites* sp. 3—*Taurocusporites* sp. 4—*Apiculatisporites vulgaris* Groot and Groot. 5—*Araucariacites* sp. 6—*Abietineaepollenites* sp. 7—*Pinuspollenites* sp. 8—*Rugubivesiculites* sp. 9—"*Granabivesiculites*" sp. 10—*Entylissa* sp. 11—*Classopollis* sp. 12—*Ephedripites?* sp. 13—*Monosulcites* sp. 14—*Tricolpopollenites* sp. 15—*Tricolpopollenites crassimurus*. 16—*Striatopollis sarstedtensis* Krutzsch. 17—*Latipollis verrucosus* Groot and Groot. 18—*Tricolpopollenites* sp. 19—*Oculopollis* sp. 20—*Sporopollis?* sp. 21—*Tricolporopollenites* sp. 22—*Tricolporopollenites* sp. 23—*Extratriporopollenites* sp.

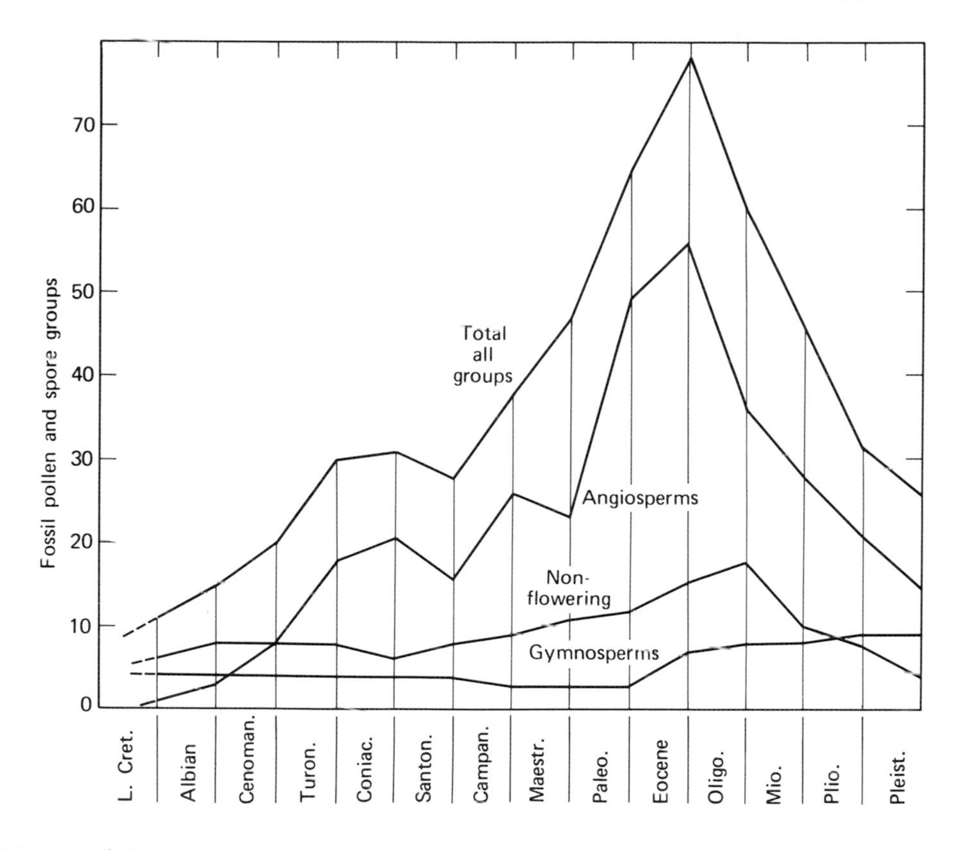

Figure 16-4. Total fossil pollen and spore groups Lower Cretaceous-Pleistocene. (After Cousminer, 1961.)

pollen floras. Present concepts of palynological characterization of Turonian sediments thus tend to be colored by Central European studies, with some confirmatory support from North American reports. As understood at the present time the Northern Hemisphere Turonian plant-microfossil record is distinguished from the Cenomanian record by two features: (a) the first clear dominance of angiosperms over pteridophytes and gymnosperms and (b) the prevalence of a morphological type of nonprolate ("Brevaxones") dicotyledonous pollen of remarkable variety, whose many form genera are usually grouped under the morphologic category Normapolles Pflug. (See chap. 2.)

Palynological Characterization of the Turonian Stage

Although some Normapolles genera (e.g., *Turonipollis*) may have evolved in late Cenomanian time, they appear in notable quantity and morphological variety only from the early Turonian upward, reaching the height of their evolution in the Campanian and Maestrichtian Stages (Krutzsch, 1960). Turonian Normapolles types—such as those classified under *Turonipollis* Krutzsch, *Monstruosipollis* Krutzsch, *Extratriporopollenites* Pflug, and others—are characterized by complex, often protruding and vestibulate, pores. (See pl. 16-1, fig. 23. Also

---

Figures 1, 4, 6, 7, 10, 12, 14, 15, 16, 17, and 18: from Cenomanian, probably upper Cenomanian, deposits near Buarcos and Tavarede, Portugal; slide from Johan Groot; photomicrographs by J. S. Penny. Figures 2, 3, 13, and 20: Tuscaloosa Group (Alabama); photomicrographs from R. H. Tschudy. Figures 5, 8, and 9: Senonian, Pond Bank deposit, near Chambersburg, Pa.; photomicrographs from R. H. Tschudy. Figures 19 and 23: Santonian of Piolenc, France; from Johan Groot; photomicrographs by J. S. Penny. Figure 11: Photomicrograph from R. H. Tschudy.

Chapter 2.) These forms give way to less lobate, more triangular "Dreieckpollen" during the succeeding Senonian Stages. The ancestry and subsequent history of the Normapolles-producing plants are of considerable interest as part of the more central issue of the origin and radiation of angiosperms. Pflug (1953), the originator of the designation, considered Normapolles to be derivatives of his *Duplosporis*-type pteridophytic spores, in which certain Y-mark features of the distal spore walls were interpreted as preangiospermous germinal structures. Kuyl, Muller, and Waterbolk (1955) and Couper (1955) questioned this interpretation, preferring instead to regard the Y-marks as folds acquired in the fossilization process. The problem of Normapolles origin still remains largely conjectural, and, in the absence of less debatable evidence, restraint here seems most appropriate. A general post-Turonian decline and extinction of inadaptive species after minor climatic deterioration during Coniacian-Santonian time may have accounted for the disappearance of some Normapolles types. Evidence that such extinction may have taken place is offered by Cousminer (1961) on the basis of his analysis of Krutzsch's 1957 data of first and last appearance of Mesozoic and Cenozoic spores and pollen. (See fig. 16-5.) The surviving Normapolles-producing dicotyledonous plants presumably were ancestral to many of the modern dicot genera appearing in the older Tertiary. Pflug (1953) saw the genealogies of the Myricales, Juglandales, Fagales, Urticales, and perhaps Salicales and Ebenales, as arising out of plants of the Normapolles complex. Thiergart (1949) had compared certain Normapolles types to pollen of the Myrtaceae, and Hofman (1948) compared them to mangrove-swamp genera such as *Rhizophora* and *Avicennia.* This latter comparison is particularly engaging in view of the possibility that widespread, lowland mudflat environments developed during the great transgressions and regressions of the Upper Cretaceous. Croizat (1952) postulated that the older Mesozoic vegetational climaxes were replaced by weedy, migrating streams ("genorheitra") of angiosperms possessing considerable genetic heterozygosity and that the majority of these "genorheitra" were mangroves or near-mangroves.

Whatever the taxonomic relations of the Turonian Normapolles may be, their record is one of considerable diversification and relatively brief geologic tenure—conditions ideally suited for zonation and correlation purposes. Sarmiento (1957) used certain triporate Normapolles and Dreieckpollen types for zonation within the Mancos group (Turonian-Senonian) of continental and brackish-water sediments in Utah. Groot, Penny, and Groot (1961) used the distribution of

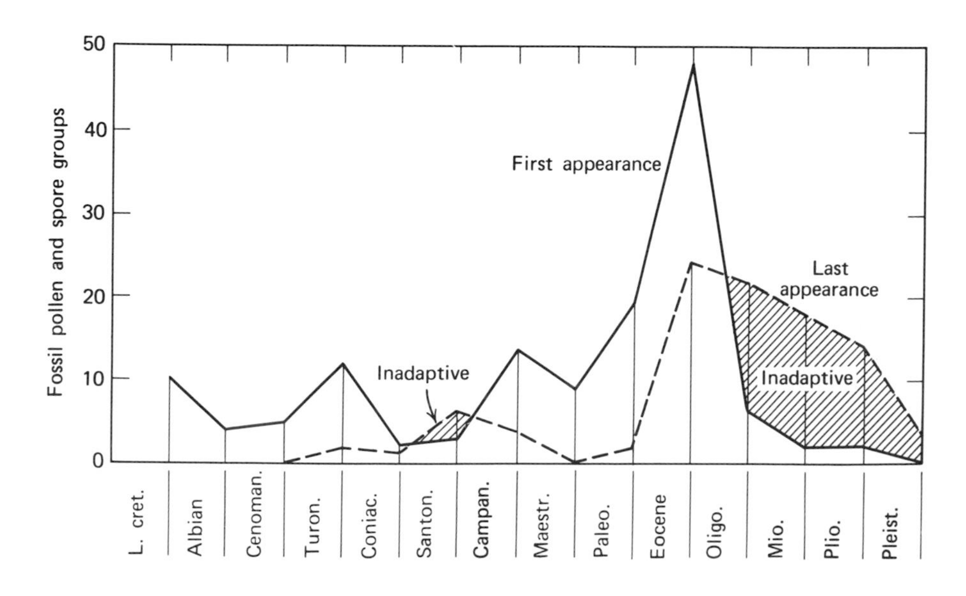

Figure 16-5. First and last appearances of Mesozoic fossil pollen and spores. (From Cousminer, 1961.)

Normapolles and Dreieckpollen as the basis of Cenomanian-Turonian, early Senonian age assignments of the Raritan, Tuscaloosa, and Magothy Formations of Eastern United States.

Senonian Floras

Although no widespread orogenic or tectonic activity marked the transition from Turonian to early Senonian, crustal oscillations continued to produce transgressions and regressions throughout the Senonian Epoch. There is no paleobotanical evidence of widespread climatic or other ecological change occurring at the onset of the Senonian, and plant microfossils of the Northern Hemisphere reflect the continued diversification and migration of the now completely dominant angiosperms. Southern Hemisphere spore-pollen floras of similar age, so far as known, remained dominated by conifer pollen; for example, the nonmarine New Zealand Paparoa beds, assigned a Senonian age on the basis of plant megafossils (McQueen, 1956) and plant microfossils (Couper, 1960b), are characterized by a distinctive assemblage of conifer pollen, including

Podocarpidites major
P. chikaesis
P. cf. ellipticus
P. marwickii
P. otagoensis
Dacrydiumites mawsonii
Araucariacites australis
Microcachryidites antaracticus
Trisaccites microsaccatus

Nothofagus and *Proteacidites*, however, make their first appearance in New Zealand during the early Senonian. The angiosperm component of the earliest Northern Hemisphere Senonian pollen floras remains characterized by Normapolles forms of uncertain botanical affinities. Some of these porate types (*Trudopollis*, *Sporopollis*, etc.), persisted with little change from the Turonian; other genera (*Emscheripollis*, *Quedlinburgipollis*, *Interporopollites*, etc.) became common for the first time. Toward the close of the Santonian Epoch, however, Normapolles types became associated with types displaying increasing morphological resemblances to pollen of modern plants. The tempo of these changes appears, on the basis of present evidence, to have been more rapid in western North America (and perhaps eastern Siberia) than in other Northern Hemisphere regions. Santonian and Campanian spore-pollen floras of central Europe remain dominated by Normapolles types such as *Trudopollis* and *Extratriporopollenites* (Weyland and Greifeld, 1953; Weyland and Krieger, 1953; Bolkhovitina, 1959; Pacltova, 1959). The Senonian-Santonian Czechoslovak flora described by Pacltova, for example, has only two species (*Palmaepollenites* sp. and *Monoporopollenites* sp.) that give any suggestion of affinities to extant angiosperm families.

By contrast, a spore-pollen flora from the western Canadian Oldman Formation of Senonian-Santonian age described by Rouse (1957), containing about 36 percent angiosperm pollen, lacks typical Normapolles grains, possessing instead pollen of modern appearance referred by Rouse to seven or more extant families.

Further evidence of a more rapid tempo of pollen evolution in western North America and in Siberia is suggested by the restricted occurrence in those provinces of an enigmatic palynomorph, *Aquilapollenites* Rouse (pl. 16-2, figs. 5–8), considered by Funkhouser (1961) and Stanley (1961) to have possible santalacean affinity. The Santalaceae is a relatively advanced family of half-parasitic plants with epigynous flowers. The possible course of its phylogeny and distributional history has invited some speculations. Lam (1948) noted the possibility of a gnetalean ancestry, and Croizat (1952) regarded the family as belonging to one of the widest and most involved migrating streams of angiosperms. Stanley has pointed out, however, that there is no confirmatory evidence from megafloral remains of santalacean genera occurring in similar-age sediments at the *Aquilapollenites* recovery sites. Chlonova (1962) has discussed additional kinship possibilities and has investigated the comparative morphology of certain pollen types called "radiata," "unica," and "oculata." *Aquilapollenites*, under this categorization, belongs to the "unica" type. Any new studies of this interesting genus should include comparison with *Pentapollenites pentangulus* (Pflug) Krutzsch and perhaps also with *Proteacidites pachypolum* Cookson and Pike, *Proteacidites polymorphus* Couper, and the pollen genus *Taurocephalus* Simpson.

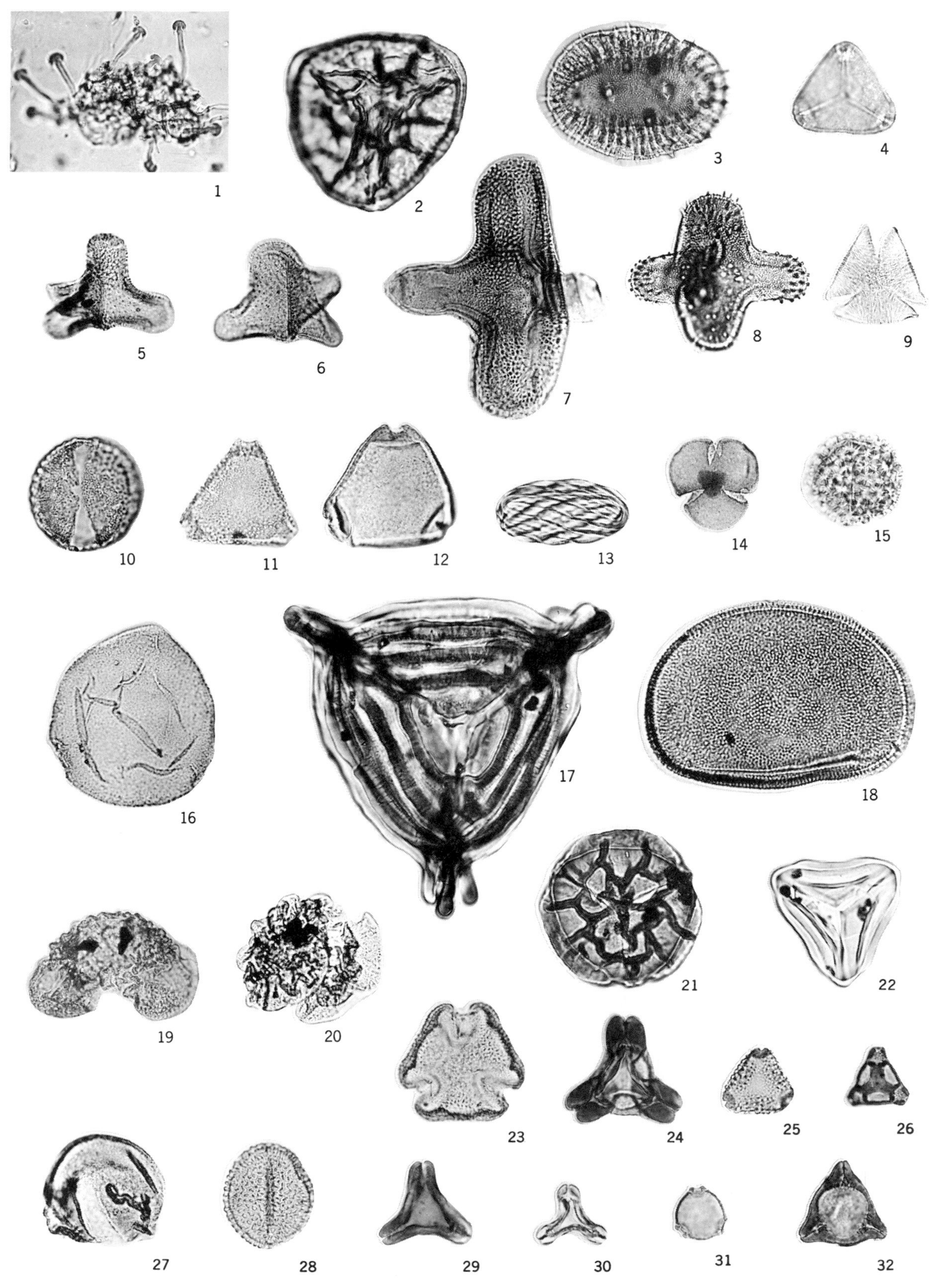
1
2
3
4
5
6
7
8
9
10
11
12
13
14
15
16
17
18
19
20
21
22
23
24
25
26
27
28
29
30
31
32

Palynological Characterization of the Latest Cretaceous

Late Senonian (Campanian-Maestrichtian) pollen assemblages tend to show a mixed character that is intermediate between Late Cretaceous and early Tertiary complexes. (See pl. 16-2.) Middle European floras—for example, of latest Cretaceous age—reflect maximum evolutionary development of such Normapolles types as *Oculopollis, Trudopollis,* and *Vacuopollis,* in company with the first appearance of pollen indicating sapotacean, nyssacean, and palm affinities. Examination by this author of sediments considered to be Campanian-Maestrichtian in age (Matawan and Monmouth groups) of New Jersey show, similarly, a mixture of Late Cretaceous Normapolles types with forms of nyssacean, juglandacean, and fagacean morphology. Southern Hemisphere floras reflect the continuing dominance by Podocarpaceae pollen, with, however, increasing invasion of angiosperm types. The Maestrichtian (Mata Series, Haumurian Stage) pollen floras from North Canterbury, New Zealand, described by Couper (1960b) show this admixture of newer types (e.g., Lilaceae, Proteaceae, Fagaceae, Olacaceae, and possibly Caryophyllaceae), in company with the older gymnosperm-fern complex inherited from earlier Senonian time. Couper's list for the Haumarian (Maestrichtian) Stage includes

Dicksonia aff. *squarrosa* Swartz
Podocarpidites otagoensis Couper
Ephedra notensis Cookson
Beaupreadites elegansiformis Cookson
Proteacidites granoratus Couper
P. retiformis Couper
P. palisadus Couper
P. scaboratus Couper
P. parvus Cookson
Nothofagus kaitangata Te Punga
Knightia aff. *excelsa* R. Br.
Caryophyllidites polyoratus Couper
Liliacidites kaitangataensis Couper
L. variegatus Couper
L. intermedius Couper

The fern-rich early Senonian floras of South America show, by Maestrichtian time, a marked influx of palm pollen in association with a variety of dicotyledonous types, presaging the more modern, and typical South American, Tertiary flora. Van Der Hammen's investigations (1954) of Late Cretaceous and early Tertiary coals of the eastern Cordillera of Colombia suggest that this replacement of pre- and early Senonian primitive species by palm and modern dicotyledonous pollen began in early Maestrichtian time and was accomplished almost completely by the end of that epoch. Western North American (and Siberian?) pollen floras of Senonian age, as noted previously, appear to show an early attainment of a modern aspect. Stanley (1960) noted that the angiosperm-rich (50 percent) Crow Butte spore-pollen flora from South Dakota of uppermost Cretaceous age, containing *Aquilapollenites* but lacking Normapolles forms, was "markedly different" from any European or Australian floras of comparable age.

From late Senonian time onward the compositions of pollen floras show, as might be expected, an increase in the number of types assignable to extant genera, and a rapidly growing literature attests to the increasing use of palynology for climatic, vegetational facies, and age studies. Martin and Gray (1962) reviewing some of the problems attending the exploitation of palynological data for Cenozoic climatological and vegetational interpretations warned against, among other pitfalls, the inadvertent selection of spurious correlations in statistical treatment of pollen counts, opining that "Palynologists seldom subject their percentages to formal correlation

Plate 16-2. Maestrichtian pollen. (All figures × ca. 500.) 1—*Azolla* sp. Massula. 2—*Zlivisporites* sp. 3—*Wodehouseia spinata* Stanley. 4—*Pemphixipollenites* sp. 5—*Aquilapollenites delicatus* Stanley. 6—*A. polaris* Funkhouser. 7—*A. trialatus* Rouse. 8—*A. attenuatus* Funkhouser. 9—*Elytranthe striatus* Couper. 10—*Monosulcites* sp. 11—*Proteacidites* sp. 12—*Triatriopollenites* sp. 13—*Ephedripites* sp. 14—*Tricolpites interangulatus* Newman. 15—*Erdtmanipollis* sp. 16—*Araucariacites* sp. 17—*Appendicisporites* sp. 18—*Schizaeoisporites* sp. 19—*Rugubivesiculites*. 20—*Rugubivesiculites*. 21—*Camarozonosporites* sp. 22—*Gleicheniidites* sp. 23—*Tricolporites* sp. 24—*Oculopollis* sp. 25—*Proteacidites* sp. 26—*Plicapollis* sp. 27—*Granabivesiculites?* sp. 28—*Monosulcites* sp. 29—*Extratriporopollenites* sp. 30—*Extratriporopollenites* sp. 31—*Triporopollenites* sp. 32—*Nudopollis* sp.

Figures 1 through 15 from Hell Creek Formation, Montana; figures 16 through 32 from McNairy Sand, Upper Mississippi Embayment.

analysis." Future use of palynomorphs of all categories for paleoecological studies in general, and facies recognition in particular, seems promising. Tschudy (1961) gave examples of the employment of a variety of palynomorph assemblages for facies inferences for some 10 Upper Cretaceous and lower Tertiary formations of the Rocky Mountain region. Assemblages most useful for indicating environmental and sedimentation situations, however, may have limited usefulness for correlation purposes. As Kuyl, Muller, and Waterbolk (1955) have pointed out, ". . . grains from plants belonging to coastal vegetation will exhibit greater quantitative fluctuations in marine sediments than pollen grains from species which occur farther inland. These are less sensitive to facies and are therefore more suitable for regional correlation." The possibility of recognizing progressive change in the generic composition of Cenozoic floras as a function of geologic age has been explored by several workers – for example, Reid, 1920; Szafer, 1946, 1954; Barghoorn, 1951; Traverse, 1955; Axelrod, 1957; Wolfe and Barghoorn, 1960. A continuing difficulty in all attempts at curvilinear representations of generic change versus geologic time attaches to the vexatious problem of reconstructing the total vegetation of a given formational unit, and many palynologists will agree with the somewhat pessimistic appraisal of Wolfe and Barghoorn (1960): "Unfortunately most fossil floras do not provide evidence from a wide range of plant parts and hence are commonly little more than tokens of the actual source vegetation of the time and place."

THE TRANSITION FROM MESOZOIC TO CENOZOIC

Toward the end of the Maestrichtian Stage the last of the great Cretaceous marine transgressions gave way to slow, worldwide episodes of regression, attended in some places by the prolonged development of swamp and mudflat environments and in other places by the onset of major orogenic disturbances. Relatively few areas of the world have records of continuous sedimentation spanning the Cretaceous-Paleogene interval, yet no dramatic geologic event seems to bisect the time boundary. The stratigraphy, moreover, of the Mesozoic-Cenozoic passage is not agreed on, particularly in regard to the stratigraphic position of the Danian. Although there seems to be growing agreement to place the boundary between the Maestrichtian and the Danian, Yanshin (1960) cautioned that differences in viewpoint will persist due to the essential transitional character of the Danian fauna and noted the occurrence of dinosauran (*Orthomerus*) remains in the Danian of the Crimea. Nevertheless, significant faunal changes – including the great extinction of dinosaurs, ammonites, and rudistid pelecypods – did occur at the Maestrichtian-Danian boundary and contributed to one of the noteworthy faunal gaps in the paleontological record (Newell, 1962). The paleobotanical record, on the other hand, seems to have no gap of comparable magnitude, so that the Cretaceous-Tertiary passage appears to have occurred without drastic vegetational change. This is not to deny, however, that floral changes reflected in stratigraphic floral breaks are encountered at the Cretaceous-Tertiary boundary; for example, Dorf (1942) estimated that only about 6 percent of the total number of megafloral species recorded from the Cretaceous type Lance Formation and the overlying Paleocene Fort Union Formation were known definitely to span the boundary. Stanley (1960) noted a major plant microfossil break between studied sections of the Hell Creek and the Fort Union Formations of South Dakota, with only 12 percent of the microfossils continuing across the passage. Leffingwell (1962) has pointed out the value of the pollen genera *Proteacidites* and *Aquilapollenites* in recognizing the Upper Cretaceous-Paleocene boundary in the type area of the Lance Formation, noting that "The Upper Cretaceous assemblages contain many species of *Proteacidites* and *Aquilapollenites* and numerous specimens of a characteristic tricolpate grain whose colpi are located between its rounded apical angles. In the lower Paleocene assemblages the characteristic tricolpate species is not present, only one species of *Aquilapollenites* can be observed." On the other hand, of the approximately 200 species of palynomorphs found by Newman (1962) in northwestern Colorado, only 16 species were restricted enough in distribution to be of value for correlative and stratigraphic studies bearing on the Cretaceous-Tertiary boundary of the area.

Data from Paleocene distribution of corals and

bauxite soils, and from oxygen-isotope paleotemperature measurements adduced in support of an inferred general cooling of the climate during that epoch, are not supported unequivocally by paleobotanical evidence. Relatively few Paleocene floras in fact have been described, and fewer still are free of suspicion of accompanying stratigraphic or paleontologic inadequacies; for example, in Western United States, a region noteworthy for the development of Paleocene leaf floras, many, perhaps most, of the floras are in need of revision (Dorf, 1955). Leaf floras of the Northern Hemisphere give evidence of extensive Paleocene temperate forests of *Populus, Platanus, Metasequoia,* and *Ginkgo,* with elements of *Cercidiphyllum, Trochodendron, Zizphus, Viburnum, Alnus, Aralia,* and *Vitis,* extending well into present arctic regions (Koch, 1959; Sachs and Strelkov, 1961). Concerning the Southern Hemisphere, Couper (1960a) expressed the view that ". . . it is almost certain that vast areas of the southern temperate and subantarctic zones were covered throughout the greater part of the upper Mesozoic and Tertiary by *Nothofagus* and podocarp forests, essentially similar to the forests of New Zealand and South America." The evidence, such as it is, bearing on the character of Paleocene tropical floras hints at strong dominance by evergreen dicotyledons and palms, with lesser representations of ferns, grass, and *Ephedra.* Pollen floras contain many kinds of unidentified dicotyledonous pollen, as well as grains showing indications of relationships within the Bombacaceae, Apocynaceae, Sapindaceae, Nymphaeaceae, and Liliaceae.

PALYNOLOGICAL CHARACTERIZATION OF THE PALEOCENE

Central Europe

In summarizing the palynological record of the earliest Tertiary floras of Central Europe Krutzsch (1957) noted that "The transition into the Tertiary takes place without any significant change. The Normapolles are somewhat shifted into the background by new form groups dominating the Paleogene." A typical Paleocene assemblage of this province might yield porate dicotyledonous pollen of suspected Juglandaceae, Myricaceae, Myrtaceae, and Haloragaceae affinities, including such form genera of Pflug as *Extratriporopollenites, Intratriporopollenites, Subtriporopollenites,* and *Stephanoporopollenites.* In company with these there may be found also *Gleicheniidites, Cicatricosisporites, Lygodiumsporites, Inaperaturopollenites, Trudopollis, Sporopollis,* and *Plicapollis*—palynomorphs of plants persisting from Senonian and earlier times, together with palm pollen and, in low percentages, pollen representing Nyssaceae, Ulmaceae, Betulaceae, Sapotaceae, Tiliaceae, and Aquifoliaceae, whose frequencies increase through the Eocene and Oligocene. Descriptions, stratigraphic ranges, and illustrations of these types are found in Thomson and Pflug (1953), Krutzsch (1957), and Durand (1962). The absence or low incidence of winged conifer pollen seems to be a characteristic feature of Central European pollen floras of Paleocene time, although several species of *Pinus* and *Cedrus* pollen were described by Zaklinskaya (1957) from Cretaceous-Paleocene sediments in the Irtysh basin of southern Siberia at approximately the same latitudinal range as the European floras. Kedves (1960, 1961) attempted a reconstruction of the flora of Sparnacian (latest Paleocene-early Eocene) time from fossil pollen obtained from borings in the Hungarian Dorog basin. Five vegetational zones were recognized, developed presumably under the altering environments of strand-line changes. *Sequoia* forests were thought to cover the highlands surrounding the marshy zones. (see fig. 16-6.) Kremp and Gerhard (1956) and Kremp, Neavel, and Starbuch (1956) earlier had detected somewhat similar floral associations in Paleocene assemblages from Midwestern United States, ostensibly reflecting changes in swamp environments. These studies bring to light the prevalence during the Paleocene of nonvesiculate conifer pollen (*Inaperturopollenites, Sequoiapollenites, Taxodiaceaepollenites*), palm pollen (*Palmaepollenites, Arecipites*), Betulaceae pollen (*Alnipollenites*), and Myricaceae pollen (*Myricipites, Triatriopollenites*). A recurring dominance of Palmae interpreted from percentage variations of monocolpate pollen, led Kedves to conclude that the climate during much of Sparnacian time was undoubtedly tropical. Chandler (1961) came to the same conclusion regarding the Sparnacian climate on the basis of leaf, fruit, and seed fossils recovered from the Woolwich and Reading beds

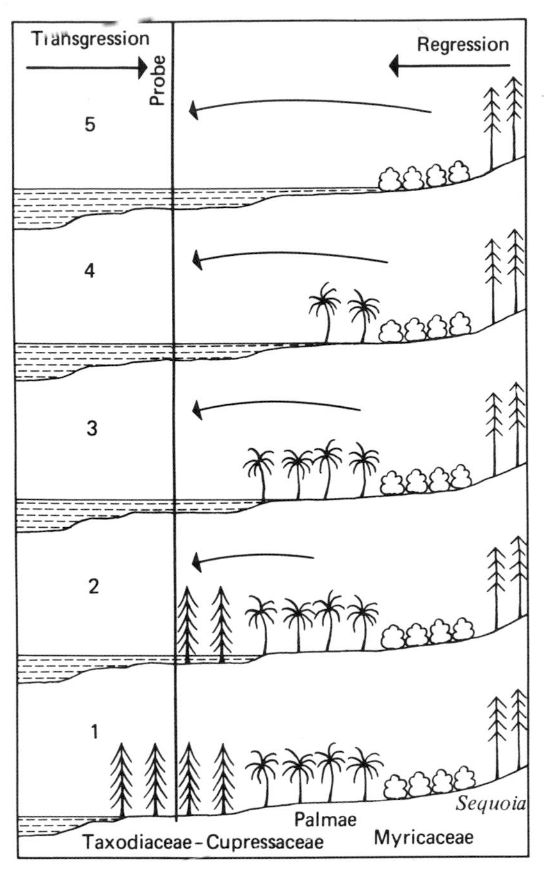

Figure 16-6. Vegetational zones developed under altering strand-line environments (after Kedres, 1960).

of the London Clay. The flora of the middle Paleocene Thanet beds, although somewhat inadequate for paleoclimatic inference, similarly is more suggestive of a warm (subtropical) rather than a temperate or boreal climate.

North America

Widespread deposits of coal and lignite in the Dakotas, Montana, and Wyoming of Western United States give evidence of the development of extensive swamp environments after the withdrawal of the Late Cretaceous interior sea. These organic sediments, especially those from the several members of the Fort Union Formation (Paleocene), tend to be rich in spores and pollen and have been used for several palynological investigations. Wilson and Webster (1946), described a small assemblage of fern and *Sphagnum* spores in company with *Sequoia, Typha, Sparganium,* hickory and oak pollen in a Fort Union coal from Montana. Gerhard (1958) recorded 68 species under 45 genera of spores and pollen from a South Dakota lignite deposit and noted some similarities between the American Paleocene pollen floras and the German Tertiary brown-coal floras. A few investigations have concentrated on floristic interpretations of Paleocene spore-pollen floras by way of inquiring into the possible relationships between coal-type and precursory-source vegetation. Kremp, Neavel, and Starbuch (1956) saw a fluctuating, if not a cyclic, change in the vegetation of certain Paleocene swamps of the Dakotas in which the dominant climaxes changed between conifer (Taxodiaceae-Cupressaceae) stands and shrubby angiosperm (Myriacaceae) cover. The conifers contributed woody anthraxylous constituents, and the angiosperms contributed cuticular and resinous constituents in the formation of these Paleocene coals. These authors suggested that oak pollen, whose percentage frequencies were not correlative with the frequencies of the coniferous and myricaceous pollen, was blown into the swamp from forests on the surrounding highlands. Kremp and Kovar (1960) amplified this suggestion, however, by pointing out that some oak species might have been able to grow in the Tertiary swamps as indeed grows *Quercus michauxii* in the southeastern swamps of the United States. Groot and Groot (1962) described a partial collection of about 30 palynomorphs from the marine Brightseat Formation of coastal plain, Maryland. On the basis of Foraminifera and megafossils the Brightseat has been assigned a Paleocene age and correlated with the Danian of Europe. Its pollen record, dominated by dicotyledonous pollen, shows a relationship with (a) the Late Cretaceous pollen floras of Europe and eastern North America in the possession of certain short-axis Normapolles species of *Triatriopollenites, Extratriporopollenites,* and *Plicapollis* and (b) typical early Tertiary Northern Hemisphere pollen floras, with the inclusion of *Ulmipollenites, Cupuliferoipollenites, Caryapollenites,* and *Tiliaepollenites.* (See pl. 16-3.) A palynological break with the Upper Cretaceous may be represented by the absence in the Brightseat floras of certain podocarpoid conifer pollen, with rugulate proximal surfaces and frilled distal surfaces, described under such names as *Abietineaepollenites ornatus* Groot, Penny, and Groot or as species of *Rugubivesicu-*

lites Pierce (pl. 16-2, figs. 19 and 20), recorded from the Cenomanian to the Senonian of North America. Groot was of the opinion that the angiosperm pollen of the Brightseat Formation indicated neither cool nor tropical but rather a temperate or warm temperate climate for eastern coastal America during the Danian Stage. Monosulcate pollen suggestive of Palmae was not recorded.

Australia and New Zealand

The Australian flora, over the long stretch of time from Late Cretaceous to mid-Tertiary, appears to have changed very little in fundamental character. Precursory elements of both the Antarctic component (such as *Dacrydium*, *Podocarpus*, and *Nothofagus*) as well as the tropical component (*Symplocos*, *Elaeocarpus*, and members of the Olacaceae and Sapindaceae) in the modern Australian flora, although apparently present in early Tertiary time, give no clear evidence of major migration into the region from any one particular direction (Burbidge, 1960). Paleocene plant-microfossil assemblages remain dominated by podocarp pollen, although frequencies of dicotyledonous pollen (mostly of unkown botanical affinities) may run as high as 64 per cent (Cookson and Dettman, 1959). A Paleocene fossil-pollen assemblage recovered from a core sample in eastern Victoria included 7 percent pteridophyte spores, 77 percent conifer pollen, and 16 percent dicotyledonous pollen (of which 3 percent was *Nothofagus*). Paleocene to lower Eocene age in Australian deposits may be indicated by the presence of *Dacrydium balmei* Cookson, *Triorites edwardsii* Cookson and Pike *Tricolpites gillii* Cookson, 1950. Burbidge (1960) has made the tentative suggestion that most subtropical climates persisted in austral regions until upper Miocene times. Couper (1960b) regarded the New Zealand Danian as Late Cretaceous, although Hornibrook (1962) indicated the intent of the New Zealand Geological Survey to classify the Danian as Tertiary. Pollen floras of both the Teurian Stage of the Mata series (Danian) and the Dannevirke Series (Paleocene-lower Eocene) remain dominated by long-ranging podocarpaceous species. *Triorites harrissii* Couper, a wide-ranging Tertiary species of possible *Casaurina* affinity, is rare in the Danian but abundant in the "Paleocene." Couper described 15 species of plant microfossils from the New Zealand Teurian Stage and the Moeraki Formation, including the following:

Podocarpidites otagoensis
Monosulcites granulatus
Proteacidites scaboratus
Nothofagus waipawaensis
Leptospermum sp.
Metrosideros sp.
Tricolpites coprosmoides
Triorites harrisii

The low frequency of *Triorites harrisii*, together with the restricted occurrence of *Nothofagus waipawaensis*, serve to identify the New Zealand Danian. As of 1960 only a small number of samples had been examined from the Dannevirke Paleocene. The small fossil-pollen assemblage obtained from the Hawkes Bay vicinity of North Island included the following:

Dacrydium aff. *cupressinum* Soland. ex Forst
Dacrydiumites mawsonii Cookson
Proteacidites annularis Cookson
Cupanieidites orthoteichus Cookson and Pike
Ascarina sp.
Anacolosidites acutullus Cookson and Pike
Triorites harrisii Couper

Ephedra notensis Cookson has been recorded from the Maestrichtian to the upper Pliocene, but its frequency is never high. Of special interest in the Australian-New Zealand Late Cretaceous-early Tertiary flora is the rare occurrence of pollen suggestive of *Anacolosa* of the pantropical family Olacaceae. Cookson and Pike (1954) pointed out that the presence of *Anacolosa* in the Indian-Malayan flora and its absence from the modern Australian flora provides another example of a migration toward the equator of plants that had a more extensive southern distribution during the Tertiary Period. The possibility of such a retreat having occurred in the Australian Tertiary *Cinnamomum* flora is discussed by Burbidge (1960).

South America

Numerous palynological investigations have been made on Mesozoic and Cenozoic sediments in South America in connection with oil geology studies, but data derived from these investigations are for the most part unpublished and remain in petroleum-company files. Present knowledge of the classification and the paleo-

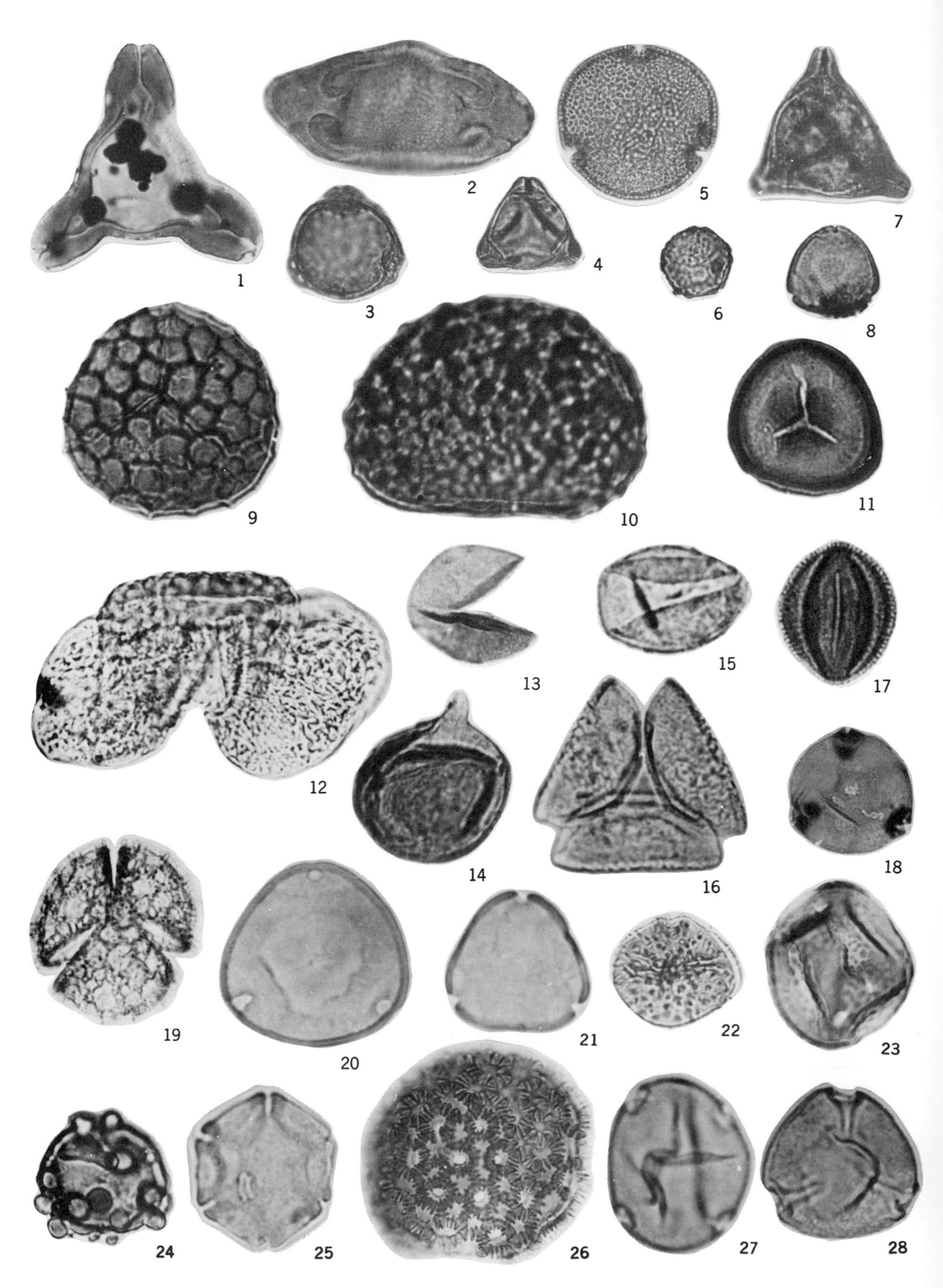
1
2
3
4
5
6
7
8
9
10
11
12
13
14
15
16
17
18
19
20
21
22
23
24
25
26
27
28

ecological and stratigraphic usefulness of Latin American spores and pollen thus rests mainly on comparatively few published papers, for example, Van Der Hammen, 1954, 1957a, 1957b; Kuyl, Muller and Waterbolk, 1955; Norem, 1955; Muller, 1959; and Van Der Hammen, Wymstra, and Leidelmeyer, 1961. According to Van Der Hammen, Paleocene age determination for the earliest Tertiary pollen flora in Colombia is made possible by floral correlations into the Hato Neuvo Formation whose equivalent in Venezuela is marine and is assigned a Paleocene age on the basis of Foraminifera. The Colombian Paleocene spore-pollen floras show marked quantitative changes among certain palm pollen from the base to the top of all sections studied. No gymnosperm pollen is recorded, and the abundance of *Psilatriletes* fern spores, which had been so high in the Maestrichtian, rarely exceeds 20 percent and usually is much less. *Proxapertites operculatus*, the Tertiary representative of the modern palm genus *Astrocaryum*, has its greatest representation in the Paleocene. The Gramineae is represented by *Monoporites lisamae*, the Bombacaceae by *Tricolporites annuae*, and the Sapindaceae possibly by *Syncolporites lisamae*. Angiosperms thus constitute practically the entire Paleocene vegetation, and the flora shows a definite trend toward the existing South American aspect. The frequency distribution of Van Der Hammen's "*Monocolpites*" *medius* group is interpreted by him as indicative of climatic change. The genera of recent palms (*Geonoma*, *Sabal*, *Attalea*, and *Acromia*) whose pollen is closely similar to that of the *M. medius* group are relatively wide ranging and, by inference, more resistant to temperature declines. These palms, and their Tertiary representatives, are thought to have attained maxima during temperature recessions capable of checking the increase of less resistant species. Pollen assemblages recovered by Van Der Hammen, Wymstra, and Leidelmeyer (1961) from borehole samples taken under the bauxite workings near Georgetown, Guyana, made possible a Paleocene age determination on the basis of comparisons with Van Der Hammen's Colombian pollen diagrams. Species occurring in both the Guyanan and Colombian sediments were

Proxaperites operculatus
"*Monocolpites*" *medius*
M. franciscoi
M. proxaperturoides
Diporites magdalenensis
Syncolporites lisamae
Polyplicadites vanegensis
Stephanocolpites lisamae

On the basis of the Tertiary distribution of "*Monocolpites*" *medius* and by extending certain data of Emiliani (1955, 1956) and Norem (1956), Van Der Hammen has suggested that for South America ". . . it seems possible that the deeper temperature minima were lower than the present-day average temperatures, and the minimum at the base of the Tertiary may even have been considerably lower."

Asia

Kuprianova (1960) noted the occurrence of pollen of *Cedrus*, *Myrica*, *Ilex*, *Nyssa*, *Liquidambar*, *Castanea*, *Platycarya*, the Rhamnaceae, Myrtaceae, and Cunoniaceae from Kazakhstan in Central Asia. Concerning the age of this pollen flora she observed that the grains were recovered from sediments, ". . . defined by some palynologists as Paleocene." The angiosperm pollen is stated to have been derived from "sclerophyl-

Plate 16-3. Paleocene pollen floras. (All figures × 500.) 1—*Extratriporopollenites* sp. 2—*Latipollis conspicuus* Groot and Groot. 3—*Triatriopollenites convenus* Groot and Groot. 4—*Plicapollis silicatus* Pflug. 5—*Tiliaepollenites reticulatus* Groot and Groot. 6—*Ulmipollenites minor* Groot and Groot. 7—*Extratriporopollenites audax* Pflug. 8—*Triatriopollenites marylandicus* Groot and Groot. 9—*Lycopodiumsporites* sp. 10—*Polypodiisporites favus* R. Potonié. 11—*Sphagnumsporites* sp. 12—*Abietineaepollenites* sp. 13—*Taxodiaceaepollenites* sp. 14—*Sequoiapollenites* sp. 15—*Monosulcites* sp. 16—*Cupaneidites* sp. 17—*Tricolpites* sp. 18—*Kurtzipites* sp. 19—*Tricolpites* sp. 20—*Caryapollenites* sp. 21—*Momipites* sp. 22—*Monoporopollenites* sp. 23—*Ulmipollenites* sp. 24—*Pistillipollenites* sp. 25—*Alnipollenites* sp. 26—*Erdtmanipollis* sp. 27—*Symplocospollenites* sp. 28—*Nyssapollenites* sp.

Figures 1 through 8: Brightseat Formation (Paleocene) Maryland; from J. J. Groot and Catharina R. Groot. Figures 9, 10, 11, 14, 17: Fort Union Formation (Paleocene) South Dakota; from J. E. Gerhard. Figures 12, 13, 15, 16, 18-28: Fort Union Formation (Paleocene) Montana; from B. D. Tschudy.

lous forms"—an assumption that, even if ascertainable, admits of more than one ecological conclusion. The inference, however, is that the Paleocene assemblage indicated a warm temperate or subtropical climate. The Kazakhstan record of *Liquidambar* (*L. thetidae* Kuprianova) is regarded as the earliest find of sweet-gum pollen. Kuprianova concluded that this genus and its Paleogene cohorts covered the coasts and islands of the Tethys. A record of *Platycarya* (Juglandaceae) in Paleocene sediments is interesting. To date *Platycara* pollen is known in the New World only from rocks of Eocene age (Hail and Leopold, 1960). The existing genus is monotypic and restricted to the temperate Orient.

The Eocene Plant World

The worldwide Eocene fossil record gives clear evidence of the modern and largely tropical to subtropical aspect of the early Tertiary vegetation. Chaney (1947) and Axelrod (1960) have summarized much of the data bearing on the distribution of early Tertiary floras, emphasizing the evidence of a widespread tropical zone ranging between latitude 45° to 50° north and south, with mild, continuously moist, temperate climates reaching into the polar regions. (See fig. 16-7.) Early Tertiary floras, like most floras, usually are indicative of lowland environments. This inherent bias must of course be weighed carefully in drawing broad paleoecological inferences. An important contribution of palynology in this regard has been the supplementation of the Eocene records of warm coastal and lagoonal vegetation with pollen indicative of extensive uplands of temperate forests (Gray, 1960; Ma Khin Sein, 1961; Jones, 1961). Climates remained broadly zoned throughout most of the Paleogene but began to modify into diverse types attending the cooling and drying of the Neogene (Axelrod, 1959). Discussion of the major hypotheses offered in explanation of these enormous climatic changes (epeirogeny, orogeny, continental drift, polar shift, crustal displacement, etc.) is not within the purview of this chapter. But if cause hypotheses remain still to be validated, there can be no disputing the profound selectional and distributional effects that followed the post-Eocene deterioration of climate.

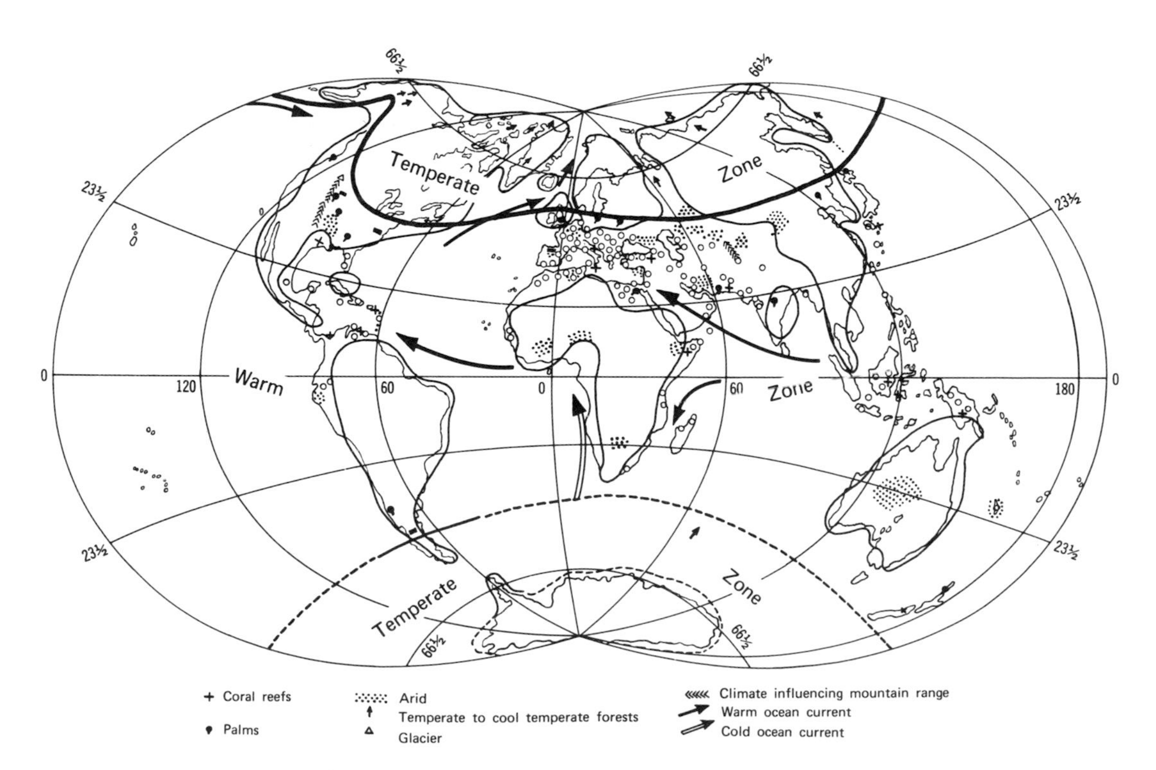

Figure 16-7. Climate and climate belts in the older Tertiary (Paleocene, Eocene, Oligocene).

PALYNOLOGICAL CHARACTERIZATION OF THE EOCENE

Many paleobotanists have observed that most well-preserved Tertiary plant megafossils, even from older Tertiary sediments, can be referred to existing genera. This observation probably does not apply with equal validity to early Tertiary pollen grains, but no reliable estimates of extinction-versus-survival percentages can be compiled until reference collections of extant-pollen slide preparations are greatly expanded in palynological laboratories throughout the world. Nevertheless, palynology offers probably more affirmative than negative evidence in support of the hypothesis that the leaves of some Tertiary plants evolved earlier than their reproductive structures in the attainment of the morphological characteristics of existing genera. Data on extinction percentages of Eocene fruit and seed genera are quite convincing in this regard; to quote but two: Reid and Chandler (1933) estimated that 75 percent of the fruit-and-seed genera of the Paleotropical lower Eocene London Clay were extinct, and Scott (1954) estimated that 66 percent of the Eocene fruit-and-seed genera of the Oregon Clarno formation were extinct. Judging from the number of palynomorph form genera described for older Tertiary pollen floras of Central Europe alone, pollen-genera extinctions might have approached these percentages. Certain major morphological classes of pollen grains are not recognized at all until Tertiary times (Couper, 1964), but the very complicated features of the pollen wall appear to have evolved fully by Eocene time, according to evidence obtained from electron-microscope studies by Ehrlich and Hall (1959) of the ultrastructure of certain lower Eocene (Wilcox) pollen grains. These authors conclude that "... no marked evolution in the basic structure of pollen exines has occurred since the Eocene." The broad steps in the morphological development of angiosperm pollen during the Cretaceous-Tertiary interval have been summarized by Couper (1964), who has selected 19 morphological groups of palynomorphs for time-stratigraphic correlations. (See pl. 16-4.) The more basic question of the origin of the angiospermous monosulcate and tricolpate morphological types from whatever may have been their pteridophytic or gymnospermous antecedents remains still in the realm of speculation.

Arcto-Tertiary Province

Early Tertiary plant megafossils from Holarctic recovery sites indicate the existence of widely developed forests of mixed deciduous hardwoods and temperate conifers. Although relatively few pollen floras of Paleogene age have been described from high-latitude northern sites, the palynological evidence in general agrees with that derived from leaf, fruit, and seed remains, and numerous genera are now known from both megafossils and microfossils, including commonly

Acer	*Fagus*
Carpinus	*Nyssa*
Juglans	*Liquidambar*
Pterocarya	*Engelhardtia*
Quercus	*Castanea*
Ulmus	*Carya*
Zelkova	*Ilex*
	Ailanthus

Tokunaga (1958) described and illustrated a flora of about 100 species of pollen and spores indicative of "warm climatic conditions" from the Ishikari (Eocene-lower Oligocene) group of Hokkaido, in which the following genera were included:

Sabal?	*Zelkova*
Cycas	*Liquidambar?*
Musa?	*Typha?*
Ginkgo?	*Johnsonia?*
Monoporopollenites	*Castanea*
Tricolpopollenites	*Nyssa*
Quercus	*Carpinus*
Salix	*Betula*
Ilex	*Myrica*
Acer?	*Engelhardtia*
Fagus	*Carya*
Ulmus	*Prunus?*
Corylus?	*Pterocarya*
Hedera?	*Juglans*
Alnus	*Carex?*

A small flora of some 52 spores and pollen species from West Spitzbergen coal seams of Paleocene-Eocene age described by Manum (1954) gives evidence of a conifer-deciduous hardwood-fern association in company with palm (*Sabal?*), *Potamogeton*, and Nymphaceae aquatics. *Acer*, *Salix*, *Alnus*, and (questionably) *Betula*

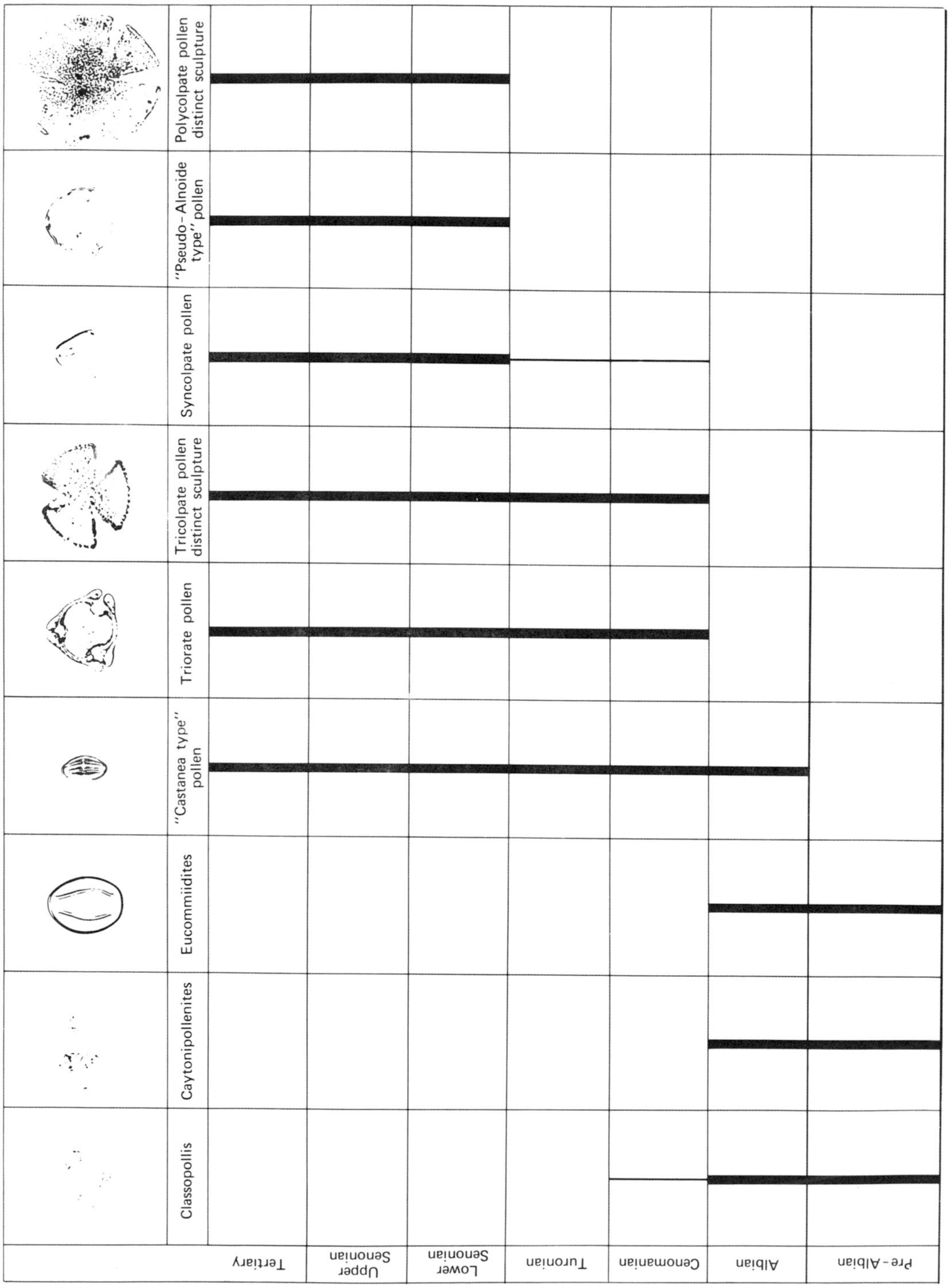

Polycolpate pollen distinct sculpture
"Pseudo-Alnoide type" pollen
Syncolpate pollen
Tricolpate pollen distinct sculpture
Triorate pollen
"Castanea type" pollen
Eucommiidites
Caytonipollenites
Classopollis

Tertiary
Upper Senonian
Lower Senonian
Turonian
Cenomanian
Albian
Pre-Albian

Plate 16-4. Stratigraphic distribution of some characteristic Cretaceous-Tertiary pollen groups. (Data from Couper, 1964.)

and *Liquidambar* were identified among the dicotylenous pollen, with Pinaceae (possibly *Abies* and *Sciadopitys*) and Taxodiaceae-Cupressaceae pollen for the conifers. Twelve species of fern spores were recorded but only by family rank (Polypodiaceae, Osmundaceae, and Schizaeaceae).

Antarcto-Tertiary Province

The fossil-pollen record from high-latitude southern localities is disappointing considering the continuing interest in the Antarctic and subantarctic landmasses as possible sources of evidence bearing on the time and site of angiosperm origin. The scarcity of spore and pollen floras is not due entirely to lack of investigative effort. Cranwell (1959) has alluded to the difficulties of Antarctic collecting and to the disappointments of barren samples. The first detailed palynological account of an Antarcto-Tertiary deposit was made by Cookson (1947) on lignites from subantarctic Kerguelen Island. Unfortunately there is no clear geological evidence of the age of the Kerguelen plant beds, and only a general "Tertiary" or at best "lower Tertiary" age is indicated by the small collection, which consists of a few species of angiosperm pollen (2 monocotyledons and 3 dicotyledons), 10 pteridophyte spores (mostly ferns) and 6 conifer pollen species (Araucariaceae and Podocarpaceae). A noteworthy feature is the absence of *Nothofagus* pollen. Bunt (1956) examined the pollen of 38 species of "grasses and herbs" growing on subantarctic Macquarie Island (latitude 54° 30′S.) and found these to be distinctly different from 17 "main fossil types" (not named) of pollen recovered from lignite. The microfossil collection lacked *Nothofagus* and gymnosperms, and showed no apparent affinities with pollen assemblages described by Couper (1960b) from New Zealand. Bunt speculated, however, that the Macquarie fossil-pollen flora might be closely related to the Tertiary floras of the Antarctic. Cranwell (1959) recovered an interesting fossil-pollen assemblage from a small rock sample obtained from Seymour Island, off the Palmer Peninsula (latitude approximately 64°S.) in the Weddell Sea. The flora is described as a mixed one, containing grains derived from possible Senonian deposits with a younger component of wind- and water-borne spores and pollen of Tertiary age. The exact age of this younger component was not determined, but a preMiocene dating is suggested by the absence of Gramineae, Cyperaceae, and Compositae pollen, which are usually present by the Miocene. Podocarp and araucarian conifer pollen, together with *Nothofagus*, dominate the flora. Pollen of Myrtaceae, Proteaceae, Loranthaceae, Oenotheraceae, Cruciferae, and possibly Winteraceae and Cunoniaceae or Elaeocarpaceae are present in low numbers. Spores of Cyatheaceae and Schizaeaceae also were recovered.

Palynomorphs from calcareous rocks collected well within the Antarctic Circle were described by Cranwell, Harrington, and Speden (1960). The samples were found amid morainic fragments at Minna Bluff and White Island, McMurdo Sound, approximately 78°S. latitude. The age of the rocks was considered as possibly Late Cretaceous-early Tertiary on the basis of resemblances to certain New Zealand sediments of that age, but no conclusive verification could be obtained from the contained palynomorphs. The assemblage is dominated by hystrichospheres and dinoflagellates. Pollen are scarce and small, although well preserved. They include *Nothofagus* and some palm and proteaceous forms. The authors expressed the opinions that the age of the flora was not likely older than Late Cretaceous nor younger than Oligocene and that pollen size and frequency, and the association with hystrichospheres and dinoflagellates, might indicate a deposition environment of offshore waters of normal salinity and low turbidity.

Neotropical Province

The first extensive flora of the Neotropical Tertiary is that of the Eocene Wilcox group of the Gulf Coastal Plain of Southern United States. Known primarily from leaf remains—most of which were described by Berry (1916b, 1930), and to a lesser extent from microfossils (Jones, 1961)—the flora serves to characterize the Neotropical early Eocene. The plant families best represented from megafossils, in order of their magnitude, include the following:

Lauraceae	Juglandaceae
Leguminosae	Araliaceae
Moraceae	Apocynaceae
Rhamnaceae	Celastraceae
Sapindaceae	Palmae

Sapotaceae
Anacardiaceae
Myrtaceae
Combretaceae
Rutaceae
Meliaceae
Sterculiaceae

Jones reported that the commonest pollen constituents were pine and oak. Berry (1937), in a somewhat overlooked paper, summarized his views on the environmental conditions prevailing at the Wilcox deposition sites by observing: "This flora is largely coastal and indicates a warm temperature climate and an abundant rainfall, more tropical in its facies than that of the late Upper Cretaceous flora which preceded it at this same region."

Later studies on the Wilcox flora (Brown, 1944; Sharp, 1951; Jones, 1961) have added additional temperate genera and have tended to interpret the Wilcox as more temperate than Berry inferred in his 1916b paper. Berry, however, had had second thoughts regarding Eocene floras and their attributed climates after a sojourn in the tropics, noting (1924): " . . . I came back from the region with the conclusion that none of the fossil floras of the temperate zone that paleobotanists have termed tropical are in the strict sense of the word 'tropical'."

Sharp (1951) saw in the mixture of subtropical and temperate species a resemblance between the Wilcox flora and the present vegetation of the Gulf coast, escarpments, and mesas of eastern Mexico, estimating that 93 genera (68 percent of the 137 Wilcox genera) still persist in these Mexican habitats. Jones' (1961) palynological investigation of Wilcox sediments from Arkansas yielded 62 spore and pollen types, comprising a mixed assemblage of tropical, subtropical, and temperate genera, including *Anacolosa*, *Manilkara*, *Engelhardtia*, *Symplocos*, *Myrica*, and *Carya*, in company with the pine and oak pollen. The Arkansas sediments were deposited under brackish-water conditions, judging from the recovery of hystrichosphaerids and dinoflagellates. Jones interpreted the pollen data as indicating a humid coastal plain with adjacent highland environments similar to those described by Sharp for Mexico. There seems, then, to be considerable evidence and agreement concerning the environmental conditions prevailing at the Wilcox deposition sites.

One lingering point of difference remains between Berry's interpretation and those of Sharp and others concerning the affinities between Eocene and living floras. Berry's 1937 view was that the early Tertiary floras of Southern and Southeastern United States more closely resembled the present flora of northern South America rather than those of Central America or the Antilles, the implication being, we may presume, that the flora of the Central American migration route[2] suffered more widespread selectional pressures (under whatever post-Laramide disturbances affected the region) than did the flora of northern South America, which retains its essential Tertiary character to this day. Sharp's estimate that 68 percent of the Wilcox genera persist in the eastern Mexican region would seem to challenge this presumption. There is, of course, the possibility that the Mexican flora is not so much a "persisting" flora as one reestablished by secondary migration, possibly from Ozarkian, Appalachian, or other centers.

The extent to which the floras of much of the Southern and Southeastern United States retain a Tertiary character is somewhat remarkable. Opinion is divided, however, concerning the interpretation of this Tertiary composition. Sharp (1951) estimated that 82 Wilcox genera are still extant in this region, and Cain (1943) estimated that 27 percent of the Cretaceous and Tertiary genera are still found in the flora of the Great Smoky Mountain National Park of North Carolina and Tennessee, a region characterized by Sharp as distinctly temperate and in part boreal. Wolfe and Barghoorn (1960) expressed doubts concerning the "relict" nature of the flora, suggesting that some of the genera had not entered the region until later Tertiary time and many others not until the Pleistocene. Future palynologic and comparative floristic studies undoubtedly will contribute valuable data bearing on the origin, radiation, and contraction of the early Neotropical Tertiary flora. The Gulf coastal region of North and Central America, and the Roraiman and Pacaraiman regions of the Guianas and Venezuela would in this regard appear to be of considerable significance. These latter regions were considered by Croizat (1952) to be of major importance as secondary centers of angiosperm distribution.

[2]Berry (1937) estimated that about 60 percent of the Wilcox genera, with over 100 species, entered Southern United States from equatorial America.

The middle Eocene Green River flora is yet another Neotropical Eocene flora known from both megafossils (about 135 recognizable species) and microfossils (43 species in the Wodehouse (1933) list). The microfossils consist overwhelmingly of pollen from anemophilous trees and shrubs indicative of a temperate assemblage. Brown (1934) noted that the association of hickory, walnut, sweetgum, *Ailanthus*, and linden suggested an original site of not more than 3,000 feet altitude (and probably considerably less). The vegetation contributing to the Green River pollen flora grew under less well watered conditions than prevailed during an earlier and later mid-Eocene stage. The species described by Wodehouse (1933) are the following:

Cycadopites sp.
"Dioonipites" sp.
Pinuspollenites (*Pinus*) *strobipites*
P. scopulipites
P. tuberculipites
Abietineaepollenites (*Picea*) *grandivescipites*
Abies concoloripites
Cedripites eocenicus
Tsugaepollenites (*Tsuga*) *viridifluminipites*
Abietipites antiquus
Taxodieaepollenites (*Taxodium*) *hiatipites*
Glyptostrobus vacuipites
Cunninghamia concedipites
Ephedripites (*Ephedra*) *eocenipites*
Potamogeton hollickipites
Arecipites punctatus
A. rugosus
Peltandripites davisii
Smilacipites molloides
S. herbaceoides
S. echinatus
Liriodendron psilopites
Caryapollenites (*Hicoria*) *viridifluminipites*
C. juxtaporipites
Juglans nigripites
Engelhardtioipollenites (*Engelhardtia*) *corylipites*
Myricipites dubius
Salix discoloripites
Alnipollenites (*Alnus*) *speciipites*
Betulaepollenites (*Betula*) *claripites*
Carpinus ancipites
Momipites coryloides
Ailanthipites berryi
Rhoipites bradleyi
Talisiipites fischeri
Vitipites dubius
Tiliaepollenites (*Tilia*) *crassipites*
T. vescipites
T. tetraforaminipites
Myriophyllum ambiguipites
Ericipites longisulcatus
E. brevisulcatus
Caprifoliipites viridifluminis

The succeeding middle Eocene Claiborne Group contains another important flora known now also from palynological studies (Gray, 1960). The group includes some of the most fossiliferous sediments in the world (Murray, 1961), but its megaflora is not so rich as that of the Wilcox. (See table 16-3.) This disparity is somewhat surprising but may be due, as Berry (1924) thought, to an "imperfection in the geologic record." The Claiborne vegetation that grew in the vicinity of the deposition sites was not markedly different in composition from the Wilcox, and Berry (1937) postulated an essentially similar environment for both, with, however, warming climates for the Claiborne Stage. One difference between Wilcox and Claiborne is the apparent absence in the latter of leaf fossils of such temperate genera as *Betula*, *Fagus*, *Sassafras*, *Staphylea*, and *Comptonia*. The pollen flora described by Gray (table 16-4) reaffirms in general the megafloral evidence concerning environmental conditions prevailing around the Claiborne deposition sites, with two important extensions: the addition of *Ephedra* and of temperate deciduous hardwoods. Gray offered two explanations for the occurrence of *Ephedra*, generally regarded as a desert or steppe plant, in a humid, coastal flora. The Tertiary *Ephedras* may not have had the "limiting adaptive relations" of recent types, or the Claiborne *Ephedras* may have existed on dunes or sandflats adjacent to the coast. The presence of pollen from a dozen or more genera of temperate deciduous plants in association with microfossils and megafossils indicative of a subtropical, broadleaf, strand, and coast flora seems at first glance a far more serious floristic discordancy. Gray's interpretation is instructive and emphasizes an important principle of palynological analysis. Avoiding the snare of seeing the coastal Claiborne flora, perforce, as considerably more temperate than the majority of its fossils proclaim, Gray noted that the total percentage of temperate genera pollen was small (about 20 percent of the total pollen flora), although most of the genera represented were anemophilous. Her conclusion favored distant transport from coeval upland vegetation; in this instance from forests on the foothills and slopes of the Eocene Appalachians.

***Table 16-3.* Middle Eocene (Claiborne) Angiosperm Genera Known from Pollen (P), Leaf-Impressions (L), or Both (+)**[1]

Alnus (P)	*Liriodendron* (P)
Arundo (L)	*Mimusops* (L)
Betula? (P)	*Myrica* (L)
Carya (+)	*Nectandra* (L)
Castanea (P)	*Nyssa* (P)
Cedrela (L)	*Octea* (L)
Celtis (+)	*Oreopanax* (L)
Coccoloba (L)	*Ostrya-Carpinus* (P)
Combretum (L)	*Persea* (L)
Conocarpus (L)	*Pisonia* (L)
Diospyros (L)	*Quercus* (P)
Engelhardtia-Alfaroa (P)	*Sapindus* (L)
Fagus (P)	*Sophora* (L)
Ficus (L)	*Sterculia* (L)
Ilex (P)	*Terminalia* (L)
Inga (L)	*Thrinax* (L)
Juglans (+)	*Tilia* (P)
Laguncularia (L)	*Ulmus* (cf. *Zelkova*) (P)
Liquidambar (P)	*Zizyphus* (L)

***Table 16-4.* Middle Eocene (Claiborne) Families of Seed Plants Known from Pollen (P), Leaf-Remains (L), or Both (+)**[1]

Ephedraceae (P)	Juglandaceae (+)
Pinaceae (+)	Ulmaceae (P)
Taxodiaceae-Cupressaceae (+)	Moraceae (+)
Gramineae (+)	Fagaceae (P)
Palmae (+)	Betulaceae (P)
Lauraceae (L)	Tiliaceae (P)
Ebenaceae (L)	Meliaceae (L)
Sapotaceae (+)	Myricaceae (+)
Leguminosae (L)	Sapindaceae (+)
Hamamelidaceae (P)	Aquifoliaceae (P)
Nyssaceae (+)	Rhamnaceae (L)
Combretaceae (+)	Nyctaginaceae (L)
Myrtaceae (+)	Polygonaceae (L)
Onagraceae (P)	Araliaceae (L)

There are few published reports of palynological studies of western North American Eocene sediments, but evidence from the major megafloras of the Pacific coastal region and the adjacent interior basins seems clearly indicative that the pre-Pliocene forests were broadleaf evergreens growing under humid, warm temperature to subtropical climates. Leaf margin analyses by Wolfe and Barghoorn (1960) of 48 selected Cenozoic floras of western North America give confirmatory evidence of such humid, mesophytic, in part subtropical, forests existing at least until the mid-Miocene.

[1] Data from Berry (1937) and Gray (1960).

The Chalk Bluffs flora, a riparian and flood plain flora from middle Eocene deposits of the west slope of the California Sierra Nevada, may be selected as representative of a Pacific Coast Eocene province (MacGinitie, 1951). The flora, consisting of about 77 species under 60 genera, shows close resemblances to the Gulf Wilcox (closer, according to MacGinitie, to that flora than to the synchronous Claiborne). On the basis of the tree-shrub ratio, leaf-margin statistics, and comparative floristics, MacGinitie concluded that the flora indicated a subtropical environment, with an average annual temperature of 60°F and an average annual rainfall of at least 60 inches at the lower altitudes. The flora's greatest resemblances lie with the floras of southeastern Asia and Southeastern United States (36 percent closely related species in common), and those of eastern Mexico and Central America (27 percent related species). Estella Leopold (unpublished report) has made generic and familial identifications of many of the pollen grains recovered from the Ione gravels of the Chalk Bluff sediments. Her list includes *Ilex*, *Nyssa*, *Alnus*, *Pinus*, *Betula*, *Carya*, *Juglans*, as well as characteristically subtropical genera, thus indicating again the same mixed assemblage of temperate and tropical genera noted for the Gulf Eocene flora. *Ephedra* pollen occurs in the Chalk Bluffs as well as in the Claiborne flora. (See pl. 16-5.) MacGinitie had noted the mixed facies in the Chalk Bluff megaflora and attributed its heterogeneity to the concurrence of three different floral components: a relict Cretaceous-Paleocene group whose modern counterparts are largely Asiatic, a migrant tropical American group, and a temperate upland group derived from the old Dakota Upper Cretaceous flora. Reviewing the composition and distribution of Eocene megafloras, MacGinitie identified the development of four botanical provinces by the latter half of that epoch in North America: (a) the Gulf Coast province, (b) the Central Rocky Mountain province, (c) the Pacific Coast province, and (d) the Alaska and Western Canada province.

Published data from South American Eocene spore-pollen floras, similarly, are disappointingly meager, and this deficiency is compounded by the incomplete state of our knowledge concerning the floras of modern tropical South America. Van Der Hammen's (1954) palynological study of late Mesozoic-early Tertiary Colombian coals and lignites—which is concerned primarily with evidence of correlative periodic vegetational changes and transgressions and tectonic movements—presents the picture of densely forested tropical climax vegetation, marked by the cyclic fluctuations and alternating dominance of ferns, palms, and unidentified dicotyledons. From his data on the age and distribution of certain palms, whose pollen seems to indicate relationship with extant South American genera, Van Der Hammen saw confirmation of Croizat's postulate of a "gate of Angiospermy" centering around the Guianan highlands, concluding that " . . . several (and probably many) important genera had their origin on, and radiated from, the old nucleus of the continent." Van Der Hammen based most of his paleofloristic calculations on certain selected spore and pollen groups occurring throughout the long interval from Maestrichtian to lower Miocene. These include the following:

- **"*Monocolpites*" *medius* group,** comprising *M. minutus, medius, heutasi, grandis* and nearly related forms, all related to some recent genera of palms.
- ***Monocolpites franciscoi* group,** comprising pollen of the "Mauritiaceae"
- ***Proxapertites operculatus* group,** comprising all the varieties of this species, and nearly related forms, all closely related to the Recent species *Astrocaryum acaule*
- ***Psilatriletes* group,** comprising all trilete spores with psilate walls
- ***Striatriletes susannae* group,** with forms belonging or closely related to *Aneimia*
- ***Verrumonoletes usmensis* group,** comprising the monolete-verrucate spores *V. usmensis* and *V. usmensoides*, and nearly related forms, probably belonging mainly to the Polypodiaceae
- ***Psilamonoletes tibui* group,** comprising all monolete-psilate spores
- **"Angiosperm group,"** including all angiosperm (mainly dicotyledonous) pollen not included in the palm groups.

Paleotropical Tertiary Province

Europe. Widespread Eocene subsidence of the continent resulted in the development of lacustrine, river-swamp, and embayment habitats. Repeated interplay of strand-line changes and luxuriant plant growth, continuing through the Miocene, produced considerable intercalations of vegetational debris, with continental sediments contributing to one of the major coal-forming periods of earth history (Brinkman, 1960). The commercially valuable brown-coal deposits, particularly in Germany, have been the object of intensive geological investigations, and since about 1930 pollen analysis has been able to make its contributions through zonation and correlation studies. Many of the important brown-coal palynologic investigations coincided with a phase in palynologic history marked by considerable preoccupation with nomenclative experiment. In consequence much of the accumulated literature is virtually a cemetery of abandoned taxonomic fashions, invalid types, quasi-trinomials, and involved new combinations. Taxonomic difficulties are expected in all pioneering efforts, but European brown-coal studies seem to have accumulated an unusually heavy burden of procedural irregularities, whose influence, unfortunately, has been worldwide (Brown, 1957). Hopefully, the critical review of original descriptions, together with a cumulative synopsis of spore and pollen genera, undertaken by Potonié (1956, 1958, and 1960) promises to rectify many of the taxonomic deviations encountered during the formative years of applied palynology. Behind the complexities of nomenclature and synonymy, however, the palynology of Eocene lignites is a record of extensive forest, thicket, and swamp communities growing under near tropical conditions. Among the first studies of European Tertiary pollen and spore floras were those made by R. Potonié (1931, 1934) on lignite beds of middle Eocene age in the Geisel valley near Halle, Germany. About 100 species were described; the spores under "Sporites" H. Potonié, and the pollen under "Pollenites" R. Potonié. In subsequent years new epithet combinations have clarified botanical affinities, and the complexion of the Geisel valley spore and pollen flora can be inferred from the following partial list:

Pteridophytes:

Reticulatisporites caelatus (R. Pot.) R. Pot.
Lygodioisporites solidus (R. Pot.) R. Pot.
Lycopodiumsporites agathoecus (R. Pot.) Thiergart
Anemiidites echinosporus (R. Pot.) R. Pot.

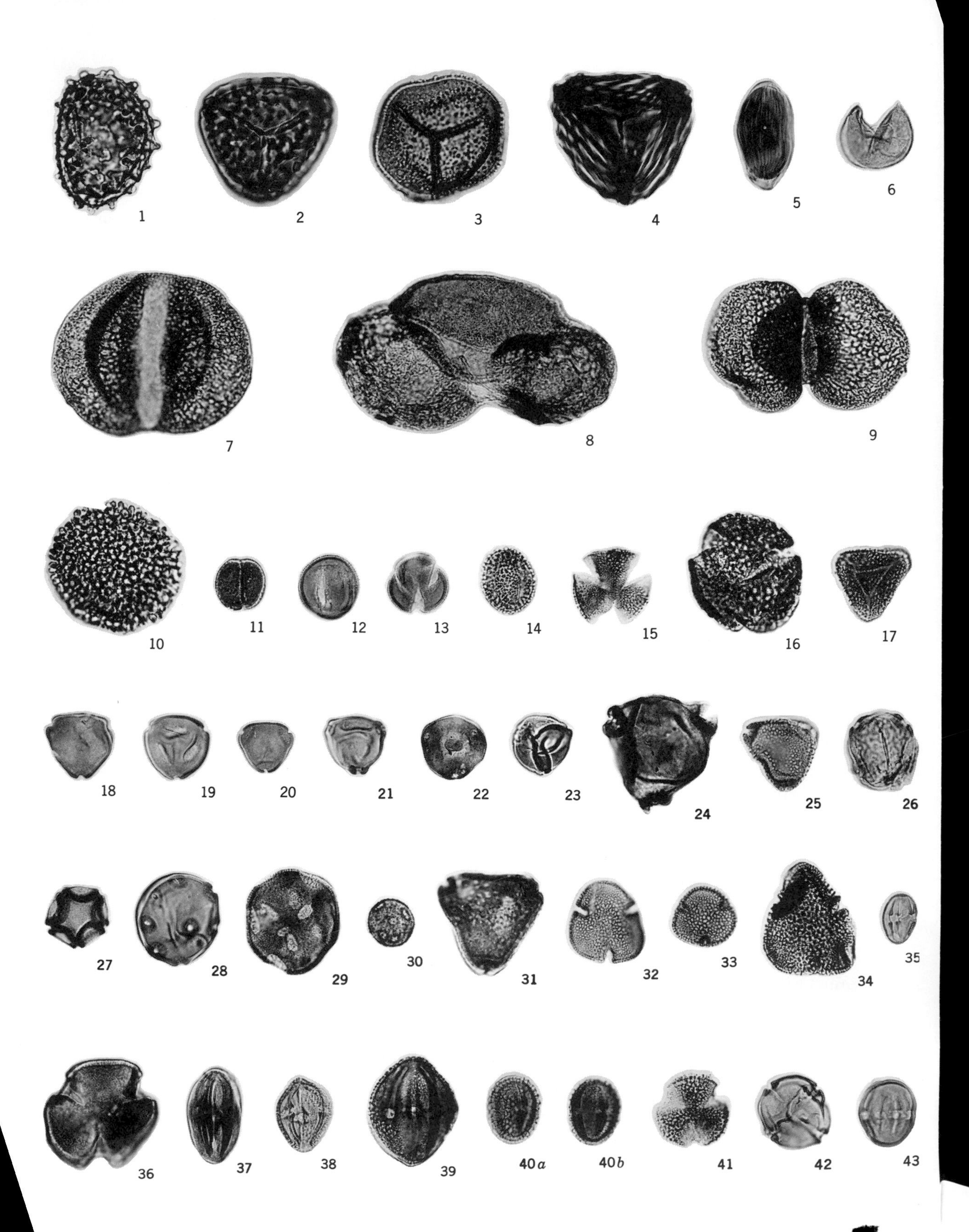

1
2
3
4
5
6
7
8
9
10
11
12
13
14
15
16
17
18
19
20
21
22
23
24
25
26
27
28
29
30
31
32
33
34
35
36
37
38
39
40a
40b
41
42
43

Conifers:

Pinuspollenites labdacus (R. Pot.) R. Pot.
Piceaepollenites alatus R. Pot.
Taxodiaceaepollenites hiatus (R. Pot.) Kremp

Monocotyledons:

Palmaepollenites tranquillus (R. Pot.) R. Pot.
Sabalpollenites areolatus (R. Pot.) R. Pot.
Smilacipites setarius (R. Pot.) R. Pot.

Dicotyledons:

Betulaepollenites microexcelsus (R. Pot.) R. Pot.
Alnipollenites verus R. Pot.
Myricaceoipollenites megagranifer (R. Pot.) R. Pot.
Sapotaceoidaepollenites megadolium (R. Pot.) R. Pot.
Anacolosidites efflatus (R. Pot.) Erdtman
Cornaceoipollenites parmularius (R. Pot.) R. Pot.
Nyssapollenites pseudocruciatus (R. Pot.) R. Pot.
Cupuliferoidaepollenites quisqualis (R. Pot.)
Ilexpollenites iliacus (R. Pot.) R. Pot.
Trudopollis pompeckji (R. Pot.) Pflug
Compositoipollenites rhizophorus (R. Pot.) R. Pot.
Rhoipites pseudocingulum (R. Pot.) R. Pot.
Araliaceoipollenites euphorii (R. Pot.) R. Pot.
Fraxinoipollenites pudicus (R. Pot.) R. Pot.
Engelhardtioipollenites punctatus (R. Pot.) R. Pot.
Tiliaepollenites indubitabilis R. Pot.

On the basis of plant-microfossil evidence alone the existence of the following major communities may be inferred for the Central European middle Eocene:

1. Swamp forests with *Taxodium* and *Nyssa*.

2. Riverbank and grove habitats of *Sabal* and other palms.

3. Shrub thickets of Myricaceae-Cyrillaceae species, Sapotaceae-Symplocaceae species, *Aralia*, *Ilex*, and polypodiaceous ferns.

4. Hardwood forests of fagaceous species, of *Fraxinus*, *Engelhardtia*, *Tilia*, *Ailanthus*, *Pterocarya*, *Carya*, and *Cornus*.

5. Conifer forests of *Sequoia* (or at least its Eocene taxodiaceous equivalent), *Pinus*, *Picea*, together with *Rhus* (vine or shrubs), and schizaeaceous ferns (e.g., *Lygodium*).

Whether the mixture of cool-climate elements such as *Picea*, *Betula*, *Tilia*, *Alnus*, and *Carya* with the more tropical forest types is to be explained by the "adjacent cool upland" hypothesis or the "drier climate, lower water" hypothesis is not known at this time.

Preliminary pollen studies by Ma Khin Sein (1961a) of the British lower Eocene London Clay, so well known for its rich tropical flora of over 230 identified fruits and seeds (Reid and Chandler, 1933), indicate remarkable compositional agreement with the megaflora in terms of the distribution of living counterparts. (See fig. 16-8.) As in the case of the Wilcox and Claiborne spore and pollen floras, the London Clay pollen flora also contains grains belonging to temperate families such as Betulaceae and Fagaceae. Ma Khin Sein (1961b) considered such pollen as possibly indicative of an upland element in the British Eocene flora, which is regarded as "underrepresented" in the fruits and seeds.

Perhaps the most interesting pollen found in the London Clay is *Nothofagus*, reported as comparing favorably with living *N. obliqua* of the Menziesii group. *Nothofagus*, together with pollen of the Restionaceae (recorded also by Erdtman, 1960), the Proteaceae, and other Southern Hemisphere types, continues to pose difficult problems in angiosperm distribution. Preliminary, of course, to the solution of distribution problems is the matter of correct taxa assignations, but here opinions concerning the validity of many identifications differ sharply; for exam-

Plate 16-5. Selected pollen and spores from the (middle Eocene) Ione gravels, Chalk Bluffs, Calif. Identifications by Estella B. Leopold; plate prepared by Bernadine Tschudy. All figures × 500. 1—*Humata* type. 2—*Lygodium* type. 3—*Osmunda*. 4—*Anemia*. 5—*Ephedra*. 6—*Taxodium* type. 7—*Picea*. 8—*Pinus*. 9—*Podocarpus*? 10—Euphorbiaceae cf. *Croton*. 11—Cunoniaceae? 12, 13—*Platanus*? 14—Ilex. 15—*Gordonia*? 16, 17—Undetermined dicotyledons. 18, 19—Betulaceae? 20—*Engelhardtia*? 21—*Betula*. 22—*Carya*. 23—*Platycarya*? 24—Onagraceae. 25—*Proteacidites*. 26—*Zelkova*, *Ulmus*, *Chaetopetala*, or *Planera*. 27—*Alnus*. 28—*Juglans*. 29—*Liquidambar*. 30—*Phytocrene*. 31—*Symplocos*. 32—Bombacaceae. 33—*Fremontodendron*. 34—Bombacaceae. 35—Undetermined dicotyledon. 36—*Ophiorhiza* type. 37, 38—Anacardiaceae? 39—Undetermined. 40a, b—*Viburnum*? 41—*Ilex*. 42—Meliaceae, *Cedrela* type. 43—Meliaceae or Sapotaceae.

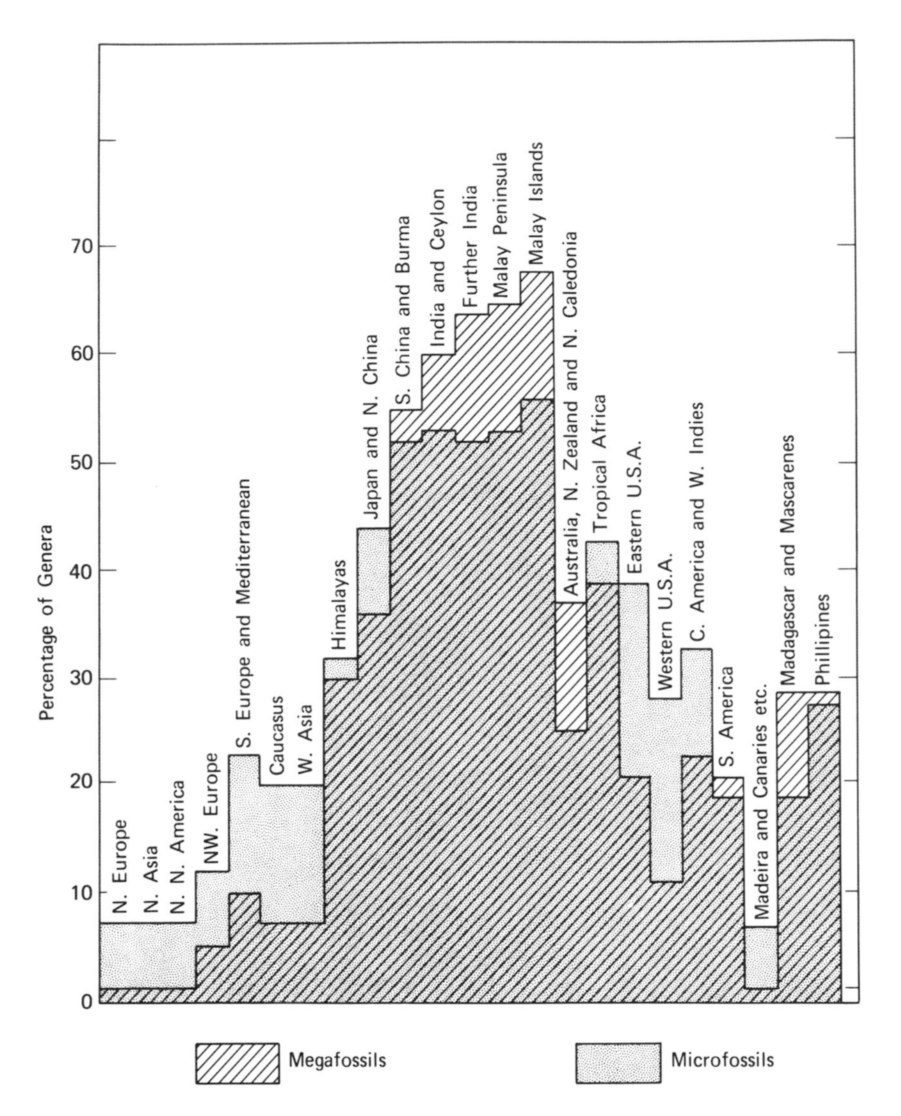

Figure 16-8. Present distribution of genera recorded in megafossil and microfossil floras of the London Clay. (From Ma Khin Sein, 1961a.)

ple, Ma Khin Sein (1961b) ruled out the possibility of misidentification of the British *Nothofagus*, but Couper (1960a) had declared categorically, "The Mesozoic and Tertiary floras of the Northern Hemisphere contain no trace whatever of *Nothofagus* . . . ."

The fossil record of the Proteaceae is similarly a matter of dispute. Cookson and Erdtman (1952) have cautioned against confusing some Proteaceae pollen types with pollen of such families as Onagraceae, Olacaceae, Rubiaceae, and Sapindaceae. Johnson and Briggs (1963) have questioned the proteaceous affinity ascribed even to fruits of *Lomatia* recorded by MacGinitie (1953) from the Oligocene Florissant beds of Colorado, observing that the fossil specimens exhibit a type of internal vascular system "unknown in the seeds of living Proteaceae." Concerning the radiation of the family, Burbidge (1960) has discussed the evidence for an "Antarctic" origin of the Proteaceae, whereas Ramsay (1963) on the basis of chromosomal studies of living species has concluded that the family originated in the Northern Hemisphere during the Cretaceous. On the assumption, however, that most of the generic determinations are correct and being mindful of the high proportion of Australian and Malasian genera in the London Clay flora, Ma Khin Sein proffered the suggestion that Eurasia served as the "bridge" for plant migration between the southern continents.

Australia and New Zealand. Although the

chronology of Australian Tertiary floral change is not known with any finality, there is evidence to suggest that the older podocarp forests gave way sometime between the Paleocene and middle Eocene to a dicotyledon-dominated "*Cinnamomum* flora," which developed with considerable uniformity across much of Australia. Some components of the new flora had a definite Australian character, others show relationships with modern New Guinean and South American counterparts. Leaf fossils indicating the assemblage of such genera as *Banksia*, *Lomatia*, *Hedycarya*, *Pittosporum*, *Eucalyptus*, *Nothofagus*, *Casuarina*, *Callitris*, and *Phyllocladus*, suggest an equable climate with uniformly distributed rainfall, comparable perhaps to conditions prevailing in mountain areas of New Guinea (Burbidge, 1960). The families Myrtaceae, Sapindaceae, Olacaceae, Haloragidaceae, Santalaceae, Casuarinaceae, and probably Proteaceae and Loranthaceae are represented by Eocene (and younger) pollen recorded and illustrated by Cookson and Pike (1954) from numerous localities in Australia and Tasmania. Prominent in their list are the pollen genera *Casuarinidites*, *Gunnerites*, *Myrtaceidites*, *Anacolosidites*, *Proteacidites*, *Santalumidites*, and *Cupanieidites*.

Couper's New Zealand studies (1960b) show that the podocarp species that had dominated the New Zealand forests into the Cenozoic were supplanted finally in late Eocene time by *Nothofagus*, especially *Nothofagus matauraensis* of the Brassi group, the group whose modern counterparts are confined to New Guinea and New Caledonia. The New Zealand pollen flora throughout the latter half of the Eocene is associated with pollen of the Bombacaceae, Sapindaceae, and Ephedraceae unknown in the Recent flora. Interestingly, Couper reports that *Elytranthe* (Loranthaceae), a mistletoelike parasite on *Nothofagus*, makes its first appearance in the pollen record contemporaneously with the attainment of dominance by *Nothofagus*.

PALYNOLOGICAL CHARACTERIZATION OF THE OLIGOCENE

Perhaps because of its transitional character, the interval between upper Eocene and lower Miocene, the 12 million or so years of mid-Tertiary time remains as one of the most troublesome epochs for paleontological studies. Much of the literature pertaining to Oligocene age determinations of invertebrate faunas and to the stratigraphic positions of vascular floras reflects an astonishing diversity of opinion (Eames and others, 1962).

North America

Two well-known floras, one Western, one Eastern, will serve to illustrate the character of North American floras considered to be of Oligocene age and to point out some of the problems attending the ascertainment of their stratigraphic positions. The Colorado Florissant Formation, an intermontane lacustrine deposit of volcanic ash and tuff, is well known for the abundance of its leaf and insect fossils preserved in paper shales. From the time of its first description nearly a century ago the floral age has been determined variously as Pliocene, Eocene, Miocene, and Oligocene. In the most thorough treatment to date of the flora MacGinitie (1953) favored an early Oligocene dating but opinion now seems inclined toward a late Oligocene dating.

Failure of agreement on age determination may of course be a transitory situation to be rectified as more information accumulates. However, the refinement of floral ages of mid-Tertiary series is not a problem entirely of collection sizes or systematics but one also of correlation, albeit indirect, with the standard European sections. For American floras considered to be of Oligocene age correlation problems are notorious; but then the Oligocene system itself was originated out of correlation difficulties. Beyrich founded the system in 1854 for marine beds in Germany and Belgium that he had earlier regarded as lower Miocene but which Lyell considered as upper Eocene. Later, when correlations between marine and continental facies were worked out, European mammalian faunas were dated, thus providing for age determinations of American Oligocene mammalian faunas independent of American marine faunas. Except perhaps for the White River beds and certain sectors of the Bridge Creek flora, American Oligocene plant beds are not intercalated with fossiliferous marine sediments, and their associated vertebrate evidence has been fragmentary (Konizeski, 1961). Although the stratigraphic positions of European Oligocene megafloras seem fairly well established, they are comparatively small and aid but little in age refinements of American Oligocene floras. Neither MacGinitie (1953) nor Becker (1961) referred to one of the better known Oligocene macrofloras, that of the Bembridge beds of the Isle of Wight, in their discussions of the age of the Florissant and Ruby floras. Thus, in the absence of external corroborative evidence, paleobotanists have had to depend on internal evidence of age determinations of American mid-Tertiary floras.

The procedures employed are mainly variations on two methods of analysis: (a) a critical floristic analysis with a view toward ascertaining what might be called a "coefficient of modernity," usually summarized by percentage or graphic representations, and (b) an interpolative comparison with other floras having relevant spatial and temporal characteristics, a method that Becker (1961) has called "paleobotanical triangulation." Basic to the first of these analyses is a calculation of geographic shift and compositional changes in the floras, reckoned on fluctuations in the number of arrivals, survivals, and extinctions of the cool-temperate (Arcto-Tertiary), warm-temperate (Neotropical), xeric (Madro-Tertiary), and Asiatic species. Thus far this method has been exploited almost entirely on megafloral evidence. Whether the newer data from palynology will alter, or merely reinforce the older conclusions remains to be seen. The validity of the second method hinges on the degree of confidence supporting age determinations of the selected reference floras; if these are not reliably dated, there is always the unfortunate possibility that paleobotanical triangulation may be scarcely more than paleobotanical circular reasoning.

The Florissant flora, as known from leaf and fruit remains, is interpreted by MacGinitie as indicative of a woody, upland flora growing under subhumid conditions at moderate elevations (1,000 to 3,000 ft). Its character is due in large measure to the strong representation of six families: Rosaceae, Leguminosae, Fagaceae, Pinaceae, Anacardiaceae, and Sapindaceae, which together account for about 45 percent of the total number of seed-plant species on MacGinitie's systematic list. According to that author's estimates, 29.7 percent of the Florissant species have living counterparts in Asia, but well over half (57.1 percent) are still native to the region within a circle of radius 400 miles centered in southwestern Coahuila State, Mexico, about 1,000 miles south of Florissant. Palynological data, as yet unpublished, may modify these ratios. Estella Leopold's preliminary studies (personal communication) meanwhile confirm the strong representation of gymnosperms, verifying the presence of *Ephedra*, *Pinus*, *Picea*, and *Abies*. She reports, however, that the taxodiaceous pollen encountered is morphologically more similar to that of modern *Taxodium* species than to *Sequoia*, despite the fact that *Taxodium* is unknown as a leaf or cone fossil whereas *Sequoia affinis* Lesq. is a common Florissant megafossil. (See pl. 16-6, fig. 4.) Even more curious is the fact that pollen removed from strobili of *Sequoia affinis* bears a closer resemblance to living *Taxodium* than to living *Sequoia*. There is much yet to be learned about the evolution of taxodiaceous pollen. (See pl. 16-5.)

Summarizing the palynology of the American Northwest, Gray (1964) noted the high proportion of undetermined angiosperm pollen and nontemperate forest genera. The Oligocene strata of that region are further characterized by the abundance of conifer pollen, presumably of taxodiaceous affinity; by herbaceous genera represented by pollen referable to the Graminae, Cyperaceae, and Typhaceae-Sparganiaceae; and by a variety of fern genera belonging to the Polypodiaceae, Schizaeaceae, and Osmundaceae. The small but well-known Brandon flora, ostensibly of mid-Tertiary age, is of renewed interest because of the well-documented palynological

Plate 16-6. Selected pollen and spores of the (upper Oligocene) Florissant Formation, Florissant, Colo. Plate prepared by and identifications by Estella B. Leopold. All figures × 500. 1 – Undetermined spore. 2, 3 – *Selaginella* cf. *S. densa*. 4 – Taxodiaceae, *Taxodium* type. 5 – *Ephedra* cf. *E. nevadensis*. 6 – *Pinus*. 7 – *Picea*. 8 – *Abies*. 9 – *Picea?* 10 – *Tsuga?* 11 – *Eucommia*. 12 – Undetermined dicotyledon. 13 – *Quercus*. 14 – *Acer*. 15 – Undetermined dicotyledon. 16*a*, *b* – *Acer*. 17 – *Salix?* 18 – *Cardiospermum*. 19 – Gramineae. 20 – *Typha* or *Sparganium*. 21 – Undetermined dicotyledon. 22*a*, *b* – Undetermined dicotyledon, Normapolles group. 23 – Onagraceae, *Xylonagra* type. 24 – *Engelhardtia?* 25 – Onagraceae, *Godetia* type. 26, 27 – *Carya*. 28, 31 – *Juglans*. 29 – *Ulmus* or *Zelkova* type, Ulmaceae. 30 – Undetermined dicotyledon. 32, 33 – Chenopodiaceae, *Sarcobatus* type. 34, 35, 37, 38 – Undetermined dicotyledons. 36 – *Eleagnus*. 39 – *Fremontia*. 40 – Ericales.

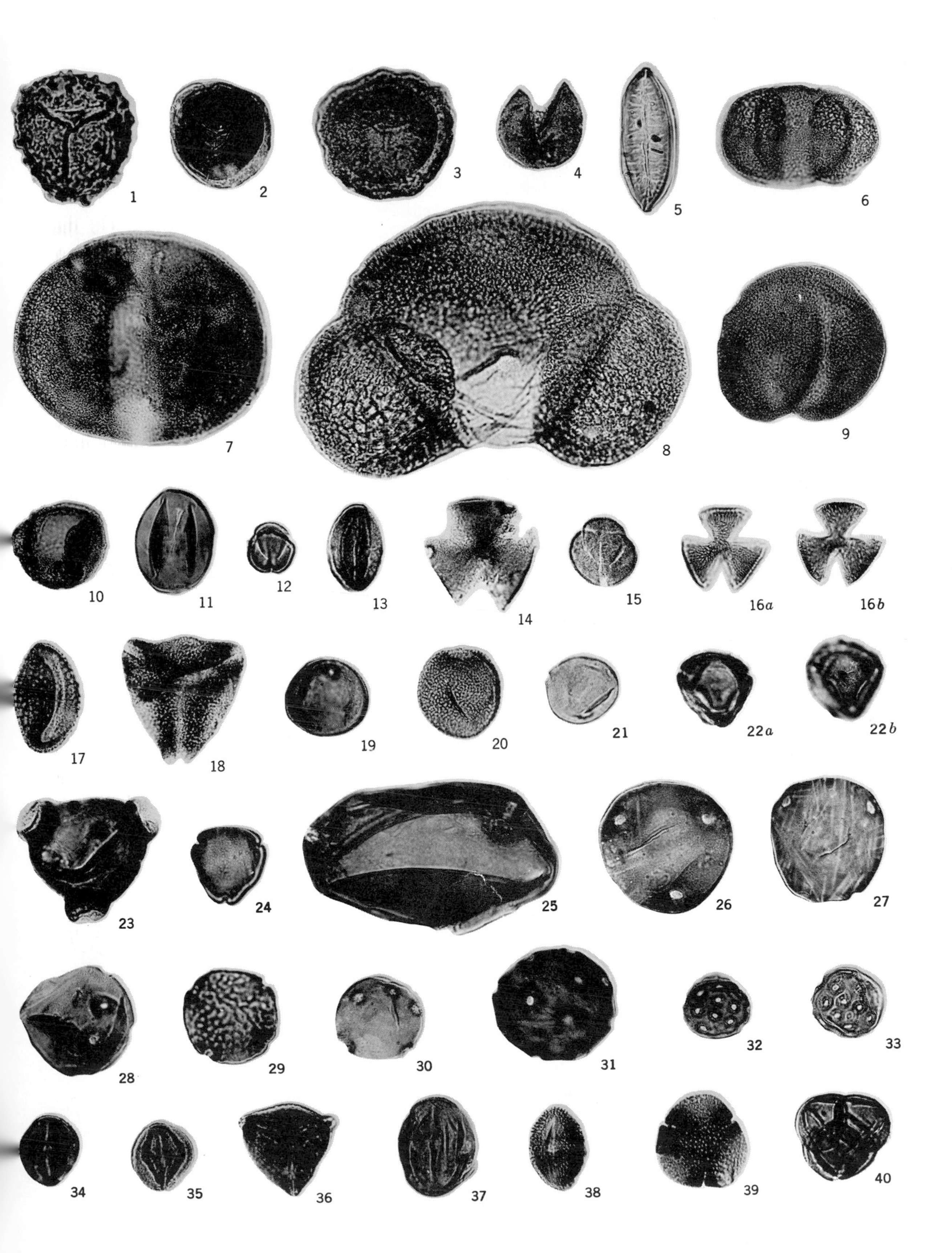

1
2
3
4
5
6
7
8
9
10
11
12
13
14
15
16a
16b
17
18
19
20
21
22a
22b
23
24
25
26
27
28
29
30
31
32
33
34
35
36
37
38
39
40

investigations of Traverse (1955). The plant remains—consisting of seeds, fruit, and wood, as well as microfossils—occur in a kind of brown coal deposited in a small Tertiary basin on the west side of the Green Mountains of Vermont, approximately 44°N. latitude. The flora is considered indicative of a forest swamp growing under warm temperate or subtropical climates (Barghoorn and Spackman, 1950). The spore and pollen flora described by Traverse consists of about 76 entities, showing an interesting overlap of Arcto-Tertiary and Neotropical genera, with a small residuum of Asiatic (*Glyptostrobus, Pterocarya*) and Paleotropical (*Alangium*) genera. Traverse calculated that 28.6 percent of the genera of the augmented Brandon flora are still native to the Vermont area. This datum, plotted on an age curve (Barghoorn, 1951; Wolfe and Barghoorn, 1960), indicated a Miocene-Oligocene age. Traverse favored an uppermost Oligocene age. The use of such curves for age determination of fossil floras remains a point of continuing dispute, and the pros and cons of the method cannot be reviewed here (see Axelrod, 1957; Wolfe and Barghoorn, 1960). Since, however, the method depends so critically on the percentage of native genera, there is some interest in noting that the purportedly coeval (i.e., upper Oligocene) flora of the Ruby basin (approximately 46°N. latitude) in Montana has twice as many (59 percent) native genera. On the assumption that in mid-Tertiary time both sites were close to, and about the same distance from, the southern border of the Arcto-Tertiary floral province (see Axelrod, 1960, p. 248), the relatively small component of native (and mostly Arcto-Tertiary) genera in the Brandon flora seems somewhat anomalous for a northern upper Oligocene site. Moreover, as Traverse points out, the presence of Neotropical and Paleotropical genera " . . . such as *Manilkara, Mimusops*, and *Alangium* as far north as Vermont should be strongly weighed . . . . " This would seem to argue for an older age for the Brandon flora, as indeed Berry (1937) earlier had argued, remarking that "My own opinion is that the Brandon deposit is of upper Eocene age."

Europe

Faunal and floral evidence both indicates the development, during the Oligocene, of two major biotic provinces, Northern and Mediterranean, but European Oligocene palynology is still mainly that of the Northern province. Here sweeping strand-line changes—beginning with the great wave of flooding at the start of the Oligocene, followed by major regressions at the close—had produced successions of marsh, lagoonal, and coastal swamp environments. (See fig. 16-9.)

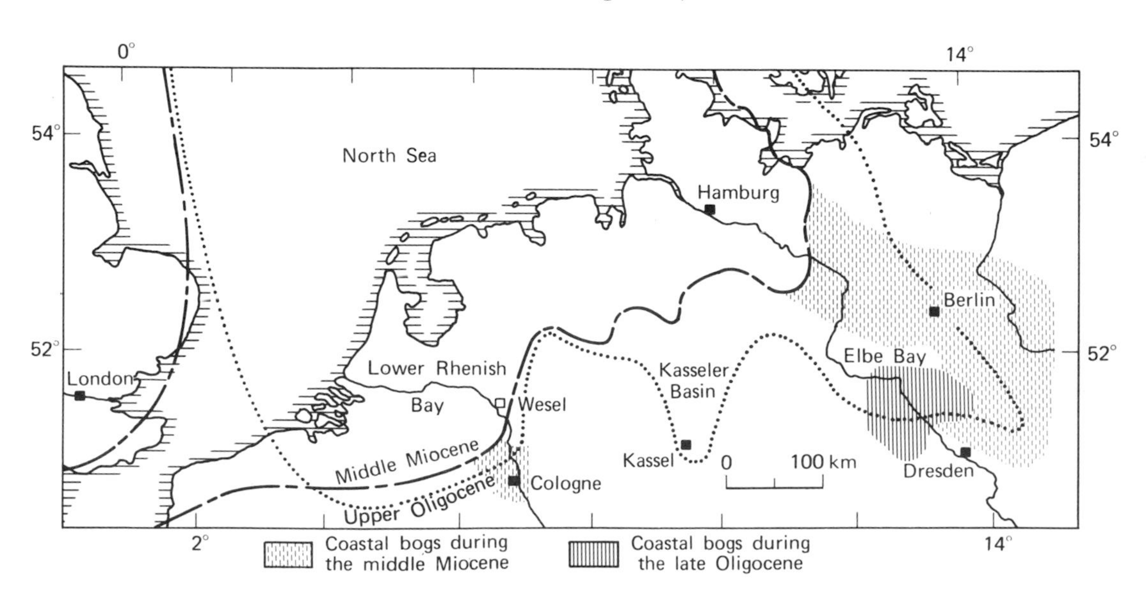

Figure 16-9. The transgressions of the North Sea into the north German lowland during mid-Miocene and late Oligocene time and the distribution of the bogs (from Teichmuller, 1958, after Bettenstaedt, 1949, Gripp, 1937, and Quitzow, 1953).

As in the case of Tertiary palynology in general, much of our foundational knowledge of the taxonomy and stratigraphic distribution of Oligocene spores and pollen has come from investigations of brown coal and the intercalated deposits associated with these facies in Central Europe, but especially in western and central Germany. Of the numerous brown-coal basins of Germany few have contributed more data for reconstructing the life and climates of Tertiary time than the lower Rhenish Basin. Although the chief deposits here are of Miocene age, Oligocene flooding of the North Sea earlier had encroached into this old arc of subsidence as far as Cologne, leaving thick sediments, of various facies, between Bonn and Wesel. One of the outstanding series of studies on the paleontology, stratigraphy, and paleoecology of brown coal is collected in the symposium edited by Ahrens (1958) on this Niederrheinische embayment. From these studies a remarkably clear picture emerges of the vegetation of Central Europe during the passage from Paleogene to Neogene times.

Under cooling, but still warm temperate conditions, with average annual temperatures declining from about 20°C in early Oligocene to about 17°C in late Oligocene time, and under relatively humid conditions, at least during the middle and late Oligocene (R. Teichmüller, 1958), corals, crocodiles, some palms and sapotaceous plants were still able to thrive. For the most part, however, the plant associations reconstructed from spore and pollen analyses reflect (a) a general decrease in Palmae and Sapotaceae, (b) a general increase in conifers, and (c) a gradual increase in Betulaceae, Fagaceae, Juglandaceae, and herbaceous genera of the Gramineae, Polygonaceae, and Chenopodiaceae (Krutzsch, 1957).

The faunal bases for separating the mid-Tertiary from older and younger Tertiary have received but little reinforcement from pollen analysis. On palynological grounds no sharp floral break is discernible between Eocene and Oligocene. In fact, for Central Europe at least, no significant floral change is detectable until mid-Oligocene time. Similarly the boundary between Paleogene and Neogene (i.e., between Oligocene and Miocene) is on palynological grounds an arbitrary division (Barbashinova, 1962). For Central Europe the floral character of the Neogene becomes first evident in sediments of Rupelian (middle Oligocene) age (Krutzsch, 1957). A clue to this change may be seen in the stratigraphic distribution of certain spores and pollen selected by Thomson and Pflug (1953); for example, the following Eocene-lowest Oligocene spores and pollen decline or disappear during the mid-Oligocene to early Miocene interval:

Concavisporites acutus Pf.
Cicatricosisporites dorogensis R. Pot. & Gell.
Triporopollenites robustus Pf.
Triatriopollenites excelsus R. Pot., subsp. *T. microturgidus* Pf.
Trudopollis pompeckji (R. Pot.) R. Pot., Pf.
Tricolpopollenites spinosus Pf.
Cupuliferoidaepollenites fallax (Th., Th. & Thierg.) R. Pot.
Intratriporopollenites rhizophorous R. Pot.

The following Neogene spores and pollen first appear, or markedly increase, during the mid-Oligocene to early Miocene interval:

Tsugaepollenites viridifluminipites (Wodeh.) R. Pot.
Sciadopityspollenites serratus (R. Pot. & Ven.) Raatz
Pinuspollenites lubdacus (R. Pot.) Raatz, R. Pot. & Ven.
Piceaepollenites alatus (R. Pot.) R. Pot.
Liquidambarpollenites stigmosus (R. Pot.) Raatz
Ulmipollenites undulosus (Wolff) R. Pot.
Caryapollenites simplex (R. Pot.) R. Pot.
Triatriopollenites rurensis Pf. & Th.

Some idea of the general character of the German Oligocene can be obtained from the following list of better known spores and pollen that continue into or first appear in the Rupelian and/or Chattian sediments of the brown-coal basins (Thomson and Pflug, 1953; Krutzsch, 1957, 1962; Thomson, 1950; Brelie and Wolters, 1958; Helal, 1958; Thiergart, 1958; R. Potonié, 1956, 1958, 1960):

Lycopodinae:
Lycopodiumsporites potoniei (Th. & Pf.) R. Pot.
Filicinae:
Osmundaceae:
Rugulatisporites quintus Th. & Pf.
Schizaeaceae:
Lygodioisporites solidus (R. Pot.) R. Pot., Th. & Thierg.
Cicatricosisporites dorogensis R. Pot. & Gell.
Divisisporites maximus Pf.

Gleicheniaceae:
Concavisporites obtusangulus (R. Pot.) Th. & Pf.
Polypodiaceae:
Baculatisporites primarius (Wolff) Th. & Pf.
Verrucatosporites alienus (R. Pot.) Th. & Pf.
Polypodiisporites secundus (R. Pot.) R. Pot.
Polypodiidites clathriformis (Murr. & Pf.) R. Pot.
Polypodiaceaesporites haardtii (R. Pot. & Ven.) Thierg.

Coniferae:
Taxodiaceae:
Taxodiaceaepollenites hiatus (R. Pot.) Kremp
Sequoiapollenites polyformosus Thierg.
Sciadopityspollenites serratus (R. Pot. & Ven.) Raatz
Pinaceae:
Laricoidites magnus (R. Pot.) R. Pot., Th. & Thierg.
Abietineaepollenites microalatus (R. Pot.) R. Pot.
Pinuspollenites labdacus (R. Pot.) Raatz, Pot. & Ven.
Tsugaepollenites viridifluminipites (Wodeh.) R. Pot.
Piceaepollenites alatus (R. Pot.) R. Pot.
Inaperturopollenites dubius (R. Pot. & Ven.) Th. & Pf.

Monocotyledoneae:
Palmae:
Sabalpollenites areolatus (R. Pot., Th. & Thierg.) R. Pot.

Dicotyledoneae:
Symplocaceae:
Symplocospollenites rotundus (R. Pot.) R. Pot., Th. & Thierg.
Symplocoipollenites vestibulum (R. Pot.) R. Pot.
Sapotaceae:
Sapotaceoidaepollenites sapotides (Th. & Pf.) R. Pot.
Ericaceae:
Ericipites ericius (R. Pot.) R. Pot.
Leguminosae:
Polyadopollenites multipartitus Pf.
Hamamelidaceae:
Liquidambarpollenites stigmosus (R. Pot.) Raatz
Elaeagnaceae:
Slowakipollis sp. Krutzsch
Nyssaceae:
Nyssoidites rodderensis (Thierg.) Th. & Pf., R. Pot.
Alangiaceae:
Alangiopollis sp., Krutzsch
Olacaceae:
Olaxipollis sp. Krutzsch
Ulmaceae:
Ulmipollenites undulosus (Wolff) R. Pot.
Fagaceae:
Quercoidites henrici (R. Pot.) R. Pot., Th. & Thierg.
Q. microhenrici (R. Pot.) R. Pot., Th. & Thierg.
Cupuliferoidaepollenites liblarensis Th.
C. fallax R. Pot.
Cupuliferoipollenites oviformis (R. Pot.) R. Pot.
Tricolporopollenites villensis (Th.) Th. & Pf.
T. asper Th. & Pf.
T. laesus (R. Pot.) Th. & Pf.
T. porasper Pf.
Faguspollenites verus Raatz
Betulaceae:
Polyporopollenites carpinoides Pf.
Alnipollenites verus R. Pot.
Betulaceoipollenites betuloides (f.) R. Pot.
Triporopollenites coryloides Pf.
T. rhenatus Th.
Tiliaceae:
Tiliaepollenites instructus (R. Pot.) Th. & Pf.
Buxaceae:
Erdtmanipollis sp. Krutzsch
Anacardiaceae:
Rhooidites pseudocingulum (R. Pot.) R. Pot., Th. & Thierg.
Simarubaceae:
Ailanthipites pacatus (Pf.) R. Pot.
Rutaceae:
Tricolporopollenites sustmanni Pf. & Th.
Juglandaceae:
Engelhardtioidites microcoryphaeus (R. Pot.) R. Pot., Th. & Pf.
E. coryphaeus punctatus (R. Pot.) Th. & Thierg.
Multiporopollenites maculosus (R. Pot.) R. Pot.
Pterocaryapollenites stellatus (R. Pot.) R. Pot.
Caryapollenites simplex (R. Pot.) Raatz
Myricaceae:
Triatriopollenites rurensis Pf. & Th.
T. excelsus (R. Pot.) Pf.
Aquifoliaceae:
Ilexpollenites margaritatus (R. Pot.) Raatz
Tricolporopollenites clavopolatus Pf. & Th.
Cyrillaceae:
Cyrillaceaepollenites megaexactus brühlensis (Th.) R. Pot.
Vitaceae:

Tricolporopollenites marcodurensis Pf.
Polygonaceae:
Persicariopollis sp., Krutzsch
Araliaceae:
Araliaceoipollenites euphorii (R. Pot.) R. Pot.
A. edmundi (R. Pot.) R. Pot.
Caprifoliaceae:
Caprifoliipites microreticulatus (Th. & Pf.) R. Pot.
Lonicerapollis sp., Krutzsch
Family alliances uncertain:
Sparganiaceae?:
Sparganiaceaepollenites sp. Thierg.
Restionaceae?:
Inaperturopollenites incertus Pf. & Th.
Salicaceae?:
Tricolporopollenites retiformis Pf. & Th.
Platanaceae?:
Platanoides gertrudae (R. Pot.) R. Pot., Th. & Thierg.
Lauraceae?:
Tricolpopollenites spinosus Pf.

The relative abundance of individual species will of course vary with facies, but, as presently understood, the spores and pollen most frequently recorded from the German Oligocene were derived from the following:

1. Haploxylon and Sylvestris pines
2. Schizaeaceous, gleicheniaceous, and polypodiaceous ferns
3. Nonvesiculate conifers
4. Oaks, elms, *Tilia, Carya, Engelhardtia, Alnus, Fagus*
5. Symplocaceae spp.; Myricaceae spp.

Helal (1958) recognized several facies represented by the total pollen flora recovered from the east Rhenish Bergish Gladbach brown-coal beds of middle Oligocene age:

1. Mixed pine-birch-myrica woodlands
2. Pine brushwoods with raised bogs
3. Mixed dicotyledonous forests
4. Pollen-rich gyttjas and clay gyttjas with high percentages of *Alnus, Taxodium,* and *Liquidambar* pollen

The source materials for the great Miocene brown-coal "Hauptflözes" were derived apparently from four major forest and bog associations. The composition of these associations has been reconstructed in considerable detail from pollen and leaf evidence by M. Teichmüller (1958). The major associations, which had undoubtedly begun to develop in late Oligocene time, consisted of the following:

1. Sequoia (or Oligocene equivalent) forests
2. Myricacean-cyrillacean shrub thickets
3. *Nyssa-Taxodium* swamp forests
4. Reed, sedge, grass everglades

Kremp and Kovar (1960) have suggested that the last-named association, the "Riedmoor" of M. Teichmüller, was not extensively developed in mid-Tertiary time. Some notion of the character of an earlier Oligocene vegetation can be reconstructed from plant microfossils of approximately 40 species recovered by Choux, Durand, and Milon (1961) from a clay deposit of latest Eocene-earliest Oligocene age from Quimper in Brittany. The assemblage is characterized by the relatively strong showing of fern spores (Osmundaceae, Cyatheaceae, Gleicheniaceae, and Schizaeaceae) and pollen of Cupressaceae (especially *Inaperturopollenites emmaensis*), Palmae (*Sabalpollenites* spp.), Fagaceae (especially *Quercoidites microhenrici*), Myricaceae (especially *Triatriopollenites rurensis*), and Juglandaceae (*Engelhardtiodites* spp.). Noteworthy is the absence or scarcity of winged conifer, elm, and *Tilia* pollen.

Simpson (1961) has described and illustrated a pollen flora of 24 species obtained from a lignite seam at the base of a Tertiary basaltic lava series at Ardnamurchan in western Scotland. On the basis of floristic analysis and calculation of the percentage of East Asian-North American species, Simpson concluded that the flora was not older than the Oligocene. Except for the presence of *Engelhardtia, Juglans,* and *Alnus,* the flora bears no resemblance to the continental Oligocene floras, and the complete absence of conifer pollen is rather surprising.

Asia

Pokrovskaya (1962) has pointed out that Upper Cretaceous and Paleogene spore and pollen complexes in the European part of the Soviet Union are very similar to coeval complexes of Germany, reflecting a broadly developed northern floral province during mid-Tertiary times extending beyond the Urals and well into the Siberian plain of Asia. The same author (1950, 1956), in summarizing the palynological characterization of the Russian Oligocene, indicated the essential similarity of German and Russian pollen floras, noting the preponderance of taxodiaceous and Haploxylon *Pinus* pollen, and the confirmed presence of *Juglans, Carya, Pterocarya, Magnolia, Fagus, Tilia, Liquidambar, Salix, Alnus, Betula, Castanea, Ulmus, Acer, Rhus*, and *Ilex*. Similarly pollen of subtropical evergreen dicotyledons and palms is scarce or absent in Russian Oligocene sediments. Pollen of herbaceous angiosperms, according to Pokrovskaya, is not encountered in sediments older than Miocene. She postulated, however, that such plants may have originated in the Oligocene to become widely dispersed only after the Miocene.

On the basis of this postulate Sung (1956) dated the Huoskaokou beds of Kansu province in northwestern China as upper Oligocene because of their herbaceous pollen component. The total pollen assemblage from the Kansu site is interpreted as indicating a rather dry climate supporting a steppe-type vegetation with nearby stands of *Ginkgo* and *Magnolia*. Families represented by the herbaceous pollen were the Gramineae, Liliaceae, Chenopodiaceae, and Compositae. Only a few species of tree pollen were recorded, representing *Betula, Ulmus, Tamarix*, the Salicaceae, and Pinaceae.

Much of the palynological investigation in Japan has concentrated on the Eocene and Oligocene sediments intercalated in the commercially important coal seams of Hokkaido and northern Honshu. Tokunaga (1964) has summarized the stratigraphic ranges of important Oligocene (and Eocene) spores and pollen. Investigations to date indicate a constant warm temperate climate for northern Japan during the Paleogene. Arboreal pollen accounts for 70 to 85 percent of the total number of grains in the Ashibetsu Formation of late Oligocene age.

South America

De Porta (1961) has described and illustrated a Miocene-Oligocene spore and pollen flora from Colombia, consisting of fern spores (Polypodiaceae, Schizaeaceae, and Cyatheaceae), podocarp pollen, and angiosperm pollen from some dozen families, all with genera still "native" to northern South America. The angiosperm component includes the following:

Ilex? sp.
Bombacaceae (unidentified species)
Euplassia sp. (Proteaceae)
Acacia sp.
Isoberlinia? sp. (Leguminosae)
Malvaceae (unidentified species)
Malpighia sp. (Malpighiaceae)
Stigmaphyllon sp. (Malpighiaceae)
Triatriopollenites rurensis? (Myricaceae)
T. coryphaeus? (Myricaceae)
Nuphar sp. (Nymphaeaceae)
Mauritia sp. (Palmae)
Palmaepollenites? sp.
Sapotaceoidaepollenites manifestus
Porocolpopollenites vestibulum (Symplocaceae)
Tiliaepollenites indubitabilis

Most interesting of these Colombian pollen are the species identified under *Quercoidites* (al. *Pollenites*) *microhenrici* (R. Potonié) R. Potonié and *Triatriopollenites rurensis?* Thomson and Pflug, so well known from the European mid-Tertiary.

Australia and New Zealand

The placement of the *Cinnamomum* flora within the Tertiary, except for the designation "early Tertiary" or "pre-Miocene," is not agreed on; moreover, evidence concerning the flora that succeeded it is scanty (Burbidge, 1960). Cookson's report (1950) of proteaceous pollen grains from mid-Tertiary (?Oligocene-Miocene) deposits of southeastern Australia suggests that the approach to the present-day flora may have begun by that time. Cookson and Duigan (1950) list the occurrences of *Callitris, Phyllocladus, Nothofagus, Cinnamomum, Tristania*, and *Causuarina* in association with the xeromorphic mid-Tertiary proteaceous types, making for a kind of mixed association that is indicative of a cool tropical climate with evenly distributed rainfall, known today in parts of southeast Queensland and certain mountainous areas of New Guinea (Burbidge, 1960). Most of the Australian mid-Tertiary proteaceous pollen has been described under the genera *Banksieaeidites* Cookson, *Beaupreaidites* (Cookson) Couper, *Proteacidites* Cookson, and *Triorites* Cookson ex Couper. Cookson (1950) has suggested affinities with such living counterparts as *Beauprea elegans* (*Beaupreaidites elegansiformis*), *Beauprea spathulaefolia* (*Beaupreadites verrucosus*), *Banksia* or *Dryandra* (*Banksieaeidites minimus* and *elon-*

gatus), *Xylomelum occidentale* or *Lambertia multiflorum* (*Proteacidites annularis*), *Isopogon anemonifolius* (*Proteacidites truncatus*), and *Adenanthos barbigera* (*Proteacidites adenanthoides*). Other species, although clearly of proteacean affinity, have no recognizable living counterparts.

The stratigraphic ranges of New Zealand plant microfossils (Couper, 1960b) show no striking differences between the flora of the upper Eocene Arnold Series and that of the lower Oligocene Landon Series. The dominant forest species continued to be *Nothofagus matauraensis* of the Brassi group. The interval is marked, however, by the first appearance of pollen representative of the Bombacaceae and Restionaceae, as well as pollen of *Podocarpus* aff. *dacrydioides* Rich. During mid-Oligocene time *Typha*, *Laurelia* (Monimiaceae), and *Pseudowintera* (Winteraceae) are recorded but with low frequencies. The first palynological evidence of climatic cooling occurs in the upper Oligocene in the strong dominance of both the Brassi and Fusca groups of southern beeches. Couper points out, however, that the common to abundant occurrence of palmaceous pollen together with pollen of certain subtropical families suggests that the late Oligocene climate was still warmer than that of present New Zealand.

THE WORLD AT THE CLOSE OF THE PALEOGENE

The passage from the Eocene to the Oligocene is regarded generally as the interval during which the Tertiary floras began to show the migrational and compositional changes that were to culminate in the complex and diversified floras of Recent times. Paleobotanists seem agreed that the direct causes of the major vegetational transformations were reduction in precipitation and lowering of minimum temperatures, but the causes, in turn, of these climatic changes are by no means agreed on. The majority opinion would infer a cause and effect relationship between orogenic and/or uplift movements and climatic modifications. Certainly, new or continuing orogenic and uplift disturbances had produced profound geomorphologic changes on much of the planet by the close of the Eocene. In North America the Laramide orogeny had run its course, producing the setting for the general physiography of most regions of the Rocky Mountains. In South America the coastal Cordillera of the northern Andes were uplifted, and the main Andean chains were to continue to rise for the remainder of the Tertiary. Central America and the Antilles were disturbed by active orogeny as well as uplift. In Europe the Pyrenean phase of folding produced maximum orogenic pressures in the Alps and Pyrenees. In Asia the old Tethyan geosyncline underwent major orogenic deformations that initiated the birth of the Himalayas. In Australia, somewhat to the contrary, crustal sinking instead of uplift marked the close of the Paleogene, and general epeirogenic disturbances appear not to have occurred until the close of the Miocene.

In survey the record is clear. Whatever the causes, major evolution within most families of flowering plants, together with their major dispersal, was accomplished during the Late Cretaceous-early Tertiary interval.

References

Ahrens, W., 1958, Die Niederrheinische Braunkohlenformation: Fortschr. Geol. Rheinld. u. Westf., v. 1 and 2.

Axelrod, D. I., 1957, Age-curve analysis of angiosperm floras: Jour. Paleontology, v. 31, p. 273-280.

______ 1959, Poleward migration of early angiosperm flora: Science, v. 130, no. 3369, p. 203-207.

______ 1960, The evolution of flowering plants, p. 227-305 *in* The evolution of life, Sol Tax, ed.: Chicago.

Barbashinova, V. N., 1962, Main features of Miocene flora in the Soviet Far East, according to spore and pollen analysis data: Pollen et Spores, v. 4, no. 2 [abs.].

Barghoorn, E. S., 1951, Age and environment: A survey of North American Tertiary floras in relation to paleoecology: Jour. Paleontology, v. 25, p. 736-744.

Barghoorn, E. S., and Spackman, W., 1950, Geological and botanical study of the Brandon lignite and its significance in coal petrology: Econ. Geol., v. 45, no. 4, p. 344-357.

Becker, H. F., 1961, Oligocene plants from the upper Ruby River basin, southwestern Montana: Geol. Soc. America, Mem. 82.

Berry, E. W., 1916a, The Upper Cretaceous floras of the world, *in* Maryland Geological Survey, Upper Cretaceous, p. 183-313.

——— 1916b, The lower Eocene floras of southeastern North America: U.S. Geol. Survey Prof. Paper 91.

——— 1924, The middle and upper Eocene floras of Southeastern North America: U.S. Geol. Survey Prof. Paper 92.

——— 1930, Revision of the lower Eocene Wilcox flora of the Southeastern States; with descriptions of new species, chiefly from Tennessee and Kentucky: U.S. Geol. Survey Prof. Paper 156.

——— 1937, Tertiary floras of Eastern North America: Bot. Rev., v. 3, no. 1, p. 31-46.

Bolkhovitina, N. A., 1953, Spores and pollen characteristic of Cretaceous deposits of Central Regions of the USSR [in Russian]: Trans. Inst. Geol. Sci., Acad. Sci., USSR, Rel. 145, Geol. Ser. no. 61.

——— 1959, Spore and pollen complexes of the Vilni beds and their value in stratigraphy [in Russian]: Akad. Nauk SSSR, Leningrad Geol. Inst., v. 24.

Brelie, G. v. d., and Wolters, R., 1958, Das Alttertiär von Gürzenich: Fortschr. Geol. Rheinld. u. Westf, v. 2, p. 473-477.

Brinkmann, R., 1960, Geologic evolution of Europe [translated by J. E. Sanders]: New York, Haffner.

Brown, C. A., 1957, Comment on European Tertiary pollen studies: Micropaleontology, v. 3, no. 1, p. 81-83.

Brown, R. W., 1934, The recognizable species in the Green River flora: U.S. Geol. Survey Prof. Paper 185-C, p. 45-77.

——— 1944, Temperate species in the Eocene flora of the Southeastern United States: Wash. Acad. Sci. Jour., v. 34, no. 11, p. 349-351.

Brown, E. W., and Pierce, R. L., 1962, Palynologic correlations in Cretaceous Eagle Ford group, northeast Texas: Amer. Assoc. Petrol. Geol. Bull., v. 46, no. 12, p. 2133-2147.

Bunt, J., 1956, Living and fossil pollen from Macquarie Island: Nature, v. 177, no. 4503, p. 339.

Burbidge, Nancy T., 1960, The phytogeography of the Australian region: Australian Jour. Bot., v. 8, no. 2, p. 75-211.

Cain, S. A., 1943, The Tertiary character of the core hardwood forests of the Great Smoky Mountain National Park: Torrey Bot. Club Bull., v. 70, p. 213-235.

Chandler, M. E. J., 1961, The lower Tertiary floras of southern England. I.—Paleocene floras, *in* London Clay flora [Supplement], Brit. Mus. (Nat. Hist.).

Chaney, R. W., 1947, Tertiary centers and migration routes: Ecol. Monogr., v. 17, no. 2, p. 139-148.

Chlonova, A. F., 1962, Some morphological types of spores and pollen grains from Upper Cretaceous of eastern part of west Siberian lowland: Pollen et Spores, v. 4, no. 2, p. 297-309.

Choux, Janine, Durand, Suzanne, and Milon, Y., 1961, Le Dépôt des Argiles Conservées au Sud de Quimper (Finistère), sous les Formations Marines Pliocènes, s'est Terminé au Début de l'Oligocène: Acad. sci. [Paris] Comptes rendus, v. 252, p. 3833-3835.

Cobban, W. A., and Reeside, J. B., Jr., 1952, Correlation of the Cretaceous formations of the western interior of the United States: Geol. Soc. America Bull., v. 63, no. 10, p. 1011-1044.

Cookson, I. C., 1947, Plant microfossils from the lignites of Kerguelen Archipelago: B. A. N. Z. Antarctic Res. Exsped., 1929-1931, Reports, ser. A, v. 2, pt. 8.

______ 1950, Fossil pollen grains of proteaceous type from Tertiary deposits in Australia: Australian Jour. Sci. Res., ser. B, Biol. Sci., v. 3, no. 2, p. 166-177.

Cookson, I. C., and Dettman, M. E., 1959, Microfloras in bore cores from Alberton West, Victoria: Royal Soc. Victoria Proc., v. 71, pt. 1, p. 31-38.

Cookson, I. C., and Duigan, S. L., 1950, Fossil Banksiae from Yallourn, Victoria, with some notes on the morphology of living species: Australian Jour. Sci., ser. B, v. 3, p. 133-164.

Cookson, I. C., and Erdtman, Gunnar, 1952, Proteaceae, p. 369 *in* Erdtman, Gunnar, Morphology and plant taxonomy. Angiosperms: Waltham, Mass., Chronica Botanica Co.

Cookson, I. C., and Pike, K. M., 1954, Some dicotyledonous pollen types from Cainozoic deposits in the Australian region: Australian Jour. Bot., v. 2, no. 2, p. 197-219.

Couper, R. A., 1955, Supposedly colpate pollen grains from the Jurassic: Geol. Mag., v. 92, no. 6, p. 471-475.

______1960a, Southern Hemisphere Mesozoic and Tertiary Podocarpaceae and Fagaceae and their paleogeographic significance: Royal Soc. [London] Proc., ser. B, v. 152, p. 491-500.

______1960b, New Zealand Mesozoic and Cainozoic plant microfossils: New Zealand Geol. Survey, Paleont. Bull. 32.

______1964, Spore-pollen correlation of the Cretaceous rocks of the Northern and Southern Hemispheres, *in* Ancient Pacific floras, 10th Pacific Sci. Cong. Ser., Univ. Hawaii Press, p. 131-142.

Cousminer, H. L., 1961, Palynology, Paleofloras, and Paleoenvironments: Micropaleontology, v. 7, no. 3, p. 365-368.

Cranwell, L. M., 1959, Fossil pollen from Seymour Island, Antarctica: Nature, v. 184, no. 4701, p. 1782-1785.

Cranwell, L. M., Harrington, H. J., and Speden, I. G., 1960, Lower Tertiary microfossils from McMurdo Sound, Antarctica: Nature, v. 186, no. 4726, p. 700-702.

Croizat, L., 1952, Manual of phytogeography: The Hague, Junk.

DePorta, N. S., 1961, Contribución al Estudio Palinológico del Terciario de Columbia: Boletin de Geologia, Univ. Ind. de Santander, no. 7, p. 55-82.

Dijkstra, S. J., 1961, Some Paleocene megaspores and other small fossils: Meded. Geol. Sticht., N.S., 13, p. 5-11.

Dorf, E., 1942, Upper Cretaceous floras of the Rocky Mountain region: Carnegie Inst. Wash. Publ. 508.

______ 1952, Critical analysis of Cretaceous stratigraphy and paleobotany of Atlantic Coastal Plain: Amer. Assoc. Petrol. Geol. Bull., v. 36, no. 11, p. 2161-2184.

______ 1955, Plants and the geologic time scale: Geol. Soc. America Spec. Paper 62, p. 575-592.

Durand, Suzanne, 1962, L'Analyse Pollinique des Formations du Paléogène Français: Colloquium on the Paleogene, Bordeaux, September 1962.
Durand, Suzanne, and Ters, M., 1958, L'Analyse Pollinique d'Argiles des Environs de Challans (Vendée) Révèle l'Existence d'une Flore Cénomanienne: Acad. sci. [Paris] Comptes rendus, v. 474, p. 684-686.
Eames, F. E., Banner, F. T., Blow, W. H., and Clarke, W. J., 1962, Fundamentals of mid-Tertiary stratigraphical correlations: Cambridge.
Ehrlich, H. G., and Hall, J. W., 1959, The ultrastructure of Eocene pollen: Grana Palynologica, v. 2, no. 1, p. 32-35.
Emiliani, C., 1955, Pleistocene temperatures: Jour. Geology, v. 63, p. 538-578.
——— 1956, Oligocene and Miocene temperatures of the equatorial and subtropical Atlantic Ocean: Jour. Geology, v. 64.
Erdtman, Gunnar, 1960, On three new genera from the lower Headon beds, Berkshire: Bot. Not., v. 113, fasc. 1, p. 46-48.
Firtion, F., 1952, Le Cénomanien Inférieur du Novion-en-Thiérache: Examen Micropaléontologique: Ann. Soc. Géol. du Nord., v. 72, p. 150-164.
Funkhouser, J. W., 1961, Pollen of the genus *Aquilapollenites*: Micropaleontology, v. 7, no. 2, p. 193-198.
Gerhard, J. E., 1958, Paleocene miospores from the Slim Buttes area, Harding Co., South Dakota: M.S. Dissertation, Pennsylvania State Univ., June 1958.
Good, R., 1953, The geography of the flowering plants, 2d ed.: London, Longmans, Green.
Gray, Jane, 1960, Temperate pollen genera in the Eocene (Claiborne) flora, Alabama: Science, v. 132, no. 3430, p. 808-810.
——— 1964, Northwest American Tertiary palynology: The emerging picture, p. 21-30 *in* Ancient Pacific floras, 10th Pacific Sci. Cong. Ser., Univ. Hawaii Press.
Groot, J. J., and Groot, C. R., 1962a, Some plant microfossils from the Brightseat formation (Paleocene) of Maryland: Palaeontographica, v. 119, sec. B, p. 161-171.
Groot, J. J., and Penny, J. S., 1960, Plant microfossils and age of nonmarine Cretaceous sediments of Maryland and Delaware: Micropaleontology, v. 6, no. 2, p. 225-236.
Groot, J. J., Penny, J. S., and Groot, C. R., 1961, Plant microfossils and age of the Raritan, Tuscaloosa, and Magothy formations of the Eastern United States: Palaeontographica, v. 108, sec. B, p. 121-140.
——— 1962b, Plant microfossils from Aptian, Albian, and Cenomanian deposits of Portugal: Com. Serv. Geol. Port., XLVI, p. 131-171.
Hail, W. J., and Leopold, E. B., 1960, Paleocene and Eocene age of the Coalmont Formation, North Park, Colorado: U.S. Geol. Survey Prof. Paper 400-B, art. 117, p. B260-261.
Hall, J. W., 1963, Megaspores and other fossils in the Dakota formation (Cenomanian) of Iowa, (USA): Pollen et Spores, v. 5, no. 2, p. 425-443.
Hammen, Th. Van Der, 1954, El Desarrollo de la Flora Colombiana en los Periodos Geológicos, I: Maestrichtiano hasta Terciario mas Inferior: Bol. Geol., v. 2, no. 1, p. 49-106.
———1957a, Climatic periodicity and evolution of South American Maestrichtian and Tertiary floras: Bol. Geol., v. 5, no. 2, p. 50-91.
———1957b, Estratigrafía Palinológica de la Sabana de Bogotá (Cordillera Oriental de Colombia): Bol. Geol., v. 5, no. 2, p. 191-203.
Hammen, Th. Van Der, Wymstra, T. A., and Leidelmeyer, P., 1961, Paleocene sediments in British Guiana and Surinam: Geol. Mijnb., v. 40, no. 6, p. 23]-23].

Helal, A. H., 1958, Das Alter und die Verbreitung der tertiären Braunkohlen bei Bergisch Gladbach östlich von Köln: Fortschr. Geol. Rheinld. u. Westf., v. 2, p. 419-435.

Hofman, Elise, 1948, Das Flyschproblem im Lichte der Pollenanalyse: Phyton, v. 1, p. 80-101.

Hornibrook, N. de B., 1962, The Cretaceous-Tertiary boundary in New Zealand: New Zealand Jour. Geol. Geophys., v. 5, no. 2, p. 295-303.

Jones, E. L., 1961, Environmental significance of palynomorphs from lower Eocene sediments of Arkansas: Science, v. 134, no. 3487, p. 1366.

Johnson, L. A. S., and Briggs, B. G., 1963, Evolution in the Proteaceae: Australian Jour. Bot., v. 11, no. 1, p. 21-61.

Kedves, M., 1960, Études Palynologiques dans le Bassin de Dorog I: Pollen et Spores, v. 2, no. 1, p. 89-118.

______ 1961, Études Palynologiques dans le Bassin de Dorog II: Pollen et Spores v. 3, no. 1, p. 101-153.

Koch, B. E., 1959, Contribution to the stratigraphy of the nonmarine Tertiary deposits on the south coast of the Nugssuaq Penninsula [northwest Greenland]: Meddelelser om Grønland, v. 162, no. 1.

Kremp, G. O. W., and Gerhard, J. E., 1956, Pollen and Spores from the lower Tertiary of North and South Dakota and some taxonomic problems concerning their designations: Origin and Constitution of Coal Conf., 3d, Nova Scotia Dept. of Mines, p. 257-269.

Kremp, G. O. W., and Kovar, A. J., 1960, The interpretation of Tertiary swamp types in brown coal: U.S. Geol. Survey Prof. Paper 400-B, p. B79-81.

Kremp, G. O. W., Neavel, R. C., and Starbuch, J. S., 1956, Coal types—a function of swamp environments: Origin and Constitution of Coal Conf., 3d, Nova Scotia Dept. of Mines, p. 270-286.

Konizeski, R. L., 1961, Paleoecology of an early Oligocene biota from Douglass Creek basin, Montana: Geol. Soc. America Bull. v. 72, no. 11, p. 1633-1642.

Krutzsch, W., 1957, Sporen- und Pollengruppen aus der Oberkreide und dem Tertiär Mitteleuropas und ihre stratigraphische Verteilung: Zeitschr. Angewandte Geol., v. 3, no. 11-12, p. 509-548.

______ 1960, Present state of spore stratigraphy of the German Mesozoic: Ann. Inst. Geol. Publ. Hungarici, v. 59, fasc. 1, p. 327-329.

______ 1962, Stratigraphisch bzw. botanisch Wichtige neue Sporen- und Pollenformen aus dem deutschen Tertiär: Geologie, v. 11, no. 3, p. 265-307.

Kummell, B., 1961, History of the earth: San Francisco, W. H. Freeman and Co.

Kuprianova, L. A., 1960, Palynological data contributing to the history of *Liquidambar:* Pollen et Spores, v. 2, no. 1, p. 71-88.

Kuyl, O. S., Muller, J., and Waterbolk, H. T., 1955, The application of palynology to oil geology with reference to western Venezuela: Geol. Mijnb., v. 17, no. 3, p. 49-76.

Lam, H. J., 1948, A new system of the Cormophyta: Blumea, v. 6, p. 282-289.

Langenheim, R. L., Jr., Smiley, C. J., and Gray, Jane, 1960, Cretaceous amber from the Arctic Coastal Plain of Alaska: Geol. Soc. America Bull., v. 71, no. 9, p. 1345-1356.

Leffingwell, H. A., 1962, Uppermost Cretaceous and lower Paleocene spore-pollen assemblages in the type area of the Lance Formation, Wyoming. Pollen et Spores, v. 4, no. 2, p. 360-361[abs.].

Lesquereux, L., 1892, The flora of the Dakota group (ed. F. H. Knowlton): U.S. Geol. Survey Mono. 17.

Loeblich, A. R., Jr., and Tappan, Helen, 1961, Cretaceous planktonic Foraminifera: pt. I—Cenomanian: Micropaleontology, v. 7, no. 3, p. 257-304.

MacGinitie, H. D., 1951, A middle Eocene flora from the central Sierra Nevada: Carnegie Inst. Wash. Pub. 534.

——— 1953, Fossil plants of the Florissant beds, Colorado: Carnegie Inst. Wash. Pub. 599.

Ma Khin Sein, 1961a, Palynology of the London Clay: Ph.D. Thesis, University College London.

——— 1961b, Nothofagus pollen in the London Clay: Nature, v. 190, no. 4780, p. 1030-1031.

McQueen, D. R., 1956, Leaves of Middle and Upper Cretaceous pteridophytes and cycads from New Zealand: Royal Soc. New Zealand Trans., v. 83 pt. 4, p. 673-685.

Manum, S., 1954, Pollen og Sporer i Tertiaere Kull fra Vestspitsbergen: Blyttia, v. 12, no. 1, p. 1-9.

Martin, P. S., and Gray, Jane, 1962, Pollen analysis and the Cenozoic: Science, v. 137, no. 3524, p. 103-111.

Muller, J., 1959, Palynology of Recent Orinoco delta and shelf sediments: Micropaleontology, v. 5, no. 1, p. 1-32.

Murray, G. E., 1961, Geology of the Atlantic and Gulf Coastal province of North America: New York, Harper.

Newell, N. D., 1962, Paleontological gaps and geochronology, Jour. Paleontology, v. 36, no. 3, p. 592-610.

Newman, K. R., 1962, Palynologic correlations of Late Cretaceous and Paleocene formations, northwestern Colorado, *in* Palynology in oil explorations—a symposium: Soc. Econ. Paleontol. Mineral. Spec. Publ. 11, p. 169-179.

Norem, W. L., 1955, Pollen, spores, and other organic microfossils from the Eocene of Venezuela: Micropaleontology, v. 1, no. 3, p. 261-267.

——— 1956, Tertiary spores and pollen related to paleoclimates and stratigraphy of California: Micropaleontology, v. 24, no. 4.

Pacltova, Blanka, 1959, On some plant microfossils from freshwater sediments of Upper Cretaceous (Senonian) in the South-Bohemian basins: Sbornik Ustredniho Ustavu Geol., v. 26, p. 1-102.

Pflug, H. D., 1953, Zur Enstehung und Entwicklung des angiospermiden Pollens in der Erdgeschichte: Paleontographia, v. 95, sec. B, p. 60-171.

Pierce, R. L., 1961, Lower Upper Cretaceous plant microfossils from Minnesota: Minn. Geol. Survey Bull. 42.

Pokrovskaya, I. M., 1950, Pollen analysis [in Russian]: Moscow, Institute of Mineralogy and Geology.

———1956, Atlas of Miocene spore-pollen complexes of several regions of the USSR [in Russian]: Trans. All-Soviet Scientific Research Geol. Inst. Ministry Geol. and Petrol. Conservation USSR, new series v. 13.

———1962, Upper Cretaceous and Paleogene spore and pollen complexes in the European part of the USSR: Pollen et Spores, v. 4, no. 2, p. 371-372 [abs.].

Popenoe, W. P., Imlay, R. W., and Murphy, M. A., 1960, Correlation of the Cretaceous formations of the Pacific Coast: Geol. Soc. America Bull., v. 71, no. 10, p. 1491-1540.

Potonié, Robert, 1931, Zur Mikroskopie der Braunkohlen: tertiäre Blütenstaubformen, [pt. 1]: Zeitschr. Braunkohlen, v. 30, no. 16, p. 325-333.

———1931, Pollenformen aus tertiären Braunkohlen [pt. 3]: Jahrb. Preuss. Geol. L. A. f. 1931, v. 52, p. 1-7.

———1931, Zur Mikroskopie der Braunkohlen: Tertiäre Sporen- und Blütenstaubformen [pt. 4]: Zeitschr. Braunkohlen, v. 30, no. 27, p. 554-556.

———1934, Zur Mikrobotanik des eocäenen Humodils des Geiseltals: Arb. Inst. Paläobot. u. Petrogr. Brennst., Preuss. Geol. L. A., v. 4, p. 25-125.

———1956, 1958, 1960, Synopsis der Gattungen der Sporae dispersae: Beihefte Geol. Jahrb., v. 23, 31, and 39.

Ramsay, H. P., 1963, Chromosome numbers in the Proteaceae: Australian Jour. Bot., v. 11, no. 1, p. 1-20.

Reid, E. M., 1920, A comparative review of Pliocene floras, based on the study of fossil seeds: Geol. Soc. [London] Quart., v. 76, p. 145-161.

Reid, E. M., and Chandler, M. E. J., 1933, The London Clay flora: Brit. Mus. (Nat. Hist.) London.

Rouse, G. E., 1957, The application of a new nomenclatural approach to Upper Cretaceous plant microfossils from western Canada: Can. Jour. Botany, v. 35, p. 349-375.

Sachs, V. N., and Strelkov, S. A., 1961, Mesozoic and Cenozoic of the Soviet Arctic: Internat. Symp. Arctic Geology, 1st, Toronto, Proc., v. 1, p. 48-67.

Sarmiento, R., 1957, Microfossil zonation of Mancos Group: Amer. Assoc. Petrol. Geol. Bull., v. 41, no. 8, p. 1683-1693.

Seward, A. C., 1931, Plant life through the ages: Cambridge.

Scott, R. A., Barghoorn, E. S., and Leopold, E. B., 1960, How old are the angiosperms?: Am. Jour. Sci., v. 258-A, p. 284-299.

Scott, R. A., 1954, Fossil fruits and seeds from the Eocene Clarno Formation of Oregon: Palaeontographica, v. 96, sec. B, p. 66-97.

Sharp, A. J., 1951, The relation of the Eocene Wilcox flora to some modern floras: Evolution, v. 5, no. 1, p. 1-5.

Simpson, J. B., 1961, The Tertiary pollen- flora of Mull and Ardnamurchan: Royal Soc. Edinburgh Trans., v. 44, no. 16, p. 421-468.

Smiley, C. J., 1961, Paleobotanical investigations in arctic Alaska: Am. Jour. Bot, v. 48, no. 6, pt. 2, p. 542 [abs.].

Stanley, E. A., 1960, Upper Cretaceous and lower Tertiary Sporomorphae from northwestern South Dakota: Ph.D. Thesis, Pennsylvania State Univ.

———1961, The fossil genus *Aquilapollenites*: Pollen et Spores, v. 3, no. 2, p. 330-353.

Stauffer, C. R., and Thiel, G. A., 1941, Paleozoic and related rocks of southeastern Minnesota: Minnesota Geol. Survey Bull. 29.

Steeves, M. W., 1959, The pollen and spores of the Raritan and Magothy formations (Cretaceous) of Long Island: Ph.D. Thesis, Radcliffe College, Cambridge, Mass.

Sung, Tze-chen, 1956, Tertiary spore and pollen complexes from the red beds of Chiuchuan, Kansu and their geological and botanical significance: Acta Palaeont. Sinica, v. 6, no. 2, p. 167 [English summary].

Szafer, W., 1946, The Pliocene flora of Kroscienko in Poland: Polish Acad. Sci., Rozprawy Wydzialu Matematyczno-Przyrodniczego, v. 72, p. 91-162 [English summary].

———1954, Pliocene flora from the vicinity of Czorsztyn (west Carpathians) and its relationship to the Pleistocene: Inst. Geol. Prace, v. 11, p. 5-238.

Teichmüller, M., 1958, Rekonstruktionen verschiedener Moortypen des Hauptflözes der niederrheinischen Braunkohle: Fortschr. Geol. Rheinld. u. Westf., v. 2, p. 599-612.

Teichmüller, R., 1958, Die niederrheinische Braunkohlenformation: Der derzeitige Stand der Untersuchungen und offene Fragen: Fortschr. Geol. Rheinld. u. Westf., v. 2, p. 473-477.

Thiergart, F., 1949, Der Stratigraphische Wert mesozoicher Pollen und Sporen: Palaeontographica, v. 89, sec. B, p. 1-32.

——1953, Über einige Sporen und Pollen der Perutzer Schichten (Bohmen): Palaeontographica, v. 95, sec. B, p. 53-59.

——1954, Einige Sporen und Pollen aus einer Cenomankohle Südfrankreichs (St. Paulet Caisson nahe Montelimar, nordlich Marseille) und Vergleiche mit gleichaltrigen Ablagerungen: Zeitschr. Geologie, v. 3, no. 5, p. 548-559.

——1958, Die Sporomorphen-Flora von Rott in Siebengebirge: Fortschr. Geol. Rheinld. u. Westf., v. 2, p. 447-456.

Thomson, P. W., 1950, Grundsatzliches zur tertiären Pollen- und Sporomikrostratigraphie auf Grund einer Untersuchung des Hauptflozes der rheinischen Braunkohle in Liblar, Neurath, Fortuna, und Bruhl: Geol. Jahrb., v. 65, p. 113-126.

Thomson, P. W., and Pflug, H., 1953, Pollen und Sporen des mitteleuropäischen Tertiärs: Palaeontographica, v. 94, sec. B, p. 1-138.

Tokunaga, S., 1958, Palynological study on Japanese coal: Geol. Survey Japan Report no. 181.

——1964, Tertiary plant records from Japan: the microfossils, *in* Ancient Pacific floras, 10th Pacific Sci. Cong. Ser., Univ. Hawaii Press, p. 13-20.

Traverse, A., 1955, Pollen analysis of the Brandon lignite of Vermont: U.S. Dept. Int., Bur. Mines, Rept. of Inves. 5151.

Tschudy, R. H., 1961, Palynomorphs as indicators of facies environments in Upper Cretaceous and lower Tertiary strata, Colorado and Wyoming: Wyoming Geol. Assoc. Guidebook, 16th Ann. Field Conf., p. 53-59.

Velenovský, J., and Viniklář, L., 1926-1931, Flora Cretacea Bohemiae: Rozpravy Statniho Geologického Ústavu Ceskoslovenské Republiky, Svazek 1,2,3,5.

Weyland, H., and Greifeld, G., 1953, Über Strukturbientende Blätter and pflanzliche Mikrofossilien aus den untersenon Tonen des Gegend von Quedlinburg: Palaeontographica v. 95, sec. B, p. 30-52.

Weyland, H., and Krieger, W., 1953, Die Sporen und Pollen der Aachener Kreide und ihre Bedeutung für die Charakterisierung des mittleren Senons: Palaeontographica, v. 95, sec. B, p. 6-29.

Wilson, L. R., and Webster, R. M., 1946, Plant microfossils from a Fort Union coal of Montana: Am. Jour. Bot., v. 33, no. 4, p. 271-278.

Wodehouse, R. P., 1933, Tertiary pollen—II. The oil shales of the Green River formation: Torrey Bot. Club Bull., v. 60, p. 479-524.

Wolfe, J. A., and Barghoorn, E. S., 1960, Generic change in Tertiary floras in relation to age: Am. Jour. Sci., v. 258-A, p. 388-399.

Yanshin, A. L., 1960, Stratigraphic position of the Danian stage and the problem of the Cretaceous-Paleocene boundary: Rept. 21st Session Norden Internat. Geol. Congr., pt. V, p. 210-215.

Zaklinskaya, E. D., 1957, Stratigraphic significance of pollen grains of gymnosperms of the Cenozoic deposits of the Irtysh basin of the northern Aral basin [in Russian]: Moscow Acad. Sci. SSSR, Geol. Inst. Contrb. 6.

17

LATE CENOZOIC PALYNOLOGY

Estella B. Leopold

Late Cenozoic floras can be compared with living plants on a more detailed taxonomic basis than can older floras. Because of this, firmly based ecological inferences can be made; these are increasingly accurate with decreasing age of the floras. The large amount of detail available from late Cenozoic floras emphasizes considerations that are not usually as apparent in older assemblages. These aspects include possible mixed geographic sources of primary pollen, reworking of older or secondary pollen, and the abundance of fossils compared to the abundance of the source plants.

Elaborate statistical techniques now utilized in the interpretation of Pleistocene floras have not yet been applied extensively to late Tertiary floras. The use of more refined methods in the interpretation of late Tertiary floras appears to be both feasible and helpful. In many instances much could be learned from the comparison of modern pollen rain with late Tertiary pollen assemblages.

The absolute age of Quaternary floras younger than 40,000 years can be determined accurately by the use of carbon-14. Many older floras can be dated, but less precisely, by potassium-argon ratios. A relatively new technique uses potassium-argon ratios to date volcanic flow rocks whose paleomagnetism can be determined (Cox, Dalrymple, and Doell, 1967). Reversals in the earth's magnetic field are recorded by the orientation of magnetic particles in volcanic rocks and deep sea sediments. The sequence of these reversals provides a time scale useful in late Cenozoic stratigraphy.

Because of the primary emphasis of this book on older floras, late Quaternary floras, ably reviewed by other authors for many parts of the world, are not treated here. Also, the large amount of information available from the north latitudes automatically slants this chapter toward the extratropical floras of the Northern Hemisphere.

Quaternary pollen chronologies have been characterized in several areas that lie well south of the glaciated areas and in mountain ranges far removed from continental glaciers, but such sequences are not included in the present work.

STAGES OF THE LATE CENOZOIC

The late Cenozoic, generally considered to be the time interval between the Oligocene and the present, includes the Miocene and Pliocene Epochs of the Tertiary Period and the Quaternary Period. The late Tertiary is often referred to as the "Neogene"[1] period in contrast to the "Paleogene" (Paleocene through Oligocene Epochs of the early Tertiary).

The term "Quaternary" was proposed by Desnoyers in 1829, and historically is recognized as encompassing all of post-Tertiary time through

Publication authorized by the Director of the U.S. Geological Survey

[1]The term "Neogene," suggested in 1903 by Hoernes to include the younger Tertiary (Miocene and Pliocene) and now accepted by the International Geological Congress (Roger, 1964), is widely used in its original sense by European geologists but has been redefined by some American sources to include the entire late Cenozoic (Fay, 1920; American Geological Institute, 1960). In the present work the first definition is used.

the present. On the other hand, the term "Pleistocene," which was first proposed by Charles Lyell in 1839, is now commonly used to refer to the ice age as separate from postglacial time (Wilmarth, 1925). Thus, as used here, the Quaternary Period is comprised of the Pleistocene plus the Holocene (Recent[2] or postglacial) Epochs.

The base of the Quaternary or Pleistocene has been defined on two main criteria, paleontology and inferred climate. Charles Lyell, in distinguishing between Pliocene and post-Pliocene sediments, emphasized paleontologic data and characterized the latter as having molluscan faunas containing more than 70 percent living species. Some more recent authors have stressed the importance of climatic change in denoting this boundary (see King, 1950; Migliorini, 1950). In 1948 at the 18th International Geological Congress (1950), a commission appointed to advise on the definition of the Pliocene-Pleistocene boundary selected the base of the marine Calabrian and the base of its supposed continental equivalent, the Villafranchian in Italy, as the type section to denote this boundary. That boundary is now known from potassium-argon dating to be 2 to 3 million years old (Evernden and Curtis, 1965). When Venzo (1965) pointed out that the base of the Villafranchian may be slightly older than that of the Calabrian, the commission then chose the base of the Calabrian as the boundary (Grichuk, Hey, and Venzo, 1965).

The late Tertiary epochs and the early Quaternary have been subdivided into formal stages based chiefly on invertebrate evidence in Europe and into informal provincial ages based on common usage of vertebrate evidence in North America. The classic European stages are based on established type sections, but the North American mammalian provincial ages are not based on type sections nor have formal stages ever been proposed for them (two exceptions are the Montediablan and Cerrotejonian (= Clarendonian) formally proposed in California. Separate Neogene stages have been set up for marine and continental sections, and foraminifer and gastropod specialists have proposed their own formal stages. Though plant fossils have been described from many of the type sections established for the classic European stages, none of these was originally defined on plant evidence.

European stage names and North American provincial age names are given in figure 17-1, showing correlations based on potassium-argon dating by Evernden and Curtis (1965). Recent invertebrate evidence from Western United States supports earlier indications that the Blancan straddles the Pliocene-Pleistocene boundary and is of late Pliocene and earliest Pleistocene age (Taylor, 1960). Not shown in Figure 17-1 are the two post-Blancan provincial ages, Irvingtonian and Rancholabrean, considered to be middle and late Pleistocene in age, respectively.

The boundary between the Miocene and Pliocene in southern Europe has been variously placed at the top or bottom of the Pontian and/or at the top of the Sarmatian; according to the isotope time scale reported by Evernden and Curtis (1965, table 1), these alternate placements would make the boundary as old as 12 million years or as young as 6 million years. The Mediterranean Neogene Committee (Roger, 1964) declared the Messinian (approximately equivalent to the Pontian) to be latest Miocene in age, implying that the boundary lies at the top of the Messinian, now dated to be 6 to 7 million years in age (Tongiorgi and Tongiorgi, 1964). For purposes of this paper the somewhat more common, older usage, the base of the Pontian, is used for this boundary.

In North America the Miocene-Pliocene boundary has been placed at either the bottom, middle, or top of the Clarendonian, or even within the Hemphillian (Wood and others, 1941; Evernden and Curtis, 1965; Charles Repenning, written communication, 1967).

The Oligocene-Miocene boundary is placed by some between the Chattian and the Aquitanian (Wood and others, 1941; Selli, 1964); that position is dated by Evernden and James (1964) as about 26 million years old. Others (e.g., Gignoux, 1960) place the top of the Oligocene at the top of the Aquitanian, which according to radiometric dating is about 25 million years old. (See fig. 17-1.) The former definition is used here.

[2]The use of Recent for the postglacial interval was formally abandoned in 1968 by the U.S. Geological Survey in favor of the term "Holocene" due to the possible confusion of the formal, Recent, with the adjective "recent."

Figure 17-1. **Correlation of European stage names and North American provincial age names for the late Cenozoic on the basis of potassium-argon dating. From Evernden and Curtis (1965).**

10^6 years	*North America*	*Europe*	10^6 years
0			0
		Villafranchian	
	Blancan, etc.		
		Astian-Plaisancian	
	Hemphillian		
		Pontian	
10			10
	Clarendonian		
		Sarmatian	
	Barstovian		
		Vindobonian	
	Hemingfordian		
20			20
		Burdigalian	
	Arikareean		
		Aquitanian	
	Whitneyan		
		Chattian	
30			30

Subdivisions of the Pleistocene and late Pliocene based on what are generally considered synchronous climatic oscillations at middle or high north latitudes, shown on table 17-1, indicate probable correlations of major Pleistocene stages and glaciations. Correlations of the last two glaciations and last interglaciation are fairly certain as they are now demonstrated by carbon-14 dating in many areas. Correlations of earliest Pleistocene stages are also well established (in part by potassium-argon dating) in northwestern Europe and in Italy, but as table 17-1 states, time relationships for events of the middle and early Pleistocene are not yet well ascertained on a worldwide basis.

LATE TERTIARY FLORAS

General Characteristics of Extratropical Floras

The best documented late Tertiary floras are from the middle and high latitudes of the Northern Hemisphere. Though little is yet known about Neogene floras and vegetational changes at low latitudes, indications are that climatic changes here were slight.

Late Cenozoic floras, whether based on pollen or on leaves, differ from earlier ones in several striking respects:

1. In any given area the Neogene flora is generally less diverse (less rich in forms) than the Paleogene flora.

2. Neogene and Quaternary floras contain a higher proportion of living genera (and sometimes species) than Paleogene floras, which may include extinct genera.

3. Miocene strata from a wide range of latitudes in the Northern Hemisphere contain the first pollen records of some plant families of herbaceous habit. Some woody groups that may have originated before the Miocene are not represented in abundance in the pollen record until Neogene time. Hence pollen of these plant groups can be useful as indicators of late Cenozoic age.

4. A comparatively high proportion of plant genera identified from Neogene fossils now live near the fossil locality; in contrast only a small proportion of Paleogene plant genera now grows locally or in the modern floristic province in which the fossil site is located.

5. Late Cenozoic floras typically are more provincial in character than most floras of Cretaceous and Paleogene age; this is particularly striking in floral assemblages younger than middle Miocene.

These characteristics of late Cenozoic floras are discussed below.

Decreasing Diversity of Flora. In any given area in the middle or high latitudes progressively younger Cenozoic floras tend to include fewer taxa.

The taxonomic diversity of modern biotas is

Table 17-1. Probable Correlative Major Subdivisions of the Pleistocene and Late Pliocene in the Middle and High North Latitudes[1]

	Venzo (1965) and Grichuk (1960)							*Hopkins and Others (1965)*		*Crandell, Mullineaux, and Waldron (1958)*	*Frye and Others (in Wright and Frey, 1965)*
	Italy (Marine Stages)	*Alps and Italy (Continental Sequence)*	*England*	*Germany*	*Netherlands*	*Poland*	*Russian Plain*	*East Siberia Chukotka Area*	*Alaska*	*Washington State*	*Central United States*
Pleistocene (Correlations Tenuous)								**Islaten** moraines	**Salmon Lake**	**Fraser**	
		Würm	**Hunstanton** boulder clay	**Weichselian**	**Taubantian**	**Baltusk**					**Wisconsinan**
							Valdai	**Vankarem** moraines		**Salmon Springs**	
		(Eemian)	(Ipswichian)	(Eem)	(Eem)	(Mazowian II)	(Mikulina)	(Konerginsk beds)	(Pelukian transgression)	(Puyallup)	(Sangamonian)
							Moskow				
		Riss									
			Gipping (Hoxnian)	**Saale** (Holstein)	**Drethnian** (Needian)	**Srodkow** (Mazowian I)	**Dneiper**	Kresta suite	**Nome River**	**Stuck**	**Illinoian** (Yarmouthian)
	Sicilian (Emilian)	**Mindel** –	**Lowestoft** (Cromerian)	**Elster** (Cromerian)	**Taxandrian**	**Krakow**		(?Pinakul' suite)		(Alderton) **Orting?**	**Kansan** (Aftonian)
	Calabrian	**Gunz** – (Villafranchian)	– (Upper Crag)	**Weybour** (Tiglian)	(Tiglian)	Mizerna	(Akchagylian)		(?Anvillian transgression)		**Nebraskan**
		Donau		**Brachtian**	**Praetiglian**				**?Iron Creek**		
Pliocene	(Astian and Plaisanscian)				(Reuverian)	(Kroscienko)	(Kimmerian)	(?Koynatkhum Suite) (?Pestov Suite)	(Beringian transgression)		

[1]Glacial stages and glaciations are shown in bold type, and nonglacial or interglacial stages and interglaciations are in parentheses. Within glaciations multiple ice advances ("stades") are known locally (e.g., Donan I, II, and III in Italy) but are not indicated here.

known to be greatest in tropical and subtropical climates and decreases progressively with higher latitudes (Fischer, 1960). In the midlatitudes of the present North Temperate Zone the change from the diverse tropical floras of Eocene time to the more simple, temperate floras of the Quaternary Period is related to climatic deterioration, particularly of the late Tertiary. Consequently this change is analogous to the modern differences between floras of low and middle latitudes. The progressive late Cenozoic loss in diversity seen in pollen floras is borne out in general by megafossil evidence from a wide range of latitudes—for example, by the late Cenozoic seed floras of Europe. (See fig. 17-2.)

Assignment to Living Taxa. A higher proportion of fossil-plant forms from late Cenozoic assemblages can be placed in living genera or species than can forms from early Cenozoic floras.

Extinct angiosperm genera are common in Eocene floras; for example, based on megafossils from England the early Eocene flora of the London Clay contains more than one-half extinct genera (Reid and Chandler, 1933, p. 46), whereas the figure drops to 43 percent in a late Eocene flora and to only 15 percent in an Oligocene flora (Chandler, 1963). Only two extinct woody genera have been reported from Neogene deposits in northwestern Europe; these are *Teschia* (Anacardiaceae) and *Jongmansia*

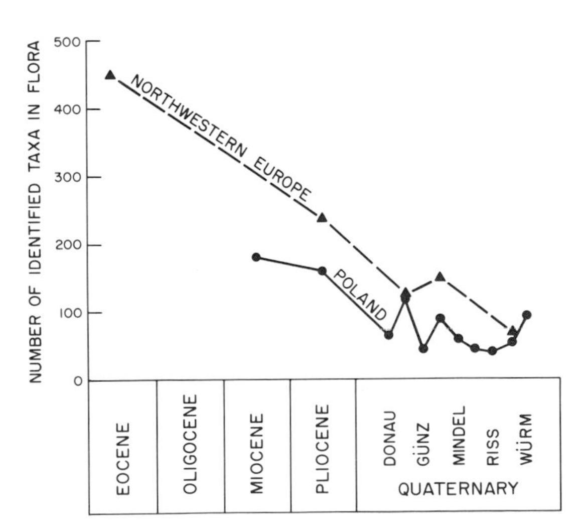

Figure 17-2. Decreasing diversity in seed floras of Europe; number of taxa identified from each flora is plotted against geologic age. (Data from Szafer, 1961; Reid and Reid, 1915.)

(Annonaceae) from the lower Pliocene (Reid and Reid, 1915). The youngest records of extinct plant genera are those from highly evolved, herbaceous angiosperm groups; some extinct grass genera—for example, *Stipidium* (Elias, 1935)—are as young as Pliocene; several extinct genera of the Lythraceae and Hydrocharyaceae are also Pliocene in age (Miki, 1959). No extinct coniferous genera have been reported from the Tertiary.

Most Neogene genera and Pleistocene species of plants are living today. On the species level many living angiosperms for which megafossil records exist in the northern middle latitudes can be traced back to the Pliocene.

Indicators of Late Cenozoic Age. Pollen of certain families first appearing or becoming abundant in the Neogene can be useful as indicators of late Cenozoic age. Such groups include highly evolved families, such as the Compositae, and certain herbaceous as well as woody angiosperm families. In some examples the late appearance of fossil pollen types may actually indicate a late Cenozoic origin for the families. In others the lack of pollen evidence in the face of megafossil evidence from the Paleogene may only indicate our ignorance concerning early Tertiary pollen floras.

The first occurrence of pollen of Compositae is a stratigraphic marker for the late Oligocene or early Miocene all over the world. In Germany Compositae pollen and fruits appear first in what are considered late Oligocene sediments (top of the Chattian) but only as rare specimens in occasional samples; in middle Miocene sediments pollen of Compositae appears more regularly but constitutes less than 1 percent of the total. By late Miocene time it composes 1 to 10 percent of pollen in all local sediment samples (Ahrens, 1958).

Potonié (1934) recorded Compositae-like pollen from Eocene and Oligocene strata of the German brown coals, but doubt surrounds these pollen records because they have not been corroborated by later work (Krutzsch, 1957, 1959; Kirchheimer, 1957).

Krutzsch (1957) in a summary of his pollen studies on the German Tertiary section states that spinose Compositae pollen first is found in the uppermost Oligocene (Flöz 3, Niederlausitz deposits) and that fenestrate Compositae types first appear in deposits of Pliocene age. He re-

ports that numbers of both types of pollen increase upward in the Neogene section; in Miocene samples occasional single grains, and in Pliocene abundant grains of Compositae occur. Kirchheimer's (1957) earliest records of Compositae seeds or fruits come from the Niederlausitz deposits, which he terms of middle or late Oligocene age.

In the Rocky Mountain region Compositae pollen first appears in the latest Oligocene or earliest Miocene and becomes quite abundant, especially the genus *Artemisia*, in the middle Miocene (Love, 1961). Occasional Pliocene Compositae may be in place from Alaskan coastal localities (Wolfe and Leopold, 1967) and from the Miocene of the Alaska Range (table 17-6). A few Compositae pollen may be in place in the late Oligocene of North Carolina (Leopold, 1959).

In the Soviet Union occasional grains of Compositae were reported from Oligocene sediments in 4 of the 12 districts studied—in the Baltic lowlands, northwestern Caucasus, western Siberia, and the Turgai lowlands (Pokrovskaya, 1956a). Of 18 districts in the Soviet Union from which early Miocene pollen has been described, Compositae pollen has been reported from only one—the Turgai Depression. In upper Miocene deposits single grains in some analyses are reported from the Don River, Moldavia, and the Ukraine districts, south of the Urals, and in the Baikal district (Pokrovskaya, 1956b). By Pliocene time Compositae pollen is reported as occurring in nearly all samples.

Pollen of some highly evolved, primarily herbaceous groups is known only from Miocene and younger sediments. Examples are Droseraceae (pl. 17-1, fig. 36; table 17-6), Polemoniaceae, and Caryophyllaceae. Pollen of Gramineae, Onagraceae, and Cyperaceae does not become abundant until Neogene time but is known from Eocene sediments.

Pollen of certain woody groups is so far known only from the late Cenozoic, although the taxa probably existed during the Paleogene; for example, Malvaceae pollen is reported from Miocene floras in Eniwetok, Palau Islands (Leopold, 1969), from the Miocene? of Nigeria (Kuyl, Muller, and Waterbolk, 1955, pl. 4, fig. 1), and from Miocene as well as from Pliocene sediments of the Rocky Mountains, but not from earlier sediments. Pollen of Sonneratiaceae—reported from the Miocene in Borneo (Muller, 1964), Eniwetok, Fiji, and Palau—is lacking in Oligocene sediments of Borneo. Sonneratiaceous fruits and flowers are reported from the Paleogene of India (Mahabale and Deshpande, 1957).

Though *Trapa*-like pollen is present in Oligocene floras of the Soviet Union, authentic pollen of the genus is first recorded in the Miocene of the Enisei Mountains (Popov, 1956). A Miocene appearance for the genus *Trapa* is also suggested by evidence from fossil fruits in the Pacific Northwest and Alaska (Jack A. Wolfe, written communication, 1968). The family Trapaceae has been recognized by megafossils in Cretaceous sediments (Vasily'ev, 1967).

Stratigraphic pollen records now available for certain plant families are summarized in figure 17-3.

Proportion of Fossil Forms Still Living Near Their Fossil Localities. A large proportion of Neogene plants or their close relatives are now living near their Neogene sites of occurrence, but typically only a small proportion of such plants or their near relatives in Paleogene floras are now a part of the local flora. This characteristic is evident on both the generic and specific levels.

Reid (1935) was perhaps the first to point out that similarity in generic composition of a fossil flora with the modern regional flora increases with decreasing age of the deposits; she expressed this change numerically and used it as a criterion of age in British floras.

In the Western United States the increasing role of local modern genera in late Cenozoic floras has been summarized by Wolfe and Barghoorn (1960) and Leopold (1965, 1967). Figure 17-4, which is based on both pollen and megafossil floras, shows that the "modernization" trend proceeded at different rates in different parts of the continent. Although the data for coastal southern Alaska and the Pacific Northwest coast show a gradual rate of change through the entire late Cenozoic, the curve for the Rocky Mountains shows a much more rapid "modernization," especially during the late Oligocene and early Miocene. Geographic factors that may have con-

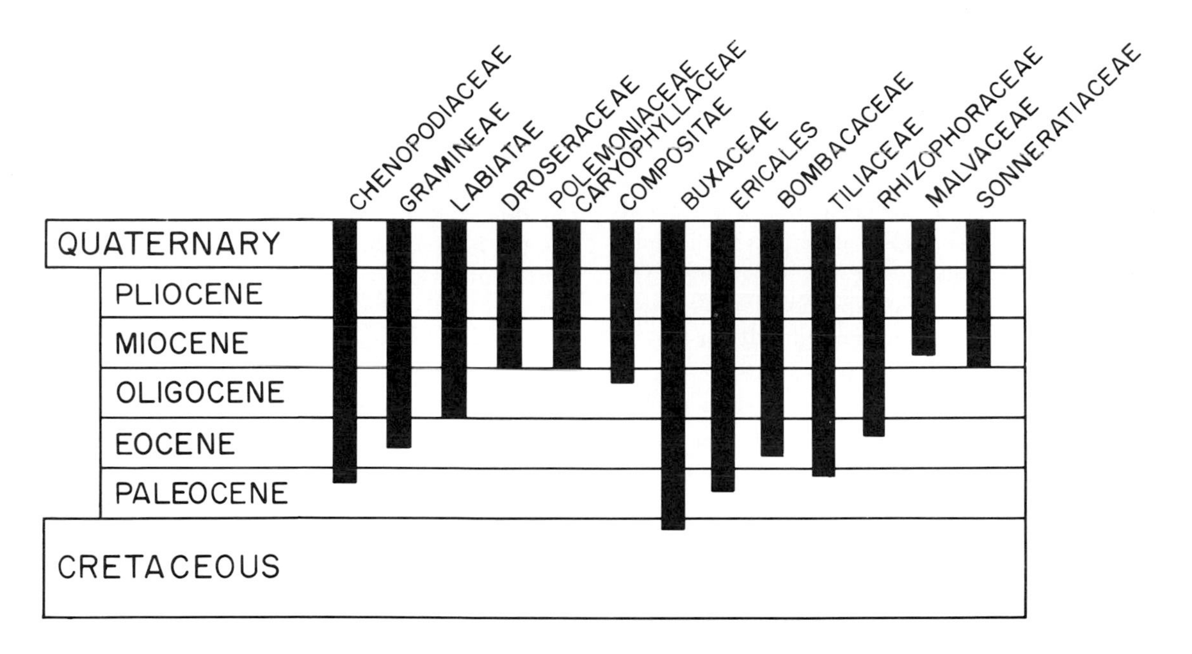

Figure 17-3. Known stratigraphic ranges for pollen of certain angiosperm groups; primarily herbaceous groups are on left, woody groups on right.

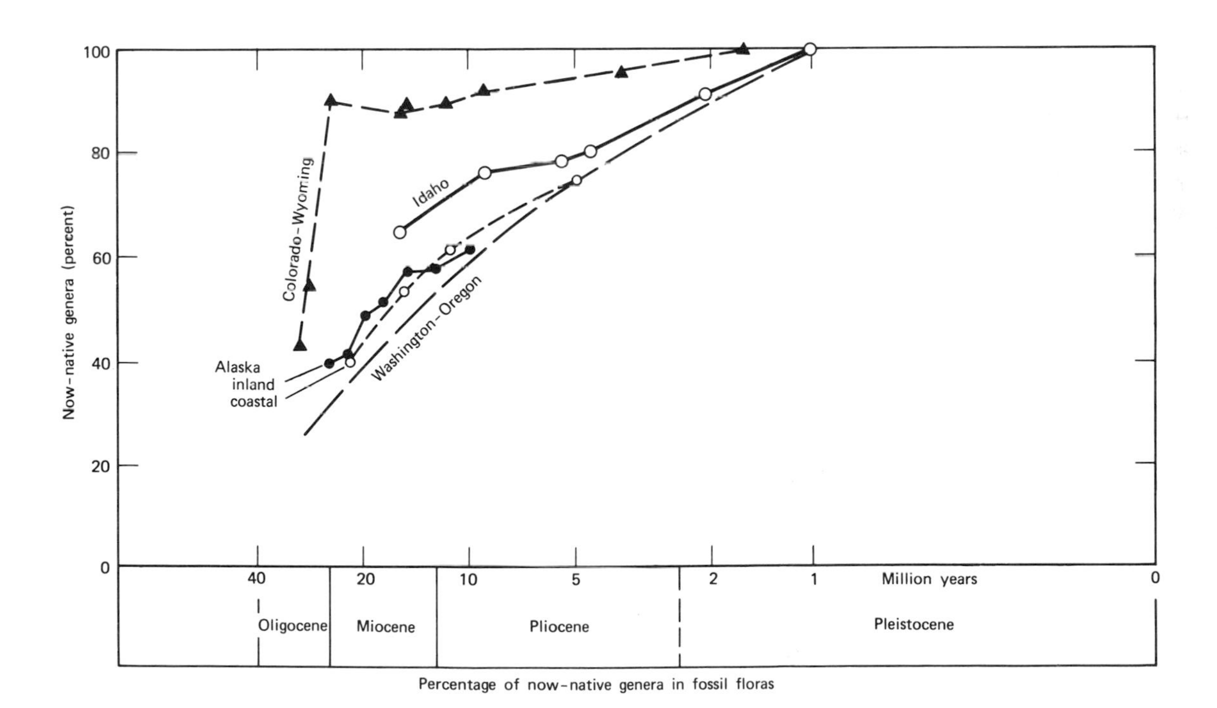

Figure 17-4. Generic age curves; percentages of now-native genera in relation to total identified genera in late Cenozoic floras of Western United States; data for Washington-Oregon curve are selected floras from Wolfe and Barghoorn (1960). Horizontal axis is a log time scale (modified from Leopold, 1967).

tributed to these patterns are discussed on pages 407-409.

The gradual modernization during the late Cenozoic of the European flora is demonstrated by Szafer's (1954, 1961) work on the seed floras of southern Poland. In figure 17-5 the percentages of various geographic elements in the floral lists are plotted according to geologic age. The data are based mainly on species identifications, but a few forms are determined only on the generic level.

The data plotted in figure 17-5 demonstrate the increasing role of plants that are now native to southern Poland. Today these plants also range into the temperate zones of Eurasia, western Europe, and elsewhere. Some species of the modern Polish flora extend back to the early Miocene, but modern species apparently did not dominate until Pleistocene time.

Increasing Provincialism in the Neogene. Unlike most floras of early Cenozoic age, late Cenozoic floras typically demonstrate marked provincialism. Assemblages of post-middle Miocene age may differ widely within small areas; this is apparently the result of latitudinal and topographic differentiation during the late Cenozoic. Because of this differentiation the distances over which floras can be correlated are lessened for Pliocene and younger assemblages.

The provincial characteristics of post-middle Miocene floras that were apparent from megafossils have been corroborated by palynologic evidence from a wide range of latitudes. The middle Miocene pollen flora of Colorado (pl. 17-2; see also p. 405) is similar to that of the Alaska Range (pl. 17-1; see also p. 400), but Pliocene and younger floras of the two regions are quite different (Wolfe, Hopkins, and Leopold, 1966).

Strong resemblances exist between the pollen floras of Paleocene age in Europe (Krutzsch, Pchalek, and Spiegler, 1960) and the Southeast Coastal Plain of the United States (Jones, 1962), and similarities still exist in the Miocene; the Miocene pollen flora from the Rhine River area of Germany (Thomson and Pflug, 1953) is strikingly similar to a Miocene flora from Alabama (Pascagoula Clay). However, the modern vegetation and the recent pollen rain in the two areas are indeed different, because Alabama is still subtropical and the Rhine River area is now temperate.

These five characteristics of late Cenozoic floras form a useful basis for interpreting and in some cases for helping to date floras in the upper part of the geologic column. Potential for recognition of living taxa among late Cenozoic forms is high, and thus detailed ecologic and phytogeographic interpretations are possible for the late Cenozoic. A "generic age curve" based on dated local floras from superposed rocks may be used to determine the general age of other local floras. The degree of provincialism and of floristic diversity can be established by regional studies of independently dated floras; the provincial character of late Cenozoic floras represents more of a limiting factor than a tool and should be considered in formulating phytogeographical and climatic interpretations. Presence of plant families or genera whose pollen is known to be limited to upper Cenozoic sediments can be useful in dating floras.

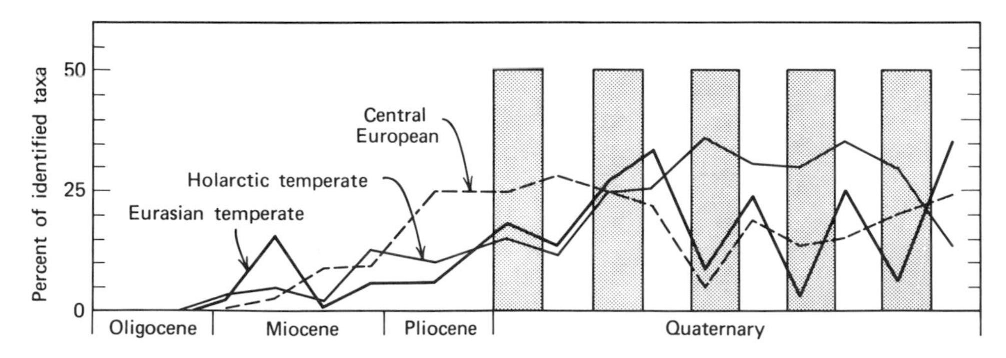

Figure 17-5. The progressive increase in representation of modern Polish taxa in late Cenozoic seed floras of southern Poland (modified from Szafer, 1961; Leopold, 1967). (Shaded bars denote times of European glaciations.)

Interpretation of Late Tertiary Floras

One of the most critical problems in evaluating the paleoecology of a fossil-pollen flora is determining what constitutes evidence of local provenance and what represents pollen drift or long-distance transport. Toward this end a familiarity with pertinent literature on modern regional pollen rain is helpful. Criteria for judging local plant sources as opposed to long-distance transport by wind or water are discussed by Faegri and Iversen (1964, p. 114-120). However, it is worth mentioning here that the palynologist should evaluate the following aspects in this regard.

1. The nature of the sediment: peats may be expected to contain more locally derived pollen than highly inorganic sediments; megafossil evidence is helpful in evaluating whether the source plants grew locally.
2. Abundance of the pollen type: this should be related to pollen productivity of the source plant.
3. Reworking: the possibility that some of the pollen is derived from older sources cannot be disregarded.

Identifying Floristic Elements. The taxa identified from a fossil flora may be classified according to their present geographic distributions or according to the distributions of their nearest relatives. This effort provides a basis for estimating paleoclimates, and it can provide leads for identifying some of the unknown elements within the flora. Stratigraphic records of pollen phenotypes are enhanced if the climatic, ecologic, and floristic connotations of the plants can be determined. But none of these objectives except pure stratigraphy can be realized unless the fossil pollen is identified with modern plants.

The more accurately and completely a fossil assemblage is compared with modern plants, the more reliable will be the resulting identifications. The design and scope of a modern pollen "herbarium" to serve as a reference collection with which to compare the fossil forms is a crucial aspect of botanical palynology. Some guidelines can be taken from Traverse's (1955) notable study of the Brandon lignite. He stressed the fact that the reference collection is far more valuable if each pollen preparation is referable to a specific herbarium sheet (voucher) and if each is prepared in duplicate permanent slides so that the sheet need not be sampled more than once. The collection should contain the following:

1. Genera and families in the modern flora of the immediate region.
2. Genera and families of woody plants from the regional floras, especially the plants that are suggested by locally available megafossil evidence and by the modern ecological associates of the megafossil taxa.
3. Genera and families that have been identified from other fossil floras of the same age in the region of the fossil locality.
4. Possibly, groups that have been found from fossil floras of about the same age in other parts of the world.

As a supplement to a modern pollen reference collection, compilations of photographs and drawings of modern pollen and spores may be helpful. Some of these are:

Geographic area:	*Groups covered:*	*Source:*
Scandinavia	ferns and higher plants	Erdtman, Berglund and Praglowski, 1961
Scandinavia	ferns and higher plants	Erdtman, Praglowski and Nilsson, 1963
Central Europe	ferns and higher plants	Beug, 1961
Japan	ferns and higher plants	Ikuse, 1956
World	angiosperms	Erdtman, 1966
World	bryophytes, pteridophytes, gymnosperms	Erdtman, 1957, 1965
New Zealand	monocotyledons	Cranwell, 1953
New Zealand	pteridophytes	Harris, 1955
Hawaii	pteridophytes	Selling, 1946
Hawaii	phanerogams	Selling, 1947
Venezuela	ferns	Tschudy and Tschudy, 1965

Geographic affinity of a fossil flora can be usefully expressed in terms of the floristic province in which the majority of the modern relatives of the identified forms live today. Additionally, the present ranges of minor elements in the flora can be of interest. Most helpful in these determinations are regional floras in which the ranges of a modern genus or its regional species are mapped. Useful sources are Hegi and others (1909-1931) for Central Europe, Hultén (1950) for Scandinavia, Faegri (1960) for coastal plants of Norway, Hultén (1941-1949, 1968) for the flora of Alaska and neighboring territories, and Hultén (1962) for circumpolar plants. Distribution maps for some woody North American forms are provided by Fowells (1965) and Munns (1938) and for amphiatlantic plants by Hultén (1958). The distributions of important plants in the Soviet Union and Asia were mapped by Gerasimov (1964). A few Malayan and South Pacific distribution maps are provided by VanSteenis (1963).

Broad floristic provinces were defined for North America by Gleason and Cronquist (1964). They recognized 10 floristic provinces that have large groups of species with similar distributions. These are listed below, together with a few representative genera or species that in North America are restricted to a single province. The diagnostic taxa with pollen structure that is distinctive on the generic level are marked with an asterisk (*); the species listed do not necessarily have distinctive pollen morphology. The genera marked with a dagger(†) also occur in the Old World. Some of the species listed belong to genera that also occur in the Old World; these are not marked.

1. Arctic or Tundra Province: *Silene acaulis, Betula glandulosa*
2. Northern Deciduous Forest Province: *Larix laricina, Abies balsamea, Picea mariana*
3. Eastern Deciduous Forest Province: *Fagus**†, *Magnolia acuminata, Pachysandra*†, *Castanea**†, *Aesculus glabra, Tsuga canadensis, Gymnocladus dioica*
4. Coastal Plain Province: *Taxodium*†, *Nyssa aquatica, Gordonia**†, *Osmanthus*
5. West Indian Province: *Rhizophora**†, Palmae†, *Sideroxylon*†, *Dipholis, Chrysophyllum*†, *Cordia*†, *Guettarda*†
6. Prairies or Grassland Province: various Gramineae, including *Buchloe*
7. Cordilleran Forest Province: *Pseudotsuga taxifolia, Libocedrus*†*, *Sequoia, Fraxinus oregona, Taxus brevifolia, Abies lasiocarpa, Picea pungens**, *P. sitchensis, Tsuga heterophylla, T. mertensiana*
8. Great Basin Province: *Sarcobatus**, *Pinus monophylla*
9. Californian or Chaparral Province: *Arbutus menziesii, Fremontia**
10. Sonoran Province: *Larrea**, *Fouquieria**, *Bursera**, *Simmondsia**

Some important eastern hardwoods occur in both provinces 3 and 4: *Carya, Tilia, Liquidambar.*

Genera now occurring in the Eastern Deciduous Forest and Coastal Plain provinces are common in the Miocene of the Western United States; for example, *Carya* is a usual element in Miocene floras of North-Central and Northwestern United States. (See fig. 17-6.) Other eastern elements—such as *Liquidambar, Nyssa, Ulmus, Ilex, Tilia, Ostrya-Carpinus, Fagus, Castanea,* and *Pachysandra*—were widespread in the Western States and Alaska during Miocene time, as were many Cordilleran elements (such as *Pseudotsuga, Sequoia,* and *Chamaecyparis*). These then grew with eastern hardwoods over a large area in western North America.

Miocene pollen documents the presence of many East Asian genera in the New World Miocene. An example is *Pterocarya*, which now has a limited distribution in China, Japan, and in the Caspian Sea region. (See fig. 17-7.) Not only was it an important Miocene forest tree in the Old World but it also was widespread in the United States, Canada, and Alaska. Other East Asian genera with a similar history include *Sciadopitys, Eucommia, Cunninghamia-Glyptostrobus,* and *Melia.* Pliocene records of such elements in western North America suggest that their ranges were shrinking rapidly during that time. So far there is no evidence that these warmth-loving elements regained much of their lost range after the Neogene.

After the taxonomic relationships are established the paleobotanist can turn to the ecology of the flora and its paleoclimate. The hypothesis that the present is the key to the past is the only basis for using fossils to determine ancient cli-

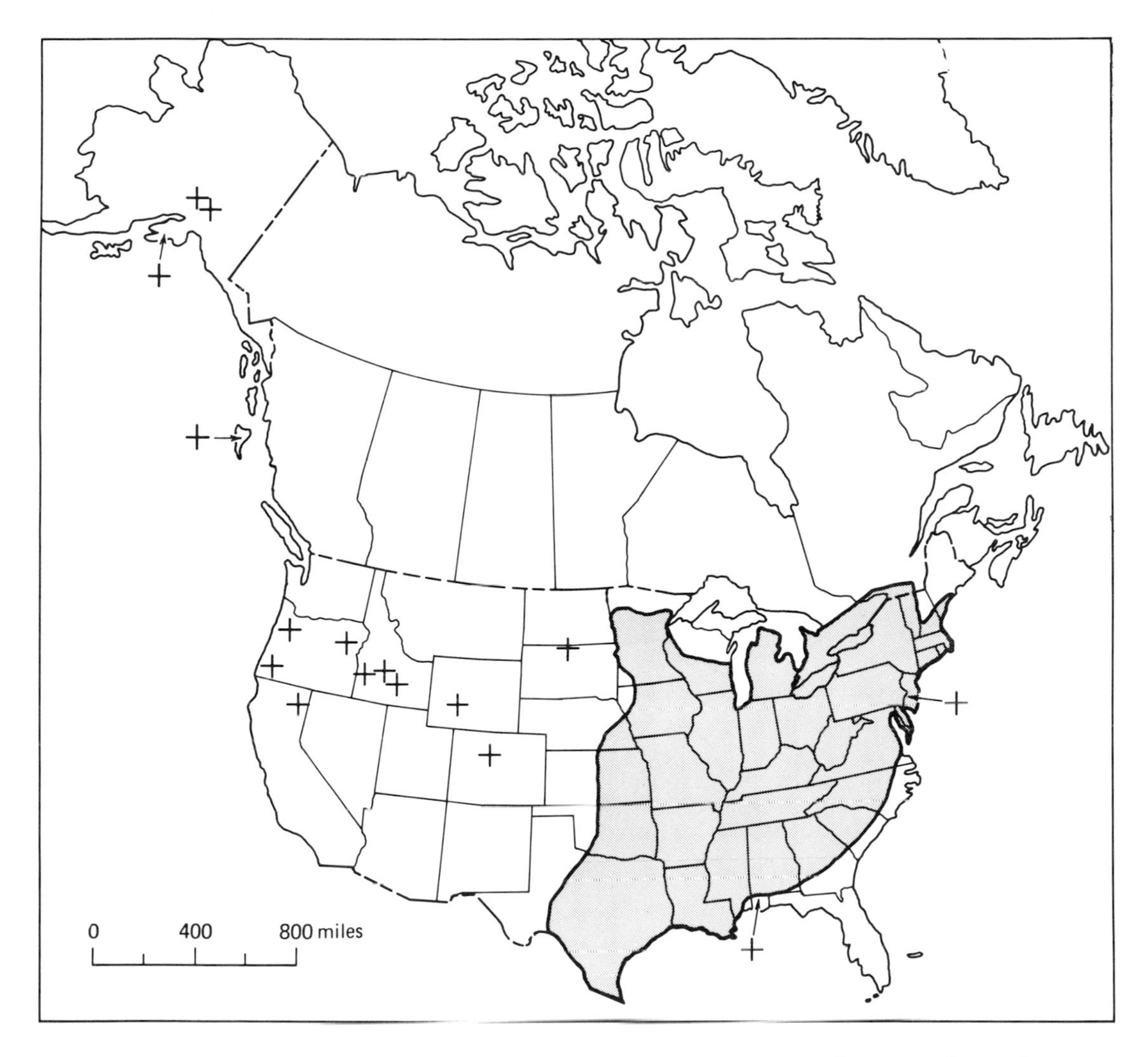

Figure 17-6. Modern and Miocene distribution of the genus *Carya* in the United States and Canada. (Modern range from Munns, 1938. Miocene records (+) from Gray, 1964; MacGinitie, 1962; Martin and Rouse, 1966; Wolfe, Hopkins, and Leopold, 1966; Wolfe and others, in press; Leopold, *in* Weber, 1965; and unpublished records.)

mate. But how far back in the fossil record can we assume that modern ecologic and climatic relationships of plants apply in detail? For an objective answer we must turn to fossil evidence on the age of extant plant species. Taxonomic analyses from the Neogene of Europe suggest that the bulk of vascular species of Pliocene age are living today. Miocene species are primarily extinct ones, and probably all Oligocene angiosperm species are extinct. However, almost all plant genera recorded in late Cenozoic floras are still living. Thus it would seem that climatic preferences and ecological relationships of modern vascular plants apply in detail to Pliocene floras and in general to those of Miocene or Oligocene age.

Comparisons of the present plant ranges with climatic maps such as those found in Gerasimov (1964), Brooks and others (1936), and in Hultén (1950) can provide a sound basis for determining the climatic significance of plant occurrences (see Dahl, 1964).

Some Late Tertiary Floras

Northwestern Europe. Northwestern Europe has been the scene of much palynological activity since the 1930's. Unfortunately a division regarding taxonomic procedures has occurred

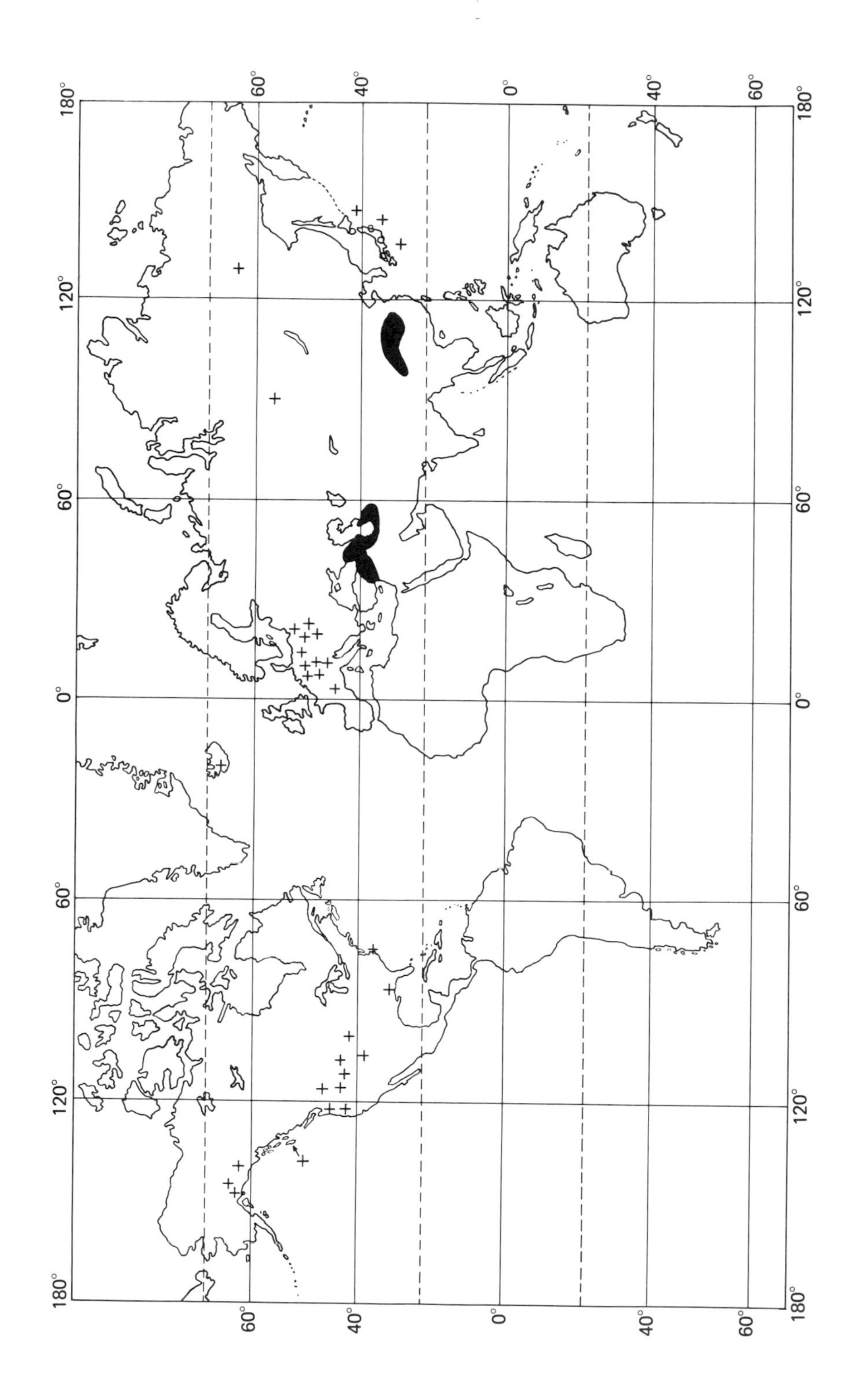

Figure 17-7. Modern and Miocene distribution of the genus *Pterocarya*. (Modern range (shaded areas and circles) from Tralau, 1963; Miocene localities (+) from Tralau, 1963; Gray, 1964; MacGinitie, 1962; Wolfe, Hopkins, and Leopold, 1966; Vangengiem, 1961; Wahrhaftig and others, in press; and E. B. Leopold and J. A. Wolfe, unpublished records.)

among European palynologists. Some workers have placed all Miocene and Pliocene taxa in form genera and species, some have used both organ and form genera, and others have assigned all possible forms to living genera. Thus comparisons between localities are made difficult by the semantic problems of taxonomy. In choosing examples of European Neogene floras I have favored works utilizing a botanical approach to taxonomy, and I have included my own interpretations of affinity for some form taxa.

Much European pollen work has dealt with the Rheinische Braunkohle from the Rhine and Elbe River deltas near Amsterdam, Cologne, and Berlin. These deposits are coastal moor and marsh sediments that range in age from middle Oligocene through early Quaternary. The sediments (including lower Oligocene beds lacking coal) were laid down in a tectonic basin whose marine shorelines had different positions in each epoch. A combination of tectonic movement and changes in sea level resulted in a sequence of marine clays and sands alternating with coaly (peaty) sediments. This section is well dated by Foraminifera and other marine fossils (Ahrens, 1958). The Oligocene and Neogene floras of these deposits are diverse: remains represented include leaves (Weyland, 1938-1948); fruits, seeds, and wood (Kircheimer, 1957; Reid and Reid, 1915); and pollen (Potonié, 1951b; Potonié, Thomson, and Thiergart, 1951; Thiergart, 1951; Thomson and Pflug, 1953; Pflug, 1953; Kremp, 1950). The pollen work by Zagwijn (1960, 1963a, b) on the Dutch deposits is especially notable.

The general sequence as inferred from pollen and megafossil evidence indicates a cooling of climate from at least middle Oligocene through the Praetiglian, or first glaciation. Tropical and some subtropical elements were eliminated during the Oligocene and early Miocene (*Mastixia*, Palmae, Sapotaceae, cf. *Anemia*). Warm temperate and subtropical forms were well represented during the early Neogene, but some were eliminated in northwestern Europe by the end of the Miocene (*Platycarya*, *Engelhardtia*, probably Cyrillaceae), and others were eliminated at the end of the Pliocene (*Meliosma*, *Liquidambar*, *Symplocos*, *Aesculus*, *Nyssa*). Strictly temperate elements were lacking in the flora of the first Pleistocene glaciation, and the main components were nontree forms and pine.

These floristic changes of the European Neogene resulted partly from secular cooling, but the changes were probably ameliorated by the presence of the Tethys sea during the Neogene. An expanded Mediterranean Sea covering parts of southern Europe and segments of the German coast during the Miocene (Termier and Termier, 1952) retreated to approximately the present shorelines by the end of the Tertiary.

Neogene pollen changes above the lower Miocene in the German brown coals include the following:

1. A general increase in abundance and kinds of saccate conifer pollen mostly of Pinaceae.
2. A general increase of Ulmaceae, *Alnus*, *Fagus* type (*Tricolporopollenites pseudocruciatus*), and *T. kruschi*.
3. A general decrease in the abundance of *Triporopollenites rurensis*, *Cupuliferoipollenites villensis*, *Quercus* types (*Quercoipollenites henrici* and *Q. microhenrici*), *Castanea* types (*Pollenites cingulum*, *P. pseudocingulum*), and *Tricolpopollenites liblarensis* (Thomson and Pflug, 1953).

Some stratigraphic features of the Neogene brown coals are summarized in table 17-2 and figure 17-8.

The palynological changes at the Oligocene boundary (at the top of the definite Chattian), are mainly quantitative, with only one or two forms terminating their ranges. However, a number of forms end (all Normapolles types) or begin (*Tilia*, various Juglandaceae, and *Liquidambar*) their ranges near the base of the Chattian within the late Oligocene (Potonié, 1951a; Thomson and Pflug, 1953). Significant changes also occur in the middle of the early Miocene; the following forms are eliminated at this level: *Symplocospollenites vestibulum*, *Araliaceoipollenites* (two spp.), *Sapotaceoipollenites (two spp.) and Cicatricosisporites*. palmaceous pollen, *Engelhardtioipollenites*, and *Platycarya* pollen become quite rare (Thomson and Pflug, 1953). It appears that floral changes during the late Oligocene and early Miocene were more spectacular than any yet recorded in these deposits at the boundary itself. (See table 17-2.)

Some important numerical changes taking place in pollen representation between the middle Oligocene and middle Miocene include the

Table 17-2. Stratigraphic Ranges of Selected Palynomorphs in the Tertiary of Germany and Probable Affinities of the Forms

Stratigraphic Ranges of Selected Sporomorphs in the Tertiary of Germany (from Potonié, 1951a)

Paleogene				Neogene			Sporomorph	Probable Affinities of Sporomorphs (by Leopold)
Paleocene	Eocene	Oligocene	Chattian	Chattian?	Miocene	Pliocene		
							Polypodiaceoisp. speciosus	*Pteris* type
							Mohrioisp. dorogensis	*Cicatricosisporites Anemia*
							Schizaeoisp. pseudodorogensis	*Schizaea*
							Lygodioisp. adriennis and *solidus*	*Lygodium* types?
							Osmundasp. primorius	*Osmunda* type
							Abietineen with wings	Pinaceae, Podocarpaceae
							Abietineaepoll. labdacus maximus	
							Tsuga spm.	*Tsuga*
							Laricolpoll. magnus	*Larix?*
							Sciadopityspoll. serratus	*Sciadopitus* type
							Sequoioipoll. polyformosus	*Sequoia* type
							Taxodioipoll. hiatus	*Taxodium* type
							Sabaliopoll. tranquillus	Palmae type
							S. areolatus	?
							Juglans spm., *Caryapoll. simplex*, and *Pterocaryapoll. stellatus*	*Pterocarya, Carya, Juglans*
							Engelhardtioipoll. microcor. and *punct.*	
							E. levis	*Engelhardtia* type
							E. quietus	*Engelhardtia* type
							Myricaceoipoll. granifer	*Carya?*
							M. megagranifer	?
							Quercoipoll. henrici	Fagaceae?
							Q. microhenrici	*Quercus*
							Poll. liblarensis	
							Poll. villensis and *cingulum*	*Castanopsis* group
							Poll. brühlensis and *megaexactus*	?
							Liquidambarpoll. stigmosus	*Liquidambar*
							Rhooipoll. dolium	
							Tilia spm., spm.	
							Nyssaceoidae and Nyssaceae	*Nyssa* and nyssoid type
							Araliaceoip. edmundi and *euphorii*	Araliaceae
							Sympolcospoll. vestibulum	Symplocaceae?
							S. clarensis and *rotundus*	*Symplocos* (in part)
							S. triangulus	*Symplocos* (in part)
							Sapotaceoip. manifestus and *megad.*	Sapotaceae
							Myrtaceoipoll. thiergarti	Normapolles type
							Poll. pompeckji	Normapolles type
							Compositoipoll. spinosus	Compositae?
							C. setarius	Compositae?

following: a considerable decrease in abundance of the species *Tricolpopollenites liblarensis* and *Triporopollenites robustus*, a general decrease in the abundance of triporate pollen, along with an increase in the abundance of *Alnus*, *Fagus* type and both winged and nonwinged conifer pollen.

Pollen changes at the Miocene-Pliocene boundary as well as at the Pliocene-Pleistocene boundary have been described by Thomson and Pflug (1953) and documented in detail by Zagwijn (1960). Figure 17-8 shows age range limits and relative abundance for key forms in the early Miocene (Thomson and Pflug, 1953) and in the late Miocene, Pliocene, and early Pleistocene based on data from 19 pollen diagrams by Zagwijn (1960, 1963a).

At or near the base of the Pliocene deposits Zagwijn observed a loss of six taxa. (see fig. 17-8.) In addition, Thomson and Pflug (1953) record the youngest local occurrence of *Triatriopollenites coryphaeus* subsp. *punctatus* (which includes *Platycarya* pollen) in the Fischbach beds of uppermost Miocene age. These changes mark a point at or near the Miocene-Pliocene boundary. The listed forms (*T. coryphaeus* excepted) are considered "Miocene elements" by Zagwijn, though they should be termed Miocene and older elements, as noted in figure 17-8.

The pollen stratigraphy of the Pliocene-Pleistocene boundary is discussed in the section on "Quaternary Floras" (p. 418).

The affinities of the Miocene brown-coal flora as identified by palynomorphs are listed in table 17-3. The Miocene flora has affinities with at least 61 living genera of seed plants, including a few that are primarily of boreal distribution (e.g., *Picea*, *Abies*), and a preponderance of forms that are now temperate (including *Juglans*, *Parthenocissus*, *Pterocarya*, *Platycarya*), a few warm temperate to subtropical genera (e.g., *Nyssa*, *Gordonia*, *Illicium*), and a few subtropical to tropical forms (*Chrysophyllum*, *Melia*).

Teichmüller (in Ahrens, 1958) has described a possible reconstruction of Miocene plant communities from the coastal marshlands. (See fig. 17-9.) This concept is based partly on dominant plant fossils and lithologic sequences in brown-coal deposits (e.g., Neuy-Stolz, 1958) and partly on inferences from environments in which many of the related groups grow today on the southeast coastal plain of the United States. According to her, *Sequoia* forest probably grew on the higher ground; low thickets of *Myrica* and Cyrillaceae were on wet but firm ground, perhaps comparable to the open marsh habitats that these groups now occupy in the southeast coastal plain. *Taxodium* and *Nyssa* might have occupied intertidal swamp areas; marshland characterized by rooted, emergent aquatics may have existed along the seaward margin. Fine-detritus gyttja and clay were interpreted as offshore or open-water deposits; because some of these are rich in quercoid pollen, Teichmüller inferred that the source plants may have grown in upland environments. Repeated sequences of these sediment and fossil types found throughout the Neogene brown-coal deposits indicate facies changes probably related to changes of the marine shoreline.

Neogene of Central and Southern Europe. A well-documented and extensive pollen record is known from Poland. Early Miocene pollen floras have been described from Silesia (Doktorowicz-Hrebnicka, 1954, 1957b, c) and from the Lausitz basin (Raatz, 1937). Late Miocene pollen (as well as seed) floras are known from Stare Gliwice in Silesia (Oszast, 1960; Szafer, 1961) and from the Konin deposits (Kremp, 1949). A study of early or middle Pliocene pollen from northern Poland was made by Doktorowicz-Hrebnicka (1957a). Pollen and seed floras from Mizerna in southern Poland, (Szafer, 1961), represent the early Quaternary section through Mindel (third European glaciation) and probably include the latest Pliocene (Venzo, 1965).

The chief differences between the Polish and German brown-coal floras of Neogene age appear to relate principally to their difference in geographic position. The Polish Miocene, like the German and Dutch sections, is rich in Tertiary relict genera. (see table 17-4.) Pollen and/or seeds of some taxa that range only to the top of the Tertiary in northwestern Europe (e.g., Symplocaceae and *Castanea*) persist in Poland into the first interglacial (Mizerna II of Tiglian age) or into the second (Mizerna III) of Cromerian age (e.g., *Nyssa*, *Liquidambar*, and *Sciadopitys*; see discussion on p. 418. However, other Tertiary relict genera (*Euryale*, *Phellodendron*, *Carya*, *Tsuga*) characterize the Tiglian both in northwestern Europe and Poland, and *Pterocarya* persists through the Cromerian interglacial in the two places.

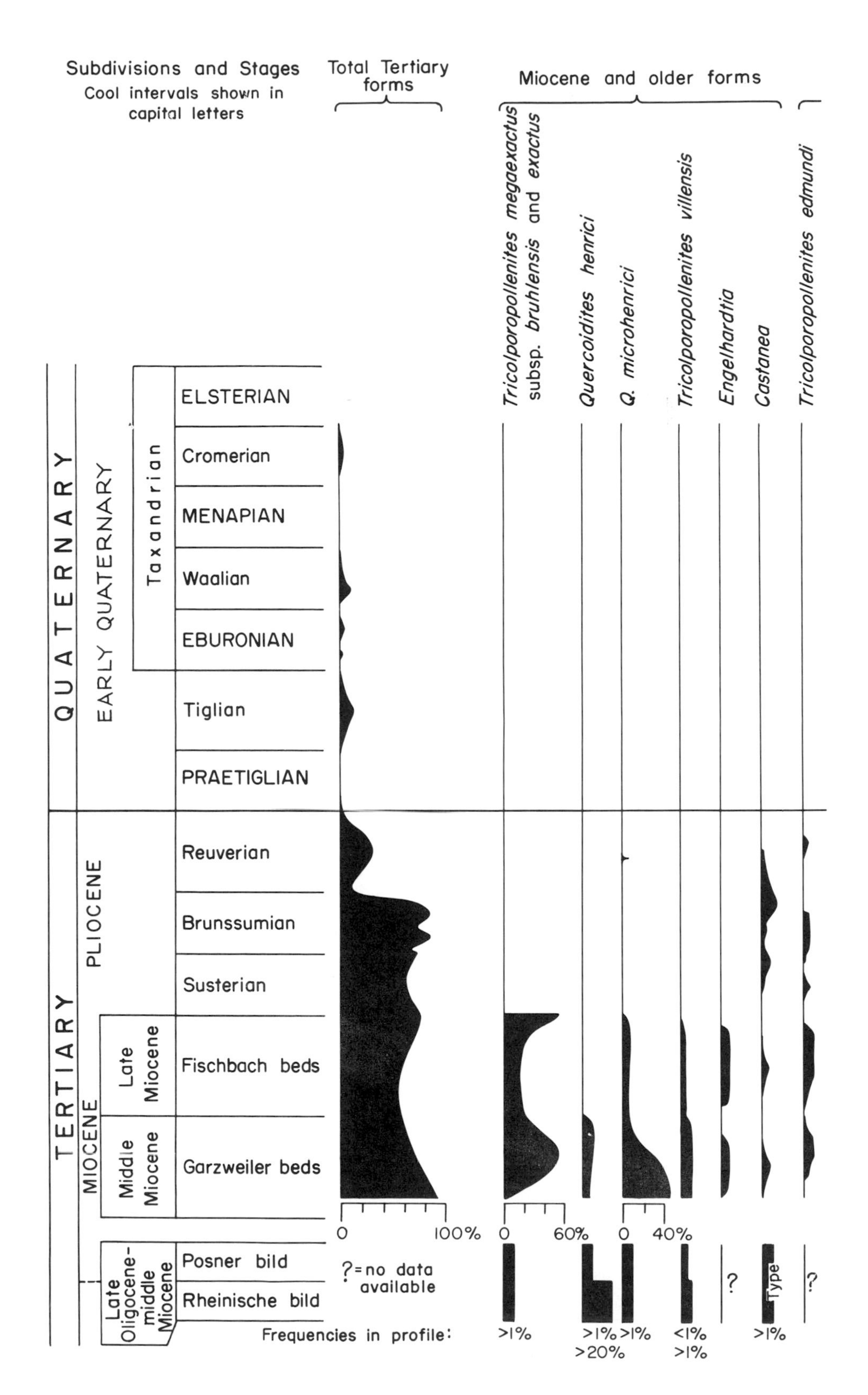

Figure 17-8. Stratigraphic chart showing general age ranges and relative abundance of key pollen forms in sediments of middle Miocene to early Quaternary age in Germany and the Netherlands. (Data summarized from

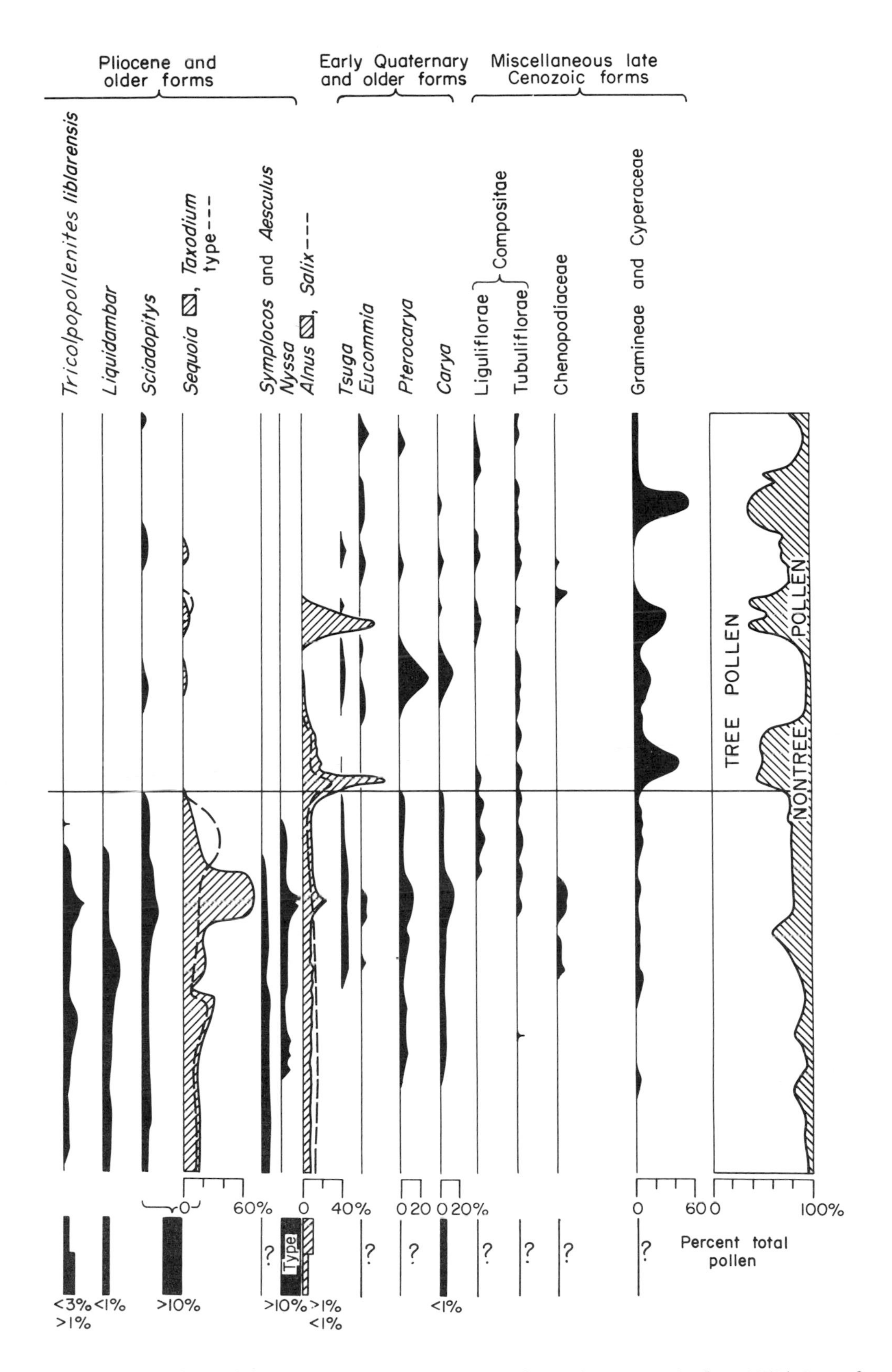

Zagwijn, 1960, 1963a; late Oligocene and early Miocene data from Thomson and Pflug, 1953.) Sum of total Tertiary forms is shown at left. Quaternary records of Taxodiaceae are considered to represent reworked pollen.

Table 17-3. Affinities of the Rheinische Braunkohle Flora of Miocene Age as Ascertained by Pollen Evidence[1]

	Gymnosperms
Pinaceae:	*Pinus, Haploxylon* and *sylvestris* types
	Picea
	Abies
	Pseudotsuga
	Pseudolarix
	Cedrus
	Tsuga
Taxodiaceae types:	*Sequoia*[2] type
	Glyptostrobus[2] type
	Taxodium type
	Sciadopitys
	Angiosperms
Magnoliaceae:	*Illicium*[2], cf. *Magnolia*[2]
Hamamelidaceae:	*Liquidambar*[2]
Platanaceae:	*Platanus*[2]
Leguminosae?[2]	
Theaceae:	*Gordonia*
Tiliaceae:	*Tilia*[2]
Rutaceae:	*Phellodendron*[2]
Hippocastanaceae:	*Aesculus*[2]
Simarubaceae:	*Ailanthus*[2]?
Anacardiaceae:	*Rhus*[2]
Cyrillaceae:	*Cyrilla* and/or *Cliftonia*
Meliaceae:	*Melia, Cedrela*
Aquilafoliaceae:	*Ilex*[2]
Vitaceae:	*Parthenocissus*
Umbelliferae	
Cornaceae	
Araliaceae:	*Hedera*
Nyssaceae:	*Nyssa*[2]
Fagaceae:	*Quercus*,[2] *Castanea*,[2] *Fagus*[2]
Betulaceae:	*Betula*,[2] *Alnus*,[2] *Carpinus*,[2] *Corylus*[2]
Myricaceae:	*Myrica*[2]
Juglandaceae:	*Juglans*,[2] *Carya*,[2] *Engelhardtia*[2] type, *Platycarya, Pterocarya*
Salicaceae:	*Salix*,[2] *Populus*[2] type
Ulmaceae:	*Ulmus*[2]-*Zelkova*[2] types, *Celtis*[2]
Loranthaceae:	*Viscum*
Polygonaceae:	*Rumex*[2]
Ericales types	
Sapotaceae:	*Bumelia*,[2] *Mimusops*,[2] *Chrysophyllum, Sideroxylon* types
Ebenaceae:	?
Symplocaceae:	*Symplocos*[2]?
Oleaceae:	*Fraxinus*[2]
Bignoniaceae:	*Catalpa*[2 3]
Caprifoliaceae:	*Sambucus*[2] type
Compositae types	
Alismataceae	
Hydrocharitaceae:	*Stratiotes*[2]
Liliaceae:	*Smilax*[2]
Cyperaceae:	*Rhynchospora, Cladium*,[2] *Cyperus, Carex*[2]
Gramineae	
Zingiberaceae	
Palmae	

[1] Data from Neuy-Stolz (1958).
[2] Taxa corroborated by megafossil evidence.
[3] I question this; the particular specimen figured by Neuy-Stolz (1958, pl. 6, figs. 48-50) is undoubtedly pollen of *Drosera*.

In both sequences the abundance and diversity of Tertiary relict genera as well as species diminish rapidly during the early Quaternary. Data compiled by Szafer (1961) from the seed and pollen floras in southern Poland demonstrate this trend. In figure 17-10 the relative importance of various geographic elements in the floras is plotted according to geologic age. The data are mainly based on species but include some forms identified only to genus.

Subtropical taxa such as *Cinnamomum* and *Alangium*, which were prominent in the Oligocene, were last recorded in Poland at Kroscienko (Pliocene). East Asiatic elements such as *Pterocarya*, which are mainly warm temperate today, were important in floras of southern Poland during the late Oligocene and Miocene but waned and died out locally by the first local (third European) glaciation (of Mindel age). North American species—such as *Dulichium vespiforme, Tsuga caroliniana*, and *Nyssa sylvatica*—rose to prominence during the Miocene and persisted as a group until the second Polish glaciation (of Riss age). Mediterranean species such as *Pinus salinarum* and *Ilex aquifolium* became abundant in the Miocene and gradually decreased, disappearing locally during the last interglacial (of Eemian age).

In Poland, as in the northwest European sequence, pollen of Gramineae and Compositae are rare or lacking in the early Miocene but become more common in younger beds. Megafossil evidence of arctic species does not appear in Poland until the Mindel, or third European glaciation.

Polish Miocene floras similar to the Miocene pollen floras of northwestern Europe have been described from Silesian brown coals by Macko (1957, 1959). However, the identifications of fossil pollen leave much to be desired; Macko claimed to have identified living genera whose pollen is not structurally distinguishable except on the family level.

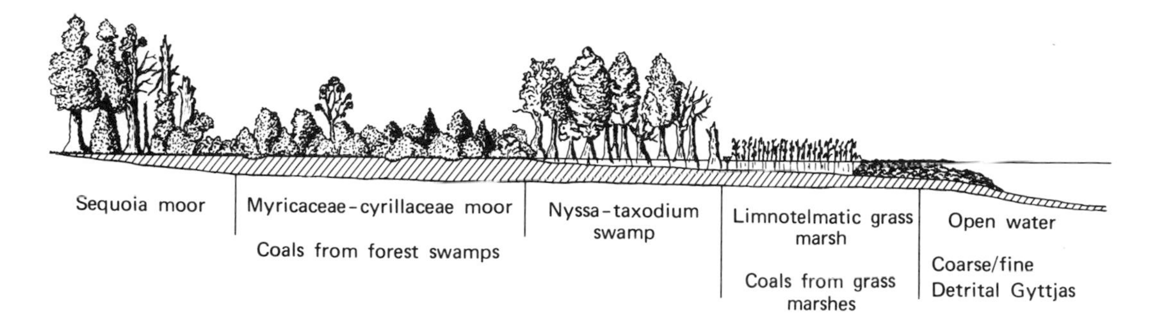

Figure 17-9. Inferred moor types of the Miocene niederreinische Braunkohle in their probable lateral succession (modified from Teichmüller in Ahrens, 1958, p. 600).

Hungarian late Miocene and early Pliocene floras described by Nagy (1958) have a general similarity to floras of similar age from northwestern Europe. Nagy assigned fossil pollen forms to living groups on a conservative basis and discussed equivalence of these identifications with German form species. Her transitional Oligocene-early Miocene flora (Nagy, 1963), however, does not seem similar to German floras of equivalent age.

Pollen, spore, and plankton floras from primarily marine deposits of late Oligocene, Miocene, and Pliocene age in Romania are summarized by Venkatachala and Baltes (1962). Each of these floras is distinctly more cool temperate than are floras of corresponding age from northwestern Europe.

Soviet Union. Extensive work on Neogene pollen floras of the Soviet Union is described in the Miocene Pollen Atlas by Pokrovskaya (1956b), and in summary articles (Tikhomirov, 1963; Zaklinskaya, 1962) available in English. The sequence of Neogene and Quaternary vegetation in the Bering Sea area is discussed by Petrov (1963, 1967) and Wolfe and Leopold (1967).

The evidence from Miocene and Plio-Pleisto-

Table 17-4. **Percentages in the Total Pollen Count of Certain Tertiary Relict Groups in the Late Cenozoic of Poland.†**

Taxon	*Early Miocene*	*Late Miocene*	*Pliocene*	*Mizerna II (=Tiglian)*	*Mizerna III (=Cromerian)*
Taxodiaceae, Taxaceae, and Cupressaceae	4-80	1-20*	0-1	0-5 (*Sciadopitys**)	– (*Sciadopitys**)
Castanea and *Castanea* type	1-43	0-1	–	0-3	–
Nyssa and *Nyssa* type	1-18	–	0-5*	0-2*	+*
Symplocaceae and Sapotaceae	1-6	–*	–	0-2	–
Tsuga	0-1	2-5	0-1*	0-3*	–
Quercus (and quercoid pollen)	0-5	1-20*	1-5	0-15*	–*
Liquidambar	+	1-3	–*	0-1*	+*
Eucommia	–	0-2*	–	–*	
Pterocarya	0-2	0-7*	0-4	0-2*	–*
Carya	0-1	0-2	0-1*	+*	–

†Sources are given on pages 391 and 419.
*Occurrence is documented by megafossil evidence.
+Percentage is less than 1.

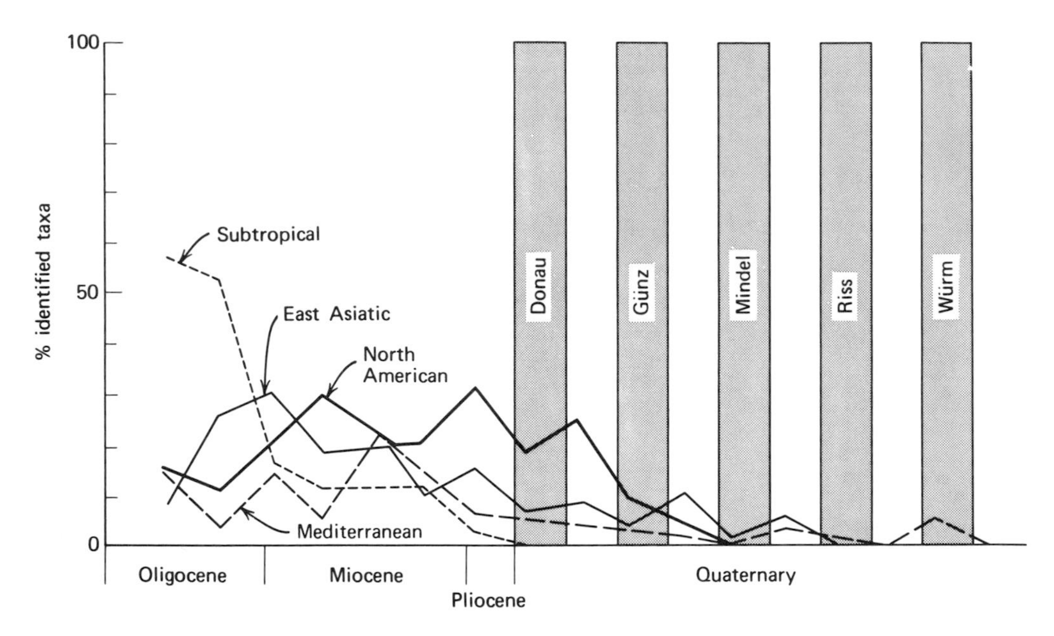

Figure 17-10. Decreasing geographic elements in late Cenozoic floras of southern Poland (modified from Szafer, 1961; Leopold, 1967); the data include generic as well as specific identifications. The present distributions of the nearest living relatives of the fossils are the basis for classifying the geographic elements. (Shaded vertical bars mark the times of European glaciations.) (Compare with fig. 17-5.)

cene pollen floras of the Soviet Union is summarized in a series of maps showing inferred vegetation patterns. (See figs. 17-11, 17-12, and 17-13.)

Early Miocene vegetation of the Soviet Union is thought to have been subtropical in eastern Europe north of the Black Sea as far north as latitude 55°; subtropical elements were present in a predominantly warm temperate forest vegetation in northern White Russia west of the Urals and in the Far Eastern province along the Pacific Coast. The remainder of the Soviet Union where records are available, which is the entire midcontinent west of Lake Baikal (longitude 110°E.), had primarily a rich, forest flora of warm temperate character.

Pokrovskaya (1956b) concluded that during the early Miocene the southern part of White Russia (zone 2, fig. 17-11) supported extensive subtropical marshes dominated by *Taxodium* and floristically diverse evergreen and broad-

Figure 17-11. Suggested scheme of plant distribution and vegetation in the territory of the Soviet Union during the first half of the Miocene Epoch (Pokrovskaya, 1956b). 1—Eastern European province, northern subprovince: diverse broad-leaved forest with pine, *Sciadopitys*, *Taxodium*, and Taxaceae (*Pinus* and Taxodiaceae pollen dominant); warm temperate, with some tropical to subtropical elements present (*Menispermum*, *Anemia*). 2—Eastern European province, southern subprovince: diverse broad-leaved forest with evergreen and subtropical plants; widespread marsh, subtropical type, covered with *Taxodium*. (Taxodiaceae consistently the dominant pollen type.) 3—Urals, west Siberian province, south Urals subprovince: diverse broad-leaved forest (*Quercus* and Juglandaceae prominent pollen types); *Taxodium* marsh widespread; warm temperate. 4—Kazakhstan province. Turgai subprovince: diverse broad-leaved forest, Juglandaceae dominant pollen types; probably widespread districts of herbaceous steppes on the watersheds and high points of relief; warm temperate. 5—(Undifferentiated Miocene) Urals-west Siberian province, northern subprovince: moderately diverse broad-leaved hardwood and pine-spruce forest with rich herbaceous flora; temperate. 6—(Undifferentiated Miocene) Urals-west Siberian province, southern subprovince: diverse broad-leaved and conifer forest (*Pinus* dominant pollen type and locally large representation of *Alnus*, *Betula*, *Pterocarya*, and Ericales); warm temperate. 7—(Early? Miocene) eastern Siberian province: diverse deciduous forest with large participation of broad-leaved genera and with spruce, fir, and hemlock (pollen of Betulaceae prominent); temperate. 8—Eastern Siberian province, Baikal subprovince: broad-leaved and spruce-hemlock forest with local prominence of *Juglans*, *Ulmus*, and Betulaceae pollen; warm temperate. 9—Far Eastern province, maritime subprovince: diverse broad-leaved, pine and spruce-hemlock forest: locally beech, pine, Taxodiaceae, *Trapa*, *Juglans*, *Ulmus* pollen abundant; subtropical to tropical forms (*Lygodium*, *Menispermum*) present; predominantly warm temperate.

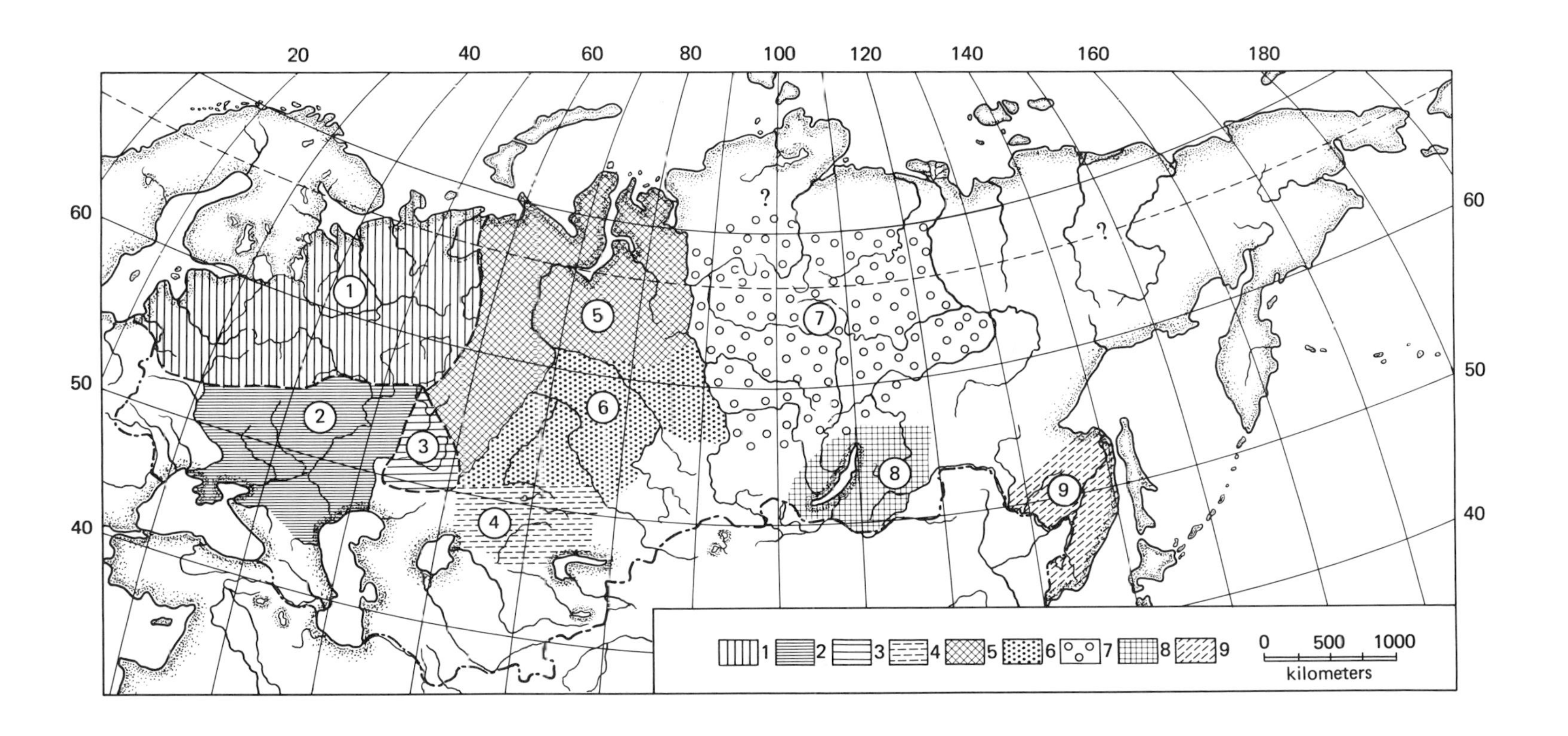

20
40
60
80
100
120
140
160
180
60
50
40
1
2
3
4
5
6
7
8
9
0 500 1000
kilometers

leaved forest. This vegetation may have been similar to that of the early Miocene in north Germany and southern Poland. In White Russia *Taxodium* swamps and a few subtropical elements occurred about as far north as the Arctic Circle (zone 1, fig. 17-11.) Perhaps the driest climate in the Soviet Union at this time was near the Sea of Aral (zone 4, fig. 17-11), where pollen evidence of extensive herbaceous steppes has been reported. (The Sea of Aral district southeast of the Urals is now in the heart of a desert.)

By late Miocene time the vegetation of the Soviet Union in areas where evidence is available was definitely temperate, and it lacked strictly subtropical taxa. In all areas reported the late Miocene floras are much less diverse than those of early Miocene age, and they contain fewer hardwood genera. Pinaceae played a more important and usually dominant role, whereas conifers of the Podocarpaceae and Taxodiaceae, which had been so important before, were rare. In general the herbaceous flora recorded here is more diverse than in earlier time.

Subtropical vegetation of early Miocene age north of the Black Sea was replaced in late Miocene time by temperate hardwood and pine forest and steppes with some elements present that occur today in both warm and cool temperate regions. Only a few subtropical to warm temperate taxa are recorded; these occur as rare pollen (*Sapindus*) or spores (Cyatheaceae) just north of the Caucasus Mountains. (See zone 1, fig. 17-12.) The Turgai basin near the Sea of Aral (zone 2, fig. 17-12) still had extensive steppe development. Siberia east of the Urals to Lake Baikal (zones 3, 4, and 5, fig. 17-12) became cool temperate in character, with spruce, fir, hemlock, or pine dominant, and with *Betula* and *Alnus* playing an important role at northern localities. The Far Eastern maritime subprovince on the Island of Sakhalin (zone 6, fig. 17-12) had a dominantly spruce forest with abundant heaths and only a few hardwoods (*Carpinus*, *Acer*, *Quercus*).

By the end of the Pliocene or the beginning of the Quaternary the hardwoods were gone from most of northern Siberia, remaining only as relicts (zone 4, fig. 17-13) on south-facing flanks and ravines of mountains near the head of the Aldan River and eastward toward the base of the Kamchatka Peninsula (areas that have since become mountain conifer forest (*Picea ajanensis*, *Larix dahurica*) or mountain scrub (Gerasimov, 1964). Pliocene relict occurrences of *Ulmus* and *Quercus* pollen are reported from western Kamchatka (Sinelnikova, Skiba, and Fotianova, 1967). Floras of "EoPleistocene" (equals late Pliocene and early Quaternary) age contain pollen of *Juglans* and fruits of *Juglans* and *Trapa* (Vangengeim, 1961) at relict sites in the Aldan Valley. Mixed conifer forest and alpine lichen and moss communities occupied the highland areas, and larch forest occurred between mountain ranges in eastern USSR. Northern Siberia was differentiated into western and eastern forest types; on the west were conifer forests of European type (zone 8, fig. 17-13) whereas on the east (zone 7, fig. 17-13) conifer forest was a strange mixture of Japanese, western American, and Russian species of Pinaceae. Alpine, "tundralike" communities were distributed only in very small highland areas, mainly in the central and eastern portions of northern USSR (Tikhomirov, 1963). Thus the late Pliocene-early Pleistocene vegetation of the Soviet Union was

Figure 17-12. Suggested scheme of plant distribution and vegetation in the territory of the Soviet Union in the second half of the Miocene Epoch (Pokrovskaya, 1956b). 1—Eastern European province, southern subprovince: broad-leaved and pine forest with rich herbaceous flora, probably widespread districts of grassy steppes on the watersheds. Pine pollen dominant; a few hardwoods present (*Fagus, Nyssa, Betula, Quercus, Castanea, Carya,* and *Pterocarya*); probably warm temperate. 2—Kazakhstan province, Turgai subprovince: broad-leaved forest with large districts of pine, alder, and birch; and probably with extensive grassy steppes (hardwood pollen as in zone 1); warm temperate. 3 and 4—(Undifferentiated Miocene only) Urals-west Siberian province: conifer woods (especially pine) with districts of broad-leaved types; rich herbaceous flora; (hardwoods more diverse than in zone 1); temperate. 5—Eastern Siberian province: coniferous forest with many large districts of spruce and hemlock, and insignificant mixture of narrow and broad-leaved types (only Betulaceae, *Fagus, Ulmus, Fraxinus, Juglans*). Rich herbaceous flora; *Trapa* present; cool temperate. 6—Far Eastern province, maritime subprovince: hemlock and broad-leaved forest; hardwoods few (*Quercus, Carpinus, Betula*) and not abundant in pollen count. *Alnus* pollen frequent; *Trapa* present; cool temperate.

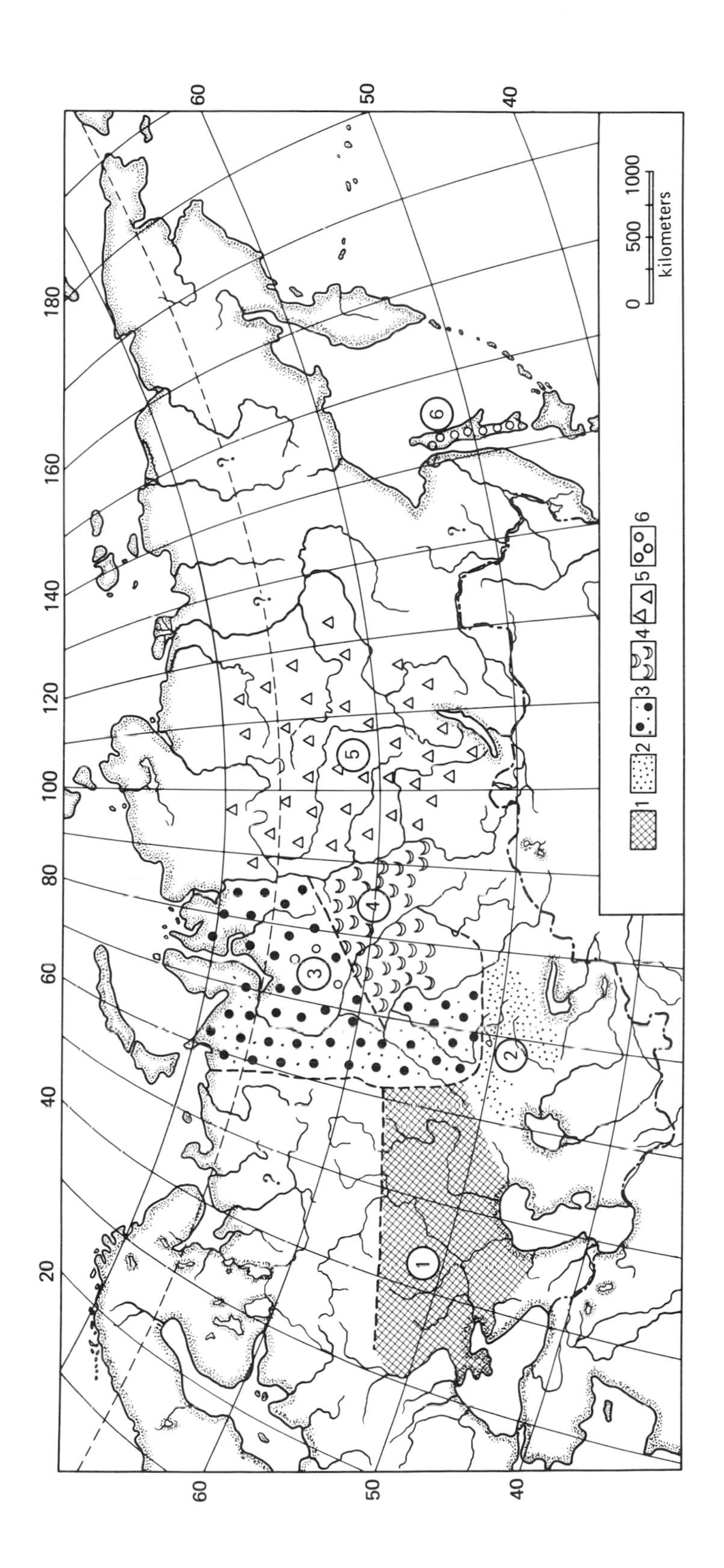

20
40
60
80
100
120
140
160
180
60
50
40
0
500
1000
kilometers
1
2
3
4
5
6

still unlike the modern vegetation and was a depleted version of the Miocene forest communities.

Northwest Pacific Basin. Floras around the North Pacific basin were similar on a generic and in many cases even a specific basis during the early and middle Miocene. Altitudinal zonation existed during the middle Miocene, when conifer forests probably dominated the highlands above 700 meters, but mixed hardwoods of a temperate character grew in the lowlands of the Pacific Northwest and in southern Alaska. The cooling of the late Miocene brought about a severe reduction of temperate woody forms in Alaska and a restriction of these forms to low elevations in northwestern conterminous United States. On a generic basis the Alaskan flora was modernized (having lost all of its exotic temperate genera) by the end of the Pliocene. (See tables 17-5 and 17-6.) The floras of high latitudes (Alaska and nearby Siberia) and those of the Pacific Northwest had few species in common by Pliocene time. Hence, through the latitudinal differentiation of climate during the Neogene the temperate floras of Old and New Worlds became isolated from each other.

A sequence of late Oligocene and Neogene leaf and pollen floras of the Cook Inlet area and of the Alaska Range in southern Alaska (Wolfe, Hopkins, and Leopold, 1966; Wahrhaftig and others, 1969; Wolfe and Leopold, 1967) spans much of the late Cenozoic. The generic changes in terms of the identified fraction of these pollen and spore floras are summarized in tables 17-5 and 17-6. The floras are dated mainly by leaves correlated with dated leaf floras in the Pacific Northwest and Japan.

The latest Oligocene (zone 1) pollen floras are characterized by the presence of extinct taxa (e.g., *Aquilapollenites, Proteacidites, Orbiculapollis*, and the gymnosperm *Gynkaletes* in table 17-5). Also present were many genera that do not now inhabit the northern latitudes: *Cedrela-Melia* (now subtropical) and many others now of warm temperate distribution, *Pachysandra-Sarcococca, Platanus, Liquidambar, Castanea.* Sixty percent of the identified late Oligocene pollen flora are now exotic to Alaska or extinct.

In Miocene and Pliocene assemblages pollen evidence of these relict groups is rare, and a progressive loss of temperate genera is recorded. The "dropout schedule" for extinct genera and genera now foreign to Alaska is summarized for the Alaska Range in table 17-5. By the end of the Miocene the percentage of "exotic" genera has dropped to 48 percent and in Pliocene time to only 35 percent. Pollen records suggest that by Quaternary time the flora of southern Alaska was modernized on a generic basis. Selected pollen types of the Suntrana Formation (zone 4 of tables 17-5 and 17-6) are shown on plate 17-1.

A feature of the Neogene pollen floras is the appearance and increase of groups now characteristic in Alaska. Their order of appearance from available pollen data is noted in table 17-6. Such groups as Cyperaceae, *Typha, Artemisia* and other Compositae, *Polygonum*, Chenopodiaceae, *Arceuthobium*, Onagraceae, Malvaceae, Droseraceae, and Gramineae first appear locally in the Miocene, and their pollen becomes increasingly abundant in younger sediments. Some of these taxa may have evolved earlier and at lower latitudes; hence their Miocene appearance in Alaska probably marks their immigration into that region. Genera of the Taxodiaceae and Pinaceae characterize the Neogene floras, but only *Picea* (and possibly pine; Hopkins, 1967, p. 471) persists into the Quaternary.

In southern Alaska during the middle and late Miocene plant evidence indicates a major deterioration of climate. On the basis of relationships of present generic distribution patterns to aver-

Figure 17-13. Plant-geographical division of northern USSR in the preglacial epoch (from Tikhomirov, 1963). 1—Arcto-alpine moss-and-lichen communities, including dwarf shrubs; 2—mountain forests, subalpine shrubs with patches of mountain forest parkland on the highland slopes; 3—larch forests with understory of *Pinus pumila* and bushy-moss complex similar to the modern one; 4—late Tertiary broad-leaved relicts preserved on the southern slopes of mountain ranges and in ravines; 5—larch and birch forests on the highland plateaus; 6—spruce-larch, pine-larch, and stone-pine-larch outside of highlands; 7—Nipponocordilleran forests as well as derivatives from 4 (above) on lowlands; 8—coniferous forests of European type lacking elements of Beringian forests on the lowlands; in the south with elements of broad-leaved forests; 9—water areas; 10—maritime communities, including salt marshes.

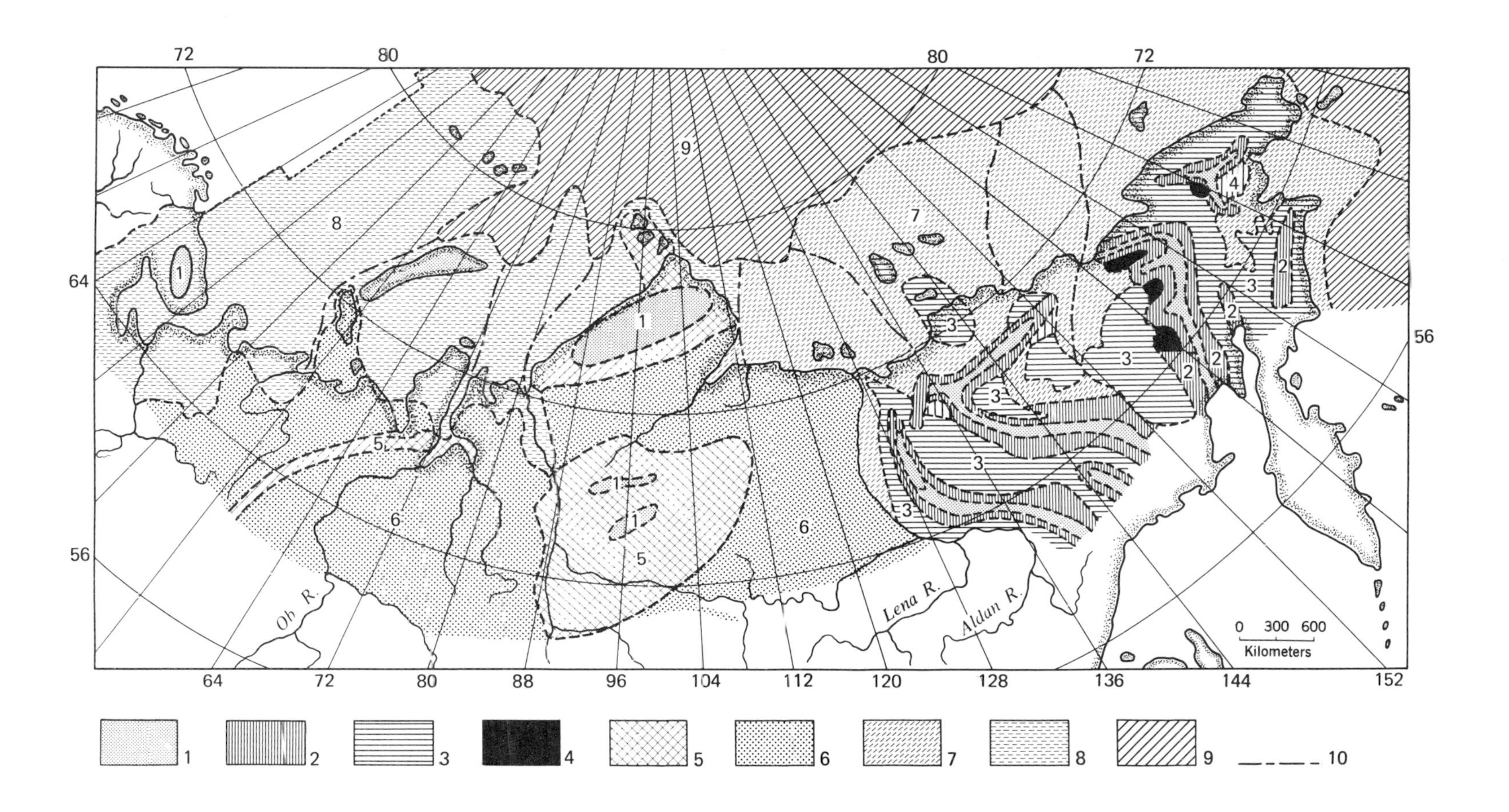

0 300 600
Kilometers
Ob R.
Lena R.
Aldan R.

Table 17-5. **Alaska Range Pollen Floras; Late Cenozoic Records of Genera Now Foreign to Alaska or Extinct[1]**

Genera Now Foreign to Alaska or Extinct (*)	Late Oligocene	Miocene				Late Miocene Early Pliocene	Pliocene	Pliocene – Pleistocene	Pleistocene
	1	2	3	4	5	6	7		
Gymnospermae:									
Taxodium-Sequoia	o	o	o	o	o	- -	o	o	
Pseudotsuga	o	o	o	- -	- -	o	o		
Libocedrus	o	o	o						
**Gynkaletes*		o	- -	o					
Glyptostrobus-Cryptomeria	o	- -	o						
Ephedra		o							
Angiospermae:									
Carya	o	o	o	o	o	o	o	o	
Juglans	o	o	o	o	o	o	o	o	
Pterocarya	o	o	o	o	o	o	o	o	
Liquidambar	o	o	o	o	o	o	- -	o ?	
Tilia	o	- -	o	o	o	o	o	o	
Ulmus-Zelkova	o	o	o	o	o	o	o	o	
Diervilla	o	- -	- -	o	- -	o	- -	o	
Ilex	o	o	o	o	o	- -	o		
Corylus	?	o	o	o	- -	o	o		
Quercus		o	o	o	o	o	o		
Ostrya-Carpinus	o	o	o	o	- -	o			
Nyssa		o	o	o	o				
Fagus	o	- -	o	o					
Platanus	o	o	- -	o					
Castanea	o	o	o						
Cedrela-Melia	o	o							
Engelhardtia-Alfaroa	o								
Pachysandra-Sarcococca	o								
* *Proteacidites globosiporus*	o								
* *Orbiculapollis*	o								
* *Aquilapollenites*	o					o?			
Itea	o					o?			
Total now-exotic genera	24	18	17	16	10	13	10	8	0
Total identified vascular genera	40	31	31	31	21	25	23	23	12
Percent now-exotic genera	60	55	52	49	48	48	41	35	0

[1]Some pollen in zones 6, 7, and "Pliocene-Pleistocene" may be redeposited.

Plate 17-1. Selected pollen and spore types of the Suntrana Formation of middle Miocene age (zone 4 of tables 17-5 and 17-6) from the Alaska Range. (All figures × 500.) 1—Polypodiaceae; 2, 3, and 4—*Sphagnum* types; 5—*Lycopodium complanatum* group; 6—Spore, undetermined; 7—*Osmunda*; 8—*Tsuga*; 9—*Sciadopitys* type; 10—*Taxodium* type; 11—*Picea*; 12 and 13—*Pinus*; 14—*Pinus*?; 15—*Abies*; 16—*Malvastrum* type, Malvaceae?; 17 and 18—*Ilex*; 19—Pericolpate, *Portulaca* type; 20 and 21—*Quercus*; 22—*Salix*; 23—Ericales; 24—*Sparganium*; 25—*Lonicera*; 26 and 27—*Betula*; 28—*Myriophyllum* type; 29, 30 and 31—*Alnus*; 32, 33, 34 and 35—*Ulmus-Zelkova*; 36—*Drosera*; 37—*Carya*; 38—*Pterocarya*; 39—*Juglans*; 40—*Liquidambar*; 41—*Polygonum*; 42 and 43—*Nyssa*; 44—*Tilia*; 45—*Ovoidites ligneolus* (cyst).

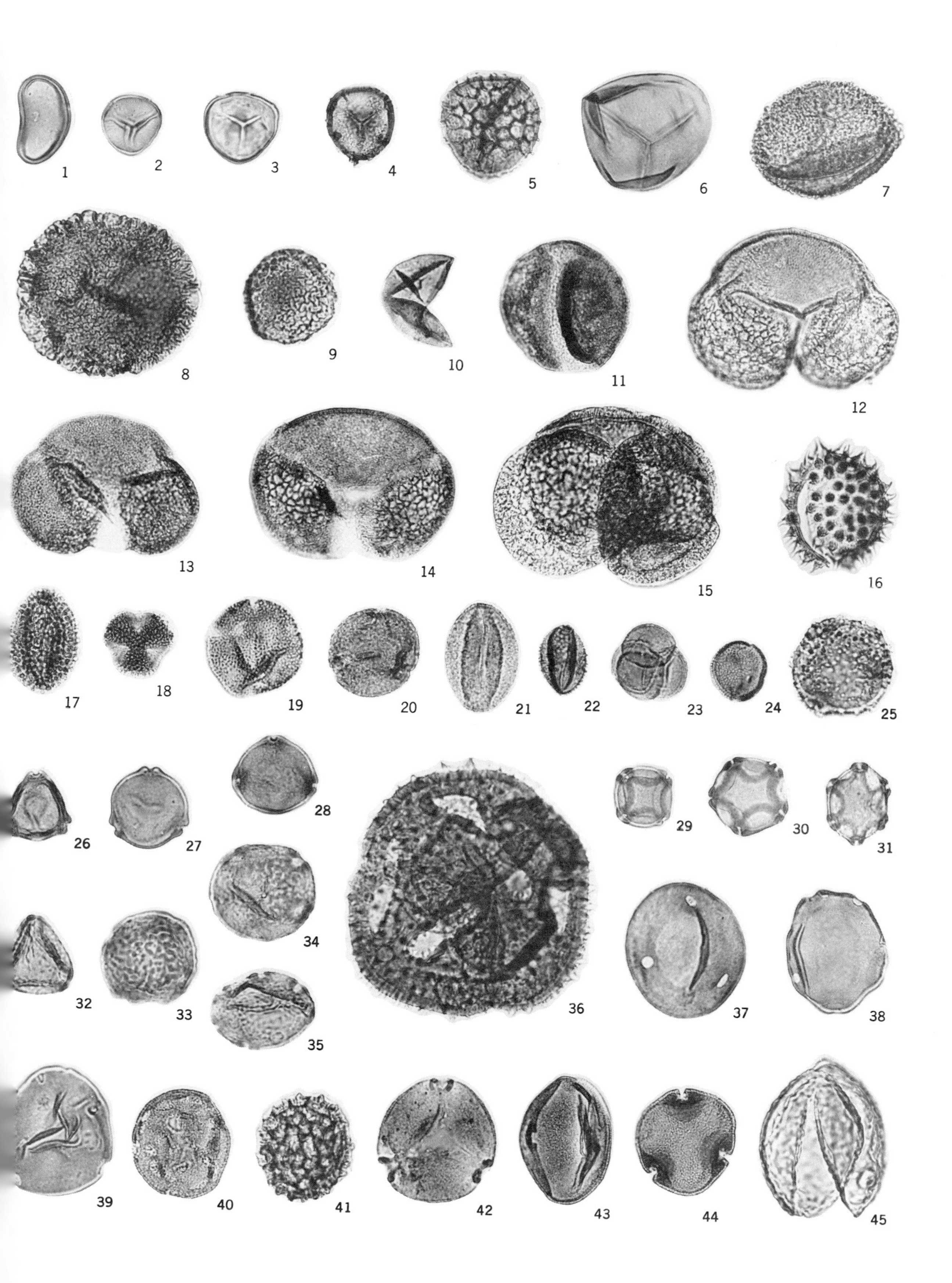
1
2
3
4
5
6
7
8
9
10
11
12
13
14
15
16
17
18
19
20
21
22
23
24
25
26
27
28
29
30
31
32
33
34
35
36
37
38
39
40
41
42
43
44
45

Table 17-6. Alaska Range Pollen Floras; Late Cenozoic Records of Genera and Families Now Native to Alaska[1]

Genera Now Native to Alaska	Late Oligocene	Miocene				Late Miocene-Early Pliocene	Pliocene	Pliocene-Pleistocene	Pleistocene
	1	2	3	4	5	6	7		
Gymnospermae:									
Picea	o	o	o	o	o	o	o	o	o
Pinus	o	o	o	o	o	o	o	o	
Tsuga	o	o	o	o	o	o	o	o	
Larix	?		o					o	
Abies	o	o	o	o	o	o	o		
Angiospermae:									
Polemonium									o
Saxifraga								o	o
Scirpus					o	o		o	o
Typha					o			o	o
Drosera				o					o
Artemisia				o					o
Polygonum			o				o	o	
Arceuthobium			o						
Nymphaea			o						
Salix	o	o	o	o	o	o	o	o	o
Betula	o	o	o	o	o	o	o	o	o
Alnus	o	o	o	o	o	o	o	o	o
Sparganium	o			o				o	o
Valeriana	o	o				o			o
Acer	o	o	o	o		?	o		
Myriophyllum	o	o	?	o		o	o		
Elaeagnus	o			o					
Viburnum	o	o							
Ferns and fern allies:									
Lycopodium	o	o	o	o	o	o	o	o	o
Osmunda	o	o	o	o	o	o	o	o	
Selaginella	o							o	
Total now-native genera	16	13	14	15	11	12	13	15	12
Total identified genera	40	31	31	31	21	25	23	23	12
Percent now-native genera	40	42	45	48	52	48	57	65	100
Selected angiosperm families:									
Compositae		×		×				×	×
Chenopodiaceae				×					×
Caryophyllaceae							×	×	×
Onagraceae						×		×	×
Malvaceae				×					
Gramineae						×	×	×	×

[1]Some grains in 6, 7, and "Pliocene-Pleistocene" may be redeposited.

age July isotherms, Wolfe and Leopold (1967) inferred from the fossil records that at the Cook Inlet the average July temperature dropped about 7°C in about 4 million years. During early Miocene the presence of *Liquidambar* and *Nyssa* indicates that the average July temperature was at least 20 to 21°C. The local presence of *Picea*, on the other hand, indicates that locally the July average was not above 20 to 21°C.

During the middle to late Miocene a group of genera (*Fagus, Liquidambar, Nyssa*) that have summer temperature requirements of about 18 to 21°C became extinct in the region. The local extinction of *Abies* (13°C) at the Cook Inlet also apparently took place in the late Miocene, indicating a rapid decline in summer temperature there.

From evidence based on the proportion of smooth-margined leaves in the floras Wolfe and Hopkins (1967) proposed a climatic scheme for the Tertiary of northwestern North America. They suggest that the Neogene climate deterioration involved a number of oscillations; these have not yet been noticed in floras of low latitudes.

The Neogene of the Pacific Northwest is summarized in the form of generic lists by Gray (1964). Her data, which are included here as table 17-7, indicate that a number of genera were eliminated during the Miocene: *Aesculus, Cercidiphyllum, Engelhardtia* type, *Itea, Liriodendron, Nyssa, Parthenocissus, Schizandra-Kadsura*.

The Neogene climate in the Pacific Northwest, as indicated by pollen evidence, was warm temperate to subtropical in the Miocene and temperate in the Pliocene. From Oregon a warm temperate to subtropical flora of Miocene age is reported by Wolfe (1962) and from British Columbia a warm temperate flora of about the same age is described by Martin and Rouse (1966).

Rather extensive work on Neogene floras in Japan is only mentioned here. Eocene and Oligocene pollen floras are summarized by Tokunaga (in Cranwell, 1964). Miocene pollen sequences are reviewed by Sato (1963) and some Miocene-Pliocene pollen floras are reported by Sato (1960) and Sohma, Jimbo, and Shimada (1959). Neogene megafloral work has been summarized by Tanai (1961).

Rocky Mountain Region. A late Oligocene or earliest Miocene flora at Creede, in southwestern Colorado in the San Juan Mountains, is severely depauperate compared to the early Oligocene Florissant flora (see chap. 16) but includes many genera that now grow in Colorado. The leaf flora from Creede is so modern in aspect that botanists have supposed it to be either Miocene or Pliocene in age (MacGinitie, 1953, p. 73). But recent potassium-argon isotope dates establish the age of the Creede flora at 26 million years (Steven, Mehnert, and Obradovich, 1967). Leaves identified by H. D. MacGinitie and R. W. Brown (written communication, 1957), plus taxa found as fossil pollen and spores (marked with an asterisk; Leopold, unpublished data), are the following:

Selaginella cf. *densa**
Potamogeton
*Pinus** *florissanti*
*Picea**
*Abies** *longirostris*
Ephedra cf. *E. torreyana**
cf. *E. nevadensis**
Juniperus
cf. *Tsuga**
cf. *Larix**
*Populus**
*Salix**
*Alnus**
*Quercus**
Jamesia (= *Edwinia*)
*Acer**
Cercocarpus myricaefolia
Oenotheraceae*
*Sarcobatus**
*Eleagnus**
Mahonia marginata
*Carya**
Ulmus and/or *Zelkova**
Crataegus
*Artemisia**

Only two of the Creede genera (or 10 percent) are now foreign to the Rocky Mountain area (*Carya* and *Ulmus-Zelkova*). Except for *Ephedra*, which now grows in very dry sites not much farther west in the pinyon pine-juniper association and on desert margins, the other genera are common in the San Juan Mountains today.

The Troublesome Formation in Middle Park, north-central Colorado (Izett and Lewis, 1963), yielded a pollen assemblage from a vertebrate horizon of middle Miocene age consisting of the

Table 17-7. **Neogene Genera Identified on the Basis of Pollen Records from the Pacific Northwest**[1]

Genus or Family	Oligocene Oregon W	Oligocene Oregon E	Oligocene Idaho	Miocene Washington	Miocene Idaho	Miocene Oregon W	Miocene Oregon E	Miocene California	Pliocene Oregon	Pliocene California
Gymnospermae:										
Abies	X	X	X	X	X	X	X	X	X	X
Cedrus	X	X	X	X	X	X	X	X	(X)[2]	
Ephedra			X	X	X	X	X	X	(X)	
Ginkgo	X		X	X	X					
Picea	X	X	X	X	X	X	X	X	(X)	X
Pinus	X	X	X	X	X	X	X	X	(X)	X
Pseudotsuga-Larix	X			X		X	X?	X?	(X)	
Taxodiaceae-Cupressaceae-Taxaceae	X	X	X	X	X	X	X	X	X	X
Tsuga	X	X	X	X	X	X	X	X	(X)	
Angiospermae:										
Acer	X	X	X	X	X	X	X	X	X	
Aesculus			X	X	X					
Alnus	X	X	X	X	X	X	X	X	(X)	X
Artemisia							X		(X)	X
Betula	X	X	X	X	X	X	X	X	(X)	X
Caprifoliaceae[3]				X		X	X	X		
Carya	X	X	X	X	X	X	X	X	(X)	
cf. *Castanea*	X			X	X	X			(X)	X
Celtis				X	X	X	X	X		X
Cercidiphyllum			X	X	X		X			
cf. *Cornus*						X				
Corylus	X?			X	X	X	X	X	(X)	X
Elaeagnaceae				X	X		X			
cf. *Engelhardtia-Alfaroa*	X		X							
Ericaceae	X		X	X	X	X	X	X	(X)	X
Fagus	X			X	X	X	X	X	(X)	
Fraxinus	X			X?	X	X	X	X	X	X
Fremontia						X?				X
Ilex	X	X		X	X	X	X	X	(X)	
Itea				X	X		X	X		
Juglans	X	X	X	X	X	X	X	X	(X)	X
Liriodendron				X		X				
Liquidambar	X	X?	X	X	X	X	X	X	X	X
Myrica				X		X				
Nyssa	X			X	X	X	X			
Ostrya-Carpinus		X	X	X	X	X	X	X	(X)	X
Parthenocissus		X		X			X			
Platanus	X	X		X	X	X	X		(X)	X
Pterocarya	X	X	X	X	X	X	X	X	X	X
Quercus	X	X		X	X	X	X	X	X	X
cf. *Rhus*				X	X	X				
Salix				X	X	X	X	X	X	
Schizandra-Kadsura						X				
Symplocos	X?			X	X	X		X	(X)	
Tilia	X	X	X	X	X	X	X	X	(X)	
Ulmus-Zelkova	X	X	X	X	X	X	X	X	X	X
cf. *Viburnum*	X			X	X					
Vitaceae-Rhamnaceae				X		X	X			

[1]From Gray (1964, table 1). (W = western; E = eastern)
[2]Genera or families new to the flora are in parentheses.
[3]*Lonicera*.

following (Leopold, in Weber, 1965, and unpublished data):

Picea cf. *engelmannii*
Pinus
Picea
Tsuga
Abies cf. *lasiocarpa*
Ephedra (three spp.)
Alnus
Acer
Salix
Ulmus-Zelkova
Arceuthobium
Carya
Juglans
Sphaeralcea
Shepherdia cf. *S. argentea*
Lonicera
Compositae cf. *Xanthium*
Artemisia
Gramineae
Symphoricarpos
Sarcobatus cf. *S. vermiculatus*
Polemoniaceae
Umbelliferae
Chenopodiaceae undet.

Selected pollen types from this flora are shown on plate 17-2.

Like the Creede flora, the assemblage is composed dominantly of pine and spruce pollen. The pollen assemblage resembles the present pollen rain in the area with the exception of *Ulmus-Zelkova*, *Carya*, and *Juglans*, which are now exotic to most western States, and *Tsuga* now of northern Idaho and the Pacific Northwest. Thus the Miocene-Pliocene floras of this region contain only a few broad-leaved trees that are late Tertiary relicts; none of these is known to persist in the Colorado flora after the Pliocene. Pliocene floras from Wyoming, Idaho, and Arizona have a generic aspect similar to those from Colorado.

Little is known of the Neogene floras in the northern Rocky Mountain region, but some evidence is now available from south-central Idaho (Axelrod, 1964; Leopold and Brown, in Mapel and Hail, 1959).

A leaf flora of probable late Miocene age at Trapper Creek (Beaverdam Formation of Axelrod, 1964) contains forms such as *Sequoia, Fraxinus, Carya, Ulmus, Zelkova, Nyssa,* and *Ilex,* now exotic to the Rocky Mountains, plus forms now characteristic of the area (e.g., *Pinus, Picea, Tsuga, Abies, Ephedra, Alnus, Quercus, Acer, Populus*). The pollen flora of the Salt Lake Formation (Pliocene; Mapel and Hail, 1959) and of the Banbury Basalt (middle Pliocene) are greatly impoverished compared with the Trapper Creek flora; for example, broad-leaved trees are represented only by *Carya, Juglans*, and *Ulmus-Zelkova* in the Salt Lake, and only by *Carya* and *Ulmus-Zelkova* in the Banbury.

A diverse pollen flora from the Glenns Ferry Formation of Blancan age (latest Pliocene and earliest Pleistocene) in the western Snake River plain represents plants now native to Idaho, except for rare pollen of *Carya* and *Ulmus-Zelkova* (see stratigraphic summary by Malde, in Wright and Frey, 1965). The flora includes various genera of Pinaceae (dominant forms), *Celtis, Populus*, and several xeric shrubs in addition to herbs and waterplants. It suggests a climate slightly wetter than today's (annual precipitation now 10 in.), which permitted low montane vegetation to grow in this now treeless plain. Pollen from the overlying Bruneau Formation (middle Pleistocene) is mainly of the Pinaceae and includes no late Tertiary relict genera. It suggests a climate somewhat cooler and/or wetter than today's.

The succession of Miocene, Pliocene, and Quaternary pollen floras from southern Idaho demonstrates gradual loss of broad-leaved tree genera that still persist in Central and Eastern United States and along the Pacific Coast. This loss may be expressed in terms of the percentages of the identified genera that were eliminated progressively with time. For Idaho the late Miocene Trapper Creek flora contains 35 percent of genera that are now foreign to the central Rocky Mountain region. The Pliocene floras contain only two or three genera that are now exotic to the region, but these genera represent about 25 percent of the total. In the Blancan (Pliocene-Pleistocene) interval these exotic genera (including only *Carya* and *Ulmus-Zelkova*) make up only 9 percent of the floral list. Younger pollen floras that have been examined contain no genera that are now foreign to the present regional flora. The generic losses (losses of Tertiary relict genera) for the Idaho and Colorado floras are indicated in figure 17-4.

The loss of broad-leaved trees from the flora of the central and northern Rocky Mountains was undoubtedly progressive, owing to gradual deterioration in regional climate and the rise of

1
2
3
4
5
6
7
8
9
11
12
13
14
15 a
15 b
16
18
19
20
21
22
23
25
26
27
28
29
30
32 a
32 b
33
34
35
36
37
39
40
41
42
43

mountains (an uplift of highlands in southern Idaho of some 5,000 to 6,000 feet occurred in post-Eocene time; Axelrod, 1968). In addition regional uplift of about 5,000 feet during the Neogene in Nevada, Colorado, and Wyoming doubtless affected the regional climate. The consequent decrease in temperate and mesic conditions permitted only a few broad-leaved trees to survive in Idaho and Colorado (e.g., *Populus, Acer grandidentatum*, and *Betula papyrifera*), though shrubby species of many dicotyledonous tree genera are also represented (*Quercus, Celtis, Acer, Alnus*, etc.).

Eastern United States. Though two or three leaf floras of Miocene age have been reported along the East Coast, no Miocene pollen work in Eastern United States has yet been published. Reconnaissance work provides a skeletal picture of common pollen types in three Miocene formations in Maryland—the Choptank and Calvert Formations, both of middle Miocene age, and the St. Marys Formation of middle and late Miocene age. (See table 17-8.)

Most of the plant groups identified are represented in the modern flora of the region. Some exceptions are *Engelhardtia-Alfaroa, Gordonia*, Sapotaceae, *Melia*, and *Ephedra*, all of these but *Ephedra* are more southern in their present distribution, having modern ranges on the southeast coastal plain or in the Old World tropics. *Ephedra* does not grow in Eastern United States.

Differences between these Miocene floras and those of Eocene age or older (Nanjemoy and Aquia Formations) from the same district are clear-cut: Normapolles types, *Platycarya, Latipollis, Sciadopitys*, and Schizaeaceae, which are represented in the Eocene floras, are lacking in the Miocene. A pollen flora from upper Oligocene sediments from North Carolina (Leopold, 1959), however, is not very different from these Miocene assemblages; many of the same groups are present in both, including rare pollen of Compositae.

Neogene of Southern and Low Latitudes

Neogene of the South Pacific Area. Miocene pollen floras from Eniwetok, Fiji, Bikini, Palau Islands, and Guam indicate that the Miocene vegetation contained Micronesian plant genera that since have been eliminated from the islands. The pollen data from Eniwetok, for example, indicate that of 18 angiosperm genera identified only four are still native to the atoll. The pollen assemblages from these localities are dominated by mangrove pollen, mainly of Rhizophoraceae but also with Sonneratiaceae, *Avicennia*, genera of the Combretaceae which are strand plants in Micronesia, and *Pisonia* (Leopold, 1969).

Early and middle Miocene pollen floras of New Zealand, according to Couper (1960), are dominated by *Nothofagus* (*N. brassi* group). Although *Nothofagus* (*N. fusca* group), Podocarpaceae, and ferns are the numerically important forms in the same beds, they become more abundant in upper Miocene and Pliocene assemblages. *Bombax* and *Cupaneidites* types (Tertiary relicts) make their last appearance in the late Miocene. Pollen of many warmth-loving plants is common in the Pliocene, including that of *Agathis, Ixerba, Quintinia, Knightia*, etc., which suggests that the climate was somewhat warmer than at present.

Neogene of New World Tropics. The Gatun Formation (Miocene) in the Panama Canal Zone furnishes evidence of Miocene vegetation in the New World tropics. The following pollen and spores types in Panama were reported by Alan Graham (written communication, 1964):

Boraginaceae	*Tournefortia*
Bombacaceae	*Bombax*
Polygalaceae	*Polygala*
Schizaeaceae	*Anemia*
Euphorbiaceae	*Hyeronima*
Polypodiaceae	*Pteris*
Meliaceae	*Trichilia*
Rhizophoraceae	*Rhizophora*

Plate 17-2. Selected pollen and spore types of middle Miocene age from the Troublesome Formation, Granby, Colo. (All figures × 500.) 1, 2, and 4—Undetermined spores; 3—*Riccia* type; 5 and 6—*Pinus*; 7 and 8—*Taxodium* type; 9, 11, 12, and 13—*Ephedra* cf. *E. nevadensis*; 10—*Ephedra* cf. *E. torreyana*; 14—*Taxodium* type; 15a, b—*Salix*; 16—Undetermined tricolpate; 17—*Quercus*; 18—Plumbaginaceae?; 19—Rosaceae; 20—*Claytonia*?; 21 and 22—*Acer*; 23—Gramineae; 24—*Carya*; 25 and 26—*Ulmus-Zelkova*; 27, 28, and 29—*Alnus*; 30—*Juglans*; 31—*Juglans*?; 32a, b—Caryophyllaceae; 33—*Sarcobatus* cf. *S. vermiculatus*; 34, 35, and 36—*Shepherdia* cf. *S. argentea*; 37 and 38—*Pleurospermum*? Umbelliferae; 39 and 40—Undetermined tricolporates; 41—*Sphaeralcea*, Malvaceae; 42—*Lonicera* cf. *L. dioica*; 43—*Symphoricarpos* cf. *S. occidentalis*; 44—Undetermined tetrad.

Table 17-8. Identified Pollen and Spores from Three Miocene Formations from Maryland[1]

Pollen and Spores	Formation: Calvert	Choptank	St. Marys
Juglans		x	
Tilia	x	x	
Carya	x	x	x
Quercus	x	x	x
Acer	x		
Cucurbitaceae cf. *Sisyos*	x	x	
Ulmus-Zelkova	x	x	
Valerianaceae	x		
Nemopanthus?	x		x
Ilex	x	x	
Castanopsis?	x		
Gordonia[2]	x		
Sapotaceae[2]	x		
Engelhardtia-Alfaroa[2]	x	x	
Morus?	x		
Melia[2]	x		x
Liquidambar	x	x	x
Tsuga	x		
Fremontodendron?	x		
Magnolia	x		
Picea	x	x	x
Pinus	x	x	x
Nyssa	x	x	
Ostrya-Carpinus	x	x	
Pterocarya?		x	
Cyrilla?	x		
Eucommiidites?	x		
Symplocos	x		
Castanea	x		
Chenopodiaceae	x	x	
Compositae	x		x
Abies			x
Alnus		x	x
Eriogonum			x
Sphagnum			x
Umbelliferae		x	x
Betula		x	
Fagus		x	
Rubiaceae, cf. *Galium*			x
Sparganium			x
Ephedra[2]	x		

[1] Leopold, unpublished data.
[2] Taxa that are now foreign to the Maryland area.

Sapindaceae	*Cupania*
Gramineae	
Bromeliaceae	*Catopsis*
Proteaceae	*Roupala*

Graham points out that in relation to higher latitude floras the Panama area may have served as a reservoir for plants for perhaps the last 50 million years; the pollen types listed above belong to plants that now grow in the area of the Panama Canal Zone and suggest that there have been few alterations or generic eliminations from the flora since Miocene time.

Summary of Late Tertiary Floras

In the Northern Hemisphere at high and middle latitudes pollen evidence records a Miocene climate that was warmer and with less seasonal variation than at present. In many areas subtropical plants, such as members of the Sapotaceae and Meliaceae, grew alongside warm temperate (Juglandaceae) and cool temperate (*Picea, Abies*) plants. These groups for the most part are not found together today, but they grow only a few miles apart in mountainous terrain of the subtropics.

Though the late Oligocene climates brought some subtropical elements as far north as latitude 63°N. in Alaska and the Soviet Union, most of these genera extended only as far north as about latitude 40°N. during the early Miocene along the Pacific Coast of North America and to about latitude 50 to 55°N. in Europe and maritime East Asia. Today subtropical elements extend northward to about latitude 25°N. in most areas. The early Miocene vegetation occupying the midlatitudes of the Northern Hemisphere was mixed warm temperate and subtropical, with the true tropics apparently restricted to relatively low latitudes (probably below latitude 35°N., judging from California leaf floras).

Middle Miocene leaf floras of Japan and the Pacific Coast of the United States indicate that the climates were warmer than early Miocene ones and that some subtropical broad-leaved evergreen elements moved northward to about latitude 45°N. during that time. Many genera now restricted to the humid Eastern United States and to temperate parts of China and Japan ranged into Western United States and Europe. Limited evidence from low latitudes suggests that Miocene floras there were not significantly different from the local floras of today.

By late Miocene time subtropical elements retreated to a position south of latitude 40°N., leaving the north latitudes a region of strictly temperate vegetation, even in Siberia, Alaska, and conterminous United States. An exception is Western Europe, where a few subtropical forms persisted as far north as latitude 50°N. until Pliocene time. During the late Miocene a temperate flora that was relatively homogeneous on the generic level occupied lowlands in the entire North Pacific Basin, though montane vegetation was more boreal in aspect. Now-desert areas of Western United States and South-Central Soviet Union showed development of steppe or subarid scrub vegetation as early as late Miocene.

In Pliocene time the widespread climatic deterioration decreased the ranges of temperate plants. The role of Pinaceae increased significantly in the high northern latitudes, replacing the earlier abundance of mixed hardwoods and Taxodiaceae. Pollen of herbaceous groups was increasingly important and more diverse than earlier. Deserts developed in the Sea of Aral area of southwestern Soviet Union and the Great Basin of Western United States, and semiarid conditions developed in the rainshadow of the Rocky Mountains in Colorado and Wyoming. By Pliocene time the mesophytic hardwood floras of Old and New Worlds were separated by the opening of the Bering Straits and by climatic barriers that limited the northern distribution of temperate plants to relict sites.

QUATERNARY FLORAS

General Characteristics

Early Quaternary floras differ from Neogene floras in that they contain few if any Tertiary relict genera or extinct species. Middle and late Quaternary floras in general resemble the equivalent modern regional floras. In some areas Tertiary relict genera have not yet been found above the Neogene, and records of extinct species are lacking (e.g., in the United States). In other areas (e.g., Europe) certain relict genera and a few extinct species are characteristic of lower Quaternary interglacial deposits.

Many Quaternary pollen assemblages, especially those of glacial origin, contain abundant and diverse nontree pollen. This composition is not usually characteristic of Neogene floras except some from regions that are now desert.

Arctic-alpine plant species are lacking in pre-Quaternary floras, are rare in early Quaternary European floras, and are characteristic in glacial floras of the middle and late Pleistocene (post-Cromerian) in the middle and high north latitudes of Eurasia.

Sequences of dominant pollen types are characteristic regionally for each interglacial, and they show both qualitative and quantitative differences from the postglacial pollen sequences.

Methods

Field and Laboratory Methods. Because sediments of Pleistocene age are usually less indurated than those of Tertiary age, different methods of collection and preparation can often be used. A general survey of palynological techniques is provided by Gray (in Kummel and Raup, 1965) and techniques particularly applicable to younger palynological samples are summarized by Faegri and Iversen (1964).

One problem in working with pollen samples from younger rocks is the recognition of reworked pollen of late Cenozoic age. Reworked pollen older than about early Miocene can in part be identified by taxonomic means and by comparison with possible source sediments (Davis, 1961a; Rohrer and Leopold, 1963; Cushing, 1964).

At least two laboratory methods have been described by which pollen or spores of widely different ages can be detected in a mixed assemblage. One of these is based on differential acceptance of safranin stain by grains of Paleozoic, Mesozoic, and Cenozoic age (Stanley, 1965). Another method relates the age of pollen and spores to their autofluorescence colors using an ultraviolet-light source. Gijzel (1963, p. 25) pointed out that autofluorescence "is based on the faculty of the grain to transpose invisible UV-light with comparatively short wavelengths into visible light of much longer wavelengths." He found that in Holocene material pollen and spores of major groups and even of species have characteristic and differing colors; these colors are dulled and modified with increasing geologic age.

It is often useful to determine the number of pollen grains per gram or per unit volume of sediment. (See "Statistical Methods.") Laboratory techniques for determining absolute number of pollen grains are of three types, based on (a) counts of pollen suspended in a measured volume of liquid, (b) pollen counts of weighed aliquots of liquid, and (c) proportion of native or fossil pollen to a measured amount of exotic modern pollen artificially added to the preparation. The volumetric techniques, which have been discussed in detail by Davis (in Kummel and Raup, 1965) are briefly reviewed here along with the weight and proportion methods.

Aliquot (measured fraction) preparations differ from ordinary preparations in that (a) the amount (volume or dry weight) of sediment utilized in the preparation is carefully determined, (b) the total yield of fossil pollen from the unit of sediment is retained, and (c) slides for counting are made up as measured fractions of the total pollen yield.

Early techniques for volumetric aliquot preparations depended on measuring volumes of glycerine jelly in which the fossil pollen were suspended (Leopold, in McKee, Chronic, and Leopold, 1959) or on the assumption that successive drops of prepared pollen slurry contain the same amounts of pollen (Muller, 1959; Sittler, 1955). Difficulties exist with both approaches. It is difficult or impossible to measure accurately the volumes of viscous media, and the number of pollen in successive drops varies greatly.

More recent volumetric techniques introduced by Davis (in Kummel and Raup, 1965; Davis, 1966) circumvent these limitations and permit easy rechecking on the counting techniques and removal of a series of aliquots of the same size from a single preparation. The refinements devised by Davis are as follows:

1. Suspending the total pollen sample in a liquid of low viscosity (e.g., tertiary butyl alcohol) that can be evaporated from the slide surface; such a liquid can be measured easily even for small volumes.
2. Counting the entire slide for each aliquot, which is easy under low power if there are only about 200 or 300 pollen per slide; this eliminates the problems created by differential distribution of pollen under the coverslip (Brookes and Thomas, 1967).

Preparation of aliquot slides by a weight method is described by Traverse and Ginsberg (1966). The number of pollen grains per gram of sample is computed according to the formula:

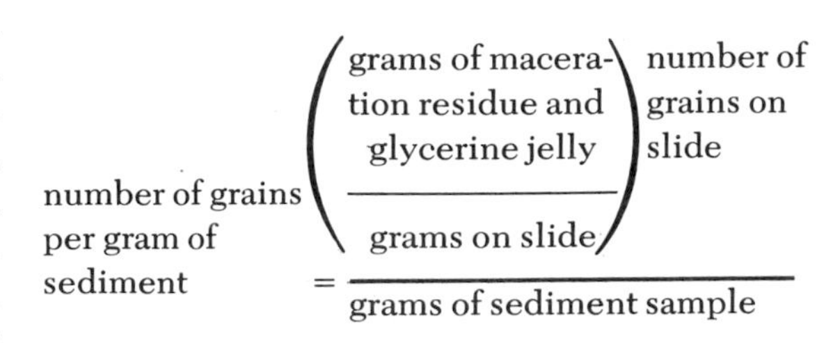

$$\text{number of grains per gram of sediment} = \frac{\left(\dfrac{\text{grams of maceration residue and glycerine jelly}}{\text{grams on slide}}\right)\text{number of grains on slide}}{\text{grams of sediment sample}}$$

This technique, which Traverse and Ginsberg applied to modern marine sediments, seems to be a shortcut to the more elaborate volumetric technique of Davis, but the accuracy of results in terms of measuring statistical variation and repeatability is as yet untested.

A method for determining absolute pollen content based on the ratio of native or fossil grains to exotic grains added to the sample in known amounts was devised by Benninghoff (1962). An improvement suggested by Kirkland (1967) involved a suspension of the exotic pollen grains (stained) in a large volume of liquid. By repeated sampling with Davis' volumetric method the number of grains per cubic centimeter in the liquid can be determined accurately. If a given volume of this liquid is then added to a measured fossil sample, the proportion of exotic to fossil grains will permit the calculation of the total number of fossil grains in the sample. This method obviates the need for preparing aliquot slides and for counting the entire slide.

Statistical Methods. Methods for determining the quantitative composition of fossil assemblages are of two broad types; one is to determine the percent composition of the pollen counts, and the other is to calculate the actual number of pollen grains. The latter may be expressed as the number in a measured volume or weight of sediment. If the amount of time represented by a sample is known, these data can be converted to the number of pollen grains deposited per unit area of surface per year. Recently determinations of pollen concentrations and pollen deposition rates have been made for the late Quaternary and for samples of modern pollen rain. There are numerous potentials for application of absolute frequency data in late Cenozoic pollen work.

Relative Frequency. When the annual rates of total pollen fallout are constant, changing percentages of pollen composition should document actual changes in the pollen rain with time. Changing percentages should reveal information on changing proportions of types in the pollen rain, although they may or may not reflect changes in the number of individual types through time. (See "Determination of Absolute Pollen Frequency.")

The determination of percentage composition of pollen assemblages, an approach utilized from 1916 (von Post) until the present, involves counting all palynomorphs encountered in traverses of a slide and determining the percentage of each entity in the total count. Percentages are then plotted in a histogram or graph for a series of superposed samples in a section.

The numerical sum for calculation of pollen percentages usually consists of one of the following categories:

1. All palynomorphs (pollen, spores, algae, etc.).
2. All pollen and spores.
3. All pollen.
4. All arboreal pollen (= AP).
5. All pollen of terrestrial plants (excluding aquatic forms).

In practice the palynomorphs that are not included in the numerical sum (i.e., nonarboreal groups in item 4 above) are calculated as percentages in relation to the numerical base; for example, in a case where NAP (nonarboreal pollen) is $1\frac{1}{2}$ times more frequent than AP (arboreal pollen), the NAP will be 150 percent of the total AP. The choice of numerical base depends chiefly on which proportional changes one wishes to study or to emphasize. A good practice is to construct first a diagram of total palynomorphs, including all forms found, and then decide which numerical bases will best present the data. Wright and Patten (1963) point out that the base is often chosen to include the plants that the worker thinks contribute to the regional pollen rain; thus pollen of plants that may be of local origin, particularly lake-margin plants, is often excluded from the sum. They consider the choice of numerical base to be a stratigraphic convenience; for example, standard European postglacial diagrams use a tree-pollen sum; nevertheless, ecological interpretation of the section often relies heavily on nontree types.

Pollen strongly representing local sources (sometimes termed "overrepresentation of local types"), such as large numbers of Cyperaceae pollen where the presence of *Carex* fruits in the sediments indicates local Cyperaceae, is often omitted from the sum. Pollen that is suspected of being reworked or redeposited from older beds is often excluded from the sum. Because not every worker will agree on which pollen types may be "overrepresented" or "redeposited," it is certainly good practice to present two kinds of plots: an objective one that includes in the sum all pollen and spores tallied, and another, based on the worker's opinion, showing a modified diagram that excludes certain groups from the sum. A good example is the pair of late glacial diagrams presented by Andersen (1954).

In a discussion of statistical pitfalls in the use of relative pollen counts Faegri and Iversen (1964) point out that the statistical significance of changes in pollen diagrams depends chiefly on the size of the counts. Sampling errors can be judged empirically, but mathematical treatment of the material is the best measure of the potential errors. Two questions that must be considered are: is the pollen count large enough to represent qualitatively the pollen flora? Is the count large enough to yield sufficiently accurate percentages for the taxa present?

By applying stepped increases in the size of the counts and by plotting the number of species against total grains counted we can decide whether the counts are representative of the total flora. In diverse floras such a graph will show a linear increase in the number of species until the total count is very large; then the number of new species begins to drop off with each increase in count size, and further counting will yield few new species. In depauperate floras incremental increases in count size will show a decrease in number of new species found when the total counts are still very small. Because diversity of a flora may provide a relative measure of the number of ecological niches and of the diversity of the landscape occupied by the flora (Odum, Cantlon, and Kornicker, 1960), it represents an important parameter of fossil data.

The effect of count size on statistical error can be studied by the same general method mentioned above. By repeating fixed sum counts on a single sample and by determining the spread of the resultant percent composition for each form we can determine the standard deviation represented by the various counts; from this figure the standard error can be calculated. Westenberg (1947) and Faegri and Ottestad (1949) showed that within a given count size the standard error may be greater for pollen forms that represent middle values (near 50 percent) than for those that are comparatively rare or abundant (Faegri and Iversen, 1964). However, if the dispersion is expressed as a coefficient of variation, the reverse is true; that is, in practice much larger counts are needed to attain the same level of error for rare or abundant items than for moderately frequent ones.

A technique now commonly used to facilitate the counting of very large sums can be called block counting. Its use is advantageous if the sample contains a few forms of high frequency and many forms of low frequency. After completing a regular count of 300 grains (the first block count) the compositional percentages for the abundant forms are calculated. For this example suppose there are 270 pine grains (90 percent) and 30 nonpine grains (10 percent). In continuing the count for the second block only the nonpine grains are tallied. After a total of 90 nonpine grains, multiplying the pine count by 3 plus the nonpine grains will show that a standard count would have had to include 900 grains to obtain the reliability and diversity of nonpine grains achieved by the combined block counts.

Pairs of counts involving "double fixed sums" may be useful if the analyst wishes to study the effects on the diagram of excluding one or more dominant forms from the count. As described by Mehringer and Haynes (1967), a first count is a fixed sum and includes total pollen, whereas the second is also a fixed sum but selected abundant forms are excluded. Pollen diagrams plotted from each type of data can then be compared.

Determination of Absolute Pollen Frequency. The difficulty with compositional percentages is that they only tell us which pollen types were represented better than others; they do not tell us which pollen types actually increased in abundance in the pollen rain over time. As Traverse and Ginsberg (1966) pointed out, relative numbers tend to exaggerate compositional differences between samples. This difficulty may be alleviated by determining actual numbers of pollen grains per gram or per unit volume of sediment if either the mineral sedimentation rate or the intensity of the pollen rain is known. If there is a constant pollen rain, absolute pollen frequency (APF) data indicate sedimentation rate. If the sedimentation rates can be determined and can be shown to have been constant within a given period, then (a) changes in the amount of total pollen rain deposited per year can be calculated, and (b) changes in actual pollen numbers of individual taxa per unit of sediment may be calculated to show "real" (as opposed to relative) changes in the amounts of individual pollen types in the pollen rain.

In a theoretical discussion of the contrast between relative and absolute pollen data Davis (1963) pointed out that, because some species produce more pollen per plant than others, a major change in abundance of a single highly productive species statistically may submerge any changes in the abundance of low pollen producers or may provide a relative pollen change that is the reverse of what actually happens in the vegetation. In hypothetical examples described by Davis (fig. 17-14) three species (*A*, *B*, and *C*) occupy a flat landscape and are homogeneously distributed. By definition species *A* produces twice as much pollen as species *B*, and species *C* produces one-fifth as much pollen as species *B*. In the first example (fig. 17-14) the parent plants represent 40, 10, and 50 percent of the vegetation, respectively. With these built-in differences in pollen production their pollen is represented in a relative frequency of 80 : 10 : 10 in the local pollen rain. These pollen percentages reflect the stated ratios, *R*, of pollen production to forest composition for the three species:

$$R_A : R_B : R_C = 2 : 1 : 0.2.$$

In the second example the forest composition has changed: species A has decreased, species *B* has increased, and species *C* has remained the same. In the resultant pollen rain the relative frequency of species *A* pollen has indeed decreased, that of species *B* pollen has increased, but that of species *C* pollen, instead of remaining the same, has shown a small relative increase. The ratios of percent pollen/percent forest composition have changed too (*R*-values, fig. 17-14), even though the ratios of the *R*-values always stand in the relationship 10 : 5 : 1. Davis showed that decreases in the abundance of high-pollen producers on the landscape (like species A) can bring about apparent but unreal changes in the relative abundance of low-pollen producers in the pollen counts. She also pointed out that the changes in relative pollen counts are not proportional to the real changes on the landscape. From the above it can be seen that simple "correction factors" relating percent composition of pollen rain and source forests as suggested by Davis and Goodlett (1960) probably would not take into account the complex relationships discussed here. Davis concluded that comparisons of pollen data with composition of standing source vegetation may be more informative when absolute pollen numbers rather than pollen percentages are used.

Where sedimentation rates can be determined and can be shown to be relatively stable pollen concentrations in sediment can be converted to number of pollen grains deposited per square centimeter per year, and in this way the actual

Species:	% Forest composition			% Pollen composition			R – values: Ratio of % Pollen / % Forest composition	Ratio of R – values
	A	*B*	*C*	*A*	*B*	*C*	*A*: *B*: *C*	$R_A : R_B : R_C$
Example 1	40	10	50	80	10	10	2 : 1 : 0.2	10 : 5 : 1
2	20	30	50	50	38	12	2.5 : 1.25 : 0.25	10 : 5 : 1

Figure 17-14. Two hypothetical examples comparing three plant species in their representation in the forest and in their relative importance in the pollen rain. On the right *R* indicates ratios for the tree species with respect to their representation in the pollen rain and in the forest. (Modified from Davis, 1963.)

fossil pollen rain can be estimated (Davis and Deevey, 1964; Davis, 1967). A disadvantage of this method is that its most reliable application is limited to the part of the section that is within reach of carbon-14 dating, or the last 40,000 years.

By assuming that modern pollen rain is relatively constant today Traverse and Ginsberg (1966) used absolute frequency of pollen per gram to infer relative sedimentation rates in modern marine sediments. Such use of pollen concentrations may have broad applications in late Cenozoic sediments, when evidence suggests that regional pollen rain was reasonably stable.

One application of pollen-concentration data to determination of fossil-pollen rain will be discussed in some detail here. The example comes from the work of Davis and Deevey (1964) and Davis (1967) on the late glacial of Rogers Lake, Conn. The authors have shown that pollen deposition rates and percentage pollen data provide different curves for major tree types.

Davis and Deevey (1964) obtained a number of carbon-14 dates within late glacial zones at Rogers Lake and established that the mineral sedimentation rate was relatively stable. With sedimentation rate known, they could convert number of pollen grains per cubic centimeter into number of pollen grains deposited per square centimeter per year. (See fig. 17-15.) These data establish that intensity of pollen rain

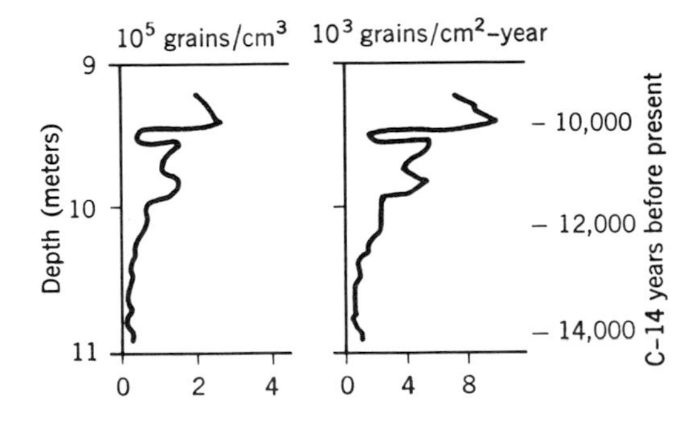

Figure 17-15. Pollen accumulation rates, Rogers Lake, Conn. (from Davis and Deevey, 1964). Comparison of pollen numbers contained per cubic centimeter of sediment with calculated pollen falling per square centimeter per year; carbon-14 time scale on right compared with depth on the left shows that deposition rates of the sediments were fairly constant over the interval.

was gradually increasing during the late glacial, though a distinct oscillation is seen during the last ice advance about 10,500 years ago. The same samples were used for relative and absolute counts, and important statistical differences emerged. (See fig. 17-16.) Although the pollen-poor sediments of the lower zone *T* were dominated by pine pollen, all pollen is rare in these sediments. At the *T/A* zone boundary the absolute numbers of spruce, pine, and oak increased greatly—an increase that is much more striking from the absolute pollen data than from the relative counts. Peak numbers from both types of counts occur at roughly the same levels, but real differences in details of changes are obtained from the two types of data. Relative counts show that during the Two Creeks interstadial "warm period" 11,000 to 12,000 years ago oak pollen rose dramatically and then fell at the end of this period; by absolute counts it rose even more dramatically but then did not fall so far. Oak pollen was still relatively abundant 10,000 years ago. Spruce pollen by relative counts shows one important peak about 11,600 years ago, but absolute pollen data show the same early peak, plus a later one that is even more impressive—one that is not even recorded in the relative counts.

Reconstruction of Climates. The potential for reconstructing climates by the use of pollen floras is greater for the Quaternary than for older periods. Because pollen identifications are usually on the generic level, additional detail can be added to the interpretation of the paleoclimate through the use of megafossils, which often can be determined on the species level. The identifications of greatest potential value will be those carried to the smallest taxonomic units, regardless of the source of the evidence. Examples of three methods that have been used by European workers for estimating paleoclimate are discussed below.

Zagwijn (1963b) estimated mean temperature in July from dominance of selected tree genera as determined by fossil-pollen tallies. He recognized four categories of tree genera:

1. Warmth-loving swamp forms mainly confined to pre-Pleistocene and earliest Pleistocene sediments (*Sequoia*, *Taxodium*, *Sciadopitys*, *Nyssa*, and *Liquidambar*).
2. Warmth-loving trees and shrubs of dry soils

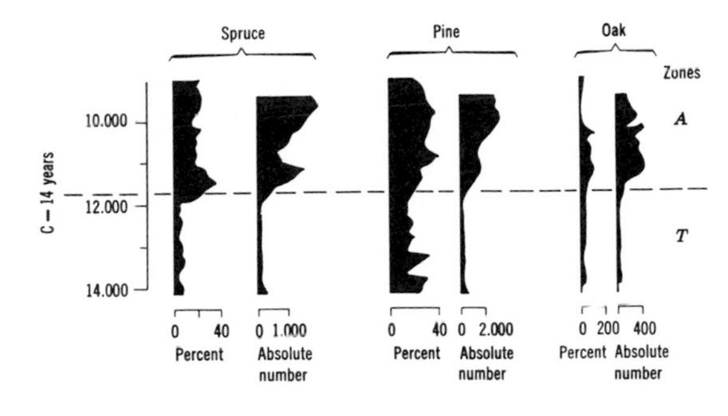

Figure 17-16. Contrasting data for the late glacial section shown in figure 17-15: percentages (standard counts) on left versus absolute numbers of pollen per square centimeter per year on right. Rogers Lake, Conn. (from Davis and Deevey, 1964).

(*Quercus, Castanea, Tilia, Carpinus, Corylus, Ostrya, Eucommia, Ulmus, Fraxinus*).

3. Warmth-loving forms of wet soils (*Alnus, Carya, Pterocarya, Vitis*).

4. Conifers (*Pinus, Picea, Abies, Tsuga*), birch and willow.

Zagwijn judged average July temperature from the characteristics of modern climates where the greatest number of the dominant genera grow together today; for example, when group 1 dominated in the Belfeld Clay (Netherlands, Tiglian) he inferred a mean July temperature of 20°C. When the conifers (group 4) dominated but small amounts of all the other groups were present, he inferred an average July temperature of 12°C. (See diagram, fig. 17-21.)

Problems derive from the use of this method, because some of the tree genera, such as the conifers, produce large numbers of pollen per tree, and others, such as *Fraxinus* or *Nyssa*, are less prolific. Thus dominance in the pollen profile is not necessarily a guarantee that the source plants dominated the vegetation. Zagwijn partly avoided this problem by including both high and low pollen producers in each of his groups.

A second method of estimating climate has been used by Szafer (1953, 1954, 1961), who judged average annual temperature from characteristics of modern climates where the greatest number of fossil-plant genera now occur. In the assemblages from Szafer's earliest Quaternary zones the component genera do not overlap in range today; for example, *Keeteleria, Corylopsis,* and *Pseudolarix* are temperate forms now of East Asia, but *Dulichium* is a temperate plant now of the United States. The fact that their megafossils are found together at Mizerna (zone II) during the first interglacial demonstrates that major changes in range have since occurred in these groups; doubtless some of these temperate plants could grow together today.

More difficult to reconcile is the fossil occurrence of *Picea* (growing season cooler than 23°C and boreal in distribution) with *Meliosma* (subtropical distribution) as megafossils in the Kroscienko flora (early Pliocene). Because the overwhelming bulk of the Kroscienko assemblage was of subtropical and warm temperate character, Szafer (1954) inferred an annual temperature of 18°C, some 10°C higher than present at the fossil locality. He estimated the average July temperature to be 25°C and the January average to be 6°C. It is difficult to imagine a climate with such a cool growing season in the subtropical belt. However, such conditions do occur very near to each other in the mountains of Formosa, where *Picea* occurs above 2,500 meters and *Meliosma* at 1,800 meters and below. In fact one species of *Meliosma* (*M. callicarpaefolia*) that is known to inhabit "high altitudes" in Formosa may actually overlap with *Picea morrisonicola* (Hui-Lin, 1963). Because the Kroscienko locality in the Dunajec Basin (elevation about 250 meters) of the Tatry Mountains is surrounded by a local and regional relief of 750 and 2,250 meters respectively, there may have been vertical zonation of montane and subtropical vegetation during Kroscienko time.

A third method utilizes a direct comparison of biogeographic range and distribution of isotherms or other climate features. If a definite relationship exists between the parameter of biogeographic range and a climatic feature, the plant is assumed to be limited by that feature; for example, in its modern North American distribution *Picea* closely follows the 10°C average July isotherm on its northern boundary and the 23°C isotherm on its southern boundary (Wolfe and Leopold, 1967). The southern limit of *Tsuga canadensis* follows (and therefore is probably related to) the 37°C isotherm of maximum summer temperature (Dahl, 1964). If fossils of *Picea* and *Tsuga canadensis* were found together, these would be a basis for determining more than one parameter of paleoclimate at the time of deposition. Various examples of plant species whose ranges have been linked with single climatic factors are discussed by Dahl (1964).

Caution is required in interpretation because of possible long-distance transport; pollen fossils cannot always be interpreted as indicating plant presence at the site or even in the region. Superposing modern pollen maps (Faegri and Iversen, 1964, p. 98) and vegetation maps, and considering data from modern-pollen-rain studies can be helpful in evaluating possible examples of drift.

The Pliocene-Pleistocene Boundary

The beginning of the Quaternary at the type Pliocene-Pleistocene boundary in Italy (see p. 378) is marked by a marine regression, by the first major late Cenozoic change from warm- to cold-water foraminifers and mollusks, and by the appearance of new vertebrate taxa.

Because the paleobotany of the Pliocene and early Pleistocene has been studied extensively in northern and central Europe, the correlation of sections from these areas with the type Pliocene-Pleistocene boundary is of importance. Such a correlation between Italy and northern Europe is discussed by Voorthuysen (1957), Zagwijn (1960; 1963b, p. 183), and Venzo (1965). Paleobotanists formerly had considered the Pliocene-Pleistocene boundary to lie at the beginning of the Günz glaciation (e.g., Szafer, 1954), but it is now clearly established at the beginning of the earlier cool period, the Donau, or Praetiglian (Venzo, 1965). This means that the well-known plant-bearing beds at Tegelen, Netherlands, which had previously been considered to be Pliocene in age (Reid and Reid, 1915; Reid, 1920), are of first interglacial Quaternary age.

At the Pliocene-Pleistocene boundary in northwestern Europe, which is represented by the boundary between the Reuverian and Praetiglian (fig. 17-8, table 17-1) dramatic numerical and qualitative floral changes take place. Pliocene and older forms—including *Tricolporopollenites edmundi, Tricolpopollenites liblarensis, Liquidambar, Sciadopitys, Sequoia, Symplocos, Castanea* type, *Aesculus, Nyssa*—which are consistently present or abundant in the Reuverian beds are eliminated from the section at the beginning of the Praetiglian (Zagwijn, 1960, 1963a). (The rare and scattered occurrences of *Taxodium* type and *Sciadopitys* pollen in the early Quaternary led Zagwijn to consider them probably reworked.) Thomson and Pflug (1953) suggest that *Tricolpopollenites liblarensis* and *Quercoidites microhenrici* occur in sediments as young as Tiglian, but Potonié (1951a) records their highest occurrence in the lower Pliocene. (See table 17-2.)

The replacement of the Reuverian temperate and subtropical Tertiary relict forms by anemophilous herbs (mainly Gramineae and Cyperaceae); shrubs, including *Salix, Alnus* (*viridis* type) and Ericales; and in some sections by *Pinus* and some *Picea* provides evidence of a distinct Praetiglian cooling. Compositae pollen, of both Liguliflorae and Tubuliflorae, appears in small but regular percentages in the late Pliocene and early Pleistocene beds, although these types are rare and irregular in occurrence in earlier Neogene strata. (See fig. 17-8.)

Correlation of the Pliocene-Pleistocene boundary from Italy to Poland is a special problem because the Polish sites are continental, not marine. In the absence of independent dating the inferred climate and nature of the floras are the only real evidence on which such a correlation can now be based. Two floras are related to the Pliocene-Pleistocene boundary in Poland: one from Kroscienko (subtropical to warm temperate in aspect), which is probably Pliocene in age; and one from Mizerna, which records two full oscillations of climate and is partly or totally of Pleistocene age. Suggested positions for the Pliocene-Pleistocene boundary are before the first or minor cooling, or before the second or major cooling at Mizerna.

Table 17-4 and figures 17-5 and 17-10 summarize the main floristic data for the Polish sequence. Table 17-9 shows the alternate placements of the boundary in the Polish deposits.

In the basal zones at Mizerna zone I contains a few warm temperate plants and zone I-II has a cool temperate flora with few Tertiary elements and abundant modern European taxa; the subtropical elements of Kroscienko are absent here. The change from the Kroscienko to the Mizerna I and I-II floras definitely indicates a climatic cooling. An overlying zone (II) at Mizerna shows a partial recovery and reversion to a warm temperate flora with an abundance of East Asiatic elements.

After the warm interval of Mizerna II the flora (zone II-III) records a second cooling, with a

Table 17-9. Correlation of the Sequence of Floras in Poland with That in Europe

Polish Sequence (pollen zones)	Correlation with Europe According to— Szafer (1954, 1961)	Venzo (1965)
Mizerna:		
IV	MINDEL=KRAKOWIAN	MINDEL
III, III-IV (warm)	"Tiglian"	Cromerian (Mizerna III)
II-III (cool)	GÜNZ	GÜNZ
	....	
II (warm)	Late Pliocene	
I-II (cool)	DONAU (pre-GÜNZ cooling)	DONAU
		....
I		
Kroscienko, Huba (warm)	Early Pliocene	Reuverian (Mizerna I)

Note: dotted lines mark placement of Pliocene-Plistocene boundary.

decreased role of East Asian species and a concomitant increase in European and Eurasian temperate elements; the significance of this change is confirmed by the megafossils and by a pollen diagram (Szafer, 1954).

Still younger zones at Mizerna (especially zone III) record evidence of a warming, with increased participation of North American taxa. The final zone (IV) records a major cooling with the appearance of arctic species as recorded at Ludwinow (Szafer, 1961).

Because of the high proportion of Tertiary elements present during the first warm interval at Mizerna (zone II) Szafer considered this to be a Tertiary flora and of Pliocene age. (See table 17-9.) He also pointed out that the cooling that follows (Mizerna II-III) is far more impressive than the earlier one (Mizerna I, I-II).

Zagwijn (1960) has suggested that the whole of the Mizerna deposits belongs in the Quaternary; he believed that the cooling marked by Mizerna I and I-II is a significant one that can be correlated with the Donau interval. Zagwijn's suggestion seems appropriate; if we accept Szafer's correlation of the first arctic interval (Mizerna IV) as relating to the first time glacial ice reached southern Poland (Krakowian of Mindel age), then the earlier cool period (Mizerna II-III) is probably of Günz age, and the earliest cool period (Mizerna I-II) should correlate with the Donau, or first European glaciation. Venzo (1965) agreed with this but drew the Pliocene-Pleistocene boundary between the warm temperate zone I and the cool zone I-II. (See table 17-9.) Although, as noted in table 17-9, Szafer places the "Tiglian" after the Günz, the Tiglian is now known to be of pre-Günz age in the Netherlands (Florschütz and Someren, 1950; Zagwijn, 1960; Venzo, 1965).

Some Quaternary Assemblages

Because the most meaningful comparison of interglacial floras depends on having a composite sequence through the Quaternary within one region, this discussion is primarily devoted to the best documented such sequence—that from northern Europe. A few comments are also included on lower Quaternary deposits from Italy and the United States. A major work by Markov and Belychko (1967) provides an excellent review of Russian literature on the Quaternary Period.

Comparison of Interglacial Floras of Northern Europe. Four criteria can be used to identify interstades of Europe:

1. The actual chronologies or successions of tree-pollen types (summarized by West, 1955; 1968); these patterns are in most cases diagnostic.

2. The total flora represented; on the basis of a taxonomic list of the definitely identified plant remains we can compute the percentage of genera that still are native to the locale. Reid (1935) used such a calculation to show the gradual increase of modern British genera and species in English floras and found that it was a good general indicator of age.

3. The presence of key fossils; for example, the occurrence of *Tsuga, Carya, Eucommia,* and abundant *Pterocarya* pollen with remains of the extinct species *Azolla tegeliensis* distinguishes the Tiglian, whereas the association of *Eucommia* with living *Azolla filiculoides* characterizes the Cromerian (Zagwijn, 1963b).

4. The present geographic affinities of forms within the flora; for example, American genera and species are present in the first three interglacials of Europe but are lacking in the last.

The Tiglian Period. The first Quaternary interglacial or nonglacial period is represented by extensive fossil beds at Tegelen in the Netherlands and derives its name, Tiglian, from them. It follows the time of the Donau glaciations in the Danube Valley but precedes the Günz glaciations of the Alps. The Upper Crag beds of England (lower Pleistocene) are also of Tiglian age. Megafossils from the Netherlands deposits were described by Reid and Reid (1907, 1910, 1915) and detailed pollen work is by Zagwijn (1960, 1963a, b; Sluys and Zagwijn, 1962). Relative abundances of key pollen forms in the Tiglian of Germany and the Netherlands are shown in figure 17-8.

The pollen sequence involving a warming followed by a cooling is characterized by Zagwijn (1963a) as follows:

Zone TC-2: thermophilous trees rather important, but *Picea* and *Pinus* also represented; forest-tree pollen abundant.

Zone TC-3: culmination of increases of thermophilous trees; minimum amounts of *Picea* and *Pinus*; forest-tree pollen abundant.

Zone TC-4 relative increase of *Picea*; *Pinus* important; decrease of thermophilous trees; forest-tree pollen abundant.

Zone TC-5: another maximum of thermophilous trees, including *Eucommia*, *Carya*, and *Pterocarya;* temporary minimum of *Pinus* and *Picea*; forest-tree pollen abundant.

Zone TC-6: *Pinus* and *Betula* dominating, strong decrease of thermophilous trees, many of them disappearing before the end of this zone; forest-tree pollen abundant and then declining.

A continuation of the late Tiglian cooling is seen in the overlying Eburonian of early Taxandrian (= Günz) age:

Zone EB-I: thermophilous trees almost absent, *Picea* of little importance; maximum values of *Betula* and herbs; park woodland.

Zone EB-II: decrease of herbs, strong increase of *Pinus*, low maximum of *Picea*; forest.

Zone EB-III: increase of herbs and shrubs, especially Ericales. Conifers, *Betula*, and *Salix* are frequent; park woodland.

Interpretation of this pollen sequence (and the herb zones in the overlying Taxandrian) is a problem because small amounts of reworked Tertiary pollen may be present. The dominant forms, however, are probably not reworked; significant amounts (20 to 40 percent total in zone TC-5) of *Carya*, *Eucommia*, and especially *Pterocarya* (fig. 17-8), which are consistently present, are thought to be in place here. Minor amounts of *Tsuga* and *Phellodendron* pollen may be in place. (*Pterocarya* and *Tsuga* are the characteristic Tertiary relicts in the English Crag deposits of Tiglian age; West, 1962). *Fagus* pollen, characteristic of the Tiglian, is absent in all later interglacials but reappears in the postglacial of northwestern Europe.

A distinguishing feature of Tiglian interglacial beds is the fact that 9 percent of the plant species are extinct as determined by diagnostic reproductive structures preserved in the sediments; for example, remains of the extinct fern *Azolla teglensis* are characteristic and common at a number of Tiglian and lower Taxandrian sections (Zagwijn, 1960).

In contrast to the Pliocene Reuver flora, which contains only 19 percent modern Dutch species, 50 percent of the Tiglian species occur in the modern Dutch flora. The Reuverian contains some subtropical elements, (e.g., *Meliosma*), but the Tiglian is strictly warm temperate in aspect.

The Cromerian Period. The Cromer forest-bed series of England, which lies between the Weybourne Crag and Lowestoft Till, represents cycles of climatic amelioration and deterioration that preceded the Elster glaciation of Germany (West, 1955) and is therefore equivalent to the interval between the Günz and Mindel glaciations of the Alps.

The Cromerian interglacial period (Duigan, 1963) is typified by the following pollen sequence (fig. 17-17): at the base of the diagram is an early *Pinus-Betula* phase; in the middle are pollen types belonging to mixed oak forest (Quercetum mixtum) reach high values, and at the end *Pinus*, *Betula*, and *Picea* become dominant. The scarcity of *Abies* and *Fagus* is an important feature of this interglacial sequence. Minor amounts (1 to 3 percent) of *Carya* and *Eucommia* pollen may represent Tertiary relict genera in the Cromerian flora. The same features (*Carya* pollen excepted) are found in Cromerian pollen sequences from the Netherlands (Zagwijn, 1960, 1963a).

According to seed evidence the Cromerian flora is much like the modern one of Britain. It is primarily composed of modern British species; only one of its known species is now extinct

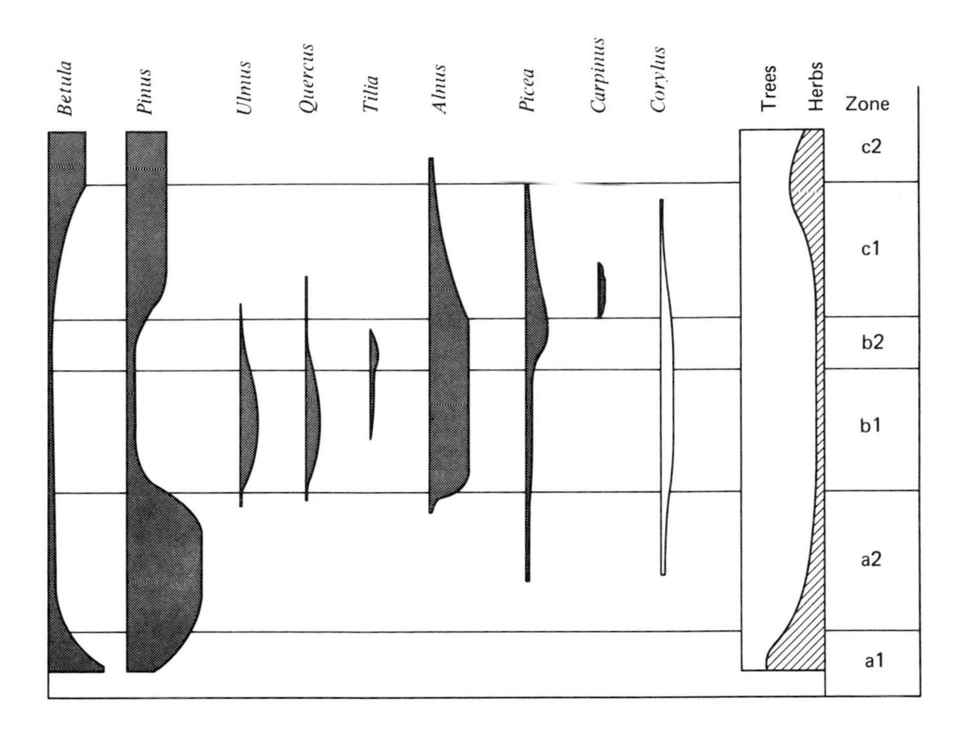

Figure 17-17. Pollen diagram from the Cromerian interglacial deposits. The pollen is represented as a percentage of the total tree pollen. (From West, 1961; Pearson, 1964; Duigan, 1963.)

(Reid, 1920; Godwin, 1956, p. 292). An extinct *Azolla* of the Tiglian is replaced by the living *A. filiculoides* in the Cromerian of the Netherlands (lower Sterksel formation; Zagwijn, 1963b).

The Hoxnian of England and Needian of the Netherlands. The Needian interglacial of the Rhine basin of Elster-Saale age (= Mindel-Riss according to the Alpine sequence) has been correlated with English deposits at Hoxne (West, 1956) and Clacton (West, 1955; Pike and Godwin, 1953; Reid and Chandler, 1923); the pollen sequences and seed floras at each are strikingly similar. The same general patterns are also seen in the Gortian (Hoxnian) of Ireland (Watts, 1967).

The pollen sequence from Hoxne (summarized in fig. 17-18) gives a representative picture of the vegetational history of this interglacial in England. An initial late-glacial stage of primarily nontree types is dominated by pollen of *Betula* and *Hippophaë*. Following this is an interval of temperate mixed hardwood and oak types (Quercetum mixtum), within which the slow rise and low values of *Corylus* are important features. These elements are partly replaced in the next stage by pollen of *Carpinus* and conifers, including both *Picea* and *Abies*. A final early glacial phase shows a rise in nonarboreal pollen types.

High frequencies of *Abies* pollen distinguish the Hoxnian from the earlier Cromerian interval. The presence of *Azolla filiculoides* (Tralau, 1959) and the small number of now-exotic vascular genera (2 out of 42) sets it apart from the more oriental Tiglian flora. Two extinct species (of *Crataegus*) reported from Clacton (Godwin, 1956) may belong to the living species *Pyracantha coccinea* (Richard A. West, written communication, 1968).

A late middle Acheulian culture is associated with the end of the Hoxnian period. A sequence of pollen changes, sudden decrease and then increase in tree pollen with temporary influx of ruderals (*Plantago*, Chenopodiaceae, and other herbs), is attributed to a temporary opening of the forest according to West (written communication, 1968) due to unknown causes.

The English records of thermophilous oceanic species now of southern distribution (species of *Hedera*, *Ilex*, and *Taxus*) indicate a climate perhaps more maritime than now during the mid-

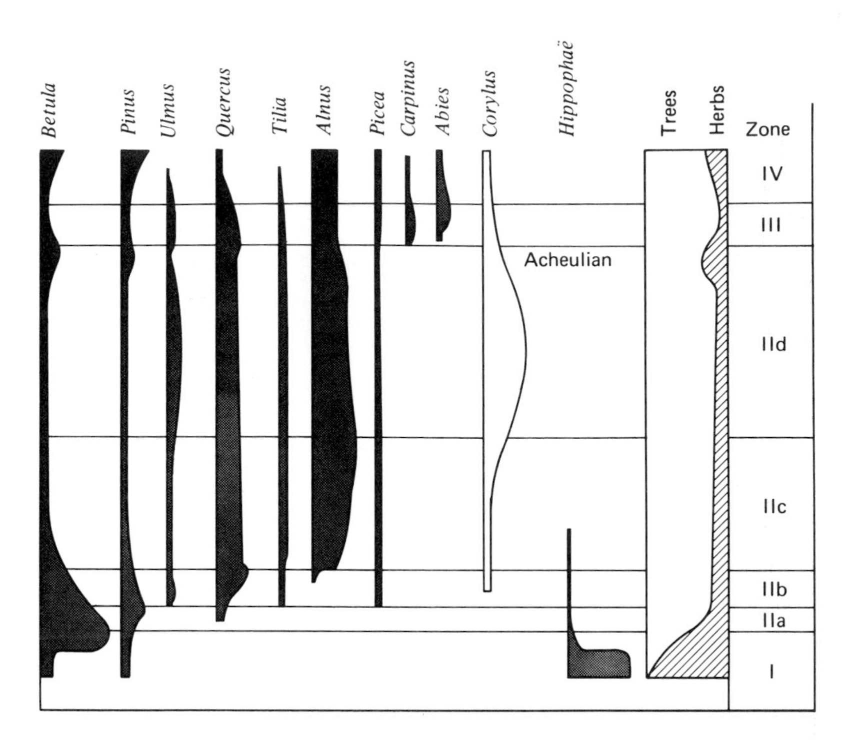

Figure 17-18. Pollen diagram from the Hoxnian interglacial deposits. (From West, 1961; Pearson, 1964.)

Hoxnian; the presence of *Hedera* indicates winter temperatures not below −1.5°C (Iversen, 1944).

The Eemian Period. Deposits of the last (Riss-Würm) interglacial in Denmark and northwest Germany were studied in detail by Jessen and Milthers (1928). Additional work was carried out in the Netherlands by Florschütz (1930) and Zagwijn (1961); in Belgium by Paepe and Vanhoorne (1967); and in England at Ipswich by West (1957) and at Histon Road, Cambridge, by Walker (1953) and West and Sparks (1960). A summary of the composite pollen sequence taken from these works is outlined below. A generalized diagram of the sequence of the Ipswichian is shown in figure 17-19.

Early arctic and subarctic zones (megafossils of *Betula nana*, *Dryas octopetala*, *Salix herbacea*, *Salix reticulata*, zones a and b) are characterized by a dominance of herb pollen and pine and *Betula*. The middle zones (c through g) show the following sequence of pollen types: a dominance of *Pinus*, succeeded by a rapid rise of *Corylus* mixed with *Quercus*, *Ulmus*, *Alnus*, and *Acer*. As *Corylus* waned *Carpinus* increased prominently at the end of the Quercetum mixtum interval. The last stage, zone k of the Eemian, shows an increase of the nontree pollen with *Pinus* and *Betula*; megafossils (*Betula nana*, *B. pubescens*) in this zone denote the return of subarctic conditions. A final oscillation after the Eemian shows the return to deciduous forest and again to subarctic flora (zones l through n).

On the basis of megafossils and pollen evidence Vlerk and Florschütz (1950) infer an initial July temperature of 12°C (zones a and b), a climatic optimum with a July temperature of 18°C, and a return to a July average of 12°C in zone k. The final oscillation repeats evidence of these July temperature changes.

The Eemian succession differs in small but important ways from that of the postglacial in Britain; though *Corylus* is important in both, it rises slowly and makes a definite oscillation during the midpostglacial, whereas it rises rapidly with no clear oscillation in the mid-Eemian. Also *Fagus*, which enters into the sequence during the late postglacial, is absent in the Eemian succession. (See diagram of British postglacial, fig. 17-20.) The postglacial differs from the Hoxnian in that it lacks *Abies* and *Picea*.

The Eemian interglacial in Europe as far as is known contains no extinct species.

Early Quaternary Floras of Southern and Central Europe. In northern Italy an early Pleistocene pollen sequence thought to span the Donau through the Günz and Mindel intervals has been described by Lona (1963) and Lona and Follieri (1958). The deposits are lacustrine and fluvioglacial terraces in the valleys of the Re River and Seriana di Vertova. The sections are dated partly by remains of *Elephas* and *Rhinoceros*. Oscillations in the abundance of "mediocratic" (warmth-loving) elements provide the basis for an inferred temperature curve for the sequence. The Cromerian warm interval was characterized by associations termed Carietum (*Carya, Pterocarya, Juglans*), Carpinetum (*Corylus, Carpinus, Alnus, Fagus, Castanea*), and Quercetum (*Quercus, Ulmus, Tilia, Zelkova*). The Carietum wanes in importance in younger warm periods; it is irregularly replaced by elements of the Quercetum and Carpinetum. Intervening cold phases are characterized by increased amounts of *Pinus, Abies, Picea,* and, at some levels, *Tsuga*. The authors state that the flora did not become impoverished so rapidly in the southern Alps as it did in northwestern Europe or in Poland; the Italian Cromerian flora, for example, contains pollen of *Carya, Pterocarya,* and *Tsuga*. However, in terrace deposits—and in the absence of corroborative megafossil evidence—the Tertiary relict types might be reworked or secondary pollen.

Pollen and seed floras of the late Pliocene and early Pleistocene, which were discussed briefly in "Pliocene-Pleistocene Boundary," are described from Mizerna in southern Poland in a major work by Szafer (1954). Using Venzo's (1965) suggested age of the section, Szafer's Mizerna zone I is of late Pliocene age and zone II is of Tiglian (pre-Günz) age; the sequence also records the Günz (zone II-III) and the Cromerian (zones III-IV). A deposit on the Bug River near Warsaw, on the basis of pollen analysis

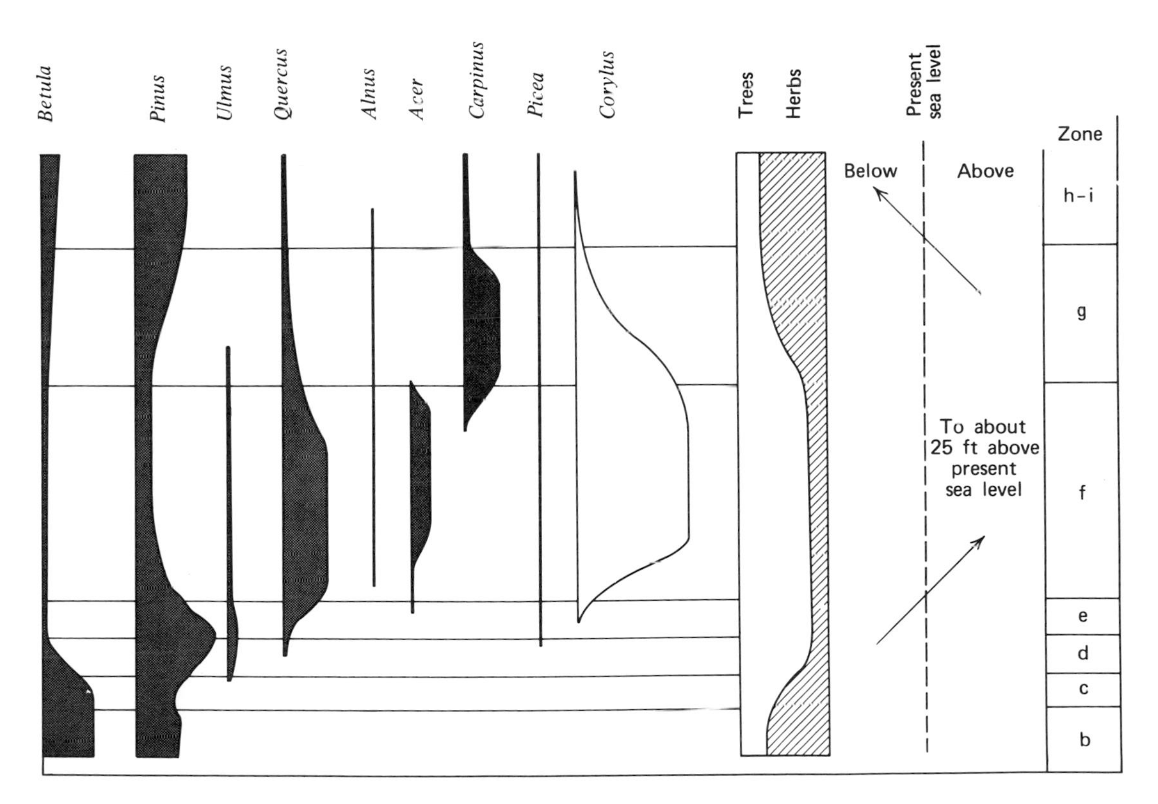

Figure 17-19. Pollen diagram from the Ipswichian (=Eemian) interglacial deposits. (From West, 1961; Pearson, 1964.)

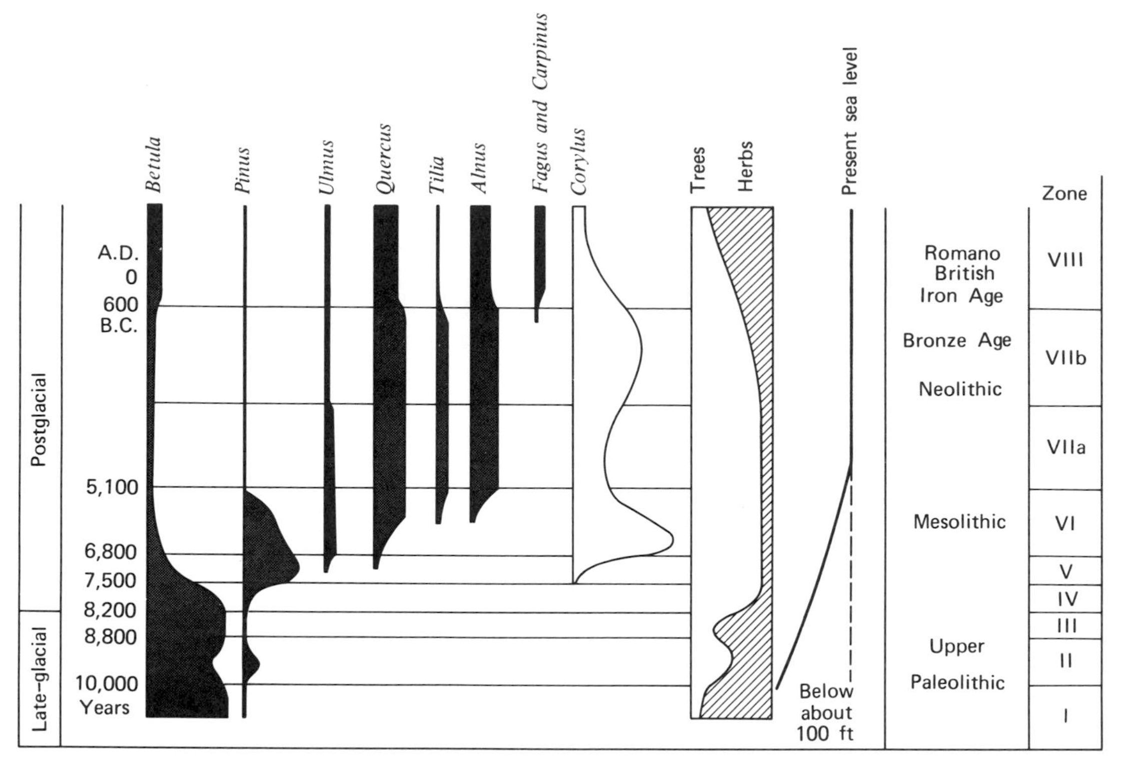

Figure 17-20. Simplified diagram of the pollen frequencies observed in deposits of late-glacial and postglacial age in the British Isles. (From Godwin, 1956; West, 1961; and Pearson, 1964.)

(Stachurska, 1961), is probably of Cromerian age; the flora contains small amounts of such hardwoods as *Carya, Juglans, Ulmus, Ilex*, and the pollen of *Keeteleria*. The scarcity of *Abies* pollen contrasts this interval with the Polish version of the Mindel-Riss interglacial at Sewerynow (Jurkiewiczowa and Mamakowa, 1960), where *Abies* rises to 50 percent of the pollen count. The Polish Eemian interglacial near Radom (central Poland) has a pollen sequence similar to that of northern Europe in that *Corylus* undergoes an early and very rapid rise to prominence and is followed by a strong peak of *Carpinus*. *Abies* plays a minor role here, but *Picea* is somewhat more important than in the north European Eemian (Tolpa, 1961).

Early and Middle Pleistocene Floras of the United States. Only a few American pollen studies of pre-Wisconsinan Pleistocene floras are as yet available; these are from isolated localities, and some are poorly dated. More is known about early and middle Pleistocene floras of California than elsewhere in the United States because of the several megafossil floras described from that state.

From the Midwest a few older pollen studies give an indication of the sequences found in early interglacials. The first recorded interglacial (type Aftonian of Iowa), thought to be equivalent to the Cromerian of Europe, yields a pollen sequence of predominantly spruce with pine, and contains in one 2-inch interval a predominance of grassland and deciduous tree forms (Union County peat beds; Lane, 1941). At other localities that may be Aftonian in age grassland elements prevail. Pollen from deposits at Quincy, Ill. (Voss, 1939), once thought to be of second interglacial (Yarmouthian) age, records spruce-fir forest and is of late Aftonian or early Kansan age (John Frye, oral communication, 1968). No pollen evidence for the Yarmouthian is now available. In the third interglacial (Sangamonian of Iowa) pollen evidence shows oscillations of pine, spruce, and grasses; supposedly Sangamonian pollen reported from Canton, Ill. (Voss, 1939), and Louise County, Iowa (Lane, 1941), are of Wisconsinan age (John Frye and Robert Ruhe, oral communications, 1968). Quaternary paleoecology based on soils from Iowa is summarized by Ruhe (in press).

More recent work on pre-Wisconsinan localities in Kansas and Oklahoma is reported by Kapp (1965). Pollen from vertebrate localities judged to be of Illinoian age indicates a flora dominated by conifers (*Picea, Pinus, Abies*, and *Juniperus*) and in some localities by grasses, *Artemisia*, and other Compositae. On the basis of rare *Pseudotsuga*-type pollen Kapp suggests that Cordilleran elements may have had outposts in the Midwest. Sangamonian sites show oscillations of *Pinus* and such nontree elements as grasses, *Artemisia*, and other Compositae; deciduous broad-leaved forest elements are rare or lacking at most of Kapp's localities. Though Kapp provides data from a large number of analyses, the sediments contain little pollen, and the pollen flora was apparently not diverse.

Pre-Wisconsinan floras from the West Coast are known mainly from megafossils and are chiefly from California. In a recent review Axelrod (1966) evaluates the age and composition of these floras and their implications for plant geography. He points out that relict occurrences in the mountains of southern California of species and plant communities that also occur in the north apparently resulted from southerly migrations during the early Pleistocene. The southern outposts are separated from their northern counterparts by low, semiarid corridors across which plant migration is not possible today. Though the late Pliocene floras of California are characterized by the presence of a few genera which are no longer native to the State and which now have their closest allies in areas far to the south, Pleistocene floras lack such relicts. Early Pleistocene pollen floras from the east slope of the Sierra Nevada and western Mojave Desert, once thought to be Pliocene (Coso florules; Axelrod and Ting, 1960), are based on few pollen grains; identifications of species are not warranted by the pollen morphology of the taxa.

Late Pleistocene floras of the Pacific Northwest, western Canada, and Alaska are reviewed by Heusser (1960). A Sangamonian or possibly older interglacial from Puyallup, Wash. contains sequences of forest-tree pollen that are distinctive and quite unlike the local postglacial forest succession (Leopold and Crandell, 1958).

Southwestern pollen floras of Pleistocene age are reviewed by Martin and Mehringer (1965). Early Pleistocene floras include records from the Sonoran desert in Arizona (Gray, 1961), southwestern New Mexico (Clisby, Foreman, and Sears, 1958), and near Channing, Tex. (Jerry Harbour, written communication, 1964). A lake sediment of possible Nebraskan age in Arizona records the lowering of vegetation zones 1,500 feet and may record the presence of northern elements as well. A long pollen sequence from the San Augustin plains in New Mexico records pollen rain for a large part of the Pleistocene but indicates that the most significant local vegetation changes occurred during the Wisconsinan. On the basis of extrapolated rates of sedimentation the authors (Clisby, Foreman, and Sears, 1958, p. 26) suggest that the lower part of the pollen sequence is of Pliocene age; independent supporting evidence is lacking, however. In Texas pollen-and-leaf evidence from a lake sediment of Blancan age establishes a western extension of *Ulmus* from its present range.

In Eastern United States a well-documented interglacial pollen flora is that from the Gardiners Clay on Long Island, N.Y., described by Donner (1964). Of Sangamonian age, this sequence shows an early development of boreal forest dominated by pine and spruce, and later a rich mesophytic forest of Quercetum mixtum followed again by boreal forest. Plant taxa identified from pollen in the clay are all represented in the present flora of the area. The inferred climate at mid-Sangamonian time was like that of today on Long Island. The Gardiners Clay on Cape Cod (Hyyppä, 1955) probably contains reworked pollen, because its flora cannot be fitted into any known pattern of vegetation development (Donner, 1964; Davis, 1961a).

The Scarborough and Don beds of Toronto, Canada, contain Sangamonian pollen sequences; the Don beds begin with a rich Quercetum mixtum assemblage followed by a boreal flora dominated by pine with fir and spruce. Pollen of the Scarborough beds of probable late Sangamonian age shows a boreal assemblage with increasing spruce, fir, and probably jack pine. Terasmae (1960, 1961) infers a warm Sangamonian climate, warmer than today's by 5°F (annual average), and a later cool Sangamonian interval that was colder than now by 10°F.

Broad Climatic Trends of the Quaternary

Inferred Quaternary climate patterns from various parts of the world are incorporated in figure 17-21; the selected curves, except two, are based primarily on plant data, but the estimated climates take into account geomorphic evidence and data from vertebrates and invertebrates. Various estimates of climate available for the late Quaternary were ably summarized by Wright (1961).

Many sources suggest that the world climates of the last interglacial (Eemian or Sangamonian) and of the thermal maximum in the midpostglacial were probably warmer than the present. Less agreement exists concerning the climates of earlier interglacials. The difference between Poland and the Netherlands in this respect may relate to the difference in timing of the first local glaciation; north Europe was first glaciated in the Günz, but southern Poland's first glaciation did not come until the Mindel (= Krakow). Apparently the cooling of the early Quaternary did not have striking effects on the vegetation in northeastern Siberia and Alaska until fairly late (probably Mindel; Petrov, 1963, 1967).

Geographic factors of many types must be considered in an evaluation of Quaternary climate. A glance at topographic and vegetation maps provides sufficient evidence that latitude, altitude and relief, proximity to continental ice sheets, and position with respect to the continental margin markedly influence the expression of local climate today. Changes of the past—such as uplifts, transgressions and regressions of the sea, and expansion of ice sheets and pluvial lakes—added many factors that complicated local climates in the late Cenozoic. With these in mind we should not automatically expect that a single secular change of climate will cause symmetrical changes in local climates on a regional basis. As more studies are completed on the ecology of the Quaternary, added detail will undoubtedly reveal a far more complex picture of prehistoric climate than is yet available from today's geologic and paleontologic facts.

Summary of Quaternary Floras

Statistical methods, potentially helpful in interpreting older late Cenozoic pollen floras, are now in use for pollen analysis of the late Quaternary. One of the most significant of these is the

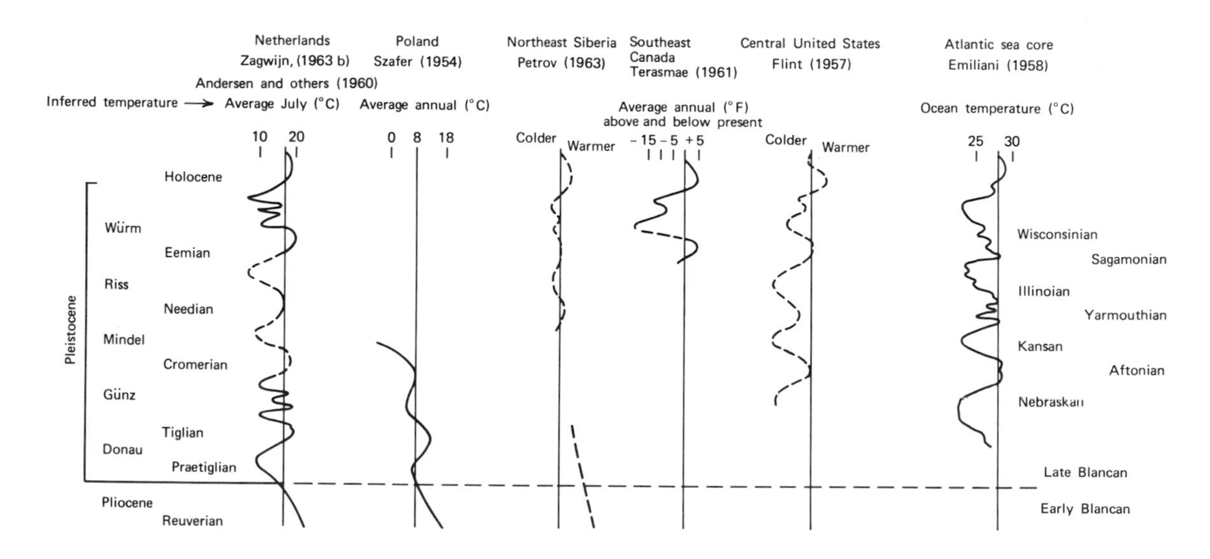

Figure 17-21. Summary of inferred climate changes of the Quaternary and late Pliocene. The first four columns are based on plant remains as well as geomorphic evidence, though Petrov (1963) also used marine mollusks and diatoms. The curve summarized from Flint (1957) utilizes evidence from vertebrates and geomorphology. Data from Emiliani's (1958) deep-sea core from the Atlantic Ocean are based on ratios of oxygen isotopes. Pre-Riss and pre-Illinoian Pleistocene correlations should be considered tentative. Vertical lines indicate modern norms.

measurement of absolute pollen concentration. Pollen deposition rates can reveal which pollen types actually increased with time in the pollen rain. Percentage composition, on the other hand, only indicates which pollen types were more abundant than others.

Early Quaternary floras of the north middle and high latitudes are notable because of the fact that not only degree of change but trends of change are widely different in various areas; for example, though the first two interglacials of the early Quaternary are marked by definite oscillations of climate in Europe, only a extended cooling occurred in Alaska and nearby Siberia.

In Europe the first recorded glacial cooling (Donau, or Praetiglian) brought boreal marine species to molluscan faunas of Italy, permanently removed subtropical taxa, and temporarily lowered the abundance of temperate plant forms in the region. Floras of the first two interglacials (Tiglian and Cromerian) contained fewer Tertiary relict genera than did Pliocene floras, and they recorded temperate climates that were warmer and more equable than the present ones in continental Europe.

Though there are a few early Quaternary records of arctic-alpine plant species, the first clear influx of such taxa across Europe was during the third (Mindel) glaciation; the Mindel is thought also to have been the first time that arctic conditions were widespread in Alaska and Siberia. In the warming of the third interglacial (Needian) that followed, the average annual temperature was cooler than that of present-day Europe, but growing-season temperatures were about like now.

American floristic records for the early and middle Quaternary either are from isolated localities, are lacking, or are not well dated; evidence from vertebrates and mollusks indicates that the Aftonian, Yarmouthian, and Sangamonian interglacial climates in Central United States (Oklahoma-Kansas region) were characterized by reduced continentality (less seasonal extremes of temperature) compared with the present; the glacial climates were also less continental but cooler than now. Exceptions are the late Wisconsinan and early postglacial, which were undoubtedly more continental than now.

The climate of the last interglacial (Eemian, Sangamonian), on the basis of paleobotanical evidence, was warmer than present in Europe as well as in the United States and southern Canada.

The third through the last glaciation brought tundra or park-tundra vegetation to Central Europe and northern Italy. In the United States pollen records of boreal forest genera (e.g., *Picea*) are found as far south as Florida, the Ozark Mountains, and Texas, but these records are not yet correlated with particular glaciations. Plant megafossils of arctic affinities have as yet been identified at only a few United States sites (in Massachussetts and Minnesota) and in Newfoundland; these sites are all of late Wisconsinan age. But pollen assemblages suggest that park-tundra conditions may have existed in a narrow zone attending the glacial border during the retreat of ice in late Wisconsinan time.

Some late Quaternary pollen localities from low latitudes and the Southern Hemisphere record partial replacement of mixed warm temperate and subtropical forest with cool temperate and boreal forest during the Wisconsinan.

Though decreases in floristic diversity are recorded for the European floras during the early Quaternary, the main changes in Quaternary vegetation of the Northern Hemisphere involved cliseral migrations of existing vegetation zones (a) in north-south directions and (b) with altitude. Few extinctions of plant species are recorded for the Quaternary interval.

References

Ahrens, Wilhelm, 1958, Die Niederrheinische Braunkohlenformation: Fortschr. Geologie Rheinland u. Westfalen, v. 1-2, 763 p.

American Geological Institute, 1960, Glossary of geology and related sciences: Am. Geol. Inst., 325 p.

Andersen, S. T., 1954, A late glacial pollen diagram from southern Michigan, USA: Danmarks Geol. Undersøgelse [Skr.] 2, Raekke no. 80, p. 140–155.

Andersen, S. T., De Vries, H., and Zagwijn, W. H., 1960, Climatic change and radiocarbon dating in the Weichselian glacial of Denmark and the Netherlands: Geologie en Mijnbouw, v. 39, no. 2, p. 38-42.

Axelrod, D. I., 1964, The Miocene Trapper Creek flora of southern Idaho: California Univ. Pubs. Geol. Sci., v. 51, 148 p.

——— 1966, The Pleistocene Soboba flora of southern California: California Univ. Pubs. Geol. Sci., v. 60, 79 p.

——— 1968, Tertiary floras and topographic history of the Snake River basin, Idaho: Geol. Soc. America Bull., v. 79, no. 6, p. 713-733.

Axelrod, D. I., and Ting, W. S., 1960, Late Pliocene floras east of the Sierra Nevada [California-Nevada]: California Univ. Pubs. Geol. Sci., v. 39, no. 1, p. 1-117, pls. 1-24.

Benninghoff, W. S., 1962, Calculation of pollen and spore density in sediments by addition of exotic pollen in known quantities [abs.]: Pollen et Spores, v. 4, no. 2, p. 332-333.

Beug, Hans-Jürgen, 1961, Leitfaden der Pollenbestimmung für Mitteleuropa und angrenzende Gebiete: Stuttgart, Gustav Fischer Verlag.

Brooks, C. F., and Connor, A. J., 1936, Climatic Maps of North America: Harvard Univ., Bluehill Meteorological Observatory, Harvard Univ. Press, Folio with 27 maps.

Brookes, D., and Thomas, K. W., 1967, The distribution of pollen grains on microscope slides; pt. 1, The nonrandomness of the distribution: Pollen et Spores, v. 9, no. 3, p. 621-629.

Chandler, M. E. J., 1963, Revision of the Oligocene floras of the Isle of Wight: British Mus. (Nat. Hist.) Bull., Geology, v. 6, no. 3, p. 321-383.

Clisby, K. H., Foreman, Frederick, and Sears, P. B., 1958, Pleistocene climatic changes in New Mexico, USA: Geobot. Inst. Rubel veröffentl., no. 34, p. 21-26.

Couper, R. A., 1960, New Zealand Mesozoic and Cainozoic plant microfossils: New Zealand Geol. Survey Paleont. Bull. 32, 87 p.

Cox, Allan, Dalrymple, G. B., and Doell, R. R., 1967, Reversals of Earth's magnetic field: Scientific American, v. 216, no. 2, p. 44-54.

Crandell, D. R., Mullineaux, D. R., and Waldron, H. H., 1958, Pleistocene sequence in southeastern part of the Puget Sound lowland, Washington: Am. Jour. Sci., v. 256, no. 6, p. 384-397.

Cranwell, L. M., 1953, New Zealand Pollen Studies: The Monocotyledons: Bull. of the Auckland Inst. and Mus. No. 3, Harvard Univ. Press Pub., 91 p.

Cranwell, L. M., ed., 1964, Ancient Pacific floras, the pollen story: Hawaii Univ. Press, 10th Pacific Sci. Cong. Ser., 113 p.

Cushing, E. J., 1964, Redeposited pollen in late-Wisconsin pollen spectra from east-central Minnesota: Am. Jour. Sci., v. 262, no. 9, p. 1075-1088.

Dahl, Eilif, 1964, Present-day distribution of plants and past climate, p. 52-61 *in* Hester, J. J., and Schoenwetter, James, compilers, The reconstruction of past environments—Fort Burgwin conf. on paleoecology, 1962, Proc.: Santa Fe, Fort Burgwin Research Center Pub. 3, 89 p.

Davis, M. B., 1961a, The problem of rebedded pollen in late-glacial sediments at Taunton, Massachusetts: Am. Jour. Sci., v. 259, p. 211-222.

——— 1961b, Pollen diagrams as evidence of late-glacial climatic change in southern New England: N.Y. Acad. Sci. Annals, v. 95, art. 1, p. 623-631.

Davis, M. B., 1963, On the theory of pollen analysis: Am. Jour. Sci., v. 261, no. 10, p. 897-912.

_____1966, Determination of absolute pollen frequency: Ecology, v. 47, no. 2, p. 310-311.

_____1967, Pollen accumulation rates at Rogers Lake, Connecticut, during late- and post-glacial time: Rev. Palaeobotany and Palynology, v. 2, nos. 1-4, p. 219-230.

Davis, M. B., and Deevey, E. S., Jr., 1964, Pollen accumulation rates—Estimates from late-glacial sediment of Rogers Lake: Science, v. 145, no. 3638, p. 1293-1295.

Davis, M. B., and Goodlett, J. C., 1960, Comparison of the present vegetation with pollen-spectra in surface samples from Browington Pond, Vermont: Ecology, v. 41, no. 2, p. 346-357.

Doktorowicz-Hrebnicka, Julia, 1954, Analiza pylkowa wegla brunatnego z [Pollen analysis of brown coal from the region of Zary (lower Silesia)] okolicy Zar na Dolnym Slasku: [Poland] Inst. Geol. Biul. 71, p. 41-108 [Polish with Russian and English summaries].

_____1957a, Wzorcowe spektra pylkowe pliocénskich osadów weglonósnych [Index pollen spectra of Pliocene coal-bearning sediments]: [Poland] Inst. Geol. Prace, v. 15, p. 87-165 [Polish with Russian and English summaries].

_____1957b, Z badan mikroflorystycznch wegla brunatnego w Miroslawicach Górnych na Dolnym Slasku [Microfloristic investigations of brown coal at Miroslawice Górne in Lower Silesia]: [Poland] Inst. Geol. Prace, v. 15, p. 167-186 [Polish with Russian and English summaries].

_____1957c, Wiek wegla brunatnego z terenu Babiny na Dolnym Slasku w swietle analizy pylkowej [The age of brown coal from the area of Babina (Lower Silesia) in the light of pollen analysis]: [Poland] Inst. Geol. Prace, v. 15, p. 187-200 [Polish with Russian and English summaries].

Donner, J. J., 1964, Pleistocene geology of eastern Long Island, New York: Am. Jour. Sci., v. 262, no. 3, p. 355-376.

Duigan, S. L., 1963, Pollen analyses of the Cromer forest bed series in East Anglia: Royal Soc. [London] Phil. Trans., ser. B, v. 246, p. 149-202.

Elias, M. K., 1935, Tertiary grasses and other prairie vegetation from High Plains of North America: Am. Jour. Sci., 5th ser., v. 29, no. 169, p. 24-33.

Emiliani, Cesare, 1958, Paleotemperature analysis of core 280 [Atlantic Ocean] Pleistocene correlations: Jour. Geology, v. 66, no. 3, p. 264-275.

Erdtman, G., 1957, Pollen and spore morphology/plant taxonomy. Gymnospermae, Pteridophyta, Bryophyta. (An introduction to palynology. II): Almqvist & Wiksell/Stockholm. The Ronald Press Company/New York, 151 p., 5 pls., 265 illus.

_____1965, Pollen and spore morphology/plant taxonomy: Gymnospermae, Bryophyta (text). (Introduction to Palynology III): Almqvist and Wiksell, Stockholm, 191 p.

_____1966, Pollen morphology and plant taxonomy, Angiosperms. (An Introduction to Palynology I): New York and London, Hafner Publishing Company, 553 p. (First published in 1952.)

Erdtman, G., Berglund, B., and Praglowski, J., 1961, An introduction to a Scandinavian pollen flora. Volume I: Stockholm, Almqvist & Wiksell, 92 p.

Erdtman, Gunnar, Praglowski, J., and Nilsson, S., 1963, An introduction to a Scandinavian pollen flora. Volume II: Stockholm, Almqvist & Wiksell, 89 p.

Evernden, J. F., and Curtis, G. H., 1965, The potassium-argon dating of late Cenozoic rocks in east Africa and Italy: Current Anthropology, v. 6, no. 4, p. 343-385.

Evernden, J. F., and James, G. T., 1964, Potassium-argon dates and the Tertiary floras of North America: Am. Jour. Sci., v. 262, no. 8, p. 945-974.

Faegri, Knut, 1960, Coast plants, *in* Maps of distribution of Norwegian vascular plants: Oslo Univ. Press., v. 1, 134 p.

Faegri, Knut, and Iversen, Johannes, 1964, Text-book of modern pollen analysis, 2d ed., revised: Copenhagen, Munksgaard, 237 p.

Faegri, Knut, and Ottestad, Per, 1949, Statistical problems in pollen analysis: Bergen Mus. Arb. 1948, Nat. Bekke, no. 3, 29 p.

Fay, A. H., 1920, A glossary of the mining and mineral industry: U.S. Bur. Mines Bull. 95, 754 p.

Fischer, A. G., 1960, Latitudinal variations in organic diversity: Evolution, v. 14, no. 1, p. 64-81.

Flint, R. F., 1957, Glacial and Pleistocene geology: New York, John Wiley & Sons, Inc., 553 p.

Florschütz, F., 1930, Fossiele overblijfselen van den plantengroei tijdens het Würm-glaciaal en het Riss-Würm-interglaciaal in Nederland: Koninkl. Nederlandse Akad. Wetensch., v. 33, p. 1043-1044.

Florschütz, F., and Someren, A. M. H. van, 1950, The palaeobotanical boundary Pliocene-Pleistocene in the Netherlands: Internat. Geol. Cong., 18th, Great Britain, Rept., pt. 9, p. 40-46.

Fowells, H. A., ed., 1965, Sylvics of forest trees of the United States: U.S. Dept. Agriculture Handb. 271, 762 p.

Gerasimov, I. P., ed., 1964, Physical geographic atlas of the world: Moscow, Akad. Nauk SSSR, 298 p.

Gijzel, P. van, 1963, Notes on autofluorescence of some Cenozoic pollen and spores from the Netherlands: Geol. Stichting Med., new ser., no. 16, p. 25-31.

Gleason, H. A., and Cronquist, Arthur, 1964, The natural geography of plants: Columbia Univ. Press, 420 p.

Gignoux, Maurice, 1960, Géologie Stratigraphique, 5th ed., Paris, Masson et Cie, 759 p.

Godwin, Harry, 1956, The history of the British flora; a faunal basis for phytogeography: Cambridge Univ. Press, 384 p.

Gray, Jane, 1961, Early Pleistocene paleoclimatic record from Sonoran Desert, Arizona: Science, v. 133, no. 3445, p. 38-39.

——— 1964, Northwest American Tertiary palynology—The emerging picture, p. 21-30 *in* Cranwell, L. M., ed., Ancient Pacific floras, the pollen story: Hawaii Univ. Press, 10th Pacific Sci. Cong. Ser.

Grichuk, V. P., 1960, Stratigraphic division of the Pleistocene on the basis of paleobotanical materials: Internat. Geol. Cong., 21st, Copenhagen, Rept., pt. 4, p. 27-35.

Grichuk, V. P., Hey, R. W., and Venzo, S., 1965, Report of the subcommission on the Plio-Pleistocene boundary: Internat. Assoc. Quaternary Research, 6th Cong. Rept., v. 1, p. 311-329.

Harris, W. F., 1955, A manual of the spores of New Zealand Pteridophyta: New Zealand Dept. of Scientific and Industrial Research Bull. 116, Wellington, 186 p.

Hegi, Gustav, and others, 1909-1931, Illustrierte flora von Mittel-Europa, 7 volumes, München, Carl Hauner.

Heusser, C. J., 1960, Late Pleistocene environments of North Pacific North America—an elaboration of late-glacial and postglacial climatic, physiographic, and biotic changes: Am. Geog. Soc. Spec. Pub. 35, 308 p.

Hopkins, D. M., ed., 1967, The Bering Land Bridge, Stanford Univ. Press, 495 p.

Hopkins, D. M., MacNeil, F. S., Merklin, R. L., and Petrov, O. M., 1965, Quaternary correlations across Bering Strait: Science, v. 147, no. 3662, p. 1107-1114.

Hui-Lin, Li, 1963, Woody flora of Taiwan: Livingston Publishing Co., 974 p.

Hultén, Eric, 1941-1949, Flora of Alaska and Yukon: Lunds Univ. Arsskr., new series, Avd. 2, v. 37-45.

———1950, Atlas of the distribution of vascular plants in northwest Europe: Swedish Gen. Staff Press, 119 p., 1,846 maps.

———1958, Amphi-atlantic plants and their phytogeographical connections: Almquist, 340 p.

———1962, The circumpolar plants, pt. 1, Vascular cryptogams, conifers, monocotyledons: Kgl. Svenska Vetenskapsakad. Handl., Fjarde Ser., v. 8, p. 3-275.

———1968, Flora of Alaska and neighboring territories: Stanford, Calif., Stanford Univ. Press, 1008 p.

Hyyppä, Esa, 1955, On the Pleistocene geology of southeastern New England: Acta Geographica, v. 14, p. 155-225.

International Geological Congress, 1950, Report of the eighteenth session of Great Britain, 1948, Pt. 9, Proceedings of section H—The Pliocene-Pleistocene boundary (K. P. Oakley, ed.), 130 p.

Ikuse, Masa, 1956, Pollen grains of Japan: Tokyo, Hirokawa Pub. Co., 303 p.

Iversen, Johannes, 1944, *Viscum, Hedera,* and *Ilex* as climate indicators; a contribution to the study of the postglacial temperature climate: Geol. Förer. Stockholm Förh., v. 66, no. 438, p. 463-483.

Izett, G. A., and Lewis, G. E., 1963, Miocene vertebrates from Middle Park, Colorado, *in* Short papers in geology and hydrology: U.S. Geol. Survey Prof. Paper 475-B, p. B120-B122.

Jessen, K., and Milthers, V., 1928, Stratigraphical and paleontological studies of interglacial fresh-water deposits in Jutland and Northwest Germany: Danmarks Geol. Undersøgelse [Skr.] 2, no. 48, p. 1-379.

Jones, E. L., 1962, Palynology of the Midway-Wilcox boundary in south-central Arkansas: Gulf Coast Assoc. Geol. Soc. Trans., v. 12, p. 285-294.

Jurkiewiczowa, Irena, and Mamakowa, Kazimiera, 1960, The interglacial at Sewerynow near Przedbôrz [Poland]: Inst. Geol., Odbitka Biul. 150, v. 9, p. 71-103.

Kapp, R. O., 1965, Illinoian and Sangamon vegetation in southwestern Kansas and adjacent Oklahoma: Michigan Univ. Mus. Paleontology Contr., v. 19, no. 14, p. 167-255.

King, W. B. R., Chm., 1950, The Pliocene-Pleistocene boundary, Introduction—Opening remarks to the discussion held on August 26th, p. 5 *in* Internat. Geol. Cong., 18th, Great Britain, 1948.

Kirchheimer, Franz, 1957, Die Laubgewächse der Braunkohlenzeit: Wilhelm Knapp, 783 p.

Kirkland, D. W., 1967, Method of calculating absolute spore and pollen frequency: Oklahoma Geology Notes, v. 27, no. 5, p. 98-100.

Kremp, G. O. W., 1949, Pollenanalytische Undersuchung des miozänen Braunkohlenlagers von Konin und der Warthe: Palaeontographica, v. 90, sec. B, no. 1-3, p. 53-89.

Kremp, G. O. W., 1950, Pollenanalytische Braunkohlenuntersuchungen im südlichen Teil Niedersachsens, insbesondere im Solling: Geol. Jahrb., v. 64, p. 489-517.

Krutzsch, Wilfried, 1957, Sporen- und Pollengruppen aus der Oberkreide und dem Tertiär Mitteleuropas und ihre stratigraphische Verteilung: Zeitschr. Angew. Geologie, v. 3, no. 11-12, p. 509-548.

——1959, Micropaleontologische Untersuchungen in der Braunkohle des Geiseltales: Geologie, no. 21-22, p. 1-425.

Krutzsch, Wilfried, Pchalek, J., and Spiegler, Dorothee, 1960, Tieferes Paläozän (?Montien) in Westbrandenburg: Internat. Geol. Cong., 21st, Copenhagen, Rept., pt. 6, p. 135-143.

Kummel, Bernhard, and Raup, David, eds., 1965, Handbook of paleontological techniques: San Francisco, W. H. Freeman & Co., 852 p.

Kuyl, O. S., Muller, J., and Waterbolk, H. T., 1955, The application of palynology to oil geology with reference to western Venezuela: Geologie en Mijnbouw, new ser., v. 17, no. 3, p. 49-76.

Lane, G. H., 1941, Pollen analysis of interglacial peats of Iowa: Iowa Geol. Survey Ann. Rept., v. 37, 1934-1939, p. 233-263.

Leopold, E. B., 1959, Pollen, spores, and marine microfossils [of the Cooper marl, late Oligocene] [of the Ladson formation, Pleistocene], p. 22-25 and 49-53 *in* Malde, H. E., 1959, Geology of the Charleston phosphate area, South Carolina: U.S. Geol. Survey Bull. 1079, 105 p.

——1963, Miocene pollen and spore flora of Eniwetok Atoll, Marshall Islands [abs.]: Am. Jour. Botany, v. 50, no. 6, pt. 2, p. 628.

——1965, Aspects of floristic change in the late Cenozoic [abs.]: Internat. Assoc. Quaternary Research, 7th Cong., p. 288.

——1967, Late Cenozoic patterns of plant extinction, p. 203-246 *in* Martin, P. S., and Wright, H. E., Jr., eds., Pleistocene extinctions, the search for a cause: New Haven, Conn., Yale Univ. Press, 453 p.

Leopold, E. B., 1969, Miocene pollen and spore flora of Eniwetok Atoll, Marshall Islands: U.S. Geol. Survey Prof. Paper 260-II, p. 1133-1182.

Leopold, E. B., and Crandell, D. K., 1958, Pre-Wisconsin interglacial pollen spectra from Washington State, USA: Geobot. Inst. Rübel Veröffent., no. 34, p. 76-79.

Livingstone, D. A., 1955, Some pollen profiles from arctic Alaska: Ecology, v. 36, no. 4, p. 587-600.

Lona, Fausto, 1963, Floristic and glaciologic sequence (from Donau to Mindel) in a complete diagram of the Leffe deposit: Berichte Geobot. Inst. Eidg. Techn. Hochschule Stiftung Rübel, no. 34, p. 64-66.

Lona, Fausto, and Follieri, Maria, 1958, Successione pollinica della serie superiore (Günz-Mindel) di Leffe (Bergamo): Geobot. Inst. Rübel Veröffent., no. 34, p. 86-98.

Love, J. D., 1961, Split Rock Formation (Miocene) and Moonstone Formation (Pliocene) in central Wyoming: U.S. Geol. Survey Bull. 1121-I, p. 11-139.

MacGinitie, H. D., 1953, Fossil plants of the Florissant beds, Colorado: Carnegie Inst. Washington Pub. 599, 198 p.

——1962, The Kilgore flora—A late Miocene flora from northern Nebraska: California Univ. Pubs. Geol. Sci., v. 35, no. 2, p. 67-158.

McKee, E. D., Chronic, John, and Leopold, E. B., 1959, Sedimentary belts in lagoon of Kapingamarangi Atoll: Am. Assoc. Petroleum Geologists Bull., v. 43, no. 3, pt. 1, p. 501-562.

Macko, Stefan, 1957, Lower Miocene pollen flora from the valley of Klondica near Gliwice (Upper Silesia): Prace Wroclawskiego Towarz. Naukowego, ser. B, no. 88, p. 5-314.

Macko, Stefan, 1959, Pollen grains and spores from Miocene brown coals in Lower Silesia. I: Prace Wroclawskiego Towarz. Naukowego, ser. B, no. 96, p. 5-177.

Mahabale, T. S., and Deshpande, J. V., 1957, The genus *Sonneratia* and its fossil allies: Palaeobotanist, v. 6, no. 2, p. 51-64.

Mapel, W. J., and Hail, W. J., Jr., 1959, Tertiary geology of the Goose Creek district, Cassia County, Idaho, Box Elder County, Utah, and Elko County, Nevada: U.S. Geol. Survey Bull. 1055-H, p. 217-254 [1960].

Markov, K. K., Glazykov, G. I., and Nikolaev, B. A., 1965, Chetvertichnyí period, v. I, II: VII International Quaternary Congress in Boulder, Colo., Moskow Univ., 371 p. and 434 p.

Markov, K. K., and Belichko, A. A., 1967, Chetvertichnyí period, III: Izdatelstvo "Nedra", Moskow, 438 p.

Martin, H. A., and Rouse, G. E., 1966, Palynology of late Tertiary sediments from Queen Charlotte Islands, British Columbia: Canadian Jour. Botany, v. 44, no. 2, p. 171-208.

Martin, P. S., and Mehringer, P. J., Jr., 1965, Pleistocene pollen analysis and biogeography of the Southwest, p. 433-451 *in* Wright, H. E., Jr., and Frey, D. G., eds., The Quaternary of the United States—A review volume for the 7th Congress of the International Association for Quaternary research: Princeton Univ. Press, 922 p.

Mehringer, P. J., Jr., and Haynes, C. V., 1965, The pollen evidence for the environment of Early Man and extinct mammals at the Lehner mammoth site, Southeastern Arizona: Am. Antiquity, v. 31, p. 17-23.

Migliorini, C. I., 1950, The Pliocene-Pleistocene boundary in Italy: Internat. Geol. Cong., 18th, Great Britain, Rept. 9, p. 66-72.

Miki, Shigeru, 1959, Evolution of *Trapa* from ancestral *Lythrum* through *Hemitrapa*: Japan Acad. Proc., v. 35, no. 6, p. 289-294.

Muller, Jan, 1959, Palynology of Recent Orinoco delta and shelf sediments: Micropaleontology, v. 5, no. 1, p. 1-32.

______1964, A palynological contribution to the history of mangrove vegetation in Borneo, p. 33-42 *in* Cranwell, L. M., ed., Ancient Pacific floras—the pollen story: Hawaii Univ. Press, 10th Pacific Sci. Cong., Ser. 2.

Munns, E. N., 1938, The distribution of important forest trees of the United States: U.S. Dept. Agriculture Misc. Pub. 287, 170 p., maps.

Nagy, Esther, 1958, Palynologische untersuchung der am Fusse des Mátra-Gebirges Gelagerten Oberpannonischen Braunkohle: Annales Instituti Geologici Publici Hungarici, v. 47, pt. 1, 294 p.

______1963, Spores et pollens nouveaux d'une coupe de la briqueterie d'eger (Hongrie): Pollen et Spores, v. 5, no. 2, p. 397-412.

Neuy-Stolz, Gertrud, 1958, Zur Flora der Niederrheinischen Bucht während der Hauptflözbildung unter besonderer Berucksichtigung der pollen und Pilzreste in den hellen Schichten, p. 503-525 *in* Ahrens, Wilhelm, Die Niederrheinische Braunkohlenformation: Fortschr. Geologie Rheinland u. Westfalen, v. 1-2.

Odum, H. T., Cantlon, J. E., and Kornicker, L. S., 1960, An organizational hierarchy postulate for the interpretation of species-individual distributions, species entropy, ecosystem evolution, and the meaning of a species-variety index: Ecology, v. 41, no. 2, p. 395-399.

Oszast, Janina, 1960, Analiza pylkowa iłów Tortońskich Starych Gliwic: Warszawa Monographiae Botanicae, v. 9, no. 1, p. 1-47.

Paepe, R., and Vanhoorne, R., 1967, The stratigraphy and paleobotany of the late Pleistocene in Belgium: Belgique Service Géol., v. 8, 96 p.

Pearson, Ronald, 1964, Animals and plants of the Cenozoic era; some aspects of the faunal and floral history of the last 60 million years: London, Butterworth & Co., Ltd., 236 p.

Petrov, O. M., 1963, The stratigraphy of the Quaternary deposits of the southern parts of the Chukotsk Peninsula: Moscow Acad. Sci. Comm. for the Study of the Quaternary Period Bull. 28 [English translation].

———1967, Paleogeography of Chukotka during late Neogene and Quaternary time, p. 144-171 *in* Hopkins, D. M., ed., The Bering Land Bridge: Stanford Univ. Press, 495 p.

Pflug, H. D., 1953, Zur Entstehung und Entwicklung des angiospermiden Pollens in der Erdgeschichte: Palaeontographica, v. 95, sec. B, no. 4-6, p. 60-171.

Pike, Kathleen, and Godwin, Harry, 1953, The interglacial at Clacton-on-Sea, Essex [with discussions]: Geol. Soc. London Quart. Jour., v. 108, pt. 3, no. 431, p. 261-272.

Pokrovskaya, I. M., ed., 1956a, Atlas of Oligocene spore-pollen complexes of various areas, USSR: Vses. Geol. Inst., Materialy, new ser., no. 16, 312 p. Moskow.

———1956b, Atlas of Miocene spore-pollen complexes of various areas, USSR: Vses. Geol. Inst., Materialy, new ser., no. 13, 460 p. Moskow.

Popov, P. A., 1956, *Trapa* L. pollen in Tertiary deposits of the Enisei Mountains: Akad. Nauk SSSR Doklady, v. 110, no. 3, p. 453-456.

Post, L. von, 1916, Om skogsträdpollen i sydsvenska torfmosselagerföljder: Geol. Fören. Stockholm Förh., v. 38, p. 384-390; diskussionsinlägg, p. 392-393.

Potonié, Robert, 1934, Zur Mikrobotanik des eocänen Humodils des Geiseltals: Preuss. Geol. Landesanst., Inst. Paläobot., Arb. Bd. 4, p. 25-125.

———1951a, Pollen und Sporenformen als Leitfossilien des Tertiärs: Mikroskopie, v. 6, no. 9-10, p. 272-283.

———1951b, Revision stratigraphisch wichtiger Sporomorphen des mitteleuropäischen Tertiärs: Palaeontographica, v. 91, sec. B, no. 5-6, p. 131-148.

Potonié, Robert, Thomson, P. W., and Thiergart, Friedrich, 1951, Zur Nomenklatur und Klassifikation der neogenen Sporomorphae (Pollen und Sporen): Geol. Jahrb., v. 65, p. 35-63, 141-144.

Raatz, G. V., 1937, Mikrobotanisch-stratigraphische Untersuchung der Braunkohle des muskauer Bogens: Preuss. Geol. Landesanst., Abh. N. F., no. 183, 48 p.

Reid, Clement, and Reid, E. M., 1907, Détermination de l'age des argiles à briques: Soc. Belge Géologie, Paléontologie, et Hydrologie Bull., v. 21, p. 583-590.

———1910, A further investigation of the Pliocene flora of Tegelen: Koninkl. Nederlandse Akad. Wetensch. Proc., pt. 19, p. 262-271.

———1915, The Pliocene floras of the Dutch-Prussian border: The Hague, Rijksopsporing van Delfstoffen Med., no. 6 in 4°, 178 p.

Reid, E. M., 1920, A comparative review of Pliocene floras, based on the study of fossil seeds: Geol. Soc. London Quart. Jour., v. 76, pt. 2, p. 145-161.

———1935, British floras antecedent to the Great Ice Age, *in* Discussion on the origin and relationship of the British flora: Royal Soc. [London] Proc., ser. B, v. 118, p. 197-202.

Reid, E. M., and Chandler, M. E. J., 1923, The fossil flora of Clacton-on-Sea: Geol. Soc. London Quart. Jour., v. 79, pt. 4, app. 1, p. 619-623.

———1933, TheLondon Clay flora: British Mus. (Nat. History), 561 p.

Rohrer, W. L., and Leopold, E. B., 1963, Fenton Pass Formation (Pleistocene), Bighorn Basin, Wyoming: U.S. Geol. Survey Prof. Paper 475-C, art. 71, p. C45-C48.

Roger, J., 1964, Réunions du Comité du Néogène méditerranéen et du Comité du Néogène nordique; Internat. Geol. Cong., 21st, Norden, 1960, Rept., 28, p. 293-299.

Walker, Donald, 1953, The interglacial deposits at Histon road, Cambridge [southern England] [with discussion]: Geol. Soc. London Quart. Jour., v. 108, pt. 3, no. 431, p. 273-282.

Watts, W. A., 1967, Interglacial deposits in Kildromin Townland near Herbertstown, Co. Limerick: Proc. Royal Irish. Acad., v. 65, sec. B, p. 339-348.

Weber, W. A., 1965, Plant geography in the southern Rocky Mountains, p. 453-468 *in* Wright, H. E., Jr., and Frey, D. G., eds., The Quaternary of the United States—A review volume for the 7th Congress of the International Association for Quaternary Research: Princeton Univ. Press, 922 p.

West, R. G., 1955, The glaciations and interglacials of East Anglia; a summary and discussion of recent research: Quaternaria, v. 2, p. 45-52 [including French and English summaries].

______1956, The Quaternary deposits at Hoxne, Suffolk: Royal Soc. [London] Phil. Trans., v. 239, ser. B, no. 655, p. 265-356.

______1957, Interglacial deposits at Bobbitshole, Ipswich: Royal Soc. [London] Phil. Trans., v. 241, ser. B, no. 676. p. 1-31.

______1961, Interglacial and interstadial vegetation in England [with discussion] Linnean Soc. [London] Proc., v. 172, pt. 1, p. 81-89.

______1962, Vegetational history of the early Pleistocene of the Royal Society borehole at Ludham, Norfolk: Royal Soc. [London] Proc., v. 155, ser. B, no. 960, p. 437-453.

______1968, Pleistocene Geology and Biology: London, Longmans Green and Co. Ltd., 377 p.

West, R. G., and Sparks, B. W., 1960, Coastal interglacial deposits of the English Channel: Royal Soc. [London] Phil. Trans., v. 243, ser. B, no. 701, p. 95-133.

Westenberg, J., 1947, Mathematics of pollen diagrams, II: Koninkl. Nederlandse Akad. Wetensch. Proc., v. 50, no. 6, p. 640-648.

Weyland, Hermann, 1938, 1941, 1943, 1948, Beiträge zur Kenntnis der Rheinischen Tertiärflora: Palaeontographica, 1938, pt. 3, v. 83B, p. 123-171; 1941, pt. 5, v. 86B, p. 79-112; 1943, pt. 6, v. 87B, p. 94-136; 1948, pt. 7, v. 88B, p. 113-188.

Wilmarth, M. G., 1925, The geologic time classification of the United States Geological Survey compared with other classifications, accompanied by the original definitions of era, period, and epoch terms, a compilation: U.S. Geol. Survey Bull. 769, 138 p.

Wolfe, J. A., 1962, A Miocene pollen sequence from the Cascade Range of northern Oregon, *in* Short papers in geology and hydrology: U.S. Geol. Survey Prof. Paper 450-C, p. C81-C84.

Wolfe, J. A., and Barghoorn, E. S., 1960, Generic change in Tertiary floras in relation to age (Bradley volume): Am. Jour. Sci., v. 258-A, p. 388-399.

Wolfe, J. A., and Hopkins, D. M., 1967, Climatic changes recorded by Tertiary land floras in northwestern North America, p. 67-76 *in* Hatai, Kotora, ed., Tertiary correlations and climatic changes in the Pacific: Sendai, Japan, Sasaki Printing & Pub. Co.

Wolfe, J. A., Hopkins, D. M., and Leopold, E. B., 1966, Tertiary stratigraphy and paleobotany of the Cook Inlet region, Alaska: U.S. Geol. Survey Prof. Paper 398-A, p. A1-A29.

Wolfe, J. A., and Leopold, E. B., 1967, Neogene and early Quaternary vegetation of northwestern North America and northeastern Asia, p. 193-206 *in* Hopkins, D. M., ed., The Bering Land Bridge: Stanford Univ. Press, 495 p.

Wood, H. E., and others, 1941, Nomenclature and correlation of the North American continental Tertiary: Geol. Soc. America Bull., v. 52, no. 1, p. 1-48.

Wright, H. E., Jr., 1961, Late Pleistocene climate of Europe—A review: Geol. Soc. America Bull., v. 72, p. 933-984.

Wright, H. E., Jr., and Frey, D. G., eds., 1965, The Quaternary of the United States—A review volume for the 7th Congress of the International Association for Quaternary Research: Princeton Univ. Press, 922 p.

Wright, H. E., Jr., and Patten, H. L., 1963, The pollen sum [with French abs.]: Pollen et Spores, v. 5, no. 2, p. 445-450.

Zagwijn, W. H., 1960, Aspects of the Pliocene and early Pleistocene vegetation in the Netherlands: Geol. Stichting Med., ser. C-3-1, no. 5, 78 p. [including Dutch summary].

——1961, Vegetation, climate, and radiocarbon datings in the late Pleistocene of the Netherlands; Pt. 1, Eemian and early Weichselian: Geol. Stichting Med., new ser., no. 14, p. 15-45.

——1963a, Pollen-analytic investigations in the Tiglian of the Netherlands: Geol. Stichting Med., new ser., no. 16, p. 49-71.

——1963b, Pleistocene stratigraphy in the Netherlands, based on changes in vegetation and climate: Koninkl. Nederlandse Akad. Wetensch. Proc., Geol. Ser., pt. 21-2, p. 173-196.

Zaklinskaya, Ye. D., 1962, Outline of the geology and paleography of the Pavlodar section of the Irtysch valley, the northern Aral region, and the Turgai lowland: Internat. Geology Rev., v. 4, no. 3, p. 310-335.

18

DINOFLAGELLATES AND OTHER ORGANISMS IN PALYNOLOGICAL PREPARATIONS

William R. Evitt

Besides spores, pollen, and finely comminuted fragments of plant tissues, palynological preparations often contain notable amounts of other microfossils of organic composition. These include morphologically diverse objects of varied natural affinities. The identities of some are still matters of uncertainty, speculation, or argument. Yet many of them are distinctive and useful fossils, as well as challenging objects for paleontological study. Most of them represent aquatic organisms that lived in waters ranging from fresh to open marine. Therefore these fossils are an important complement to those derived from land plants, which often occur in the same samples.

This chapter chiefly concerns three groups of fossils: the dinoflagellates, the acritarchs, and the chitinozoans. A few additional types are mentioned briefly in a concluding section. Since recent evidence has shown that the fossils heretofore commonly attributed to the order Hystrichosphaeridea are not a homogeneous group and are in part dinoflagellates, this order has been abandoned. These microfossils are treated here in part as dinoflagellates (the hystrichospheres sensu stricto), and in part as acritarchs, a heterogeneous assemblage of fossils of unknown and varied affinities.

Among the authors who have been concerned with these microfossils five have made especially significant contributions. Ehrenberg (1838) was the first investigator to identify and illustrate fossil dinoflagellates (including hystrichospheres), having found them in Cretaceous samples from Germany and Poland. Otto Wetzel (1933) launched the modern study of fossil microplankton preserved as tests of organic composition with his major work on Cretaceous cherts from the glacial deposits of northern Germany. Georges Deflandre and Alfred Eisenack, in a series of works, large and small, spanning nearly 40 years have written extensively on dinoflagellates, acritarchs, and chitinozoans. They have dealt with fossils of all ages, but specially noteworthy are the papers of both on Jurassic and Cretaceous dinoflagellates, and those of Eisenack on Ordovician and Silurian chitinozoans and acritarchs. Finally some of the most thorough observations and meticulous descriptions of the morphology of Cretaceous and lower Tertiary dinoflagellates (mainly hystrichospheres) are recorded in the exemplary series of short papers by Maria Lejeune-Carpentier published from 1935 to 1941.

The important early investigations by Wetzel and Deflandre were based on specimens studied while still embedded in transparent slivers of chert deftly chipped from larger pieces. An outline of the technique used is given by Deflandre (1936), and a more complete discussion will be found in an earlier paper (Deflandre, 1935). In contrast with these methods, most recent studies have utilized isolated specimens recovered by acid treatment and prepared finally as either single-specimen mounts or strew preparations, each type having its special advocates and advantages. Many of the techniques applicable to spores and pollen (Gray, 1965) are equally useful for the fossils under discussion here.

The number of publications concerning fossil microplankters with organic tests has increased rapidly since 1955. Although these works include much that is new and valuable, they rest on the foundations laid mainly by Deflandre and Eisenack. Since the descriptive literature pertaining to the fossil microplankton is large and diffuse, use of the portion that treats of dinoflagellates, acritarchs, and chitinozoans (as well as several other groups not discussed in this chapter) is facilitated by the bibliographies and generic indices compiled by G. and M. Deflandre (1943-1967), Downie and Sarjeant (dated 1964, distributed 1965), Eisenack (1965), Norris and Sarjeant (1965), Loeblich and Loeblich (1966), and Combaz ed. (1967). The contents of the first five of these works are described and compared by Evitt (1967). (See last section chapter 4).

Germinal Openings in Dinoflagellates and Acritarchs

Operculate openings in many fossil dinoflagellates and some acritarchs have been widely referred to as pylomes. Some authors (e.g., Eisenack, 1962, 1963a,b) have held that the common occurrence of pylomes among the microfossils that were formerly assigned to the Hystrichosphaeridea is evidence of a close interrelationship of the organisms that left these fossils. However, this suggestion ignores a critical distinction that can be made between two different types of openings. Accordingly in this discussion the term *archeopyle* (Evitt, 1967) is applied to openings whose shape or position (commonly both) may be correlated with the arrangement of plates in a dinoflagellate theca. Most archeopyles are operculate and basically polygonal, but they may also be slitlike and of irregular shape. Archeopyles, first known in a wide variety of fossil dinoflagellates, have now been observed in resting cysts of modern species (Evitt and Davidson, 1964; Wall and Dale, 1967a; Evitt, 1967; Evitt and Wall, 1968). The term *pylome* is reserved for openings among acritarchs. They are most often approximately circular and operculate, more rarely polygonal or slitlike, and cannot be clearly correlated with a pattern of plate arrangement as in dinoflagellates.

Chemical Composition

There has been much experimentation to determine the composition of the organic remains of dinoflagellates, acritarchs, and chitinozoans, but the exact nature of the compounds involved remains obscure. The extreme chemical resistance of the materials is their most striking feature, and this of course makes accurate analysis difficult. Microchemical and X-ray tests commonly applied to organic compounds have given mostly inconclusive results. Informative reviews of the experimental work and summaries of results are given by Deflandre (1938), Pastiels (1945), Collinson and Schwalb (1955), and Eisenack (1963a). Significant observations are the cutinoid composition of some acritarchs and the variable silica content in some dinoflagellates reported by Eisenack (1936), the inconsistent and inconclusive reactions to microchemical and optical tests reported by Deflandre, and the similar composition of fossil dinoflagellates and fossil pollen reported by Pastiels. The only generalization that seems safe is hardly helpful: the somewhat varied and generally undetermined composition of the fossils today reflects an unknown postdepositional modification of unknown original organic compounds.

Dinoflagellates with siliceous or calcareous external tests have been described (e.g., Deflandre, 1933, 1949; Lefevre, 1933). In addition stellate siliceous structures like those that occur within *Actiniscus*, a modern unarmored dinoflagellate, are frequently encountered in Tertiary diatomites and were given the name *Actiniscus* by Ehrenberg (Deflandre, 1952). These fossil dinoflagellates with fully mineralized remains are not common constituents of palynological preparations and will receive no further mention here.

DINOFLAGELLATES

Summary. The fossil record of dinoflagellates extends from Silurian to Recent, but a single Silurian occurrence is the only pre-Permian one yet reported, and specimens are rare before the Middle Jurassic. Beginning then dinoflagellates are common constituents of marine assemblages, although fossil freshwater dinoflagellates are rare. The tests are morphologically diverse and reasonably complex. They are thought to be cysts (or structures that housed cysts) rather than the thecae of organisms in the actively swimming stage. Although many of the fossils do resemble thecae, others (the hystrichospheres) are of quite different aspect, and intermediate types occur. Characterization of genera and species is on the basis of shape, number, and position of major projections (horns) or lesser projections (processes and septa), character of a distinctive opening (the archeopyle) through which the contents escaped, wall structure (especially the number and relationship of layers, presence or absence of an inner body), and a variety of features that reflect the plate pattern (tabulation) of the now-vanished theca. Local and cosmopolitan species occur. Extensive geographic ranges, combined with rapid evolutionary changes, render many types excellent tools for long-range correlation as well as for local zonation.

Modern Dinoflagellates

Dinoflagellates are unicellular aquatic organisms generally treated today as a class (Dinophyceae) within the division Pyrrhophyta among the algae. They commonly range from about 10 to 100 microns in size, with occasional giants up to 1.5 millimeters. The majority are free-living elements of the oceanic plankton; but the group also includes bottom dwellers as well as symbiotic and parasitic types, and their habitat extends to brackish estuaries and freshwater rivers, lakes, and ponds. Some of the free-living dinoflagellates are heterotrophic, but the majority are autotrophic. Characteristic pigments are chlorophylls *a* and *c*, beta-carotene, and four xanthophylls. An identical combination of chlorophylls and beta-carotene (but with different xanthophylls) occurs in the brown algae.

Diagnostic of dinoflagellates is an actively swimming, or motile, stage during which the cell is propelled by two flagella, one extended longitudinally and the other encircling the longitudinal axis (fig. 18-1*a*). The forward movement of the cell is often combined with a distinctive spiral motion. The parasitic and symbiotic species may pass through this stage quickly, but it constitutes the major part of the life cycle of free-swimming species. The longitudinal flagellum is typically whiplike and extends behind the cell. It lies partly in a furrow on the ventral surface called the *sulcus*. The transverse flagellum is ribbonlike and undulates in a furrow called the *cingulum* (also called the girdle), which may be equatorial in position or lie closer to the anterior (apical) extremity of the cell. The girdle may encircle a cell in a plane or its course may be conspicuously spiral. It is interrupted on the ventral surface by the sulcus.

Dinoflagellate life cycles are poorly known. In almost all cases reproduction appears to be exclusively of the vegetative type involving either a single or multiple division that results in two or more daughter cells, each capable of developing to maturity. The daughter cells may look like miniature editions of the mother cell, or they may first progress through one or more different stages before attaining the characters of the parent. Incomplete separation of daughter cells in some species results in chains of individuals. Immobile encysted stages are discussed in a separate section below.

Dinoflagellates of many types are naked cells in the motile stage (the so-called unarmored dinoflagellates), but in two large groups (the armored dinoflagellates) the motile cell is enclosed in a *theca* consisting chiefly of cellulose. In one of these groups (Dinophysidales, fig. 18-1*b-d*) the theca has a sagittal suture and comes apart into two roughly symmetrical halves. In the other group (Peridiniales, fig. 18-1*e-r*) the theca consists of a number of polygonal plates arranged in several latitudinal series. The plates are separated by sutures that are often marked by narrow elevations (low ridges or higher membranous walls) and the plates themselves may be smooth, perforate, or ornamented with various short projecting structures. The sutures are closed, and the plates remain tightly juxtaposed except under certain circumstances; for example, when the living protoplast emerges from the theca or after the death of the cell. Then the plates – all of them or, at first, only certain ones – separate along the sutures. The number and arrangement of the thecal plates differ among taxa and can be described by standard terms and symbols, as explained in figure 18-2. Both the cingulum and sulcus are floored with separable plates, but those in the sulcus are especially small and difficult to observe. The cingulum di-

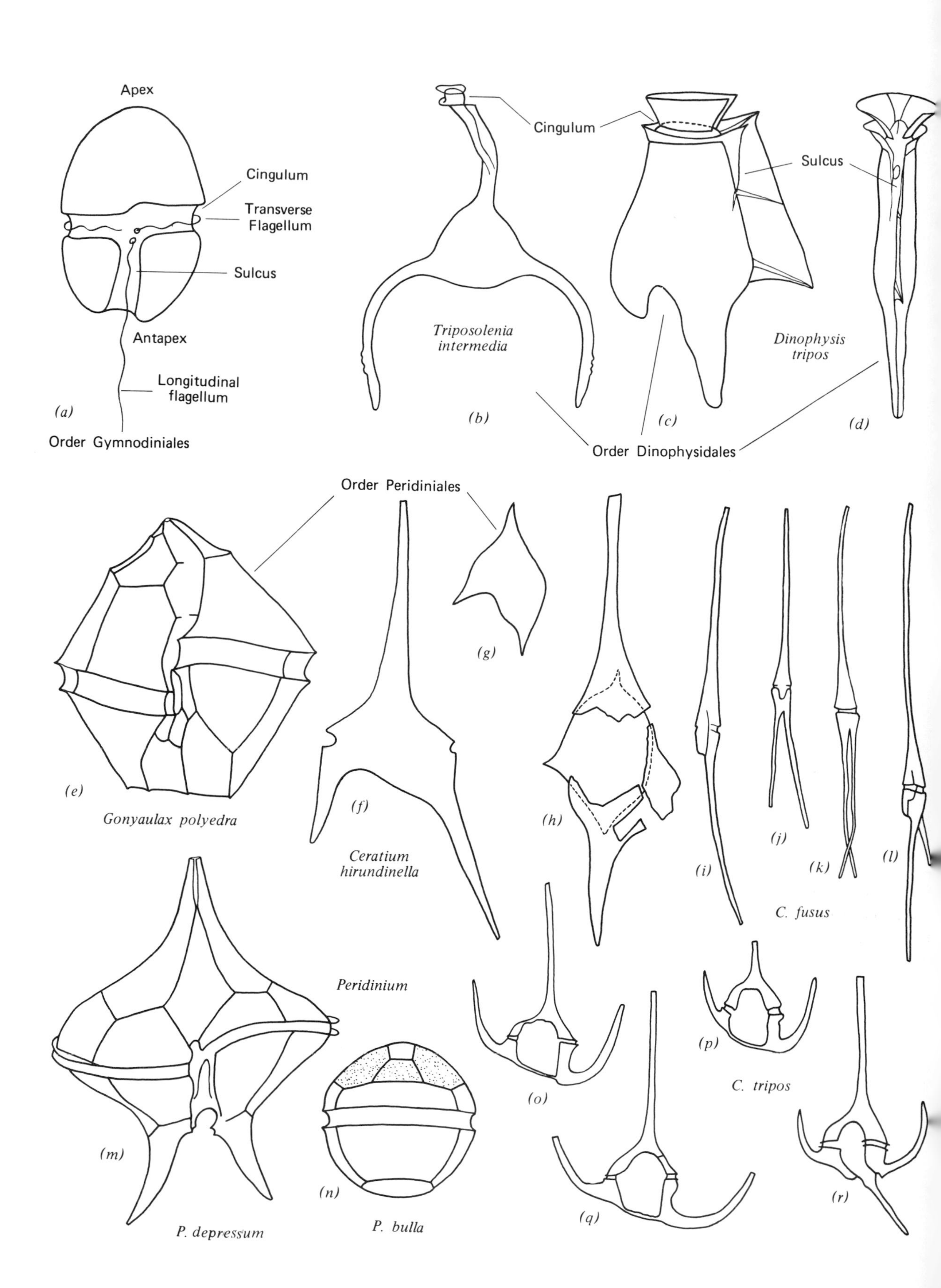
Apex
Cingulum
Transverse
Flagellum
Sulcus
Antapex
Longitudinal
flagellum
(a)
Order Gymnodiniales
Cingulum
Triposolenia
intermedia
(b)
(c)
Sulcus
Dinophysis
tripos
(d)
Order Dinophysidales
Order Peridiniales
(e)
Gonyaulax polyedra
(f)
Ceratium
hirundinella
(g)
(h)
(i)
(j)
(k)
(l)
C. fusus
Peridinium
(m)
P. depressum
(n)
P. bulla
(o)
(p)
C. tripos
(q)
(r)

vides the theca into two regions—the epitheca (including the apical, anterior intercalary, and precingular plates) and the hypotheca (including the postcingular, posterior intercalary, and antapical plates).

Armored dinoflagellates exhibit a great variety of outline shapes with nearly circular to elliptical shapes dominating. A single apical horn and one or two antapical horns are common (fig. 18-1*b*, *c*,*d*,*f*, and *m*); one or two postcingular horns are characteristic of a few genera (notably *Ceratium*). (See fig. 18-1*f*-*l*, *o*-*r*.)

Four important living armored dinoflagellates will serve to illustrate some of the significant and variable features of shape and thecal patterns. *Dinophysis* (fig. 18-1*c*, *d*) represents the exclusively marine Dinophysidales. The theca consists of two laterally borne parts joined along a sagittal suture. The forward position of the girdle—at or nearly at the apex and often on a necklike extension—is distinctive of the Dinophysidales. The sulcus is not strongly impressed, but the two flagellar furrows are flanked in many species by high septa that rise from the thecal plates along the sutures. Other projecting septa or spines are also conspicuous. The theca is nearly right-left symmetrical and laterally compressed, and if two antapical horns are developed they are dorsal and ventral in position. The Dinophysidales, almost unknown in the fossil record, appear to be represented by the Jurassic genus *Nannoceratopsis* (pl. 18-1, fig. 2) and the Permian *Nannoceratopsinella*.

The other three genera are *Peridinium*, *Ceratium*, and *Gonyaulax*. They represent the Peridiniales and are abundantly represented today by both freshwater and marine species. Their thecae are composed of several plates whose number and arrangement are partly illustrated. (See fig. 18-1.) Variation among the species of each genus is so great that no one species can be taken as representative of the genus. However, in combination the selected examples illustrate many of the important morphological features of thecae in the Peridiniales. Salient points are given below.

Many *Peridinium* species (fig. 18-1*m*, *n*) are nearly bilaterally symmetrical, and a prominent group of anterior intercalary plates in median position is characteristic. Some species show striate accretionary growth bands between the rough surfaces of original plates (cf. fig. 18-4*b*). Two antapical horns and dorso-ventral flattening are common.

Gonyaulax (fig. 18-1*e*) exhibits a strongly asymmetrical plate arrangement: a spiral girdle, one large posterior intercalary plate, and small, asymmetrical anterior intercalary plates (if present). Prominent horns are uncommon.

Ceratium species (fig. 18-1*f*-*l*, *o*-*r*) characteristically possess conspicuous horns, including one antapical horn and one that rises from the postcingular region. The plate arrangement is asymmetrical and often obscure, and the ventral surface is highly modified, possessing a thin covering of special plates not present in other genera.

Many living dinoflagellates exhibit a large intraspecific variability in thecal shape. *Ceratium fusus* (fig. 18-1*i*-*l*) and *C. tripos* (fig. 18-1*o*-*r*) are striking examples. Temperature, salinity, and dissolved nutrients, either singly or in combination, seem to affect the form assumed by some species, but the precise role and interrelationships of the several factors are unknown (Hasle and Nordli, 1951).

Dinoflagellates are small but fundamentally important organisms in the sea today where, to-

Figure 18-1. Examples of Recent dinoflagellates. Gymnodiniales and gymnodinioid stage (*a*): diagram of simple, unarmored cell characteristic of the Gymnodiniales and of the so-called gymnodinioid stage in the life history of dinoflagellates from other orders. Dinophysidales: (*b*) *Triposolenia intermedia*, lateral view (× 375); (*c*) and (*d*) *Dinophysis tripos*, lateral and ventral views (× 350). Peridiniales: (*e*) *Gonyaulax polyedra*, ventral view of theca of a marine species associated with some outbreaks of "red water"; (*f*-*h*) *Ceratium hirundinella*, a freshwater species; (*f*) outline of theca (× 200), (*g*) cyst (× 500) (note similarity to *Pseudoceratium*, fig. 18-9*g*), (*h*) theca in process of dissociating as cyst breaks out (× 300); (*i*-*l*) intraspecific variation in *Ceratium fusus* from Oslo Fjord (× 250); (*m*) ventral view of marine species of *Peridinium depressum* with prominent horns; (*n*) dorsal view of hornless and nearly spherical *Peridinium bulla*; three anterior intercalary plates stippled; (*o*-*r*) intraspecific variation in *Ceratium tripos* from Oslo Fjord; note supernumerary horn in (*r*) (× 200). (*b*-*e* after Schiller, 1931-1937; *f*-*h* after Entz, 1925; *i*-*l*, *o*-*r* after Hasle and Nordli, 1951; *m*-*n* after Meunier, 1910.)

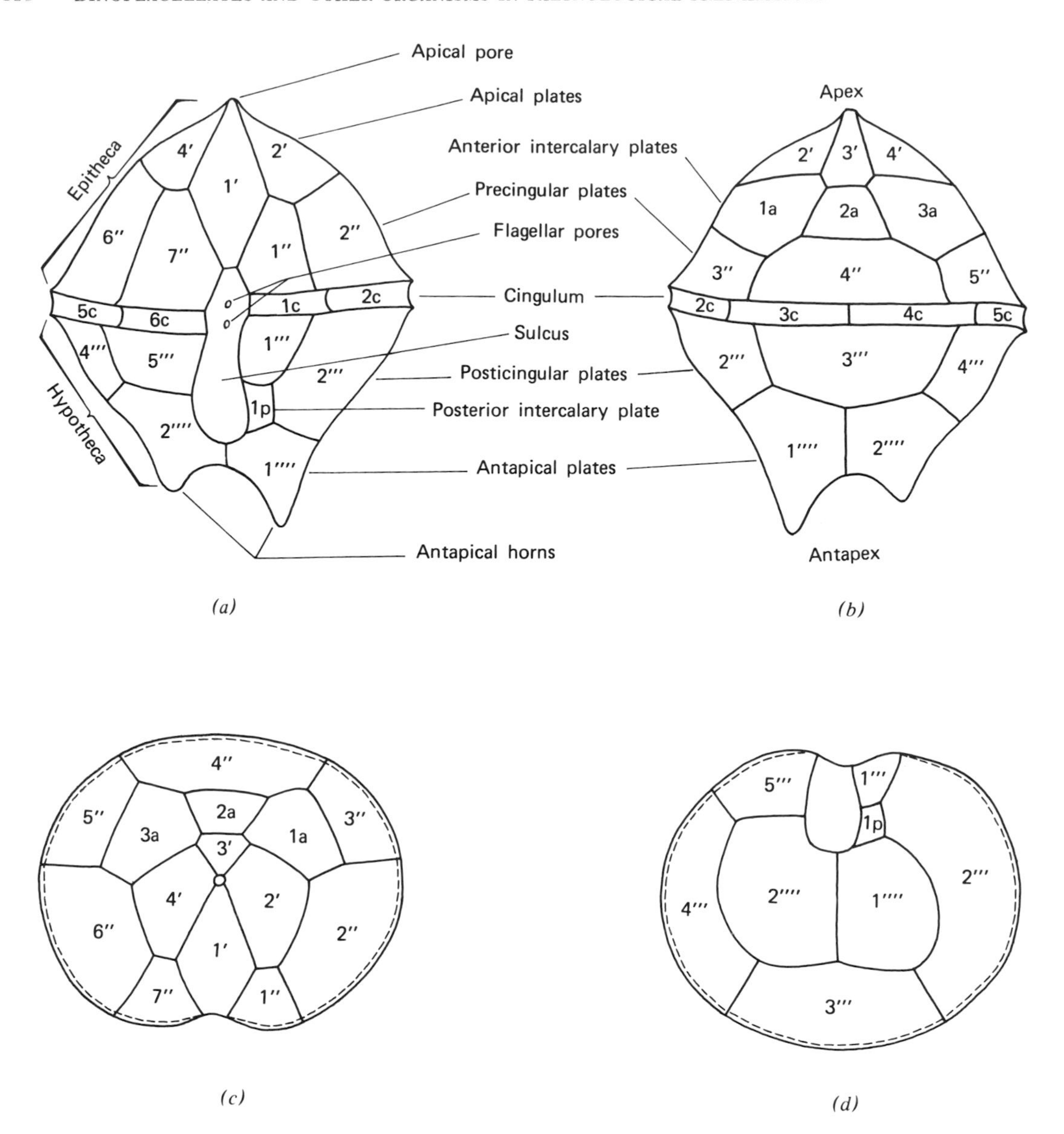

Figure 18-2. Dinoflagellate terminology and tabulation. Diagrams of a hypothetical dinoflagellate theca to illustrate the terminology for the principal thecal structures and the symbols used to designate individual plates (slightly modified from Kofoid, 1909): (*a*) ventral view; (*b*) dorsal view; (*c*) apical view; (*d*) antapical view. The complete tabulation of this example can be described in an abbreviated "formula" as follows: 4′, 3a, 7″, 4*c*, 5‴, 1*p*, 2⁗, meaning that the theca is composed of four apical plates, three anterior intercalary plates, seven precingular plates, four cingular plates, five postcingular plates, one posterior intercalary plate, and two antapical plates. A similar terminology and the same symbols can be used in describing the distribution of structures on the tests of fossil dinoflagellates.

gether with the diatoms, they are a basic link of the food chain. Besides their role as food organisms, they play some lesser, but still spectacular, parts in the economy of the sea; for example, the appropriately named genus *Noctiluca*, among others, is luminescent and a major cause of the "phosphorescence" of the sea at night. Species of other genera are associated with outbursts of "red water" or "poisonous tides" that periodically take severe toll of life in local regions of the sea. Brongersma-Sanders (1957) gives a thorough review and bibliography of this sort of mass mortality. The luminescence and toxicity of these microscopic organisms would of course pass unnoticed were it not for their presence at certain times and places in astronomically large numbers. When individuals number in the millions per liter, the water may turn milky, reddish, purplish, or yellowish and acquire a syrupy consistency.

Dinoflagellates are most varied and abundant in modern tropical seas. However, some species thrive in the seas of high latitudes, whereas others are abundant in intermontane lakes. Some of the same genera (notably *Peridinium*, *Ceratium*, and *Gonyaulax* among the armored forms) are represented by both freshwater and marine species. Both locally restricted species and cosmopolitan ones with virtually worldwide distribution are known.

For a more comprehensive discussion of modern dinoflagellates consult Lebour (1925), Chatton (1952), Hutner and McLaughlin (1958), Schiller (1931-1937), Paulsen (1949), and Loeblich and Loeblich (1966). Brunel (1962) provides some unusually fine photomicrographs of modern species.

Dinoflagellate Cysts

Many dinoflagellates pass through an encysted stage in addition to the mobile stage, but encysted stages of modern dinoflagellates have been relatively little studied. Because dinoflagellate fossils appear to be the remains of cysts, rather than of once-mobile thecae, a better understanding of cysts and encystment in modern dinoflagellates is needed for a fuller interpretation of the fossil types.

Depending on the species and the circumstances, cyst formation in modern dinoflagellates apparently may be associated with (a) the onset of unfavorable environmental conditions, (b) a resting period in the life cycle, (c) a part of the reproductive phase, or (d) a period of "digestion" of solid food. The resting cysts, which are seemingly represented by fossil dinoflagellates, may be simply spherical, with or without spines —or they may look more or less like the mobile cell.

Among the small number of modern resting cysts whose specific identity has been established, two will serve as useful examples here. They are shown inside their thecae in the sectional views of figure 18-3. One, the cyst of *Gonyaulax digitalis*, is a spiny, ellipsoidal type, with a two-layered wall, altogether like the fossils that have been referred to the genus *Hystrichosphaera* (fig. 18-4*u*). This cyst has been found inside its theca in the position that had been hypothesized from the study of the fossils, and living cysts have been observed to excyst in culture (Wall, 1966), yielding free-swimming gymnodinioid flagellate individuals that soon secrete a small replica of the parent theca. In contrast the second example, the cyst of *Peridinium leonis*, is peridinioid in outline, lacks spinelike projections, has a single-layered wall, and lies close to the surrounding theca. Very different from *Hystrichosphaera*, this type of cyst resembles the more obviously dinoflagellate-like fossils (such as fig. 18-4*b*), appearing as only a somewhat abbreviated version of the theca.

Until a few years ago only a few such objects as these among modern planktonic organisms had been identified as dinoflagellate cysts. Now increasing numbers are being associated with the thecate stages that give rise to them (Wall and Dale, 1967). The walls of these cysts are chemically more resistant than the thecae, and they possess excystment apertures (archeopyles) comparable to the openings known in many fossil dinoflagellates.

Entz (1925) presents an especially illuminating account of cysts and cyst formation in modern freshwater species of *Ceratium*. It is worthwhile to summarize here some of his observations that are pertinent to the study of fossil dinoflagellates.

The shape and size of cysts in *Ceratium*, characteristic for each of the several species and formae studied, vary much less than do the thecae of the motile stage. However, variation from one species to another is great: in one the cyst may approximate the theca in appearance (fig. 18-1*f*–*h*); in another it may be simply ellipsoidal. The cyst forms inside the theca, but sooner or later it is liberated when the theca splits into pieces. Encystment involves a reorganization of the cell contents, accompanied by a decrease in volume through loss of water. The continuous cyst membrane begins to form as the protoplasm pulls away from the inner surface of the theca. The

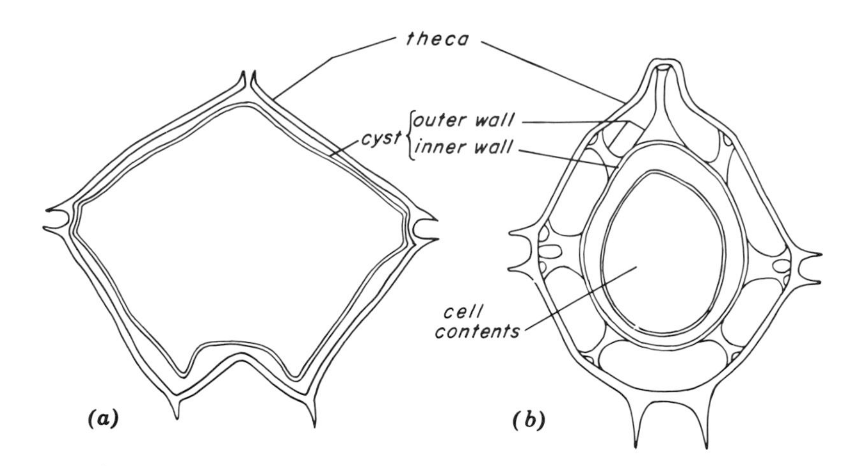

Figure 18-3. Diagrammatic sections of Recent dinoflagellate cysts inside thecae: (*a*) *Peridinium leonis* with its dinoflagellate-like cyst (the cyst is shown somewhat contracted, as it appears after mounting in glycerine jelly; when fresh the cyst wall is close to the theca except at the apical and antapical extremities); (*b*) *Gonyaulax digitalis* (the cyst is a typical *Hystrichosphaera*).

membrane is thin at the start but gradually thickens. It may be homogeneous or layered in different species. The horns characteristic of the *Ceratium* theca are noticeably abbreviated if present in the cyst, and the cingulum, prominent in the theca, is not discernible in the cyst. Excystment may be by several means: general dissolution of the membrane, local thinning of the membrane and eventual rupture, or tearing of the membrane along a line that has the same position in all specimens.

Although much remains to be learned about the cysts of modern dinoflagellates of the order Peridiniales, we are now certain of the following points:

1. Resting cysts, which form inside the theca of the motile cell and in some examples are supported against the theca by spinelike projections, are liberated when the thecal plates separate at the sutures and fall away.

2. Resting cysts of some species are morphologically like fossil hystrichospheres; those of other species more closely resemble the thecae in which they formed.

3. Species with similar thecae may produce cysts that are strikingly different.

4. Archeopyles like those in fossil dinoflagellates also occur in resting cysts of modern species. These openings function as excystment apertures through which emerge flagellate cells that develop into new thecate individuals.

5. The resistant wall material of the cysts is not visibly degraded by chemical treatments that cause the immediate dissociation and subsequent complete destruction of the plates of the thecae.

FOSSIL DINOFLAGELLATES

We have learned much from intensified study of these fossils in the last few years. At the same time new and perplexing problems have been uncovered. One new observation brings enlightenment, but another renders absurd an interpretation that seemed reasonable only yesterday. With rich and previously untouched material currently being studied by several workers in distant parts of the world, it is understandable that communication is not altogether abreast of observations and that new interpretations are not at once accepted by all workers. So it happens that the validity of a new idea presented by one worker may be documented by the independent observations of a second at the same time it is being attacked by a third.

These remarks suggest why the following discussion of fossil dinoflagellates must be read in the perspective of the changed circumstances created by each new discovery in this rapidly growing field. The interpretations offered are largely the author's. They are based on evidence of varying firmness and varying familiarity to other workers. The major points either have been or shortly will be presented with suitable documentation and discussion in appropriate journals. If some of these interpretations prove correct, progress will have been made. If others prove incorrect—as some of them certainly will—a demonstration of the degree and nature of their error will also be progress. Many of our ideas about fossil dinoflagellates ultimately will be proved or disproved by a penetrating study of certain aspects of the morphology of living dinoflagellates (especially the structures and mechanisms of encystment in armored types). Only the first steps in such a study have been taken. Here is an opportunity for exciting cooperation between biologist and paleontologist, each alternately leading and following toward a better understanding of dinoflagellates.

A comprehensive synopsis of families and genera of fossil dinoflagellates must await revisions of many taxa, although the summaries of some categories presented by Eisenack (1961a) and Evitt (1963a) may be useful, and the more recent papers by Sarjeant and Downie (1966) and Davey and others (1966) have this objective. In this chapter some of the important and frequently encountered fossil types are introduced to illustrate the salient morphological features discussed in the text.

General Morphology

Dinoflagellates are represented by fossils that vary between wide extremes in morphology (fig. 18-4). At one extreme are tests that look superficially like the thecae[1] of modern dinoflagellates. Such fossils have been unhesitatingly called dinoflagellates and were thought to represent fossilized thecae of organisms in the motile stage. At the other extreme are tests of quite a different sort (fig. 18-4*r–y*). They are spherical, ellipsoidal, or discoidal in shape and bear radiating projections of varied types. At first glance almost nothing about them suggests possible relationship to the dinoflagellates, and such objects (along with a diverse array of other small, hollow fossils of organic composition) were long placed in a separate group of microfossils, the order Hystrichosphaeridea—or hystrichospheres, as they were widely known.

Before considering the morphology of fossil dinoflagellates in detail we must first consider briefly the history of concepts relating to hystrichospheres and to the precise nature of dinoflagellate fossils. The affinities of the hystrichospheres have been argued extensively (Sarjeant, 1961; Eisenack, 1963a,b, give reviews and references). One school, headed by Deflandre, held the group to be polyphyletic and to include fossils representing various stages in the life cycles of both plant and animal groups. Another school, headed by Eisenack, argued for the biological integrity of the hystrichospheres, considering them a group of mutually related organisms somewhere among the algae. Definition of the order Hystrichosphaeridea was clearly restated by Eisenack (1962). All parties recognized that a few hystrichospheres exhibit dinoflagellate-like features (exemplified by the conspicuous platelike areas on tests of the genus *Hystrichosphaera*, fig. 18-4*u*), but these were difficult to explain satisfactorily.

Evitt (1961c) presented evidence that suggested to him that some of the so-called hystrichospheres were dinoflagellate cysts formed within thecae that have subsequently disintegrated or disappeared. He viewed the radiating processes typical of many hystrichospheres as supports between the theca and the contracted mass of encysted protoplasm, and he recognized various traces of dinoflagellate-like characters in the form and distribution of surficial structures on these hystrichospheres. Other so-called hystrichospheres lacking these structures were considered nondinoflagellates. Later Evitt (1963a) proposed to use the term "acritarch" for the nondinoflagellate ex-hystrichospheres and to include the hystrichospheres having dinoflagellate-like characters with the dinoflagellates. This is the procedure followed here. Although this proposal made the ordinal name Hystrichosphaeridea unnecessary, the casual term "hystrichosphere" remains convenient for fossils close to that extreme of the dinoflagellate morphological "spectrum" characterized by more or less

[1]In this discussion the term "theca" is used exclusively for the covering, composed of separable plates, that characterizes a motile armored dinoflagellate. The noncommittal term "test" is used in referring to the fossils except where it is intended to emphasize that the fossils are probably the resistant remains of resting cysts that formed inside now-vanished thecae. Authors of many works published up to now have used "theca" for the fossils, believing them to be the remains of true thecae of motile individuals.

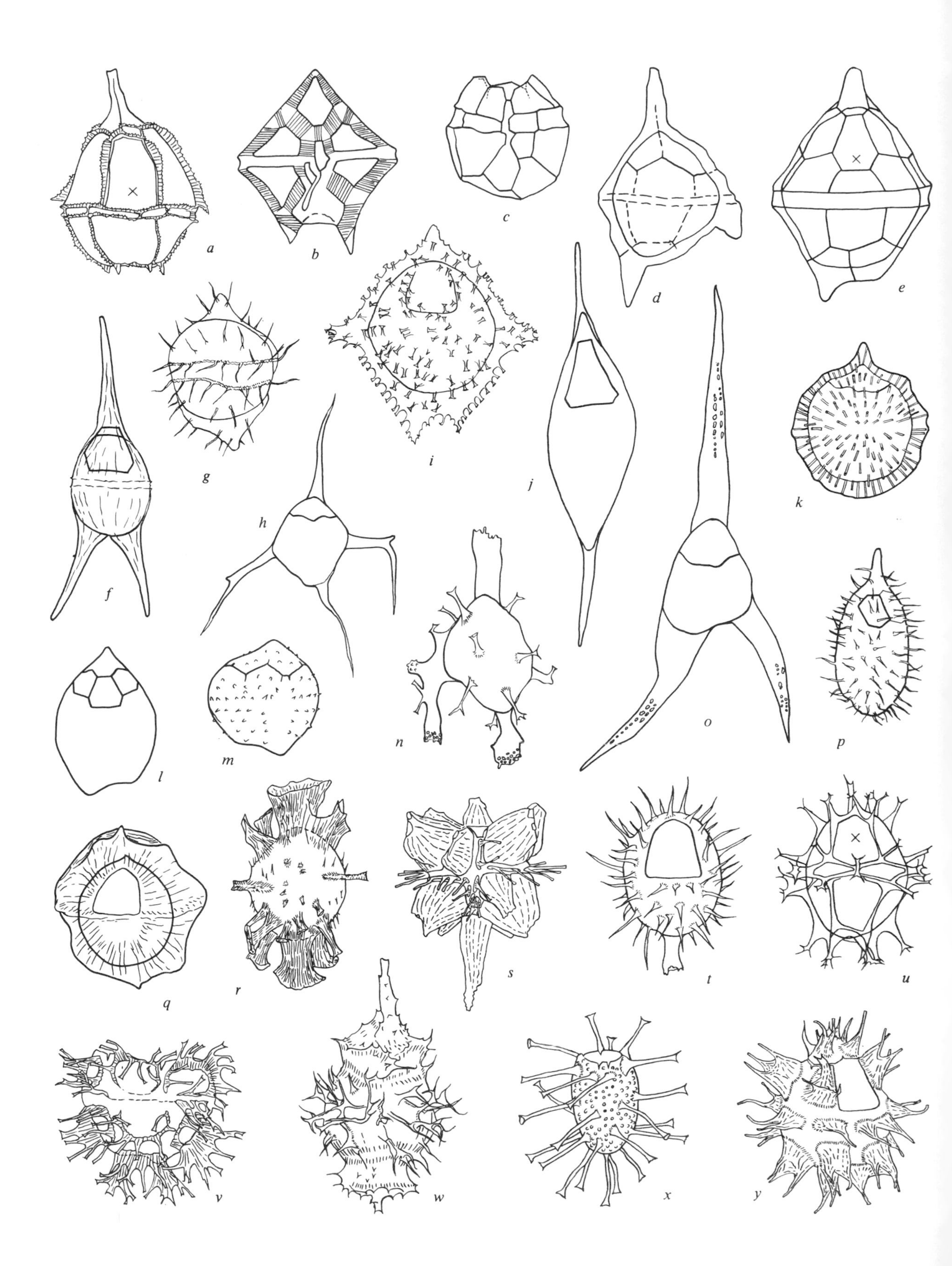
a
b
c
d
e
f
g
h
i
j
k
l
m
n
o
p
q
r
s
t
u
v
w
x
y

spherical main bodies with radiating processes. However, we must remember that the limits of the term in this sense are vague, because a continuum of gradations connects the extreme morphological types that might be labeled "obvious dinoflagellates" and "typical hystrichospheres."

Like Deflandre and Eisenack, Evitt originally considered many fossils with obviously dinoflagellate-like features to be the thecae of motile individuals. He suggested criteria, first, for recognizing fossil dinoflagellates and, second, for distinguishing fossil-dinoflagellate cysts from fossil thecae. The former distinction seems soundly based, and the criteria are enlarged on here; but the alleged distinction between fossil cysts and fossil thecae does not appear to be real. It now appears that possibly no dinoflagellate fossils are true thecae. Instead the typical fossil dinoflagellate seems to preserve a structure which is chemically more resistant than the theca and which formed inside the theca for protection of the protoplast in an encysted state (resting cyst). Deflandre (1962) expresses general accord with this view, although he considers a small number of dinoflagellate fossils, especially those with conspicuous growth bands (e.g., pl. 18-1, fig. 3) to be true thecae. In addition, as he notes, the exact nature of some extinct forms will very likely remain uncertain. He stresses that the presence of an archeopyle is a determinative character of the encysted state.

The possible relationships of cyst wall to theca and of "obvious fossil dinoflagellate" to "typical hystrichosphere" are summarized by the following picture of the process of cyst formation. The picture is generalized, partly hypothetical, and is based to a large degree on observations of fossil dinoflagellates. However, the few observations of resting cysts of modern dinoflagellates that have been made support this picture, and many features of the fossils that it readily explains are otherwise difficult to interpret and harmonize. The picture is presented at this point to help provide a perspective for viewing the varied morphological features that will be discussed later.

The basic steps of the process of cyst formation are common to all species. A thecate dinoflagellate in the actively swimming stage begins to encyst. The flagella are cast off and the cell content contracts, forming about itself a thin covering of chemically resistant organic material. The plates of the less resistant theca dissociate or disintegrate. Eventual excystment is accomplished by rupture of the cyst wall along definite lines or, in some examples perhaps, by general dissolution of the wall.

Many principal morphological characters of the cyst wall are plausibly explained as resulting from the particular spatial relationships of the cyst and theca when the cyst is formed. If the outer surface of the cyst wall forms in relatively close contact with the theca, not only does the cyst acquire roughly the shape and configuration

---

Figure 18-4. A spectrum of fossil dinoflagellates. The arrangement emphasizes the transition from "obvious" dinoflagellates at the top to "typical" hystrichospheres at the bottom. Symbols in parentheses refer to archeopyle type (see fig. 18-8). Other symbols: dv – dorsal view, vv – ventral view, odv – oblique dorsal view. A small "x" (figs. *a*, *e*, *u*) identifies the archeopyle where it is not obvious. (*a*) *Gonyaulacysta jurassica*, Upper Jurassic, dv, × 400 (P). (*b*) *Palaeoperidinium pyrophorum*, Upper Cretaceous, vv, × 300; archeopyle never observed. (*c*) ***Meiourogonyaulax*** sp., Lower Cretaceous, vv, × 1,000 (A) (operculum not present). (*d*) ***Pseudoceratium ludbrooki***, Lower Cretaceous, dv, × 300 (A). (*e*) ***Deflandrea*** sp., Upper Cretaceous, dv, × 500 (Ia). (*f*) *Deflandrea* cf. *D. diebeli*, Upper Cretaceous, dv × 400 (I). (*g*) *Palaeohystrichophora infusorioides*, Upper Cretaceous, dv × 600; archeopyle never observed. (*h*) *Muderongia tetracantha*, Lower Cretaceous, vv, × 250 (A). (*i*) *Wetzeliella articulata*, Oligocene, dv, × 200 (Ia). (*j*) *Svalbardella* sp., Upper Cretaceous, dv, × 400 (I). (*k*) ***Chlamydophorella nyei***, Upper Cretaceous, dv, × 500 (A). (*l*) ***Trithyrodinium*** sp., Upper Cretaceous, dv, × 300 (3I). (*m*) *Tenua* sp., Upper Jurassic, vv, × 750 (A). (*n*) Undescribed genus, Upper Cretaceous, vv, × 300 (A). (*o*) *Odontochitina operculata*, Upper Cretaceous, dv, × 400 (A). (*p*) *Imbatodinium villosum*, Upper Jurassic, odv, × 750 (2I). (*q*) *Thalassiphora pelagica*, Oligocene, dv, × 200 (P). (*r*) *Cordosphaeridium filosum*, Eocene, vv, × 300 (P). (*s*) *Hystrichokolpoma cinctum*, Oligocene, vv, × 400 (A). (*t*) *Coronifera* sp., Upper Cretaceous, dv, × 400 (P). (*u*) ***Spiniferites*** sp., (=*Hystrichosphaera*), Upper Cretaceous, dv, × 600 (P). (*v*) ***Areoligera*** cf. ***A. senonensis***, Upper Cretaceous, dv, × 400 (A) (operculum not present). (*w*) ***Triblastula*** cf. ***T. utinensis***, Upper Cretaceous, vv, × 500 (P). (*x*) ***Tanyosphaeridium***, Upper Cretaceous, vv, × 750 (A). (*y*) *"Hystrichosphaeridium" ferox*, Upper Cretaceous, odv, × 500 (Aa + P).

of the theca but it may also preserve traces of suture lines or even minute details of the inner plate surfaces, such as impressions of pores, reticulate markings, or growth bands. If contact with the theca is somewhat less intimate, short projections of the cyst wall may serve to hold the encysting cell in position within the theca. If the cyst wall forms chiefly about a compact body very much smaller than the motile cell, then still longer projections may extend to the theca as spinelike processes or high thin walls. In some species the arrangement of these projections gives no clear evidence of the tabulation; in others it definitely does. If the projections form only beneath the sutures between plates, then they record the positions of these sutures on the test and outline polygonal areas resembling thecal plates. If, on the other hand, the projections are restricted to positions beneath the central portions of plates, one or a group of projections represents each thecal plate—still reflecting the tabulation but in quite a different way, without trace of sutures. Concordant tips of processes, at times expanded or elaborately interconnected by membranes or trabeculae, directly record the position of the inner surface of the theca. Ultimately rupture of the cyst wall, mainly along lines that reflect sutures between thecal plates, produces an opening (the archeopyle) whose shape and position are also closely related to the tabulation of the theca. Excystment without formation of an archeopyle may proceed in some species by general dissolution of the cyst wall.

In other words, the fidelity with which a fossil-dinoflagellate cyst reveals the features of the theca and therefore the extent to which the fossil is "dinoflagellate-like" in appearance depends partly on the proximity of the main surface of the cyst to the theca.

As noted earlier in this chapter, Entz (1925) reported on the formation of cysts in modern *Ceratium* that, to a degree, resemble the thecae within which they developed (fig. 18-1g, h). Figure 18-5 presents the contrasting example of a hystrichosphere forming within a theca of grossly different appearance. When first constructed this figure was entirely hypothetical. However, as discussed earlier, we now know that cysts of some modern dinoflagellates are typical hystrichospheres supported by processes within the theca as shown here (fig. 18-3*b*; Evitt and Davidson, 1964).

Criteria for Recognition

In view of the morphological extremes manifested by fossil dinoflagellates, a reasonable and fundamental question is: How can a fossil be recognized as a dinoflagellate? Flagellar furrows, tabulation, and, to a much lesser extent, shape are the important criteria; but a gathering of experienced workers would not quickly agree on all the details of the manner and combination

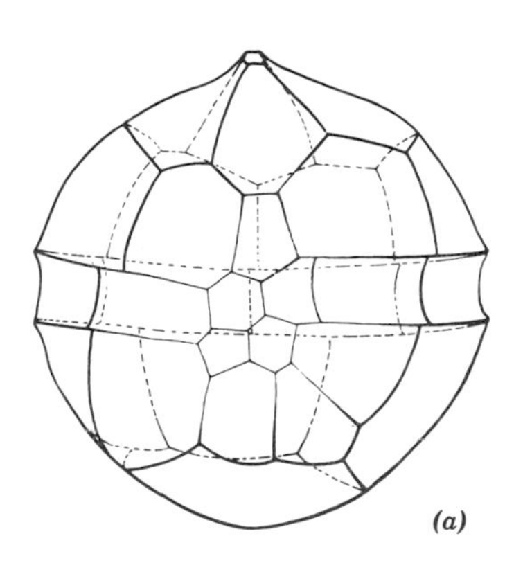

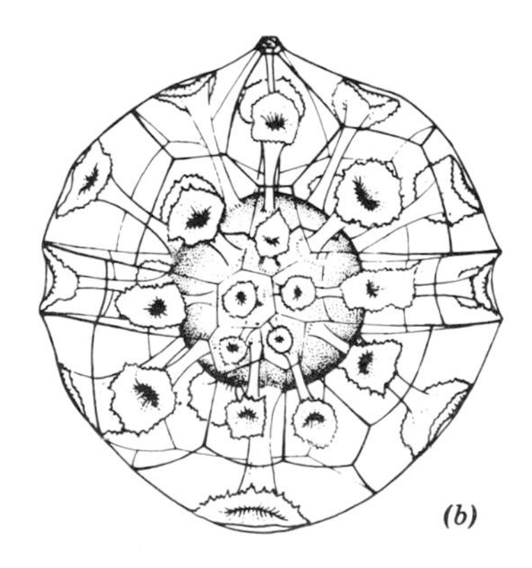

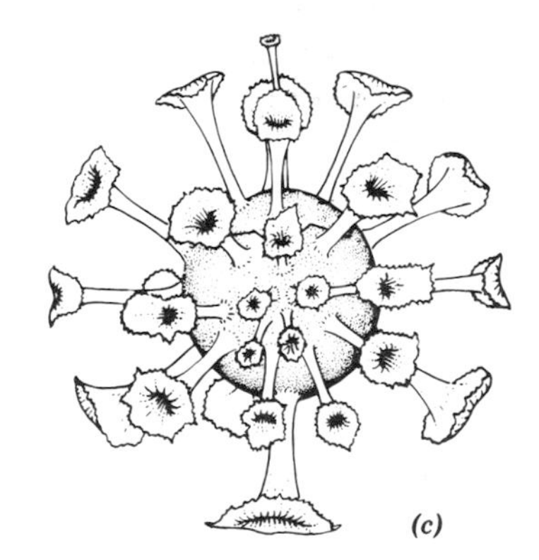

Figure 18-5. Test development in a typical hystrichosphere. Three stages, all shown in ventral view, to illustrate the inferred manner of formation of the test of *Hystrichosphaeridium tubiferum* (Upper Cretaceous) (× ca. 600): (*a*) tabulate dinoflagellate theca before test (cyst) begins to form; (*b*) test supported within theca by trumpet-shaped processes whose number and position correspond to thecal tabulation; (*c*) test alone (as found in fossil state) after disappearance of original theca. Relative sizes and arrangement of processes are based on actual specimens; relation of test to theca in this species is inferred from studies of many fossil species and from the observed relation in modern *Gonyaulax digitalis*.

in which these features must be displayed for a fossil to be called a dinoflagellate. The following discussion of criteria assumes the general accuracy of the explanation of cyst-theca relationships just reviewed.

Flagellar Furrows. Flagellar furrows alone are conclusively diagnostic of dinoflagellate affinity if they can be reliably identified, as, for example, in *Dinogymnium* (pl. 18-1, fig. 1). So much the better if the furrows are combined with other meaningful features, as in certain *Deflandrea* species (fig. 18-4*f*) with a clear cingulum and a distinctively dinoflagellate-like shape or in a specimen (fig. 18-4*c*) whose cingular plates are conspicuous in a general tabulation. If other dinoflagellate characters are lacking and mere faint depressions are suspected of representing sulcus and cingulum, special care must be exercised. Simple folds caused by compression of a thin-walled fossil may create features that simulate these furrows in objects that are not dinoflagellates. On the other hand, furrows need not be present as furrows for a fossil to be a dinoflagellate. Certain structures on the test that are not at all furrowlike may still be reasonably accepted as reflecting the furrows in the vanished theca (see later discussion of projecting structures). Even if all traces of cingulum or sulcus are wanting, a convincing combination of other dinoflagellate characters may be present.

Tabulation. Tabulation is the pattern of plate arrangement in a dinoflagellate theca. There may be no trace of tabulation on the test of a fossil dinoflagellate—or it may be reflected by various structures and may range from obvious, through obscure but demonstrable, to faint and uncertain. This reflected tabulation may be revealed in one or both of two ways: by polygonal areas that at least superficially look like true plates (fig. 18-4*a*,*b*,*c*) or by a wide variety of other structures that are not at all platelike. (See figs. 18-4*r*,*s*,*y*; 18-5.)

A very important, if only partial, indication of tabulation is the archeopyle, which is treated in a separate section. This opening in the test usually corresponds in shape and position to a recognizable plate or group of plates in the theca. Like well-formed and easily recognizable flagellar furrows, a distinct archeopyle of identifiable type is sufficient by itself to establish the dinoflagellate nature of an unknown fossil. Note, however, the distinction made earlier in this chapter between the dinoflagellate archeopyle and the pylome that occurs in certain acritarchs.

Shape. The overall shape of dinoflagellates ranges through gradational stages from nearly spherical to conspicuously three-pointed (or rarely having more than three points), with many variations along the way. Near the spherical end of this range shape in itself has no significance whatever as an indicator of affinities. Toward the other extreme, however, shape becomes increasingly meaningful. A distinctive outline with a single prominence at the apical pole and two, usually somewhat unequal, prominences at the antapical end is so typical of certain dinoflagellates that it has been called "peridinioid" from its characteristic development in some members of the modern genus *Peridinium* (fig. 18-1*m*). An extreme peridinioid outline among fossils (fig. 18-4*b*,*f*,*i*), although highly suggestive of dinoflagellate affinities, is probably not reliably conclusive without some sort of supporting evidence. Less extreme examples are correspondingly less convincing and require correspondingly greater supporting evidence.

Detailed Morphology

Shape and Projecting Structures. Fossil dinoflagellates exhibit a variety of shapes. (See fig. 18-4.) Although some are virtually spherical or equidimensional, an original dorso-ventral flattening (occasionally extreme) and longitudinal extension are common, whereas lateral and longitudinal compression are rare. Original flattening, a significant character in itself, has important mechanical effects also: it determines the dominant direction of secondary compression in specimens from severely compacted strata as well as the common orientation of specimens in prepared slides. Therefore it can be a troublesome factor when critical details are disadvantageously located. In some species shape is relatively constant, in others it may be highly variable.

The outer surface of the test may be smooth and featureless, but it is often modified by various projecting structures. These may range from large spinelike projections or broad bulges to scarcely perceptible granules. The *main body* is the central portion of the test from which the projections extend. Very often a single feature or

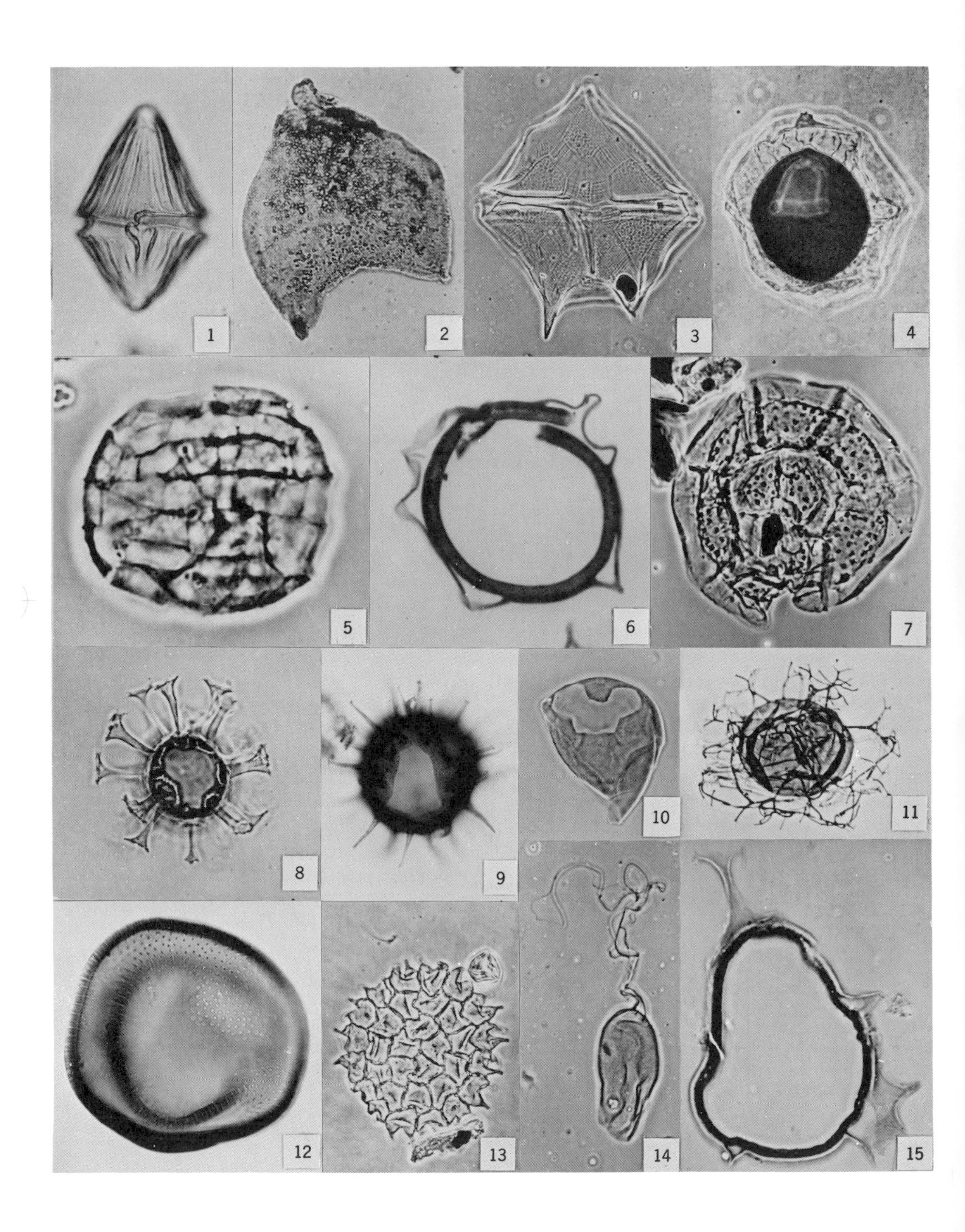
1
2
3
4
5
6
7
8
9
10
11
12
13
14
15

a line of features can be recognized as the cyst equivalent of the cingulum. This separates two general regions of the cyst: a more apical portion called the *epitract* and a more antapical portion, the *hypotract*. As the terms suggest, these regions are equivalent to the epitheca and hypotheca, respectively.

Horns are major projections of the test. They seldom number more than five and appear to be abbreviated versions of major projections (also called horns) that characterize thecae of many modern dinoflagellates. Figure 18-6 shows the "cardinal" positions that may be occupied by horns. *Apical* and *antapical horns* occur in a wide assortment of modern and fossil types; *cingular horns* are relatively rare (fig. 18-4*i*). *Postcingular horns* in modern types are especially characteristic of *Ceratium* and its close allies, and they occur in tests of several fossil forms (fig. 18-4*d, o, n*) that have other traits also suggesting alliance with the Ceratiaceae. Horns in fossil dinoflagellates may be solid or hollow. Their size and prominence may vary within a species.

In contrast to horns, *processes* and *septa* seem not to represent structures that projected from the theca but to be unique features of the cyst that formed within the theca. They are more numerous than horns and less restricted in position. *Processes* are essentially columnar or spine-like projections and range from simple to intricately branched and interconnected. *Septa* are essentially planar structures (varying from membranous to relatively thick and stiff) that rise more or less perpendicularly from the main body. They have also been called wings, ledges, crests, lists, and membranes. Processes and septa may occur separately or in combination, merging into one another. They also grade downward in size into lesser structures—processes into features such as spinules, denticles, or granules; and septa into simple ridges or the low muri of a surface reticulum.

The contrast in appearance between "obvious" dinoflagellates and "typical" hystrichospheres is determined largely by the presence of processes or septa in the latter. The prominence of these structures depends on the ratio of the diameter of the main body to the total diameter of the cyst. If it is useful to distinguish variations in this ratio by names, then cysts in which the ratio exceeds 0.8 may be described as *proximate* in reference to the supposed proximity of the main cyst wall to the theca at the time of encystment. In contrast, in *chorate* cysts this ratio is 0.6 or less, whereas in *proximochorate* cysts the ratio lies between 0.6 and 0.8. "Obvious" dinoflagellates like *Gonyaulacysta* (fig. 18-4*a*) and some others (fig. 18-4*c*) are proximate cysts, in contrast to "typical" hystrichospheres such as *Hystrichokolpoma* (fig. 18-4*s*) and *Hystrichosphaera* (fig. 18-4*u*), which are chorate cysts.

The form and arrangement of the projecting structures are important taxonomic features. Some of the more common types are illustrated

Plate 18-1. Dinoflagellates and miscellaneous organic microfossils. 1—Order Gymnodiniales: *Dinogymnium acuminatum*, Upper Cretaceous (× 400). 2—Order Dinophysidales: *Nannoceratopsis gracilis*, Lower Jurassic (× 600). 3-11 and 15—Order Peridiniales: 3—*Palaeoperidinium pyrophorum*, Upper Cretaceous; ventral surface showing reticulate intratabular areas with striate growth zones between (× 400); 4—*Thalassiphora pelagica*, Oligocene (× 250); 5—*Dapcodinium priscum*, Lower Jurassic; oldest known dinoflagellate with clear tabulation (× 1,200); 6—*Pentadinium laticinctum*, Oligocene; microtome section showing relation of thin outer-wall layer to thick capsule (× 500); 7—*Dinopterygium* sp., Lower Cretaceous; apical view, showing wide peripheral septum along girdle and polygonal fields with intratabular markings (× 600); 8—*Hystrichosphaeridium tubiferum*, Upper Cretaceous; apical view, showing open apical archeopyle (type A) and distinctive trumpetlike processes (× 400); 9—*Exochosphaeridium* sp., Upper Cretaceous, with distinct precingular archeopyle (type P) as the only conspicuous dinoflagellate character (× 400); 10—deformed thin-walled sphere, Upper Cretaceous, whose single but diagnostic dinoflagellate character is a well-formed intercalary archeopyle (type 3I) compare fig. 18-Im, 18-3, 18-7 (× 400); 11—*Cannosphaeropsis* cf. *C. utinensis*, Upper Cretaceous; a hystrichosphere with elaborate trabecular connections between the tips of its few processes (× 400); 15—*Hystrichosphaera tertiaria*, Oligocene; phase-contrast photomicrograph of microtome section showing capsule (dark) surrounded by outer-wall layer (light) that forms the single open-tipped process at top, the more massive process complex at lower right, and the sutural ridge between process bases seen in cross section at bottom left (Compare Fig. 18-3u), (× 500). 12, 13, 14—Miscellaneous microfossils of organic composition: 12—*Tasmanites* sp., Miocene (× 250); 13—*Pediastrum* sp., Upper Cretaceous (× 600). 14—*Ophiobolus lapidaris*, Upper Cretaceous (× 600).

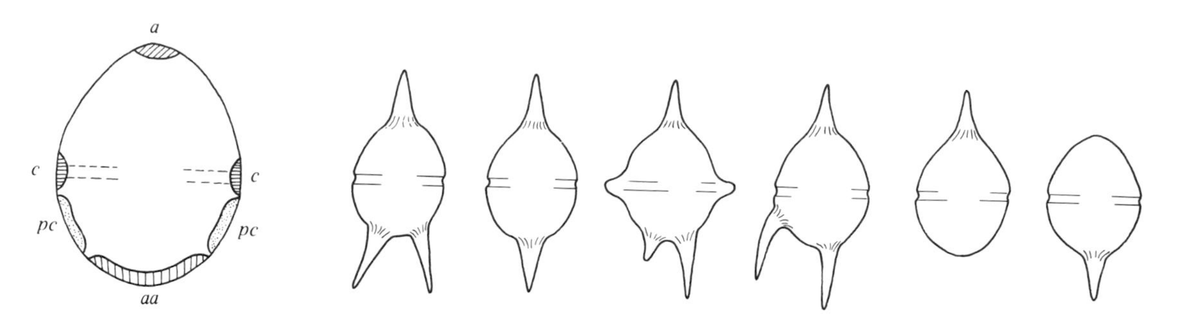

Figure 18-6. Horns in fossil dinoflagellates. At left, shaded areas on a simple, oval theca as seen in dorsal or ventral view show the "cardinal" positions in which horns are most common: *a*—apical, *aa*—antapical, *c*—cingular, *pc*—postcingular. The other diagrams show examples of single and combined horn types.

in figure 18-7. Their distribution is *tabular* if it reveals the pattern of thecal tabulation, *nontabular* if it does not. Tabular projections are either *sutural* or *intratabular*; nontabular ones may be disposed regularly or irregularly. Sutural features correspond in position to the sutures between thecal plates and may include processes, septa, or other projecting structures. If sutural projections in the position of suture junctions (i.e., corresponding to the corners of thecal plates) are different from those in between, they may be distinguished as *gonal* and *intergonal* projections, respectively. *Intratabular* features occupy positions on the test corresponding to more or less of the central portions of thecal plates rather than to the lines of separation between them.

Processes may be solid or hollow, open or closed at their tips or bases, in substance fibrous or hyaline. They may stand entirely free and separate from one another—or the bases may be joined by ridges or low septa, and the tips by trabeculae (pl. 18-1, fig. 11) or by a membranous layer (fig. 18-4*k*). The tips may be unbranched or furcate, acute or variously enlarged. Summits of adjacent processes may be conspicuously concordant. A detailed descriptive terminology for processes of different shapes and arrangement is presented by Downie and Sarjeant (in Davey and others, 1966).

A single specimen may exhibit one or more styles of projecting structure. Often an apical process or one or two antapical processes are distinct from the rest (fig. 18-4*t*). Equatorial processes, which apparently correspond to cingular plates, and a longitudinally elongate group of processes, corresponding to sulcal plates, are often distinguishable and may differ in size or structure from processes of the precingular and postcingular series (fig. 18-4*s*, *y*). Processes asso-

Figure 18-7. Surface features of fossil dinoflagellates. (*a*) Diagrams of several different styles of processes. (*b-d*) Three types of processes that occur in *Hystrichosphaera*. Prongs of trifurcate tips alternate with folds in outer wall layer that radiate from bases. Processes vary from solid (*b*) to hollow but closed at the tip (*c*) to hollow and open at the tip (*d*). (*e*) Diagram of small portion of dinoflagellate test and inset diagram of tabulate theca to show relationships of areas portrayed in perspective diagrams (*f-p*). Dashed lines on test correspond in position to suture lines on theca; shaded areas correspond to inner areas of thecal plates. (*f-o*) are tabular patterns; (*p*) is nontabular. (*f-h*) Surface features in sutural arrangement, corresponding to dashed lines in (*e*): (*f*) large granules; (*g*) trifurcate gonal and bifurcate intergonal processes joined by basal ridges (as in *Hystrichosphaera*); (*h*) thin septa forming a reticulum, as in *Pterodinium*. (*i-o*) Surface features in intratabular arrangement, corresponding to all or part of shaded areas in (*e*): (*i*) apiculate granules or spines; (*j*) trumpet-shaped hollow processes (as in some *Hystrichosphaeridium*); (*k*) thin septa around isolated areas, not forming a reticulum (as in *Schematophora*); (*l*) like (*k*), but septa deeply dissected and incomplete (as in *Areoligera*); (*m*) large process with constricted base, polygonal tip and wall of strands (as in some "*Hystrichosphaeridium*"); (*n*) ring of processes rising from a connecting basal ridge, free at tips (as in *Systematophora*); (*o*) like (*n*), but process tips joined by ringlike trabeculum (as in *Polystephanephorus*). (*p*) Coarse granules in nontabular arrangement. Distribution pattern of projections does not reveal either a pattern of sutural lines or a pattern of rounded to polygonal platelike areas.

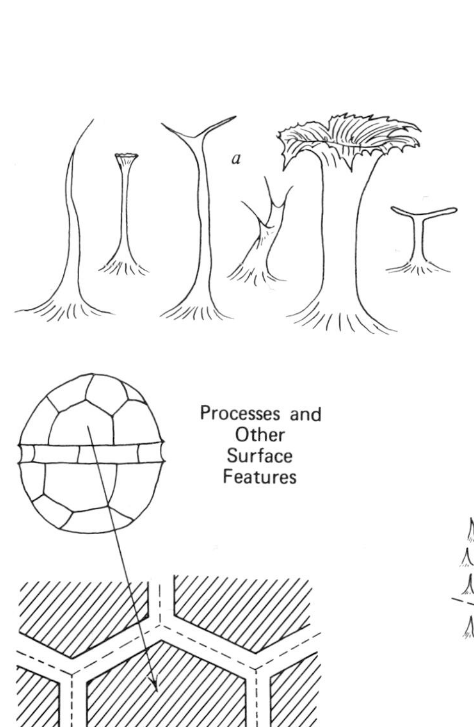

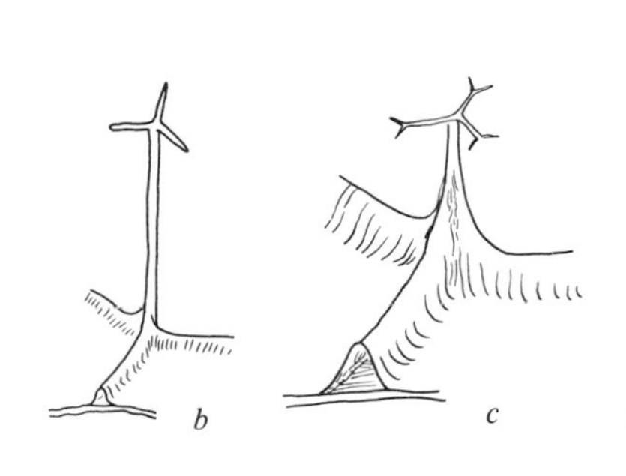

Processes and Other Surface Features

i–o Intratabular

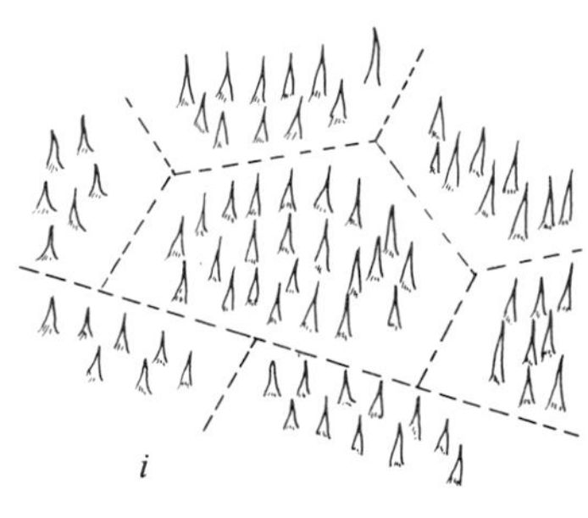

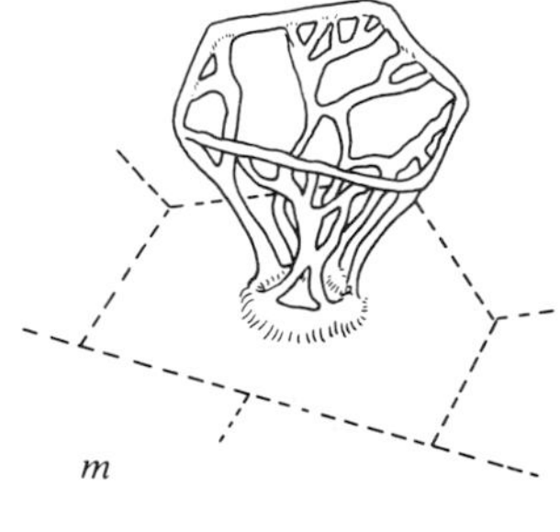

f–h Sutural

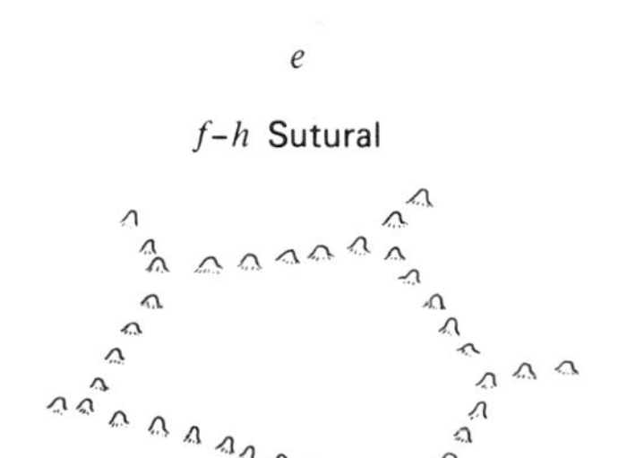

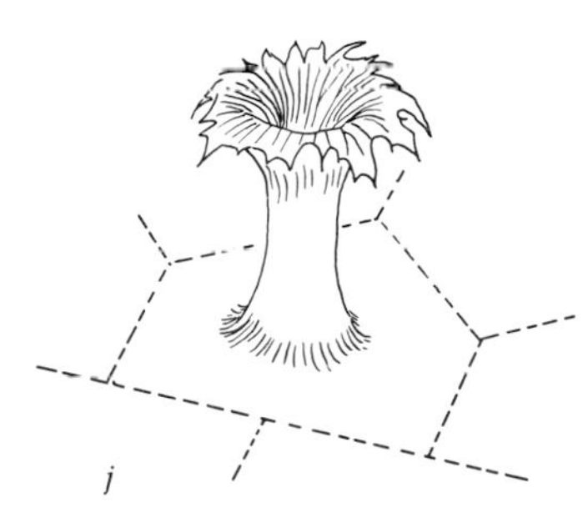

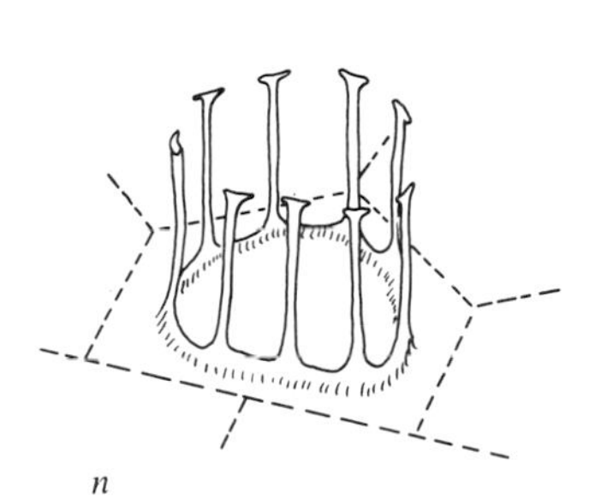

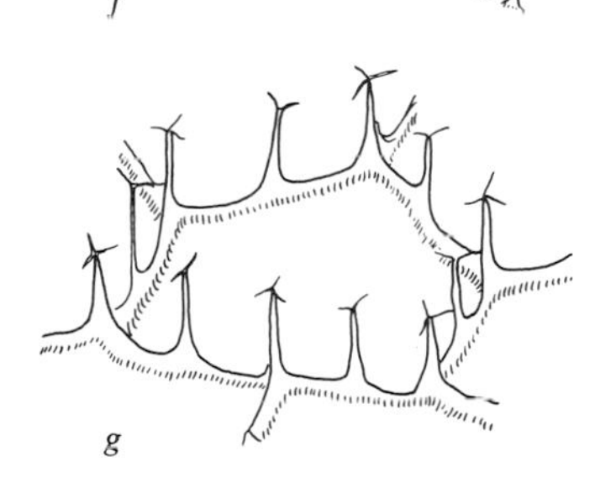

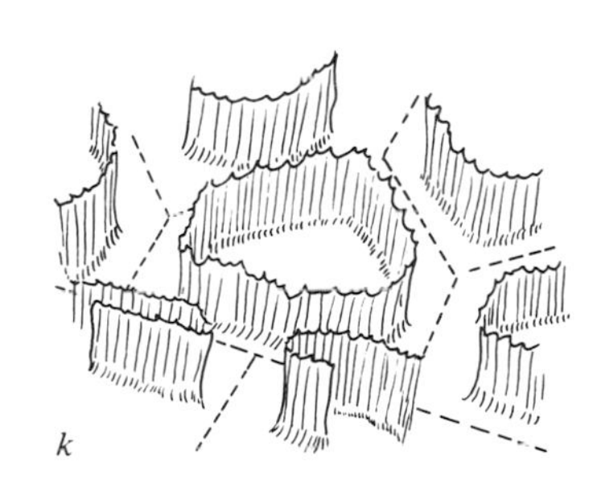

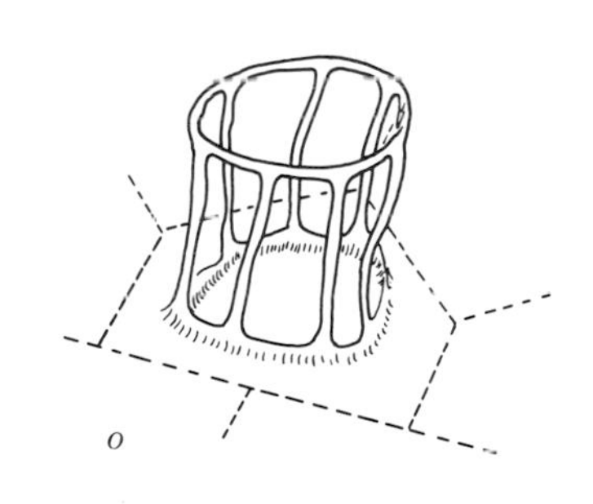

p Nontabular

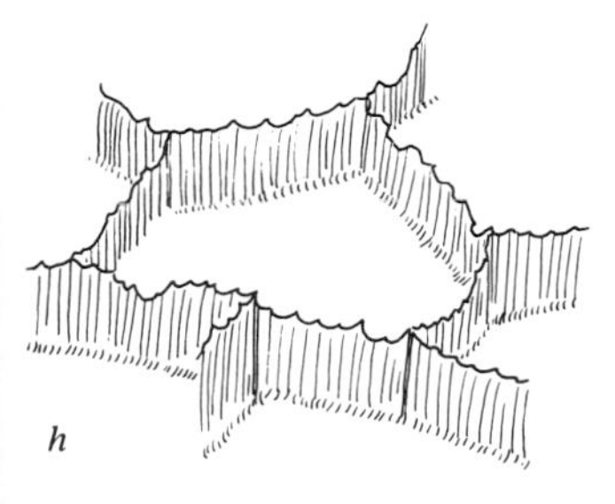

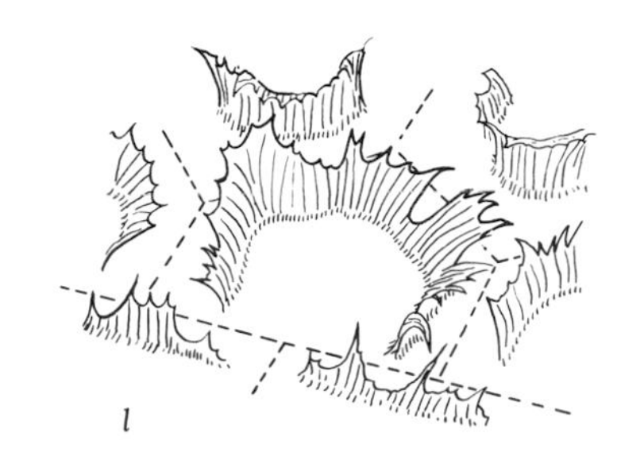

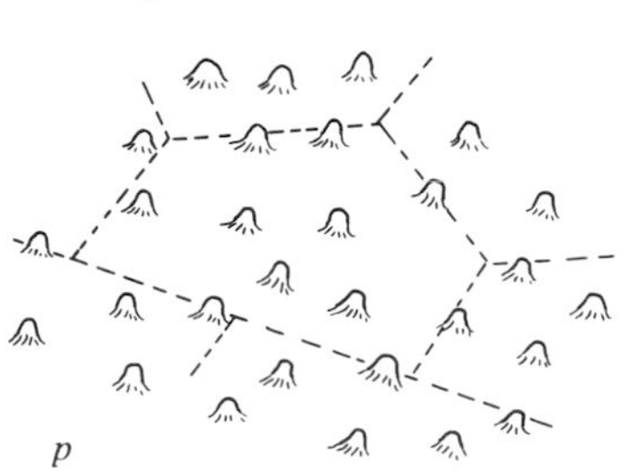

ciated with the cingulum may be recognizable by their alignment even when other processes on the test are nontabular in distribution. (See fig. 18-4*k*, *t*.)

In many but not all hystrichospheres with processes in intratabular arrangement it is possible to recognize which thecal plate is reflected by each process or process group and thus to determine the formula of the complete thecal tabulation. If some of the plates are not represented by processes (e.g., the cingular and sulcal plates), as happens in some species, then the tabulation may be only partially determinable. In dinoflagellates with sutural processes (including some hystrichospheres) the tabulation formulas can be determined from the pattern of sutural lines. Only a few slightly different tabulation patterns seem to be represented by the many genera of hystrichospheres for which formulas have now been determined. The tabulation of *Hystrichosphaeridium* is the most common type: 4′,0*a*,6″, 6*c*, 5‴, 1*p*, 1⁗; that of *Hystrichosphaera* differs but slightly: 3′, 0*a*, 5(6)″, 6*c*, 5‴, 1*p*, 1⁗.

The Archeopyle. The *archeopyle* (fig. 18-8) is the excystment aperture in the resistant wall of a dinoflagellate resting cyst. Its shape and position are closely related to a pattern of thecal tabulation.[2] Rarely the archeopyle is a slitlike separation in the wall, but the opening is usually formed by the partial or complete release of a portion of the wall, the operculum, which corresponds to one or more thecal plates. The operculum is freed by parting of the wall along the *principal archeopyle suture*. The operculum is *simple* if it is a single unit or *compound* if it is divided into two or more separate pieces by *accessory archeopyle sutures*. The operculum (or part of it, if it is compound) is described as *free* if the primary archeopyle suture surrounds it completely and *attached* if a connection with the rest of the wall is maintained because the principal archeopyle suture is incomplete.

The shape and position of the archeopyle generally suffice to identify the thecal plates that are reflected by the portion of the test wall involved in its formation. It is typically more or less polygonal, but in some fossils the archeopyle suture runs roughly parallel to, instead of exactly along, a reflected suture line. In such cases the size of the opening differs somewhat from that of the reflected plates to which it corresponds, and its corners may be rounded.

In the past the archeopyle has often been overlooked or only superficially described, but it is now proving to be a feature of unusual value in discriminating between fossil-dinoflagellate taxa. A description that does not adequately treat the archeopyle cannot be considered complete. About a dozen main archeopyle types and several minor ones are now known. Details of archeopyle form and structure, together with examples, are given by Evitt (1967); a few are shown in figure 18-8.

In some species an archeopyle has never been observed; in others it is consistently present (although possibly concealed in distorted or disadvantageously oriented specimens), in still other species the operculum is in place in some specimens and missing from others. Species with two wall layers (i.e., outer wall and inner body) usually have archeopyles of the same type in both layers, but there are exceptions. In most species the archeopyle suture is evidently preformed for it can be opened simply by chemical treatment of unruptured modern specimens retaining protoplasts. Therefore fossil-dinoflagellate cysts with open archeopyles probably include both cysts from which the contents had emerged during the life of the organism and cysts of organisms that died before excysting. The constant absence of an archeopyle in tests of some species may mean that general dissolution of the tests, rather than formation of a local opening, provided exit for the contents of the living cell.

Opercula found separated from the rest of the test can often be identified with the species they represent on the basis of details of surface structures, processes, size, or shape. Not infrequently, especially in species with precingular or intercalary archeopyles, the displaced operculum lies inside the test. The reasons for this are not altogether clear. Rossignol's (1963) observations on *Hystrichosphaera* suggest that the operculum adheres to the thick layer about the enclosed protoplast, is pulled inward, and is rotated away from the opening in the course of the emergence of the protoplast.

[2]The term "pylome" has been widely used for this opening in fossil dinoflagellates and for circular openings in certain acritarchs. Here it is limited to openings in acritarchs. An archeopyle thus differs from a pylome by its obvious relations to a system of tabulation, as revealed by its shape and position.

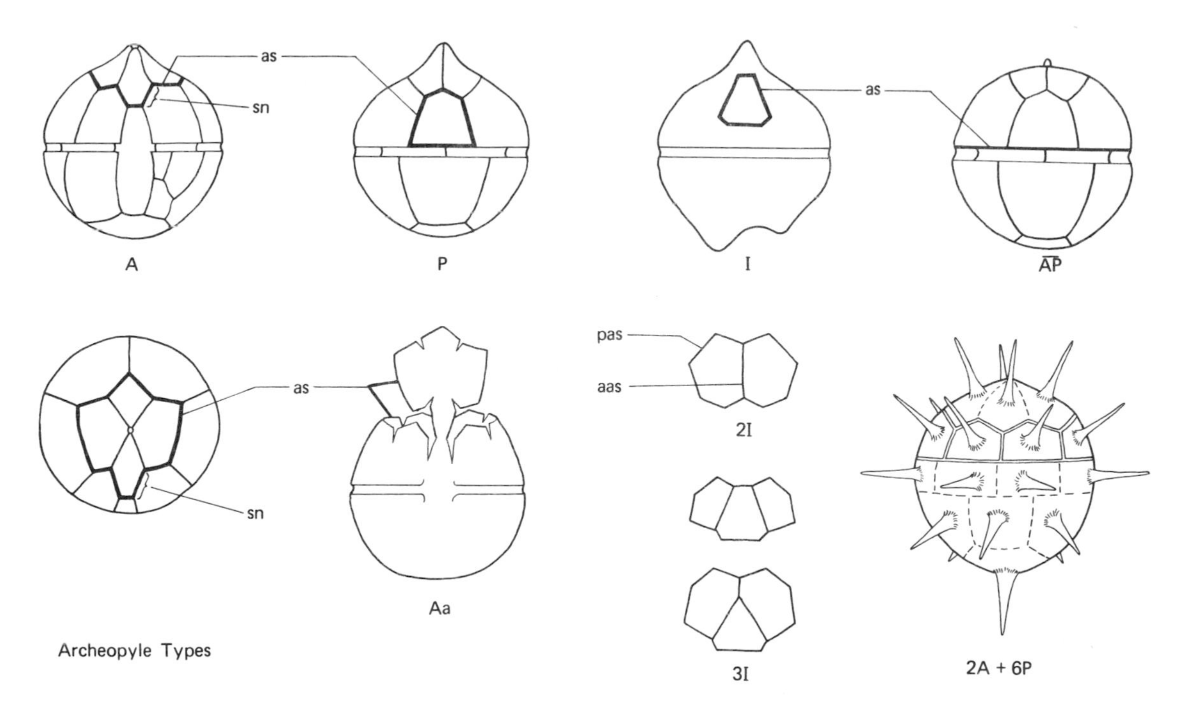

Figure 18-8. Archeopyle types in fossil dinoflagellates. (Symbols: as—archeopyle suture, pas—principal archeopyle suture, aas—accessory archeopyle suture, sn—sulcal notch. The four best known types are shown in the upper diagrams; selected details and variations are shown below. The symbols, fully explained by Evitt (1967) identify the series of thecal plates reflected by the portion of the cyst wall that is released to form the opening and give additional information about the operculum as follows: "A," "P," and "I" stand for apical, precingular, and intercalary series, respectively. A single capital letter (e.g., P), or two capitals without spacing, indicates a simple operculum, and a bar above the letter (e.g., $\overline{A}$ or $\overline{AP}$) indicates that the simple operculum corresponds to two or more thecal plates. The operculum is known to be attached only when the symbol includes a small "a" (e.g., Aa). A numeral or a plus sign indicates a compound operculum, the numeral showing the number of pieces to the operculum (e.g., 3I) and the plus sign indicating coincidence of archeopyle sutures with boundaries between two plate series (e.g., 2A + 6P). In figure at lower right dashed lines are not represented by any visible features or specimen, but are drawn in order to show arrangement of processes with respect to thecal tabulation; only some of the secondary archeopyle sutures are visible. Figure at lower left is apical view of one above.

Symmetry. A tendency toward bilateral symmetry (never perfect) is recognizable in most species, even when shape or projecting structures suggest a radial or axial symmetry at first glance. In most species with a precingular or intercalary archeopyle both the opening on the dorsal surface and the sulcus on the ventral surface lie astride what would be the sagittal plane. In species with an apical archeopyle, however, the sulcus (as indicated by the sulcal notch, *sn* in fig. 18-8) may lie astride the sagittal plane or noticeably to one side, as in some dorso-ventrally flattened species with bilaterally symmetrical outline or process arrangement. (See fig. 18-4*m*.) The significance of the position of the sulcus with respect to the sagittal plane is unknown, but the feature is useful in characterizing taxa.

Wall Structure. The walls of fossil dinoflagellates are as varied as other features of their morphology. The substance that forms the walls and the projections that extend from them may be hyaline or fibrous. The quality seems constant within species—possibly within some genera—and is easy to determine under the microscope. Minute inclusions occur in the hyaline walls of some species and may be numerous enough to make the wall appear granular.

The majority of fossil dinoflagellates have two walls, but among the minority are to be found cysts with one, three, or four walls. Eisenack (1961a), considering only the "obvious" dinoflagellates, emphasized the significance of wall structure and hypothesized an evolutionary progression from two separate walls, through two walls closely spaced, to one wall resulting from ultimate fusion of the original two. This appears to be, at best, a gross oversimplification. A few other authors have also drawn attention to two walls in certain hystrichospheres. However, in most works wall structure has received much less attention than it seems to warrant. An exception is the recent study by Downie and Sarjeant (in Davey and others, 1966) that serves as the basis for the somewhat expanded terminology presented here. The terms employed incorporate the suffixes *-phragm* to refer to the wall itself, *-coel* to refer to the cavity between a given wall and the next inner one, and *-blast* to refer to the entire body formed by the wall. (See fig. 18-9.)

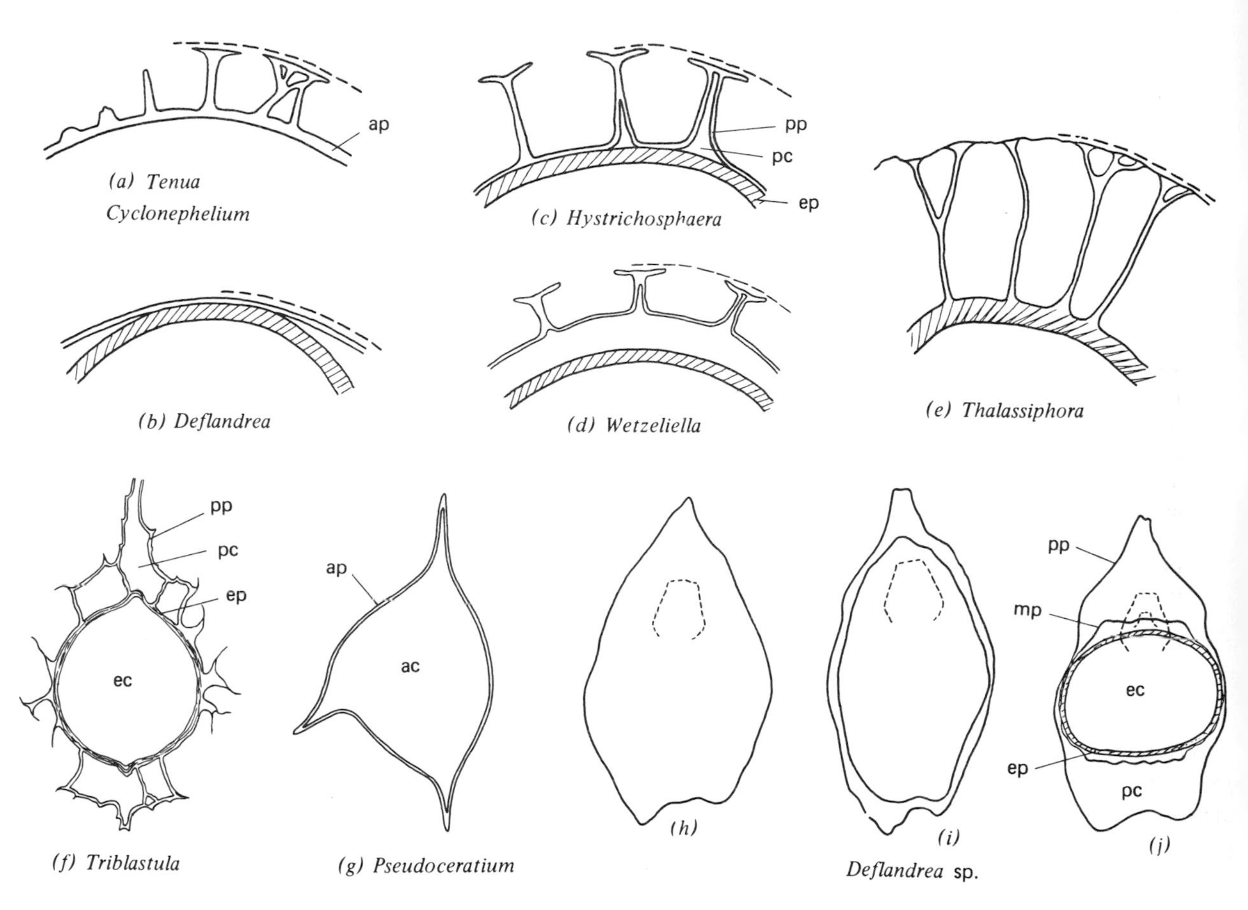

Figure 18-9. Wall structure in fossil dinoflagellates. (All specimens as seen in optical or cut sections; ac—autocoel, ap—autophragm, ec—endocoel, ep—endophragm, mp—mesophragm, pc—pericoel, pp—periphragm.) (*a-e*) Diagrams of some typical wall relations (shaded layer is the structure that has generally been called the inner body or capsule; dashed line shows inferred position of original theca): (*a*) wall of a single layer (autophragm) bearing a composite of processes that might occur in different species; (*b*) thick capsule (endophragm) and thinner periphragm (the two walls here are comparable to the inner and outer layers of (*j*); (*c*) and (*d*) diagrams to different scale to stress similarities of basic layers and structures in the two-layered walls of *Hystrichosphaera* (Pl. 18-1 fig.15) and *Wetzeliella*; each diagram shows several process types that might be found on one or different specimens; (*e*) processes supporting a membranous outer wall (Pl. 18-1 fig. 4). (*f*) *Triblastula utinensis*: outer layer (periphragm) and capsule (endophragm) closely appressed near equator, separated (by pericoels) at poles, with occasional strands spanning the separation (× 500). (*g*) Single-layered wall (autophragm) of *Pseudoceratium pelliferum* (abundant short processes not shown) (× 300). Compare cyst of *Ceratium hirundinella*, shown in figure 18-1g. (*h-j*) Progressive formation of inner walls in one species of *Deflandrea* (dashed line shows suture of intercalary archeopyle (type I) of outer wall): (*h*) outer wall only is present; (*i*) second wall has appeared but shares archeopyle with outer wall; (*j*) third wall is present but unruptured, whereas intermediate wall is shrunken and almost fully separated from outer wall (small dashed outline shows shriveled archeopyle operculum of intermediate wall).

Another example is provided by the hystrichospheres in which one tabulation pattern (4′, 0*a*, 6″, 6*g*, 5‴, 1*p*, 1⁗, as evidenced by number and arrangement of processes) recurs in many fossils assignable to a dozen or more different genera (including *Hystrichosphaeridium*) distinguished by the kinds of processes or interconnections between them. Similarly *Hystrichosphaera* typifies another group of genera with projecting structures in sutural arrangement that indicate a thecal tabulation of 3′, 0*a*, 6′, 6g, 5‴, 1*p*, 1⁗. All these fossils could be referred to the single genus *Gonyaulax* if thecal tabulation alone were considered. Indeed, we have already seen that the cyst of modern *Gonyaulax digitalis* is identical to *Hystrichosphaera*.

What is now known of both fossil and modern dinoflagellates seems to justify the conjecture that the evolution of the cyst in the peridinialean dinoflagellates has been rapid and varied relative to the more conservative evolution of the theca. As we learn more about modern cysts it may become both desirable and feasible to effect a comprehensive reclassification of modern peridinialeans to take full account of cyst morphology. It is likely that fossil-dinoflagellate cysts could be accommodated satisfactorily in such a classification. But that is for the future. However, at least temporarily, disparate classifications will persist at the levels of genus and species. At the family level it may prove easier to work out a single generally acceptable classification, but this must certainly be different from any yet proposed for fossils. Pending more information and more widespread agreement, it may be best simply not to refer fossil dinoflagellates to families, or to refer them only to living families insofar as that is possible. Either of these actions would avoid expanding pointlessly the array of alternative family assignments already possible in the contrasting classifications of Eisenack (1961a, as extended by himself and others in later publications), Vozzhennikova (1965), and Sarjeant and Downie (1966).

The following synopsis of the classification of dinoflagellates above the level of family attempts to put the fossil examples in some perspective with the more diversified modern types.

Division Pyrrhophyta Pascher 1914. The unicellular algae that constitute this division are characterized chiefly by their pigments: chlorophylls *a* and *c*, beta-carotene, and four xanthophylls. There are two classes, the Dinophyceae, or dinoflagellates, and the Desmokontae.

Class Desmokontae. The Desmokontae (in some classifications also included in the dinoflagellates) comprise free-swimming cells with two nearly equal flagella at the apex of the cell. They are naked or have a two-valved cellulose envelope with a sagittal suture. *Prorocentrum* and *Exuviella* are modern examples; no fossils are known.

Class Dinophyceae Pascher 1914. Pyrrhophyta with two unequal flagella inserted laterally. Naked or with a theca.

Order Dinophysidales *Stein 1883.* Dinophyceae with a theca of two symmetrical valves joined along a sagittal suture. Furrows margined by ridges, crests or erect membranes. Girdle far forward in most genera, often subpolar.

Remarks. The Jurassic genus *Nannoceratopsis* probably includes cysts referable to this order (Evitt, 1961a). *Nannoceratopsiella* Tasch 1963 from the Permian appears similar. No other fossil representatives have been recognized. Resistant resting cysts of modern species have not been described. Apparently all recent and fossil species are marine.

Order Gymnodiniales *Schütt 1896.* Naked Dinophyceae without either a pellicle or theca separable from the protoplasm.

Remarks. Fossil representatives of this order, it would seem, should attest extraordinary conditions of preservation and therefore should be rare. Actually many fossil species appear best referred here (formerly to the genus *Gymnodinium,* now to *Dinogymnium*—pl. 18-1, fig. 1; Evitt, Clarke, and Verdier, 1967). They are widespread in Upper Cretaceous rocks and form perhaps the single most distinctive dinoflagellate element of rocks of that age. Otherwise, fossils similar to modern *Gymnodinium* are virtually unknown. Are these Cretaceous fossils closely related to the living *Gymnodinium*? And are the fossils cysts or the resistant walls of motile individuals? Whatever the fossils are, they represent dinoflagellates that differed from modern species of Gymnodiniales by having a chemically resistant covering of considerable morphologic complexity that recorded the configuration of the fragile cell. Possibly the similarity of the fossils to modern *Gymnodinium* is misleading, and they are merely another manifestation of the generalized gymnodinioid cell that appears in the life cycles of many distantly related dinoflagellates. The order Gymnodiniales is represented today by many living species, but chemically resistant resting cysts have not been described. Apparently all recent and fossil species are marine.

Order Peridiniales *Schütt 1896.* These are Dinophyceae with a motile stage armored with thecal plates joined along sutures. Known fossil representatives are probably exclusively resting cysts that formed inside thecae that were not fossilized. The fossils have a resistant, often two-layered wall and may or may not resemble thecae of living types. Processes or other surface features on fossils may reflect thecal tabulation. An archeopyle (excystment aperture) of distinctive form is present in cysts of many species, both living and fossil. Walls of fossils are mostly organic, but calcareous and siliceous examples occur. Recent species inhabit marine, brackish, and fresh water.

Remarks. The boundary between unarmored Gymnodiniales and armored Peridiniales is not sharp. The latter is believed to have been derived from the former; some modern genera have been referred first to one order, then to the other, depending on interpretations of platelike markings on their thin walls. Eisenack (1961a) has emphasized that what he calls the theca varies gradationally in fossil dinoflagellates from thick and clearly tabulate to tenuous with obscure tabulation. He implies that this may record transition between the two orders. However, it now appears that the fossils are remains of cysts with resistant walls secreted within the original thecae. So the gradation noted by Eisenack may simply record variations in the thickness of the cyst wall and in the fidelity with which it reflects the plates of the true theca.

Almost all the genera of fossils described until now (June 1969) that qualify as dinoflagellates by the criteria discussed early in this chapter, (exceptions include *Dinogymnium, Nannoceratopsis,* and *Nannoceratopsiella*) are here considered to belong to the Peridiniales because their features seem more likely to reflect structures typical of modern Peridiniales than of any other living group of dinoflagellates.

Other Orders. From one to five additional orders are recognized by different authors. They comprise free-living, parasitic, and symbiotic dinoflagellates of specialized form, including amoeboid, coccoid, and filamentous types. The motile stage is transitory, gymnodinioid, and naked. No fossil representatives are known, and the cysts that occur in some types would be difficult to recognize if fossilized. Being simple in shape (e.g., tetragonal, spherical, lunate) and without trace of structures (such as flagellar furrows, tabulation, or archeopyle) that are critical for positive identification of fossil dinoflagellates, these cysts would be referred to the acritarchs. *Palaeotetradinium* from the Cretaceous is perhaps an example.

Stratigraphic Distribution and Significance

Occurrences of fossil dinoflagellates before the mid-Jurassic are rare. They include probable representatives of the Dinophysidales from the Lower Permian of Kansas (Tasch, 1963) and the Lower Jurassic of Denmark (Evitt, 1961a); probable Peridiniales without clear indications of tabulation from the Lower Permian of Kansas (Tasch, 1963) and the Upper Permian of Alberta (Jansonius, 1962); Peridiniales with sutural ridges clearly showing tabulation from the Upper Triassic of Britain (Sarjeant, 1963) and the Lower Jurassic of Denmark (Evitt, 1961b); and Silurian specimens from Tunisia (Calandra, 1964), whose sole but convincing dinoflagellate character is a distinct, probably precingular archeopyle. Almost nothing is yet known of dinoflagellates from the strata between the horizons represented by these occurrences, but, beginning with the Middle Jurassic, dinoflagellates are conspicuous elements of many marine assemblages. Upper Jurassic and younger deposits contain a vast array of forms only in part referable to the 200 or so genera described until now.

Many species were geographically as widespread in the past as at present, and the evolution of some stocks within the group was rapid. As a result, some dinoflagellates are good horizon markers and highly useful tools for zonation and correlation over both short and long distances. This is illustrated by published regional studies like those by Klement (1960) and Alberti (1961) and by numerous unpublished studies carried out in oil company laboratories. However, range charts of genera and species that might be prepared now from the literature would, with some exceptions, be relatively meaningless. There are three chief reasons for this. First, our current knowledge provides too inadequate a sample, as shown by the many new taxa and extensions of known ranges that are the products of studying each new assemblage rich in species. Second, too many published records of species, and even genera, are unreliable because identifications were based on inadequate morphological criteria. Third, many published descriptions of fossil microplankton (especially in the earlier works) contain only broad stratigraphic datings and inexact geographic details. Therefore a reliable and meaningful picture of the stratigraphic distribution of dinoflagellates requires more precise data on the stratigraphic position and the detailed morphology of many of the described taxa, as well as extensive studies of undescribed assemblages.

Little can yet be said about the paleoecological significance of fossil dinoflagellates. Here is an open door for future research. Since the physicochemical requirements of living species are imperfectly known and appear to vary greatly from one species to another, any detailed interpretation of environment from fossil species, whose degrees of affinity with living forms are still speculative, is at present out of the question.

Living Gymnodiniales and Dinophysidales are limited to marine and brackish waters. Fossils that appear to be referable to these orders are also from known or supposed marine sediments. The order Peridiniales, on the other hand, is represented by both marine and freshwater species today, and there is no obvious reason why this salinity range may not be of long standing. Nevertheless, the majority of fossil assemblages with dinoflagellates that are not plausibly believed to be reworked seem to be of marine origin, as is often evidenced by associated microfossils and megafossils. We should not, however, accept the occurrence of fossil dinoflagellates—certainly not of those belonging to previously undescribed species—as prima facie evidence of marine conditions, because a few fossil freshwater dinoflagellates are now known (Traverse, 1955; Krutzsch, 1962; Churchill and Sarjeant, 1963) and some living freshwater species produce chemically resistant cysts strikingly similar to many fossil examples (Evitt and Wall, 1968).

ACRITARCHS

Summary. The acritarchs (Downie, Evitt, and Sarjeant, 1963) comprise unicellular or apparently unicellular microfossils that consist of a test, composed of organic substances, enclosing a central cavity. Shape, symmetry, structure, and ornamentation are varied. An inner body may be present or not; where present it may be connected to the outer wall by varied means or it may lack such connection. The test may be unruptured or may open by formation of a pylome of varied design. Acritarchs include many of the fossils formerly known as hystrichospheres, especially those from the Paleozoic. Many types bear spinelike processes and superficially resemble the dinoflagellate hystrichospheres. Acritarchs are widespread in carbonates, cherts, and fine clastic sediments from Proterozoic and younger horizons.

General

The acritarchs (ακριτος, uncertain and ἀρχή, origin), defined in the terms quoted in the summary, constitute a "catch-all" utilitarian category of organic microfossils. They are morphologically varied and often abundant microfossils whose affinities cannot at present be precisely determined. The group Acritarcha, which is not a formal taxon under the Code, replaces the order Hystrichosphaeridea in the nondinoflagellate sense; but the fossils assembled here are thought to represent a highly polyphyletic association of organisms. At its inception the group included all those former Hystrichosphaeridea (sensu latissimo) that could not be identified as dinoflagellates; that is, a varied assortment of microfossils from Proterozoic and younger geologic periods, ranging in size from a few microns to 150 or more.

General affinities of these microfossils with the algae have been suggested by Eisenack (1938, 1962), but other possibilities exist (Deflandre, 1947). Some post-Paleozoic forms are probably dinoflagellates that lack the minimum of morphological features required for positive recognition. The rest probably represent a variety of life stages (e.g., eggs, cysts, mature tests) of assorted one-celled and higher organisms, both plant and animal. However, characters adequate to justify referral to a particular class of organisms have not been recognized.

Downie, Williams, and Sarjeant (1961) recommended that the dinoflagellates, hystrichospheres, and other organic microfossils of uncertain affinities (including acritarchs in the sense used here) be treated under the Botanical Code. This recommendation was followed by Evitt in proposing recognition of the acritarchs. When the biological affinities of an acritarch genus are determined, that genus should cease to be referred to as an acritarch and should be assigned to its proper place in the taxonomic hierarchy under the appropriate nomenclatural code.

As most authors have not distinguished between the hystrichospheres and the acritarchs, the comprehensive reviews of hystrichospheres by Deflandre (1947) and Sarjeant (1961), as well as remarks by Eisenack (e.g., 1954, 1962, 1963a), contain much that is applicable to the acritarchs. Useful indexes to the literature will be found in Downie and Sarjeant (1964(1965)) and Norris and Sarjeant (1965). (See Chap. 10.)

These fossils occur in rocks of many lithologies, shales and limestones having yielded the richest assemblages. Little is known about the paleoecology of the organisms they represent. Most appear to have been elements of the marine plankton, although freshwater examples have recently been reported (Churchill and Sarjeant, 1963). Eisenack (1961b) suggests some acritarchs were food of lower Paleozoic foraminifers. A discussion of acritarch distribution in relation to sedimentary facies is presented by Staplin (1961) as part of a study of fossil microplankton from the Upper Devonian Woodbend Formation in Alberta. He found acritarchs to be most abundant in the offreef strata and certain morphological types to occur closer to the reefs than others. In general thin- and thick-spined types were limited to areas more distant from the reefs than the more ubiquitous smooth and papillate ones.

Morphology and Classification

The essential morphological feature of an acritarch is a central cavity closed off from the exterior by a wall of primarily organic composition. Restriction of the name to fossils that consist of a single such hollow body is implicit in the definition; that is, fossils that might be described as "colonies" of similar units arranged into a larger compact structure are excluded. Variations in overall shape of the test are best comprehended by reference to the selected examples discussed below.

Spines or other projecting structures occur on many acritarchs. They commonly vary in number within a single species and may also vary in length on a single specimen. However, in some acritarchs the summits of adjacent processes are conspicuously concordant, and expansions of process tips are constructed as they would be if the fossils had formed as it is suggested the hystrichospheres did; namely, inside a surrounding wall, the processes serving as supports between that wall and the central body. At least acritarchs of this type may be resting spores or cysts of organisms with more active stages.

Surface structures that project appreciably from the central body fall generally into two categories. *Processes* are spinelike to columnar projections and may have simple to elaborately branched tips, free or interconnected. *Septa* are membranous structures that rise more or less at right angles to the surface of the central body. Minor ornamentation of varied sorts (e.g., granules, verrucae, spinules, pilae) also occurs. The wall of most acritarchs is single layered, but two closely spaced wall layers have been reported and several subgroups are characterized by a discrete inner body within an outer wall. No openings (other than apparently accidental ones) have been observed in the majority of acritarchs, but some possess a pylome as an operculate opening, which is usually circular (rarely polygonal), or as a simple slitlike aperture. If the pylome has not yet actually opened, traces of it may be difficult or impossible to recognize.

The variety of form and structure evidenced by the acritarchs seems virtually limitless. It is therefore understandable that different workers, especially when dealing with a group having wholly unknown and probably mixed biological affinities, may evaluate differently the criteria that might be used as the basis for classification of these fossils. In reviewing here some of the more common morphological types represented among the acritarchs the morphological classification of Downie, Evitt, and Sarjeant (1963) is followed. In this classification acritarch genera are considered to be form-genera and are arranged in subgroups that are named and defined outside the jurisdiction of the Botanical Code, in a manner analogous to the turmae, etc., of the widely used form classification of sporomorphs developed by Potonié and others. Names of acritarch subgroups have stems descriptive of principal morphological features and have a standard ending, *-omorphitae*. To facilitate comparison with earlier literature family names previously used are noted where appropriate.

An extreme morphological variation is often evident among acritarchs from a single sample, thus presenting an acute problem of defining species limits. This has been solved by the "splitters" with narrowly defined, but generally similar species, and by the "lumpers" with broadly defined, highly variable species. Similarly contrasting attitudes toward definition of genera are also apparent from published descriptions. The group requires extensive study to establish affinities where possible and to understand better the underlying morphological features that can be most effectively used as a basis for circumscribing taxa.

Subgroup Acanthomorphitae. This classification is given to acritarchs with a spherical to ellipsoidal main body and radiating spinelike processes. (See fig. 18-10*a–d*.) These are the acritarchs that are most likely to be confused with hystrichospheres (s.s.). *Baltisphaeridium* (fig. 18-10 *a,b*) and *Micrhystridium* (fig. 18-10 *d*) are the most important genera. They differ in being, respectively, greater than and less than 20 microns in diameter. Justification of a generic distinction on size alone has been debated (e.g., Staplin, 1961; Downie and Sarjeant, 1963), but the separation appears to correspond to a discontinuity in size distribution (although some species overlap the boundary) and to a difference in distribution in the geologic record. *Micrhystridium* ranges from lower Paleozoic to Tertiary and is notably common in some Mesozoic samples. *Baltisphaeridium* species are most common and diversified in Upper Cambrian to Devonian assemblages. The variety and elaborateness of morphologic structures appear greater in *Baltisphaeridium* than in *Micrhystridium*.

Many species of *Baltisphaeridium* were attributed originally to *Hystrichosphaeridium* and became well known under that name before *Hystrichosphaeridium* was restricted by emendation (Eisenack, 1958; Evitt, 1963a). Acanthomorph acritarchs, although most common in the Paleozoic, have post-Paleozoic representatives as well. Some of the latter are probably dinoflagellates.

The wall is usually single, although a few examples of two-layered walls have been reported. Some species exhibit a distinct circular pylome with an operculum, but these openings are relatively rare, and assemblages of numerous individuals and species occur in which the opening has not been seen in a single specimen.

Many Acanthomorphitae superficially resemble hystrichospheres among the dinoflagellates because similar types of processes (fig. 18-7*a* and fig. 18-10*c*) occur in both groups. However, Acanthormorphitae and other acritarchs differ from the hystrichospheres by (a) lacking an arrangement of surface structures comparable to the arrangement of plates in a dinoflagellate theca, and (b) having a circular or slitlike (rarely polygonal) pylome—if any opening is present—in contrast to an archeopyle whose shape or position is correlatable with a dinoflagellate tabulation.

Subgroup Polygonomorphitae. *Veryhachium* (fig. 18-10*g*, *i*), characterized by a polygonal or subpolygonal body from whose angles extend a small number (3 to 8) of hollow spines, is the best known representative of this subgroup. Genera with somewhat similar characters include *Wilsonastrum*, *Estiastra* (fig. 18-10*f*), *Pulvinosphaeridium* (fig. 18-10*e*), and *Evittia* (fig. 18-10*h*). In all of them the wall is single and smooth to granular, but the massiveness of the projections and the size of the tests is variable. An opening consisting of a straight or arcuate slit along one edge is known in a few species.

Acritarchs of this sort are most typical of Paleozoic rocks, but the simple triangular types range at least through the Cretaceous. A high intraspecific variability in shape and spine length among specimens from a single sample is characteristic (Wall and Downie, 1963).

An algal affinity for *Wilsonastrum* has been suggested by Jansonius (1962) on the basis of a bristlelike structure and the occurrence of matted masses of individuals reminiscent of certain living Chlorococcales.

Subgroup Herkomorphitae. *Cymatiosphaera* (fig. 18-10*p*) comprises tests with a spherical main body bearing an external reticulum of thin septa. In similar genera the reticulum may be formed of ridges (*Dictyotidium*) or rows of processes (*Cymatiogalea*) with simple or furcate concordant summits. The polygonal lumina of the reticulum may be regular or irregular in

a
b
c
d
e
f
g
h
i
j
k
l
m
n
o
p
q
r
s
t
u
v
w
x
y
z

shape; but their arrangement does not define a dinoflagellate-like tabulation with girdle, sulcus, and latitudinal series of other polygons. A large polygonal to circular pylome occurs in *Cymatiogalea*.

Tests of this sort are distributed in rocks that range from lower Paleozoic to at least Cretaceous in age. Several types are common in Silurian-Devonian strata, and others, mostly smaller and with a conspicuously regular reticulum of low septa, occur in the Jurassic and Cretaceous.

Subgroup Sphaeromorphitae. In this subgroup are assembled simple spherical to ellipsoidal tests with a single wall. They occur throughout the fossiliferous geologic section, although a majority of described species are from the Ordovician and Silurian. The fossils range in size from 10 to several hundred microns. The wall is smooth to minutely spinulate, and a circular pylome has been observed in some species. *Leiosphaeridia* (fig. 18-10*q*) is a representative genus; about 20 others have been proposed (family Leiosphaeridae). Eisenack (1956) interprets small spherical bodies within some *Leiosphaeridia* as young individuals. Wall (1962) suggests that the affinities of *Leiosphaeridia* lie with *Halosphaera*, a modern planktonic green alga. Since it is difficult to derive cogent morphological evidence for or against such a possibility from so simple a structure as a thin-walled sphere, fossils of that character are here retained in the acritarchs. On the other hand, three other genera (*Tasmanites*, *Tytthodiscus*, and *Crassosphaera*) sometimes treated with *Leiosphaeridia* are discussed here in the section on miscellaneous organic microfossils.

Subgroup Netromorphite. In *Leiofusa* (fig. 18-10*m*) the test is elongate with rounded or pointed poles. The wall is one-layered, smooth to granulate, and may be thrown into minute longitudinal folds. Eisenack proposed the family Leiofusidae for the genus. *Lunulidia* (fig. 18-10*o*) includes arcuate tests of similar design. These genera are characteristic of the lower and middle Paleozoic.

Domasia (fig. 18-10*j-l*) has an elongate test with a smooth surface, bearing two processes at one pole and one at the other. The wall is single and smooth; the processes are solid or hollow at least at the base. No opening has been observed. Somewhat similar species are referred to the genus *Deunffia* (fig. 18-10*n*), and undescribed types that are transitional between these genera occur. Some forms show a distinct bilateral symmetry and compression at right angles to this symmetry plane; others are more or less axially symmetrical. Acritarchs of *Domasia-Deunffia* type appear to be widespread in Silurian strata. Some exhibit a striking variability, making separation into distinguishable species difficult. Downie (1964) discusses the stratigraphic distribution and possible evolutionary relationships of these two genera.

The outline of three-spined *Domasia* suggests the three-horned peridinioid outline of certain dinoflagellates, but no other character points to any affinity with dinoflagellates. Cell shapes analogous to some of the forms shown in figure 18-10 are to be found among the modern freshwater chlorophyte algae (e.g., *Schroederia*).

Subgroup Pteromorphitae. The genus *Pterospermopsis* is characterized by a globular to len-

Figure 18-10. Representative acritarchs. (*a-d*) Acanthomorphitae: (*a*) *Baltisphaeridium longispinosum*, Silurian (× 300); (*b*) *B. brevispinosum*, Silurian (× 300); (*c*) a variety of process types that occur in the Acanthomorphitae; (*d*) *Micrhystridium inconspicuum*, Silurian (× 1,250). (*e-i*) Polygonomorphitae: (*e*) *Pulvinosphaeridium pulvinellum*, Silurian (× 150); (*f*) *Estiastra magna*, Silurian (× 100); (*g*) and (*i*) *Veryhachium* spp., Silurian (× 750); (*h*) *Evittia* sp., Silurian (× 750). (*j-o*) Netromorphitae: (*j-l*) *Domasia* sp., three variants from a single highly variable population, Silurian (× 750); (*m*) *Leiofusa* sp., Silurian (× 250); (*n*) *Deunffia* sp., Silurian (× 750); (*o*) *Lunulidia lunula*, Ordovician (× 50). (*p*) Herkomorphitae: *Cymatiosphaera* sp., Eocene (× 750). (*q*) Sphaeromorphitae: *Leiosphaeridia ovalis*, Ordovician (× 100). (*r*) Pteromorphitae: *Helios aranaides*, Silurian (× 800). (*s-u*) Prismatomorphitae: (*s*) *Polyedryxium* sp., Devonian (× 400); (*t*) *Polyedryxium pharaonis*, Devonian (× 500); (*u*) *Staplinium hexaeder*, Triassic (× 1,500). (*v-w*) Diacromorphitae: (*v*) *Acanthodiacrodium barbullatum*, Cambrian (× 600); (*w*) *Lophodiacrodium bubnoffi*, Cambrian (× 750). (*x-z*) Acritarchs with a capsule: (*x*) *Trigonopyxidia ginella*, Cretaceous (× 500); (*y*) *Cirrifera unilateralis*, Cretaceous (× 600); (*z*) *Wallodinium krutzschi*, Cretaceous (× 500). (*a-f, o, q* after Eisenack; *d* after Valensi; *s, t* after Deunff; *u*, after Jansonius; *v, w* after Timofeyev; *x, y* after Cookson and Eisenack.)

ticular main body provided with a thin equatorial flange that may be smooth or undulating and that often bears conspicuous radial folds. Species have been described from strata of Jurassic to middle Tertiary age, but morphologically similar types have been observed in Devonian samples and virtually indistinguishable objects occur in present seas.

A pylome has not been seen. Wetzel proposed the family Pterospermopsidae for the genus. The generic name alludes to resemblance to *Pterosperma* Pouchet, a component of living marine plankton of uncertain systematic position.

Duvernaysphaera (fig. 18-10*r*) from the Silurian and Devonian is a somewhat similar morphological type, with the equatorial flange supported by 10 to 20 spoke like extensions from the main body. The spokes may be hollow at the base with an extension of the cavity of the main body, and the whole structure is saucer shaped.

The distinctive and widespread Eocene-Oligocene species *Thalassiphora pelagica*, originally referred to *Pterospermopsis*, has an archeopyle and further characters indicative of a dinoflagellate; dinoflagellate-like characters in other species referred to *Pterospermopsis* are unknown.

Subgroup Prismatomorphitae. Tests of *Polyedryxium* (fig. 18-10*s*, *t*) are basically polyhedral, with angular or thickened edges. Spinelike processes or chimneylike processes with flared tips rise from these edges at or between corners. Expanded tips of processes between corners may be conspicuously concordant.

Staplinium (fig. 18-10*u*) is similarly polyhedral but lacks the projections from edges and corners. The general surface is smooth to granular or minutely spinose.

No pylome has been observed in these genera, which are known sparingly from Devonian and Triassic rocks.

Subgroup Diacromorphitae. The family Diacrodiaceae was introduced by Timofeyev (1958) for distinctive Cambrian microfossils that he thought were unicellular algae. They are characterized by a hollow test with an equatorial smooth zone separating two polar areas that may be simply roughened or, more commonly, granulose or spinose. The equatorial zone may be straight or slightly concave or convex, so that the overall shape ranges from nearly spherical to ellipsoidal or obesely dumbbell shaped. No pylome is known. (See fig. 18-10*v*, *w*.)

Deflandre and Deflandre-Rigaud (1961) after reviewing the nomenclatural validity and taxonomic justification of a whole suite of genera named by Timofeyev, accepted four genera (*Trachydiacrodium, Lophodiacrodium, Acanthodiacrodium*, and *Dasydiacrodium*), which they referred to a new zoological family Trachydiacrodiidae, equivalent to Timofeyev's botanical family Diacrodiaceae. (See Chap. 10.)

The forms described by Timofeyev are from the Cambrian of Baltic Russia. Other species are known from the Tremadocian of England and North Africa. The Cretaceous *Palaeostomocystis echinulata* Deflandre is possibly similar, but more strikingly similar are the objects from modern arctic plankton described as *Echinum* by Meunier (1910). The affinities of *Echinum* are uncertain, and, in the absence of intervening fossil representatives, it seems doubtful that *Echinum* and *Palaeostomocystis* have anything to do with the Cambrian Diacromorphitae, which probably represent a closely knit group of organisms, whatever their precise affinities.

Acritarchs with an Inner Body. Most of the acritarchs discussed in the foregoing sections clearly have a wall composed of one principal layer. However, an assortment of organic microfossils, especially from the Jurassic and Cretaceous, consist of an outer wall about a distinct inner body. They may be further categorized by differences in shape (subgroups Disphaeromorphitae, spherical; Dinetromorphitae, fusiform; Platymorphitae, flattened). (See fig. 18-10*x*,*y*,*z*).

Although critical identifying features are lacking, distinctive external shapes and traces of openings, which may prove identifiable as archeopyles on closer study, suggest that some of these fossils may be dinoflagellates. Several genera have been proposed but most are known from a single species each. In contrast, *Wallodinium* (fig. 18-10*z*) is represented by several Jurassic species in Europe and Australia. It is cylindrical (straight or curved) and is truncated by an opening (archeopyle?) at one end. A large inner body of somewhat similar shape is enclosed.

CHITINOZOA

Summary. The chiefly vase-shaped tests of Chitinozoa range from 30 to 1,500 microns in length and resemble pseudochitin in composition. They are widespread in Ordovician to Devonian marine sediments and have proved highly useful for stratigraphic zonation in some areas. Genera and species are distinguished chiefly by differences in the shape of the test, presence and structure of spines and other projections, and structures associated with a single terminal aperture. Recent studies illuminate wall structures and internal bodies commonly concealed by opaque tests as well as associations of individuals in linear series and cylindrical masses. The affinities of the Chitinozoa remain uncertain.

General

The Chitinozoa, named by Eisenack (1931), comprise an assortment of essentially vaselike, commonly dark-colored, organic microfossils in lower Paleozoic marine sediments. They are readily distinguished from associated fossils and seem to constitute a closely interrelated group. The fossils are composed of a substance whose exact composition is unknown, similar to pseudochitin (not to chitin as the name would suggest). Specimens usually appear black or dark brown except in the thinnest areas, but, with best preservation, they are transparent and reveal internal structures.

Chitinozoan tests range in maximum dimensions from 30 to 1,500 microns, with the majority under 500 microns. Whole individuals of some of the smaller species and test fragments from larger species fall in the size range of the spores and acritarchs that commonly occur in the same samples. The larger specimens can be separated easily from fine debris by differential settling or heavy liquid treatment and then picked individually from concentrated residues with a fine pipet under a low-power (× 50) microscope. Specimens may be studied and illustrated dry or embedded. Dark-field illumination, chemical treatment to clear the wall, and photography with infrared light, to which the wall is more transparent, are useful in detecting structures otherwise difficult to observe.

The comprehensive discussion of these fossils by Taugourdeau (1966) includes chapters on methods of study, morphology and nomenclature, systematic position, and classification, along with an extensive bibliography and tables showing ranges of 34 genera and about 260 species. The great potential of these fossils as tools in stratigraphic paleontology was demonstrated by Taugourdeau and Jekhowsky (1960), who studied their distribution in lower Paleozoic subsurface strata in North Africa.

A series of important papers on Chitinozoa by several authors—beginning with Combaz and Poumot (1962) and including Kozlowski (1963), Van Oyen and Calandra (1963), Jansonius (1964), Taugourdeau (1964), and Taugourdeau and Magloire (1964)—have illuminated hitherto unsuspected complexities in these fossils. Much of this new information has come from exceptionally well-preserved material from Tremadocian to Upper Devonian strata in North Africa. The tests described in earlier papers as simple, opaque, vaselike bodies are now known to possess varied internal structures, to exhibit dimorphism in size, to evidence possible spore or cyst formation, to occur commonly in chains of individuals whose successive members may show differences that suggest progressive maturation, and to occur occasionally in masses completely enclosed in a cocoonlike membrane.

These advances in knowledge of chitinozoan morphology so far have only increased the uncertainty as to their biological affinities. At times in the past they have been considered ciliate, rhizopodous, or flagellate protozoans (Eisenack, 1931; Collinson and Schwalb, 1955) and chrysophyte algae (Staplin, 1961). Kozlowski's observations (1963) led him to suggest that they are the eggs or egg cases of an as yet unidentified metazoan. No theory of affinity yet advanced seems to account satisfactorily for all the morphological features that have now been observed.

Morphology

The morphological features of Chitinozoa are shown in figure 18-11. The typically vaselike tests range from nearly spherical to irregularly cylindrical. The cavity of the *chamber* may open directly through the terminal *aperture* (pseudostome), or a distinct *neck* may separate the two. Externally the *base* (aboral surface) and sides of the test may merge along a smoothly convex surface. Alternatively the contact may be marked by a distinct angulation or ridge, the *carina,* or by a row of basal horns, or appendages. The carina is rarely extended into an elaborate

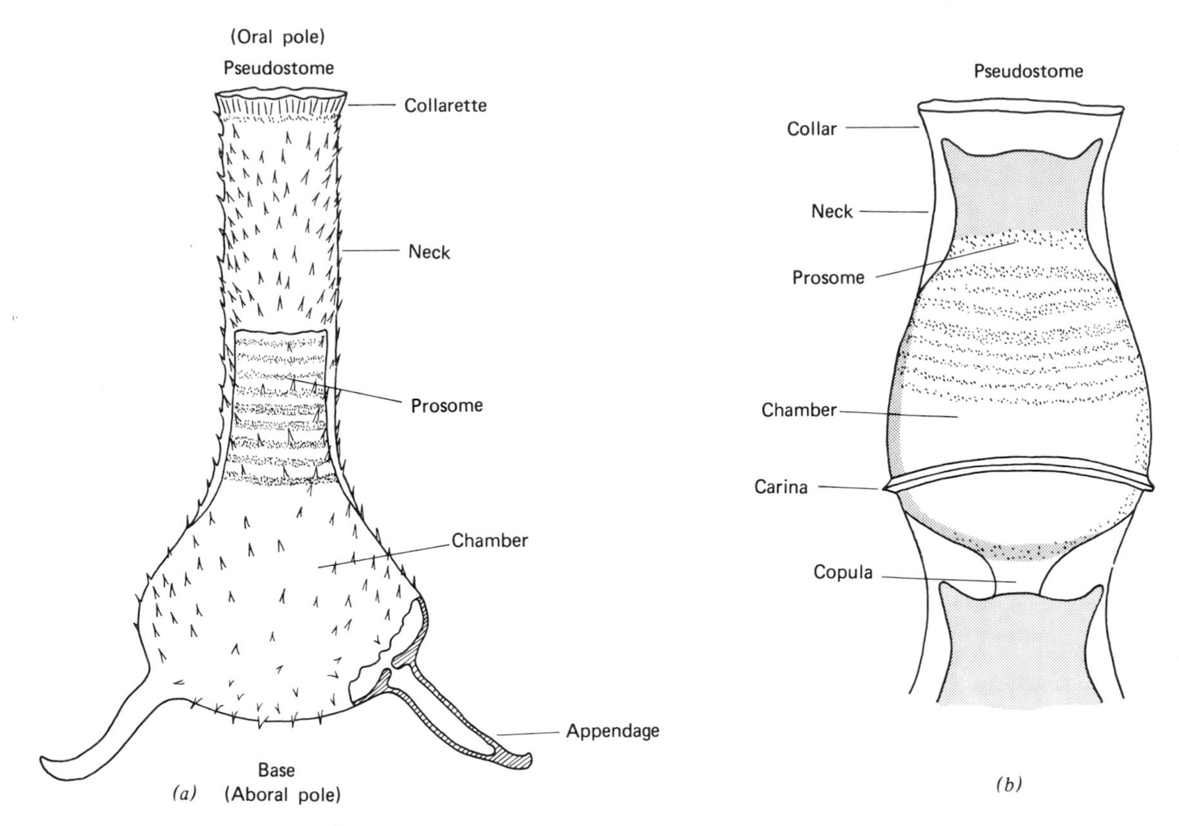

Figure 18-11. Terminology of Chitinozoa: (*a*) chitinozoan of the solitary type, with a portion of the wall at the lower right cut away to show the hollow appendage; (*b*) an individual of the "colonial" type, with a portion of a second individual attached below. Terminology and figures after Combaz and Poumot.

network. The basal horns are variable in size, number, and structure. If hollow, their interiors may open by a pore into the chamber. The basal horns are distinct from the spinules or larger projections that occur on the outer surface of the neck and upper parts of the chamber in some species.

The base commonly bears at its center either a small *mucron*, which is a nipplelike elevation usually with a central perforation, or a larger, tubular structure called the *copula*. The latter may merge with the base to form a tunnel-shaped aboral termination. The chief internal structure is a more or less cylindrical, and apparently contractile, lining of the neck called the *prosome*, which may expand toward the aperture or be closed by an *operculum*. The prosome is frequently torn or displaced so that it lies deeply within, or partly to wholly outside, the test. The supposed existence of a second internal structure, the opisthosome, in the aboral part of the body, is possibly based on misinterpretation of inturned portions of a fragmented base.

For many years only isolated individuals of most types of chitinozoans were known, and so-called "colonial" forms, with chains of individual tests attached end to end, were recorded from a single genus, *Desmochitina*. It is now clear that similar linear arrangements occur in several genera. Kozlowski (1963) described for the first time aggregates of a different sort, in which many individuals are arranged in cylindrical masses. Some of these masses are enclosed in a thin, faintly striate covering of apparently the same composition as the contained units. Clear evidence of attachment to a substrate is preserved in some of the linear series and larger masses observed by Kozlowski.

The internal structures of chitinozoans and the devices that appear to have linked together successive members of chainlike "colonies" are relatively new discoveries. The character of the preservation of many specimens often wholly precludes the observation of these features, and only under the most fortunate circumstances can their details be closely studied. Therefore it is not surprising that opinions differ as to the nature and possible function of these often delicate structures. For the most thorough discussion published to date the reader should consult Taugourdeau (1966). Although the significant studies published since 1961 have contributed greatly to our understanding of the chitinozoans, many questions concerning morphological details and especially functions and affinities invite increased attention to these problematic microfossils.

Occurrence and Ecology

First described from the Ordovician and Silurian of Baltic Europe, Chitinozoa have now been reported also from Cambrian-Ordovician (Tremadocian) to Upper Devonian rocks of France, North Africa, and North America. They are probably to be expected wherever marine calcareous or fine clastic strata of these ages occur. One specimen has been reported from the Mississippian (Wilson and Clarke, 1960).

The most comprehensive study of chitinozoan distribution yet made is by Taugourdeau and Jekhowsky (1960). They analyzed nearly 300 Silurian-Devonian core samples from wells over a large area of the Algerian Sahara, examining some 7,000 specimens of Chitinozoa, half of them identified as to species. In addition to descriptions and illustrations of many species, new and old, they presented numerous charts to show the interrelations of species, genera, rock types, and stratigraphic position. Distinctive ranges of both genera and species permitted them to divide the sequence studied into 10 zones, some of which were independently dated (Taugourdeau, 1962). Half the species were limited to two zones or less. Benoit and Taugourdeau (1961) have also described Ordovician chitinozoans from the Sahara.

Clear evolutionary trends in the Chitinozoa have not been recognized. Some of the morphologically simple types are represented throughout the range of the order, but these are particularly characteristic of the Ordovician. Sharp differentiation of cylindrical neck from inflated body is more typical of Silurian and Devonian forms. Stout basal horns occur in many Silurian and Lower Devonian species, whereas a general covering of shorter spines is more common in the Upper Devonian. Chains of bulblike individuals especially characterize the Upper Silurian and Devonian.

Chitinozoans have been widely accepted as free-swimming or floating members of the plankton, but their relatively large size, thick walls, aggregation into series and masses, and definite indications of attachment argue effectively against this. They were perhaps attached to floating objects if not themselves truly benthic.

Report of a seemingly dominant association of Chitinozoa with sediments deposited in oxygen-poor waters (Collinson and Schwalb, 1955) was based on relatively few occurrences and may be illusory. The tests are now known from a wide variety of rock types. Taugourdeau and Jekhowsky (1960) found specimens particularly abundant in fine clastics with cephalopods, graptolites, acritarchs, clams, and scolecodonts, and rare in coarse clastics and carbonates. In contrast, Dunn (1959) described a rich and varied assemblage from limestones with brachiopods and corals.

Classification

Eisenack (1931) originally divided the Chitinozoa into three families (Conchitinidae, Lagaenochitinidae, Desmochitinidae) and six genera. Taugourdeau (1966) lists 34 genera. He divides the order Chitinozoa into two suborders. The Copulida comprises forms with a perforated base and most often with a connecting device indicating association in linear "colonies." The Acopulida includes forms with an imperforate base, the individuals having been free or in nonlinear "colonies." Two families are distinguished within each suborder on the basis of the nature of the connecting device (Copulida) or the presence or absence of special ornamentation around the basal periphery (Acopulida). Examples of selected genera are shown in figure 18-12.

MISCELLANEOUS ORGANIC MICROFOSSILS

Other organic microfossils besides the types discussed in the preceding sections of this chapter

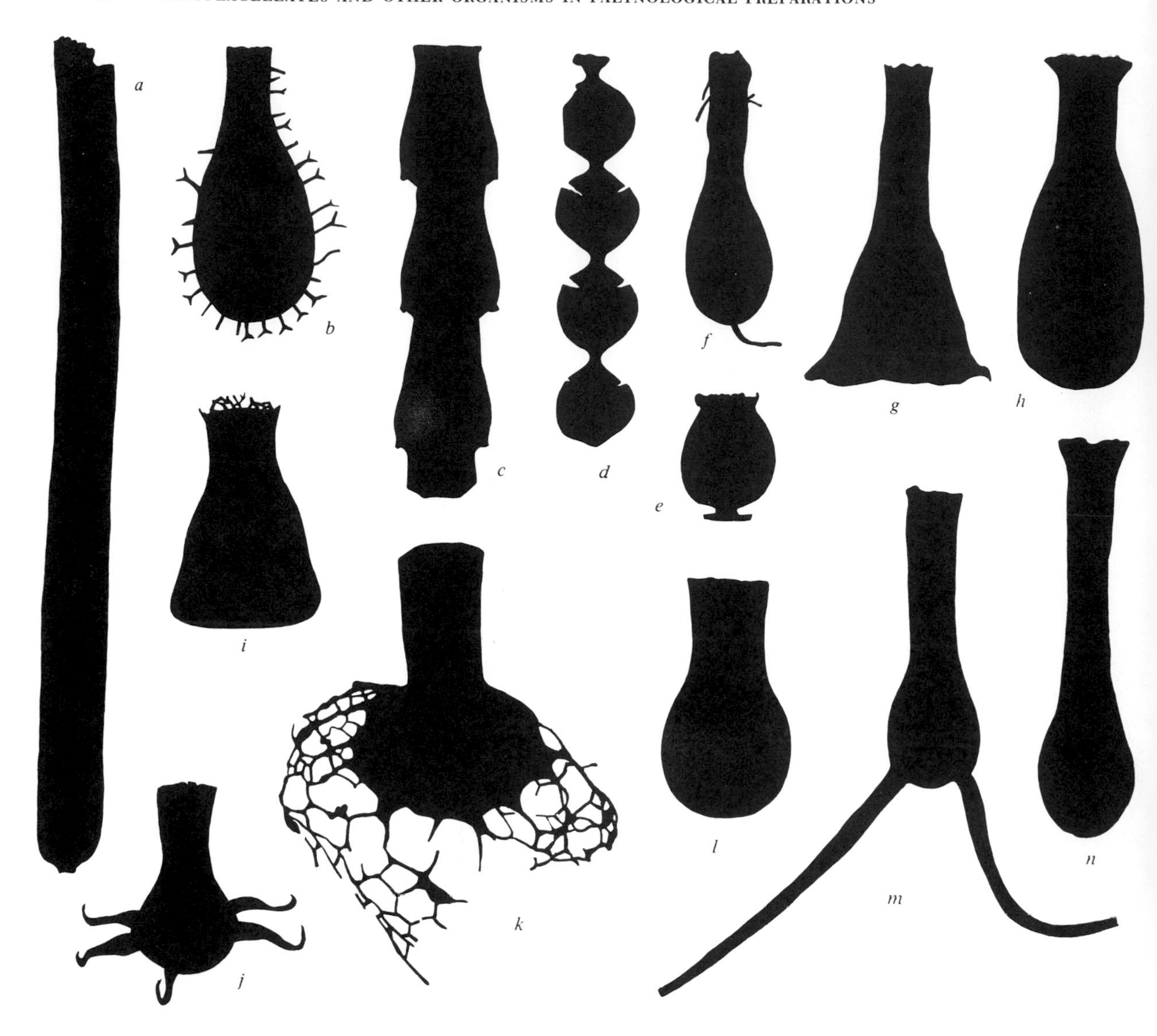

Figure 18-12. Representative chitinozoans (× 200 except as noted): (*a*) *Conochitina* cf. *C. minnesotensis*, Silurian (× 150); (*b*) *Angochitina bifurcata*, Devonian; (*c*) *Eremochitina? cingulata*, Silurian-Lower Devonian; (*d*) *Desmochitina margaritana*, Silurian-Middle Devonian; (*e*) *D. minor*, Silurian; (*f*) *Urochitina simplex*, Devonian; (*g*) *Cyathochitina kukersiana*, Silurian (× 150); (*h*) *Lagenochitina esthonica*, Ordovician (× 75); (*i*) *Euconochitina conulus*, Ordovician (× 275); (*j*) *Ancyrochitina fragilis*, Devonian; (*k*) *Clathrochitina retifera*, Ordovician; (*l*) *Sphaerochitina cuvillieri*, Devonian (× 150); (*m*) *Plectochitina longicornis*, Silurian-Lower Devonian; (*n*) *Lagenochitina* sp., Silurian-Devonian. (After Taugourdeau and Jekhowsky, except: (*b*) after Collinson and Schwalb, (*e*) and (*f-h*) after Eisenack; (*l*) after Taugourdeau.)

occur in palynological preparations. Their morphology is varied and their identities are only partly known. Many are encountered only occasionally, others more commonly. Since it is impossible to include a complete catalogue of described types here, only a few are discussed.

Tasmanites and Possible Allies

Tasmanites (pl. 18-1, fig. 12) includes hollow, spherical to lenticular fossils with a thick organic wall partly or completely penetrated by many fine canals; a pylome is present in some specimens (Schopf, Wilson and Bentall, 1944; Eisenack, 1958a). The objects range from less than 100 to 600 microns in diameter. Their known occurrence is Ordovician-Tertiary. Specimens are abundant in some Devonian-Mississippian dark shales. (See Chapter 10).

The wall structure of *Tasmanites* appears identical with that of living species of *Pachysphaera*, a planktonic green alga (Wall, 1962). Two other genera, *Tytthodiscus* and *Crassosphaera* (Jurassic-middle Tertiary), resemble *Tasmanites* in having perforate walls, but the structure and spacing of pores are different. The perforate walls set all three apart from simple Sphaeromorphitae. No living counterparts of *Tytthodiscus* and *Crassosphaera* are known.

Pediastrum

The distinctive coenobia of this colonial green alga are easily recognized when encountered. Living *Pediastrum* is exclusively a freshwater organism, but fossils of *Pediastrum* occur in both freshwater and marine deposits. Abundant fossil *Pediastrum* in marine deposits apparently records either transport of the resistant coenobia from freshwater lakes or streams, or, possibly, local derivation from fresh surface waters above marine bottoms. (See Evitt, 1963b, for discussion and references.) The oldest known examples of the genus are from the Early Cretaceous. A Devonian genus, *Deflandrastrum*, (Combaz, 1962) is strikingly similar to *Pediastrum*. (See pl. 18-1, fig. 13.)

Ophiobolus

This genus includes hollow ellipsoidal bodies, about 15 by 40 microns, with a smooth to minutely granulate wall and one to several long, threadlike appendages continuous with the wall. They were originally observed in flakes of Cretaceous chert by O. Wetzel (1933) and Deflandre (1937), who suggested they might be remarkably preserved protozoan cells with fossilized flagella. Alberti (1961) and Evitt (1968), finding well-preserved specimens in insoluble residues, view them as cysts or eggs with filamentous holdfasts or entangling devices. Specimens are abundant in some Upper Cretaceous deposits. Deflandre recognized a second genus and a family, the Ophiobolidae. (See pl. 18-1, fig. 14.)

Acknowledgement. For helpful criticisms of an early draft of this chapter I am especially indebted to Georges Deflandre and W. A. S. Sarjeant.

ADDENDUM

In the considerable period of time between the final revision of the manuscript for this chapter and the correction of the first proof (June 1969) significant progress has been made in studying dinoflagellates, acritarchs, and chitinozoans. Although most results of this work could not be integrated into the text, a few contributions of particular note should be pointed out.

A wealth of varied information on all these fossils, including a summary of their geologic history (Downie, 1968), appears among the published papers of the Second International Conference on Palynology, which constitute the initial volumes of the *Review of Paleobotany and Palynology.*

Dinoflagellates

Wall and Dale (1967b, 1968a, b, c) have reported investigations that effectively bridge the gap between the world of living dinoflagellates and the prolific dinoflagellate fossil record from Middle Tertiary and older strata. They have described Pleistocene marine assemblages, in part with ecological emphasis, and have identified as to species many kinds of resting cysts after first inducing them to develop in culture into thecate individuals. As a result, some important questions about the biologic identity of fossil dinoflagellates in terms of the accepted classification of living ones have been answered, and new questions have been raised. These authors have also shown that calcareous dinoflagellates like the fossils reviewed by Deflandre (1949) still exist in present seas and that these calcareous bodies pertain to resting cysts from which thecate cells can be cultured.

A major contribution to the knowledge of dinoflagellates from Mesozoic and early Cenozoic deposits of the Soviet Union has appeared in a comprehensive work by Vozzhennikova (1967). An overlooked but important work on the living freshwater dinoflagellate *Ceratium hirundinella* by Huber and Nipkow (1922, 1923) has come to light and apparently represents the first exhaustive investigation of excystment in a living species.

Chitinozoa

A monograph edited by Combaz (1967), which will long remain a vital reference, presents a comprehensive synthesis of information on the morphology, taxonomy, and stratigraphic distribution of these fossils, as well as a complete bibliography. Jenkins (1968 unpubl.) presented evidence which, although largely circumstantial, would suggest that a genetic or ecological relationship exists between chitinozoans and graptolites. Jenkins also reports (1969) on an evolutionary lineage of chitinozoans from the Ordovician of Oklahoma and on the feasibility of correlation between northern Europe and North America on the basis of the many chitinozoan species common to both regions in the Late Ordovician.

References

Alberti, Gerhard, 1961, Zur Kenntnis mesozoischer und alttertiärer Dinoflagellaten und Hystrichosphaerideen von Nord- und Mitteldeutschland sowie einigen anderen europäischen Gebieten: Palaeontographica, v. 116, sec. A, p. 1-58, pls. 1-12.

Benoit, A., and Taugourdeau, Ph., 1961, Sur quelques chitinozoaires de l'-Ordovicien du Sahara: Inst. Français du Pétrole Rev., v. 16, p. 1403-1421, 5 pls.

Bouché, P. M., 1964, Revue bibliographique des hystrichosphères (acritarches) du Paléozoïque et du Permo-Trias: Inst. Français du Pétrole Rev., v. 19, p. 3-31.

Brongersma-Sanders, Margaretha, 1957, Mass mortality in the sea, p. 941-1010 *in* Hedgepeth, J. W., and Ladd, H. S., Treatise on marine ecology and paleoecology: Geol. Soc. America, Mem. 67, v. 1.

Brunel, Jules, 1962, Le phytoplancton de la Baie des Chaleurs: Univ. Montreal Press, 365 p., 66 pls.

Calandra, François, 1964, Sur un présumé Dinoflagellé, *Arpylorus* nov. gen. du Gothlandien de Tunisie: Acad. sci. [Paris] Comptes rendus, v. 258, p. 4112-4114.

Chatton, Éduard, 1952, Classe des Dinoflagellés ou Péridiens, *in* Gassé, P.-P., Traité de Zoologie, v. 1, p. 309-406, pl. 1, figs. 2-4. Paris, Masson & Cie.

Churchill, D. M., and Sarjeant, W. A. S., 1963, Freshwater microplankton from Flandrian (Holocene) peats of southwestern Australia: Grana Palynologica, v. 3, no. 3, p. 29-53.

Collinson, Charles, and Schwalb, Howard, 1955, North American Paleozoic Chitinozoa: Illinois Geol. Survey Rept. Invest. No. 186, p. 1-33, 2 pls.

Combaz, A. 1962, Sur un nouveau type de microplanctonte cénobial fossile du Gothlandien de Libye, *Deflandrastrum* nov. gen.: Acad. sci. [Paris] Comptes rendus, v. 255, p. 1977-1979.

———(ed.), 1967, Microfossiles organiques du Paléozoique. Les Chitinozoaries: v. 1, 96 p., 11 pls.; v. 2, 43 p., 5 pls. Paris Centre Nat. Recherche Sci.

Combaz, A., and Poumot, Cl., 1962, Observations sur la structure des chitinozoaires: Rev. Micropaléontologie, v. 5, p. 147-160, 5 pls.

Cookson, I. C., and Eisenack, Alfred, 1960, Microplankton from Australian Cretaceous sediments: Micropaleontology, v. 6, p. 1-18, pls. 1-3.

Davey, R. J., Downie, C., Sarjeant, W. A. S., and Williams, G. L., 1966, Studies on Mesozoic and Cainozoic dinoflagellate cysts: British Mus. (Nat. Hist.) Bull., Geol. Supplement 3, 248 p., 26 pls.

Deflandre, Georges, 1933, Note préliminaire sur un péridinien fossile *Lithoperidinium oamaruense* n. gen., n. sp.: Soc. Géol. France Bull., v. 58, p. 265-273.

______1935, considérations biologiques sur les microorganismes d'origine planctonique conservés dans les silex de la craie: Bull. Biol. France Belgique, v. 69, p. 213-244, pls. 5-9.

______1936, Microfossiles des silex cretacés, Première partie, Généralités, Flagellés: Ann. Paléontologie, v. 25, p. 149-191, pls 1-10.

______1937, Microfossiles des silex cretacés, Deuxième partie, Flagellés incertae sedis, Hystrichosphaeridés, Sarcodinés, organismes divers: Ann. Paléontologie, v. 26, p. 49-103, pls. 11-18.

______1947, Le problème des hystrichosphères: Inst. Océanographique [Monaco] Bull. no. 918, p. 1-23.

______1949, Les Calciodinellidés, Dinoflagellés fossiles à thèque calcaire: Le Botaniste, v. 34, p. 191-219.

______1952, Dinoflagellés fossiles, *in* Grassé, P.-P., Traité de Zoologie, v. 1, p. 391-406.

______1964, Quelques observations sur la systématique et la nomenclature des Dinoflagellés fossiles: Paris, Lab. Micropaléont. E.P.H.E., Inst. Paléont. du Muséum, 8 p.

Deflandre, Georges, and Deflandre, Marthe, 1943-1967, Fichier micropaléontologique [a continuing series of cards]: Paris, Centre de Documentation du C.N.R.S.

Deflandre, Georges, and Deflandre-Rigaud, Marthe, 1961, Nomenclature et systématique des hystrichosphères (s. l.). Observations et rectifications: Paris, Lab. Micropaléont., Inst. Paléont. de Museum, 15 p.

Downie, Charles, 1963, "Hystrichospheres" (acritarchs) and spores of the Wenlock Shales (Silurian) of Wenlock, England: Palaeontology, v. 6, p. 625-652, pl. 91-92.

______1967, Geologic history of the microplankton: Rev. Palaeobotany Palynology, v. 1, p. 269-281.

Downie, Charles, Evitt, W. R., and Sarjeant, W. A. S., 1963, Dinoflagellates, hystrichospheres, and the classification of the acritarchs: Stanford Univ. Publ., Geol Sci., v. 7, no. 3, 16 p.

Downie, Charles, and Sarjeant, W. A. S., 1963, On the interpretation and status of some hystrichosphere genera: Palaeontology, v. 6, p. 83-96.

______1964 (1965), Bibliography and index of fossil dinoflagellates and acritarchs: Geol. Soc. America Mem. 94, 180 p.

Downie, Charles, Williams, G. L., and Sarjeant, W. A. S., 1961, Classification of fossil microplankton: Nature, v. 192, no. 4801, p. 471.

Dunn, D. L., 1959, Devonian chitinozoans from the Cedar Valley formation in Iowa: Jour. Paleontology, v. 33, p. 1001-1017, pls. 125-126.

Ehrenberg, C. G., 1838, Über das Massenverhältniss der jetzt lebenden Kiesel-Infusorien und über ein neues Infusorien-Conglomerat als Polierschiefer von Jastraba in Ungarn: König. Akad. Wissensch. Berlin, Abh. 1836, p. 109-135, 2 pls.

Eisenack, Alfred, 1931, Neue Mikrofossilien des baltischen Silurs, I.: Paläont. Zeitschr., v. 12, p. 74-118, pls. 1-5.

______1936, Dinoflagellaten aus dem Jura: Ann. Protistologie, v. 4, p. 59-63, pl. 4.

______1938, Hystrichosphaerideen und verwandte Formen im baltischen Silur: Zeitschr. Geschiebeforschung, v. 14, p. 1-30.

______1954, Mikrofossilien aus Phosphoriten des samländischen Unteroligozäns und über die Einheitlichkeit der Hystrichosphaerideen: Palaeontographica, v. 105, sec. A, p. 49-95, pls. 7-12.

——1956, Probleme der Vermehrung und des Lebensraumes bei der Gattung *Leiosphaera* (Hystrichosphaeridea): Neues Jahrb. Geol. Paläont., Abh., v. 102, p. 402-408.

——1958a, *Tasmanites* Newton 1875 and *Leiosphaeridia* n. g. als Gattungen der Hystrichosphaeridea: Palaeontographica, v. 110, sec. A, p. 1-17, pls. 1-2.

——1958b, Mikroplankton aus dem norddeutschen Apt, nebst einigen Bemerkungen über fossile Dinoflagellaten: Neues Jahrb. Geol. Paläont., Abh., v. 106, p. 383-422, pls. 21-27.

——1961a, Einige Erörterungen über fossile Dinoflagellaten nebst Übersicht über die zur Zeit bekannten Gattungen: Neues Jahrb, Geol. Paläont., Abh., v. 112, p. 281-324.

——1961b, Hystrichosphären als Nahrung ordovizischer Foraminiferen: Neues Jahrb. Geol. Paläont., Mh., p. 15-19.

——1962, Mitteilungen über Leiosphären und über das Pylom bei Hystrichosphären: Neues Jahrb. Geol. Paläont. Abh., v. 114, p. 58-80, pls, 2-4.

——1963a, Hystrichosphären; Bio. Reviews, v. 38, p. 107-139, pls. 2-4.

——1963b, Sind die Hystrichosphären Zysten von Dinoflagellaten?: Neues Jahrb. Geol. Paläont., Mh., p. 225-231.

——1965, Katalog der fossilen Dinoflagellaten, Hystrichosphären, und verwandten Mikrofossilien, Band 1, Dinoflagellaten: Stuttgart, Schweizerbart'sche Verlags., 888 p., 9 pls.

Entz, Géza, 1925, Über Cysten und Encystierung der Süsswasser-Ceratien: Archiv Protistenkunde, v. 51, p. 131-183.

Evitt, W. R., 1961a, The dinoflagellate *Nannoceratopsis* Deflandre: morphology, affinities and infraspecific variability: Micropaleontology, v. 7, p. 305-316, pls. 1-2.

——1961b, *Dapcodinium priscum,* n. gen., n. sp., a dinoflagellate from the lower Lias of Denmark: Jour. Paleontology, v. 35, p. 996-1002, pl. 119.

——1961c, Observations on the morphology of fossil dinoflagellates: Micropaleontology, v. 7, p. 385-420, pls. 1-9.

——1963a, A discussion and proposals concerning fossil dinoflagellates, hystrichospheres, and acritarchs: Nat. Acad. Sci. [U.S.] Proc., v. 49, p. 158-164, 298-302.

——1963b, Occurrence of freshwater alga *Pediastrum* in Cretaceous marine sediments: Amer. Jour. Sci., v. 261, p. 890-893, 1 pl.

——1967, Dinoflagellate studies. II. The archeopyle: Stanford Univ. Pub. Geol. Sci., v. 10, no. 3, 82 p., 9 pls.

——1968, The Cretaceous microfossil *Ophiobolus lapidaris* O. Wetzel and its flagellum-like filaments: Stanford Univ. Publ., Geol. Sci., v. 12, no. 3, 9 p., 1 pl.

Evitt, W. R., Clarke, R. F. A., and Verdier, J. P., 1967, Dinoflagellate Studies. III. *Dinogymnium acuminatum* n. gen., n. sp. (Maastrichtian), and other fossils formerly referable to *Gymnodinium* Stein: Stanford Univ. Publ., Geol. Sci., v. 10, no. 4, 27 p., 3 pl.

Evitt, W. R., and Davidson, S. E., 1964, Dinoflagellate studies. I. Dinoflagellate cysts and thecae: Stanford Univ. Publ., Geol. Sci., v. 10, no. 1, 12 p., 1 pl.

Evitt, W. R., and Wall, D., 1968, Dinoflagellate Studies. IV. Theca and cyst of Recent freshwater *Peridinium limbatum* (Stokes) Lemmermann: Stanford Univ. Publ., Geol. Sci., v. 12, no. 2, 15 p., 4 pls.

Gray, Jane (coord.), 1965, Techniques in palynology: Part III, p. 469-706 *in* Kummel, B., and Raup, D. M., Handbook of paleontological techniques: San Francisco, W. H. Freeman and Co.

Hasle, G. R., and Nordli, Erling, 1951, Form variation in *Ceratium fusus* and *tripos* populations in cultures and from the sea: Norsk Videnskaps-Akademi, Avhandl., I. Mat.-Naturv. Klasse 1951, no. 4, p. 1-25.

Hutner, S. H., and McLaughlin, J. J. A., 1958, Poisonous tides: Scientific American, v. 199, no. 2, p. 92-98.

Jenkins, W. A. M., 1968 unpubl., Chitinozoa, a talk presented at First Annual Meeting, American Assoc. Stratigraphic Palynologists, Baton Rouge, October 1968.

———1969, Chitinozoa from the Ordovician Viola and Fernvale Limestones of the Arbuckle Mountains, Oklahoma: Palaeont. Assoc., Spec. Papers in Palaeontology, no. 5.

Jansonius, Jan, 1962, Palynology of Permian and Triassic sediments, Peace River area, western Canada: Palaeontographica, v. 110, sec. B, p. 35-98, pls. 11-16.

———1964, Morphology and classification of some Chitinozoa: Bull. Canadian Petroleum Geology, v. 12, p. 901-918.

Klement, K. W., 1960, Dinoflagellaten und Hystrichosphaerideen aus dem unteren und mittleren Malm Südwestdeutschlands: Palaeontographica, v. 114, sec. B, p. 1-104, pls. 1-10.

Kofoid, C. A., 1909, On *Peridinium steini* Jörgensen, with a note on the nomenclature of the skeleton of the Peridinidae: Arch. Protistenkunde, v. 16, p. 25-47.

Kozlowski, Roman, 1963, Sur la nature des Chitinozoaires: Acta Paleont. Polonica, v. 8, p. 425-449.

Krutzsch, Wilfried, 1962, Die Mikroflora der Geiseltalbraunkohle. Teil III. Süsswasserdinoflagellaten aus subaquatisch gebildeten Blätterkohlenlagen des mittleren Geiseltales: Hall. Jahrb. Mitteldeutsch. Erdgesch., v. 4, p. 40-45, pls. 10-11.

Lebour, M. V., 1925, The dinoflagellates of northern seas: Plymouth, Marine Biol. Assoc. of United Kingdom, 250 p., 35 pls.

Lefevre, M. M., 1933, Sur la structure de la thèque chez les Peridinites: Acad. sci. [Paris], Comptes rendus, v. 197, p. 81-83.

Loeblich, A. R., Jr., and Leoblich, A. R., III, 1966, Index to the genera, subgenera, and sections of the Pyrrhophyta: Univ. Miami Inst. Marine Sci., Studies of tropical oceanography, No. 3, 94 p.

Meunier, A., 1910, Microplankton des mers de Barentz et de Kara: Brussels, Camp. arctique Duc d'Orleans, p. 1-355, atlas of 30 pls.

Norris, Geoffrey, and Sarjeant, W. A. S., 1965, A descriptive index of genera of fossil Dinophyceae and Acritarcha: New Zealand Geol. Survey Paleont. Bull. 40, 72 p.

Pastiels, André, 1945, Étude histochimique des coques d'Hystrichosphères: Mus. Roy. d'Hist. Nat. Belgique Bull., v. 21, p. 1-20.

Paulsen, Ove, 1949, Observations on dinoflagellates: Kong. Danske Videnskab. Selskab. Biol. Skrifter, v. 6, no. 4, p. 1-67.

Rossignol, Martine, 1963, Aperçus sur le développement des hystrichosphères: Mus. Nat. d'Hist. Nat., Bull., Ser. 2, v. 35, p. 207-212.

Sarjeant, W. A. S., 1961, The hystrichospheres-A review and discussion: Grana Palynologica, v. 2, no. 3, p. 101-111.

——1963, Fossil dinoflagellates from Upper Triassic sediments: Nature, v. 199, p. 353-354.

Sarjeant, W. A. S., and Downie, Charles, 1966, The classification of dinoflagellate cysts above generic level: Grana Palynologica, v. 6, p. 503-527.

Schiller, Josef, 1931-1937, Dinoflagellatae (Peridineae), v. 10, sec. 3 *of* Rabenhorst's Kryptogamen-Flora: pt. 1 (1931-1933), p. 1-590, pt. 2 (1935-1937), p. 1-604. Leipzig, Akademische Verlagsgesell.

Schopf, J. M., Wilson, L. R., and Bentall, Ray, 1944, An annotated synopsis of Paleozoic fossil spores and the definition of generic groups: Illinois State Geol. Survey, Rept. Invest. No. 91, p. 5-73, pls. 1-3.

Staplin, F. L., 1961, Reef controlled distribution of Devonian microplankton in Alberta: Palaeontology, v. 4, p. 392-424, pls. 48-51.

Tasch, Paul, 1963, Hystrichosphaerids and dinoflagellates from the Permian of Kansas: Micropaleontology, v. 9, p. 332-364, 1 pl.

Taugourdeau, Ph., 1962, Associations de chitinozoaires dans quelques sondages de la region d'Édjelé (Sahara): Rev. Micropaléontologie, v. 4, p. 229-236, 1 pl.

——1964, Sporulation ou enkystement chez un *Ancyrochitina* (chitinozoaire): Soc. Géol. France Comptes rendus Somm., 1964, p. 238.

——1966, Les Chitinozoaires. Techniques d'études, morphologie, et classification: Soc. Géol. France Mem., v. 45 (Mem. 104), p. 1-64, pl. 1-4.

Taugourdeau, Ph. and Jekhowsky, B. de, 1960, Répartition et description des chitinozoaires Siluro-Dévoniens de quelques sondages de la C.R.E.P.S., de la C.F.P.A. et de la S.N. Repal au Sahara: Inst. Français du Pétrole Rev., v. 15, p. 1199-1260.

Taugourdeau, Ph. and Magloire, L., 1964, Le dimorphisme chez les Chitinozoaires: Soc. Géol. France Bull., v. 6, p. 674-677.

Timofeyev, B. V., 1958, Über das Alter sächsischer Grauwacken. Micropaläophytologische Unterzuchungen von Proben aus der Weesensteiner und Lausitzer Grauwacke: Geologie, v. 7, p. 826-845, 3 pls.

Van Oyen, F. H., and Calandra, F., 1963, Note sur les chitinozoaires: Rev. Micropaléontologie, v. 6, p. 13-18, 1 pl.

Vozzhennikova, T. F., 1965, Vvedenie v izuchenie iskopaemykh peridineevykh vodorosley: Akad. Nauk SSSR, Sibirskoe Otdelenie, Inst. Geol. Geofys., 156 p. Moscow, Isdatelstvo "Nauka."

——1967, Iskopaemye peridinei yurskikh, melovykh, i paleogenovykh otlozheniy SSSR: Akad. Nauk SSSR, Sibirskoe Otdelenie, Inst. Geol. Geofys., 224 p., 121 pls. Moscow, Isdatelstvo "Nauka."

Wall, David, 1962, Evidence from Recent plankton regarding the biological affinities of *Tasmanites* Newton 1875 and *Leiosphaeridia* Eisenack 1958: Geol. Mag., v. 99, p. 353-362, pl. 17.

——1966, Modern hystrichospheres and dinoflagellate cysts from the Woods Hole region: Grana Palynologica, v. 6, p. 297-314.

Wall, David, and Dale, Barrie, 1967a, "Living fossils" in western Atlantic plankton: Nature, v. 211, p. 1025-1026.

——1967b, The resting cysts of modern marine dinoflagellates and their paleontolgical significance: Rev. Palaeobotany Palynology, v. 3, p. 349-354, 1 pl.

——1968a, Modern dinoflagellate cysts and evolution of the Peridiniales: Micropaleontology, v. 14, p. 265-304, 1 pl.

——1968b, Early Pleistocene dinoflagellates from the Royal Society borehole at Ludham, Norfolk: New Phytology, v. 67, p. 315-326, 1 pl.

______1968c, Quaternary calcareous dinoflagellates (Calciodinellidae) and their natural affinities: Jour. Paleontology, v. 42, p. 1395-1408, pl. 172.

Wall, David, and Downie, Charles, 1963, Permian hystrichospheres from Britain: Palaeontology, v. 5, p. 770-784, pls. 112-114.

Wetzel, Otto, 1933, Die in organischer Substanz erhaltenen Mikrofossilien des baltischen Kreidefeuersteins, mit einem sedimentpetrographischen und stratigraphischen Anhang: Palaeontographica, v. 77, p. 147-186; ibid., v. 78, p. 1-110, pls. 1-7.

Wilson, L. R., and Clarke, R. T., 1960, A Mississippian chitinozoan from Oklahoma: Okla. Geol. Notes, v. 20, p. 148-150.

AUTHOR INDEX

SUBJECT INDEX

Page numbers in bold face type refer to illustrative material, figures, tables and plate legends; those marked with an asterisk refer to descriptions of taxa.